ENCYCLOPAEDIA OF NARCOTIC DRUGS *AND* PSYCHOTROPIC SUBSTANCES

ENCYCLOPAEDIA OF
NARCOTIC DRUGS
AND
PSYCHOTROPIC SUBSTANCES

(Vol. II—DRUGS' DANGEROUS DIMENSIONS AND TOBACCO SMOKING)

Giri Raj Shah

Foreword by:
Kiran Bedi

GYAN

GYAN PUBLISHING HOUSE

New Delhi-110002

Encyclopaedia of Narcotic Drugs and Psychotropic Substances
(Botany, Chemistry, Economics, Health, International Relations, Law, Police/Crime, Pharmacy, Sociology, Sports)

ISBN : 81-212-0545-X (Set)
ISBN : 81-212-0541-7 (Vol. II)

Published in 1998 in India by
Gyan Publishing House
5, Ansari Road,
New Delhi-110 002

Laser Typesetting at Manas Typesetter, New Delhi
Printed at Mehra Offset Press, New Delhi.

Contents

List of Illustrations

List of Tables

Preface

The drug abuse was confined to the opium, cannabis, cocaine and a few other drugs of vegetative origin but the progress in the field of bio-chemistry and pharmacology led to abuse of entirely new groups of drugs of synthetic origin. The drugs of synthetic origin were generally more potent and very easy to transport. These drugs are not manufactured for intoxication or for change of mood but for the treatment of diseases. Therefore, the experimentation and later on dependence also started among the doctors and some of the chronic patients. These developments gave new group of climate to the industry and in turn a new dimension was added to the illicit trade far more difficult to control because of its potentialities to influence society.

The increase in the number of substances available for misuse also encouraged multiple drug abuse, frequently in combination with alcohol and tobacco. The social acceptability of these substances tends to increase its misuse among the group of tender age, thus causing more harm to the society and adding more complexities to the problem of detoxification and treatment. Social acceptability of certain chemicals, depending on historical and geographical factors, is one of the important determinants in its likely misuse. The vegetative drugs or synthetic substances *per se* are not imbued with moral values but the severity of intoxication and potentialities to some physical and psychological impairment capable of causing harm to the individual and to the society are the factors which determine the social sanction to lack of it. A highly potent drug is heroin which is misused by a small minority of a community but invites more attention of society than tobacco which is used by a large majority of people and commands so much acceptability that a non-user at times invites social embarrassment. Social attitude towards the use of alcohol is also conditioned by hordes of social and geographical factors and differs from one cultural pocket to another. The psychoactive substances, widely used within a society, are integrated in the way of life and harmonized in its value system. Many centuries are required to weave a psychoactive substance into the cultural warp and woof, and to lend it individual and social symbolism. Social contracts were exerted over intoxication by religious strictures. While religious proscription has been very effective in controlling the use of intoxicants in many oriental societies, the move towards secularism and away from organized religion has weakened the moderating effect of religious sanction on uncontrolled use of intoxicants, hence the need for more concerted efforts to control the misuse of drugs to save society from its ill-effects.

The use of intoxicants by a group of people for a long time makes the abuse of the substance as the centre of their lives, consequently alienating them from their families, friends, co-workers and increasingly compelling them to spend more and more time in society, using and experiencing the use of intoxicant, thus leading to formation of a sub-culture of drop-outs, skid-row drinkers, drug peddlers, opium smokers and others who indulge, abet and gain from such drug abuse. These affiliations have sub-group norms and values, meet at specific times and places and more often than not they have their own vocabulary of slangs and dress.

The struggle against the drug abuse is the first and foremost fight against degradation and debasement of human values. The stakes are quite high and the failure to arrest the spread of drug use means total change in the value system and chaotic socio-economic conditions which may imperil the very existence of humanity. Under the circumstances it becomes imperative on our part to fight determined with all the resources at our command. In the wake of unprecedented progress in the field of science and technology and consequential rapid occurrence of world events, has come the multiplicity of relations between human beings from different geographical units, differing in terms of race, language, religion and stage of economic development attained. The change in the field of science and technology could not be absorbed in the blood vessels of national and international life, leading to myopic considerations even while we plan our strategy against the universal problem. The discussions on international fora are replete with examples when world bodies fail to arrive at unanimous decisions on the ways and means to face such universal problems like drug abuse and narrow political considerations cloud our vision to draw a well-knit plan to counter the spread of drug dependence. Now the time has come when we should give a go-by to tentativity and ad hocism and formulate a plan based on best possible data and scientific aids with in-built mechanism to correct the declivity creeping during the course of execution in different geographical and ethno-cultural units and unitedly start working for the whole of the world as one family.

1.1.1998 **Dr. Giri Raj Shah**

Volume II

DRUGS' DANGEROUS DIMENSIONS AND TOBACCO SMOKING

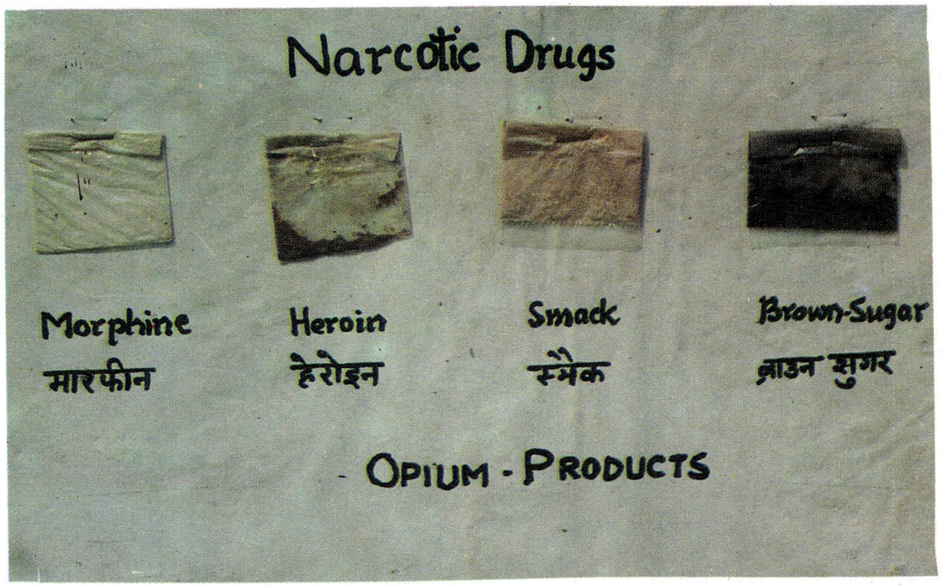

Opium-Products (Narcotic Drugs).

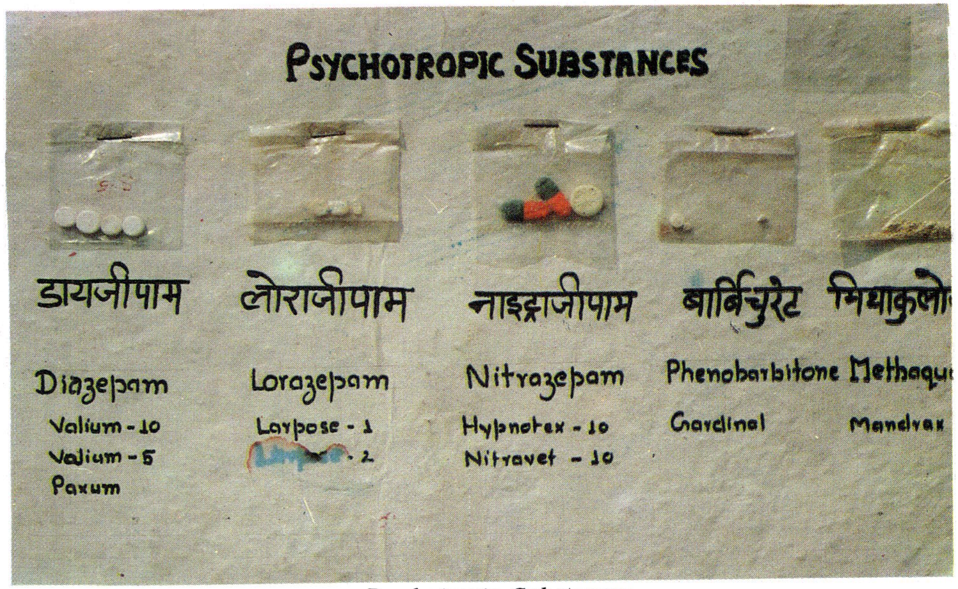

Psychotropic Substances.

Bugs to Stop Drugs.

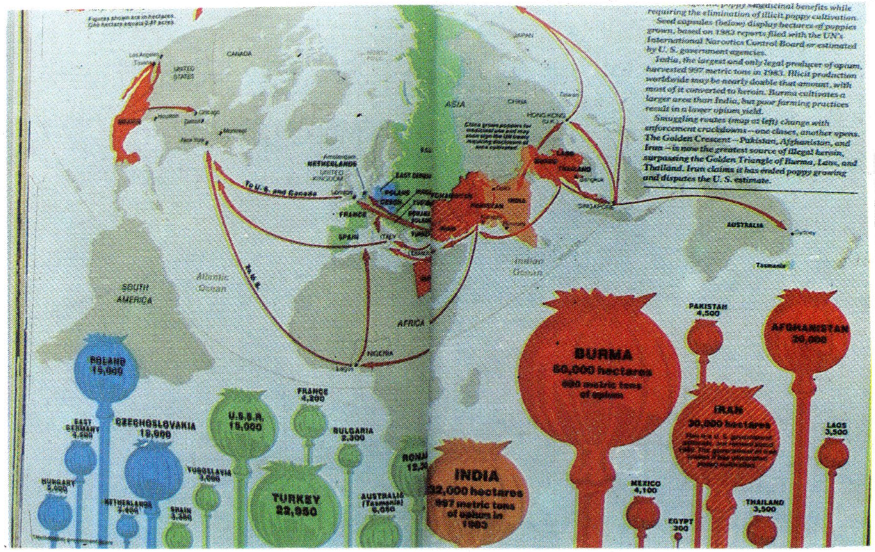

Golden Angle.

First Opium War (1839-1842).

Opium War between two Nations (U.K. and China).

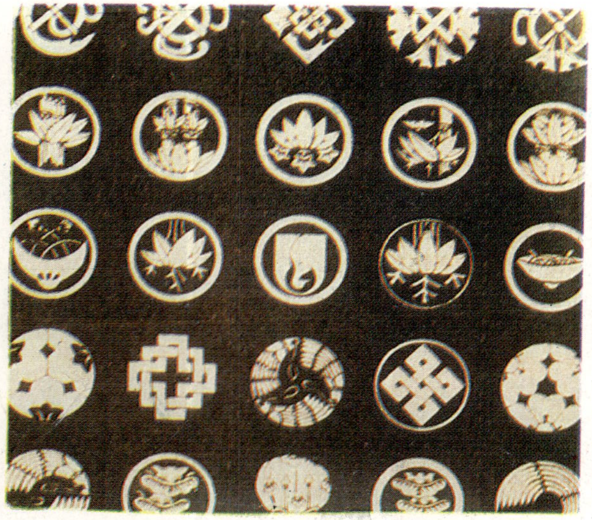

LSD Blotter Paper.

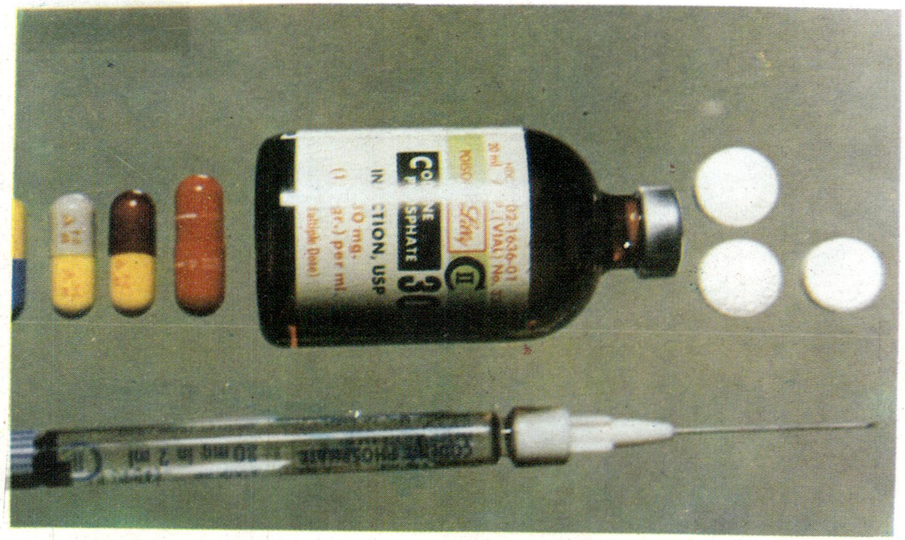

Codeine; as a Liquid for Injection, as a Capsule and as a Tablet.

Illicit Production of Narcotics.

Cocaine Hydrochloride.

Hallucinogens (a substance which produces hallucination).

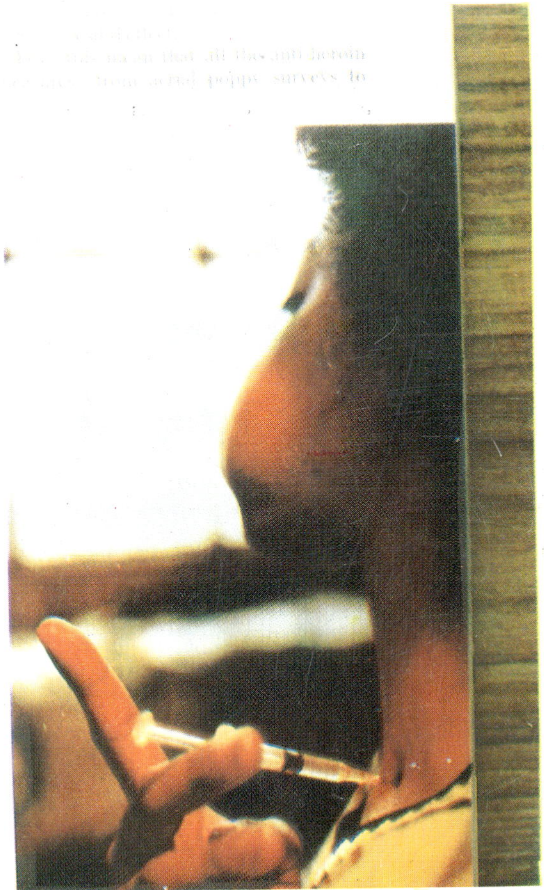

The Needle of Death.

Stimulants (temporarily quickens activity of some parts or organs).

Enjoying his Last Puff.

Smoking of Charas.

Chinese Smokers using Narcotic.

38

THE GOLDEN TRIANGLE AND CRESCENT

Golden Triangle and Golden Crescent

For decades the region known as the Golden Triangle, which includes parts of Thailand, Laos and Myanmar and the Golden Crescent, which cuts across Pakistan, Afghanistan and Iran, has supplied the bulk of the opium for the world's supply of heroin. But in the past, the poppy fields provided only raw materials. Asians left the task of refining and consumption to the drug kingpins of Europe and the United States.

Today Asia has itself triggered off into a big market where thousands of youth spend hours searching for the next fix. Across the Asian continent, drug abuse may no longer be considered only a matter of concern as it is now an epidemic. "It is in every town and every social strata", says Toaha Quereshi, Director of Pakistan Narcotics Control Board, "It is completely out of hand."

For years, the subterranean world of international drug smuggling operated much like a private club. Heroin traffickers protected their interests from the prying hands of outsiders and were even suspicious from each other as well. It was an exclusive group, largely dominated by the notorious, Marseilles based French connection.

Clans of ethnic Chinese spread across the European continent and in the United States primarily—through the Italian-American Mafia. Each faction used its own well-established channels to take raw poppy from the rich fields of the Golden Triangle in South-East Asia and the Golden Crescent in South-West Asia, refine it into heroin and smuggle it into Europe and the United States, where it would be shot into the arms of thousands of addicts.

Tougher anti-drug enforcement and Turkey's 1971 ban on poppy cultivation, shut down the critical French connection and the police have been crushing down traditional smuggling routes ever since. In recent years, Asian poppy growers also realised that they could make more money by refining the opium themselves. Once these growers and other small town middlemen came into possession of the readily transportable drug, it immediately opened the way practically for anyone to smuggle heroin into the West. And that's what exactly happened. As a result even while old time smuggling syndicates continue to make use of traditional routes through Thailand and Hong Kong and the West through Turkey, a new wave of traffickers, dominated by small bands of Asian drug runners, is opening up a labyrinthine system of routes through India, Pakistan, Afghanistan, the Gulf states of the Middle East and even China. In addition to promoting an epidemic of drug use in Asia itself, the new network has complicated the drug control, efforts of Western nations who are unable to cope up with the profusion of traffickers.

The most active of the new smuggling route goes directly through India. Recently, Indian authorities have seized more than 2.5 tons of heroin either leaving or entering the country, more than triple the total for all of 1985. Much of the heroin enters the state of Punjab from the poppy growing areas of northern Pakistan, but drug runners from Myanmar also smuggle shipments through Bangladesh or across the Bay of Bengal. The Indian route began as an emergency.

Interpol sources pointed out that India is sandwiched between the Golden Triangle in the east and the Golden Crescent in the west. The Islamic revolution in Iran and the former Soviet incursion in Afghanistan and their stringent drug laws have deteriorated the drug bases there. They have now shifted to Pakistan and India which have become big transit-centres.

With the increase in heroin trafficking along the Nepal-Myanmr borders. The Opium Act (1878), The Excise Act (Charas) 1914 and the Dangerous Drugs Act (1930), have become obsolete and ineffective. The draft bill based on the Gopalan Committee recommendations incorporates all drug related laws. The comprehensive Bill is to be finalised after report studies of recent trends and pattern of drug abuse in close consultation with all concerned, including, Interpol and intelligence outfits. The problem needs a co-ordinated and integrated approach—a regular system for evaluation and monitoring of drug abuse—specific grounds and areas to be identified by chief attention.

Establishment of de-addiction centres in selected hospitals and psychiatry department on Gopalan Committee lines, has been taken up under the Health Ministry. Publicity campaigns against drug abuse should not arouse undue curiosity but the focus should be on the families and groups of drug addicts rather than singled-out cases. The programmes for educational campaigns should be reviewed from time to time in view of their relevance to the specific requirements of the target groups. The newly established Social Awareness and Information Division under the Ministry of Social and Womens' Welfare will launch publicity campaigns and collaborate with the Ministry of Youth and Sports in formulating both training and education campaigns. The Nehru Yuvak Kendras will take up preventive actions at their centres in Imphal, Andhra Pradesh, Goa, Rajasthan, Cochin and Varanasi was headed. Miss Rama Majumdar, Secretary, Ministry of Social and Womens' Welfare headed high-level inter-ministerial group to tackle the problem. India, over the past few years, has emerged like Afghanistan and Pakistan, as a refiner and not merely a exporter of opium. In fact, the 'Golden Crescent' formed by the three countries is said to be rapidly overtaking the established 'Golden Triangle', comprising Myanmar, Laos and Thailand as a principal source of heroin supplier to the international market.

The Men Behind Trafficking

The drug traffickers have a large group of determined, experienced, and, in many cases, very intelligent people. Drug traffickers are in good position to ship a relatively small amount of easily concealable powder across borders with almost unlimited financial back behind them.

The Iran-Iraq war and the fight in Afghanistan have forced many poppy growers in these countries to abandon these traditional shipping routes and to seek safer outlets. According to US officials most of the heroin that comes via the United States from the Golden Crescent each year is refined and processed in Pakistan.

Search for New Drug Routes

The search for new and safe drug routes has produced some innovative solutions. Perhaps the most unlikely twist in the smuggling game is the so-called 'China connection'. In fact, these days there are more new routes, more new tricks, So, of course, new players. The latest in this dirty game is involved the Israeli organised crime syndicate. In Japan, Yakuza gangs are widely suspected to be involved in both importing drugs and acting as middlemen, in shipments into Hawaii and the US Mainland.

Finding a way to thwart the new breed of smugglers promises to be a Herculean task. But Malaysian government has shown the path by adopting an extremely hard line, which includes imposition of death penalty. One can earn much money in less time in this business, therefore, everyone is easily attracted towards this lucrative business which provides thrills and money both.

Golden Triangle

The Golden Triangle of South-East Asia, from which the flow of drugs, the United States has been trying to control for 25 years, is once again the single largest source of heroin coming into America. In 1987, only 18 per cent of the heroin in the US originated in South-East Asia; today the figure is at least 45 per cent. Six years ago, 5 per cent of the heroin in New York city was Southeast Asian; today, according to the US Drug Enforcement Administration, it is 80 per cent. This increase is a result of many factors. According to western diplomats and law enforcement officials, unrest in Myanmar, decision by some drug smugglers to bypass Latin America, which has got increasing attention from governmental authorities, and to seek safer routes are some of these factors.

Thailand at the Centre of Golden Triangle

The changing pattern of heroin supply has brought new and unwanted attention to Thailand as the main smuggling and shipment centre for heroin from the Golden Triangle. Although Thailand is praised by western diplomats for having reduced the amount of opium-poppy cultivation on its territory sharply, the rapidly developing communications and transport systems of its booming economy, are being used for shipping to Europe and America, increasing amounts of the heroin produced in Laos and Myanmar. The growth in Thai exports and tourism provides more options for smuggling, ranging from individual couriers to large shipments hidden deep in legitimate exports moving through containerised ports.

The amount of money at stake, makes the problem of corruption among ill-paid Thai police, army and custom officials, a growing concern even for a well-intentioned Thai Government that co-operates with Western law enforcement authorities. Thai sensitivity to public suggestions of corruption and complicity is very high, but the tone of even private discussions about drugs between US and Thai officials is becoming more "forthright" and "even, at times, angry." According to a western diplomat, "We are trying to get across to the Thais that drugs are an international problem and that Thailand is a target too despite extensive cooperation between the two governments."

The Drug Boom in Myanmar and Laos

The major increase in opium cultivation is taking place in Myanmar and Laos, two of the world's most isolated countries. The Laotian Government, which the US has in the past, accused of involvement in drug production, agreed only in January to start a crop substitution programme, with an initial US commitment of some $ 200,000 in aid. A $ 6 million to $ 8 million-a-year programme to aid Myanmar's military Government in crop eradication, was suspended along with all other US aid and loans in September 1988, after a bloody crackdown on pro-democracy demonstrators there. But some members of the US Congress argue that with opium cultivation in Myanmar "out of control", the aid should now be renewed.

Based on 1988-reports western experts said that "depending on the growing season which runs through at least 2,600 tonnes of raw opium will be produced in the Golden Triangle as much as 3000 tonnes, p.a. Even the lower figure represents an increase of 44 per cent over two years ago, much of it coming from Myanmar. Nearly 2,200 tonnes will be produced in Myanmar, compared with 2,000 tonnes last year and about 1,600 in 1988. Thailand will produce about 50 tonnes, about the same as in the last two years, but only a third of that produced in 1970. The estimated figure for Laos is between 300 and 400 tonnes up from 200 to 300 tonnes last year and 100-200 tonnes in 1988.

With the increase in heroin production, Thailand's role as regional entrepot is drawing more criticism. A member of a delegation of US Congressmen said in Bangkok recently that Congress was "concerned about the explosion of opium and heroin coming through Thailand. We are losing the war on heroin, and we are losing this war in South-East Asia.

Another commentator said that like Columbia, Thailand risked becoming "stigmatised" internationally as a major drug centre and a focus for well-armed and well-financed drug syndicates that might threaten the stability of the state.

According to the US Government's "International Narcotics and Controlled Substances Review," heroin production and refining in Thailand has been increasing, unmatched by "corresponding interdiction success" by the Thais; "corruption and lack of conspiracy law hinder effective enforcement."[1]

Drug Menace in East

Malaysia with its per capita income of about 1800 US dollars and Singapore with about 3,500 US dollars are providing a good market for the peddlers of heroin and other drugs. It has been estimated that while Singapore addict population increased by 40% what is worse, about 65% of drug takers fell within the 14-29 age group. The most common drug is heroin which has led to broken homes, parental neglect, sheer hedonism in a consumer society and no worthwhile objective in life.

The proximity of Malaysia and Singapore to Golden Triangle ...that is the world's largest poppy growing area has exposed it to frequent inroads of narcotic mafia operations. Thailand is full of mafias specialising in this trade. Drug traffickers are cold and calculated businessmen who are not dissuaded even by massive hauls of narcotics.

Relationship between Thailand and USA. is very closed which is the final destination of a good part of these drugs. Therefore, partly as a result of US pressure, the Thai government is quite keen to stamp out this evil. The Prime Minister, Prem Tinsu Larveda, told his cabinet that "the government was determined to vigorously clamp down on narcotic trafficking but the difficulties are stupendous—livelihood for thousands who live in the 'Golden Triangle', corrupt enforcement agencies and politicians."

In Singapore, the mandatory punishment is death and in Malaysia the courts have option between life imprisonment and death. The courts in Malaysia are ruthless in awarding death sentences to drug traffickers.

The poppy fields of the Far East are situated in "jungles darker than night" in the border region of Myanmar, Thailand and Cambodia. These regions are so inaccessible that national borders through those hills are purely national. It is impossible to build military fronts in these forests. Of these countries in the Golden Triangle, Thailand has, for decades, been the primary concern of International enforcement agencies. The closest markets are Australia and New Zealand, and the immediate outlets are Philippines, Hong Kong, Singapore and Malaysia.

The Drug Ship from Myanmar (Burma)

It is the silky route at Paona Bazar where young men loiter around or stand in groups as if waiting for something to happen. Suddenly, a man appears on the scene and is surrounded. "Has the ship come?", the youths ask. "Not yet, but you can soon expect it, very soon," says the man.

The real thing is heroin for which the market seems to be growing enormously with youth in the metropolitan cities waiting anxiously for the drug to be transported from Manipur. This magic powder comes to India essentially from Burma through the small Manipuri border hamlet of Moren, 100 kms from Imphal.

KMT Army in Golden Triangle

In remote villages of the Golden Triangle—the meeting point of Myanmar, Thailand and Laos—scattered remnants of the late Chiang Kai-Shek's anti-communist Chinese army still plans the long march to Beijing. At Kuomintang (KMT), headquarters at La-ton just inside the Laos border, the first question directed at any Westerner arriving there is: "Are you American; Are you CIA agent ?" If your answer is "No", they are heartbroken. "Are you sure, you are not a secret agent?" they ask, eager for a cheque from the Central Intelligence Agency.

The Chinese National Army flag flutters above the shack and Chiang Kai-Shek beams down from faded posters adorned with dried orchids. "We are still fighting for our home-land", said KMT commander Cheng Tsu, "It has been a very long fight". Cheng Tsu has been "at war with communism" since 1949. His troops, fled Mao-ze Dong's revolution and took refuge in Myanmar. Now, in this mountainous no-man's land, they try to keep alive "the true spirit of China". "The CIA has abandoned us, said Chiang Chi, another KMT officer", "We protected their interests, grew their heroin and fought against China. Why did they go away? Now we have many problems."

Chinese Dependence on Opium

Foremost of these problems is Chinese dependence on opium. In the villages around La-ton addiction is chronic, running at about 60-70 per cent among adults. Many children also use opium as a hunger suppressant. Some are only seven years old. Said one Chinese: "At Kui village, Khun Sa's men burned the houses to the ground. They took the rice and the opium for the next season. They raped women and captured men to work in their fields."

In the Myanmar districts of Putao and Myitkha, Chinese work as slaves to powerful Kachin and Shan ethnic group tribes waging war against Burma's government in Rangoon. Kept by force, the slaves refine and pack heroin on great opium plantations. Conditions on the plantations are described as brutal by one farmer who escaped from there : "Work never stops and for each worker the only pay is a weekly packet of heroin and a small amount of food. You are not allowed to have children and if you try to leave, the guards will shoot you."

Among Thailand's ethnic Lahu, Shan and Karen, peoples' resentment against Chinese refugees is growing and anti-KMT programmes are common.

Drug Smuggling in N-E States

The current political turmoil in Myanmar has already begun to have an impact on the bordering North-Eastern states amidst reports that various insurgent groups operating from across the border are re-grouping and that there has been a sharp increase in the smuggling of narcotics and drugs, according to well-informed sources.

The almost total withdrawal of the Myanmar Army, from near the border with India and its induction into the power struggle in Myanmar, has provided a breather to the insurgents and smugglers who are reported to be having a field day. Sources have warned that the delay in solution of the political crisis, would have serious repercussions in the bordering states while any outcome of the crisis would be important from the point of view of India's security. Informed sources have also reported that an attempt is being made to set up a "joint front" of the various insurgent organisations operating from the Burmese territory. These include the NSCN, the United National Liberation Front (UNLF), the Peoples' Revolutionary Party of Kangleipak (Prepak), the United Liberation Front of Assam (ULFA) and the People's Liberation Army (PLA). These organisations, the reports add, have decided to co-operate with each other and coordinate their "operation".

Besides the insurgents, the drug peddlers and smugglers are believed to be having a free-hand and have stepped up their activities. It is a well-known fact that the smuggling of narcotics from the "Golden Triangle" is mainly carried out through the border with Manipur and Mizoram. The addiction to drugs in the two North-Eastern states has already reached a serious level. The poor record of the Customs does not reflect the magnitude of the problems during the last couple of years. The requisition of the Myanmar army from the border has reportedly eased the pressure on the smugglers. On the other hand, no special steps have been taken by India to check the illicit trade."

Some of the assorted insurgent groups, including those pitted against the Burmese government, survive on the smuggling of narcotics and jade which abounds in the northern parts of Myanmar. At the time of weak government in Burma, these groups are expected to increase their illicit activities.

The Drug Trail

Fighting between Opium King Khun Sa and the Myanmar Communist Party insurgents for the control of the opium crop on the Thai-Myanmar border continues. Several hundred Myanmarese were thrown out of Thailand. Some 800 men of the "Crack zone 6 special force" on 40 trucks, 6 armoured vehicles and 30 motorbikes were sent under Col Thong Aun Ban Hun Tach and encircled the village located close to the Frontier. The Colonel grabbed the loudspeaker and called upon Khun Sa to surrender but Khun Sa opened fire with his Shan force. Colonel's troops swung into action and with the help of reinforcements from the "Zone 5 BPP Force" and armoured helicopters, the opium king's head quarter was ultimately overrun, but Khun Sa made a safe escape through an underground passage into the jungle. Since then Khun Sa and his 4,000 men have determined to stay in hiding.

The opium war has produced about 400 victims. Ban Hun Tach, meaning the "place of the Broken Stone", has been renamed Ban Thod Thai, as "the Place of Upheld Freedom", nothing changed. The Thais need American weapons to counter "Vietnamese Imperialists" and to keep in check the rebellious Meo tribes in the North and Red guerillas in the South, who are supported by Thailand's communist neighbours. As a price for American weapons, Bangkok has been asked to dry up the swamps of the "Golden Triangle" (on the Upper Streams of the Mehong) which is flooding the US drug market with heroin. But the Thai government Chief Gen Prem Tinsulanand also needed opium king Khun Sa and his military forces—splintered remnants of Chiang-Kai shek's defeated Kuomin Tang army and its offspring, who control the opium trade in the "Golden Triangle".

Khun Sa, whose influence extended to the largest parts of North-East Myanmar, want to enforce the creation of an independent Shan country on Myanmar territory. It puts pressure on Rangoon and eases the pressure on Thailand, which feals threatened by Myanmarese expansionism. Simultaneously, Khun Sa's authority keeps in their place the militants Meo and Karen population in the North, which too are a great headache for Bangkok: the Thai believe that it is better to see Khun Sa controlled, who are strenuously trying to seize control of the drug trade. To maintain order in this area, Khun Sa needed the proceeds from the opium market. Even the local daily, Bangkok Post, not known as a sharp critic of the regime, admitted : "The Government found it always useful to cultivate his (Khun Sa's) friendship."

The largest portion of the harvest in Thailand is destined for domestic consumption. According to UN drug experts, there are 500,000 opium addicts in Thailand. In seizing power in 1976, the Bangkok Junta promised to do away with three basic evils of the western civilisation, "drugs, prostitution and communism", but the achievements of these objectives to put it mildly is hopelessly falling behind schedule. Even if Gen. Prem tried his best to get at the root of drug menace he could not achieve the desired success. The opium kitchen in neighbouring Laos produces at least as much as the Thai one, in Myanmar 4 to 5 times as much; and of the Chinese province of Yunnan, whose produce reaches the world market via Hong Kong or Macao, even more.

The US Role

How the Yankees solve the problem of their junkies causes no sleepless nights to the authorities in Peking, Rangoon or Vietnam. US Ambassador John Gunther Dean praised the operation against Khun Sa as a "forceful and brave act". Dean knows that, during the Vietnamese war, the CIA's own Air American planes kept flying with heroin to Saigon and Vientiane, since the mountain tribes in the producing regions supported the American cause but the Pathet Lao insurgents had endangered the trade routes for big dealers. The 700 tonnes of opium which, according to the estimates of the US "Drug Enforcement Administration (DEA)", is a record crop. The refining process is now quietly under way. A DEA agent says in Tshiang Mai, "heroin price will go down in downtown Manhattan".

Posed Problems in Buah Tshan

In the Thai province of Tshiang Rai, the prices of unrefined opium have temporarily risen. Lao Isu, the village elder from Buah Tshan in the Meo country, complains bitterly that the opium war has disorganised things all round. Never have so many chickens been stolen in Buah Tshan. The native opium smokers are stealing the neighbour's fowl to meet the higher cost of opium. Opium smoking is not considered a social problem in Buah Shan, just like beer drinking is adopted in Western countries. The place reeks with opium and burned dog meat. Men lie in huts or in the streets among wandering pigs and chicken. Gazing absent-mindedly in the distance, they keep pulling at the arm thick bamboo ducts of their porcelin pipes—silent and hollow cheeked.They smoke 20 to 30 pipes a day.

A few families in the village have joined DEA sponsored crop replacement programme, that replaces poppy with useful plants, such as strawberries, salad, cauliflower or coffee. The U.S. programme is lavishly backed with subsidies. The King Bhumipol, himself tilled the land to give a lead to his people.

Khun Sa's Offer

Khun Sa made his own contribution to the problem of drug flow from South-East Asia about two years ago. He offered to Washington, the entire harvest of the "Golden Triangle" for 6 million dollars. But the US Embassy commented that it would not get involved in bargaining with dope peddlers. In fact, the Carter Administration turned the offer down in full knowledge of the drug market mechanisms—neutralising some 300 to 400 tons would have driven prices of opium up, created a powerful rise in demand and, ultimately, an increase in the planting area—thus producing more opium than before. The Americans knew more than they admit. Richard Nixon once concluded such deal and sparked an unprecedented debate.

Early in February 1982, Khun Sa renewed his offer. Three American tourists, who had been taken prisoner, were set free by Khun Sa and sent to Washington with a letter to President Reagan in which the Opium King had offered his services as an independence fighter and suggested to help America fight against the drug problem. White House did not give any reaction to this offer.

Thailand being one of the major world tourist centres, there are number of ways in which smugglers are attempting to make a fortune by taking drugs out of the country. Once, the police stopped a tourist who swallowed 59 preservatives filled with 'Snow' or 'heroin powder'. He was found to take laxatives to dislodge from the bowl the contraband drugs while police watched on in an airport rest room. The Frontier office in South Thailand arrested several Malayese who tried to smuggle heroin stuffed into the still warm body of a dead baby. However, the hunt fails to touch the "distinguished smugglers" using often, their diplomatic status, and the high officials and politicians like those of the formerly ex-Premier Kittikashosn who protected them, plus the so-called Triods—group of big dealers from Hong Kong, Singapore and Amsterdam. All these circles have woven a tight net of powerful political protection throughout South-East Asia.

Thai soldiers had thrown the dead man's body before the door of Kamdaeny Duangdi because he was the only one in the village who knew the whereabouts of opium King Khun Sa. The fact that the corpse had been marked with huge cuts was according to Pantshang, an old Thai military habit, brought back to life in Opium War.

In fact, the Thais have little reason to hate the Shans. The opium war waged in the "Golden Triangle" enveloping Thailand, Laos and Myanmar was not conceived originally as a war but as a public relations action stimulating a war, launched by Thailand to appease the US protectors. But things, for unexplained reasons, went wrong, producing wide-spread killings at Ban Hin Tach.

Golden Crescent

In Turkey poppy is known as 'has-has', pronounced hash-hash, and opium is 'afyon', plentiful here since antiquity. The Turkish farmers dry the stem, the capsules so harvested from a Turkish field are known as poppy straw and can yield no opium, only poppy straw concentrate (FSC). Turkey banned the poppy growing in 1971 under US pressure. In order to satisfy the unhappy farmers, it began licensing growers in 1974, but only to produce poppy straw for sale to a government factory where technology is required to convert it to PSC and then Morphine and Codeine. Before the dried capsules are pressed into PSC, the seed is shaken out and then pressed for cooking oil and the remaining high residue cake is fed to the cows. Some poppies produce white seeds, a favourite in Scandinavia, but blue poppy seeds bring the highest price on the international spice market. Turkey, Netherland, and the Australian state of Tasmania are the top suppliers of seed.

However, Turkey is still a conduit for heroin traffic. Officials cannot inspect the thousands of vehicles that each month cross the Bosporus bridge linking Asia and Europe.

Until the early 1970s Turkish black-market opium was refined in Marseilles for heroin of the infamous French connection. But now, "No afyon, not even a gram", says the Director of TMO, the Government Soil Products Organisation which buys all agricultural products—wheat, barley, rye and dry poppy capsules. When capsules ripen without incisions, morphine stays in the capsule walls. It can be efficiently extracted only in a sophisticated factory. Turkey has built one at enormous expense, "We have done all this for humanitarian reasons, for our friends in the Western world."

At the huge new alkaloid factory at Bolvadia in Afyon Province on the eastern Autolian plateau, dry poppy capsules—called poppy straw—are crushed and dissolved in stainless steel tanks. The outcome is a darkish liquid. More husks, more acids; outcome net crystals for drying in a centrifuge, then in ovens. Now it is ivory coloured powder. More dissolving, filtering, centrifuging, and then it is white poppy straw concentrate, or PSC, crude morphine. Armoured cars will take it under heavy guard to Izmir for export. Most of it is processed into Codeine.

A worldwide notion for a long time was that heroin is mainly an American problem. But that has been changing. Addiction is rising not only in Thailand and Myanmar, Malaysia and Pakistan, but also in Western Europe; and Turkey is on a transit route. Kintey, the official Bulgarian export-import agency, has been blamed by US authorities for trading weapons to assorted militants in the Middle East for heroin to be sent on to Western Europe and the United States.

Iran's War on Drugs

In the Islamic Republic of Iran, the government is engaged in fighting against opium and heroin but smugglers from Afghanistan and Pakistan bring the stuff over mountains at night, (so the gendarme helicopters cannot detect them), and bury it in the desert until a deal is made. They come along with jeeps or with camels fed on opium and conditioned to follow a special route. Thus, they keep on running by themselves. The smugglers have binoculars and are heavily armed.

There are estimated 150,000 heroin addicts in the country and 450,000 opium addicts. In fact, Iran has a centuries long history of opium—until the fall of the Shah, opium smoking after dinner was acceptable in circles high and low; registered addicts could get legally produced opium cheap. The revolution made opium unlawful, sinful and forbidden. According to Khomeini the spread of narcotic drugs is a plot, a conspiracy of the East and West to addict Iranian youth so they become physically unfit and useless for the economy. Therefore, a massive programme for curing the addicts and educating them has been taken up behind the steel doors at Teheran's Bahar Addict Centre. Inmates from 13 to 60 stay for two months—no medicine, plenty of pep-talk, "If you could topple a regime supported by the super powers, you can also defeat your habit, with determination."

In Iran there are Revolutionary Courts to deal with Narcotic crimes. The judicial procedure is simple: if they've done it before, and the evidence is clear, the High Judicial Court approves, their execution.

Shah's Brother a Jail Bird

Yamid Reza Pahlavi, younger brother of the late Shah of Iran, grinned a toothless smile in the cell he shares with 23 other convicts in the notorious Evin prison and said, "This could be worse". Speaking to a group of foreign reporters on a government-guided tour of the prison in north Teheran, Pahlavi said, "He was serving a 10-year sentence for a family reason", but declined to be more specific. He said he had seven more years to serve. Prison officials said the prisoner had been disowned as a drug addict by Shah Mohammed Reza Pahlavi before the February 1979 Islamic Revolution that toppled the monarchy. The officials said Yamid Reza Pahlavi was arrested as a vagrant and imprisoned three years ago.

Drug Traffickers Hanged

The Iranian authorities hanged a total of 56 drug traffickers throughout the country in one day, the official IRNA news agency said in a despatch. The executions underlined the authorities,"firm determination" to stamp out drug trafficking. It said that the hangings took place in Teheran and 16 other towns. A total of 56 kg. of illicit substances had been found with the people executed.

Those executed included four foreigners: three Afghans and a Pakistani. Under a new drug law, traffickers found in possession of more than five kgs. of opium or hashish, or 30 gms of heroin or morphine, risk death sentence. The law also gives the country's drug addicts a grace period of six months, in which to kick their habit, after which time they risk prosecution.[2]

Pakistan at the Centre of Drug Peddling in Golden Crescent

Pakistan was emerging as a major opium producer and its drug producing area was hardly known to Delhi, until the DEA of United States sounded India in May 1984. Golden Crescent was bringing 40 pounds of heroin a year. In Delhi few were aware that Pakistan was fast emerging on the world's heroin map in the last quarter of 70s. It began with President Ayub Khan's reluctance to import opium from India. Opium trade had begun to flourish in NWFP. The growers had initially begged Ayub for just 3,600 acres of land and an opium processing factory at Lahore. Once granted, this confederation of 30 families turned into Pakistan's heroin mafia. In 1974, when Zulfikar Ali Bhutto fell out with Ayub, he returned to Sind, his home town, to organise Pakistan People's Party. He was in hard need of money. In search of easy money Bhutto fell back on the trade that once flourished under Ayub in NWFP.

America was watching the events in Pakistan very closely. The Drug Enforcement Administration (DEA) of the USA had been repeatedly warning the Presidency of the emerging Golden Crescent—the confederation of Muslim countries like Iran, Afghanistan and Pakistan—as a rising threat to the traditional heroin producers in South-East Asia—the Golden Triangle. In 1979, Pakistan emerged as the world's largest exporter of heroin; just a decade earlier Pakistan did not even produce its own opium and had to depend upon India.

In 1979, the poppy kingdom of the NWFP had expanded to 80,000 acres. Pakistan, that year, was bragging for a record scoop of 800 metric tonnes of opium and it was being sold in 33 government licensed shops, for which there were 300,000 registered addicts. Most of it was being turned into heroin in clandestine laboratories of Landi Kotal, Khyber which was fast replacing Bar Han Tiek, the heroin fortress of the drug baron Khun Sa, set along the Thailand border of the Golden Triangle.

Pakistan has become one of the world's biggest suppliers of "White death" from the Golden Crescent which consists of border areas of Pakistan, Iran and Afghanistan. More than 1,000 tons of opium is smuggled out to world markets. Now India has become a major conduit for this smuggling from Karachi to Amritsar in Punjab or the towns of Rajasthan that is the usual drug route used by smugglers. From here the foreigners fly the billion dollar cargo out of India. A number of Pakistanis are held up in India on charges of smuggling drugs. The Narcotic branch of the Central Bureau of Investigation seized 20 kgs. heroin, worth twenty million dollars. In 1984, Pakistan authorities too seized heroin, opium and hashish worth $ 510 million on January 16 1984.

The Pakistan International Airlines PIA has been accused of smuggling heroin to other countries. Very recently, a haul worth £ 125,000 of heroin hidden in a PIA plane was found at London Airport.

Pakistanis bring 5 to 20 kgs. of brown sugar, a low grade heroin, into India. They are of two types: No. 3 which is very inferior and No. 4 a high pure drug. In Baluchistan and Pakistan the tribes have been producing No. 3 quality drug for quite some time. Landi Kotal in the Khyber Pass has been turned into Pakistan's most notorious drug market. One can smuggle it in Bombay through Indo-Pak border so it is easily available at Dongri, Shuklaji and Colaba.

There does not seem to be any immediate solution, unless and until India and Pak governments take strict measures and severe action. A strict vigil should be kept on the Indo-Pak border, particularly at the NWFP from where the drug enters India, particularly into Punjab.

The fact is that the laws are lax in Pakistan as in

India and the courts are also very lenient. This has helped the growth of illicit cultivation of poppy and hemp, unlicensed processing into opium, morphine, heroin and ultimately illegal peddling. Interpol reports have shown that the main drugs consumed by drug users have been processed by traffickers, in laboratories located where drugs are grown or in countries particularly well situated from the geographical point of view. Interpol reports also say that it needs experienced specialists to detect laboratories.

Intelligence sources say that one of the biggest dealers in opium and heroin in Pakistan is a man named Nek Khan Pathan, of Barg, a small town five miles from Peshawar. Khan's role in large scale drug running came into open when a consignment of opium worth 1.5 crore was intercepted by the Pakistan Police.

Smuggling on Indo-Pak Border

The border is long and it is difficult to find exactly where it commences. In many places there are fields on either side and only a small dirt track separates the two countries. Says a BSF official, "The people on either side look the same, speak the same language and wear the same clothes, so it becomes difficult to distinguish between them and the smugglers." Besides this the opium networks are well organised. Smugglers usually travel in groups of 10 to 15 and the main party is led by escorts who are well armed. The consignments are brought over in polythene bags by carriers. Animals are seldom used as they are easily spotted.

Chitral Hills area of Pakistan, which forms North West Frontier Province with Afghanistan, is the main source of opium and heroin. The poppy fields of the Golden Triangle and cocaine farms of South America have some common characteristics, such as inaccessibility due to dense jungles in hilly regions. Moreover, the tribal area of Chitral is semi-autonomous where local administration is run by fierce independent tribals and it is difficult for government agencies to pierce through. Moreover, the Afghan Mujahideen also found shelter there and took to drug peddling for arms and ammunitions.

On the western front, Barmer-Jaisalmer district has become an increasing threat to India because of heroin smuggling from Pakistan. The work here goes on as planned by the international mafia-men. Men and women are equally adept in this trade. In these two districts, Rivadi, Negdada, Bambara, Bali ki Basti, Antiya, Bijota, Deoda villages have become focal points. Economically backward area has been flooded with Jonga jeeps now and surprisingly all of them are registered in Gujarat. Heroin is stealthily brought on camels to such spots in the desert, as the jeeps can easily reach and pick up the smuggled material. Each camel, for the purpose, is hired on Rs. 40,000 in the smuggling market. The gang, working on the border, has links with the smugglers of Gujarat and Bombay. According to a survey on the Sri Ganganagar-Barmer border belt, almost 5 to 6 thousands have made their homes who are citizens of both the countries. They hold bank accounts and ration-cards in both the nations and move freely across. Pakistani currency is freely available in the market. In these border districts, some villages which till recently were desolate, are now thickly populated and mainly Pak citizens have occupied them. Such persons also travel in India to invest money in drug-trade.

Effective patrolling and seizures have not been possible for reasons of corruption among the low rank officials. Another important factor is that Sikhs and Muslims have come up against "Vishnoi" dominance in the field. The new community seeks shelter easily and finds substantial patronage from the border villages.

India—The Growing Menace

The Assistant General Secretary of United Nations Fund for Drug Abuse Control (UNFDAC), Guaspe-de-Gennaro, expressed his serious concern at the growing trend of trafficking narcotics through India as a transit centre to other countries of the world. He quoted statistics to substantiate the data of seizure—in 1983—138 kgs, 1984—202 kgs, 1985—760 kgs, and the first eight months of 1986—2,500 kgs of heroin were channelled through India. He revealed that 95 per cent of the total seized heroin enters the Western countries via India from Afghanistan and Pakistan. He felt that narcotic-trade spreads rapidly in the disturbed areas as revealed by analysis of last ten years of Latin America and Asia. It is felt that international criminal organisations are more powerful and faster than local criminals, therefore, they find India as their paradise, in establishing and deep-rooting themselves on the wings of terrorism. He revealed that such organisations work like big companies with sophisticated and up-to-date facilities, employing highly efficient, educated persons. One of the main criminal organisations in Florida owned 3,000 aircrafts and they owned many banks and ships too. He advocated a social set-up, mentally prepared to fight the menace of narcotics effectively, otherwise severe punishments like the capital punishment, cannot deter them. If society stands against terrorism, smuggling is automatically rooted out.

Mr. Gennaro felt that Indian Panchayat system could be an effective method of mobilising public opinion against this fast spreading menace. Coordinated efforts at international level were necessary to check this. The seizures recorded an increase of 300 per cent and were made by Indian enforcement agencies without any international assistance. India is the only country growing opium for medicinal purposes but other countries did it for export. India has been compelled to reduce the poppy-growing area from 63,000 hectares to 33,000 hectares due to poor demands, thanks to exports falling in drug importing nations.

Smuggling of Narcotic Drugs and Psychotropic substances at Indo-Pakistan border via desert route is also very common. The camel rider chooses the route some where from Jaisalmer Rajasthan to collect and handover the goods of either side. They used to dump the concerning packets and bags at that point where motor vehicle can come. A few hours later trucks roll out of Jaisalmar, Rajasthan to retrieve the cargo—hundreds of packets—kilo after kilo—of Golden crystal heroin, from the traffickers border strongholds. The deadly powder is distributed to India's teeming cities. There are an estimated 1,10,000 heroin addicts in New Delhi alone; in India, as a whole, perhaps as many as half a million. Neighbouring Pakistan has upto 450,000. Hong Kong 40,000. In most major cities in Asia, thousands of people spend their crores searching for their next fix.

Turkish-French Connections

For many years opium poppies grown in the desolate, windswept Anatolian plateau of Central and Western Turkey were the source of most of the heroin smuggled into the United States. Each year, Turkish peasants held back a portion of their opium crop from the official government marketing system to sell at a far better price to illicit drug traffickers. The traffickers, in turn, smuggled it across the border into Syria, where it was converted into less bulky and more easily concealed morphine base. The base was then smuggled either by sea or across the border into Europe to one of the many heroin laboratories located in Southern France. For many years drug traffickers used the sea route. The base could be easily hidden in the vessels that regularly sail through the waters between such ports as Istanbul or Beirut and Marseilles. Later, many of the traffickers switched to the overland route from the Middle East to Europe. One of their favourite tactics was to conceal morphine base in the thousands of tin trucks that carry meat and other produce from Turkey to neighbouring countries, such as Lebanon, Syria and Europe.

Entry into USA

A variety of routes were used to smuggle narcotics into the United States. Some was sent directly from France. One favourite *modus operandi* was to conceal it in automobiles that were being shipped. Other routes led from Europe to Canada and ultimately to the United States. Still other traffickers preferred to route the drug via Mexico or other Latin American countries such as Brazil and Panama.

France-Turkish Narcotics Combine Smashed

During the early 1970s a series of events combined to effectively smash the Turkish-French nexus. Intense international enforcement efforts resulted in immobilising a large number of important French and Corsican drug traffickers. Key drug figures throughout Europe, Latin America and in this country were successfully put out of business. At the same time, intense diplomatic pressure was brought to compel the Turkish government to ban the opium poppy cultivation. Turkey finally banned it in 1972.

The Arab States on the Gulf Route

The Arabian Peninsula consists of Bahrain, Kuwait, Oman, the People's Democratic Republic of Yemen, Qatar, Saudi Arabia, the United Arab Emirates and the Yemen Arab Republic. The Arabian Peninsula has the gulf on the North East and the Red Sea on the West, holding two vital gateways from Asia to Europe and the Mediterranean region. With oil resources in most of the countries stimulating an expanding economy, the growing international trade, and the induction of modern technology in fields of communication, power and construction, the region has emerged as an important landmark on the global scene.

According to the International Air Traffic Association nearly 25 international airlines transit through most of the Gulf airports almost daily, carrying heavy passenger traffic and air freight East and West bound. With growing international trade and commerce in the region, the movement of bulky export and import cargo, has led to the expansion and modernization of ports. Finally, rail and road links within the region have expanded over the past six or seven years.

Briefly, these recent developments in the region have brought the Arab sector much closer to the rest of the world, which in turn has involved the Arab States along the drug trafficking route, hence increasing their drug abuse problems.

References

1. Reproduced from *New York Times*.
2. *The Pioneer*, Lucknow, January 18, 1980, p. 12.

39

ASIA'S DRUG EPIDEMIC

The plague of drug addiction has taken a toll in Asia that goes far beyond our estimates. In Thailand, drug related offences now account for 57 per cent of all crimes. Everywhere the scourge of addiction flourishes under umbrella provided by bureaucratic corruption. Asia's war on narcotics is equally tough. The main area of poppy production regions along the Thai and Myanmar borders, the desolate bad lands of Pakistan's Northwest Frontier Province, Afghanistan's highland plains—are effectively outside the control of any central government. When authorities do manage to destroy a field or shut down a route, the gangs simply move their operations elsewhere.

Indian Drug Scene

In 1986, India reported 31,714 registered opium dependents, mainly the elderly working classes, low income labourers, students and the upper income youths. A rise in heroin abuse has been noted in New Delhi, Bombay, Calcutta and Madras. Unconfirmed reports indicate Bombay has over 100,000 heroin dependents.

Asia : Largest Drug Addict Concentration

There has been an alarming increase in the number of drug users in Asia with India alone accounting for about five million addicts, according to International Labour Organisation (ILO) estimates. From a continent regarded in the early 1970s as being relatively free of the drug menace, Asia today has the largest concentration of drug addicts, especially in India, Pakistan, Sri Lanka, Nepal, Indonesia, Malaysia and Thailand.

The ILO has urged Asian countries to adopt fresh measures to check the spread of drug abuse and is expanding its programme of rehabilitation and reintegration of addicts into society. In India the number of addicts has been roughly estimated to be between three and five million, a majority of them being teenagers and under 30 adults.

With only a few hundred drug addicts six years ago Pakistan has now about 1.3 million people regularly using drugs like cannabis or opium. Heroin addiction has also become a serious problem in Pakistan, an ILO release said. Stringent punishment against drug abusers has not solved the problem either. In Malaysia, despite the mandatory death penalty for drug offence, there were an estimated 120,000 addicts.

Sri Lanka which had virtually no heroin addicts in 1980 has an estimated 30,000 addicts today. According to the ILO' Bangladesh has about 2.5 million cannabis users and another 10,000 heroin addicts.

Nepal with some 20,000 drug abusers, Indonesia with 130,000 and Thailand with 500,000 drug addicts are all faced with a growing problem.

Much of Asia's drug problem stems from the Golden Triangle and Golden Crescent which between them supply a significant share of the world's illicit narcotic drugs. The Golden Triangle, covering an area of 75,000 square miles spread over Laos, Thailand and Myanmar, produces between 500 and 800 tonnes of opium each year. The Golden Crescent covers the border regions of Pakistan, Iran and Afghanistan and most of its 600 tonnes of opium find their way into the illicit market of West Europe and the USA.

India and Australia are the only countries of the Asia-Pacific region which produce opium legally and between them they provide about 60 per cent of the world's medicinal needs of the drug.

Realising the threat posed by the spread of drug addiction, most Asian countries have taken stringent measures to curb the menace. The ILO's efforts have concentrated on the rehabilitation and reintegration of addicts into society and helping them to stay off drugs through training, transitional work, family counselling and income-generating activities. According to ILO inter-regional drug adviser, Mr. Behrouz Shahandeh, cases of relapse, with rehabilitated addicts going back to the drug habit should not necessarily be seen as a failure of the programme "but as a manageable step towards a drug-free life style."

The global illegal trade is today an enormously profitable racket with an estimated turnover of $ 300 billion and according to the ILO it is this enormous profit which makes it worth the risk for the drug barons to violate the most stringent anti-drug smuggling laws.

The elimination of drug abuse cannot be achieved overnight and the abundant availability of drugs is a major reason why many nations are shifting the emphasis to demand reduction initiatives such as prevention, treatment and rehabilitation. The ILO as part of its programme of rehabilitating drug addicts has focussed on the escalation of drug and alcohol abuse problem at the workplace, according to Mr. Shahandeh.[1]

Golden Crescent at the Hub of International Trafficking

Lundy Kotal in the Khyber Pass witnessed a spate of violence following the arrest of the gang-leader Sheikh-Zumer who had a dispute with other leaders of the 'Kabila' on the prevention of drug in the Afghan-border land area. The place is notoriously known for its sale of toxic liquids in Pakistan.

Pakistan, it is believed, supplies 90 per cent of heroin used in America. The custom officials seized 15 kgs of cocaine worth about £ 3 million ($ 4.5 million) concealed inside bottles of champagne at Heathrow airport. It was the largest ever seizure of cocaine in Britain. The police, keeping surveillance for 6 months, arrested four women from Paris, along with two men of British nationality all coming from Pakistan.

Drug Addiction Grips Bangladesh

There are some 10,000 heroin dependents. The extent of drug alone is not known as the country freely allows marijuana use and distribution. About 2.5 million people use marijuana often in combination with opium. One million are thought to use sedatives of which 700,000 do on daily basis. Trapped in heroin-taking, thousands of youths in Bangladesh are destroying themselves, apart from ruining their families. They are heroin addicts, feared by everybody wanting to distance himself from these forlon youths, occasionally—and sometimes more frequently—resorting to anti-social activities, including thefts to collect the money for heroin.

Heroin smuggling is continuing unabated, even though it attracts death punishment in Bangladesh. According to reports, the number of heroin addicts in northern districts of Bangladesh might have exceeded ten lakh. But what would be their number in Dhaka and other parts of Bangladesh would be difficult to guess. One can draw an inference only when one takes it for an epidemic throwing the youth into the hallucinated world. A press report said about 4,700 heroin addicts are languishing in 13 jails of northern districts, with 2,448 cases registered against 1,226 dealers, including three women.

Police report said heroin addicts and illegal narcotic traders are hardly caught by them except in cases of thefts and other anti-social activities, in which the arrested admit to have committed these vices to collect heroin money. Even fathers unable to correct their sons give information to the police. Reports said addicts steal rods, grills of the houses only to sell them to buy heroin. The police said in Natore, Bogra, Rajshahi and Rangpur Sadar police stations 3,000 heroin addicts were arrested in the last six months. For money to satiate their heroin hunger, they do everything from thefts and dacoities to house-breaking.

The police version from Pabna, Natorea, Rajshahi and Naogaon districts say it is one of the main causes for deterioration of law and order. It was believed that 80 per cent of the present volume of crimes was being committed by addicts. Recently, a daring dacoity on a night coach was committed at Goyeshar on Pabna-Nagarbari highway. During escape, six dacoits, later proved heroin addicts, were killed as the coach they were travelling by was overturned.

Even the hospitals do not want to admit heroin patients. Doctors in four modern hospitals in north Bengal said when the heroin addicts were taken to hospitals in unconscious state, admissions cannot be refused. They said there was no use treating since after their departure from hospitals they start drug-taking again. Many heroin addicts are to be kept in hospitals for a long time, much to the suffering of the normal patients.

Doctors said they are voracious eaters, one taking the meal of five persons. They said 5,000 heroin addicts in jail consume the food for 25,000 people. As they resort to fighting in jails, it had become a great problem for jail authorities. De-addiction centres in northern areas are mostly absent. Unless checked in time with socio-economic measures, it might pose in due course a hydra-headed monster for the country.[2]

India, over the past few years, has emerged like Afghanistan and Pakistan, as a refiner and not a mere exporter of opium. In fact, the 'golden crescent' formed by these countries is said to be rapidly overtaking the established 'golden triangle', comprising Myanmar, Laos and Thailand as the principal source of supply to the international market.

Smuggling from Golden Triangle to North-Eastern Region of India

A bumper crop of poppy was expected in Golden

Triangle area of Myanmar reportedly controlled by the rebel Shan United Army. As a result of which the total production rose to 1,100 tonnes and there was increased inflow of narcotics through its North-eastern border, particularly, through Manipur which is facing increased drug addiction problem. Unfortunately, Mizoram, Nagaland, Meghalaya and Assam are facing the same problem. The police and excise departments of those states have requested the customs and Central Excise Training Institute, Calcutta, to conduct courses in Narcotics Laws and enforcement in order to train personnel, to combat narcotics addiction and smuggling.

In 1986 India reported 31,714 registered opium dependents.

Flow of Narcotics from Golden Triangle

The agencies concerned admit that smuggling of high grade heroin and other narcotics from the Golden Triangle and Kachin areas into India from Tamu in Myanmar, to Imphal via Morek, is on the rise. But they have not been able to quantify the amount smuggled in. The explanation put forward is that in the absence of big seizures, it is difficult to arrive at any estimates. Most of the seizures so far have been of small amounts, some as little as five miligrams.

The two major seizures of opium, till now, were made in 1986 and 1987 by the Shillong Customs Collectorate. In 1986 it was 490 kgs. and in 1987 630 kgs.

Problem Areas—Myanmar Border

On the 1,000 km India-Myanmar border, the problem areas are south Manipur (Churachandpur and Mopeh). The common routes that have been identified are Tamce-Moreh-Imphal. Mandaley-Tiddim-Churachandpur-north Manipur, Mandaley-Tiddim-Champa-Aizawl and Imphal. Churachandpur is said to have a motorable road with Myanmar.

According to intelligence reports, there are at least 22 major heroin refineries in Myanmar. Seventeen of these are along the Myanmar-Thailand border. Besides, there are 60 clandestine laboratories controlled by the Shan United Army. Heroin is manufactured in the laboratories with the aid of acetic anhydride, smuggled out in jerry cans of five gallons, tied to floating rafts, along the Chindwai river upto Mandlay in Myanmar. Some of the heroin is locally consumed in the North-East, while the rest is sent out to Nepal, New Delhi and Bombay.

Heroin from Myanmar is highly rated in India because of its over 90 per cent morphine content. One kg. of Myanmarese heroin fetches as much as Rs. 2.5 to Rs. 3 lakh compared to the inferior Indian variety which sells for Rs. 50,000 to 10,0000 Lakh per kg. Besides high grade heroin, other narcotics opium and cannabis too are smuggled in. According to intelligence reports, illicit cultivation of poppy for the manufacture of opium is carried on in 8,000 to 10,000 hectares in Myanmar. The annual production is 800 to 1,000 tonnes.

Students as Couriers

It is suspected that sometimes students are used as couriers. However, narcotics control agencies have not yet nabbed any students acting as couriers. Some tribes in Arunachal Pradesh and a sizeable section of tea garden labourers in Assam are said to be addicted to opium. When the local produce and that from Myanmar fails to meet the demand, some opium is smuggled in from Rajasthan, Uttar Pradesh and Madhya Pradesh. Unfortunately, amongst younger generation, psychotropic drugs like mandrax and methagolon are said to be in great demand in the north-eastern states, particularly in Manipur. According to the narcotics control agencies, almost all north-eastern states have complained about the misuse of the codein-based cough syrup phenedril.

Efforts to check narcotics smuggling fail because of insurgency situation in North-east states and the nature of the border.

One half of the country's 160,000 drug dependents in Myanmar use about 100 tons of opium annually. Of about 40000 registered drug dependents, 8000 are on heroin although, it is believed, there could be upto 17,000 abusers of tranquillizers and 3,193 of sedative hypydics.

The drug 'ship' from Myanmar comes to Paona Bazar in Imphal with loads of smuggled goods and magic powder. The ship begins its journey from Thailand and has little trouble making its way to the Indo-Myanmar Border because the authorities have no good hold in the countryside of Myanmar where Khun-Shah and others are actively engaged in narcotic trade.

Sri Lanka

Sri Lanka is one of the latest victims of the heroin invasion in the region. In 1984, some 30,000 heroin dependents, mostly 16 to 30 years old, consumed about 500 miligrams of heroin a day.

Nepal

Heroin abuse was first noted in 1980 when the problem moved to some non-urban centres. There are now some 15,000 to 20,000 drug abusers, mostly on heroin, of which 3 per cent are women. A serious problem is the increasing involvement of the Nepalese in international marijuana and heroin trafficking. Nepalese youth are increasingly taking to drugs use.

Pakistan

Pakistan stands at the centre of drug smuggling and is considered as the biggest wholesale supplier of drugs, collected from Chitral area and Afghanistan. Its youth are drug addicts on a large scale.

There are over 500,000 heroin dependents in Pakistan and a larger number using opium and marijuana. Karachi and Bombay have been pin-pointed as the two main sources for heroin smuggling in the international trade.

In 1986 alone, 72 Pakistanis and 62 Indian nationals were arrested for drug trafficking in Europe.

Singapore

It is among the few countries to have successfully contained its drug-dependence problem. When it launched its "Operation Ferret" anti-drug programme in 1982, there were about 15,000 to 20,000 drug dependents in the island state. Today, the number has gone down to 6,500 hard core drug addicts mostly using heroin. Singapore is now concerned more with the problem of volatile substance abuse, or glue sniffing of 3,110 known victims. 80 per cent are below 26 years.

Malaysia

It has identified about 1,22,329 drug dependents, up from 711 victims in 1970. Officials believe the number could be as high as 4,00,000 to 5,00,000 and 73 per cent are on heroin; about 70 per cent of drug dependents are in the 19-23 age group.

Malaysia has dealt with the drug problems with a heavy hand. It has identified about 1,28,000 drug dependents, 90 per cent of whom are young. Reports show a persistent pattern to shift from marijuana use to other drugs, like heroin and multiple drug abuse.

Indonesia

Indonesia has some 123,000 dependents on opiates, marijuana and other psychotropic substance. Almost 70-80 per cent are on heroin.

The famous resort of Bali has the largest number of drug addicts both foreign and local.

The problem of increasing abuse of volatile substances is on the rise.

Thailand

Thailand has about 5,00,000 dependents on opiates, marijuana and other psychotropic substance. About 70-80 per cent are on heroin. Cocaine has also spread in and around Bangkok.

Fiji and Papua New Guinea

Cases of opiate abused are small due to the fact that marijuana and its derivatives are easily available for a growing illicit demand.

Drug Menace in Malaysia and Singapore

In Singapore, if anyone is found in possession of more than a fixed quantity of heroin the mandatory punishment is death; the courts have no discretion. In Malaysia, the courts have the option between life imprisonment and death. The Federal Court of Malaysia has laid down strict guidelines for deciding between life imprisonment and death.

The real problem is to destroy the source. Unfortunately, poppy growers are not in the firm grip of any of the three governments—Laos, Thailand and Myanmar. One of their overlords is Khun Sa, who is rich enough to maintain a private para-military force of 3000-4000 and to arm them with sophisticated weapons like rifles, machine guns, mortars, grenade launchers, etc. President Reagan made a personal appeal to the Thai Prime Minister to intensify the efforts to nab the Mafia don Khun Sa, who quickly shifts from one country to another, knowing fully that due to the political difficulties, a joint operation by the three governments involved—Myanmar, Thailand and Laos—appears remote; moreover, Khun Sa keeps several political and military leaders on his pay roll.

Indo-Pakistan Borders

All day the desert haze screens the plodding camel caravan. As night falls the caravan crosses from Pakistan into India. Reaching the caravan, the drivers dump their bags and turn back toward home, sending a lone messenger on to report their delivery. A few hours later, the smuggler's trucks which were lying in wait, roll out of Jaisalmer, and other places to retrieve the cargo, which is subsequently distributed to India's teeming cities, where it reaches the hands of hundreds and thousands of drug addicts.

Heroin Concealed in Straws from Pakistan

Information received indicates that heroin is being smuggled into United States from Pakistan in straws, stitched in clothing worn by passengers. The straws, which are of unknown composition, are flattened and inserted into the lapels or side panels of blazers or sportcoats worn by male couriers. This method of concealment is said to have been used successfully on three occasions this year to introduce heroin into the New York City area via Pakistan International Airlines. The concealment method has potential for use in baggage, cargo, postal items and various forms of apparel and accessories.

Strange Methods and Remarkable Recoveries

Some of the strange methods employed by the smugglers and equally effective methods employed by the police are described below :

(1) Police arrested one person near Chitranjan Park, Varanasi with heroin when he was to contact foreign citizens involved in the trade. At his information, house-search in Shitla Gali brought to the police discovery a number of instruments used in heroin processing, acid flasks, tubes and syringes.

(2) The seizure of a can-seaming machine from a canning factory at Palluruthi has strengthened the Cochin Customs officials belief of the existence of a multi-crore heroin racket in the city. The heroin was seized by the British Customs Officials at Manchester International airport in May, packed in cans carrying the labels of "Crab Meat". Mr. V.A.N. Pillai, manager of the factory, and Mr. Y.M. Ibrahim of Brite International of Malipuram were arrested as their factory premises had been used by the smugglers for packing the cans containing heroin. The Customs officials are actively investigating unauthorised export of 6 kgs. of heroin worth Rs. 26 lakh to the UK two months ago.

(3) From a Departmental Store at Tanjung in Yugoslavia a woman purchased a talcum pack but after using it, she felt strange. She reported it to the Assistant of the shop who found the matter suspicious. At his behest, the police investigated and discovered that in the thin layer of the talcum pack heroin was kept. From all the packs of the store, the police seized 4 kgs. of heroin worth 3.5 lac pounds. The store had purchased these Talcum packs from the Jesenis Custom-house at the Yugoslav-Austria boundary in an auction.

(4) According to Al-Usefi, the Egyptian and the Greek police surrounded the ship Alexandus which was waiting to cross the Suez Canal and seized drugs worth Rs. 12 crore (10 lac dollars). The arrests included 7 Greeks and one Italian.

(5) The Crime Branch, Delhi Police, arrested a Sri Lankan national from a hotel-room in Ballimaran and seized 1 kg. of heroin valued at about Rs. 1.5 lakhs and 20 electronic watches. Mohammed Nalim, 23, is believed to have smuggled it from Peshawar. He was charged under the Dangerous Drugs Act.

Illicit Manufacture of Methaqualone

1. On 21.9.88, the Bombay Zonal Unit of the DIR, on information, recovered 5 cardboard cartons from the godown of *M/s Siddharth Trotters Pvt. Ltd*, Kothari Mansion, 337, S.V.P. Road, Bombay-4. Examination of the cartons showed that they contained a total of 251 Polythene packets—each containing 1,000 tablets bearing makes "MX" on one side and "R/L" on the other indicating that they are Mandrax tablets (Methaqualone preparations). In all 2,51,000 tablets weighing around 125.5 kgs. were seized. While the search operations were being conducted one Manoj Kumar Dayanand Jain, R/o Room No. 138, 1st Floor, 52 Amarwadi Gali, S.V.P. Road, Bombay-4 entered the search premises. Since spot enquiries revealed that he used to get telephonic instructions about the receipt and disposal of the Mandrax tablets from one S.K. Chelawat of Indore, message was flashed to the Central Excise Collectorate at Indore, who organised a raid on the premises connected with the said person. Search of the residential premises of Suganchand Chelawat and his father Kanhaialal Chelawat at 157 Textile Clerk Colony, Indore resulted in seizure of the following:

(i) 71,700 kgs. Mandrax tablets.

(ii) 148,300 kgs. Methaqualone powder.

Two machines used for making the tablets were also seized. Since two cartons of mandrax tablets were found in a car belonging to Chelawats, the same was also seized. Dharmendra Chelawat, S/o Sugan Chand, who is involved in the clandestine manufacture and disposal of the mandrax tablets made an attempt to commit suicide by consuming an insecticide and was admitted to a hospital. In this operation, the officers and staff of the Narcotics Commissioner of India who are based at Indore also rendered assistance. The disposal channel at Bombay was identified as one ABDU of Dongri.

2. Based on specific information, the officers of NCB, Bombay seized 1,06,000 nos. Mandrax Tablets from a deserted bungalow at 19, Juhu Vile-Parle Development Scheme, North South Road, No. 5, Bombay 49 on 29.10.88. Mr. Ratan Singh Bahadur, Mr. Arifbhai, Mr. Usman and Mr. Ramswaroop K. Agarwal have been apprehended on 30.10.88.

3. On 19th October, 1988, the Customs officers at Bombay Airport checked the baggage of Mr. KENNEDY DJUEKA MAIBARA Passport No. A-077669 issued at Nairobi on 15.6.86 (Kenyan national) and seized 28,000 Mandrax Tablets. He was to travel to Dar-es-Salam via Abu Dhabi. The tablets were found concealed in the false bottom of the suitcase. He was arrested.

Insurgency and Drugs in Myanmar

Nearly 50 per cent of the area in Myanmar has always remained under the control of various insurgent groups within the country. The most powerful of such groups, the Kachin Independent Army (KIA), virtually rules the entire northern part of the country. While getting tacit support from China, the KIA has also been helping some of the recessionist organisations against India. The Burmese Communist party is well-entrenched in the North-Eastern parts of the country and defence experts do not rule out the possibility of China stepping in the area by making use of the current political stability in Burma. Experts stress that such an event would seriously endanger the security of the country which shares the second largest and potentially very sensitive border with Myanmar.

A former Indian ambassador to Myanmar, Prof. S.G. Swell, who is also a former Deputy Speaker of the Lok Sabha, has warned that "the border could be used for large scale infiltration intended to destabilise the country". Talking to ENS, Prof. Swell said the current crisis in Burma could lead to a civil war and a final disintegration of that country. The extension of Chinese interests in such a situation could prove disastrous for India. He pointed out that unlike the Indo-China border along the Himalayas, the border along Manipur and Mizoram was open and more suited for invasions. The country has already been invaded thrice from this area, he added.

Prof. Swell severely criticised the Indian foreign policy in regard to the developments in Burma.[3]

References

1. *The Times of India*, New Delhi/Lucknow, December 21, 1989, p. 2.
2. *The Hindustan Times*, New Delhi, July 30, 1989, p. 1.
3. *Indian Express*, September 25, 1988.

Note : Burma is now called Myanmar. "Burma" has been retained wherever it suits the context.

40

INTERNATIONAL PHENOMENON OF NARCOTICS SMUGGLING

Crime draws all criminals under one cover and in case of narcotics peddling different nationalities join hands. They forge their passports ingeniously. This fact came to light when Sri Lankan national P. Nihal Chanelera Sirpiris joined hands with Nurul Haq Nuri, an Afghan national.

The procuring of these drugs is not always very easy. This fact, in turn, gives rise to the widely prevalent smuggling trade, which is believed to be highly profitable. The narcotic traffic is manipulated by resourceful people who operate on a global scale. They are professionals with lots of money and powerful allies in the right places. When someone is caught, it is usually the dope pusher or peddler and never the organiser behind the scenes.

Most countries have laws prohibiting the production, processing, distribution and consumption of 'Cannabis Indica' and their end products and connected derivatives. The Golden Triangle countries too have their laws but traffickers have thrown all laws to the wind.

The Motivation

The main purpose to take drug is an attempt to escape from reality. Sometimes, it starts as an adventure, or among modern youth, as a passport "to be in with the crowd", and at other times as just a means of relief from pain or frustration.

The new induction is invariably among the youth of well-to-do families of all countries, including India. The initiation into drug taking is accompanied by a pleasure sensation which facilitates an easy and encouraging impact on a widening circle of votaries. But this pleasure sensation is transitory. It weakens the will-power and develops an overpowering desire to go on taking the drug in increasing quantity.

The Mechanics of Drug Smuggling

The bulk of the traffic comes through the Indo-Pak border. The drug largely produced in Pakistan, Afghanistan and Nepal is responsible for marijuana and hashish. From Pakistan, it is brought through deserts in Gujarat and Rajasthan from where it is passed on to agencies at Bombay and Delhi. In the recent past, the Border Security Force has seized some big consignments of illicit drug.

The trafficking of narcotics is much more organised, and functions as clockwork operations from the poppy fields to retail outlets in Amsterdam or Los Angeles, that is trans-global circuits. The *modus operandi* by and large is that a sub-agent of the American or European Drug Syndicate, flies to Bombay or Delhi and contacts the local supplier, places the order and flies out clear. It is now left to the local supplier to have the contraband sent out to Bombay to a pre-arranged rendezvous abroad. They may do this either by engaging 'runners' who will carry it on their own persons or in their luggage and catch a flight out, he may conceal the contraband in cargo, despatched to benami firms registered abroad. According to Ivan Fera, "Apart from the range of narcotics from Mandrax to heroin, Bombay offers a package of associated services including storage facilities for a consignment booked (pending transhipment), instant havala banking and the most crucial and specialised skills of all—repackaging and concealment in any size and form that the importer desires.[1]

The most vital factor of all, on which success of the entire operation depends, is the method employed to conceal narcotics. One of the most ingenious methods recently employed was to conceal the hashish in between several layers of diesel engine pump mountings. The mountings are sheets of steel, covered with rubber, that provide diesel pumps with a foundation and serve to absorb engine vibrations. In this case hashish was sandwiched between the steel plates, and the insides of the packing case were coated with a lead based paint—to avoid X-ray detection. However, the nature of the cargo aroused the suspicion of the air intelligence unit. It seemed odd that a firm in Canada would want to import only engine mountings and not any pumps at all. The cargo was detained and opened which revealed the whole concealed contraband.

In another interesting case, a large consignment of automobile spares was booked. The customs personnel checked the consignment and cleared. Just before loading on to the flight, the crates were substituted with other identical goods in every detail, including the customs markings. In place of auto spares they contained 900 kgs of hashish. The customs markings are easily replicated, since they are often no more than chalk strokes. However, the changeover of the crates would be impossible without the connivance of the airport employees handling cargo. The closest links that smugglers and narcotics dealers have with airport personnel are with the field staff of the International Airports Authority of India. The IAAI regulates the movement of vehicles and individuals all over the airport, and as such its employees have immediate access to any point.

Bombay is, however, a well established bazar and the most international agencies have agreed that the narcotics traffic crosses the Indo-Pak border to three Indian states—Punjab, Rajasthan and Gujarat. The traffic through Punjab finds its way through Delhi but the drugs going through Rajasthan and Gujarat are routed through Bombay which has the additional advantage of a sea port. The largest seizures, so far have been made at the Sahar International Airport.

The Canadian Connection

The international airport at Toronto has the busiest drug section in the whole of Canada and the volume of business being transacted through this route appears to be on the increase. According to Staff Sgt Fred Foster of the Royal Canadian Mounted Police (RCMP), head of the Drug Squad at the airport, some sixty million dollar worth of illegal drugs had been seized which is almost triple the value of such seizures in 1982. The drugs seized in the first eleven months of 1983 include heroin, cocaine, hashish, hash oil and marijuana.

Although the RCMP Squad at the Toronto Airport has been increased from ten to eighteen alongwith the services of a handler, but considering the volume of traffic, the staff is too inadequate. According to a modest estimate nearly 4 million passengers pour in Toronto Airport through 45,000 international flights. For instance, between 3 p.m. and 7 p.m. as many as 2,000 passengers an hour arrive from Europe, where South American Cocaine, is usually diluted before being flown to Toronto.

A Plot that Failed: Hashish Haul

Had it succeeded, it would have been the stuff of a Hollywood crime thriller, replete with mid-ocean air drops, high speed boats and clandestine flights in and out of high mountain valleys. As it turned out, the Derrick hashish case became the proudest feather in the cap of India's narcotics sleuths, who cracked it open after a year of painstaking effort, bagging more than one ton of hashish and 4 of the suspected 10 drug runners involved. This, happened between mid 1976 and May 1, 1977.

It all began with a tip-off. In the summer of 1976, Western intelligence agencies passed word to Indian Directorate of Revenue Intelligence (DRI) that an international narcotics syndicate led by a Canadian citizen Donald Mervyn Derrick was planning a major drug shipment from Nepal to Canada. Derrick had drawn an audacious plan to pull of this scheme. Squatting on a tarmac in Singapore was. DC 3 Dakota, the old work horn of most of the World's airports, which Derrick had bought equipped with modern avionics. Sailing at the Pacific at that prime time was the light weight ocean going Thrasher which carried on board a crew of four with latest electronic equipment. The investigators later found that the idea was for Derrick to fly the plane to Nepal, pick up hashish loaded in watertight tin containers and drop the consignment on board Thrasher at a secret Bay of Bengal rendezvous. The boat would then directly sail to Canada without touching any port minimising the risk of interception.

Unfortunately for Derrick, the scheme did not work as planned and instead he had to make changes because the Thrasher developed trouble. He hired an air-conditioned coach registered QM 9137 which was being driven by a Britisher Paul Gray. The coach visited Nepal and returned later with a load of the contraband. The DRI sleuths tracked it down and to their utter joy recovered the jackpot 1008 kgs of hashish separately concealed in a chamber. The Thrasher and its crew were taken into custody the next day.

Soft Laws

So much of effort and so little punishment. The local Visakhapatnam court fixed bail in August 1977 at Rs.10,000 each. Wiggnis jumped bail and vanished. Each of the others were sentenced for six months and fined Rs. 200 on each count. They served their sentences and left India in mid 1978.

Drug Trafficking in Goa Beaches

Large scale drug-running and trafficking in narcotics is flourishing unchecked behind the tranquil facade of the picturesque coastal villages of Bondez taluka in Goa. Such is the intensity of the spell cast by drugs that an increasing number of youths and children are dropping out of schools and colleges and are being steadily sucked into the murky world of narcotics. The growing drugs menace is tearing apart the social and moral fabric of the village society but the local residents prefer to bear their pains silently because the drugs trade has brought to the villages lots of money and superficial prosperity.

At Anjuna, believed to be the nerve-centre of the drugs trade and an international transit point for narcotics smuggling, houses have been rented out to foreigners at fancy prices. So these foreigners who are involved in drug trafficking retain their places on rent for long periods even while they shuttle in and out of the country.

Most of the foreigners in the narcotics trade carried fake passports and travel documents on their person. In case they were arrested, only the documents on their person were seized, as a result once they were bailed out, they could always escape with the original documents hidden elsewhere. The police are trying to tighten their vigil on the Bondez coast to effectively check the drugs trade. The police has restricted the extension of Visas to foreigners and visas are now being extended only in exceptional cases.

Drug Use and Abuse in America

The US was the first country to use heroin for medical purposes, and also the first to be hit by addiction. In the twenties there was a substantial number of heroin addicts. The number went up slowly. The real explosion came wih the Vietnam war. Heroin addiction became a menace to US troops in Vietnam, and America soon became the most lucrative market for heroin.

The US like India woke up to the menace quite late. Interpol had started sending warning signals to the police forces of the member-states, very soon after the United Nations Single Convention in Narcotic Drugs in 1961. India signed the International Treaty and enforced it on December 13, 1964.

A study has confirmed that the US capital Washington and its suburbs are the worst affected by narcotics problem.

United States Concern over Drug Peddling

America wanted that the two legitimate growers of poppy—India and Turkey—should give up the cultivation of Poppy Somniverum. America was able to pursuade Turkey to give up the cultivation by means of a soft loan of Rs. 32 crore which was sequel to President Nixon's famous Presidential speech when he took a vow to stamp out the poppy from the face of earth. India refused to yield to US pressure tactics and it was emphasized that the cultivation and processing of opium was efficiently controlled in India and it was impossible for any pilferage from the farm or factories.

According to an intelligence assessment, most of the narcotic traffic, which passes through Bombay, is headed for Africa, Spain, Portugal and Switzerland. From there the Nigerians carry the drug to Europe and USA who mainly resort to bodily concealment. Packed in condoms, between 100 to 150 cocaine condoms are either swallowed or stuffed in the anus. This poses an extremely grave health hazard—if the condom bursts, death is almost instantaneous and moreover, the heroin can be so carried for a maximum of between six to seven hours in the stomach without food, otherwise the bodily juices would react and make the condoms move down to intestines. One such carrier, a woman died instantly at Kennedy International Airport where the condoms she swallowed burst. US officals found her body being cut open by drug operators to find out the concealed heroin pouch.

It is impossible to tell how much of the heroin flowing through Bombay comes from the Golden Triangle and how much from the Golden Crescent. Narcotics from Golden Triangle has other conduit points through Hong Kong Singapore, Malaysia and Sri Lanka. The Golden Crescent route is through Delhi.

A Survey of Drug Related Crimes in America

According to the Bureau of Justice Statistics Survey State prison inmates in 1979; were asked a series of questions about their drug use, beginning with whether they had used drugs and ending with whether they were under the influence of drugs at the time of crime for which imprisoned. The procedure was adopted to minimize attempts by inmates to cite drugs as the reason for their crimes. The responses were classified and it was discovered that—

(a) one-half of all drug offences, were committed under the influence of a drug which was heroin in 20 per cent of the cases;

(b) about 25 per cent of all burglaries and 20 per cent of all robberies and all drug offences were committed under the influence of marijuana;

(c) approximately 12 per cent of all robberies and 10% of all larcenies, were done under the influence of heroin; and

(d) 60 per cent of inmates had used drugs the month before their crimes, compared to just over 40 per cent of those with no prior convictions.

Most of the users of marijuana had almost at one time or another used heroin daily and about 10 per cent had used cocaine daily. About 3/5th of drug users convicted of drug offences were in prison for selling drug than for mere possession and those convicted of drug offences were the heaviest users of drugs before incarceration.[2]

Asian Mafia Control US Narcotics Smuggling

It is causing great concern in America that Chinese,

Pakistanis, Indians and other ethnic gangs are increasingly replacing the mafia gangs of New York's multi-dollar heroin trafficking industry. According to New York Times, the new gangs include Israelies, Afghans, Nigerians and Thais. According to a reliable report, last year, the American officials had to scour the city for eight translators of the African language to assess the wire-traps of a ring of heroin smugglers from Ghana.

More than any other ethnic group, overseas Chinese with contacts in Bangkok, Hong Kong and throughout the United States have developed large sophisticated, but little understood international drug trafficking organizations. Law enforcement officials of America say, "The Chinese groups are now responsible for more than 30 per cent of heroin entering the New York city".

For decades the Sicilian mafia dominated the racket which involved the racket of buying Turkish opium, processing it in laboratories in France or Italy and selling it through distribution in poor inner city neighbourhoods in the United States. In the early 1970s, a crackdown on opium production in Turkey shifted the source of heroin increasingly to Mexico, South-east Asia, Pakistan, Afghanistan and Iran. Today the scene is dominated by the Chinese.

Anti-drug Indo-US Co-operation

United States Attorney General, Mr. Edwin Mese, told that a working group of representatives consisting of officers from the Ministry of Home Affairs and other departments such as Enforcement and Narcotics Bureau from India and such similar representatives from the United States will make it an on-going body for intelligence sharing and co-operating in pursuing the overall objectives of a joint policy. Mr. Messe said he had met the then Prime Minister, Mr. Rajiv Gandhi, to apprise him of the US Government's intention to fight in co-operation with India, the menace of drug and terrorism and to hear from him his country's commitment to the same objectives. Referring to the concerted efforts in the US against drug trafficking in the last three years, he said the drive has paid rich dividends, with about 90 per cent convictions and reduction in trafficking. Drastic steps such as the confiscation of property purchased with illegal funds from drug trafficking had been further supported by a nation wide effort to educate young people so that they did not fall prey to this habit. According to Mr. Messe, 7,600 traffickers had been convicted between 1983 and 1986.

US Anti-drug Campaign

US President Ronald Reagan and Mrs. Nancy Reagan in an unusal joint television address, hoped to intensify Washington's widening war on drugs to the level of a national crusade, stressing the need for what Mrs. Reagan planned to call "outspoken intolerance" toward drug abuse of all sorts. "When we all come together...for this cause", her husband agreed in his prepared remarks, "then those who are killing America and terrorizing it with slow but sure chemical destruction will see that they are up against the mightiest force for good. They will have no dash alleyways to hide in". Behind the vintage rhetoric, Washington's election year enthusiasm for quick action against drugs had reached a feverish pitch. By an overwhelming 392 to 14 votes, the House approved a sweeping omnibus bill, that called for the immediate employment of enough US force, to halt the flow of narcotics into America. The legislation included the death penalty for drug racketeers charged with murder, mandatory life sentences for those convicted of selling drugs to children for the second time and a controversial provision to ease the rules of evidence in all federal criminal prosecutions. It also increased federal spending on virtually every aspect of the drug war—from stepped-up enforcement activities to prison construction, to expanded treatment and drug abuse programmes.

Indo-Nepal Border

On the Indo-Nepal border, the Nepalese 'ganja' is being illegally smuggled on a large scale. In August 1985, police netted 22 quintals of 'ganja (Nepalese) which was to be sent to Varanasi. The eight persons arrested were handed over to the Customs and they seem to have links with an MP of the area. Smuggling seems to have picked up on the Nepal-Deoria border, as their secret stations in the district are finding ways to send it via Bhagalpur and Kapanbar Ghat, to Varanasi and other directed places. Nepal is a major source of Cannabis Indica' supply.

Tasmania

New-comer to poppy growing, Tasmania, now holds the record yield per hectare for morphine, meeting perhaps 20 per cent of the world's codeine needs. Experimental fields of poppy straw in the mid-1960s, have blossomed into 6,050 hectares, largely on the island's north coast. Agricultural research and sophisticated cultivation boost the region's natural advantage in poppy production—Pharmaceutical companies have contracted the crop with the farmers. The crop is regularly supervised by the police.

Tasmania's Poppy Advisory and Control Board is careful to call them oil poppies; after all, they were hybridized mainly from European varieties primarily grown for oil. Besides, no opium is produced here, only poppy straw. Here it is a high technology crop; precision seeding—a machine inserts anti-fungus coated seed and fertilizer simultaneously, seven inches apart, two and a half inches down; computer controlled herbiside, spraying precisely 200 litres per hectare; harvesting with a newly designed comb and knife on a combine to pull the dry poppy capsules down and slice them off.

The question arises why while other countries are

curtailing their poppy production, Tasmania is increasing it? Because, it is politically stable and climatically, it is far from the equator, meaning many daylight hours in the growing season. Further supplemented by harmons applications, genetic engineering, controlled cross pollination and scientific cultivation through sprinklers morphine yield is higher. In the last five years, morphine yield has doubled to ten kilos per hectare.

Mexico

Poppy growing in Mexico started when Chinese labourers planted some in Sinaloa, in the 1920s, to get opium for smoking. During World War II, American gangsters, fearing a cut-off of the morphine and heroin they were smuggling from overseas, promoted large scale planting. By 1951 the smuggling of heroin from Mexico to America touched a new high. Even today, according to a US intelligence estimate, about a third of the heroin entering the market of USA is from illicitly planted Mexican poppies. In order to destroy this illegal cultivation of poppies, USA supplies most of the air craft and Mexican pilots and navigators to do the job. It's the most strenuous anti-poppy campaign. The helicopters work in pairs—one low spraying and the other higher, directing "A little more to the left...do the part near the stream again...climb a little, there's another field...

Nigerian Heroin Traffickers

Of late, Nigerian nationals are increasingly engaged in heroin trafficking all over the world. Intelligence indicates that a majority of Nigerians act as freelance operators, more interested in quick profits than long-term organisational objectives. They recruit couriers not only from their own country but also from other African and European countries. The couriers leave India primarily from the airports in Bombay and New Delhi.

Nigerian traffickers in India tend to gather into small ethnic local groups and within these groups, members usually distribute heroin independently to other outside. Eventually, a co-ordinator or 'boss' usually emerges and he is the one who arranges for heroin to be supplied by traffickers. In New Delhi one such co-ordinator was using Nigerian students, studying in Delhi University, as couriers. It appears that there are multiple loosely associated groups operating within several cities, primarily Bombay and New Delhi. Co-operation enhances the chances of their success. As per intelligence, Nigerian trafficking organisations outside India are slowly developing a net-work in the USA.

Heroin in Groin

Nigerian heroin smuggling organisations mainly operate in major metropolitan cities like. Bombay and New Delhi. Through a network of relatives or trusted tribal members, arrangements are made for exportation of multi-kilograms of heroin from India which is received in Nigeria and thereafter distributed to the Western consuming countries. A large majority of these couriers conceal heroin in their body cavities. In recent months, all over the world, there has been an increase in the number of Nigerian/African female couriers, concealing heroin in body cavities and on their bodies near the groin area. Heroin is also concealed in small bags which are further sewn into Nigerian/African native-attire and in the shoulder pads of other clothing. In addition, these couriers use traditional methods of concealment—hair, wigs, false bottom of suitcases/briefcases, soles of shoes, etc. Once the heroin is cleverly concealed, it is smuggled out of the source countries (usually India and Pakistan) to Nigeria and onward to the United States and Europe via London and other European airports.

Mechanics of Smuggling

Intelligence indicates that a majority of the Narcotic couriers is uneducated and between 25 and 40 years of age. They carry very little luggage, have made frequent and regular trips into and out of India and sometimes provide bogus addresses of their stay while in India. On several occasions, the addresses given are just contact locations to ensure that the courier has successfuly made it to his desired destination. Couriers generally carry a small amount and pose as legitimate businessmen, but have very little knowledge about the business they claim to represent. The couriers' passports usually do not show frequent visits to India/Pakistan which indicate that they have access to two or more passports or that different couriers are being used to smuggle heroin from the source countries to Africa and then onwards to other destinations. Passports of most of the couriers show extensive travel between Lagos and/or London to United States.

Other indicators associated with the couriers are—

(i) tickets purchased by cash;

(ii) possession of additional ticket, showing a different travel route;

(iii) lost tickets;

(iv) returning passengers with one-way tickets or open return;

(v) possession of additional passports;

(vi) passports revealing travel to heroin source countries or major transit countries;

(vii) re-routing travel;

(viii) indirect travelling routes, e.g., instead of routing a trip from Lagos to New York, they may travel Lagos-London-Montreal-New York;

(ix) no apparent reason for travel;

(x) short duration of stay;

(xi) tourists and students frequently recruited as couriers; and

(xii) unescorted children.

Noted appearance and behaviour associated with Nigerian/African couriers are—

(i) appearance does not conform to employment income claimed by the subject:

(ii) unusual/awkward walk which may indicate concealment of narcotics in rectum;

(iii) nervous manner; and

(iv) answer questions in a vague manner or as if rehearsed.

Items which are sometimes found in the possession of couriers are—

(i) rubber contraceptives;

(ii) rectal lubricants;

(iii) large quantities of capsules in which the medication has been replaced with heroin; and

(iv) medicines used as pain killer.

Some known methods of concealment used by couriers/ traffickers are—

(i) body cavity;

(ii) on person;

(iii) hand bags, carry-on luggage, attache-cases, false-side bottom of suitcases;

(iv) soles and heels of shoes;

(v) drugs sewn in clothings;

(vi) electronic equipments;

(vii) hand-carry water coolers;

(viii) coat hangers (both wooden and plastic);

(ix) baby pillows;

(x) marlboro cigarette packs;

(xi) wall clocks; and

(xii) inside jars of commercial products like Vicks, Ponds, face cream, etc.

References

1. "The great Narcotics Bazar"—*The Illustrated Weekly of India*, February 16, 1986.

2. *Justice Assistant News*, April 1983, Vol. 4 No. 3, Many Inmates Committed Crimes while Abusing Drugs, Survey finds, p. 3.

41

WEST-WARD HO !

Far away, in Western countries, drug enforcement agents have never before impounded more contraband drugs than in recent months—yet the flow is unabated. Many experts believe that success of the fight against drug abuse is not in sight. Heroin, they say, has become a world power which has taken hold of large segments of the Western world. In the emotionally depressed West, no important industrial branch has posted as impressive advances as international heroin smuggling. US citizens alone spent last year a staggering 100 billion dollars on drugs—as much as on tobacco and alcohol put together.

Every 20th American citizen, every 4th soldier, takes regularly or occassionally a chard' trip.

The United States is ahead of the rest of the world in the field of drug abuse. Americans now-a-days go increasingly for cocaine, which the Paris News Magazine Le point called "The champagne of drugs" In Miami the customs seized, a shipment of 1.8 tonne of cocaine from South America worth 1 billion dollars in the retail trade in a single operation.

The drug industry, pays no taxes, makes more profits than the entire "Top ten" of the U.S. Industry from Exxon to IBM rolled together. H. Rose Perot, Chief of the Texas War on Drugs Organisation, believes that the drug prices, sensitive to shifts in market conditions generates a far greater economy than the OPEC countries.

Serious Situation

Even in Europe, according to the statement made by Jaqises Franque, head of the French anti-drug police, in the Le Point, the situation has "never been so serious". According to Franque, the police and the customs succeeded in capturing only 60th ton of contraband drugs. In Britain the number of addicts has doubled. Italian Interior Minister Virgimio Regnoin believes that drug abuse in his country is "more dangerous than terrorism". Regnoin says that drug abuse has reached the state of national emergency and has "poisoned the entire life of Italian society", and led to the "build up to power mechanism which could be directed against the national community". In Rome alone, according to the police, there are around 70,000 drug addicts—about as many as in the entire West German Republic. Any prominent figure in Italy's underworld is involved, one way or another, in the drug trade—from ordinary gangsters to Red Brigades, Subversive Palestinian factories, Libyan and Lebanon agents with diplomatic passes, KGB and members of the "Propaganda Two" secret society accused of planning a coup.

The drug boom has not spared even prudish Switzerland. The Confederation reported that 107 persons—mostly youth—died last year from "The Golden Shot" a lethal overdose of heroin.

Addiction in Amsterdam

Amsterdam is one of the world drug centres Like Italy, the Netherlands has a wide organised retail in dope which the authorites find hard to destroy. The trade is dominated by "Trials" from South Asia. The Trials—which in Chinese philosophy represented the triple units of heaven, earth and man were originally secret political societies, which wanted to free China in the middle of the 17th century from Mandshu domination. They degenerated over the past century into Mafia style gangster groupings, who organised the opium and later, the heroin trade between the Far East and the West. The Trials firmly control the wholesale of Dutch drug, although the government, in a crackdown on the China towns in Rotterdam and Amsterdam shipped back to the Far East 1,800 illegal Chinese immigrants. The Trial bosses in Hong Kong and Singapore have long filled in the holes in their foreign subsidiaries and have decentralised their related structures so that raids no longer lead to collapse of their trade chains.

Dramatic battles with pistols and knives take place between rival class of drug peddlers in Amsterdam's red light districts, with the Chinese trying to fend off inroads into their drug market domination by exiles from former Dutch Indonesia.

Sources Dried Up

The Soviet invasion of Afghanistan has largely dried up supplies of drugs from that near eastern nation. The reign of terror imposed on Iran by the Islamic Mullahs has thinned out drastically crops of Iranian opium producers. Above all, since the military seized power in Ankara, the Turkish opium kitchens, which in 1979 supplied half of West Germany's opium alone, are working at reduced speed. The Turkish Generals have strictly forbidden any further growing of poppies. The former Soviet Bloc countries had signed the 1961 "Single Convention" calling for international co-ordination in containing drug smuggling. But, since the problem of drug abuse hardly exists in these countries, their custom officials had to close their eyes and let smugglers go through their territory with impunity.

Smuggling of Opium in Golden Triangle

Thai government and international agencies promise aid if hill villages switch over to crops such as coffee and mushrooms from poppies. But official pressure is lax in this region, where insurgent groups often hold sway. Surprise visits by Thai Rangers found opium, harvesting despite an agreement to stop commercial poppy growing.

Jungle refineries are quite common in Thailand and quite often refineries are demolished by the Thai Border Patrol Police. Their crude factory consists of a few crude pots; simple chemicals, and a source of fresh water are all that is needed to produce this substance from which illicit heroin is made. About ten kilograms of opium makes one kg. of base, which is compacked into a cake in a press. The lid has the number 999, once the trade mark of Golden Triangle dealer. Heroin is made by treating morphine base with a chemical, acetic anhydride. More potent than morphine, heroin fetches more money for the same amount of powder. Heroin was first marketed in 1828 as a cough and diarrhoea remedy and falsely hailed as a morphine substitute less likely to cause addiction. Interestingly, once inside the body heroin converts back to morphine.

Spectacular treatment of addicts at Tham Krabok monastery north of Bangkok results in a 90 per cent success rate, according to its records. The ten-day free treatment begins with a vow to Buddha never to use narcotics again. Then patients are given herbal medicine that makes them vomit immediately. The monks who have treated 80,000 addicts since 1959, claim the medicine clears poisonous drug residue from the body and helps eliminate "the physical desire for drugs". Herbal steam bath, milder tonics and herbal pills, alongwith continuous counselling are part of the regimen. Majority of the 1,00,000 to 3,00,000 heroin addicts in Thailand inject drug.

Malaysia, one of the largest importers of Golden Triangle heroin, now counts 3,50,000 addicts. Herbal tea is also part of the cure dispensed by a Malay healer called a Bomoh, who paints Islamic verses on the chests of the patients. Some 25 Bomohs treat addiction in Malaysia, and authorities say the cure rate is about 60 per cent.

According to an estimate the Myanmar-Thailand-Laos areas of South-east Asia, supply 20 per cent of the heroin consumed. This area consists of nearly 3,000 villages scattered in hilly terrain, inhabited by mountain people, called the Yao, Henoy, Akhai, the Lahu and Lisu. They produce rice to eat and opium for medicine and for cash. Extensive poppy growing began here in the 19th century. Poppy growing is a flourishing business because you have not to look for customers. They come and knock at your doorstep.

Smuggling Wars

Smuggling of opium, on carvans of mules, for turning it into morphine at clandestine laboratories is a continuous process. This is a hallmark of Golden Triangle Traffic and involves a sizable force of armed outlaws, who are from minorities in revolt against the government, such as SUA, the Shan United Army, and the KMT or Kumintang Soldiers—remnants of anti-communist armies that left China after Mao's victory, now settled in northern Thailand.

According to a report published in *National Geographic*, written by T. White, "One day, in 1967, at the place where Myanmar, Thailand and Laos meet—a conjunction that gives the Golden Triangle its name—a SUA caravan with 16 tons of opium, coming from South Myanmar, tried to avoid paying the customary tax to KMT. So the SUA crossed the Mekong river into Laos; the KMT gave chase. At the height of the fight, at the village of Barkhwan, Laos, war planes swooped in, dropping bombs, Then came Lao paratroopers, seizing the opium, it went to the Commander-in-Chief of the Royal Lao Army, who was in the opium business too." The SUA leader in that miniature opium war, Khun Sa, is now the area's top trafficker and has had skirmishes with Thai forces and Myanmar soldiers. He is fighting for the liberation of the land of Shan. Modern warfare costs lots of money, and opium and heroin are the only way to get it. He repeatedly offered to sell all his opium to U.S. for 36 million dollars but President Carter rejected.

Thailand's annual opium production, as high as 180 tonnes in the 1960s, declined as the Thai government stepped up economic development to counter the threat of insurgency; it's now about 40 tonnes. Laos produces about 50 tonnes. The main problem is Myanmar—where about 600 tonnes of opium is produced. The government there plans to eradicate poppies but can't do much in areas out of its control. And so, heroin continues to flow south to Bangkok, to Malaysia, to India and to Hong Kong; then to Europe and the US, Canada and Australia. It is bound for destination Europe and U.S.A.

Drug Addiction in United States

Opium preparations were used medicinally in the United States from colonial days, but there was a large increase in consumption in the latter half of the nineteenth century. When opiates were prescribed for the relief of symptoms caused by a variety of disorders especially diarrhoea and gynaecologic conditions, patent medicines, including soothing syrups for infants, contained opium and later morphine, codeine, and even heroin. When the original disease was treated the patient often found himself addicted to opiates.

Opium smoking was introduced in the United States by Chinese labourers, imported to work on railroads construction, taken by the gamblers and criminals on the West Coast, and rapidly spread to delinquent individuals all over the country. By the 1870s people began to connect the use of opium with criminal behaviour.

Dr. Sichtenstein's Study—1913-14

Dr. Sichtenstein was the physician of the city prison in New York in 1913-14. His description shows that in some ways the addict of those days resembled today's preponderant males, the onset of addiction at an early age, introduction to narcotics by friends rather than in the course of medical treatment, and skating to get money to pay for narcotics. Differences, however, may also be noted. Many of the addicts were minority group members, but in 1913 they were Chinese, not Negro or Puerto Rican. The most frequent methods of taking the narcotics were by sniffing, subcuteneous shots, smoking and by mouth; today most American addicts quickly learn, and continue to use the intravenous—the "mainline"--rouete. Mr. Bronac, the State Food and Drugs Commissioner of Tennessee in 1914, described the narcotic addicts in that State. At the time women addicts outnumbered men two to one in Tennessee, while they were in minority in New York.

1964—U.S. Drug Addiction Study

In 1964, an Associate Editor of Life magazine, James Mills, and a photographer, Bile Eppridge, entered the addict world in New York City and literally lived with addicts for several months. Their word and picture portrait of that world is an accurate and vivid description of what it means and how it feels to be an addict in New York today.

"From legal poppy fields in Turkey, by camel across the sands of Syria to the not so legal laboratories in Lebanon, then by ship to southern France for final refining, back to Italy and, country of the mafias to New York docks and airports—heroin comes in Harlem. And from Harlem, the drug moves swiftly through the city of New York, as efficiently and regularly as from New Jersey or first from Fulton Street. As it moves, the illicit stream swells into pools from which addicts in various parts of the city draw their daily needs. Addicts—and the police—are as aware of the selling locations as the housewife is of her neighbourhood shopping centre.

In the rush and hub-bub of 96th Street and Broadway, addicts gather on the corner to meet the pushers and buy their drugs—ignored by crowds of New Yorkers on their way to work. On the south-west corner of 82nd Street and Columbus Avenue, addicts spend thousands of dollars a day for heroin. In front of the drug store at 47th Street and Broadway, within the chaotic glow of Times Square, unacquainted tourists brush shoulders with barbiturate addicts waiting for their connections. It's the same just two blocks to south, among the honky-touch bars and night clubs, or down in Greenwich village, where heroin and marijuana pass from hand to hand on the benches of Washington Square. 'Needle Park', like the rest of the world of the big city narcotics addicts, is crowded by a conglomeration of individuals of different backgrounds, who prefer different combination of drugs. Simply by hanging around 'Needle Park'... you can meet a whole spectrum of addiction: Irene, a slight crispy lesbian addicted to "Goofballs", barbiturates. Goofballs, usually produce a quiet drowsiness, but also at times a tense aggressiveness that can be frighteningly unpredictable...

When addicts meet, they talk continuously about drugs, which is better, heroin mixed with a bombita or cocaine. Both cocaine and bombita are stimulants, and either one mixed wih heroin, which is a depressant, produced a more pleasurable high flight than heroin alone. The mixture is called "Speedball".

Junkies hang around "Needle Park" because it is surrounded by cheap hotels, needed by addict prostitutes; because three blocks away, a short walk for a junkie, are respectable neighbourhoods which are good for burglary and 'cracking shots'—breaking into cars, and because, probably, a long time ago someone started selling dope there and the area just got to be known as a good place to make a connection to "score".

The addict is habitually dirty, his clothes filthy, and he stands slackly, as if his body were without muscles waiting for a connection, he is nervous and intently stares for minutes in the direction from which the pusher is expected. Once the addict has drugs on him, he keeps moving. He is about to achieve the one thing for which he lives. His shoulders are hunched, his head is down, and he strikes out with what some detectives call a "tearing—the set—walk", as if he had just won a jack-pot where a million dollars were hidden. When the heroin addict is high, his pupils are "pinned", constricted, and if the shot was sufficiently powerful he goes in a nod—his head drooping, eyelids heavy.

But though he appears terribly drowsy, he speaks coherently. His mind wanders, he day-dreams, and everything he does, he does with maddening slowness. He can take half an hour to tie his shoelaces. But he always resists admitting that he is on a nod. He is very sleepy, he says, and if he stops talking in mid-sentence, he argues that he is not nodding, only trying to frame the sentence properly. Once the addict has used his shot and is "straight", he may become admirably, though briefly, industrious, suddenly deciding to shine shoes, brush his coat, comb his hair—all the while scolding himself bitterly for having slipped so far.

Heroin depresses sexual desire—men may become impotent, women often do not mensurate. Almost all heroin addicts are childishly immature; full of demands, empty of offerings. When they want something, they want it immediately and effortlessly. Perhaps the dominant emotional characteristic of the addict is his enormous compulsion to abdicate all responsibility for his own life. He craves to be told what to do. If he is persuaded to go to a hospital by someone he trusts, he will go; but soon, when he finds the hospital not to his liking, he will leave, and then blame the failure not on himself but on the person who had exhorted him to go. An addict will walk along a street openly displaying a container of drugs, all but asking to be arrested. He thus purges himself of the responsibility of choosing between "addiction and abstinance, or continued addiction on the street".[1]

Media-legal Addicts

The number of persons addicted to narcotics or using them with some regularity in the United States is not known. There is one type of addict not reported to the Bureau of Narcotics, and probably underrepresented in other samples. There are addicts who obtain their narcotics from legal medical sources and who do not come into conflict with the law. Then there is also an unknown number of persons who are technically ill, and who are needing narcotics to an extent which produces physical dependence.

The Opium Habit in Michigan

The Secretary of State, Board of Health, Michigan State, projected these facts about opium habit in his State,[2] "At a meeting of the State Board of Health in January, 1877, by a written communication, I called its attention to the large number of opium eaters in the vicinity of North Lansdy. From the nature of this habit, complete and perfectly reliable information about addiction is impossible. Those who have become addicted resort to almost every possible means which human ingenuity can devise. To such an extent are fabulousness and deception practised by the opium eaters that they seem to be almost a part of the habit itself."[2]

The Secretary of State defined the drug habit in these words, "In this State, the drug is used principally in three forms: first, crude or gum opium; second, sulphate or morphia, called morphine; third, tincture of opium called Landanum. Opium is concrete juice of the poppy (Papaver Somniferum)". *Landanum* (to praise or praised) is the alcoholic solution of opium, made by saturating opium in diluted alcohol. It is used by opium eaters when the additional kick of alcohol is desired. When *landanum's* use is continued for long the system becomes accustomed to it, and larger doses are tolerated and required to produce the desired effect; the attempt to leave it off is then followed by depression and suffering so that some one may have the courage or endurance to discontinue its use by the opium eater. Charles Lamb describes thus—the pleasures of opium smoking—"The drug is resorted to on account of the pleasurable sensation which comes of it. The mental faculties are said to be enlarged; a state of mind is felt which is described as perfect happiness; the ordinary capacities of enjoyment are refined and amplified. Dreams and hallucinations of a wonderous but agreeable kind present themselves to the thinker—but there is no intoxication like that produced by alcohol. To his friends, the opium eater may seem a grave, silent, and distracted person, or his flow of conversation and expression of his ideas to others may be more remarkable and more vivid than theirs. At length, he sleeps, still dreaming, sleeps long; and when awoke, still feels drowsy. Now his headaches, his frame is usually weary, his tongue is coated, his appetite is diminished his eyes dull and listless he longs for sleep again. Soon there comes a depression of mind which is described as intense. To cure this, the wretched man knows but one way. Death would be better than its continuance. He takes opium again, so he goes on, till the daily stimulant must be added to produce its usual effect. From one grain of opium, or ounces of *landanum*, he goes on along a terrible and slippery incline which in this world has no end. It produces, it is said, a disinclination to all business, and yet, while its influence lasts, a singular ability and aptitude for employment mental or physical.

Opium eating once contracted as a habit, is far more difficult to give up than the practice of alcoholic intoxication. The torments of an addict deprived of his accustomed stimulus, is described so graphically by Charles Lamb.

Drug Distribution and Marketing

Once the heroin arrives in the United States, it flows into a multilateral extremely profitable illegal distribution and marketing system. At the top of this system is one of the major drug importers upon whose orders large shipments of heroin cross over borders. At the bottom are the street dealers, many of whom are themselves addicts.

The marketing process begins when a major drug dealer in this country arranges with a foreign supplier for a multi-kilo shipment of heroin at going prices, the dealer will pay in the vicinity of $ 20,000 per kilo for heroin that is brought in bulk—say 50 or 100 kilos at a time. After the heroin enters America it changes hands several times and by the time the individual dealers take delivery for distribution, the price will have shot to about $ 40,000 per kilo.

Adolescent Narcotic Addiction

The use of drugs by young people to alter their psychological behaviour is not a new phenomenon. Over 100 years ago, before the introduction of ether as an anaesthetic agent, Godman and Colman reported that medical students and other young people had experimented with the drugs by holding ether parties or "jags" as they were called. Doctor Crowford Larg often attended such ether "frolics" before he had a friend inhale ether after which he removed a small tremor from the neck, thus introducing a new era in anaesthesia.

Pescor, in the first statistical analysis of the information contained in the clinical records of 1,036 patients admitted to the US Public Health Service Hospital, Laxington, Kentucky, during the period from July 1, 1936 to June 30, 1937, points out:

"While no age is exempt from drug addiction, there is, nevertheless, a heavy concentration of cases in the age group 20 to 29 years, more than half of the patients being victimised during this period. A substantial percentage of adolescents 19 years of age or less also yielded to the temptation of using drugs. On the other hand, very few individuals became addicted after the age of 80. Therefore, drug addiction seems to be a greater potential danger to youthful individuals, than to older men. Youth characteristically seeks adventure, excitement, more thrills, anything but settling down with one sexual partner, to the humdrum business of making a living and rearing a family. Drugs offer new thrills and an escape from uninteresting reality. Furthermore, drugs act as a balance for those unfortunate individuals who, after they are cut adrift from parental protection, are unable to decide upon a career or to carve a niche for themselves in social structure. The state of indecision and dissatisfaction is bound to be reflected in emotional upsets which are relieved, at least temporarily, by alcohol or drugs".[3]

Drug Abuse Scene in the United States

Twenty-five years ago fewer than 2 per cent of US High School seniors had experimented with an illicit drug at some time. Now more than two thirds have. Perhaps 10 per cent of adolescents experiment with drugs on a regular basis. Around 10,000 young people die in auto mishaps each year, primarily because alcohol is a major factor in the case of youthful accidents.

An alarming number of crimes are drug related and influenced. New York police estimate that nearly a quarter of the city's homicides are drug related. In San Francisco, California, dope is believed to be an element in 7 of every 10 crimes.

In the United States there is illicit drug trading, estimated by some to be more than $ 10,00,00,000—all untaxed.

On the streets of some major US cities, illicit drug is so common that police can't cope with. Some prisons are jammed and urban court systems are in gridlock because of drug arrests. Drug pushers are aware of the fact that the chances are good they won't even be indicted, or they can place bargain for a minor penalty. "We're chasing the same people over and over", laments one district attorney. For years, the United States and, upto some extent, other Western nations have borne the stigma and shame of having the highest rates of drug and alcohol abuse in the world. But no longer are these Western problems alone. Now the alarming tentacles of illicit drug trafficking and alcohol abuse and the resulting corruption at many levels are seriously undermining whole regions politically, economically and socially.

Lobby for Legalising Drugs

The old school dates back to the hippy culture and argues that drug use should be allowed: "As a matter of individual liberty and that drugs aren't that dangerous", or there is compelling medical need for illicit drugs; or that drug laws are evil plot by big brother government; or that people use drugs only because they are illegal. However, with public awareness gaining ground against the use of drugs, this group appears to have lost ground. Mr. Robert E. Peterson, deputy attorney-general in the Commonwealth of Pennsylvania observed: "America has grown increasingly intolerant of illicit drugs. Ironically, in the midst of this intolerance, a renewed cry for legalisation is being heard. Many of us feel that this call is primarily based on fear and frustration. "The inability of the law enforcement department to single-handedly solve America's drug problem is cited as evidence of the futility of imposing legal prohibition on drugs. "The cost of enforcement, it is argued, outweighs the benefit society derives from such enforcements. Such faulty economic theories are being applied tc bolster the legalisation position", Mr. Peterson added. "They are faulty because they overlook historical experience, neglect economic realities, ignore the biological and psychological effects of illicit drugs", he elaborates.

The advocates of drug legalisation claim that the war on drugs cannot be won and "we must withdraw now and cut out our losses". In support of this claim they cite the factual position of drug availability, drug consumption in spite of stronger law enforcement efforts and tougher drug laws.

History and experience of other nations, including Great Britain, Netherlands, China, Japan, Sweden, Norway and Spain reveal that perhaps the most effective way of solving the drug problem is by launching a vigorous campaign for public education and rehabilitation. Mr. Johnson felt that drug has to be stopped at the very source so it doesn't reach anyone.

There are areas where the anti-drug campaign seemed to be gaining ground. Annual statistics show that use of all types of drugs, except alcohol, has come down among the high school students and young adults. It brought about 62 per cent drop in drug use in the US Navy.

Legalisation proponents say that if drugs were legal and made less-expensive, then, trafficking would cease and related crimes would fall. They also claim that various schemes of the state regulation of drugs could be tried. In defence, they analogise the current drug laws with alcohol prohibition. Supporters of legalisation of drugs think that drug users commit crimes solely to support expensive drug habits. But, lowering the price through legalisation of the drug may not solve the problem, as it never happened in Sweden, Great Britain, Spain or the Netherlands, where the issue of legalisation always remained an unhappy experience.

Drug users will commit crimes to obtain drugs regardless of drug price, as it is evident from the fact that crime levels have risen where drug prices fell in recent times. In areas of Washington DC where crack now sells for only 3 $ per dose, viloent crime has increased sharply. According to the Washington DC metropolitan police out of the 256 murders committed this year so far, 90 per cent were drug related. Most of the lobbyists for legalisation admit that there would be at least a temporary rise in drug use, if legal prohibition were lifted. However, they cherish an ill-founded belief that some natural level of use would be reached and suggest a proposal to utilise the money saved on law enforcement for education to prevent drug use and treatment. Mr. Peterson says that over 50 per cent of the high school seniors had smoked marijuana, and 1 per cent got stoned on pot every day during the period in which some states decriminalised the drug and support for anti-marijuana laws was at a low ebb. Today, criminal penalties for pot are strongly backed and the use of marijuana among high school seniors has dropped, with the daily use going down to 4 per cent.

"Our experience with alcohol proves that a drug that is legal for adults cannot be kept from children", Dr. Mitchell Rosenthal of the Phoenix House Treatment Centre said and he also estimates that 75 per cent or more of all illicit drug users become addicted, while 10 per cent of the nation's drinkers are alcoholics. "Almost every American has been personally affected by this problem. But none has suffered as much as our youth. Whether it's a baby born addicted, a child murdered by drug-crazy parents, or a teenage abuser destroying his mind and future, the young have paid dearly, concluded Mr. Starling Johnson.

The argument, that the only solution to the drug problem is to legalise illicit drugs has resurfaced with a vengeance. This time it is not just the usual group of drug lobbyists calling for marijuana decriminalisation. Legalisation proponents now include Conservatives such as William F. Buckley, Jr., Commentator Andy Rooney, and the Mayors of Baltimore and Washington DC. They say the use of drug is prompted by the fact that they are illegal and argue that "if crack was legal, there would be no crack epidemic".

Following the appeal of the Mayor of Baltimore, Mr. Kurt Schmoke, for a national debate on legalisation in the US conference of Mayors, the media swiftly brought back the issue into public domain. Even though the majority of Americans oppose drug legalisation, the debate is very much back here. Simply the question "Should drug be legalised?" is confusing particularly to the young people. One class of 14 and 15-year-olds wrote a letter to the 'New York Times' saying "America is in a 'moral rut' and drug abuse is one of the main causes, "and concluded that" legalising drugs is not the answer. The chief prosecutor for narcotics in New York, Mr. Starling Johnson asked : Don't these children have enough pressure trying to resist drugs and say 'no' to their peers and pushers everyday?

There is both a new and old school of thought on why drugs should be legalised.

War on Drugs

The then Vice-President of America George Bush declared in 1984, "I want a drug-free America and this will not be easy to achieve. ...To night I challenge the young people of our country to shut down the drug dealers around the world... My administration will be telling the dealers 'whatever we have to do we'll do, but your day is over, you're history'."

United States Operation against Drug Smugglers

For quite sometime, the underworld grapevine had been ripe with rumours and it was not long before US Drug Enforcement Officials got the reward for persistently tracking them. In April 1984, the US Drug Enforcement officials seized heroin worth Rs. 94 crore from five passengers at New York's JFK International airport. Drug enforcers the world over sat up and took notice. All the five persons, including a woman, caught with what is considered one of history's largest heroin seizures, had flown out from India, confirming that the country has emerged as one of the major transit points for drug traffickers.

The story of this biggest ever catch started with the arrival at Kennedy airport on April 9, on Pan American Flight 73 from New Delhi via Frankfurt, of Shabin Hassan Siddiqui, and his brother Syed Salim Ahmed, an electronics technician were residents of Chicago and were returning after visiting relatives in India. Custom officials discovered three suitcases with false bottoms containing 32 pounds of heroin which was neatly packed in lining and frames. The Siddiqui and Ahmed were arrested and handed over to DEA who then tracked down their distributor contact in Chicago, an Indian-born Amercian industrialist and Indian citizen Raghvinder Mahal, a bus driver.

Next day, in a separate and correlated incident, Kublay Wangdi, 41, a consular officer with the Royal Bhutan foreign service, who had just resigned his post in New Delhi, also landed at Kennedy airport, arriving with a business vacation visa but without diplomatic immunity. He was carrying 56 pounds of heroin, scattered in 23 separate bags in his two suitcases. Wangdi was permitted to check into the International Hotel at the airport. There Wangdi made three phone calls to Delhi, which were monitored by the authorities. Kumar, the person he called in Delhi, advised Wangdi to await his New York contact. The latter, an Indian passport holder by the name of Satya Paul Arora, arrived on schedule at 10.30 A.M. and left Wangdi's room five minutes later with the smuggled heroin. Drug enforcement officials closed in and arrested both Wangdi and Arora—in *flagrante delicto*.

Washington Mayor Held for Drug Abuse

Just hours before Washington Mayor Marion Barry was arrested on charges of smoking crack cocaine, he preached an anti-drug sermon to high school students mourning an 18 year-old football player killed in a street shooting. "He came to our school talking about drugs and not to use them, and then later I saw him on TV getting arrested for cocaine", said John Williams, 16, a junior at McKinley High School. "I take him serious as a Mayor, but his personal life is something to talk about. He says not to do something and then he does it", Williams said.

Williams and three buddies were eating breakfast on Friday in the Florida Avenue Grill, a northwest Washington landmark so popular that people stand and wait for seats all day long. Mr. Barry often ate breakfast there himself. He grins out of several photos in a row of celebrity pictures over the counter. And his name was in practically every conversation. "They think young kids sell drugs, and that's not the case", said Andre Simmons, 17, also a student at McKinley High School, "Look at our Mayor".

Mr. Barry, 53, was arrested with a woman at night at a swank Washington Hotel in an undercover operation by FBI and Washington DC police. He was all allegedly Videotaped smoking crack cocaine. The sting team took Barry to FBI headquarters where he was charged with possession of cocaine. He was then released and allowed to go to his south-east Washington home. "It's a disgrace", said Dick Brown, a political consultant who has lived 20 years in Washington. He was eating pancakes in the front booth of the Florida Avenue Grill.[2]

US Drug Dealers More Sophisticated

There are some of the newest ways drug dealers are transporting drugs, money or message into and around the United States in a cat-and-mouse game with authorities, government prosecutors say. In Boston last year, authorities, found 190 pounds (86 kgs) of Asian heroin concealed in a bean sprout washing machine shipped from Hong Kong. Officials in Philadelphia reported the seizure of more than 2,400 pounds (1,088 kgs) of cocaine hidden in a shipment of anchovies from Argentina. The cocaine was weighed and packaged "so as to be virtually indistinguishable from the surrounding cans which actually contained anchovies", said a report by the Philadelphia-based US attorney Mr. Michael Baylson.

"Dealers are using increasingly clever ways to deliver drugs and collect money", Mr. Baylson said. In addition to the use of cars, taxies, trucks and pedestrian, there are reports from Philadelphia city council man, Mr. Angel Oritz, that dealers are now using carrier pigeons. Fuel tanks are popular with drug dealers, as is strapping drugs to carrier bodies. Los Angeles gang members strap-crack their clothes or send it on greyhound or united parcel service for transport. In Hong Kong, authorities found spare fuel tanks loaded with 250 pounds (113 kgs) of heroin base on a trawler in 1984. The same Asian drug organisation that used the fuel tanks also shipped heroin in ice buckets, wall plaques and in a shipment of T-shirts, headed for New York, but seized in Thailand in December 1986, according to the US attorney, Mr. Andrew Maloney of New York's Eastern district.[5]

The Search for Effective Tactics

It has been a long year for the United States as it has wrestled fitfully with what to do about its severe drug abuse problem. While members of the Federal Government have debated what Uncle Sam should be doing, there have been visible failures in dealing with the drug incursion and largely unseen but basic successes also.

The Bush administration's anti-drug strategy, formulated earlier in 1989 by the Drug Policy Director, Mr. William Bennett, has drawn applause from those experts and politicians who seek additional emphasis on punishing drug law violators, coupled with a go-slow approach to increased spending on drug education and treatment until society can agree on which tactics really work. But the strategy has been criticised by other experts and politicians—those who want much more money spent now on education programmes and treatment facilities, or both, to reduce demand for drugs and aid drug addicts in overcoming their habits.

After months of discussion Congress passed last November or moved to the brink of passage a number of anti-drug laws in the annual flurry to adjourn. Included are provisions for—

— Building $. 1 billion worth of new prisons.

— Spending $ 1 billion more for drug treatment, prevention, and education programmes.

— Additional funds to Bolivia, Colombia and Peru—chief sources of cocaine—to combat drug production.

— Forcing the states to put drug treatment plans into effect before they can receive Federal anti-drug funds.

What the Federal Government is doing "is pretty much what we have done in the past, only more so", says Mr. Norman Omstein, a resident scholar at the American Enterprise Institute. "There isn't a heck of a lot that the Government can do".

The government at all levels is now confronted by cheap and deadly crack cocaine that has turned largely minority inner cities into war zones. Cocaine, marijuana, and their attendant problems have spread their tentacles into sleepy towns from Georgia to Texas to New Hampshire. Despite highly publicised seizures of large drug shipments, the flow of illegal narcotics continues to ovewhelm the ability of the US Government to curtail it.

Yet experts say America is making progress in fundamental areas. One is the fast-growing public awareness of the seriousness of the problem.

"Over the last year there has been a dramatic increase in the percentage of the public saying drug is the No. 1 problem facing the nation". says Mr. Everett C. Ladd, executive director of the Roper Institute. This public concern is significant in a society that requires public support for major expenditure, such as the anti-drug efforts demand.[4]

Also on the rise is an awareness that simple, broad-brush approaches are not enough in dealing with drug treatment and education programmes. Combating different kinds of drugs often requires separate types of treatment and new types of drugs are coming along all the time. There will be a continuing need for research. Further, different people taking the same drug may need different aproaches: for example, chemical approaches to negating the body's craving for cocaine appear promising, but cannot be used by women who are pregnant—crucial exception since many children are now born with an addiction acquired from cocaine-using mothers. Similarly, in drug education, "You need specific kinds of education", "You need to know a lot more about what works and what doesn't" with specific groups. Among many high school and college students—especially those with aspirations and a pattern of academic success—drug education already is working, with a 37 per cent drop reported in recent months in the number of students who use drugs. This is encouraging, but the educational programmes that have helped produce this decline have not shown similar success among low-achieving inner-city minority youth.

Also encouraging is the way American business has moved ahead in drug programmes. In theft, absenteeism, accidents and deaths, business pays approximately two-thirds of the nearly $150 billion estimated annual cost in America stemming from drug and alcohol abuse, according to Mr. Delattre. Most large American firms now have programmes to identify early and aid employees who use drugs or abuse alcohol. Many smaller firms, too, are installing such programmes.

The Challenge

Despite these areas of progress, much remains to be done before the anti-drug effort can be deemed successful. Some experts say the States spend too much time beseeching the Federal Government for aid, and not enough time helping themselves. "The States are doing far too little to fight drug abuse", says Mr. Jeffrey Eisenach of the Heritage Foundation. He says America's "greatest single failure" in combating drugs has been the inadequate effort of the States to enforce their own laws against drug possession and even dealing.

But if huge sums of Federal money were found today and poured into treatment and education programmes, much would be wasted, Mr. Delattre and others say, because more must be learned about which programmes work. One point of which everyone agrees is the staying power to combat it for years. "If ever the phrase 'the price of freedom is eternal vigilance' was true", says Mr. Delattre, "It is true about drugs".[7]

References

1. "The Drug Takers"—a Special Time-Life Report-1963-Inc.
2. Annual Report-Michigan State Board of Health, Vol. 6-1878, pp. 63-73.
3. Adolescent drug dependence. C.W.M. Wilson, p. 11.
4. *The Hindustan Times*, New Delhi-January 21, 1990, p. 15
5. *The Times of India*, Lucknow October 20, 1990, p. 15
6. *Indian Express*, New Delhi, August 10, 1989, p. 7.
7. *The Hindu*, Sunday, March 11, 1990, p. 18.

42

NOREIGA TRIAL AND CUBAN SCANDAL

Panamanian strongman Gen Manuel Antonio Noreiga turned out to have been the ultimate spy, purveying intelligence information for payment to a surprisingly varied clientele. According to intelligence officials, among those who considered Noreiga an asset were the CIA, the Defence Intelligence Agency, the Drug Enforcement Administraton (DEA), Cuba's DGI, Nicaragua's secret service, the Soviet KGB, Israel's Mossad, and Colombia's Medellin drug cartel. "We paid him, but we never owned him," a former high-ranking intelligence official was quoted as saying in the *Washington Times*.

CIA Link

Noreiga's CIA link began in the late 1960s when he was a low level officer in the Panama national guard. He was then paid several hundred dollars a month. The association ended in 1986 when the CIA halted his annual stipend of $ 250,000 following public disclosure that he was involved in large scale drug operations.

Noreiga's payments from the Medellin cartel were estimated at more than 4 million dollars. At the time of his arrest by DEA agents, Noreiga had amassed a fortune estimated at nearly one million dollars in bank accounts and real estate around the world. According to sources familiar with a secret CIA biography of Noreiga, the wealth was spread out in Spain, the French Riviera, Britain, Switzerland, Hong Kong, Cuba, the Dominican Republic and the United States.

Noreiga was regarded as one of the most valuable US intelligence assets, albeit a dubious one, who helped the CIA spy on Cuba and Nicaragua. He also assisted covert US operations, by permitting the training in Panama, of Nicaraguan resistance forces and by assisting in the secret Contra arms pipeline operated by former White House aide, Oliver North.

The National Securtiy Agency and the army intelligence and security command also were permitted, under the Noreiga regime, to operate two listening posts in Panama, near each end of the Panama Canal.

Double Agent

At the time he was working for the CIA, Noreiga also provided assistance to the Nicaraguan Sandinistas, before and after they took power in 1979, and to Cuba's DGI intelligence service, known to be a surrogate of the KGB. Noreiga's penchant for doing everyone down was apparently limitless. In the late 1970s, he conducted effective intelligence operations against the US.

According to intelligence officials, Noreiga found out in 1976 that the army and the NSA were eavesdropping on Panamanian officials involved in negotiations that led to the 1977 Panama Canal treaty. He then successfully "turned" several army enlisted men who were conducting the wiretaps and telephone surveillance and bribed them to provide valuable information about GMS, Electronic spying capabilities. US officials believe that Noreiga sold or traded the information to the Cuban DGI. The Soviets are believed to have obtained the information, causing serious damage to US national security.

Drug Dealings

Noreiga displayed another aspect of his fertile brain in the early 1970s, when he tipped off the DEA about the activities of his rivals in the drug trade. "He used the DEA to burn his rivals", an intelligence official said. As late as 1987, the DEA sent Noreiga a letter of commendation for his work in countering narcotics trafficking. What the DEA did not know was that Noreiga was exposing his rivals, so that he could conduct the same business without competition.

Noreiga provided sanctuary for drug kingpins of the Medellin cartel. In 1984, top cartel leaders stayed in Panama, during a period of Colombian, government's anti-narcotics operations. That did not mean, however, that the drug lords could do whatever they wanted: he broke up a cartel of cocaine-processing laboratory in Panama that was operating without his permission. Later, Cuban President Fidel Castro served as a mediator in reconciling Noreiga and the cartel leadership during a meeting in Havana in 1984.

Ties with Mossad

It was in the early 1980s that ties developed between Noreiga and Israeli intelligence. One of Noreiga's closest advisers until the US Military crackdown, was former Mossad operative Mike Harari, who profited from business deals in Panama and who also helped arm and train security forces for the Noreiga regime. According to US Military officials, Harari found out about the US Military operation eight hours in advance and fled to Israel. The Israel Government claimed that Harari is no longer linked to Mossad, but US intelligence believes that he was closely linked to the Israeli Government and that Noreiga received payments from Israel. The Israeli interest in getting close to Noreiga was based on the fact that agents of the PLO and Libya were operating out of Panama city, according to officials.

KGB Connections

The Soviet Union and Panama do not have diplomatic relations, but the KGB presence in Panama city had been increasing since 1986. According to intelligence officials, Aeroflot, the Soviet airline, had planned to open an office in Panama city to be used as a cover for KGB operations.

According to former DGI Major Florentino Aspillaga, who defected to the US, Noreiga permitted the DGI to establish numerous front companies in Panama during the 1970s so that they could be used to smuggle high technology and other scarce goods into Cuba.[1]

Noreiga Trial

Deposed Panamanian strongman Manuel Antonio Noreiga, dressed in a neatly pressed five-star general's uniform, came face to face with justice in a Florida court. American style and decided that he wanted no part of it: he refused to enter a plea on drug charges and would not recognise the court's right to try him. During a brief hearing, the dimunitive general—many—people were surprised that he was only five feet three—listened to the case against him through a headset that provided a Spanish translation. He spoke briefly to answer some questions from US district judge William Hoeveler. When the judge asked him if he had used drugs under psychiatric care, Noreiga snapped "absolutely not".

Earlier the judge over-ruled defence efforts to prevent Noreiga from making a court appearance and subsequently entered a "not guilty" plea for him when the general refused.

The 15-minute preliminary hearing marks the first step in what promised to be a protracted legal battle. Defence attorney Frank Rubino estimated that it would take nine months to prepare the defence. He indicated the defence team would subpoena a broad range of federal documents showing Noreiga's ties to several US intelligence agencies. According to defence lawyers, the formal defence of Noreiga will begin in earnest with a motion to have the case thrown out for lack of jurisdiction, because Noreiga was a head of State and therefore immune to prosecution, because he cannot get a fair trial, and because the arrest occurred in coercion.

However, the US Attorney in Miami, Dexter Lehtinen, dismissed those claims saying, "everything the US did was legal under national and international law". Defence attorney alleged that the general was duped into surrendering when officials at the Vatican Embassy in Panama said they would move the embassy elsewhere if Noreiga stayed past a deadline. That frightened the general into surrendering.

According to a Drug Enforcement Agency source, the general stayed true to his usual superstitions and was wearing the red underwear he believed would ward off evil. Noreiga was housed in a special security cell inside the Miami federal court system.

Mr. Noreiga, who took refuge in the Vatican Embassy in Panama City on Christmas eve, left on Wednesday night and surrendered to US Drug Enforcement Agency officers at Howard Air Force base in Panama. He was immediately flown on a US military plane. Meanwhile, a report from Panama City said, Noreiga's wife left the Cuban Embassy and headed for political asylum in the Dominican Republic.[2]

Noreiga Refuses to Submit to US Court

The fallen Panamanian dictator, Gen Manuel Noreiga, told a federal judge that he is a political prisoner and refused to enter a plea on charges that he took $ 4.6 million to turn his nation into a way station for the cocaine trade. Gen Noreiga's attorney, Frank Rubindo, told US District Judge, William Hoeveler, that "General Noreiga refuses to submit to the jurisdiction of this court because he is a political prisoner brought to this country illegally". The judge entered an innocent plea for Gen Noreiga.

Gen. Noreiga who listened to his attorney through a translator, was dressed in olive green trousers and a khaki uniform shirt with four stars of a general on his epaulets. When questioned by the judge, he stood erect with hands clasped tightly behind him in a military bearing. The 51-year-old Panamanian answered softly in Spanish and stared directly at the judge, showing no emotion.

The defence said it could take nine months to prepare the trial. US Attorney Dexter Lehtinen said he thought the trial would take only two months. An indictment accuses Gen. Noreiga of accepting $ 4.6 million in bribes from Colombian drug cartels to aid and protect their smuggling and processing-charges that carry maximum penalties of 145 years in prison and $1.1 million in fines.

He is not coverd by the new US law that allows execution of drug kingpins, an assurance Gen Noreiga reportedly demanded before turning himself in. He also faces drug charges in another indictment. US prosecutors said that once ousted Gen. Noreiga gets through fighting cocaine-smuggling charges in Miami, marijuana-smuggling charges await him in Tampa. "The plan is to try him in Miami and then bring him to Tampa", said Tampa-based US Attorney Robert Genzman, "No matter what happens with the Miami case, he will come to stand trial".

Although the two drug indictments were filed on the same day in February 1988, they were aimed at different parts of Gen Noreiga's alleged drug dealings. Kalish, who pleaded guilty to federal drug-smuggling charges for a reduced 14- year sentence, was one of the Tampa smugglers who said, "He entered into a partnership of convenience with Gen Noreiga in 1982".

US Denial

The commander of the United States who invaded Panama, denied that American forces had killed large numbers of civilians during the operation to overthrow Gen Noreiga. "There is absolutely nothing to that", Lieutenant General Carl Stiner told a group of 29 visiting US Congressmen, responding to what he said was undue criticism in Washington over civilian casualties, caused by US forces. According to US officials at least 250 civilians were killed in the December 20 invasion. Felicidad Siero De Noreiga, wife of deposed Gen. Noreiga along with her three daughters, brother and son-in-law took refuge in the Cuban embassy the day after the December 20 US invasion and headed for political asylum in the Dominican Republic.

Paris Justice Minister, Mr. Pierre Arpaillange, said that France had frozen a bank account in the name of General Noreiga's wife. Mr. Arpaillange, who was interviewed by the French Europe-I radio station, did not name the bank involved to put a figure on the amount. He said the account was frozen on a request from either the United States or Panama.

Bonn West Germany has granted a United States request to seize bank accounts held by Gen. Noreiga in West German institutions, officials here said. US authorities filed an official request with the West German institutions, officials here said.[3]

Noreiga Won't Plead Guilty

Attorneys for the ousted Panamanian ruler, Gen Manuel Antonio Noreiga, said they aren't seeking a bargain plea and will move ahead with plans, to bombard a federal court with motions to dismiss drug-trafficking charges.

Gen. Noreiga remained hidden from the public eye, reportedly in subterranean room at the Federal Court House where he met several times with defence attorneys. "A man who is innocent is not receptive to any plea bargain" the defence lawyer, Mr. Steven Kollic, said after a two-hour conference with Gen. Noreiga, "Gen. Noreiga is not guilty of these charges and if we can find a jury of 12 people who will have an open mind...General Noreiga will be acquitted".

Under the US legal system, a plea bargain is a deal worked out between prosecutors and defence attorneys, under which a criminal defendant agrees to plead guilty to some of the charges and co-operate with investigators. In exchange, prosecutors dismiss the remainder of the charges and often pledge to recommend a lighter sentence to the judge hearing the case. Mr. Killin said he has "absolutely" ruled out a plea bargain with the government. The US justice department officials have been quoted in published reports as saying they might be interested in striking a deal in exchange for information about Latin American drug networks. Federal prosecutors also said they were gearing up for a trial.

No Deal

President George Bush also denied that the Administration was seeking a deal with Gen Noreiga. Mr. Bush refused to rule out a plea bargain, however. He said Gen Noreiga has the same right as any federal prisoner to seek a deal. Speculation on a plea bargain is linked to the possibility Gen Noreiga's trial exposing embarrassing details of his long time connections with the CIA. Mr. Bush declined to comment on the controversy about his dealings, first as CIA chief in 1976 and then as vice-president, with Gen. Noreiga, who headed his country's intelligence service. He told reporters that he had made some statements on the subject in the past and "I will have nothing more to say about it".

General in Tears

The Washington Times reported that Gen Noreiga was in tears aboard the transport plane that brought him to the US. "I really put my foot in it this time," Gen Noreiga was quoted by the US drug agents aboard the plane as saying." I should not have left the nunciature. A source familiar with the flight said Gen Noreiga carried a Bible in one hand and three rosaries in another when he was brought to the plane.

In Washington, US intelligence and law enforcement officials began pouring over tonnes of documents captured by the military in Panama, seeking information on subjects ranging from possible double agents to arms and drug trafficking, an intelligence source said. According to a Vienna report, Felicidad Fiero Noreiga, wife of Gen. Noreiga has $ 1.1 million stashed away in an Austrian bank. The mass circulation daily *Kurier* and the political magazine *Profil* said Mrs. Noreiga had opened the account in a Vienna branch of the Erste Oesterrejchische (First Austrian) Bank. Mrs. Noreiga was holed up in the Cuban Embassy in Panama.

Support to Invasion

A report from New York says Panamanians almost unanimously support the US invasion that toppled the General, despite casualties that personally touched nearly a quarter of the population. A poll there found more than six in 10 of 794 Panamanians "polled expect the invading US troops to remain there for six months or longer, and nearly eight in ten said they want the soldiers to stay that long or "as long as necessary."[4]

Castro to Testify

Mr. Rubino told the newspaper that he flew to Cuba last month where Mr. Castro gave him documents which will "greatly help" Noreiga's defence. Mr. Castro is eager to assist the defence and may well testify for Noreiga in a video-taped deposition, he was quoted as saying. Noreiga, whose trial is set to start next year, has already been granted prisoner of war status. As POW, he is regularly visited by an American army officer and is allowed to wear uniform.

The defence lawyers asked for dismissal of the case against Noreiga after the United States government froze all accounts of Noreiga claiming that they contained profits from drug sales. The lawyers are arguing that in the absence of funds, Noreiga would not be able to defend himself and, therefore, the case be dismissed. The defence had subpoenaed almost all major American intellligence, defence and foreign policy agencies and Mr. Rubino said he intends to subpoena present and former intelligence officials and documents.[5]

A video-taped testimony by Castro, appearance of Americn intelligence officials and perusal of hundreds of classified documents by the judge and lawyers is likely to mark the trial of the former Panamanian leader, Gen Manuel Antonio Noreiga, the New York Times reported yesterday, reports PTI.

Noreiga was seized by American forces during invasion of Panama and brought to the US to stand trial on charge of drug trafficking. The Washington government may also disclose the amount it paid to Noreiga during the time he worked for its intelligence agencies. But this could itself become a bone of contention as American officials were quoted as saying that the amount paid was much less than claimed by Noreiga. "Fidel said he wants to help and is willing to do whatever he can to show Noreiga is innocent", the paper quoted Noreiga's chief lawyer, Mr. Frank A. Rubino as saying.

Cuba Drug Scandal

The drugs scandal that is rocking the Cuban government and Communist Party is more than the show trial of a former comrade who made a pact with the infamous Medellin cocaine cartel of Colombia. It suggests that some important changes may be taking place in Cuba, and also in Central America.

Gen Arnaldo Ochoa Sanchez, number three in Cuba's hierarchy after President Fidel Castro and his brother Raul, the Defence Minister and armed forces chief, has shocked his country's revolutionary establishment. He was found guilty this week by a military tribunal of 47 Cuban generals of accepting 3 million over a period of two and a half years for assisting in the smuggling of six tonnes of cocaine into the US. With him have been arrested some six other senior officers from the armed forces and Interior Ministry. The Interior Minister, Gen Jose Abrantes, was obliged to resign on Thursday. For Cuban society, in which the armed forces and Interior Ministry have symbolised revolutionary purity and dedication, this fall from grace is like a blow to the very soul of the revolution.

It is not the first scandal to hit the government. But President Castro has always been a clever politician. Whether the scandals were economic and planning failures, cases of bribery and corruption, or military desertion, the Cuban leader until now has been able to lay the blame for the revolution's shortcomings on individuals rather than on the revoultion itself and thereby head off political reform.

If Gen Ochoa is awarded the death penalty, as requested by the military tribunal, the severity probably has as much to do with the spreading problem of drug abuse in Cuba as with the now publicly acknowledged drug trafficking to the US.

United States Complaint

The US has been complaining of a Cuban government drugs connection since 1982. Cuba lies in the direct flight line of light aircraft carrying drugs from the Columbian coast to southern Florida. Significantly, the accusations of Ochoa's drug involvement date back to 1987, when the Medellin cartel began to encounter difficulties in refuelling through its traditional conduit, the Bahamas. It is not inconceivable that President Castro might have turned a blind eye as long as Cuba itself was not affected, in order to weaken his arch enemy, the US. But now there are many reasons for wanting a show trial and exemplary punishment of Gen Ochoa.

Severe economic problems are undermining the authority of the Cuban regime, which is now concerned to improve relations with the US. Furthermore, Cuba, a member of Comecon, the eastern bloc economic grouping, was looking more isolated than ever in the Gorbachev era. It was under great pressure from the erstwhile Soviet Union to adopt political and economic reforms along the lines of glasnost and perestroika. President Castro has resisted this pressure, choosing to clean up the bureaucracy rather than reform it.

This, however, was the most far-reaching purge since the 1959 revolution. Scandal had never reached so high in the system. And like Iran-Contra affair in the US, a serious doubt existed as to whether the chief executive was entirely ignorant of the goings-on of his subordinates. Officially there was no drugs problem in Cuba. Segments of Cuban society nevertheless do have access to cocaine and marijuana. The problem had spread its tentacles beyond artists and bohemians to youth, blacks and now, it seemed, the armed forces.

But the unresolved mystery of the Ochoa case lies in his motivation. As a decorated hero of the revolution, the former commander of Cuba's troops in Angola and Ethopia, he had reached the peak of his career. He was part of the revolutionary elite, with access to privileges denied the average Cuban. Would the risk of losing these be worth a share in $ 3 million which he could not in any case spend in Cuba without arousing suspicion? The answer may prove to be more complex than a simple desire to secure a nest egg for exile and retirement. Some observers of Cuba's closed ruling group believe that Gen. Ochoa might have had own political ambitions, for which he was amassing a war chest.

There could be wider implications. President Castro was reportedly approached by the US to mediate in the long-running Panama crisis, precipitated when US authorities accused Panama's military strongman Gen. Manuel Antonio Noreiga of drugs trafficking. If there was a drugs connection between the Panamanian and Cuban military, then Gen. Ochoa's arrest would be a logical first step to break it. President Castro's best efforts to secure Gen Noreiga's withdrawal from power could, some analysts now believe, be traded for an easing of US sanctions against Cuba.

Nicaragua Connection

In Nicaragua too there seemed to be a connection. Cuba and Nicaragua were the only countries in Latin America that continued to express support for Gen. Noreiga. Col. Antonio de la Guardia, also accused of trafficking, stated that Gen. Ochoa, by his own admission, had drugs "connections" in Nicaragua. Gen. Ochoa was for two years the chief of the Cuban team of military advisers to Nicaragua. For years the US had fired drugs accusations at both countries. But these had been largely ignored as rhetorical petards in the intense anti-communist crusade Washington had been waging against Havana and Managua.

The Ochoa scandal will revived these accusations and put pressure on the ruling Sandinistas in Nicaragua to make their own investigations public. The defections to Costa Rica of a senior Nicaraguan officer, Lt. Col. Douglas Zuniga, together with his brother, an army lieutenant, set off speculation that they were linked to the Ochoa affair Nicaraguan military patrols had been stepped up along the Costa Rican border and restrictions had been tightened on light aircraft flying near that frontier even though there had been almost no military activity there by the largely defeated US-backed Contra guerillas. The reason could be to impede further defections before the net closed, either on the real offenders or on scape-goats.

If the Sandinistas had their own house-cleaning to do, now was a good time, allowing them to treat it as an adjunct to more sensational Ochoa affair rather than in an electoral campaign and under the international spotlight.[6]

Crackdown on Drug Dealers in Cuba

Cuba had been rife with rumours about the real reasons for the execution of Maj. Gen. Arnaldo Ochoa for drug smuggling. As the scandal widened, observers also wondered how many other high-level officials were involved in shipping illegal drugs through Cuba.

Fidel Castro had denied any knowledge of the narcotics schemes, but a lawyer for convicted cocaine smuggler Reinaldo Ruiz said that his client had directly implicated Castro and his brother Raul in the trafficking. Maj. Gen. Arnaldo Ochoa Sanchez was arrested in June and charged with being part of a group that conspired with Colombia's Medellin cartel to ship more than six tons of cocaine into the United States between 1987 and 1989. The scandal resulted in the court-martial of 14 army officers and senior state security personnel and the execution of Ochoa and three others.

In the wake of the trial, other officials were purged from sensitive government posts. Gen. Jose Abrantes Fernandez, director of the Interior Ministry, was fired, and two Ministry officials were executed. Also named in the investigations were a top secret department of Cuban intelligence and an agency controlling private and commercial air traffic over Cuban territory.

Throughout the trials, it was claimed that Fidel Castro knew nothing about the drug trafficking. However, rumours had spread like wild fire, both about a hidden motive for Ochoa's execution and about Castro's own knowledge of the smuggling. The charismatic and popular General Ochoa, Cuba's most decorated officer, was a much admired hero of the war in Angola. He also attained high distinction in Ethiopia and Nicaragua and earlier fought alongside Castro in the revolution. Several observers had pointed out that the real reason for Ochoa's execution was that he could have become a threat to Raul Castro's hopes to succeed Fidel as President of Cuba. They noted that Raul had personally announced Ochao's arrest and, in his speech for the prosecution, accused Ochoa of "unrestrained populism". Ochao was also castigated for his "popular character" and "wit". Others believe that the execution was ordered so that Ochoa would not implicate other officials whom Castro wanted to protect. These "others" might include even Castro himself. In addition, Ochoa may have alienated himself from Castro by being too outspoken. In the weeks before the trial, he was said to have been openly critical of Castro's rejection of glasnost and perestroika.

It is widely believed that if Ochoa did traffick in drugs he did so on behalf of the Government. Cuba has allegedly used the proceeds of drug smuggling to obtain illegally high-technology items form the United States and to support revolutionary activity in other Latin nations. Cuba needed the injection of cash badly after the money flow from the Soviet Union slowed down, forcing the country into drug dealing on a large scale.

The Cuban Government continues to be extremely cautious in saying anything about the four executions. It has been widely reported that the relatives were not permitted to see the bodies or learn where the men were buried. In fact, shortly after the trial, three human rights leaders were arrested, allegedly for telling foreign journalists that Ochoa had been drugged by Cuban authorities before his trial appearance.

But Cuban involvement in illegal drug trafficking has long been known. In 1982, four Cuban officials were indicted in Florida along with Jaime Guillot Lara, a Colombian drug trafficker, for running a secret drugs—for arms network since 1975. That trial revealed the involvement of the Cuban Navy, the America's Department of the Cuban Communist Party and a senior Cuban diplomat in providing official Cuban protection for drug-laden vessels. Testimony indicated that Castro received huge payoffs, up to US $ 500,000 per shipment, which were used to support revolutionary activity in Latin America.

References

1. *The Hindustan Times* New Delhi, 9 January, 1990, p. 13
2. *The Hindustan Times*, New Delhi, Jan. 6, 1990, p. I
3. National Herald, New Delhi, January 9, 1990, p. 9
4. *The Times of India*, New Delhi, Sunday, January 7, 1990, p. 7
5. *The Times of India*, New Delhi, May 22, 1990, p. 16
6. *The Economic Times*, New Delhi, July 6, 1989, P. 10

43

LAND OF DRUG DONS : COLOMBIA

A chance encounter with a band of rebels or a hunt for gems in a jungle river can mean life or death in Colombia, a dangerous land long before the drug lords began their latest war of terror. The government's war with narcotics traffickers has captured the world's attention while partially obscuring the fact that Colombia, for many years, has been a violent place to live. More than 75,000 people were killed in political violence in 25 years, most of them in attempts by half-dozen leftist guerilla groups to overthrow the Colombian Government. Between 1949 and 1953, 250,000 people died in a war between the nation's two biggest political parties. Such was the hatred in that conflict that entire villages, including elderly people and children, were wiped out when dump trucks would pour screaming prisoners over steep cliffs.

Today, guerillas execute their own men for treason, drug traffickers wage war on other drug traffickers, bandits battle for control of the nation's emerald mines, and right-wing death squads issue periodic reminders that they are still around. Additionally, about 12,000 people are killed every year by criminals, the national police said in a recent communique.

Colombian Violence

The enormity and intensity of Columbian violence makes Colombians scoff at foreign news describing them as panicky and terror stricken, a population huddled indoors while the drug war is played in the streets.

Colombia's emerald mines, for example, are regular scenes of violence. Dozens have been killed as bandits struggle for control of the mines in the hot, steamy parts of the country. An army colonel once told an Associated Press reporter who visited one of the mines that there is an average of one murder a day among the people who look for emeralds. Emerald buyers who go to the mines travel in convoys and carry machine guns and hand-grenades. Many times in a year, the convoys coming from the mines are ambushed by bandits. Rightwing death squads have taken a heavy toll in Colombia, killing more than 1,000 people in the past four years.

The Leftist Patriotic Union Party has been the hardest hit with 751 of its members gunned down or blown up. Colombia's Justice Department said recently that it suspects the death squads are operated by drug traffickers. Guerillas have long extorted money from cattle ranchers in the form of a so-called revolution tax. In recent years drug traffickers have bought up hundreds of ranches. When the traffickers refused to pay the revolution tax, the guerillas began kidnapping ranchers for ransom. The death squads are believed to be the durg traffickers method of settling the score with the rebels.[1]

Rival Colombian Cocaine Cartels Join Hands

Rival Colombian cocaine cartels may be joining forces to keep smuggling lines open during their government's crackdown, according to a US Federal agent investigating a record cocaine seizure.

Drug agents were surprised to find cocaine from both the Medellin and Cali cartels in a Los Angeles warehouse where 20 tonnes of cocaine were seized, Joun M Zienter, the drug enforcement administration's special agent for Los Angeles said : "The cartels are named after the Colombian cities where they are headquartered. The cartels are trying to get much of the cocaine out of the country and to maintain distribution channels".

Two Drug Lords Offer To End War, Share Profits

Colombia's top two cocaine lords want to negotiate an end to their war with the government and have offered to pump some of their illicit profits into the nation's industrial sector. Meanwhile, two men confessed to killing Senator Luis Carlos Galan, the popular presidential candidate whose assassination touched off the war between the government and Colombia's drug lords.

The president of Colombia's House of Representatives, Mr Norberto Morales, said at a news conference in Medellin that he was telephoned by Pablo Escobar and Gonzalt Rodriguez Gacha, kingpins of the infamous Medellin cartel. A tape recording of the news conference was broadcast by Bogota radio stations. There was no answer at the Morales office in Medellin. Mr. Morales said Escobar and Rodriguez Gacha offered to invest millions of dollars in Colombian industry if a truce was reached. He said he had passed the information to the President, Mr. Virgilio Barco. This would not be the first time that Escobar, Rodriguez and other suspected members of the Medellin cartel have tried to buy amnesty. In 1984, the President, Mr. Belisario Betancur, even sent the former President, Mr. Alfonso Lopez Michelsen, to Panama to meet with members of the cartel. The cartel offered to pay off Colombia's foreign debt, then about 10 billion, if Colombia would end the war declared on durg traffickers after the Justice Minister, Mr. Rodrigo Lara Bonilla, was murdered. The government rejected the offer, but a few weeks later the army stopped confiscating property of suspected traffickers.

The Medellin mayor, Mr. Juan Gomez Martinez, has been the biggest advocate of negotiations with drug traffickers. A US embassy official said earlier this month that the traffickers' specific demands have included full pardons in exchange for promoting all of their drug operations from Colombia, perhaps to Bolivia and Peru. The officials also said traffickers offered to invest part of their fortunes in helping government security forces rid the country of Leftist guerillas. Mr. Barco refused to conduct negotiations with the traffickers. The Forbes magazine, US business publication, ranked Escobar as having between $ 2 billion and $ 3 billion. Colombian media have reported that all five members of the Medellin cartel are billionaires.

The US Justice Department had on priority the capturing and extradition of 12 Colombians facing drug charge in the United States, including Escobar and Rodriguez. None of the 12 had been captured. The Medellin cartel is responsible for 80 per cent of the cocaine that is smuggled into the United States.

The newspapers quoted General Ramon Nicoles as saying the army knews who hired the two men who confessed to killing Senator Galan, the leading presidential candidate when he was assassinated, at a Bogota rally. But the General did not say who ordered the killing. The two men who confessed were part of a band called "the blackies", The general was quoted as saying that the two other men in the band thought to have set off a 227 kg bomb that seriously damaged the Bogota daily EL Espectador.

The army in the past was accused by Amnesty International and Colombian Human Rights Committee of forcing confessions by torture. Colombia's armed forces have the authority to arrest and detain people under a state of siege that had existed in Columbia for all but a few months of 40 years.[2]

Outraged Colombia Fights Drug Lords

A society outraged by the assassination of its most dynamic young statesman is displaying dramatic, if still tentative, signs of a widespread commitment of battling the drug underworld despite terrifying risks. When five suspects in the slaying of Senator Luis Carlos Galan were paraded before 15 witnesses, the witnesses did not hesitate in identifying four of them, the police said. They also said special telephone numbers set-up to receive tips provided information that enabled the police to raid several of the hundreds of drug lords' estates and businesses.

"We've had a tremendous cooperation from the people", the National Police Force's press chief, Col. Alfredo Salgado, said in an interview. "The people have always wanted to cooperate with the police, but they were afraid", He said "they have left aside their fear and are cooperating." He said the cooperation was inspired by citizens' witnessing the police, for the first time, engaged in a virtual all-out war with the country' drug cartels. The crackdown was sparked by the assassination of the senator, a police colonel and a magistrate within a 48-hour period.

Colombia's drug lords stepped up a campaign of intimidation and violence, killing one person in a bomb attack and declaring "absolute war" in response to a crackdown on the narcotics trade. Drug gangs bombed two political party offices and burned the homes of two politicians in Medellin, the police said.

Emergency Powers

President Virgilio Barco issued new emergency decrees requiring suspected traffickers to come forward or give up $ 200 million worth of seized property, including hundreds of mansions, ranches and airplanes. Invoking emergency powers under a four-year-old state of siege, Mr. Barco decreed that the suspects must make personal claims of restitution for property seized by the police and the military. The Defence Ministry estimated the total value of that property at approximately $ 200 million and said it also included restaurants and shops, office buildings, helicopters, cars and trucks.

The crackdown was prompted by the assassination of Luis Carlos Galan, leading presidential candidate who promised to combat drug trafficking. Mr. Galan, a critic of Colombia's dependence on the drug trade, was gunned down just before he was to deliver a campaign speech. The police said drug traffickers had placed a 5,000,000 dollar price on his head. His death, coupled with the slayings of a national police official and a magistrate, triggered an outpouring of public anger as many Colombians hoped the battle against the cocaine barons could be won this time.

The police said one person was killed in the bombing of Liberal Party offices in Medellin, an industrial city that serves as a base for a crime cartel. The conservative party's head quarters also was bombed, the police said. Politicians said the wave of drug sponsored and political violence jeopardized the country's electoral process. According to the former president, Mr. Michael Pastrana Borrero, leader of the Opposition Conservative Party, "The electoral process is badly compromised" by killings committed by drug dealers and leftist guerillas.

According to a Washington report the US government responding to bomb attacks in Colombia, offered a 2.5 million-dollar aid package to help train Colombian court guards, buy armoured cars and protect judges involved in the drug crack down there.

Colombia Extradite 4 Drug Peddlers to US

On August 24, 1989, Colombian government announced that it would extradite four suspected drug traffickers to the United States and urged Washington to do its part by cracking down on consumers of narcotics. The four included Edurado Martinez Romero, believed to be a top money-launderer and treasurer for the Medellin cocaine cartel, and Rafael Oriandes Gamboa, both wanted by the US Justice Department. The identity of the two other suspects to be extradited was not revealed, but none of the four were on the department's "most wanted" list of a dozen alleged top drug traffickers. The four suspected drug traffickers were among some 11,000 people arrested in an unprecedented crack-down on drug traffickers, sparked by the assassination of Mr. Luis Carlos Galan.

The police also reported the arrest of drug suspect Helena Beatriz Rodriquez in the coastal resort city of Cartagena. They said she was wanted in the United States on drug-related charges, but US officials in Washington said they were unfamiliar with her. The arrests came during an anti-drug campaign by police and the military prompted by last week's assassinations.

The President, Mr. Virgilio Barco, established emergency procedures for extraditing traffickers wanted in the US. The Supreme Court nullified the extradition treaty with Washington in June 1987, after years of intimidation of the judiciary. At least 220 judges and other officials have been killed. Mr. Barco also authorised seizures of traffickers' possessions. The police and soldiers continued raids on mansions, farms, office buildings, restaurants and other property believed to be owned by the drugs bosses.

Mr. Carlos Lemos Simonds, the Communications Minister, declared "there's a firm decision" to extradite suspects, but the US bureaucracy must act quickly. He said Mr. Barco's emergency measures empower the police to hold suspected traffickers for up to seven days without charges, but then they must be freed. Mr. Lemos also asked Washington to provide more aid for the antinarcotics campaign.

At his vacation home in Maine, US President, Mr. George Bush, said: "they might need certain technical assistance...They can use training for some of their forces—police, for example". Mr. Bush said Mr. Barco made clear "that he was not requesting United States troops...I know enough about this hemisphere, and I h've had enough experience in dealing with countries in Latin America...to know the constraints...in terms of despatching troops."

Meanwhile, the Bolivian government reinforced border patrols to head off Colombian drug runners from entering the country to dodge Colombian offensive against drugs. Police helicopters were placed on alert and additional guards were sent to Bolivia's northern border with Brazil. Mr. Gonzalo Torrico, the official responsible for drug interdiction, said the government was concerned that Colombia's war against cocaine would "generate a stampede of traffickers looking for refuge in areas they have contacts."

There are dozens of cocaine processing labs in northern Bolivia's Beni jungle region, which borders Brazil. From there cocaine is flown to Colombia and then distributed to the US or Europe.[3]

Cocaine Financier Extradited to US

The government put a $ 250,000 price tag on the heads of Colombia's top two reputed cocaine cartel bosses, and a US official said a reputed drug financier had been extradited to the US. Eduardo Martinez Romero, the accused financier was flown on a US government plane, said the drug enforcement administration official, Mr. Frank Shults in Washington. It was the first reported extradition to the US of a reputed cocaine cartel member since Colombia declared war on drug traffickers. The US officials in Bogota, had predicted even greater bloodshed in this violence-rocked country if anyone was extradited. The powerful Medellin cartel, the world's largest cocaine smuggling gang, had warned it would kill 10 Colombian judges for each person extradited. The last Colombian drug boss to be extradited was Carlos Lehder a leader of the Medellin cartel who was sent to the US in early 1987 and is serving a life sentence in prison there.

Earlier the Colombian Justice Ministry announced on national television the $ 250,000 reward on notorious cartel chiefs Pablo Escobar and Gonzalo Rodriguez Gacha The TV broadcast showed "wanted" posters of the two men. Escobar and Rodriguez Gacha, now in hiding, also wanted in the US. Both were on the list of the 12 most wanted Colombians issued by the US government. Martinez was not on that list, but was reported to be a financier for the cartel.[4]

Drug Lords Seek Amnesty, Offer to end Trafficking

The Medellin drug cartel offered to dismantle its trafficking empire in exchange for amnesty, but a spokesman for President, Mr. Virgilio Barco, said the traffickers would have to take their chances in the courts.

The cartel released two kidnap victims to back up its offer, which was similar to previous proposals rejected by the government, except for its acceptance of defeat. Also, a Colombian brother and sister were flown to the United States to face cocaine-trafficking charges. The drug cartel's communique followed Mr. Barco's comment, in response to newsmen's questions that his government is not "inflexible" regarding a negotiated settlement with drug lords. Earlier, the president had ruled out talks.

US Send Troops in Colombia

The United States was ready to send combat troops to fight the deadly war against the drug lords of Colombia, President Bush's Chief of Staff John H. Sununu said in a television interview on CBS. Sununu was asked on the "Face the Nation" programme if Bush "is willing to risk American servicemen in this war on drugs." Sununu said: "Well, if that request ever came in, I'm sure the President would recognise that the American public thinks that there is a great feeling among American citizens that we probably would take even that risk".

The private armies of the drug lords have been trained by tough Israeli and other mercenaries. With a large pool of international thugs of various nationalities, willing to fight for anyone for a consideration, the drug lords will not lack the means to fight a guerilla war, according to observers.[5]

The US Government is spending millions to help Peru, Colombia and Bolivia fight the drug cartels. Mr. Bush decided to fly to Colombia-despite security concerns of the US secret service—to provide support for President Barco's armed struggle with the Medellin and Cali drug cartels producers of about 80 per cent of the cocaine that floods the streets of Washington and other American cities.[6]

Assets of Colombia Drug Baron Frozen

Over 61 million dollars in assets, belonging to accused Medellin cartel drug kingpin Jose Gonzalo Rodriguez Gacha, had been frozen in his international bank accounts, US Attorney General Dick Thornburgh said. The Attorney General said that records found in a raid in Colombia and turned over to the US led to the seizure of 61.8 million dollars in cash and securities kept by Rodriguez Gacha in five countries. Action against the assets of the Colombian, reportedly of the cocaine cartel's ruling triumvirate, began with the freezing in Luxembourg of 10 bank accounts in which 39.4 million dollars in assets had been deposited. The following day, the US Justice Department initiated similar action that blocked 10.3 million dollars in Swiss banks, 5.9 million dollars in Austrian banks, 4.2 million dollars in British banks and 2 million dollars in US institutions.

Officials declined to estimate how much of the money in foreign banks would be forfeited. Mr. Thornburgh said the law of the four other nations would determine how authorities in those countries dispose of the money. "While Mr. Rodriguez Gacha remains at large, he has been divested of substantial assets provided by the illegal activity with which he is charged", Mr. Thornburgh said.

The seizures stemmed from a raid by Colombian authorities on property of Alfonso Rodriguez Munoz, a nephew of Rodriguez Gacha. Papers confiscated during the raid led officials to accounts managed by Colombian financial expert Mauricio Vives. The records indicated that Vives had been entrusted with investments of 81 million dollars. Before all the assets could be frozen, however, 20 million dollars were moved from other countries to banks in Panama, which was not very eager to cooperate with the US.

Mr. Thornburgh pointed out that "it is important to recognise these are not funds in the process of being laundered. They were long term investments in most cases. I believe that represented the stash of the Gacha organisation".

Drug enforcement agency administrator Jack Lawn revealed that the records seized in the raid might enable the DEA to trace the assets of other cartel figures.[7]

The Name that Spells Death

Pablo Escobar, the world's most dreaded cocaine trafficker, is one man whose presence is felt on Colombia's presidential election even though his name does not appear on any ballot paper. Whoever wins the election will have to decide how to handle the elusive drug baron, accused by the Government of helping mastermind a bloody campaign of bombings and assassinations in a bid to scare people away from voting.

The Government accused the Colombian cocaine barons of murdering three presidential aspirants. Escobar explicitly denied involvement in at least one of the killings. Escobar, who rose from obscurity to head a cocaine cartel with a turnover the size of a multinational corporation, evaded arrest even though he was in his home town, Medellin.

Escobar acted with impunity, building up an aeroplane fleet and buying hotels and apartments at home as also abroad until Justice Minister Rodrigo Lara Bonilla exposed the activities of the Medellin cartel and ordered the seizure of Escobar's assets.

Most Hunted Criminals

Lara Bonilla was murdered in April 1984 and Escobar went underground to escape prosecution. But it was not until the government cracked down on the drug trade after the murder of Liberal presidential candidate Luis Carlos Galan that Escobar became one of the world's most hunted criminals. Escobar and other drug lords declared an all-out war on the state, allegedly ordering bombings and assassinations which killed more than 300 people.

Catching Escobar had become a key test of Government credibility after a nine-month war with the drug barons which had won down the public's resilience and prompted many Colombians to clamour for peace at any price. Remarks by two Colombians as they left polling stations summed up the dilemma facing Escobar and other drug barons. One man, said he had voted for Mr. Gaviria because of his hard line on drugs. "He is young and has a young man's ideas and he is against drugs", the man said. But other said they voted for Mr Gaviria's opponents because they feared the retaliatory action to a continued crackdown against Escobar and other Colombian drug lords.

Gaviria Voted Colombian President

Voters ignored threats by drug traffickers and elected Mr. Cesar Gaviria with an overwhelming majority as their next President—the candidate who took the hardest line against the traffickers. Mr. Gaviria, 43, told supporters in a nationally televised speech that his victory represents a triumph for demorcracy in this violence-wrecked country.

"All those who have sought to threaten our democracy, listen closely. The people have spoken with courage and clarity, and we will triumph," the ruling Liberal Party candidate told a cheering crowd at a Bogota hotel.

Mr. Gaviria had been sworn in almost one year after the previous Liberal party candidate, Mr. Carlos Galan, was slain by a drug cartel hit-man. Mr. Gaviria, who had been Mr. Galan's campaign manager, then became the candidate. Two other candidates were killed in attacks allegedly by the traffickers.

The Caracol radio network estimated that just 45.5 per cent of the 13 million exercised their voting right. About six million votes were cast, compared with nearly eight million in the 1986 presidential election.

Kingpin of Cocaine Network Arrested

Rafael Munoz Talavera was away on business that day when US Drug Enforcement Administration agents trooped into his palatial 3 million estate in suburban El Paso. Only the caretaker was at home as DEA staff cars pulled into the Munoz family's circular driveway. Munoz's wife and three children were gone, too, leaving behind spacious closets filled with finely tailored clothes and a few scattered pieces of jewellery. Rifling through the closets, Federal agents found purses and wallets containing petty cash. The business matter that caused Munoz's absence that day was the seizure by Los Angeles area police and drug agents of 21.4 tons of Colombian cocaine in a Los Angeles warehouse—the largest cache of narcotics ever found in a single location. As the alleged head of the Mexican smuggling operation hired by the Medellin and Cali drug cartels to ship 3 tons of cocaine a week into the United States, Munoz lost a considerable portion of his inventory, according to US and Mexican law enforcement authorities.

With the heat on in the days after the raid, Munoz disappeared into the underworld of Ciudad Juarez, a bustling border town of 1 million people across the Rio Grande from El Paso. His capture a month later, in a cramped motel room less than a mile from a border checkpoint, proved to be the most significant development in the investigation. "Next to the Colombians, you can't get a bigger guy than Munoz", said Phil Jordan, the DEA's special agent in charge of the Dallas Division.

American and Mexican law enforcement officials said that Munoz's arrest—and a subsequent "confession" narcotics kingpin made to investigators, provided the most definitive links to the Colombian drug cartels. Mexican federal authorities said they had issued a warrant for a Colombian associate of Munoz who allegedly worked with him in Juarez, supervising the smuggling of 3-ton shipments of cocaine flown in each week from the Colombian interior.

Jose Luis Alcalde, a prosecutor with the Ministerio Publico Federal (the Mexican Justice Department) said that the Colombian was believed to be a member of the Medellin cartel.[8]

To Crush the Cartel

Colombia, the South American nation blessed with the world's finest coffee crop and exotic tropical fruits and flowers and at the same time notorious as the home of the international cocaine cartel, re-emerged as a resurgent democracy after going through a bloodbath caused by narcotics terrorists. Its President, Cesar Gaviria Trujill, who took over on August 7, 1987 had vowed he would fight to the finish the war on drug traffickers, began by the outgoing President, Virgilio Barco.

The people of Colombia, who ignored threats of an attack on polling day held by the drug cartel, voted in their millions—the country has a population of 31 million, of which 18 millions are voters—and elected the 43-year-old Gaviria in the hope that he would pick up the threads from where Barco left and crush the cartel dug in the city of Medellin. "We have to break terrorism, we have to defeat terrorism", said Gaviria who was heavily guarded against any assasssination attempt by the goon squads of the cartel. "We don't have many alternatives. We cannot allow those practising terrorism to believe that they can change government policies by planting bombs."

At the same time, Gaviria reminded the United States, the world's biggest consumer of cocaine, that mere words will not suffice but active financial assistance was necessary, on Colombia's own terms, to continue the fight. He called for vigorous steps on the part of the US to check drug consumption in that country. "If they fail, the tremendous effort being carried out by Colombia, and our great sacrifices, will be lost in the face of rising demand", he said. Gaviria said Colombia was in the present crisis because of the ability of criminal organisations to kill or intimidate any judge in the country who interfered with their interests and went away scotfree. During the election campaign, three presidential candidates were assassinated. A respected Senator, Frederico Estrada Velez, who had supported Gaviria's candidacy, was gunned down while travelling in his car to his home.

A week before polling, there were more killings by gunmen riding motorcycles. Had a powerful bomb not been defused in time by a vigilant guard, it would have spelt tragedy for 500 children, including two granddaughters of President Barco, at a school. A dozen car bombs claimed the lives of 40 people in the cities. Some 320 persons were hospitalised with injuries sustained in explosions.

While the terrorists' attempt this time was aimed at preventing large turnouts at the polling centres, their targets of assassination and intimidation earlier had been judges and magistrates. They killed the judicial officers if they were not successful in compelling them to resign. The result was the demoralisation of Colombia's judges and magistrates many of whom resigned or sought premature retirement. The cartel was fighting both for its survival and for retaining its ill gotten wealth. It did not lack weapons or manpower, nor money. The assessment is that the drug cartel, including the Medellin group, had more money at its command than some of the Latin American Governments had in their treasuries.

The international drug economy runs on cash says a US Government paper. Worldwide drug profits, from about 80 per cent of the total narcotics trade, were estimated as high a figure as $ 240 billions (Rs. 4,08,000 crore) in 1987. While the cartel spent huge sums on intermediaries and on buying influence, the main beneficiaries were a small coterie of drug overlords who have acquired vast real estate and corporate interests and political power throughout the hemisphere.

Barco believed that the cocaine traffickers, facing the worst challenge of their lives from the Government forces, exploited the election process to widen their goal and were determined to destroy democracy to protect their interests. Barco, during the campaign, had stressed that his aim was to defeat their designs by mustering a 50 per cent turnout on polling day to drive home the message that the masses were not intimidated by them. "There would have been 500 children dead", he said, alluding to the attempted bombing of a school in Bogota. "It just shows the perversity of these people."

The elections did produce a 50 per cent turnout, Barco's message went home. Ever since Barco hit world headlines by declaring war on the drug traffickers, the bulk of the Colombian armed forces and the police were on a round-the-clock vigil to prevent any mischief by hit men. The Government spent annually about $ 1 billion (Rs. 1,700 crore) on the war—an extremely high commitment for a developing nation. It come out with innovative ideas to raise additional funds, the latest being "war bonds." Surprisingly, the election day was singularly calm. Earlier, the Government had seized some four tonnes of explosives and rounded up a number of persons.

But one top man who was said to have masterminded the rash of pre-election bombings still eluded the security dragnet. He was identified by the news media as Pablo Escobar, leader of the Medellin group. "Pablo Escobar is on the run in this country, apparently not staying in the same place for more than a day", Barco said, "We are following him closely. Sooner or later, we will find him."

All eyes were now on Gaviria. He is a bright economist, he became prominent as Finance Minister.

The manner in which Gaviria came to be chosen as presidential candidate is an emotion-swathed story. He was a protege of an assassinated presidential candidate of the Liberal Party, Luis Carlos Galan. At the funeral of the much revered Galan, the 17-year-old son of the deceased leader embraced Gaviria and said, "We entrust in you the banner of my father. You can count on our support to become the President that Colombia needs. Save Colombia."

American Initiative

Nearly a year's relentless pursuit of the cartel's rank and file had resulted in some extraditions to the US and seizures of substantial cocaine stocks, but the cocaine mills produced faster and ringleaders such as Escobar were still at large.The pace of drug traffic did not show any signs of slowdown either.

Experts in the US, to which much of the Colombian cocaine goes, say the law of supply and demand remains unaffected and cocaine consumers get the "stuff" they want without any difficulty. All narco-rich South American leaders believe the drug problem could be brought under control only if the US as a major consumer nation is able to bring down the present unconscionably high rate of drug use.

President George Bush, on his part, agreed. "Let's face it: Americans cannot blame the Andean nations for our voracious appetite for drugs", he said. Instead of blaming these countries, the US should work with them. But the difficulty comes in financing enforcement operations and in providing economic assistance to help coca-producing countries to end such production. The US provides some assistance but this is not considered adequate by the recipients in the Andean coca belt.

Future

Gaviria gone on record saying. "Drug trafficking is a multilateral problem, and industrialised countries aren't doing much to confront the problem of drug consumption. If they don't do this, it will become increasingly difficult for the Colombian people to understand that their sacrifice isn't being matched."

Colombians were, elated that they went through the traditional democratic process of elections once again for choosing their President despite threats of disruption by the drug cartel.[9]

Under Siege

February 15, was a bad day in Colombia—typically bad. On Bogota's Avenida Decima, in a district of flossy boutiques and expensive hotels, two car bombs exploded within five minutes of each other, showing flame, shattered glass and ragged steel in all directions. Four people were killed, 117 wounded. The bombs were believed to have been planted by goons working for drug baron Pablo Escobar Gaviria. Later the same day, in Madellin, the world cocaine capital, assassins killed two policemen, two off-duty soldiers and six teenagers in a billiard hall; seven were injured. Every day like that—and there were many—was another blow to the standing of President, Caser Gaviria Trujillo, 45, who took office in August 1990, with the promise that he would put an end to "narco-terrorism" within a year. Once compared by many Colombians to John F. Kennedy for his vigor and flair, Gaviria was now denounced in many quarters as a failure, largely because, despite some gains in the drug war—he had not been able to fully deliver on his pledge.

Narco-terrorists Spread Terror

In late 1991, the Colombian government took-out full-page advertisements in the Washington Post and the New York Times boasting that leaders of murderous Medellin cocaine cartel were "either dead or in jail", including the king-pin Escobar. Escobar had the last word, however. Several months after the ads ran, he escaped from a well-appointed jail in Envigado, outside Medellin, when he had agreed to be imprisoned after the government promised it would not extradite him to the US to face drug charges. Ever since, Escobar and his team have been on a rampage. Not long after Escobar's escape, a judge, who was about to convict him in absentia of a 1986 murder, was shot dead. In the seven months since the jail break, 70 Medellin policemen were gunned down, as were countless members of competing narcotics gangs. While most of the major drug lords, who dominated the Medellin cocaine trade in the 1980s were indeed in jail or dead, a new generation had replaced them.

The situation was, in fact, worse than before. The random violence terrorised daytime living. Smothered night life in Bogota and Medellin and terrorised daytime living. Bogota Mayor, Jaime Castro banned streetside parking in down-town areas. People left work early to avoid potential trouble after dark, A telephone intimidation campaign aimed at foreign corporations, schools and embassies left the recipients jumpy. Some firms like US-owned Occidental Petroleum, decided to repatriate all non-Colombian executives and technicians. We are under siege", said General Oscar Pelaez, the Bogota Metropolitan Police Chief.

Money Laundering

Clearly, new bosses among the cocaine, and increasingly heroin, traffickers were trying to tighten their clamp on Colombian Society. Fernando Corrales, the government's agrarian reform director, said that they had one-third of the country's best agricultural land, both to launder money and to establish a political base. In Washington, Craig Passio, an analyst for the US Drug Enforcement Administration, claimed that the cartels were buying private companies by the dozen, though the sales never showed up in official records. The mobsters, simply made offers that owners could not refuse, then used the firms to launder drug profits. "Colombia is poisoned to the core by the influence of drug money", said a diplomat in Bogota. "Politicians are bought by the mafias and criminal funds are infiltrating the government on all levels".

Critics link the spread of the drug lords' influence to some of Gaviria's most touted economic policies. The government has dismentalled barriers to foreign investment and privatised two dozen state-owned companies. Under its few questions asked policy on repatriation of Colombian capital from abroad, foreign reserves shot up from $ 3.9 billion in December 1990 to $ 8.1 billion this year. Economist Jorge Child suggests that at least $ 1.5 billion of that flow consists of drug profits.

Gaviria rejected such assumptions. "Our programmes are a great success", he told Time in an interview, "We have dismantled what was left of the Escober organisation. We are getting traffickers surrendered not just from the Medellin organisations but from other cartels too. No one comes close to our marks on confiscation of drugs—60 tons in 1991 and 42 tons in 1992. Never before has Colombia been so united in solidarity against drug trafficking." Members of his administration claimed that the government had greatly strengthened the judicial system in its fight against narcotics traffickers, made major strides in professionalising the police and struck a number of severe blows against other drug networks, like the notorious Cali Cartel. Moreover, Gaviria pointed out an estimated $ 1.796 billion of Colombia's $ 50 billion economy, one of the most diversified in Latin America, was based on the drug trade.

Soft Peddling

The government's opponents claimed that its victories in the narcotics war were due to "soft policies" that accommodate rather than attack the cartels. US officials, who persistently lobbied for tougher measures, were fuming over the provision in Colombia's revised 1991 constitution ending an extradition agreement with Washington that the traffickers truly feared. There was also criticism of the administration's "surrender programme", under which those in Colombia involved in the drug trade, like Escobar, can have prospective sentences reduced by upto three fourths if they turn themselves in and plead guilty, as a result, few will serve longer than 10 years in jail, no matter how serious their crimes.

Gaviria's presidency had been beset by other problems as well. Two Marxist guerrilla groups tried to force the administration's hands at the negotiating table with attacks on government troops and facilities; the rebels killed more than 1,000 people and disrupted oil exports. A drought severely affected the hydro-electric power system, forcing the government to ration electricity. A plan to finance repairs on roads, bridges and other infrastructure projects resulted in a highly unpopular tax increase for the middle class.

To no one's surprise, Gaviria's approval rating plummeted. From a peak of 73% after Escobar's surrender, it slummed to 12.7% and rose slightly to 22%. The uptick was linked to policies Gaviria adopted after a spate of bombings in October and November of 1991, by narco-terrorists and guerrillas. He declared a state of emergency that put a clamp on press reports about those two groups, established new reporting requirements for banks that will allow officials to trace suspicious money movements and gave extra-ordinary powers of arrest and detention to the military and police. The crack-down paid off: since the emergency was declared, no cocaine cartel members who escaped from jails with Escobar turned themselves in. Police killed two of Escobar's deputies, Brance (Tyson) Munoz and Johnny (the Pigeon) Rivera. Another key Escobar aide, Carlos Alzate Urquijo, surrendered in Medellin.

Gaviria said it was "only a matter of time" before Escobar himself was arrested. The government doubled the reward for his capture to $ 7 million. US authorities were offering an additional $ 4 million. Interpol and several European governments $ 3 million. The Colombian Weekly 'SEMANA' claimed that a rival drug gang known as the Itagui group and an organisation of prominent Medellin businessmen who called themselves Colombian Libre put up a total of $ 4 million. Other Escobar enemies had taken to using his own methods against him: shortly after the car-bomb blast in Bogota, armed men set off a similar explosion in suburbun Medellin outside a house allegedly owned by Escobar's sister-in-law. Later, about 60 armed men drove upto a dwelling said to be owned by Escobar himself and set fire to the garage, destroying a collection of antiques including a 1933 Pontiac that supposedly once belonged to Chicago crime legend Al Capone. The incidents forced Escobar's wife, two children and sister to try to leave Colombia for Miami. They were turned back by authorities at Medellin airport because the children lacked an important document: a letter of permission from their father to travel abroad. Escobar was believed to be hiding out in the Medellin area.[10]

Mafia Don Testimony

A former member of the Sicilian mafia told an US senate panel that "money, drugs and greed" have corrupted that organisation. Testifying behind a screen in the Senate's most extensive organised-crime hearings in a quarter-century, Tommaso Buscetta said the mafia had changed so much in Italy because of the drug trade that he no longer felt bound by the code of 'omerta'—or silence. FBI officials, who also testified before the Senate permanent sub-committee on investigations, said the drug involvement of the US organised crime families is less extensive, although many individual mobsters are linked to narcotics. "Before the mafia would have nothing to do with drugs", Buscetta testified. "Drugs brought too much attention from the authorities, too much heat. Now, drugs were the main part of the mafia's business and everyone in the Sicilian mafia was rich because of it".

The sub-committee, headed by Sen Sam Nunn, held the first of four hearings to assess the state of U.S. organised crime in the 25 years since the mafia figure, Joseph Valachi, gave dramatic testimony before the same panel. In those 1963 hearings Valachi introduced the phrase "la cosa nostra" to describe the organised crime families. Buscetta said the drug involvement is much deeper in the Sicilian mafia—and has changed the face of crime families there. "Over the years, I have seen our organisation change from within", he said. "I have seen money, drugs and greed corrupt and destroy the cosa nostra code of honour and loyalty to the families". He said he had left Italy in 1981 for Brazil and while there, much of his family was wiped out by gang violence in his native land. He was arrested in Brazil in October 1983 and extradited to Italy in 1984 on an Italian warrant for "Mafia association". There, he decided to cooperate with Italian and US authorities "because I felt that mafia that I had known no longer existed".

Mr. Sam Nunn said Buscetta's testimony in Italy and the US has helped convict nearly 500 mob members, mostly in Italy.[11]

Futile War

The military effort to combat drugs is a waste of money. The war on drugs should be fought in the minds and souls of American consumers—not by destroying the livelihood of poor people in under developed countries who will trade on whatever gives them a chance to a better subsistence, legal or not. Cutting consumption is the best way to assure a reduction in production, not the other way around.

The most effective way to flight the drug mafias in Colombia, Peru and Bolivia would be to send in the U.S. Marines to occupy the drug-producing areas and publicly execute the cocaine colonels and mafia leaders. Since this is unthinkable, the only other way would be to more stringently control the U.S. coastline and other ports of entry, with the support of reliable undercover intelligence at the source of shipment. All the money and training of local forces in these Latin countries only helps to strengthen the "bad guys". What's worse is that since the traditional points of shipping are being tightly sealed, the drugs now seep across the borders into Brazil and Venezuela for reshipment elsewhere, bringing organized crime and a rapidly growing local drug market.[12]

Drug Cartels Steal US Aeroplanes

Hundreds of stolen American planes have ended up in the hands of the drug cartels, according to Robert J. Collins, president of the Maryland based Airplane Crime Prevention Institute. In 10 years, 1,400 planes worth over $ 150 million had been stolen in the United States with 70 to 90 per cent of the thefts involving drug dealers, Collins said.

The volume of laundered drug money, once estimated at $ 85 billion in the US and Europe has remained high in spite of the efforts of a task force set up to combat it by the leading industrialised nations at their 1989 economic summit. "I cannot imagine that the figure has dropped since we started", Mr. Alexis Lautenberg, a Swiss foreign ministry official, said. Mr. Lautenberg, who chaired the task force in the past year, said that even though many countries, including Switzerland, had introduced measures to prevent drug money laundering, the activity had merely shifted "geographically and institutionally".

Money launderers were moving to eastern European and far eastern countries and using unregulated institutions and more circutous paths. The report also revealed that many of the 27 governments participating in the task force had not yet implemented many of its recommendations, especially those requiring legal action. Only 10 have ratified the Vienna Convention criminalising drug money laundering.

Progress will be accelerated when member countries of the European community and European Economic Area will comply with the EC money laundering directive. It will oblige all financial institutions to know the identity of their customers, to keep full records and to report any suspicious transactions.[13]

Nigeria's Growing Role in Drug Traffic

Nigerians have become important cogs in the world heroin machine. More than 45 per cent of heroin seizures at US ports of entry last year involved Nigerian couriers, according to a US State department report.

More Nigerians were arrested trying to smuggle drugs into Britain in 1990 than any other nationality and 527 are in British jails on drug-related offences, the British embassy has reported. "Nigerian organisations are a new and aggressive heroin-trafficking threat", said Mr. Robert Bonner, administrator of the US Drug Enforcement Administration. The arrests in September of 28 Nigerians and nine Americans on charges of smuggling heroin through Baltimore highlights Nigeria's growing role in the international drug traffic. A 15 month investigation of four Nigerian drug-smuggling rings accused of supplying more than $ 200 million worth of heroin to dealers in Washington, New York, Chicago and Los Angeles has produced seven grand jury indictments.

Gen. Ibrahim Babangida's government works with the DEA against the traffic, but it is an uphill battle in a country with endemic corruption. New Nigerian laws sentence citizens convicted abroad to an additional five years in prison on their return for "bringing the country's name into disrepute". The Central Bank has organised seminars to track money laundering and a proposed law would compel banks and other financial institutions to reveal details of suspect deposits. Photographs of convicted traffickers appear regularly on posters and in newspapers. The campaigns against drug-smuggling appeal to Nigerian pride. The poster shows a man covering his face as customs officers lead him away from the Lagos airport. "Why do something you would be ashamed of?" it asks, "Drug trafficking will end in shame for you and the family". Since 1985, DEA agents have trained anti-drug personnel, provided X-ray equipment, communication sets and information on suspected couriers, and worked with the US embassy to review visa applications.

Three-fourths of all heroin seizures in the United States are made at John F. Kennedy airport in New York and more than half the people arrested there are Nigerians. According to DEA figures, 96 per cent of the 851 Nigerians arrested in the United States on drug charges during 1984-89 were involved with heroin. Still, young unemployed Nigerians continue to be attracted by quick money to be made in a single trip carrying heroin swallowed inside condoms or hidden in the heels of their shoes.[14]

Noriega Faces 120 Years in Prison

The former Panamanian leader, General Manuel Noreiga, the only ever foreign leader, brought to trial in the United States faces life imprisonment, following his conviction in a Miami Court on Thursday on drug trafficking, racketeering and money laundering charges. The one-time self-styled "maximum leader" of Panama, dressed in a four-star military uniform, showed no emotions as a jury of nine women and three men that had earlier remained deadlocked, found him guilty on eight of ten counts. He was acquitted of two minor drug trafficking charges. The trial lasted seven months.

President Bush hailed the verdict as "a major victory against the drug lords" and said it should "send a lesson to drug lords here and around the world that they will pay a price if they continue to poison the lives of our kids". President Bush, who had ordered the invasion of Panama in December 1989, justified the controversial American military action, saying it was "worth bringing Noreiga to justice...it is certainly worth it when you protect the lives of American citizens and when part of the result of that is democracy in a country, it makes it doubly worth it".

But General Noreiga's Chief Attorney, Frank Rubino, who had argued that the defendent was a dupe for the Central Intelligence Agent and the Bush administration, denounced the United States for acting as the "world's policeman" in invading Panama and capturing Noreiga. "This in our opinion, is the modern day working the United States will now trample across the entire world, imposing its will upon so-called independent, sovereign nations", Mr. Rubino said, "Unless the foreign governments are willing to kneel once a day and face Washington and give grace to George Bush they too may be in the same posture as General Noreiga". Mr. Rubino indicated he would appeal the verdict right upto the Supreme Court. General Noreiga faces 120 years in prison and that means that the former Panamanian leader will have to spend rest of his life in prison. The sentencing was set for July 10, 1992.

Some legal experts maintain that General Noreiga's abduction—even for the purpose of trial—violates international law. The U.S. Supreme Court is expected to rule shortly on a case in which the government solicitor-general argued that US agents may kidnap alleged criminals to bring them to trial in the United States. "General Noreiga is not a very sympathetic character, but there are still some real problems with the way the case was brought against him", said Michael O' Kane, attorney representing a co-defendant in the Noreiga trial.

US Attorney-General William Barr, welcoming the verdict said, "it sent an important message to the drug lords—there are no safe havens—their wealth and their free power cannot protect them". Mr. Barr said when General Noreiga was indicted nearly 4 years ago, "few observers believed that this day would ever come." Reminded about the long-term US co-operation with General Noreiga before the invasion of Panama, Mr. Barr said: "as came out in the court clearly, this was a case of corrupt cop. The full extent of his corruption of his involvement in drug trafficking was not evident to the United States at the time there was co-operation".

Panamanians Jubilant

Panamanians who campaigned for an end to Noreiga's harsh rule cheered guilty verdict, saying it would serve as example for future generations, reported AP from Panama city. "Justice has been done", declared Carlos Abadia, a leader of the Civic Rusade, a broad-based organisation that led anti-Noreiga protests before the military ruler was ousted and carried off by the US troops. "this verdict is a cause for delight for the country's youth," he said; it would help assure that "this type of man never exists again in Panama". Vice-president Guillermo Ford said Panamanians "recognise that (Noreiga) deserves that kind of sentence". Charges are pending against the former general in Panama, he said but the country does not want to jeopardize proceedings in the United States. On the sweltering streets of the northern port city of Colon, 80 kms. from the capital, people tended to agree.[15]

Little Headway at Drug War Summit

The presidents of the USA, Mexico, Venezuela, Colombia, Peru and Bolivia held a summit meeting at San Antonio in Texas to review and follow up the earlier one held at Cartagena two years ago on fighting the drug war. The presidents represented the largest consumer of drugs (the USA), the producers (Peru, Colombia and Bolivia) and the two countries of transit (Mexico and Venezuela). At the end of the summit there was the expected declaration to pursue the drug war with determination but there was little else by way of specific steps. President Alberto Fujimori of Peru bitterly complained about the cut in US aid to Peru to enable it to fight the drug war and received no consolation but only an explanation that it was due to US budgetary constraints. The US proposal to have a multinational force to be stationed in the Latin American countries to monitor the drug production and flow was firmly rejected by the Latin American presidents. If USA had made a proposal to have a multinational force stationed in all the countries, including the USA to monitor not only production and transit of drugs but also its consumption it might have been accepted but no one expects the USA to treat itself on a par with other nations.[16]

Some of the international squabbles between Peru and Colombia also surfaced and the latter complained about the former not doing enough to prevent the leaves produced in the country reaching the processing centres in the latter. One of the issues that has attracted a lot of attention in the outside world in recent times is the changeover from Coca to poppy by the cultivators in Colombia which now boasts of producing not only the highest quality coffee but also the highest quality heroin and its derivatives. It is stated that heroin is less bulky and cost effective compared to cocaine and is easier to handle and process in laboratories. This is not surprising to observers of the drug scene who have been predicting such changeover for years. The drug war, as all other wars, cannot be successfully fought piecemeal and the US strategy of focussing exclusively on cocaine instead of both cocaine and heroin was fatally flawed from the beginning.

In spite of reports of capture of increased quantities of drugs the flow into the USA continues unabated. The US authorities should be able to understand the predicament of India in respect of flow of drugs, arms and militants across the Indo-Pakistan border. So long as there are powerful interests on both sides of the border—whether US-Mexican or Indo-Pakistan who gain enormously out of drug flow there are enormous difficulties in stemming such flows.

One of the biggest money launderers of drug money, the Pakistani Bank of Credit and Commerce International (BCCI) has been allowed to enter into a plea bargain arrangement. It is not clear at this stage whether this step will stave off more intensive and intrusive investigations of its drug related criminal activities over the western hemisphere. There have already been complaints of agencies of the US government having not been equally enthusiastic on the investigation into the BCCI.

The USA is not exercising the same amount of pressure as it does in Latin America on other two global centres of narcotics generation—the golden crescent of Afghanistan-Pakistan and the golden triangle of Myanmar Thailand. That would mean applying pressures on Pakistan and Myanmar. The US narcotics policy has many similarities to its nuclear policy obsessed with single dimensional consideration. The USA tolerates nuclear proliferation by Israel and Pakistan and also trafficking and money laundering. Noreiga, BCCI, Zia-ul-Haq and the Thai generals were all American clients at some time or the other in the past. One does not see evidence of a total and comprehensive drug war strategy on the part of the country which has become the greatest victim of drug menace at least partly because of its own permissiveness.

References

1. *The Pioneer*, Lucknow, September 3, 1989, p. 4.
2. *The Hindustan Times*, New Delhi, September 25, 1989, p. 1.
3. *The Times of India*, New Delhi, August 25, 1989, p. 14.
4. *The Times of India*, New Delhi, August 25, 1989, p. 1, sec. 2.
5. *The Times of India*, New Delhi, Friday, September 8, 1989, pp. 12.
6. *The Hindustan Times*, New Delhi, September 6, 1989, p. 1
7. *The Hindustan Times*, New Delhi, December 29, 1989, p. 13.
8. *The Hindustan Times*, New Delhi, December 8, 1989.
9. *The Hindustan Times*, New Delhi, November 5, 1989, p. 13.
10. *Frontline*, June 23-July 6, 1990, p. 44-47
11. *TIME*, March 1, 1993, Reported by Tom Quinn/Bogota.
12. *The Times of India*, New Delhi, April 13, 1988, p. 8.
13. *Newsweek*, February 17, 1992.
14. *The Economic Times*, July 1, 1992.
15. *The Times of India*, December 31, 1991.
16. *Indian Express*, April 11, 1992.

44

PAKISTAN NEXUS AND NARCOTICS NUCLEUS

Pakistan has emerged in recent years as the major producer-cum-supplier of drugs like heroin, morphine, opium, charas, ganja and bhang in the region, contributing to the sudden escalation of drug trafficking in India. The extent of drug trafficking from Pakistan can be gauged by the seizures in India in 1983, which was 138.7 kg. of heroin, of which 98 kg. was from Pakistan, the figure went up in 1984 to 202 kg. and in 1985 to 534.4 kg. and in 1986 to 1300 kg. out of which 99 per cent was from Pakistan.

Steps to Seal the Transit Points

Realising the increasing enormity of the problem, the government of India and Pakistan are working towards bilateral co-operation for narcotics control. While earlier, the known routes identified by Pakistan law enforcement authorities were by road from Lahore via the Wagah border and again Lahore to India via Ganga Singhwala, present routes could be anywhere along the 2100 km. long border with Pakistan.

While Pakistan drug enforcement officials talk about fighting traffic of narcotics across its borders, the problem of cultivation of drug based flowers and plants continues to gather momentum. The International Narcotics Control Board of the UN reported that between 70 and 80 tonnes of opium are estimated to have been produced from illicit cultivation of some 9000 hectares of poppy there in 1987.

The problem has increased also due to the changed political situation in Afghanistan and Iran. The traditional narcotics route from this region to the western world via Turkey has been affected and the traffic diverted through India, particularly through Delhi (via Peshawar and Punjab) and Bombay (via Peshawar-Quetta-Karachi). India seized 2,270 kgs. of heroin that originated in Pakistan as evinced from such markings as—"555" and two swords crossed. This was 84% of the total seizures made the previous year. In 1986, Pakistan heroin constituted 88% of the total amount seized here. Likewise, 9,911 kgs. of hashish from Pakistan was seized in India, 69% of the total amount seized. The opium-heroin seized from Pakistan in 1987 amounted to 1,740 kgs., mostly from the Punjab and Rajasthan. The remaining 590 kgs. was seized at Delhi and Bombay (Mumbai), which happen to be the exit points westward for traffic.

Indian Point of View

Submitting these facts before the Indo-US Working Group on Narcotics, the head of the Indian delegation, Mr. B.V. Kumar, member of the Central Board of Customs and Excise, and also the Director-General of the Narcotics Control Bureau (India) further stated that the success of "Indian narcotics in the West where India was believed to be the originating point have significantly come down". Quoting excerpts from studies conducted by INTERPOL, Mr. Kumar said "Heroin seizures in Europe where India was believed to be the originating point have declined from 35 per cent in 1985 to eight per cent in 1987. Similarly, in the UK where until 1986, 80 per cent of the heroin was coming from India, the figure last year was brought to only 28 per cent."

On April 16, 1987, Kazi Abdul Aojib Abid, Pakistan's Minister for Information and Broadcasting, who was chairman of a special committee submitted a report before the National Assembly Secretariat of Pakistan. He wrote : "First reported in Quetta in 1980, before which heroin was neither processed nor abused in the country". He attributed the emergence of heroin in Pakistan to three factors in 1979. These were—

(a) The Soviet occupation of Afghanistan disturbing the opium trafficking route to west, and putting an end to the French and Scicilian connections;

(b) The Iranian revolution which led to a closure of borders and penalties of death to traffickers in Iran, and

(c) Promulgation of Prohibition (enforcement of Hodd) Order 1979, and subsequent closure of 330 opium vends and sale points for opium, blocking the marketing channel for the opium produced in the North-Western Frontier Province (NWFP).

The Special committee further reported that "heroin that escaped the notice of law enforcement agencies in Pakistan in 1986 were three times the seizures made in Pakistan-London-Amsterdam-New York route where 55 metric tonnes were seized at a total value of US $ 1.1 billion". Thus the value of escaped heroin was estimated at US $ 33 billion or approximately equal to the entire export earnings of Pakistan in that financial year."

However, according to Indian sources the problem in Pakistan has accelerated to the "inability" of the Pak agencies to capture the big fish, which not only include retired members of the Pak Military establishments but also serving ones. "This was further confirmed by the Pakistan Special Committee's Report:" From the early 80's the governor of the NWFP and the deputy CMCA: H. General Fazal-e-Haq developed the reputation of patronising the herion manufacturers and smugglers in and around the Khyber Agency. The President should not have allowed the General to continue (in office) with such reputation...The 24 laboratories destroyed in the Khyber Agency later were done under an arrangement. No heroin was recovered. It seemed to be a personal vendatta and battle over who controls the heroin trade...Army officers too either escaped with detention or were given very light terms wherever convicted despite being caught with over 174 tonnes of heroin in crude form. Since his (the General's) removal as the governor of the NWFP; smuggling had been on the decline to some extent. [Sic] The Committee observed further that the Pakistan's Narcotic Control Bureau (PNCB) created in 1973 had not been adequately supported by the government of Pakistan.

Opium Smuggling

Opium smuggling from Pakistan to India has some other aspects also concerning the security of the country. It has come to the government's notice that it enables Pakistan Intelligence agencies find facilities and contacts in India under the garb of smuggling. What is happening is that Pakistani intelligence agents establish contact with Indian smugglers and then develop and expand their intelligence network.

The smuggling racket also provides the Pakistani intelligence agents with a convenient cover, if caught. During their interrogation, they confess to their smuggling activities and explain their presence in India on this account only, thereby hiding the true purpose of their trip.

Deliberate Plan

There is also reportedly a deliberate plan by Pakistan to introduce more consumption of opium in the Indian part of Punjab. Pakistan authorities feel that consumption of opium by the inhabitants of Punjab will damage the health of the youth in this region which is in the forefront of providing front rank soldiers to the Indian Army.

Pakistan's biggest city Karachi is now the world's largest market for cannabis (charas). Nearly 62,000 kgs. of cannabis worth about Rs. 900 million destined for the United States market were seized in or near Karachi during 10 weeks. According to a Pakistani drug detective, as quoted in 'The Muslim', an Islamabad daily, only 40 per cent of the stuff dumped in Karachi for shipment abroad is intercepted by the drug enforcement authorities. "The amount of cannabis seized in Karachi during the three months ending March 18 was more than the combined seizures made all over the world", according to an official of the Pakistan Narcotic Control Board. But in spite of such huge seizures, drug enforcement agencies failed to smash any of the syndicates running there multi-tonne drug smuggling operations.

Most of the cannabis dealers ship their contraband through branches which take these consignments to cargo ships for their onward journey to U. S., Belgium, Mexico , the Netherlands and the U. K. The Pakistan Coast Guards, mainly responsible for thwarting these smuggling attempts at high seas, did not make any significant drug seizure during the past six months.

Pakistan Authorities Hand in Gloves with Smugglers

In a two series Article in the Times of India[1] Lawrence Lifschultz described the inside story of Pakistan's dirty hands in the narcotics trade. The revelations are shocking and startling.

"When Raza Quereshi' a young Pakistani' was arrested at the Oslo airport in December 1983 with 3.5 kgs. of heroin, the Norwegian authorities had no idea of what they had laid their hands upon, because at a time when Mr. George Bush and Gen. Zia walked among the pines at Muree' in Oslo prison, heroin smuggler Raza Quereshi sat speaking into a tape recorder, describing the details of a syndicate which would ultimately lead to the doorstep of the President of Pakistan, Gen. Zia-ul-Haq. Ironically enough, even as the US Government was congratulating Gen. Zia-ul-Haq for helping control narcotic traffic, the police in Pakistan, under Norwegian pressure arrested Hamid Hussain—the"adopted son" of General Zia who turned out to be the kingpin in the drug running mafia. In Hussain's possession were found books and bank statements of the Pakistani President and his family.

Quereshi had been arrested in December 1973 at Oslo's Farnebu airport entering the country with 3.5 kg. high grade heroin. Following his arrest, Quereshi was told the hard facts of Norwegian prison life. He faced a 15-year prison sentence without remission. If he cooperated with the police to uncover the entire operation behind the heroin syndicate he worked for he would be sentenced for 10 years with the prospect of a release after 5 to 8 years. Quereshi began to cooperate. He led the police to a Norwegian contact George Trohr, who had been working with several Pakistani-born syndicates. Over the next year Quereshi set up several courier runs from Pakistan under police surveillance. Phone calls back and forth to Pakistan making arrangements and transfers were taped by the Norwegian police. In the summer of 1984, the Norwegian and Swedish police arrested three more Pakistanis carrying heroin for the Nordic and German markets. But the Norwegians were after bigger fish. They wanted not the minions but those in Pakistan who stood behind the couriers.

In the summer of 1984, a quiet but determined Norwegian police officer, Oyvind Olsen, arrived in Islamabad. For the next year he worked to identify and break the syndicate behind Quereshi's operation in Oslo. Back in Norway an equally determined young journalist Mild Jonasson of the newspaper 'After Posten', began a relentless pursuit of what the Norwegian press came to term "the Pakistan League".

In May 1985, Jonasson despatched a telex directly to Gen. Zia, describing the convictions of Quereshi in Norway and the fact that during Quershi's trial three leaders of the Pakistan based heroin syndicate were identified. In his telex to Gen. Zia, Jonasson asked why such individuals were still living as free persons with good position in the community. Jonasson stated that Norwegian prosecutors could prove that the men named were the 'top leaders in this organisation'. The Norwegian authorities had through their drug liaison officer provided documentation nine months earlier to the Pakistan foreign office, identifying at least three main figures in the syndicate. Jonasson described how plans had been unearthed in Norway to kill two of the Norwegian police officers working on the case. He received no reply to his telex for his request for an interview with the General.

Yet another interesting case is of a Japanese courier named Hiseyoshi Maruyama who was arrested in Amsterdam in May in 1983 carrying 17 kilos of high grade heroin. Serving a ten years sentence in a Dutch prison he spoke 'in camera' describing the organisation in Pakistan with which he had worked as a courier for several years. Maruyama identified the head of the syndicate as Mirza Iqbal Baig who based his smuggling operation out of the syndicate at Plaza and Capital cinemas in Lahore. A BBC team travelled to Pakistan and managed to briefly meet Baig in his office at the Plaza cinema. Carrying a disguised camera and microphone, as soon as the BBC reporter started asking questions about his involvement in narcotic smuggling; the two British reporters were promptly dragged out of Baig's office and beaten and their equipment smashed on the road in front of the cinema opposite to which is the police station. Baig has not so far been arrested because he enjoys a high place in Pakistan administration and political circles. In spite of the fact that American and other European Narcotics police and the US Drug Enforcement Agency had attempted several times to compel him to leave for abroad in order to make an arrest but Baig has not fallen in the trap. Americans have not pressed for his arrest because it could lead to embarassing revelations at the highest levels of a government so closely allied with the United States.

Army Involvement

There is definitely a syndicate within Pakistan Army which is deeply involved in the smuggling of narcotics. This suspicion has been confirmed by the tension between Pakistan police and Army because the latter appears determined to keep the police out of millitary affairs, especially where it concerns narcotics. However, the police was successful to lay its hand on Army personnel on two separate occasions. The first was in July 1986 when an Army Major, Zahoor-uddin-Afridi was arrested while driving to Karachi from Peshawar. Major Afridi, was captured with 220 kilos of high grade heroin. It was the largest consignment ever intercepted in Pakistan. Exactly two months later police arrested an Air Force Officer, Flight Lieutinent, Karim Rahman. This officer was also intercepted with a consignment of 220 kilos of high grade heroin. Rahman confessed to police that it was his fifth "mission".

Put together, total value of the catch from Afridi and Rahman was worth $ 800 million dollars on arrival in Europe. Once "cut" or diluted for sale, it would ultimately be worth upto $ 4 (four) billion, or more than the total merchandise exports of Pakistan in a single year. The sum was absolutely staggering. Just one shipment equalled the entire defence budget for the Afghan war for eight years.

Army managed to get both the officers from police custody and detained them under Army's Main Cantonment outside Karachi. After preliminary investigation, both officers were to face court martial proceedings. However, before an enquiry even began, both men escaped under what Pakistan's 'Defence Journal' (August 1983) has generously termed "mystifying circumstances". The police service which made the arrest was again set out to trap its prey once again. European police sources allege the escape was arranged for nearly $. 100,000 per head, considered small charge to protect a major net work.

Afghan War—A Drug War

Allegations involving the Pakistan Army have come closer and closer to that select cadre which had been most intimately involved in the 'Covert effort' in support of Afghan war and the drug pipeline which has supplied the resistance to Soviets. But the Afghan war is clouded with an enormous illicit trade in arms and narcotics. The Pakistan press led by a remarkably courageous English language monthly called—The Herald, had repeatedly noted the widely held belief that the principal conduit by which weapons reach the Afghan resistanc in the north is, in fact, one of the main organised routes by which heroin reaches Karachi for transhipment to Europe and the United States. "It is really very simple," wrote The Herald, January 1987, "If you control the poppy fields, Karachi and the road which links the two; you will be so rich that you will control Pakistan."

National Logistic Cell (NLC) is the largest transport organisation in the country and is wholly manned by Pakistan Army personnel and security is tight at its main installation. According to reports which have appeared in The Herald and other newspapers, NLC trucks have been used repeatedly in the shipment of heroin from the Frontier province to Karachi port. In its September 1985 issue The Herald gave the following eyewithness report : "The drug is carried in NLC trucks, which come sealed from the NWFP and are never checked by the police. They come down from Peshawar where they deliver their cargo, sacks of grain, to government godown. Some of the sacks contain packets of heroin...This has been going on for about three and a half years." The Herald went on to describe an incident, again quotiing an eye witness in support of the allegations against the NLC : "A few months ago, an NLC truck was involved in an accident near Tjama Bula Khan. The driver was thrown out of the cab and lost consciousness. When he became conscious and found a number of people gathered around him he became anxious. Not surprisingly—because when the trailer had overturned, a white powder spilled out all over the road."

The figure discussed most frequently in the country press and referred to most often by European police sources as having fostered an environment in the North-West Frontier Province within which the heroin trade could flourish is the former Governnor and current Chief Minister of the Province, H. Gen. Fazale Haq. The General, however, has categorically denied the allegations against him. Would the US base tolerate for eight years a governor involved in drug trafficking", asked Gen. Haq in an interview. He further claimed that opium production substantially dropped during his tenure. But from an overall perspective of opium and heroin production, the Afghan and Pakistan border regions must be considered a single confederal unit. Almost the entire production of opium in Afghanistan is now controlled by Pakistan based syndicates with powerful link to liberated areas of Afghan countryside.

Drug Syndicates of Pakistan

According to European police sources who have worked closely with the American Drug Enforcement Agency, US Narcotics agents have identified nearly 40 significant Narcotics syndicates functioning in Pakistan. The DEA when recently approached declined to comment obviously because drug trade is closely linked with CIA and ISI (Inter Services Intelligence-Pakistan) operations, particularly the transfer of areas and money to the rebels. The Far Eastern Economic Review (5 March, 1987) reported that in its testimony before the US Senate, Andrew Eiva, an American Advisor to Afghan Mujahideen groups and an official of the Federation for American Afghan Action FAAA, alleged that US $ 700 million out of $ 1.09 billion in aid earmarked for the Afghan rebels had disappeared and charged that "Pakistani Military officers had become overnight millionaires."

A number of prominent figures in the Afghan resistance have for many years been highly critical of the links which have existed between the heroin trade, senior figures in the Pakistan government and the use of Afghan exiles as intermediaries. According to the American Drug Enforcement Agency report "much of the opium and heroin originating in Afghanistan is transported to Peshawar via the tribal areas adjacent to the Afghan border...large amounts of opium and heroin are smuggled into Pakistan...some Afghan refugees are involved in the heroin trade as opium poppy growers, opium stockists, manufacturers, middle-men or international traffickers."

Drug Addiction Growing in Pakistan

The heroin trade in Pakistan is flourishing. Diplomatic sources and law enforcement sources say heavily-armed poppy producers are expanding their field into remote corners of the lawless North-West Frontier Province. Heroin addiction in Pakistan cities is also growing swiftly. According to conservative estimates, this nation of 107 million people now has 700,000 heroin addicts, more than double the number four years ago, said Sajjad Zahid, Regional Director of Pakistan Narcotics Control Board in North-West Frontier Province. According to Dilshad Majimuddin, chairman of Pakistan's Narcotic Control Board, "We are as prone to corruption as any law enforcement agency. There's a lot of money involved. We get cases of corruption on occasions, and we deal with them on a case by case basis."

Moreover, drug enforcement is hampered in the North-West by the location of the poppy fields in tribal areas that are only marginally under the control of the Central government. Tribal people, who police themselves, do not believe poppy cultivation is a crime. Few are willing to give up the lucrative cash crop.

Western Concern and Anxiety

Several West European countries and the United States have expressed concern about drugs from Pakistan, which make their way to Western Europe through Iran and Turkey and also to North America. According to sources 40 per cent of the heroin sold on US streets originates in Pakistan or crosses into the country from neighbouring Afghanistan.

In 1982, some 300 passengers, 120 of them women, travelling by Pakistan International Airlines, were held at Heathrow airport after customs men discovered £ 125,000 worth of heroin hidden aboard the plane. None was arrested but the British Government lodged a strong protest with the Pakistan Embassy. The London based Pakistan newspaper "Mussawat", in its September 14, 1981, issue reported how Gen. Zia's family was allegedly involved in illegal drug operation during his visit to the United Nations in 1980. According to "Mussawat", Gen. Zia had ordered several marble lamps to be carried in his presidential plane, to be given as gifts to western leaders. Somehow, the lamps were left behind. Then unwittingly, Pakistan Federal Investigating Agency officials seized the lamps and found on examination that they were stuffed with heroin. Several persons, including an ADC to the president, Capt. Qamar, were arrested later, when the military leader returned home, he not only saw to it that his "adopted son" was freed but also ensured that nothing was published about the incident in the press.

According to a Drug Officer of the US Drug Enforcement at the American Embassy in Pakistan, "Pakistan which seized less than 10 kgs. of heroin in 1980 but netted 2000 kgs. in 1984, while no other country of the world has even recorded a seizure of 1000 kgs. in a year".

In the past few years, heroin has been flowing across the Indo-Pak border into India in steadily increasing quantities. With prices in Western markets at around Rs. 5 lakh per kg. it is a tempting lure for drug peddlers out for a kill.

Perhaps the most daring case was uncovered when the Central Bureau of Investigation seized 20 kgs. of heroin worth twenty million dollars! Another interesting case was crab meat consignment which when opened in Manchester, England was found filled with heroin in 48 tins of crab meat. This was worth 25 lakhs in India and it would have fetched a whopping 2.5 crore. Further investigation revealed that this consignment was brought from Karachi to Cochin and then sent to Bombay (Mumbai).

Prohibition has Failed to Affect Pakistan's Spirits

It is somewhat odd that no Indian journalist based in Pakistan has so far written about the prohibition policy in the country. "Financial Times" had carried a feature by Christina Lamb in which she wrote more about the production woes of Bhandara's beer factory and less about liquor comsumption in general in Pakistan.

Prohibition in Pakistan has been as successful as in our own Gujarat, and successive regimes—of Ayub, Bhutto, Zia and now Benazir—have felt it convenient to pay lip service to prohibition to placate the Mullahs while doing littler to enforce it. In fact, it was a big surprise when under Z.A. Bhutto total prohibition was promulgated even for non-muslims in 1977. The media and elite had jeered this act of legislation from a man who did not spend a day without Scotch. Also in focus was the religious and *Wakf* affairs minister, Maulana Kausar Niazi, who was widely known as "Maulana Whisky". But Bhutto had to soon modify the policy under pressure from the tourist trade and the hotel lobby. Non-Muslim guests at five-star hotels were allowed to buy foreign liquor on permit and this still continues. The permits are issued by hotel managements themselves and later approved by the prohibition officer, who makes a daily round of such hotels. But many guests do not want liquor. Their quota is served to others, including Muslims, at enormous profits. "Invited guests", businessmen, executives, journalists, government officials—are allowed an extra quota for entertainment purposes, which is also acquired by the hotel if not utilised by the guest.

Enormous quantities of liquor are being smuggled into Pakistan. Except for the Bhandara beer factory there is no source of legally-sanctioned liquor for home consumption. Gen. Zia did, however, allow the setting up of two factories in Punjab some years ago for the production of industrial alcohol, after the Islamic Ideology Council ruled that the Islamic injunction is against "intoxicants for human consumption" and not against liquor for production purposes. But Zia insisted that these factories be owned only by non-Muslims.

Spread of Alcoholism

It must be noted that alcohol and drugs go together. The reported alarming increase in the number of drug addicts in Pakistan should also be viewed as an indicator of the spread of alcoholism in the country. In 1986, the Zia regime, acting under the American inspired anti-narcotics policy, ordered the total evacuation of Pathan-dominated Sohrab Goth area of Karachi as it was serving as a centre for collection, processing and distribution of drugs—hashish and heroin—alongwith a wide variety of arms and hooch. Army and police evacuation squads were not surprised to see almost half the houses in Sohrab Goth equipped with "Sharab bhatties" which were earning for the Pathans million.

A Press Correspondents Report

I visited Karachi in 1986 and 1987 as an invitee journalist from the Gulf. The first time, I stayed at a five-star hotel owned by a leading non-Muslim family. During the second visit, I stayed at a hotel which is part of an international chain. For the four-day duration of each visit, both hotel managements drew upto 20 bottles of Black Label on my "invited guest" permit. I asked no questions because my hosts, who were paying for my stay, were present each time I signed the form and I assumed they knew what was happening. During the course of my stay, I also found out that all five-star hotels in Karachi, though apparently booked, keep five per cent of the rooms vacant at all times to be given at short notice for a "private" lunch or dinner where liquor is freely served. It is certainly expensive but that presumably does not matter when it is an expense-paid account.

For the poor and middle-class, Karachi abounds in illegal hooch joints, which in appearance, production and price are no different from those found in the bylanes of Surat, Ahmedabad or Dharavi. I made it a point to ask my journalist colleagues to take me to some such joints and visited many of them in Saddar, Clifton and Gulshan-e-Iqbal, all of which are known as the "posh" areas of Karachi. In each of the areas, the police were present but were apparently unconcerned about the goings-on. They were also making no move to apprehend those emerging drunk. What distinguishes the situation in Pakistan from the Gujarat scenario is that the Pakistani press does not report any liquor-related crimes, there is no public airing of demands for the scrapping of prohibition and there is not even any estimate of how much the government is losing by way of duty and excise because of prohibition."[1]

Pakistan Drug Barons Offer

With the count down for the no-confidence motion against the former Prime Minister Ms Benazir Bhutto, having commenced powerful drug barons jumped into the fray assuring support of national assembly members to ruling Pakistan Peoples' Party (PPP) and also the opposition.[2]

The minister of state for narcotics, Mr. Muzaffar Shah, claimed that a drug baron had approached the PPP with an offer of support of 10 national assembly members provided another drug baron, Anwar Khattack was released from the Karachi jail. The minister claimed the offer was declined and that action would be taken against the drug baron (whom he did not identify) as soon as sufficient material was collected against him.

Ms Bhutto had also accused the Opposition of using drug money to browbeat and purchase assembly members soon after the national assembly admitted the motion of no-confidence against her. The drug mafia has its tentacles in virtually every aspect of life in Pakistan. Take for instance the VVIP treatment accorded to Khattack in the Karachi jail. He was permitted to meet visitors and got other privileges denied to other jail inmates. He was convicted by a military court for smuggling 13 tonnes of hashish to Belgium and Holland. Two political leaders met him to seek his help in gathering support for the crucial no-confidence motion debate.

Speaking to a press correspondent the Baluchistan chief minister, Mr. Akbar Khan Bugti, referred to the power of drug barons and claimed that some important political leaders were actively involved in drug trafficking. There were reports of links of some parliament members with Haji Iqbal, a notorious drug smuggler, who was released for want of evidence. The government had blamed some foreign countries, including India, for not handing over evidence against the Haji.

Benazir Government Faces Threat from Drug Mafia

Ms Benazir Bhutto, the Pakistan Prime Minister, was forced to cancel most of her engagements for several days running and instead she held important conferences on the security of the State, and democracy, showing that she was already facing perhaps the most serious crisis of her government's short stint of eight months in office. She had also been forced to request the President to undertake the tour of the SAARC countries, which she should have herself undertaken as the chairperson of the SAARC. The threat, according to Prime Minister's mother and the senior minister in the Federal Cabinet, Begum Nusrat Bhutto, came from the drug mafia that is said to be urging the army to declare martial law yet again. The President, for his part, continued to behave in public with utmost dignity and suavity. But he also received all manner of conservative and sometimes questionable personalities, most of whom were demanding his intervention against the PPP government. It has been an ongoing process for some time. The same obscure Deputies of either House, mostly belonging to the Muslim League, Frontier Province, or tribal areas, were making a beeline for the President's House in Islamabad to say.

The PPP was emphasising that the drug mafia was out to destroy democracy and preferred army rule under which it grew so prosperous and so powerful. Most of the high gentry has been in the forefront of Pakistani establishment during the long night of General Zia's martial law and the drug mafia is supposed to have close links with many of them (What is the nature of the links and how close they are is naturally a matter of hearsay and speculation).

Pakistan Heroin Boom

Despite the ruling Pakistan People's Party (PPP) claim that the elimination of the narcotics menace was one of its priorities, the heroin boom in Pakistan continued. Each day, reports appeared in the local press about heroin seizures. And, a till-unknown quality of heroin found its way through checkpoints and past customs officers to the West. Most of Pakistan's own poppy crop is grown in the remote tribal areas of the North-West Frontier Province (NWFP) and Baluchistan.

During the Afghan war, production fell from 800 tonnes in 1979 to just 130 tonnes in 1988—partly due to government crackdowns on growers, but mainly because of the staggering drop in the price of opium—the traditional product—from 3,000 Pakistani rupees to 500 rupees a kg. This was due to a glut of opium coming into Pakistan from Afghanistan, where poppy-cultivation boomed during the war years. The result was the emergence of a network of distilleries to convert the opium into far more profitable heroin in Pakistan's tribal areas. Most of these heroin laboratories are mobile. Houses are rented out at 1,200 rupees ($ 66) a night—good money for those who are prepared to take the risks.

Meanwhile, the government's failure to offer farmers here an alternative to poppy cultivation is driving more of them back to it. Attempts to introduce onion cultivation have failed due to low prices for onions and the government's inability to buy up the crop.

The Farmers' Welfare Organisation of Dir district, a major poppy growing area in NWFP, has complained that they suffered financial losses by switching over because the massive onion crop perished for want of demand. They said most projects meant to provide farmers with alternative sources of livelihood did not recognise the genuine needs of poppy-cultivators.

Meanwhile, law-enforcement officers have alienated locals. Mohammad Iqbal, president of the Khakodel Farmers Association in Swat district threatened to retaliate with force if the administration continued to arrest poppy farmers. Adding that farmers would not give up poppy cultivation until they were provided with alternative means of income he disclosed that the people of Khakodel had grown poppies on 3,000 acres of land during the current season, with an estimated yield of 30 tonnes.

Meanwhile, the tribals of the Tangi Badinzai area in South Waziristan decided to grow more poppies and have said the money they earn from the crop would be spent on development projects. Foreign agencies have so far invested $. 118 million in projects meant as an alternative to poppy cultivation but the results have not been satisfactory. The people of the remote villages of the NWFP do need roads, schools and dispensaries, but what they also need is an alternative source of subsistence.

The setting-up of industrial estates—like one alredy installed in Gadoon Amazai in Swabi district—is considered an ideal solution, but industrialists from Karachi and Punjab do not generally look favourably on projects in these areas, despite government incentives. Another problem faced by the government is that the old link between the central administration and the 'maliks'—men of power and influence in the tribal areas appears to be fast eroding. In the past, including British times, the maliks tended to carry out government orders unquestioningly.

Many observers say the 'maliks' have become obsolete as an institution and corrupted by the massive windfalls from the heroin trade. During a military clean-up operation in Dharra Adamkhel tribal district, opium and heroin were recovered from shops and godowns in the area along with heavy weapons. Retired Lt. Gen. Fazle Haq, an opposition legislator from NWFP whom the ruling party suspected of heroin smuggling, had condemned the operation, describing it as a witchhunt meant to embarass the National Assembly member for the area, who voted against Ms. Bhutto. This had done little to boost confidence in the government's ability to fight the heroin trade and control poppy production.

Karachi—Capital of Cannabis

All multi-tonne cannabis leave the port of Karachi through launches which take these consignments to cargo ships for their onward journey to the US, Belgium, Mexico, Netherlands and the UK. At least 20,000 kgs. of hash leave the Karachi port and airport every month, according to unconfirmed intelligence information. Despite consistent reports that urge consignments of hash leave Karachi's long coastal area at regular intervals, Pakistan Coast Guards, mainly responsible for thwarting these smuggling attempts at the high seas, did not make any significant drug seizure during the past six months, the paper says. Drug officials attribute the tremendous rise in hash production to the opening of more cannabis production units in the federal-administered tribal areas of NWFP.

"We have concrete information that more hash and heroin producing units have been opened in the tribal areas recently and the drugs start their journey from the tribal areas through the NWFP and Punjab province", a Daily says quoting a senior customs official in Karachi.

"In making Karachi the world capital of charas, thousands of pople are operating from this port city to the self-governed tribal areas".[3]

Drug Used in Zia's Killing

A top-level American warfare expert has suggested that narcotics, particularly psycotropic substance, were used in a plot to assassinate former Pakistan President, Zia-ul-Haq. The drug *3-methyl-pentanyl*, which is 3,000 times more powerful than heroin, was probably placed on the controls and headsets of pilots flying the late President's aircraft that crashed killing all occupants on board, according to the US Defence and Foreign Affairs Weekly. The skin-permeating drug was placed in a dose sufficiently diluted to allow an elapse of time before disorienting the aircrew. US officials are known to be concerned that such "drug assassination" techniques could be used again. The chemicals used in the assassination plot are similar to a psychoactive drug known as MDMA, which employees of a West German Company have been charged with selling illegally in the United States narcotics market.

It now appears that similar chemicals in the same psychoactive drug category were responsible for disorienting the aircrew of the Pakistan Air Force C-130, the weekly says. A full recording of all conversation in the cockpit after takeoff has revealed that aircrew chatter stopped soon after takeoff and then the captain failed to respond when the tower called "Pak one, Pak one".[4]

Soon after and just before the crash, a Pakistani brigadier who was travelling with the President in the VIP capsule and who managed to clamber to the flight deck was heared to say, "what is the matter, what is going on?" The aircraft crashed almost immediately and the evidence clearly indicates that the sabotage was confined to the flight deck and the crew was disoriented, although apparently alive, when the brigadier reached the flight deck.

The German company has also been charged by the West German government with helping to construct a chemical weapons facility in Libya. The Libyan plant has the capacity to make such designer drugs as MDMA—also known as "ecstasy"—but there is no evidence that the drugs used to assassinate President Zia came from that facility. There was sufficient time for the aircraft to have been contaminated with the substance. Workmen were inside the aircraft for what the guard said was a "long time", repairing a damaged C-130 door. It is estimated that the contamination of the flight deck could have been achieved in less than a minute, the weekly says.

The development highlights the first use of narcotics especially psychoactive drugs as tools of assassination and throws new light on the possible uses planned for the Libyan "pharmaceutical" facility. It is possible that the facility could be used to create chemicals narcotics for sale abroad. Drug experts estimate that at least 30 per cent of all heroin currently sold on the US streets is in fact synthetic.[5]

Pak 'Prince' Hired Man to Kill Zia

A Pakistani man charged with heroin smuggling has testified that he paid a million dollar for the assassination of former Pakistani President Gen. Mohammad Zia-ul-Haq. U.S. prosecutors allege that Mushtaq Malik, 41, paid Samir N. Houchaimi of Amman, Jordan, to smuggle ten million dollars worth of heroin into the United States. Malik testified in a US district court that he had hired Samir to kill former Pakistan President Gen. Mohammad Zia-ul-Haq, rather than to smuggle drugs. Malik said Houchaimi was a former Colonel in the Jordanian army. Gen. Zia died on August 17, 1988 in a mysterious plane crash, along with U.S. Ambassador Arnold Raphel and 28 other people. Malik was questioned by his attorney, Stephen Ferrarone.

Malik who claimed to be a Pakistani prince, said he gave Houchaimi a million dollars to secure personnel and equipment, if necessary, for the assassination operation. But Houchaimi, 51, testified that he had been given two kilograms of heroin in Nicosia, Cyprus to sell in United States for Malik. Houchaimi was arrested with a nylon handbag filled with heroin on February 6, 1988, near a bus station in downtown Springfield. Prosecutors say Malik is known as "the black prince", and they allege he is one of the biggest heroin traffickers in the world.

After Houchaimi was arrested, he arranged a meeting in Rio De Janiero, Brazil, between Malik and a special agent with the federal drug enforcement administration. The two met in March 1988, supposedly to fly to Costa Rica together but Malik was arrested and sent to the United States, where he was held on suspicion. Malik said he met Houchaimi while in Karachi jail and later conducted a background check on him. After his credentials were checked, Houchaimi was given a contract by Malik's superiors in Pakistan People's Party to kill Zia. Malik said that on October 5, 1987, Houchaimi left Pakistan, and in February 1988, he learned that Houchaimi had used diplomatic channels to bring 18 kilograms of heroin from Pakistan to Cyprus. Malik said he had lost face among his compatriots because he recommended a man who failed to carry out the task of killing Gen. Zia.

"I don't know anything about money for heroin in America," Malik testified, "I wanted to have that money back for my people and save my face."

References

1. *The Times of India*, October 24-25, 1984.
2. Dara Kadva—*The Times of India*, New Delhi, 1989, p. 3.
3. *The Hindustan Times*, New Delhi-April 4, 1989, p. 3
4. *Indian Express*, New Delhi, November 8, 1989, p. 12.
5. *Hindustan Times*, March 25, 1989, p. 12.

45

INDIAN CONNECTION

A few years ago, international anti-narcotics sleuths would probably have put a not-so-important yellow flag on the map of India. For them, it was just a country wedged between two gigantic sources of illegal, opium-based drugs, "the Golden Triangle" and the "Golden Crescent" formed by Afghanistan, Pakistan and Iran. Few years ago drug traffickers were across its borders and it was an acknowledged fact that for the drop-out generation of the West, they came to India in search of spiritualism sex and stimulant with the passage of time India grew the largest illicit opium for medicinal use. India, due to its geographical position, has become to-day the centre of trafficking in narcotics and drugs and the situation is causing serious concern not only to authorities here but also in western countries.

Today, India as a modest transit country and a relatively harmless fun time destination for soft drug addicts is fast burgeoning into a major transit country and even a possible source of opium based narcotics to the drug-dozed western world. The recent hauls of heroin, morphine and cocaine were described by one narcotics official as a 'the tip of the iceberg'. Coming in quick succession, those record breaking seizures with intelligence reports which clearly indicate that the country has climbed several notches higher in the twilight world of the multifarious ill-equipped and under-manned outfits which have to combat this cancerous and quite often insurmountable evil of modern civilisation. The salient features of this drug trading phenomenon are as follows:

(i) Interpol's latest confidential "Drugs Intelligence Bulletin", a basic source of intelligence on narcotic smuggling, states quite explicity that "Pakistan and even India are now said to be major exporters" of opium and drugs. The bulletin elaborates, "It is believed by many that considerable amount of India/Pakistani heroin, through Nepal is reaching Australia and Newzealand. Though described as a "theory" this was the first time that Interpol or anybody else had said that India might be making its own mark in the world's estimated 100 billion (Rs. 80,000 crore) traffic in opium derivatives, a sudden development considering that the UN International Narcotics Control Board reports for 1979 and 1980 makes no mention of India in their discussion of illicit traffic.

(ii) India's delegation rang warning bells at the 29th session of the U.N. Commission on Narcotic Drugs in Vienna in February 1981, when it sounded to the collected delegates that drug smugglers may well be trying to convert the country into a vast staging ground between Asian source countries and the end of the pipeline in North America and Western Europe.

(iii) Traditionally, the opium trail has found its deadly way to the West starting in two mountainous, highly policed and fiercely independent regions of Asia. To India's east, the "Golden Triangle" has pumped an estimated 400 to 600 tonnes of illicit raw opium—worth upto Rs. 2000 crore on the streets of Hong Kong today—into the world 1980—compared with a paltry 8 kg. in 1979. Part of the better policing on the border had shown its positive impact. According to Interpol "the bulk of Indian and Pakistani output of opium and heroin may very well be of Afghanistan origin". There is also evidence that opium from poppy fields in Pakistan's North West Frontier Province is finding its way to India. Burmese and Thai origin drugs are also increasingly transmiting in India.

(iv) What bores smugglers more than anything else is the laxity of Indian laws. While Iran shoots drug traffickers after its summary trials, South-East Asian countries achieve the same results by death sentence but in India the culprits are let off with just six months in jail and a paltry fine of Rs. 1000. In the famous Derrick case four smugglers caught with 60 kg. hashish in 1977 were punished with a mere six months sentence. Bail is granted generously.

The statistics show that the doubts are not misplaced. Till 1980, most Indian officials were not even familiar with heroin, and even a seizure of a few grammes was considered alarming. The trickle began in 1981 and when 9 kg. of heroin was seized. And in no time it became a flood with 29 kg. seized in 1982 and 118 kg. in 1983, a figure that came as an eye operner, to the Indian drug enforcement officials. As the latest report of the United Nations International Narcotics Central Board records: "India, situated between two illicit supply areas—South-East Asia and the Near and Middle-East is mainly a transit country for heroin and cannabis resin. Reports available to the board show that in 50 cases involving a total of 123 kg. of heroin, the traffickers apprehended in Europe during the first eight months of 1983 began their journeys in Bombay or New Delhi".

While Indian officials still do not take the threat seriously, arguing that being a transit point it does not make India a heroin producing and exporting country like Pakistan, experts feel the sense of security may be rather misplaced. According to Jagjit Singh, Vice President of the International Narcotics Control Board and a former Chairman of India's Central Board of Excise and Customs, "Our officials should learn to live with reality. The drug menace is real now. There should be no shame in admitting this when the world still faces rampant infiltration and political instability. The infusion of drug money in the economy has created a monster that devoures the livelihood of the humble peasant and fills the pockets of the narcotic barons."

How India has Come in the Centre Stage of Narcotics Trade ?

With the Russian occupation of Afghanistan and Vietnamese presence in Cambodia, the flow of opium had been cut off at source both in the Golden Crescent and Golden Triangle. Particularly, with the Russian occupation of Afghanistan, and the American presence on Pakistan to turn off the exports from Karachi, the only access from Pakistan to the West is through Delhi and Bombay—with nearly 40 international flights a day and a major sea port, east of Suez. New centres for drug trafficking have emerged for the world's narcotics trade at Bombay and Delhi.

The largest and the most revealing seizures of narcotics were recorded in the three years since Bombay has become known world wide as a major transit centre for the drug traffic to Europe. The Directorate of Revenue succeeded in raiding two large consignments, one after another, in the year 1983 and again in 1986. On January 9, in a raid on a farm house in Talasari, a tribal belt 150 kms off Bombay, the DRI obtained what is almost certainly the largest seizure of heroin effected by an enforcement agency 615 kilos, which is a pointer to the increase in traffic. The source of this big haul was Pakistan and Afghanistan.

Top-level drug enforcement officials from all over the world, the UN and Interpol sat at long sessions of serious discussions at the Capital's Vigyan Bhawan, the point was being repeatedly made, much to the discomfiture of Indian authorities, who had hosted the international seminar-cum-training programme on the control of narcotics.

Informally, however, Indian officials admit that the international drug scene was changed suddenly over the past two years and brought India into the very heart of the narcotics business. While junior officials shied away from admitting this, the most candid admission came from M.L. Wadhawan, Director of Revenue Intelligence, who told delegates that the problem had now come to India and called for a highly intensified effort. The statistics show that the fears are not misplaced.

The Heroin Boom

The heroin boom in India coincided, as is apparent from the figures, with the change in the geopolitical scene towards the West, where the Islamic revolution in Iran and Soviet invasion of Afghanistan cut off the traditional "Golden Crescent" route to Europe via Iran and Turkey, for peddlers buying the contraband from the Pakistani and Afghan growers. Moreover, the influx of Afghan refugees into Pakistan further strengthened the drug mafias. As production rose and the traditional route to Europe became increasingly inaccessible, the traffickers looked for new conduits—and the vast land border between India and Pakistan beckoned. Says Doug Wankel, the US drug enforcement officer posted at the American Embassy in Islamabad: "Our estimates based on intelligence operations show that 20 to 40 kg. of heroin crosses into India from the Wagah border every week".

The Indian official position is that while transit heroin traffic is a problem, narcotics based on Indian opium produce have never found their way into the underworld markets. M.L. Bhatnagar, Chief of India's Central Bureau of Narcotics, said, "We are victims of heroin smuggling from Pakistan. In the last few years there has been glut of over production of heroin on the Near and Middle Eastern region. This unleashes a centripetal force which effects us". He asserted that western fears that the widespread licit opium cultivation in India is a permanent threat on the world scene, are unjustified. He says : "In the past seven years our area under opium has been reduced to 25,000 hectares from 60,000. At the same time our collection of opium per hectare has gone up from 34.8 kg. to 40.85 kg. proving the fact that there has in fact been a substantial reduction in the leakage from the licity produced opium stocks".

The UN and some foreign governments, however still nurse fears about the stocks. The International Narcotics Control Board observed in its 1983 report: "Two factors continue to cast a shadow over the market in opiates: the existence of stocks of opium in India and poppy straw in Turkey which are capable of meeting demand for several years as well as an excess capacity for processing of raw materials".

The Indian Position

The fact is that with Australia and some other countries going in for poppy straw production in a big way against the spirit of the 1961 Single Convention, the traditional domination of India and Turkey in world markets for medicinal scientific use of opiates has disappeared. Against the UN resolutions many countries including the UK have switched over the new sources, leaving stocks of over 2,500 tonnes of unsold, legally produced opium in India. Says an Indian official: "This is a problem. But why must people go on suspecting that pilferages will take place? And if the western powers are so worried about the stocks why don't they buy it".

UN officials are now pressuring the buyers-countries to increase purchases. Said Stephen Walsh, a top-level coordinator from the Vienna-based UN Division of Narcotic Drugs: "There has been a suggestion that the user countries should buy off these stocks and build buffers". Adds Jagjit Singh: "Eight years ago I, as an Indian representative at the UN Commission on Narcotic Drugs, had warned the world of the consequences of allowing proliferation of opium cultivation in other countries which was bound to upset the supply-demand balance in India, a traditional producer. I said, this way you will have the moral responsibility for making a country with excellent controls a problem country".

But Walsh, a New Zealander policeman with a long career in drug law enforcement and Interpol, agrees the real problem at the moment is of transit traffic. "Traffickers are skipping old routes which have become difficult now and they must devise new routes", says he, repeatedly stating that he is not criticising the Indian authorities who "are aware and doing their best".

Indian intelligence authorities have also been sounded out by the Interpol office in Delhi. At a recent meeting of the European Division of Interpol at Paris it was discovered that a very substantial number of drug peddlers caught in the recent past were of Indian or Sri Lankan origin. As Jagjit Singh explain: "The French connection via Turkey has vanished. With traffickers looking for new routes Bombay and Delhi are becoming new focal points. But these are experimental channels which seem to be succeeding and once they entrench it will be difficult to remove them".

The Directorate of Revenue Intelligence too was in the process of devising a new strategy for more co-ordenation among various countries. As a senior official equipped: "There seems a lot more cooperation between smugglers across the borders than between the governments". In North Block, Finance Ministry officers were weighing their options. Said one: "We are evaluating the threat. But the basic point is there will be no eradication of the Indian connection till the Pakistanis, who are responsible for the supplies, co-ordinate". According to the firm indications the Indian authorities have, in a bid to check the illicit drug traffic now approached Pakistan for co-ordinated action. But while co-ordinated moves are crucial, Indian officials realise that such multinational action is often difficult to matter how serious the problem is in various countries. India will, therefore, have to gear up its own enforcement machinery for, as Jagjit Singh summed up: "There is no point harping on old postures now. You have to build up and re-orient the system here and now. You cannot fight traffickers here with contacts abroad".[1]

Opium Cultivation-Apprehensions

This, paradoxically, is one crop the failure of which causes world-wide jubilation particularly among those struggling to fight the drug menace. For them, this has been a happy year. Extended, severe winter has destroyed over 90 per cent of the poppy crop in the districts of Kota, Jhalawar, Bhilwara and Chittorgarh in Rajasthan and adjoining Madhya Pradesh districts of Neemuch, Ratlam, Ujjain and Mandsaur. The region produces nearly 80 per cent of Indian opium which, in turn, accounts for the bulk of legal opium production of the world. Says a foreign drug official: "It is a terrible thing to say. But the crop failure is a god-send". Many drug enforcement experts in the West think that the loss of crop in India will help get rid of some of the excess stocks and "lessen the dangerous state of oversupply".

Indian sources, however, point out that jubilation at crop failure is short-sighted, if not misplaced. As one official says: "While for the moment it takes some of the supplies out of stores, it will generate pressures among the farmers to grow more next time to make up for losses, and if we do not allow them to do so, it will be invitation for illicit cultivation. Moreover, this may make us slow down the efforts to gradually reduce the area under poppy".

Indian apprehensions seem more realistic than the jubilation in the West. Officials of the Central Bureau of Narcotics, spread all over the growing areas are having a tough time placating farmers who have suffered heavy losses on what is virtually their only cash crop from terrain that produces more stone out of its quarries than grain from the patches of cultivable land. "Every year, they impose a substantial cut on the area, on which we can plant poppy. From 10 acres in 1976, I now have only one", complains Moti Das, surveying his ruined patch of poppy near Bhowani Mandi in Rajasthan's

Jhalawar district, adding, "if they take even this away now we will have to resist. We have no choice".

Along the network of narrow roads in central India's poppy country most farmers echo similar sentiments. For Gobardhan Lal and Shaitani Bair of Jhalawar, the destruction of their half-acre crop means the loss of practically all the surplus they had hoped to generate from their five-acre farm and if even this is taken away now, they say, it will be an unkind cut. For decades, the region's agricultural economy has depended on opium, all produced legally and under strict controls of the Government. For the farmer, an acre under opium means over Rs. 1,000 in terms of opium alone and also a substantial amount for poppy seed which fetches upto Rs. 2,000 a quintal and is sold freely for use as an ingredient of certain medicines and as a condiment. The farmer in the region has obviously reacted adversely to the Government's continuing efforts to curtail the area under poppy and the reaction could be violent this year, if the process of reduction continues even after the crop failure. Only two years ago a Deputy Commissioner of the Narcotics Control Bureau was killed, allegedly by irate farmers, in Bhowani Mandi. Said Haresh Chand, a Bhowani Mandi cultivator: "It will be in the Government's own interest to let us continue growing opium. Or there will be no stopping the smugglers from taking over". He points out that unofficial buyers pay Rs. 900 a kg. while the Government purchases for less than Rs. 200.

Restoration of Supply Demand

Many experts agree and say the best way of the problem is not a haphazard eradication of the cultivation but restoration of supply-demand balance. Indian officials dismiss the fears of western observers about this traditional cultivation and assert that their control measures are absolutely foolproof. Said M.L. Bhatnagar, Chief of the Narcotics Control Bureau: "Our controls are absolutely effective and have won praise all over the world. I can say with absolute certainty that none of our opium is pilfered out to the smuggling market".

Under the system followed by the bureau, each individual farmer is given the licence to grow poppy on a certain area of land. At the end of the season the farmer has to give a fixed opium of something like 7 kg. per acre which, according to officials leave no surplus. The failure to give the stipulated volume of opium per acre to the Government invites penal action.

Besides, the bureau maintains watchman in each village in the opium-growing area to keep an eye on the crop round the clock. Frequently, bureau offcials tour the region to ensure that the controls are not violated. This surveillance is intensified towards the end of March as poppy pods ripen and farmers begin to lance them early in the morning for sap collection. Said a bureau official: "Our stringency is obvious from the fact that it was after weeks of extensive surveying that we decided farmers shouldn't turn in any produce this year due to crop destruction even though they cursed. There is no casualness in this business".

Indian officials, in fact impute motives to the "exaggerated" western concern about the licit poppy cultivation in India which, they feel, could be part of an effort to divert attention from the fact of Pakistan's emergence as the world's dominant drug-supplier. The problem, they say, will not be solved by voicing fears or threats that do not yet exist, but by striking at the nucleus of the contraband drug trade. And that lies in the highlands around the Khyber Pass on the Pakistan-Afghan border and not in the arid central Indian plains.[2]

Mumbai—Busiest Bazar of Narcotics

A new supply centre has emerged for the narcotics trade—Mumbai. This has triggered a rare concern among the authorities not only in India but in other countries as well where it is marketed. The fact of Bombay becoming a transit centre in the drug trade dates back to at least 1980.

Bombay has become international menace as a major transit centre for the drug traffic to Europe and the U.S. occurred in the very first days of 1986, within a week of one another, one of the seizure was the largest known for a single operation anywhere in the world.

Hashish Haul

Likewise, on January 1, 1986 DRI seized 2.95 tonnes of hashish from a container at a Bombay dock to be loaded on a ship bound for Rotterdam via Singapore. Apart from its being one of the largest consignments ever seized, its real significance lay in the fact of its being seized at the docks. Such seizures of hashish largely came from the airport or from godown, seizures at the docks were rare and negligible. Thus the seizures at Falasawad docks define the entire spectrum of the traffic.

The earlier trade routes to the West had been severely disrupted by the Russian occupation of Afghanistan, and the Vietnamese presence in Cambodia. In other words, the flow had been cut off at source—in both the Golden Crescent: Iran, Afghanistan, Pakistan and the Golden Triangle: Burma, Thailand and Laos. Particularly with the Russian occupation and subsequent exit from Afghanistan, and the American pressure on Pakistan to turn off, the exports from Karachi, the only access from Pakistan to the West is through Delhi and Bombay, particularly with more than 40 international flights a day and a Sea port, Bombay which is well suited as a conduit point.

Bombay now caters to India's growing internal consumption of narcotics. In Bombay alone, over a lakh of people are addicts of brown sugar, which means at an average consumption of 10 kg. heroin per day.

The CIA Operation

The first ever batch of narcotics to be prepared for clandestine export from Bombay to the West was by the CIA in 1972—800 kgs of opium to the U.S.A. itself, whose weakness for the opiates was by then rampant. Packed in tins, ready for shipment, the opium was intercepted by the Bombay police at a desolate beach on the city's shoreline. The police had no clue as to the CIA's involvement, and took it as a first instance of the city's underworld trying its hand at hard drugs instead of gold. The seizure made headlines—'no'—we had ever heard of a traffic in narcotics from Bombay but it was a redity now. On that fateful day, Narcotics Commissioner was in Bombay. He was most surprised and embarrassed because it was his responsibility to ensure that none of the 800 to 1000 tons of opium grown annually in the country was smuggled out, because all this legalised produce was meant for international pharmaceutical industry.

Bombay provided the facility of a page of services such as storage facilities, instant banking, repurchasing and concealment in any shape, size and form.

The Gradual Transformation of Mumbai into a Narcotic Wholesale Bazar

How exactly does Bombay functions as conduit? Have the city's underworld operators been recruited into the international drug syndicate? Or has Mumbai developed a narcotics mafia of its own, capable of financing, planning and executing trans-global operations exporting opiates to the West? According to the DRI, which is in constant touch with drug enforcement agencies all over the world, Bombay functions virtually as a narcotics bazar, a trading centre. The city's drug operators, in other words, are middlemen who procure narcotics from Pakistan and sell them wholesale to agents or sub-agents of the American and European mafia. In short, they are traders not financers or entrepreneurs.

According to D.R.I. officials the Californian and Italian mafia do not have their agents or representatives based in Bombay or recruited from its underworld. The trafficking of narcotics is much more loosely organised, and does not function in terms of clockwork operations from the poppy fields to retail outlets. In Amsterdam or Los Angeles that is in transglobal circuits. The practice by and large is that a sub-agent of the American, Italian or European drug syndicate flies to Bombay and contacts a local supplier here. The agent rarely carries back the stuff with him; he flies out clean—as clean as he came in. It is normally left to the local supplier to have the contraband sent out of Mumbai to a pre-arranged rendezvous abroad.

The Men Behind Trafficking

The reasons are apparent for all to see the profit margins jump at least twelve times at every stage. Ten kg. of raw opium brought from a cultivator for Rs. 5000 or more produces at least one kg. of heroin. With the wholesale price of three lakhs after a simple laboratory process which can easily fetch anything around 25 lakhs and if it finds its way outside India, it can fetch anywhere 16 million dollars or more. This has lured smugglers and motion of all hues to undertake drug peddling in an effort to get quick. The prominent names here.

Birendra Kumar Rai

Birendra Kumar Rai was arrested by me when I was Superintendent of Police, Ghazipur and was detained under MISA. He was again arrested by the officer of the Narcotics Bureau, Varanasi on 2.8.88 at Babatpur, Varanasi. He has been a jail end.

Born on August 15, the day India attained freedom, Birendra Rai hailed from village Malsa in district Ghazipur (U.P.). He was born a with a silver spoon in his mouth in the respectable and land owning Rai family of village Malsa. But it is said that during his student days, he fell in bad company and when in the later 70's heroin appeared on the narcotic scene around Varanasi, Birendra Rai was quick to move into it. Opium could be procured from illicit channels in Eastern Uttar-Pradesh. There was demand for heroin at Varanasi. Birendra Kumar Rai thus found the stage set for him and found the illicit trade tempting.

Several factors helped Birendra Kumar Rai to grew and attain almost an aura around him that he could not be touched by any government agency. He belonged to a relatively reputed family. His activities during his student days created an image that made people mortally afraid of his good connections. He had people who could be trusted, and he could organize the things. It is because of these factors that though he was talked about as the biggest durg trafficker of Varanasi, he was never caught carrying the drug and went on growing in stature. From Varanasi, he graduated to Bombay and started running a heroin refining laboratory there. This laboratory came to surface on 18.5.86 when Bombay Customs raided a flat in Andheri (West) Bombay. The extent of operations of Sri Birendra Kumar Rai could be gauged from the fact that 14.700 kgs of refined heroin along 40 litres of liquid solution containing heroin under process was seized from the flat belonging to Birendra Kumar Rai. Also seized was an auto pistol made in Czechoslovakia along with 20 rounds. People were justifiable afraid of him. While his younger brother and three other accomplices were arrested, Birendra Rai himself gave the slip and absconded. Maharashtra government issued a detention order under COFEPOSA in October, 1986 against Birendra Kumar Rai. Prosecution was launched against him by Bombay Customs in the court of Additional Chief Metropolitan Magistrate, Bombay. Despite several attempts, the detention order under COFEPOSA remained unserved on him. He also did not attend the court at Bombay. Ultimately, N. C. B.., Varanasi located him at Varanasi when he was about to emplane for Bombay and produced him in the court. COFEPOSA detention order was served on him and at present he is undergoing the detention.

Birendra Kumar Rai and his family have amassed considerable wealth. The family have a house in Birdopur, Varanasi. He owns flat No. 503, 5th Floor, Sinchan Building, New Link Road, Extension, Versova, Bombay. Flat No. 21 Seaside Building Versova Beach, Bombay is also said to be belonging to the family. The family have a Silk Emporium in Nadesar locality of Varanasi, and own two trucks with Regn. Nos. UGZ-774 and UGZ-775. Birendra owns a Maruti car bearing registration No. MPP—8846. He also has an orchard in Pandaypur locality of Varanasi. Besides this, there may be several benami acquisitions.

Trideep Banerjee

Trideep Banerjee was arrested by the officers of Delhi Zonal Unit of the Narcotics Control Bureau on 14-10-1988 in New Delhi while he was caught red-handed supplying 400 grams of white heroin to a Dutchwoman of Surinamese origin staying in Hotel Janpath.

Born on 28.5.1952 in a mediocre Bengali family, Trideep Banerjee was like any other boy of his age. He passed his Intermediate Examination in the year 1970 from a Government School of Delhi and started to support his father financially by doing clerical jobs in some private firms in Delhi for 3 years after which he drove auto-rickshaw for about six months and also worked as a travel agent during which he came across an Austrian couple. In order to become rich overnight he and the Austrian couple planned to smuggle out 800 gms. of heroin to Austria but got caught by Italian Customs at Rome where he was sentenced to three years' imprisonment. The jail term proved to be "blessing in disguise" for Trideep Banerjee as he spent the whole term in making contacts with other international drug traffickers lodged in that prison and before coming back to India in October, 1981 (on release) from the Italian prison, he told his jail colleagues that he could supply any quantity of pure white heroin to them from India at reasobable rate on a very short notice.

After returning to India, for a short period of about two years, he worked with his father-in-law in his brassware business at Paharganj, New Delhi but once again the dream of becoming multi-millionaire overnight compelled him to utilise the contacts made by him with other international drug traffickers in the Italian jail. First of all, he successfully traced out the source of white heroin from somewhere in Ghazipur district of Uttar Pradesh. He had already started receiving orders for the supply of white heroin from some of the drug traffickers who were lodged with him in Italian jail for similar offences. After being assured for the supply of pure heroin he agreed with the international buyers for supplying the heroin at a fixed rate of US $ 10,000 per kg. (approximately Rs. 1.5 lac per kg.) which was being acquired by him at a price ranging between Rs. 50,000 and Rs. 75,000 per kg. He was very cautious in his illicit trade and all his buyers were foreigners. He never supplied heroin to any Indian buyer and that was the reason he was doing his heroin trade totally unnoticed by any of the drug law enforcing agencies. Whenever he went to supply heroin to his buyers, he always took his four years old innocent son with him to avoid suspicion about his movements.

Ultimately, like a number of other major drug traffickers, Trideep Banerjee too walked into the trap laid by the Delhi branch of Zonal Narcotics Control Bureau on 13.10.1988. He alongwith his son entered a room of Hotel Janpath, New Delhi to supply 400 gms. of pure white heroin to the Dutch women to whom he had already supplied 2.800 kgs. of pure white heroin on 11.10.1988 for US $ 28000. It was Banerjee's bad luck that his buyer-the Dutch woman was intercepted by the officers of Delhi Zonal Unit of NCB in the early hours of 12.10.1988 while she was leaving India for Amsterdam and 2.800 kgs of white heroin was recovered from her and seized. The Dutch woman disclosed the name of Trideep Banerjee as the supplier of the heroin. The Dutch woman was made to check-in-again at Hotel Janpath, New Delhi on 13.10.1988 morning where she had received the supply of 2.800 kgs of heroin from Banerjee on 11.10.1988. She was asked to contact Banerjee at his residential phone for placing an additional order of 400 gms of white heroin. After receiving the message, Banerjee took his son with him and reached the hotel for supplying the desired quantity of heroin where he was successfully nabbed by the waiting NCB officers.

In the follow-up, the NCB officers also searched the residence of Trideep Banerjee situated at E-45, Jawahar Park, Laxmi Nagar, Delhi-92 on 13.10.1988 from where an additional quantity of heroin weighing 11.150 kgs., Indian currency amounting to Rs. 80,17,000/- foreign currency (US $ 1700 and £ 785 and some bottles of foreign liquor wee recovered and seized.

Trideep Banerjee in his statements admitted that, so far, he had supplied more than 200 kgs of white heroin to his international buyers earning a profit of about Rs. 1.5 crore approximately out of which he had got his Laxmi Nagar house constructed in 1987, besides purchasing a flat in Bombay in 1986 and other banami properties. Thus, with the arrest of Trideep Banerjee NCB made a big dent in the circle responsible for supplying white heroin to the international drug traffickers.

Hottest Name

The hottest name on the lips of senior narcotics investigator in Bombay now is that an underworld don who has been known to deal in other kinds of contraband, including gold. The other names from the DRI list include of Karim Lala, whose men were caught recently. Amongst the other well known operators is Som Bhargani, who was once arrested trafficking in Goa. Satya Prakash Behl, known as trafficker was recently arrested by the Rotterdam police with 12 kgs. of heroin on his hands, worth Rs. 12.5 corres in international market.

Ibrahim Patel, Machchliand Satish Sain were suspected to be mandrax trafficker to African countries.

The big drug traffickers are assisted by agents, subagents and retailers.

Smugglers Gaining Legitimacy

Since 1974 or so, most of the big time smugglers have moved iinto legitimate business and they have used corruption, wheeling and dealing and muscle power to amass immense wealth. They are Iranians with strong Indian links, Abdul Rahim, 55, Abdul Wahab, 53, Abdul Latif, 49. Galadharies soon became among Dubai's leading exporters and set-up a sophisticated net-work of henchman in India who received the consignments and smuggled them past the authorities. In 1977, the Galadharies were declared by India court as 'offenders' and remained on the wanted list till 1981. They mainly concentrated their smuggling from three points: (Bombay's Cuffe Parade, Beach Candy and a village near Ratnagiri). Mrs. Gandhi's government, changed its tone towards them and treated them like international celebrities. After the oil-price hike of 1973, Dubai had been transformed from a smugglers' haven to a flourishing trade centre. The Galadharies builts up an enormous industrial and financial empire that took in two newspapers (The Khaleej Times and Gulf News), a cement plant in Karachi, hotels all over the gulf (including the Dubai Inter-continental) a trading company in London's Wood Mews, the Maxims casino also in London, and even an ice-cream factory in Dubai. In 1978, Abdul Wahab, separating from his brothers, set up his own Galadhari Enterprises and his own bank, the Union Bank of the Middle East by becoming a speculator on gold and silver markets. The Galadharies using their political links had ensured India withdrawing all cases against them. The Wahab Galadhari, because of his refusal to sign over his US $ 400 million Hyatt Regency—Galadharis, Gallevia Complex at a throw-away price to the ruling family that helped him amass wealth, had his passport confiscated and accounts seized and put under house arrest in early 1984. Under the U. A. E. law, directors may not borrow more than 5% of the bank's paid-up capital and this provided a convenient stick to beat Galadhari with and stepped of the Chairmanship of his bank as the bank lent Wahab's companies US $ 380 million and issued him guarantees for his another US $ 270 million against the paid-up capital $ million only. His two brothers pumped US $ 400 million into Wahab's companies to secure his release and stand as his guarantee but now they are perceived as world-wise legit businessmen.[3]

Galadhari-Antulay Connections

In past, A. R. Antulay's alleged smuggling connections derive from the time, a decade ago, when a smuggler was arrested by customs in the Konkan region. Antulay-Galadhari connections have remained a subject of speculation. One story had it that Wahab routed his contribution to Mrs. Gandhi's 1979 campaign through Antulay. His connections with other politicians are well established. In 1977, he asked lawyer politician Rajni Patel to see if he could get the cases against him withdrawn. It failed because of the then Finance Minister H. M. Patel's determination, then they tried to work through the Jana Sangh component of the Janata Party met Satish Agarwal (State Minister of Finance) and A. B. Bajpai who are said to have promised to help out and lead into the matter but the government fell before deal could get throgh.

Romesh Bhandari, the then Secretary (West Asia and North Africa) in the External Affairs Ministry, recommended to the government that the Galadharis be welcomed to India on grounds that this would please the Dubai Gove. Abdul Latif during his visit to Delhi, met through Bhandari, the then Finance Minister R. Venkataraman; then Commerce Minister, Pranab Mukherjee and even Mrs. Gandhi.

The Haji Mastan connection with Sanjay Gandhi was initiated by Rukhsana Sultana and later through Akbar Ahmad Dumpy. Yusuf Patel used Rajni Pater's legal services for 20 years and remained an adviser till his death in May 1982. After that, like other smugglers, he shifted to Ram Jethmalani. Yusuf Patel, never got to meet A. R. Antulay, the Chief Minister, but retained close links with Shalini-tai-Patel, the Revenue Minister with Rama Adik and Chandrakant Tripathi.

Political patronage

The CBI enquiry into a drug-running case against one ex-Chief Minister is nearing finalisation. The report states that the ex-Chief Minister flew with three trunk loads of raw opium to Kathmandu on a state visit on June 29, 1978 and returned with refined opium and hashish. He earned a profit of Rs. 65/ lakh of which Rs. 31 lakh were deposited in a Kathmandu Bank. This case is one of the 84 important cases including those against public servants.[4]

The case against Gulab Singh, ex-Minister registered by Chanchal Singh, a patwari of Chakrata Tehsil in Dehra- dun district under section 5/11 of Dangerous Drugs Act 1930 and under section 4/9 of the Opium

Act 1878, for growing poppy without licence and on the land which was not notified for the purpose, is likely to be dropped because his supporters find it as a conspiracy planted against him for a political murder and the land, no doubt was owned by Gulab Singh but it was leased for cultivation to someone else and Mr. Singh never knew what was being grown on this land. The raiding parties found that the lancing of poppy plants to extract latext which, when dried, becomes opium, was already done.

Involvement of M.L. A's

Fifty-one years old, MLA Nazir Ali from Shahjahanpur was trapped with 20 kgs. heroin at his room no. 50 at Darulsafa which was supposed to have been purchased from a Bareilly hideout shop. The interrogation revealed the relationship with an International gang with its headquarters at Calcutta but also the patronage of the political bosses. The heroin packed in bags polythene twenty weighed 19 kgs. 540 gms., which could have been easily sold at Rs. 2 crore in international market. An instrument to test the purity of heroin was also discovered at the MLA's official residence. When asked about his involvement, the man showed his ignorance to the whole affair and turned it as the image-tarnish movement.

Ghazipur police hauled up ehammad Zakir for uranium smuggling charges, found him to be the Block President, Congress-I, President of the Crime Prevention of his party and the President, Ghazipur Youth Congress-I.

Mr. Nazir Ali, previously a Lok Dal member, joined Congress I in December 1981. Before him, another Minister Gulab Singh was charged with growing opium on his farm. The resultant furore forced the Chief Minister to drop him but Sripat Misra inducted him again in his Cabinet. In 1970, Ram Murthy Anchal, a Congress MLA from Badrol of Shahjahanpur was also detained for smuggling activities. The then opposition leaders like Mulayam Singh Yadav, questioned, when the Govt. servents depend on smugglers and criminals what happens to the public?[5]

International Traffickers

Gurdav Singh Sangha, who went to UK as a labourer, took his Ph.D. in jail, became an international drug syndicate boss. He eluded the Interpol and figured topmost on the wanted list as the files credit him with building an alternative to the traditional Laos-Burma-Thailand Triangle. The Sangha organisation with "Henri Claude Regholi, shipped multi-kilograms of hashish concealed in rolls of drappery materials, four feet long and eleven inches in diametre, with outer wrappings of white burlap from Orient Exports Co., New Delhi to Michael Fashions through a customs broker in Toronto".[6]

Four persons including one Afghan of an international gang were arrested by the Narcotics Branch of CBI and recovered heroin and hashish worth Rs. 63/ lakh. Three cars and a motor-cycle were also taken into possession (used for the purpose). This arrest led to a major-gang busting and a wide arrests of many persons involved.

International Racketeer

Ashok Solomon, was first caught in 1969 by Delhi Police for forging cheques and defrauding a bank and again in 1973 at Palam Airport with large quantity of foreign currency, detained under MISA. He secured his release and was wanted under COFEPOSA when he left country in 1973 to U. S. where the Federal Bureau of Investigations caught him for trafficking in Narcotics and sentenced to ten years imprisonment. After 6 years in jail, he secured his release in 1982 to return to India in 1983. In these three years, he amassed so much of wealth that he, now, owns a palatial house in Golf Links, a beauty clinic and a boutique in Connaught Circus. An Electronic factory, a cosmetic factory in Okhla and other business interests in reputed companies. According to Mr. Pereira, a close-watch of police on Solomon brought to light his shrewd operations from behind the scene. His arrest was dramatic and that too mainly because of his driver's being on leave that he had to drive his car himself and surface. Solomon's *modus operandi* was to export charas in the garb of antiques, ready-made garments and other exportable items to U. S. where his contact, former jail inmate Robert Evens sold it to other parties.[7]

Stray Smuggler

There are a few stray smugglers who are arrested occasionally like—Dr. Hari Bhajan Nath Gulati, a lecturer in the Political Science department of Dayal Singh College, Delhi was caught with one kg. sealed pack of heroin at Panchsheel Park selling to the decoy customers employed by the Delhi Police. He revealed that the packet was handed over to him by a Bombay-based drug chemist. He was in need of money to get his daughter married well-off and had lent previously some money to the said chemist.[8]

Ajit Mohan Balse—Jet plane pilot of IAF sent for training in 1964, settled with an American girl in Santiago and gave up the job. He was successfully trapped and during his second visit to Mumbai he was arrested with 8 kg. of heroin worth Rs. 3.2 crore.

Political Cover-up for Smuggling

An interesting smuggling feature came to light on December 27, 1985 when four members of parliament from Mauritius were found at Amsterdam with heroin worth one million US dollars in their bags. Ironically enough, they had been cleared by Bombay Airport from where they boarded the plane for Amsterdam, under diplomatic cover, hence escaped customs scrutiny but at Amsterdam their passports did not give them clearance from x-ray screening.

Clues to Drug Trafficking Provided by the Arrested Persons

According to Kanpur ex-army man arrested after a wild chase near Chachoda in Guna district, vital clues to international racket in opium smuggling were eis-closed. In those last 10 days arrest and opium seizure, S. N. Tewari, a culprit and other persons divulged the names of 10 persons involved in trafficking spread over Rajasthan, Uttar Pradesh, Punjab and West Bengal provided clues to the two main bases in the country and the kingpins who operate from Calcutta and Taran Tarn (Punjab). The intermediaries based at Lucknow and Kanpur seem to assume the role of 'clearing houses' for the dealers in Calcutta and Taran Tarn. These intermediaries purchase opium at generally rate of Rs. 1200/- per kilogram from the villagers of Kota and Jhalawar districts of Rajasthan and pass it on to the main dealers with huge percentage of profit at a rate of Rs. 5000/- per kg. It is mainly from these two places that the country product opium gathered at different centres in India, finds its entry into foreign countries, mainly Pakistan, Afghanistan and Dubai where it is processed into heroin and other drugs. The intermediaries own or have access to a fleet of vehicles to Punjab. It is mainly transported in truks to Calcutta in specially built-in-secret compartments of jongas. Is Rajasthan camels are also used for the purpose to inter-state functioning. They establish their strategic outposts on borders as M. P.-Rajasthan border on Agra-Bombay road between Shivpuri and Guna where they exchange hands of pre-determined places so as to escape the clutches of law.

Barabanki

Every fourth person is allegedly involved in the smuggling of 'black gold'. In Barabanki over 80,000 kgs. of narcotics (worth Rs. 8 crore) is smuggled out of this remote region every year. The inhabitants of Barabanki district are largely dependent on poppy production. The narcotics department reveals that the region has 45 per cent of total licensing for poppy cultivation.

The smugglers face no problems in managing the drug in huge bulks from the farmers. On an average an opium producer sells over of his yield to the smuggler who pays two or three times more price then the government rate of Rs. 120 and Rs. 220 per kg. The reason for this can be that during harvesting, the fields are not guarded as it is normally don in madhya Pradesh and Rajasthan and the farmers easily conceal a major share. The racketeers have their own tailors for making special suits used for peddling. They also stitch 'burqas' used for similar purposes. Local agents including women peddle small quantity in the district and stores the rest in secret godowns that become fine factories for processing. When a considerable quantity is collected, it changes hands once or twice in a year on the Indian border or the docks in catching them. The police know but hardly succeed.

Ghazipur

An international gang of smugglers in narcotic drugs had been smashed in a tiny Ghazipur village following recovery of a large amount of contraband narcotics valuing over Rs. 35 lakh in India and Rs. one crore abroad and the arrest of two branded smugglers B. K. Rai and Sunil Kumar Rai. A posse of over 150 police personnel led by the then Superintendent of Police, Ghazipur Sri Giri Raj Shah surrounded a cluster of ten villages with Malsa as a nucleous and the searches stretching over twelve hours resulted in the recovery of eleven kgs. of morphine, one kg. of heroin, ½ kg. of hashish and 20 kgs. of chemical liquids. The raiding party recovered some equipment used for conversion of different contraband narcotics through a mechanised system, three foreign-made fire-arms including a Pakistani revolver, a double-barrel gun and one American rifle. The police also seized accounts-books showing a transaction of more than 1,52,00,000 of narcotics through various means. The register contained the names of such foreign countries where smuggling had been on a continued spree. Mr. B. S. Bedi, the then Dy. Inspector General of Police, Varanasi Range distributed cash reward of Rs. 5000/- for those three prosonnel including one sub-inspector of the Ghazipur police who worked in disguise for ten days to collect informations leading to the successful crack-down on international narcotics smugglers.

Ghazipur police also seriously enquired into a matter that led to 18 months old case wherein, the local opium factory, sent a sealed parcel of opium worth Rs. 4900/- and the person in Calcutta found it packed with stones.[7] Officials of Central Bureau of Narcotics based at Varanasi, nabbed two jawans of Central Industrial Protection Force with govt. sealed polythene bags of morphine 450 gms. worth Rs. 8000/- in India and Rs. 40,000/- in foreign market.

The Killing Fields of UP

Around noon on 22, November 1989, a white Ambassador bearing the State insignia and pioleted by motorcyclists drove up to the district hospital in Ghazipur, where notorious criminal and hitman of the Mukhtar Kamlesh gang, Sadhu Saran Singh was convalescing. The hospital authorities were informed that the DIG wanted to interrogate the criminal. A man, dressed as an officer, entered the room, whipped out a pistol and shot Sadhu Saran dead. The killer, it was later discovered, was not policeman but a member of rival gang led by Brijesh. The gang took revenge by shooting Sadhu Saran involved in more than two dozen murder cases.

This incident market the first of a continuing series of gang war and bloody mafia-like shoot out in eastern UP, On 8 June 1990 when two of those involved in the Sadhu Saran murder case stepped out of the police van before the court, gunfire cracked from a neighbouring high rise building. One of the accused was kilied and the other seriously injured. At least 20 persons including five police constables, have been killed in these shootouts as heavily armed drug traffickers continue their bloody war to control the underworld drug trade. Lethal weapons such as the AK-47 assult rifles and Israeli Uzi machine pistol are frequently used by the gangstes who are thriving on quick money from smuggliing narcotics—especially heroin and brown sugar.

Political forces of different hues have aligned themselves with these gangsters to use their money and muscle power for electoral gains. While the National Front—Left parties combine sought the support of the Mukhtar-Kamlesh gang during last election in Ghazipur, a BJPMP from Gonda is known to be providing shelor to the Brijesh gang. The brazenness with which these gangs are operating, increases every day, Recently, desperadoes armed with automatic weapons attacked Maya Shanker, a hitman of Mukhtar gang, iinside the district court premises, though Maya Shanker escaped, two constables escorting the criminal to the court were injured in the shoot-out. One of them later succumbed to his injuries in hospital.

What imparts seriousness and urgency to the entire issue is the fact that these fafia groups are now trying to subvert State authority by using mney and muscle power. These groups are steadily recruiting police constables into their gangs. According to the SSP, Varanasi, a onstable recruited by Mukhtar gang had been involved in almost all recent shoot outs. Similar tactics employed another constable as a hitman. Varanasi, Ghazipur and adjoining districts have witnessed a reign of terror unleashed by these gangs. But all the police has done is to remain mute spector.

Nabbing Narco Smuggler

Most of NCB's efforts are directed towards stopping drugs coming across the Pakistan border. Heroin originating in South-West Asia is brought into India through aresa like Firozpur and Amritsar districts in Punjab, and Jaisalmer and Barmer in Rajasthan.

The terrain is difficult to survey particularly with tall elephant grass on the Pakistani side. Effective sealing of the borders is difficult since the barbed wire is not continuous. Matters are not helped by the lack of mutual co-operation between Indian and Pakistani security agencies. It was only recently that Pakistan started worrying about increased drug addiction amongst its populace.

So far there has been lack of intelligence about activities of drug smugglers operating from Pakistan. The US drug enforcement which has 20 agents in Pakistan was extremely miserly on sharing information, says an NCB official. The US was quite prepared to give information on other areas but the reason for reticence over Pakistan, says the official, was because drug money had for long financed the Mujahideen.

However, NCB pressure had forced operators to open other routes through the Balkans, for sending the heroin to Europe, said Dalbir Singh, Dy. Director General of NCB. Quoting Interpol figures he said that in 1985 thirty-five per cent of the heroin reaching Europe came through India. In 1986 the figure came down to 29 per cent, and in 1988 it was just 7 percent.

NCB Handicapped

The NCB has been handicapped by organisational weaknesses which have reduced the efficiency of its operations. It is desperately understaffed with just 160 personnel. Considering the country's size and the extent of drug operations, this is sorely inadequate. Of these just 50 are field operatives. In co-ordinating anti-drug efforts and gathering intelligence on the big timers, NCB officials have to depend on local police for much information of pushers and peddlers. Here the problem of frequent collusion between local forces and petty criminals arises, as X[1]s story above, illustrates. Another handicap the NCB suffers from is the lack of computerised data management. Often proceedings in courts are delayed as NCB puts together various evidence. Lack of equipment also delays laboratory tests. The intelligence is still stored in old-fashioned files with diagrams showing the connections between gang members.

NCB investigators say that they have yet to uncover a large nexus between politicians and drug cartels. Local politicians accept drug money for campaigning, particularly in Bombay, and in turn arrange favours like prompt bail. But care is taken to see that the connections are always indirect.

However, there has not been any political effort to pressurise NCB investigations, says Dalbir Singh Kiran Bedi, a Deputy Director with the bureau, pointed out that a few people in power in the north-east have been arrested for durg running. Officials, however, admit that associates of drug smugglers are known to frequently meet certain politicians but the evidence is at best circumstantial. They, however, claim major successes in Delhi where they say that the big fish have been wiped out. (But what about the man profiled above?) Straining the NCB's thin resources further is the lack of enforcement in neighbouring countries. In Myanmar widespread insurgency has given rise to a flourishing drug trade. Nepal is the main source of the 'ganja' that finds its way into India since 'ganja' cultivation in India has largely been restricted. Heroin is also brought in by many of the hundreds of Pakistanis who come on visas now handed out liberally, Records show that half of them just stay on. They are known to stay in certain specific hotels in cities like Bombay and Delhi. The NCB does not believe in letting a medium rank operative exist too long after he is identified, in the hope of uncovering more links. Even letting just one kg of heroin into the market may lead to 10,000 new addicts, says Singh. The price for the waiting game is too high, he says. Getting the 'Mr. Big'is often intelligence work and nabbing him puts 200 to 300 persons out of business.

The NCB has also initiated international collaboration. In a recent example, a group of Nigerians carrying heroin were arrested at Amsterdam following a tip-off by the NCB. Nigerians, officials say, are the most common couriers of the drug. They are now even taking over drug peddling in Delhi. In some cases they have also become organisers.

Portrait of a Drug Smuggler

This story about a drug king published in the Indian Express reads like a story out of some Indian action cinema-He is 27 years old and is so innocent-looking that you would not credit him with the ability to do even small-time racketeering. He is five feet seven inches, stockily built, and dresses so ordinarily that he would, in a street attract no attention at all. And yet stepping out casually is something he would neven dream of doing. Several policemen guard the entrance to the gali which leads to his house, and a clutch of hatchet faced bodyguards dodge his steps.

The Portrait

He lives in the Ajmere Gate area, behind Lal Kuan, in what it looks like a crumbling old house. Entry is through a narrow winding lane, lined by smelly ditches. Through the front door you enter a shabby room with creaky divans, faded paint on walls, and cracks in the mosaic on the floor. But this too is just a front. You pass into the next room that tells the real story. The furniture is as luxurious as that in any five star deluxe hotel, and apart from the usual VCR there is a lacer disc player, there are large four way speakers, and the liquor cabinet is well stocked with the best that money can buy. The bedroom too, has every conceivable luxury and then some.

He lives with his mother, brother, brother's wife and child. The brother closed down the small business he had because going out every day to tend it was too risky. All of them live under constant threat of abduction and do not step out without bodyguards. His mother detests what her son does for a living, but like him she knows that now there is no turning back. As her son puts in, "If I quit, I lose my protection and get bumped off in a month". They have few friends. Their neighbours fear them, and keep their distance. They all want to live a little longer.

Q. How did you get into this racket?

A. It was while I was in my IInd year at college studying engineering. I was 19 at the time. A friend of mine used to spend money indirectly, though he was from a middle class family. I asked him how he managed to do that, after a lot of persuasion he confided that he was pushing drugs and acting as a sort of courier. He would pick up drugs from the supplier and give them to all the pushers. It was quite safe as he wasn't doing the actual peddling on the streets. I jumped at the chance when he offered to take me to his boss. I want, and after a long questioning session was 'in'. Initially they gave me small packages and I want about distributing them to pushers. With time, when they found me reliable and trust—worthy, I was given bigger jobs like collecting stuff from outstation and getting it to Delhi. It was quite easy. Our trucks would come and be ushered across the checkposts (most policemen are on our payroll). We just informed them when our rival groups truckloads were passing and the cops got a big cut when these were caught (an ordinary policeman pays as much as Rs. 40,000 to get stationed for a month at these checkposts!). I made a lot of contacts. I got to know where the stuff comes from, and who supplies it.

Q. Did you have any brush with law that you particularly recall?

A. Oh yes—once I really felt I had pushed my luck too far. I was assigned the job of getting 20 kg. of heroin from Hong Kong. How I got it across was left to me. I decided to bypass the normal method of getting it by sea. It takes a lot of time and hence there are greater chances of getting caught. To top it, coast guards are a problem. I simply dissolved the entire powder in water and soaked it all in numerous blotting papers. When dry, I just took these blotting papers to a contact of ours who fronts as a stationer, and got them bounded to make some 50 odd note pads. These I crammed into two suite cases along with some clothes, and came by air. The customs officer was suspicious of my luggage and detained me while he got it checked. I knew I was caught. What saved me then was another customs man who recognised and wangled me through. (He got handsomely paid afterwards).

The stuff was worth Rs. 10 crore. I got it through single-handed. That was when I decided to break off from the gang and work for myself. (The paper was later cut into small squares and any 'consumer' had to put it in his mouth, suck on it for a while, and spit out the paper and that's it. He got his high). Initially I worked alone but now I have got almost a 100 people working directly under me and must be a thousand indirectly.

Q. How much do you make in a month?

A. Nothing fixed. Depends on the number of consignments we get across. On an average, I figure I make about Rs. 40-50 lakh in a month.

Q. How do you spend this kind of money?

A. I cannot, Nobody can my needs are still the same as they were in my college days. Occasionally when our foreign counterparts come, we throw lavish parties for them, but the expenses are negligible. I have bought over some chit fund companies, I own DDA flats and farmland in Mehrauli, I have bonds. A lot of it of course is in cash. All our transactions are in cash.

Q. What sort of a network do you have?

A. My operation has become so big, that now, I am supplying their entire requirements to my earlier boss and many others like him. I don't sell to pushers. I sell to smaller set-ups who in turn sells to purshers, who in turn sell to the addicts. I get most of my stuff from abroad and shipments arrive about once a month. A single shipment constitutes almost three full truck-loads of the stuff. There are small time farmers down south who occasionally sell me their produce but mostly it is from abroad. A lot of stuff comes from the Nepal border and Afghanistan. My subordinates report to me every transaction they carry out. May be once in every three months, all trusted guys, get together and we have broad meeting to decide whom to work with, whom not to work with, whom to bribe or whatever. I keep juggling my men around so that they don't become conspicuous. One has to also travel constantly, to Singapore, Bangkok and Hong Kong, also to Afghanistan, not to bring the stuff but of work on links abroad.

Q. For a racket of this magnitude to survive and flourish, you must be having lots of contacts and connections. Do you pay money to keep them quiet and their tails wagging?

A. Of course, without protection, we can't survive a single day. Most big politicians get huge amounts from me to keep mum and help us across barriers. It is because of them that we manage to get our trucks across frontiers unchecked. The politician only has to make a phone call and our vehicle is let through. Of course the cops at the barriers are also paid. Why, last Diwali, I covered the entire route of our consignments by car. All along the way, I distributed sweet boxes to all the check post officials and cops. Of course, the boxes had no sweets in them, they had cold hard cash.

Q. But do things always work out so smoothly for you? There must be some hitches and setbacks, some times.

A. Well yes. Sometimes the scene does go sour. Someone somewhere along the line gets greedy and demands more money. A year back the top cop in the Narcotics Division asked for a huge amount—Rs. 50 lakh to get one of my trucks across the state border check posts. I offered 25 lakhs. The truck was standing about 2 kms. from the border. I had to move fast. We could not come to a compromise and he threatened to get my consignment caught. I told him to go ahead if he dared. He did and I lost that consignment worth one crore. The very next day, he got transferred to a remote area. That's how deep my political connections go. But mostly we manage to get through without problems. I have found out that every man has his price. It just varies from person to person.

Q. Any operation of this size can't function without violence. What about yours?

A. Yes, that's true. Violence does come into play. And it is the best way to get co-operation. Where money does not work violence and threats do. Knowledge of the consequences in case of squealing, persuades a man to remain loyal to us. Every now and then there is a guy who wants to quit or rat on us. They either get severely beaten up or bumped off. There are no half measures possible, in this business. Stay loyal and work well, you reap mammoth rewards, step out of line, you are maimed or dead.

Anyway, killing is not something you can be squeamish about. (And some of the killings has nothing to do with business. He gives the example of a visit to Nainital with a girlfriend when a man in another car said something provocative to her. Following the altercation that resulted, he told one of the bodyguards in the car behind, "Is pe petrol daal do". They did, and set fire to the man and drove off).

Q. Do you ever go on vacations; to hill stations or tourist spots?

A. Almost every month I go abroad for 'business'. In this trade, you can mix pleasure with business. There isn't too much of a chance to go anywhere just for a holiday. I keep going to hill stations too, but that also for prearranged meetings with my associates.

Q. Lastly, how come you agreed to give this interview?

A. The main reason is that I know that this thing can't really hurt me. It's not as if nobody know what is going on. Damnit. I've got police protection because 'some people' know everything.

Secondly, you helped a friend of mine when he was in dire straits. I am just returning the favour. I stick to my friends.[1]

However, this story was blasted by Addl. Commissioner of Delhi Police as pure concoction and the paper had to print, the contradiction. But the story has some grains of truth, though details have been exaggerated to heighten the impect.

References

1. *Indian Today*, May 15, 1984, pp. 147-149.
2. *India Today*, May 15, 1984, p. 149.

1. Imprint, July 1983
2. *The Times of India*—January 18, 1983
3. *Times of India*—April 23, 1984
4. *India Today*—June 15, 1984
5. *Indian Express*—July 23, 1988
6. *Dainik Jagran*—August 1, 1985
7. *Indian Express*, Delhi—Saturday, March 3, 1990, Rajeev Narayan.

46

STRANGE SMUGGLERS AND DUBIOUS DESTINATION

"No investigating agency anywhere in India", says a DRI Director, "can name one individual as the brain behind a narcotics syndicate. The reason is the manner in which the trade operates, which is why we call it a system of cut-outs, which ensures that in case of an arrest and a seizure, nothing can be traced back."[1] The DRI has recounted a story of an aged couple between 58 and 60 years old, who were on a pilgrimage to Rome. One of their suitcases was found to have had a false bottom lined with pouches of narcotics. They had been contacted by a man they did not know, who asked them to carry the suitcase to Amsterdam. He gave them his name, but it was of course fictitious. Over and above, the price of their tickets by air to and fro, and the cost of their stay and travel tickets in Europe, which he provided, they had money to spend. Their instructions were to check into any hotel. They lived in Amsterdam and to book a trunk-call to a certain telephone number of Mumbai, ask for a particular name and inform him of the telephone number and the room number of their hotel in Amsterdam. It turned out that they were given the telephone number of a hotel in Mumbai, and on that the individual name was a guest, who had checked into a room. The name was, of course, false.

Men Behind the Scene

As per scheme, if the call had come through from Amsterdam, the man in Mumbai would have called his contact in Amsterdam, who would approach the couple, give them a pre-arranged code word, and arrange for them to travel to Rome by train. The suitcase was to have been picked by someone en route. To have got the couple and the stuff, the DRI points out, is not to have got very far, because they genuinely do not know real names of the individuals who contacted them, much less who they were really looking for.

The situation is not very different, with the narcotics operators in Mumbai, who need to know nothing more than the names of their suppliers in Pakistan. But these are only wholesale dealers, not the men who make the drugs.

"The state of affairs has crucial implications for the drug enforcement agencies in India. Based on an intuition, developed on day-to-day experience, and piecing together discreet list of intelligence, they know who, in fact, the chief narcotics operators in Mumbai are but have little evidence on them directly. All that the seizures at the airport and elsewhere yield, are the drugs themselves, and smaller minions-drivers or clerks who are in collusion. Further investigations may lead to the arrest of agents and sub-agents, but to get the syndicate heads themselves is a one in a million chances because of the way the traffic is organised. It is unlikely that any of the real dons would make that big a slip.[2]

The hottest name on the lips of senior narcotic investigators in Bombay is that of an underworld don who has been known to deal in other kinds of contraband, including gold.

Clandestine Laboratories

The mathematics of cultivation, processing and subsequent clandestine transformation into morphine and heroin became known to me only when I got posted in Ghazipur in June, 1882. When I joined there, the city was still rocking from the after-effects of large-scale recoveries of the drugs by the police. This immediately attracted my attention. In my previous postings, I was engaged and deeply involved in anti-dacoity operations which developed in me deep insight into the working of criminal gangs. I also made in-depth study of criminal's behaviour and their insidious ways. Therefore, I at once started unearthing the drug racket with the help of an S.I. Shiv Agyna Yadav and two-three constables who played their role with adroit skill. A *pan* and tea shop was opened opposite village Malsa so as to keep watch over the activities of smugglers. The setting was so perfect that nobody could guess their identity. They noted all the details of the ongoings. On a given day, I decided to launch a massive raid in the homes of Virendra Rai and Raj Kumar Rai. At about 4 AM the police party surrounded the house and entered into it by scaling the walls. The police party succeeded in recovering a clandestine factory with a lot of raw material and equipment along with 24 kgs. of heroin and imported transistor sets. Though the real culprits Virendra Rai and Raj Kumar Rai escaped, through careful planning they were ultimately nabbed under Maintenance of Internal Security Act. After this biggest ever haul of heroin various other recoveries of contraband goods followed.

The first clandestine factory to be cracked in Mumbai on 1 August, 1985, was set up in a room of a company guest house in Worli. Company officials were in fact staying there in the last week of July, when heroin was being processed, but didn't suspect a thing till the police broke in. The Worli police were tipped off by other occupants of the building, who were perturbed by the strange odours. As it turned out, the police arrived when the batch was in final stages of processing and were able to seize the drug—white and fresh as snow. The entire apparatus was captured intact—bell jars, pipes, test-tubes, a laboratory thermometer, two domestic masala mixtures and several jars of chemicals, including acetic anhydride, which is reacted with morphine to yield heroin.

Kohire's product was chemically analysed. It was found to contain 67 per cent of heroin—a percentage which is remarkably high for a crude lab—only achieved according to drug experts by laboratories abroad. A highly placed Customs official, who doubts that any crude lab can yield a product with such high purity, says that it is actually on this basis that the Customs officials distinguish between drugs produced in India and those imported from abroad. Police also recovered a jar of lactose, commonly used to adulterate heroin. By the time the drug reaches the addicts on the streets of American or European cities, the heroin content is reduced to between five and two per cent.

Global Trade that Brooks no Barriers

The narcotic—that is at present the primary target of drug enforcement officials in the USA, the world's largest market—is not heroin but cocaine. According to the US Customs, the consumption of heroin has more or less stabilised, but addiction to cocaine is rife and rampant. From airstrips in the deep jungles and exclusive coca islands in South America, the narcotic is flown into the USA in small, single or twin engined craft and landed on private runways. American agencies are now engaged in an all-out war with the Columbian mafia, on their terms. An attorney-general and several policemen as well as Drug Enforcement officials have been shot dead. A recent indication of their financial clout is their offer to pay off the national debt of their countries in an attempt to negotiate a reprieve with their governments.

Mumbai is a clearing house for West Asian and Far Eastern heroin bound for USA and Europe. It is not the major concern of international drug enforcement agencies. To begin with, it is only a transit zone, not a source area. No cocaine is grown in India, and most agencies are satisfied that Indian opium, cultivated legally under government supervision, does not feed the narcotics trade. Secondly, Mumbai is not the only transit channel for narcotics from West Asia and the Far-East; Turkey and Hong Kong are the other conduits to Europe and the USA.

According to recent intelligence reports between international and Indian drug enforcement authorities, most of the narcotics traffic through Mumbai is bound for the African countries as well as Spain, Portugal and Switzerland, en route to the rest of Europe and the USA Most of the heroin is crossed into the USA by Nigerians who resort mainly to body concealment.

Cannabis Disguised as Fish Cakes

Canadian authorities had come across a novel method of concealment of Herbal Cannabis. This method adopted in Jamaica is that the drug is compressed into circular shapes of about 3 cm thick and then wrapped in foil paper. This was then shaped to resemble "fish steaks" and "fish pellets" and then dipped in butter and fried. Real fish patties and fried onions were mixed with them that contained the drug.

The Canadian Drug Enforcement authorities had come across an unusual concealment method which was discovered during a cocaine smuggling investigation. In this case the accused had taken an old car battery and removed the bottom and emptied the inside components. He then placed an empty beer can which was mounted on a piece of wood and fastened to the frame of the vehicle. Approximately 1 kg of cocaine was found in the beer can. The battery was then connected to an elevator to give the look of a second battery.

Air-route—Quickest Possible Mode of Transhipment

One of the ingenious methods recently employed was to conceal the hashish in between several layers of diesel engine pump mountings. The packaging was almost detection-proof. The mountings are sheets of steel, covered with rubber, that provide diesel pumps with a foundation and serve to absorb engine vibrations during operation. In this case, the hashish was sandwiched between the sheet plates, and the insides of the packing case were coated with a lead bond paint—to avoid X-ray detection. However, it attracted the suspicion of the air intelligence sleuths, who doubted that a firm in Canada would want to import only engine mountings and not pumps at all. Moreover, the cost of freighting the mountings to Canada was far greater than the value of sheets in the first place. The crates were ordered to be opened. The contraband was recovered. However, the name of the exporter could not be identified because various hands had been independently employed to prepare the consignment for transhipment. The man who procured the pump mounting was not the one who booked the cargo for export through a clearing house agent. None of those who were arrested knew the others The papers were perfect, including the RBI clearance.

Even more sophisticated methods have now been developed to send large quantities of narcotics across, which cannot be detected even when the crates are opened, or the walls drilled through. A large consignment of automobile spares was booked for export in 1985. The Customs men found nothing objectionable or suspicious hence the goods were cleared for transhipment. But just when they were loaded on to the flight, the crates were substituted with others identical in every detail, including the Customs marking. In place of auto spares they contained 900 kgs. of hashish.

In one case last year, recorded by the Air Intelligence team, IAAI staff whisked away a consignment of narcotics with the 11 labels straight from the tarmac. The contraband can be easily transported into the airport simply by using an IAAI vehicle. There are instances of vehicles bearing the sign—Follow me—meant for guiding aircraft to the hangar, used to carry contraband. A year ago, a theft was reported from one IAAI warehouse situated outside the airport, of a consignment that had been lying uncleared for weeks. Investigations revealed the IAAI staff in close collusion with the importer, had stage-managed the burglary.

Clearly, the largest consignments of narcotics are smuggled out concealed in cargo rather than as passenger luggage. Whether crates or suitcases are used, depends on the drug. Heroin, a kilo of which costs a crore, can be sent out either way, but hashish is normally sent out in crates since it would take at least 25 kgs of the drug to make the effort worthwhile.

The commonest mode of concealment of narcotics is in false bottoms of the suitcases. But this can easily be detected by tapping the suitcases for a hollow sound, or piercing it through. More ingenious methods are consequently resorted to by the experienced. The drug, for instance, may be packed in a hollowed out transistor battery which can nevertheless still be shown to work. The trick is to use a 1.5 Volt cell to which its terminals are connected. More intimate though excruciating means are also known. There are instances of the pouches being pushed inside the anus of the passengers or packed in condoms and swallowed.

An exporting firm in Cochin was caught canning heroin along with shrimps; an 8 kg. heroin consignment was nabbed just as the last nails to a mango crate (meant for export) were being hammered in. In another case, Toshiba Anand battery cells were cunningly packed with the drugs.

Heroin Worth 88 m Seized in USA

Customs officials at the JFK International airport seized on two consecutive days heroin having an estimated street value of 88 million from four passengers arriving from India. In two apparently unrelated incidents a routine baggage check yielded 88 pounds of heroin. The seizures, which were announced by the US attorney for the eastern district of New York, were said to be the largest ever made in the city. In all, six persons including a woman were arrested.

The *New York Times*, in a report from New Delhi said, quoting authorities in India and Pakistan, that Delhi and Bombay were rapidly becoming the main staging, shipping and trafficking centres for heroin smuggled to the USA and Western Europe from the Khyber Pass area of Pakistan and Afghanistan. The paper said some Indian officials disputed this contention. "I think it would be more accurate to say that it is an important and convenient staging area, an official of the Indian Central Bureau of Intelligence was quoted as saying.

Army Vehicles used for Smuggling

Drug smugglers and insurgents in the north-east, who work closely for more reasons than one, have taken to one of the safest modes of transporting narcotics from areas near the Indo-Myanmar border to Calcutta: using vehicles of the army and paramilitary forces. Smuggling of drugs in army vehicles has been going on for some time in the South-Asian countries like Thailand but their use in India is a recent phenomenon. The army and the paramilitary forces have a strong presence on the Indo-Myanmar border, particularly in Manipur and Mizoram, through which most of the drugs are smuggled. Recent seizures have revealed that the vehicles of the Assam Rifles are used the most by the smugglers. The Assam Rifles is under the control of the Army. Sources said the Border Roads Force was also allegedly involved.

In July, Customs officials were surprised when as many as 300 kg. of heroin was recovered from a Bihar-bound truck of the Central Reserve Police Force (CRPF) at Nagium in Manipur. Besides, 100 kg. of charas was seized from a truck of the Border Roads Task Force at Kangpoppi also in Manipur. The sources said the heroin smuggled from Moreh in Manipur and Dimapur in Nagaland was of high quality and was a craze.

Smuggling on Indo-Myanmar Border

The magnitude of drug smuggling from the Indo-Burma border could be considerably higher than what the seizures have indicated. The matter has reportedly reached the Union Home Secretary who discussed it several times with concerned officials.

The drug traffic from Myanmar has increased considerably ever since political turmoil erupted in that country. During the initial days of the trouble in Myanmar, the smugglers and insurgents had a field day due to lack of vigil from Myanmar side. But since the first week of October the flow of Myanmar heroin has turned into a trickle at Moreh, due to the virtual sealing of the Myanmar border at Tamu. Still, smuggling continues due to the long, porous border along Manipur and Mizoram.

The Myanmar heroin fetched a high premium wherever it was taken in India. At Moreh, one kg. of heroin cost nearly Rs. 1.7 lakhs but by the time it reached Imphal Rs. 50,000 was added to the cost and the price virtually doubled when it reached the metropolitan cities. When safer modes are not available for transporting the heroin, it is brought into India through numerous couriers in small amounts and then accumulated at Imphal or Aizawal.

In Manipur also, nearly one kg. of heroin is consumed everyday, primarily in Churachandpur area in south Manipur. Though many youths have been arrested under the Narcotic Drugs and Psychotropic Substances Act 1985, yet effective curb has not been possible. It is apparent that with so much money and power at stake, the contraband smuggling enjoys political patronage in the area. On 5 August, a Congress councillor of Moreh, Mr. Hminga, was arrested when Customs officials found 120 gm of heroin from his possession. It was later revealed that he himself was an addict and was actively involved in smuggling heroin and cocaine from across the border. He has since been expelled from the Congress.

Drug smuggling is further facilitated by the rules allowing the entry up to 20 km on either side of the Indo-Myanmar border by tribals living in the area. Such tribals need not carry documents. For them, movement across the border is a routine affair. Sources at Moreh told this correspondent recently that insurgents on the Myanmar side controlled over 15 heroin refineries from where it was supplied to the Indian drug market. Besides some mobile laboratories near the border produce heroin.

A motion moved by Gauri Shankar Bhaiya in UP Legislative Assembly, disclosed to the shock of the House that the heroin processed from opium was being supplied not only to the hostels and hotels, but to the foreign countries from the Central Drug Research Institute Lucknow. The Institute is controlled by the CSIR. In this institute, opium was mainly being used to check unconsciousness and shivering cold in the pilot project of Analgesic and CNA group. In this Research Institute, heroin under the supervision of the scientists is processed in 20 minutes. Under this method, opium is heated with acetic hydride (C-2) to be converted into brown sugar and then white sugar which is easily processed into heroin. Gauri Shankar accused that some high officials and researchers are involved in supplying not only heroin to the hostels and cantonments but heroin packs are sent with research journals and magazines to institutes abroad as well.

Smuggling and Seizures

Anti-narcotic agencies all over the world are coming with a heavy hand over the smugglers. The Interpol, an international agency has stepped in to nab narcotics dealers using Gujarat as a route for their drug-trafficking. Interpol has traced three Gujaratis suspected to have been involved in smuggling narcotics from Borsad taluka in Kaira district to the United States and Britain. The customs authorities busted the narcotics racket at Bodal village of Borsad taluka run by two travel agents, one from Borsad and another from Bombay (Mumbai).

The customs authorities have already seized 867 grams of brown sugar from the bound cover of two books on wildlife meant for despatching to Britain. The international value of the seized quantity of brown sugar is Rs. 90 lakh. However, the value of the total quantity of narcotics involved in the racket ran into crores. Interpol telephonically asked the Narcotic Bureau of Bombay to direct the customs authorities here to investigate the case thoroughly. Interpol authorities also rang up the customs authorities to find out how much quantity of brown sugar was recovered from the books seized by the officials from the residential premises of Bipinbhai of Patel Travels at Bodal village. They also inquired whether the officer involved in busting the racket had received any threat or had any difficulty in dealing with the case.

Meanwhile the interrogation of Bipinbhai has revealed that his counterpart travel agent in Bombay was involved in this racket for the past two years. Every week he used to send ten books, each containing narcotics ranging from 500 grams to 800 grams to U.K. and the U.S.A. A 52 years old woman had received two such books at the Chicago airport last month. Interpol sent a message that they had started investigation of the cases at the international airport of UK and the US Further, it was revealed that the Borsad agent had earlier sent five books, each containing 500 grams of brown sugar to the US.

The *modus operandi* of Mr. Patel revealed that he used to approach various passengers at Bombay airport and was helping them in one or the other for luggage shifting. Later he requested them to carry the books on wildlife which contained narcotics. He used to take down the addresses and telephone numbers of the passengers and informed his men in the UK and the USA.

The Customs Preventive Collectorate established a close nexus between Air India and narcotics smuggling and had fined the airline Rs. five lakh for attempting to smuggle out 125 kgs. of hashish valued at over Rs. 60 crore. If the airline had not brought the contraband back to India from New York, where it was bound, then the aircraft could have been impounded by the US authorities.

Additional Collector (Preventive), L.D. Arora, who submitted his findings to the court said that the contention of the airline that they had no knowledge about the contraband cannot be believed. The findings stated that the Air India staff brought back the five packages from New York by their flight only to safeguard their own interest because the customs authorities had already been on the lookout for the packages and they were aware that the packages contained narcotics.

The findings revealed that Air India could face the legal consequences of carrying narcotics to the US and that too as unmanifested cargo.

Airports in neutral countries are now being included in the narcotics network as decoy centres. This is because the established routes are well known and are subjected to constant surveillance. This had led narcotics operators to re-route their traffic. In a recent case, a consignment from Sri Lanka to West Germany was diverted through the Maldives and transferred on to a flight from Malaysia to West Germany. Drug enforcement officials point out that any airport can be used as a decoy.

What to Look for on an Air Way Bill

When considering airfreight as a smuggling method, almost any country or geographical area is a potential source or transit point for drugs which can easily be concealed in a freight consignment.

UK Customs officers—responsible for freight controls at London/Heathrow airport—have prepared the following information which should assist generally with the profiling of freight consignments which are more likely to contain drugs. A copy of an original air way bill with manuscript numbers 1 to 13 representing the various boxes required to be completed on the document is annexed for reference purposes.

Box 1 *Shipper's name and address*

Is the address vague-possibly mis-typed, misspelt or incomplete (PO Box, hotel, etc.).? Is the address proper to the broker or freight agent?

Is the shipper and consignee one and the same? Is it possibly connected with a previous case (local records)?

Box 2 *Consignee's name and address*

Is this a legitimate company? (Check local directories).

Is it a first time importer? New local registering number. Does the address appear to be in a private dwelling?

Is the consignee and importer the same person?

Is the address vague, possibly mis-typed, mis-spelt or incomplete?

Does the telephone number quoted on the documents correspond with the area for the address.

Is the company or the individual the subject of a local lookout?

Box 3 Issuing Carrier's Agent, Name and City Check that this information is not the same as in Box 1. If it is, then try to ascertain who the shipper is.

Box 4 *Airport departure*

Is it a source country, or is it a major transshipment point?

Have they tried to hide the source country?

Box 5 *Routing*

Is a specific day of flight demanded? (e.g., goods must travel as booked)

Check the unusual routing.

Box 6 *Handling information*

Look out for the following special instructions. Hold at destination—pending further instructions. Shipper will make contact with further instructions.

Contact immediate-possible suspicion if the goods in question are not perishable—or "rush" (blood serum, etc.)

Must travel as booked.

Hold for pick.

Box 7 No. of pieces, weight, commodity description, etc. Excessive weight for the commodity.

High risk goods? (e.g., handicrafts, personal effects, returned goods and furniture).

Does this type of consignment fit the current trends?

Shipment of low value goods where the shipping cost exceeds the declared value.

Description of goods is vague, insufficient

or evasive.

Goods are not logical (for type of business) or economical.

Have the goods the potential for concealment?

Box 8/9 *Accounting a payment (Valuation charge)*

Prepaird and/or cash payment is suspect.

N.B.—Most of the regular exporters/importers have accounts with freight transporters and they deal on a "collect" basis.

Paying by credit—has not been used to date at London/Heathrow by a drug trafficker.

Box 10/11 *Declared Value Insurance*

Where there are no insurance details, declared values or a value for Customs—Check the importation further.

Box 12 *Other changes*

Any change quoted should be reviewed for necessity.

Box 13 *Shipper's signature*
Is there a signature?

Use of an "X", swirl or other mark in place of a signature—possibility of persons trying to maintain anonymity.

Heroin Worth Rs. 90 crore Seized by BSF

The BSF intelligence unit and customs seized 90 kgs. of heroin valued at Rs. 90 crore in the international market from the Khem Karan area of the Indo-Pakistan border. The D.I.G., B.S.F., Mr. V.K. Rajagopalan, told reporters that the heroin was smuggled from Pakistan for onward transport to Delhi and Bombay and was hidden in the oilseed field of Puran Singh. The joint raid by troops of 45 and 69 BSF battalions and Customs was carried out on prior information. However, nobody was arrested. This is the fourth seizure of narcotics from the Indo-Pakistan border since 1989.[3]

1,200 kgs. of Hashish Seized

A truck carrying 1200 kgs. of fine hashish was stopped at the East Delhi border and five persons were arrested in this connection. According to the police, this was the largest haul of hashish made in Delhi. They said that information leading to the seizure was first obtained by the Vigilance Branch about a month ago, and was secretly developed. The police came to know that a large consignment of hashish, originating from the Nepal-U.P. border, was to come to Delhi for some international destinations via Bombay.

A truck run by a Delhi-based transporter, Narender Singh, arrived in the Capital a shortwhile after midnight. It was stopped at the check-point at the G.T. Road-Loni border contagious to Ghaziabad, and the police asked the driver what was inside. The truck driver, Talvinder Singh, a resident of Amritsar, Punjab, said that he was carrying high-quality bone china crockery. He was asked to unload the consignment, which was packed in 46 wooden crates. He refused, saying it could be easily damaged.

CBI sleuths were able to identify the six men well before the date of the voyage and maintained a constant watch on their movements. They had brought over 13 kgs. of high-purity opium from Moradabad, while 24 kgs. of hashish were procured from Bombay. The kingpin, Syed Ahmed Shaukat Hussein, was travelling under the name of Akhtar Hussain. His associates were Mehboob-Akhtar, Haroon Bhonde Khan, Mohammed Hafiz Rabban, Abbas Salam and Mohammed Hanif Shafi, all of whom described themselves as agriculturists from Moradabad in their travel permits. At 7.30 pm., they arrived at Indira dock closely followed by the CBI officers. The dock presented a chaotic scene with over 4,500 people gathered there. They were nabbed just as they collected their heavy luggage, which had arrived separately by a truck. Examination of the trunks and boxes revealed that the narcotics had been concealed in huge quantities of chilly powder, garlic, onions, mango chips and dry fish. CBI sources surmise that besides concealing the contraband, the etables were also packed to cover the strong odour emanating from opium.

The six men were produced before a magistrate and remanded to police custody for 15 days. Investigators believed that they were not just drug carriers, but part of an organised international syndicate. The CBI expected to follow up the seizure by raiding the places in Bombay from where the gang had procured hashish.

The Haj Committee in Delhi was understood to have cooperated in the investigations. The police unloaded the crockery, and found that under the floor of the truck lay 1,200 kgs. of fine, soft, hashish, rolled up into balls. The driver and his three assistants tried to flee, but were nabbed.

The culprits were identified as Talvinder, his father Tanwant Singh, who is a retired head constable of Punjab Police; Mashvinder Singh of Amritsar who was the second driver and the cleaner Gugan Ram of Mohindergarh, Haryana. Tanwant Singh is said to be the organiser. A fifth person Narinder Singh, the transporter, was arrested from his house on Roshanara Road, later in the morning. The police said that the hashish had come from a Nepalese district bordering Gorakhpur in U.P. It was to go to Bombay via Kanpur, Hardwar and Delhi. The police said that some of the consignment was for domestic use in the aforementioned cities. The extent of network, the man in Bombay and the international destinations were still to be ascertained by the police. They said that the transporter had been paid Rs. 1,20,000 to transport the goods.[4]

Opium Seized from 6 Haj Pilgrims

The Central Bureau of Investigation (CBI) detained six Haj pilgrims at Indira dock and seized opium and hashish, valued internationally at Rs. five crore from them. The six were arrested moments before they were to board M.V. Akbar to Jeddah. The vessel set sail, on its third Haj voyage in the year 1989 with over 1,500 pilgrims on board. All the narcotic smugglers were arrested.[5]

Smuggling Via Toronto

The heavy volume of the traffic apparently helps smugglers. For instance, between 3 pm. and 7 pm. as many as 2,000 passengers an hour can arrive from Europe, where South American cocaine is usually diverted before being flown to Toronto. Heroin can come from anywhere.

To hoodwink the customs and police the smugglers go to great lengths in concealing their imports. Some hide the drugs in custom made body harnesses, bury them in hidden luggage compartments or swallow them wrapped in condoms. Being panicked on the plane some smugglers abandoned their bodypacks in the washrooms. But because they fitted the smugglers profile they were put through a strip search revealing marks on their torso left by the tight fitting harness. Searches of the airplane reveal the abandoned drugs, with finger-prints sometimes left on the harness.

Four years ago, customs officials made an unsuccessful bid to use trained dogs for searches. The animals were hidden in cages at check-in points and tipped off inspectors when they smelled drugs by pushing an alarm. The animals, however, proved to be over-sensitive sending too many innocent passengers off for full scale searches.

One smuggling case pertaining to India involved 671 pounds of hashish found in a cargo shipment. It was labelled as research materials destined for Toronto General Hospital.[6]

NCB Pressure Forces Drug Mafia to Reopen Balkan

It could go down in the history of anti-drug trafficking as a remarkable feat. The two-year-old Narcotics Control Bureau (NCB) of India has forced the drug lords of the Golden Crescent to virtually stop using India as their favourite route for heroin trafficking to Western Europe and the United States. A recent Interpol report states that the traffickers are again being forced to reopen the Balkan route to Europe.

The meek and humble Mr. Madan M. Bhatnagar, strikes the casual observer as a quiet academic, most comfortable in an armchair with his favourite book. It is difficult to associate him with the illicit trade of automatic weapons and heroin. But the sharp intelligent eyes of the Director General (DG) of the Narcotics Control Bureau (NCB) cannot be deceived. Mr. Bhatnagar as the DG of the two-year old NCB has perhaps the least envied job in the country. With little over 160 dedicated officers of the NCB he has the difficult task of battling out the international drug traffickers operating from within the country and across the borders. In an interview with The Economic Times, Delhi, Mr. Bhatnagar explained the current drug trafficking situation in the country and talked about the achievements of the NCB, that the bureau has made remarkable recoveries despite manpower constraints. This is borne out in a recent Interpol report which pays tribute to the NCB saying that it is the only example in the world where a nascent agency has succeeded in checking drug trafficking in a substantial manner.

Some of the major drug lords operating from the country have been neutralised effectively. Armed with the recently amended Narcotics and Psychotropic Substances Act of 1989, which has teethed it with the provisions for death sentence in certain cases for trafficking offences, NCB is closing on its net over some other drug mafias still operating. The coming months will see application of the Act in full vigour and there are likely to be some cases of forfeiture of properties owned by the drug mafias.

NCB's success is substantiated by Interpol reports stating that drug trafficking in the Golden Crescent, comprising Pakistan, Afghanistan and Iran, have been forced to reopen the Balkan route across Turkey and the Mediterranean to Western Europe. India as a transit country to Europe and North America has become unsafe for the drug mafias of the Golden Crescent, because of the NCB. From January till the end of September, NCB has seized 1,807 kgs of heroin, of which 1,338 kgs were seized in the western borders coming from the Golden Crescent. During the same period about seven kgs. of heroin was seized in the eastern borders coming from the Golden Triangle, comprising Myanmar, Thailand and Laos. The heroin seizures in 1988 totalled 3,029 kgs, 84 per cent of which were seized in the western border.

India, Mr. Bhatnagar said, has in the last couple of years emerged as a transit country in respect of trafficking in heroin and hashish as it is sandwiched geographically between two of the major producing areas—the Golden Crescent on the west and the Golden Triangle in the east. Through constant vigilance the bureau could effect large drug seizures and this is one of the major reasons for a marked decline in the flow of drugs in Western Europe, which is believed to have come from India, and the declining trend continues. But NCR's major worry now is the large volume of drug seized for which there is no known source of origin.[7]

The Comparative Figures

The comparative figures for the seizure of narcotic for the last five years are as given below:

National Drug Enforcement Statistics

As on 31.8.1994

1. Quantity of Various Drugs Seized in kgs with No. of Cases.

		1990	1991	1992	1993	1994
OPIUM	SEIZURES	2114	2145	1918	3011	1058
	CASES	506	566	1286	1679	592
MORPHINE	SEIZURES	6	6	3536+110 Amps		36
	CASES	27	21	158	105	52
HEROIN	SEIZURES	2193	622	1153	1088	464
	CASES	764	1158	2779	3383	1453
GANJA	SEIZURES	39090	52633	64341	98867	83921
	CASES	1782	3140	5839	5214	2844
HASHISH	SEIZURES	6388	4413	6621	8238	3668
	CASES	753	335	2516	2827	1184
COCAINE	SEIZURES	1	0.008	0.420	2	0.38
	CASES	2	1	4	4	5
METHAQUALONE	SEIZURES	2141	4415	7475	15004	27261
	CASES	60	78	167	283	175
AMPHETAMINE	SEIZURES	—	—	—	—	—
	CASES	—	—	—	—	—
PHENOBARBITAL	SEIZURES	—	—118020 Tabs.		—	—
	CASES	—	—	2	—	—
L.S.D.	SEIZURES	—	—	50 Sq. Papers	164 sq. Papers	157 sq. Papers
	CASES	—	—	—	1	—
ACETIC ANHYDRIDE	SEIZURES	—	—	—	19758 ltrs	19020 ltrs
	CASES	—	—	—	22	31

2. Persons Arrested.

	1990	1991	1992	1993	1994
(a) No. of Persons Arrested including foreigners	4302	5300	12850	13723	6528
(b) No. of foreigners arrested	234	93	116	144	59

3. Action Taken Against Persons Involved in Drug Trafficking

	1990	1991	1992	1993	1994
(a) No. of persons prosecuted	4299	5546	7172	9964	2845
(b) No. of persons convicted	551	855	761	1488	336
(c) No. of persons acquitted	1160	1940	1762	1633	1239

4. Action Taken Under COFEPOSA/PITNDPS (NDPS) Act, 1988

	1990	1991	1992	1993	1994
(a) No. of detention orders issued under PITNDPS Act, 1988 with no. of detained in Bracket	211 (190)	109 (80)	97 (80)	116 (92)	50 (17)

5. Destruction of Narcotic Drug Yielding Plants

	1990	1991	1992	1993	1994
(a) Poppy plant area (in acres) with potential yield in bracket (in kgs.)	59 (1616)	46 (671)	19 (135)	67 (754)	— —
(b) Cannabis plant area (in acres) with potential yield in bracket (in kgs.)	1907 (1896034)	1013 (1497284)	1219 (1230209)	2587 (3270661)	758 (948770)

6. Destruction of Manufacturing Facilities

(a) Facilities detected and quantity of finished drugs (in kgs.) in bracket.

	1990	1991	1992	1993	1994
HEROIN	3 (0.090)	2 (1.060)	5 (1.010)	4 (1)	1 (0.600)
HASHISH	—	—	—	—	—
METHAQUALONE	1 (525.000)	—3 (3651.000)		2 (1710)	1 (4820)
MORPHINE	—	—	1 (0.760)	—	—
(b) No. of persons arrested	2	7	7	3	5

(c) Incriminating materials seized in kgs

	1990	1991	1992	1993	1994
OPIUM	—	3.150+5 ltrs Opium solution	37.950	2.750 Opium Solution	—
GANJA	—	—	Acetic Anhydride Sodium Carbonate Ammonium Chloride Methanol Diethyl ether Acetic Chloride	Acetyl Chloride Acetyl Chloride Sodium Carbonate Ammonia Chloride	
PRECURSOR CHEMICALS AND EQUIPMENTS	Lab. Equipments			Lab. Eqpt.	

7. Disposal of Seized Narcotic Drugs and Psychotropic Substances

	1990	1991	1992	1993	1994
OPIUM	43	101	254	449	19.5
MORPHINE	—	—	1	—	—
HEROIN	1708	1578	807	240	61
GANJA	4780	4046	5962	10699	6159
HASHISH	1810	2386	1117	1115	2165
COCAINE	—	—	2	—	—
METHAQUALONE	724	1615	114	17345	—
AMPHETAMINE	—	—	—	—	
OTHER DRUGS					

8. Forfeiture of property

	1990	1991	1992	1993	1994
(a) Value of Property Forfeited	—	603000	888337	—	—
No. of cases in bracket	—	(2)	(5)	—	—
(b) Approx Value (in Rs.) of the frozen property of Drug Traffickers	17985040	7023158	37026070	904044	—
No. of Cases in bracket.	(18)	(8)	(4)	(7)	

9. Agency-wise Number of Seizures/Cases

NAME OF THE AGENCY	1990	1991	1992	1993	1994
ALL INDIA	3903	5298	12751	13518	6329
NARCOTICS CONTROL BUREAU	134	68	88	116	40
DIRECTORATE OF REVENUE INTELLIGENCE	3	5	5	18	24
CUSTOMS & CENTRAL EXCISE	676	186	518	474	169
NARCOTIC COMMISSIONER'S ORGANISATION	46	81	114	79	35
B.S.F.	57	34	42	125	97
C.B.I. —	—	—	—	12	2
STATE AGENCIES (TOTAL)	2987	4924	11984	12694	5962
POLICE	—	—	—	12497	5774
EXCISE —	—	—	—	197	188

10. Agency wise Break up of Quantity Seized (in kgs)

	1990	1991	1992	1993	1994
(A) Narcotics Control Bureau					
OPIUM	—	5	50	166	—
MORPHINE	1588 Amps.	—	—	—	—
HEROIN	337	137	85	69	69.1
GANJA	1979	4965	359	3349	160
HASHISH	814	52	2	347	45
COCAINE	—	—	—	0.06	—
METHAQUALONE	1238	1496	1491	5776	575
AMPHETAMINE	—	—	—	—	—
PHENOBARBITAL	—	—	63590 Tabs.	—	—
(B) Directorate of Revenue Intelligence					
OPIUM	—	—	243	393	0.04
MORPHINE	—	—	0.190	—	—
HEROIN	15	1	0.035	140	6
GANJA	5	—	—	703	1170
HASHISH	—	935	74	—	1340
COCAINE	—	—	—	—	—
METHAQUALONE	75	399	—	372	10394
AMPHETAMINE	—	—	—	—	—
PHENOBARBITAL	—	—	—	—	—
ACETIC ANHYDRIDE	—	—	—	2000 Ltrs.	—
(C) Customs and Central Excise					
OPIUM	20	35	68	111	6
MORPHINE	1	3	2	—	30
HEROIN	384	78	83	114	72
GANJA	11212	6394	12221	22146	8321
HASHISH	1512	1107	115	1369	487
COCAINE	1	—	0.310	1	—
METHAQUALONE	796	2251	1638	4140	1523
AMPHETAMINE	—	—	—	—	—
PHENOBARBITAL	—	—	54430 Tabs.	—	—
ACETIC ANHYDRIDE	—	—	—	8000 Ltrs.	3605 Ltrs.
(D) Narcotic Commissioner's Organisations					
OPIUM	305	596	533	127	376
MORPHINE	—	1	2	2	—
HEROIN	11	40	10	25	7
GANJA	17	21	53	152	135
HASHISH	1	—	—	19	—
COCAINE	—	—	—	—	—
METHAQUALONE	—	—	—	—	—
ACETIC ANHYDRIDE					
(E) B.S.F.					
OPIUM	—	74	—	12	6
MORPHINE	—	—	—	0.295	0.5
HEROIN	362	13	29	77	47
GANJA	351	363	281	1156	524
HASHISH	150	408	373	813	346
COCAINE	—	—	—	—	0.19
METHAQUALONE	—	—	—	—	—
AMPHETAMINE	—	—	—	—	—
PHENOBARBITAL	—	—	—	—	—
ACETIC ANHYDRIDE	—	—	—	7501 Ltrs.	695 Ltrs.
(F) C.B.I.					
OPIUM	106	150	—	3	2
MORPHINE	—	—	—	—	—
HEROIN	7	—	—	13	4
GANJA	—	—	—	2	—
HASHISH	5	218	—	7	—
COCAINE	—	—	—	—	—
METHAQUALONE	—	—	—	—	—
AMPHETAMINE	—	—	—	—	—
PHENOBARBITAL	—	—	—	—	—
ACETIC ANHYDRIDE	—	—	—	—	—
(G) State Police					
OPIUM	1675	1285	1017	2196	668
MORPHINE	4	2	30	34+41 Amps.	5.3
HEROIN	1027	347	938	638	253
GANJA	12074	37042	49374	68310	68873
HASHISH	3877	1692	2057	5680	1446
COCAINE	—	0.008	0.110	1	0.2

COCAINE	—	0.008	0.110	1	0.2
METHAQUALONE	32	269	4346	4736	14769
AMPHETAMINE	—	—	—	—	—
LSD	—	—	50 sq. Paper	164 sq. Paper	157sq. Paper
ACETIC ANHYDRIDE	—	—	—	2257 Ltrs.	14720 Ltrs.
(H) State Excise					
OPIUM	7	—	7	3	—
MORPHINE	—	—	1	69 Amps.	—

HEROIN	50	6	8	12	6
GANJA	4472	3848	2053	3049	4738
HASHISH	29	1	—	3	4
COCAINE	—	—	—	—	—
METHAQUALONE	—	—	—	—	—
AMPHETAMINE	—	—	—	—	—
PHENOBARBITAL	—	—	—	—	—
ACETIC ANHYDRIDE	—	—	—	—	—

References

1. *The Illustrated Weekly of India:* The Great Narcotics Bazar—Ivan Fera.
2. *The Illustrated Weekly of India*, Feb. 16, 1986.
3. *The Times of India*, November 15, 1988.
4. *The Times of India*, September 27, 1989.
5. *The Times of India*, New Delhi, June 25, 1989.
6. *Northern India Patrika*, January 13, 1984.
7. *The Economics Times*, New Delhi, October 23, 1989.

47

METHODS AND TECHNIQUES OF NARCOTIC SMUGGLING

Violating the laws which are not so stringent, Narcotic smuggler gangs manage to operate freely. Attracted by enormous profits involved in this clandestine trade, they resort to methods bordering on dare-devil. Intensive checks and preventive raids have a temporary effect on the underworld. The smugglers hardly bother about the law which can put them behind the bars for only a period not more than three years.

According to narcotics officials, the price shoots up as the contraband is transported across the borders of the state or closer to the national borders. Even pure opium for which government pays the cultivators anything between Rs. 130 and Rs. 300 per kg. (depending on the quality) could net much more in any clandestine deal. Some recent seizures made by the state unit of the Central Bureau of Narcotics (CBN) have revealed that there has been a sudden trend to convert opium into morphine and in turn morphine into heroin because one kg. of morphine could at least fetch Rs. 2 lakh in the international market, while the same quantity of heroin could provide a return of nearly Rs. 1 crore. Once outside the state, its price shoots three-fold and the escalation continues as it passes from hand to hand.

The small number of possibilities involved and the common features found in most of the cases encountered have made it possible for us to offer the following analysis and classification of cases.

Methods of Concealing Drugs

Under this heading we shall consider only methods which may necessitate intervention by a doctor.

(a) Concealment of Drugs in Corpore

Experience has enabled us to divide people who conceal drugs within their bodies into two categories: "Swallowers" and "Stuffers".

(i) *"Swallowers"*—To avoid customs checks, the courier swallows the drug in the form of small pellets weighing 1 to 5 grams, wrapped in several layers of latex (finger tips of surgical gloves) and tied up. Once the pellets have been swallowed, they make their way along the digestive system (the stomach, the small intestine and colon, and the rectal cavity). The total weight swallowed can be as much as 600 grams.

In the producer country, the courier fasts for 48 hours and then, over a period of varying from 12 to 16 hours, swallows from one hundred to two hundred pellets between mouthfuls soup or jam. Once a pellet has got past the pharyngo-oesophageal sphincter, the swallower can no longer control its course until it reaches the anal expulsion state. The motor impulses which carry the swallowed item between the oesophagus and the anus are automatic. To slow the process down, the courier takes intestinal poultices, anti-spasmodics or anti-diarrhoea medication.

He then boards his plane and throughout the flight usually refuses any kind of solid food, consuming only cold drinks. Once he arrives at his destination, he eliminates the pellets in the normal way. The pellets swallowed last are in different coloured wrappings from the rest, so that he can tell when elimination is complete. It is possible to retain the pellets in situation for 48 hours after swallowing the last of them. This method is used by couriers (both men and women) bringing cocaine from South America. However, since mid December 1985, Nepalese couriers have been found using this method to smuggle heroin: the first were intercepted at Roissy—Charles de Gaulle Airport on arrival from Kathmandu. The latex tube containing the drugs are much simpler than those used for cocaine.

In January 1986, two Swiss nationals were stopped an qd found to have swallowed about twenty pellets of cannabis each weighing 5 grams. The pellets had been wrapped in cellophane which had then been thickly coated with bees wax.

In March 1986, a German passenger on an Air France flight from Bombay was found, on arrival, to be in a state of coma. He was rushed to hospital for an operation, and 292 pellets of heroin, protected only by a wax coating, were found in his body.

(ii) *"Stuffers"*—The courier introduces suppositories containing drugs into rectal cavity and/or vagina. The drug to be smuggled is compressed and given form of a long, 1 cm. in diametre and weighing 50 grams cylinder. These small cylinders are first wrapped in cellophane, then covered by several layers of adhesive plastic tape, and finally put into one or two condoms to facilitate insertion. Among both men and women, the rectal cavity is the one most commonly used. Since the vagina has no shincter, any drugs inserted have to be kept in place with a sanitary towel. Insertion is made easier by the use of some form of grease (vaseline, haemorrhoid cream, cocoa butter, etc.).

In this way, a courier can carry upto 550 grams of a drug. A woman—who was 7 months pregnant—was intercepted at Roissy—Charles de Gaulle Airport in January 1985 and was found to be carrying 100 grams of heroin in her vagina. This method has been used for a variety of drugs.

(b) Concealment of Drugs Para Corpore

Again, there seems to be no limit what smugglers can imagine. Possible hiding places which a doctor may be required to investigate include plaster casts, bulky dressing, orthopaedic devices, etc.

Observations by Customs Officers or Doctors giving Grounds for Suspicion or Guilt

We are not able to divulge the criteria governing a Customs Officer's decision to make a passenger undergo a medical examination, however the medical factors giving grounds for suspicion or establishing the presence of drugs in corpore are described below:

(A) "Swallowers"

— Baggage found to contain medication to slow down the passage of substances through the intestine, but no intestinal disinfectant: Obviously, a person with diarrhoea contracted while travelling is less suspect.

— Breath smelling of acetone, which confirms that the suspect has been fasting: passengers tend to compensate for the boredom of long-haul flights by eating and drinking a great deal, with the full encouragement of the airline companies. It is, therefore, not normal for a passenger to refuse all the meals and snacks offered to him, unless of course the digestive tract is full! Refusal of hot drinks or solid food can also be motivated by the fear that a pellet might burst under the effect of digestive juices.

— The medical examination; the doctor must:

(1) Check that the suspect has signed the consent form presented by the Customs Officer.

(2) Carry out the examination with the suspect naked, lying on his or her back, with an empty bladder.

(a) The examination begins with thorough palpation of the abdomen to detect the presence of any small lumps, which move under pressure, in the projection area of the right, transverse and left colon (If the doctor finds small lumps, the suspect will immediately claim to have swallowed ganglions!).

(b) It ends with pelvic touch examination (rectal and/or vaginal). Where pellets have been swallowed, they will have passed easily through the digestive tract, and are bound to be in the rectal cavity.

If a suspect is thin, the pellets are more easily detected by palpation. The presence of swallowed pellets can only be confirmed by pelvic touch examination.

(B) "Stuffers"

— Tube of some form of cream or grease in baggage.

— A large number of condoms.

— Surgical thread for stitching, or dental floss.

— Discrepancy between a woman's supply of contraceptive pills and her claim to be mensurating in order to avoid a body search by a female Customs Officer.

— Claims to have some of venereal disease or haemorrhoids.

— Grease marks on the crotch of underpants.

— The medical examination—The doctor must:

(1) Check that the suspect has signed the consent form presented by the Customs Officer.

(2) Carry out the examination with the suspect naked and with an empty bladder.

(a) Have the suspect get up and sit down on a hard seat several times: a person not concealing drugs will sit down normally on both buttocks, whereas a "stuffer" will place one buttock on the seat, sliding the other buttock after it.

(b) Spread open the groove between the buttocks to see if there is any sign of a thread attached to a suppository.

(c) With the suspect lying on his or her back, thoroughly palpate the abdomen, feeling for the presence of any lump in the projection area on the left colon and of the sigmoid flexure. Lumps are rarely found unless the suspect is very thin.

(d) Pelvic palpation is the vital part of the examination.

Rectal palpation immediately gives the answer, easily disclosing the presence of a lump or lumps. The examination is carried out with the suspect lying on the back or on the left side, but never with the knees pulled up to the chest, as that position would make the suppository drop down under its own weight into the sigmoid flexure, thus escaping detection. With women, vaginal palpation immediately indicates whether or not there are suppositories in the rectal cavity—they can be left through the recto-vaginal wall. Rectal palpation should then follow.

The rectal cavity always contains some foetal matter; if the rectal cavity is absolutely empty (in which case, the finger-staff will emerge clean and greased), this can mean that the suspect had time to eliminate the suppositories before or during the customs baggage check. It will, therefore, be necessary to look for them either in the baggage checking area itself or between that area and the medical centre.

(C) Subsequent Action

Once the diagnosis has been made, the drugs must be eliminated either digitally or spontaneously. It is important for the doctor to obtain a specimen pellet or suppository in order to:

— Check the state of wrapping and have the suspect taken to hospital immediately if it looks as if the wrappings may burst spontaneously.

— Provide physical evidence.

— Reassure the Customs Officer that his intuition was right.

(D) Other Examination Techniques

For Customs checks at Roissy—Charles de Gaulle Airport, the palpation techniques has the advantage of being rapid and requiring only simple equipment which is generally available with every doctor. Use of X-rays is not appropriate, either socially (the passenger is held up) or as far as diagnosis is concerned—lumps of the same density as the soft tissues do not always show up on an abdomen X-ray without prior preparation.

Ultrasonic scans do not come up to our expectations; the expansion of gas in the hollow viscera caused by the plane journey prevents the passage of the waves. Consequently, this method is no longer used.

Results of Medical Examinations Made at Roissy—Charles de Gaulle Airport Record Following Figures

(A) Seizures in Corpore

	Heroin	*Cocaine*
1980	2,363 g.	—
1981	3,747 g.	—
1982	1,125 g.	2,280 g.
1983	3,659 g.	7,160 g.
1984	570 g.	11,015 g.
1985	8,486 g.	3,046 g.
1986	18,100 g.	19,100 g.

(B) Number of Couriers

	Men	*Women*	*Total*
1980			18
1981			35
1982			12
1983	40	8	48
1984	22	4	26
1985	68	8	76
1986			87

(C) Types of Courier

	Swallowers	Stuffers
1980		100%
1981		100%
1982	40%	60%
1983	83%	17%
1984	79%	21%
1985	25%	75%
1986	30%	70%

(D) Quantities Concealed in Corpore

	Minimum		Maximum	
	Swallowers	Stuffers	Swallowers	Stuffers
1983	60 pellets (Cocaine)	100 g. (Heroin)	140 pellets (Cocaine)	420 g. (Heroin)
1984	3 pellets (Cocaine)	70 g. (Heroin)	118 pellets (Cocaine)	210 g. (Heroin)
1985	20 pellets (Cocaine)	2 g. (Heroin)	80 pellets (Cocaine)	260 g. (Heroin in vagina)
1986			150 pellets	550 g. (Heroin in vagina)

These figures correspond to the seizures made at Medical Emergency and Treatment Centre at Roissy—Charles de Gaulle Airport. However, the actual total weight of drugs seized is large, since the couriers may be handed over to the police with some pellets or suppositories still remaining in corpore.

Conclusions

Medical examinations by doctors at the Medical Emergency and Treatment Centre at Roissy—Charles de Gaulle Airport are carried out at the request of Customs Officers, who are responsible for preliminary screening of passengers, and with the consent of the people concerned. These medical examinations are an effective method of detecting drugs being transported in corpore and have made it possible to define two types of courier; "Swallowers" and "Stuffers".

The examination techniques—palpation and pelvic touch—are purely clinical. For reasons of speed and reliability, this type of examination is preferred to other types such as X-rays or scanning. The results obtained from these medical examinations are highly significant, bearing in mind the fact that hiding places under consideration are few in number and restricted in scale.[1]

Containers and Drug Smuggling

Containers have now been used by commercial freight carriers for a little over two decades. In that time, the industry has become highly mechanised to handle about 55 million container movements annually. Special ports have been constructed to accommodate container ships. Such is the streamlining process that a fully laden container ship can now be turned around at a modern port in about eight hours. As examples of the high level activity at major ports, in 1987, Le Harve in France—handled 600,000 containers, Rotterdam in Netherlands, about 2.25 million and Hong Kong about of 2.25 million.

It is, therefore, not surprising that drug traffickers have begun using this port to port service to transport their products. Traffickers, ever on the alert to discover new and innovate methods of concealing and transporting their drugs, could easily see the advantage provided by containers. The sheer numbers of them in constant use has almost guaranteed the undetected shipment of extremely large consignments of the various drug types. The cocaine barons of South America have used containers on a regular basis to ship their products to the United States and Europe. Cannabis producers in Southeast Asia and the Middle East transport cannabis and its derivatives in containers to the world's illicit drug centres. Southeast and Southwest Asian produced heroin has also been shipped in the same manner.

Cannabis products are often despatched in multi-tonne quantities; cocaine and heroin shipments can range from a few kilogram to a tonne or more depending on the method of concealment. The least complicated method is to include the drug in the middle of the regular cargo: for example, bags of coffee beans considering that the awkwardness of the cargo will deter close inspection. More sophisticated methods of concealment include the creation of special cavity spaces on the roof, and wall, floors or framework of the container. It is the last type which are being encountered on a large scale. They are also a cause for alarm, as many of the conversions are carried out with a high degree of workmanship and can be difficult to detect unless closely examined.

Under normal circumstances, police have very little to do with the seizure of drugs carried in containers. However, with this method of transportation seemingly increasing, it is considered that it maybe helpful for police officers to know something about that which, one day, they may confront.

Container Identification

The size and shape of regular container can vary considerably according to its function but, in the main, the majority are the "box" shape. The standard container by which all others are measured is 20 feet or 6,000 mm in length and is described as a "twenty equivalent unit" (T.E.U.) Therefore, a single 40 feet or 12,000 mm container is described as two twenty equivalent units. There are approximately 4.7 million T.E.U. containers currently in use around the world. Of this figure, the majority are registered with the Bureau International des Containers (B.I.C.) which has its headquarters in Paris, France. This organisation is an independent, non-profit body which was set up in 1933, originally under the auspices of the International Chamber of Commerce. Its principal function has been to create a register of internationally accepted identification codes relating to container ownership. It has grouped all identification markings required by customs safety and other regulatory bodies into a single document known as the BIC-CODE. When containers are built, the manufacturer can affix the various identification numbers and owner's code which are registered with the B.I.C.

Heroin Disguised in Ceramics

The Swedish Drug Enforcement authorities have learnt about a sophisticated method of conealment of heroin in the guise of ceramic tiles. The September issue (at page 10 S. No. 4) of this bulletin mentioned a method of making cocaine in the form of commemorative plaques. A similar technique can be adopted in the case of heroin also, whereby the heroin of very high purity (95 to 100 per cent) would first be dissolved in distilled water and then the water allowed to be evaporated. The resultant product would be a glass-like sheet of heroin which can be moulded into different shapes. After moulding the substance can be painted and at the receiving-end the same can be powdered for use as heroin. In the instant case in Sweden, an information source indicated that the heroin disguised in the form of *Decorative Tiles* was being smuggled from Thailand by a courier who posed himself as a sales representative for these tiles. The National Laboratory of Forensic Sciences in Sweden have not so far been able to identify the technology followed at the stage of moulding. But interpol reports show this method of smuggling heroin had been used successfully in France, Holland, U.K. etc.

Use of Live Animals to Ferry Drugs

Again, the Swedish authorities have reported use of animals like dogs to ferry drugs from one country to another. The dogs have been induced to swallow small parcels containing the drugs before transporting them. The packets are recovered at the destination by giving enema. Dogs of Dobermann breed have been used for this purpose in Belgium and Sweeden. (News from Narcontrol Oct. '88).

Nigerian Traffickers—New Smuggling Techniques

The CIS has received reports of two new smuggling methods currently being used by Nigerian traffickers. The first involves items such as Nigerian soups. Heroin has been discovered sewn inside animal intestines used in making soups. The heroin filled intestine is securely wrapped and submerged in the soup juices. Information on methods employed to transport the soup is unavailable but devising a variety of methods is possible.

The second method is an internal body carrying technique in which an incision is made below the navel in human carriers by a doctor or other person with some medical knowledge. The heroin is placed in the fatty tissues, covered by the skin of the abdomen and then stuffed. It is reported that up to 1 kg. of heroin can be concealed within the body by this method. It is believed that retrievel of the heroin is made in the United States by, as yet, unidentified Nigerian doctors, or other doctors of unknown nationality. The smugglers believe that they will have no problem passing through Customs, because the smuggler has only to *indicate that an operation was recently performed.* No information is presently available regarding X-ray examination.

Categories of Containers

Containers are grouped into one of three categories. The first category are those which are purchased new, registered with BIC and owned and operated generally by the major container leasing companies and shipping lines. They have life expectancy of between 10 and 15 years. After this time they are sold and become second category, the new owner usually being a small leasing company. They become third category containers when their second owner feels that they have reached the end of their useful life and either sells or disposes of them in some other way. At this stage they may be put to a variety of uses; instant housing in some Third World countries, garages and on-site workshops. A large number of them will also be purchased by very small container firms, many of which will be based in the drug producing countries. It is those containers which customs term "high risk", as they are unregistered, difficult to trace as far as owners are concerned and easily converted to carry drugs. The United States authorities recently experienced a number of importations involving this type of container, the hollow bearer of which has been packed with bags of cocaine. The owners of these containers were shipping legitimate cargo consigned by reputable firms in South America to reputable companies in North America. After the goods had been delivered, the empty containers were taken to the shipper's container yard and relieved of their illicit cargo. They were then shipped back to their country of origin with or without a cargo, but in either case, they could be used to carry United States currency instead of cocaine. Each time a new shipment of cocaine was loaded, the ends of the bearer beams had to be cut and re-welded and required repainting. This meant that the "identification" number and various markings had to be re-applied. Various discrepancies such as numbers being affixed vertically instead of horizontally, multi-layers of paint work and scorching of floor timbers by welding equipment attracted the attention of alert Customs officers.

Another more complicated method of smuggling drugs in "empty" category three containers has been encountered by customs in Europe. It requires dubious container operators to be based in at least three countries; the source country, an intermediary country and the country of final destination, for example Lebanon, Italy and Spain. Operator "A" in Spain ships a legitimate cargo through operator "B" in Lebanon. After the container has been unloaded and "B" has retrieved it, he modifies it by adding, for example, a false end wall, the cavity created by which is packed with perhaps a tonne of hashish. "B" then ships the "empty" container to "C" in Italy. "C" has a client, probably a legitimate company which has a cargo for delivery to Spain, which is where "C" wishes to send the hashish. "A", the owner of the container collects it when it has been emptied and removes the hashish. The container is then ready to repeat the round trip.

The advantages of this method for the traffickers are, that Italian customs would not normally pay close attention to an empty container, and the Spanish customs would not be duly concerned about a container "originating from Italy".

It will be appreciated that it is essential to have the co-operation of the dubious operators at all re-direction points to ensure that the illicit shipment arrives at its intended destination. Quite obviously, first category containers could not be used as the traffickers would have no control over their routing.

Controlled Delivery as a Tool of Investigation

In the last few years, the controlled delivery technique has proven as an important enforcement tool in identifying the Principals involved in drug trafficking and other major smuggling offences. The United Nations Conference for the adoption of a Draft Convention against Illicit Traffic in Narcotics Drugs and Psychotropic Substances has defined Controlled Delivery as follows:

1. "Controlled delivery" means the investigative technique of allowing illicit consignments of narcotic drugs and psychotropic substances (or substances in List A and List B) (or consignments which are intercepted by competent authorities and allowed to continue with the narcotic drugs or psychotropic substances replaced by innocuous substances), to pass out, through or into the territory of one or more countries with the knowledge and under surveillance of their competent authorities, with a view to identifying persons involved in illicit traffic and to take legal action against them (if permitted by the national laws and regulations of a State party)."

2. Further, Article 7 of the proposed convention covering controlled delivery reads as follows:

(i) Consistent with the basic principles of their respective nation's, legal system, the parties shall take the necessary measures to allow for the appropriate use of controlled delivery at the international level, on the basis of agreements or arrangements mutually consented to, with a view to identify persons involved in illicit traffic and to taking legal action against them.

(ii) Decisions to use controlled delivery shall be made on a case-to-case basis and may, when necessary, take into consideration financial arrangements and understandings with respect to the exercise of jurisdiction by the parties concerned.

3. The Secretariat of the Customs Co-operation Council conducted a survey of Member-countries and it was found that :-

— Most countries do not have legislative or regulatory provisions explicitly governing or authorizing the controlled delivery techniques;

— Despite this lack of legal "authority", controlled delivery is widely used, particularly against illicit traffic;

— Co-operation in this field between the Customs and other national services is necessary to prevent on-going operations from breaking down;

— The practice of substituting a harmless substance of the drugs for well known but is not widespread because of the difficulties encountered by the Customs (or police) in the case of consignments whose forwarding falls within the competence of other services (e.g., postal consignments);

— Countries are unanimous in recognising that controlled delivery could prove extremely effective in the context of international co-operation;

— Such international co-operation stems primarily from direct and personal contacts between the officials responsible for repressing drug trafficking rather than from close adherence to the provisions of international legal instruments.

Advantages

(i) It enabled the capture of the organisers and principals involved in drug trafficking;

(ii) It entitled as a rule the application of normal Customs and Police investigative procedures;

(iii) It could be applied in all countries of the world;

(iv) It fostered co-operation between enforcement agencies, both national and international;

(v) It assisted in the development of communications between countries;

(vi) It led ultimately to sharing of intelligence between countries.

Disadvantages

(i) There was always the risk of losing the drugs;

(ii) The costs involved in terms of manpower, etc. were often very high;

(iii) The legal situation did not always warrant its application;

(iv) Safeguarding the security of the operation was a demanding task.

Controlled Delivery System in India

Under Section 4 of the Narcotic Drugs and Psychotropic Substances Act, 1985—This clause enables the drug law enforcement agencies in India to use the controlled delivery techniques as an effective tool of investigation against the organised syndicates, nationally and internationally. However, before adopting this technique, the investigative agencies of the originating country and the recipient country should discuss in detail the entire operation, maintain surveillance simultaneously in both the countries, keep close surveillance on the movement of drugs either through cargo or through couriers and synchronise the final strike operation in both the countries for achieving optimum results. It will also be necessary to authenticate the evidence gathered in either country for successfully prosecuting the traffickers involved.

A typical case usually develops as follows :—

(i) The informant in a source country is approached by a source of supply to act as a courier to take drugs to a destination country.

(ii) The informant reports the approach and the identity of the source of supply to the drug law enforcement / police / customs agency.

(iii) The informant learns the contact point of the customer.

(iv) The drug law enforcement agency verifies its dossiers for the previous trafficking activities of the source of supply.

(v) Discrete enquiries are made to identify the customer and his activities, location, possible associates, places where drugs are received and source, etc.

(vi) With permissibility under the laws of the particular country, surveillance is ordered to know more about the customer / recipient.

Truly, unique are the smugglers' methods and strange are the destinations! But equally determined are the steps taken by anti-smuggling organisation which succeed in hunting and countermanding the ingenious tricks of dope traffickers.

Reference

1. This article is a reproduction from the International Criminal Police Review—January-February, 1988.

48

MONEY LAUNDERING

The term "money laundering" has taken on many different definitions, all of which describe the generic term of changing the form of money. There are as many ways to launder money as there are people doing it. Any criminal enterprise that deals in cash generally will be doing some form of money laundering. The idea is to change the form and ownership of monies generated from illegal sources. The primary focus here is narcotics financing, but it applies to any crimimal group that deals with currency as a product of illegal activities.

A clear way to see how money is laundered is to look at the problem from the narcotics dealer's perspective. There are a number of problems that the drug dealer must deal with. First of all, security is a large problem. Volumes of cash present an inviting temptation to those within the criminal organisation as well as to those from the outside. Secondly, the dealer must accurately account for the drugs bought and sold. Thirdly, he must ensure that sufficient money supplies remain liquid to allow for quick purchases and payment of drugs, transportation, manufacturing and distribution costs. Fourthly, the ownership of the money must be concealed from possible investigative agencies in order to avoid or minimize taxation on income generated by illegal sources. Finally, the criminal organisation must separate the financial transactions from the drug sales. Generally, drug dealers will have one employee dealing with money matters, or will have an independent money launderer conducting the affairs for many trafficking groups.

There are two constants within the money laundering rationale. First, the drug dealer must conceal the true ownership of the money.

He will want to conceal his involvement in the transaction while controlling the cash. He is apprehensive of being identified, targeted, and prosecuted because of being associated with large volumes of cash generated from illegal activities. He will want to create a false paper trail that will not identify him as the true owner of the money. Second, the drug dealer must change the form of the money. He will need to shrink the huge volumes of money generated from the street sales of narcotics. Since these large quantities of currency are difficult to handle, money laundering is needed to reduce the volume of cash to be more manageable.

Variation of Money Laundering

While there are many variations of money laundering, there are some transactions or instruments that are common throughout the criminal financial community. These include: (1) small bill for large bill exchange; (2) the cashier's cheques exchange; (3) the dummy account, individual or corporate; (4) the wire transfer; (5) the offshore jump; and (6) the physical transport.

The exchange of small denominations of United States currency to larger denominations is the simplest form of money laundering. Here, the launderer wants to exchange his small bills for larger ones, usually one hundred dollar bills. If currency must be used, the trafficker normally will prefer to have one hundred dollar bills as his standard. This will shrink the volume of currency he must deal with and also simplify the time spent counting the money and accounting for it. Large volumes of one hundred dollar bills being exchanged at a financial institution is often an indicator that money laundering is taking place.

The favourite method of money exchange is through the use of cashier's cheques. Here, huge amounts of currency can be reduced to one cheques or series of cheques. The check can be in any name and for any amount. The use of the cashier's cheques greatly simplifies the money launder's problems. First, the cheques is very secure and easy to handle and transport. Second, it is not covered under current Treasury Bank Secrecy Act regulations. Third, it can be negotiated many times in any alias name before presentment for payment is made. Fourth, banks sometimes favour this type of transaction because if the checks are taken out of the country, the turn around time is extended, normally 21 to 45 days, and the bank can use the float on the money for its own investment.

Another commonly used method of laundering money is setting up a dummy corporate account or personal account in a financial institution and depositing large amounts of cash into that account. If the bank begins questioning the amount of cash being deposited or authorities begin examining the account, the account becomes dormant and the money launderer simply goes to another bank and sets up another account under a different name. The money deposited into such an account can be withdrawn in the form of cheques to anyone, real or otherwise, or wire transferred offshore.

The wire transfer is often utilized by money launderers. The wire transfer requires one financial institution to send money and another to receive it. Normally, there will be some account relationship at the bank prior to money being wired, as the money representing the wire must be physically in the hands of the sender. Funds are sent off shore to a country that has a good Bank Secrecy and Business Act. The favourite countries that are utilized by narcotics traffickers are Panama, The Bahamas and the Cayman Island.

Any good money laundering scheme will usually involve an offshore bank secrecy jurisdiction. A number of countries have strict bank secrecy law which forbid the release of banking information to investigative agencies. Once criminal money arrives in one of these jurisdictions it becomes very difficult, if not impossible, to trace the money further.

The least attractive, but sometimes necessary method of laundering money is to actually remove it from the United States. It is unattractive for security reasons, but can be very effective if successful. Flying cash out of the United States to a bank secrecy jurisdiction would be an example. Once in that jurisdiction, money is then deposited to a bank and the trafficker can safely utilize the money in any capacity he so desires.

A typical variation of the above methods is the wire transfer of funds to an offshore Bank Secrecy Act jurisdiction, where the money is transferred to a trust or corporate account within the jurisdiction. The money then returns to the United States narcotics source as the proceeds of a loan or finders fee. This process has great United States Tax advantages for the trafficker and gives the money a legitimate look.

The advantages of Bank Secrecy Act jurisdictions to criminal elements are enormous. These countries shield banking and, in many cases, business transactions completely from investigation, Dummy companies or trusts can be created to invent any necessary documentation to support a loan, fee, or trust proceeds. The government's inability to examine the actual records prevents any meaningful challange to the validity of these types of transactions.

The Scope of the Problem

The Treasury Department currently estimates that approximately 40 billion dollars is generated in the United States from the illegal sale of narcotics annually. Many areas of the United States have become money laundering centres. The Federal Reserve Banking system keeps track of currency going out of its district banks and the currency coming in from its member banks. Large unexplained volumes of currency coming in from banks in a particular region may indicate money laundering from illegalsources.

The first question law enforcement personnel in Florida began asking in the late 1970s was "where is all of this excess cash coming from?" There certainly was not enough traditional business to support such huge amounts of currency being reported from that area, and flight capital from Latin America could account for only a small portion of the money. The seemingly logical answer was that a large amount of the 8.7 billion dollar excess cash total in the State of Florida was being generated by illegal sources, principally narcotics traffickers. It was evident to law enforcement personnel in the field that large scale narcotics transactions were being structured in the marijuana and cocaine traffic. Money was being moved in boxes and suitcases and deposited in banks who were very happy to help their customers and themselves.

With all of this money came a series of problems associated with narcotics trade. Public corruption increased and extended itself to the judges, prosecutors, and law enforcement personnel. There were problems in the management of the local economy.

Once the illegal currency source of the problem was identified, the next question was what to do about it. The Department of the Treasury and the Department of Justice set up a financial task force to develop methodologies to target and destroy narcotic money dealers. Operation Greenback was formed in 1980 in South Florida to find how narcotics financing was taking place, where the money was going, and what could be done to stop it. The concept also entailed looking at a narcotics organization as a financial organization and identifying its members through their illegal currency transactions by tracing the money to the drug dealers. The concepts that were developed proved that such an approach was sound and cost effective. The experiment was fruitful in that it provided a vehicle to test the criminal side of the Bank Secrecy Act.

The United States Government's Response to Illegal Money Laundering Aspect of the Bank Secrecy Act

The Bank Secrecy Act was passed by Congress in 1970 (hereinafter the Act). This Act requires private individuals, banks and other financial institutions to report certain domestic and foreign transactions to the Federal Government. The primary purpose of the reporting requirements of the Act is to identify the sources, volumes, and movements of United States currency being transported into or out of the United States or being deposited in domestic financial institutions for criminal, tax and regulatory investigation purposes.

The Act was also a reaction to the problem of foreign bank secrecy jurisdiction. Congress passed a records keeping statute which would track financial deals involving currency transaction regardless of whether foreign governments assisted investigators in tracing illegal money transaction. This statute then created a system of records which could be utilized in criminal investigations. Shortly after the Act passed, the foreign and domestic reporting statutes were challenged on First, Fourth, and Fifth Amendment grounds as violating the United States Constitution. The Supreme Court held that the reporting requirements did not violate the Fourth Amendment. While the Supreme Court has not ruled on the First and Fifth Amendment question, a number of the Circuit Courts of Appeals have addressed those issues and found the Act constitutional. The Act did not begin to be used as a law enforcement tool until the late 1970s. It took the Treasury Department about five years to implement the reporting requirements. Many banks looked at the reporting requirement as another bothersome Federal regulation and initial compliance with the terms of the Act was poor.

Bank Secrecy Act Regulations

The Bank Secrecy Act requires individuals, as well as certain financial institutions, to report to the Federal Government designated currency transactions. The Act covers three basic financial areas including: (1) domestic currency transaction of a specified amount, (2) foreign transactions of a specified amount, and (3) record of maintaining a foreign account. The Act also deals with records that financial institution must maintain and the period for which to be maintained. The Act consists of seven main parts. These parts include: (1) definitions by which the Act is enforced, (2) reporting provision, (3) records keeping provision, (4) criminal penalty provision, (5) civil remedy provisions, (6) exemption provisions, and (7) provisions dealing with the dissemination of financial information for law enforcement purpose.

The Act makes it a crime to fail to report certain financial transaction. It is necessary to understand the definitions of the Act when determining whether an event under ivestigation conforms to a criminal violation as defined by the Act. The definitions are very broad but can apply only in specific factual situations. The term "financial institution" is broad enough to include any individual who regularly receives over $ 10,000.00 and fails to report it. Under a literal reading of the Act, that individual as a "domestic financial institution" must file currency transaction reports. In addition, there is a requirement that the individual had been informed about the Act and chose to violate it. This broad construction has been used in money laundering cases where the individual handled money or "laundered it for narcotics interests".

The Reporting Provisions

The heart of the Act is the reporting provisions. The purpose of the provisions was to create an internal record keeping system that monitors money flow and is helpful in criminal investigations. An explanation of the three main reports will demonstrate the type of information that the government can use in criminal, tax and regulatory investigations.

All of the reports that are discussed here are computerized. This capability can actually develop targets for investigation of illegal money laundering and can lead to the narcotics dealers themselves.

Domestic Currency Transactions

The first major area covered by the Act is the reporting of certain domestic currency transaction. The current regulations require that financial institutions report currency transaction above $ 10,000.00.

This means that the financial institution must report every individual or corporation depositing currency above this amount to the Federal Government within 15 days. The report that is generated is called a "Currency Transaction Report" or "CTR". The CTR contains a wealth of valuable investigative information concerning the person or persons conducting the transactions. There are criminal penalties for failing to file a CTR and for supplying false information on a CTR.

Information of the CTR can be a very valuable investigative tool without alerting a potential defendant that he is under investigation. The CTR form requires the financial institution to provide detailed information concerning the currency transactions. The person making the transaction must identify himself; he must also identify the real owner of the money if the individual making the transaction is acting as a nominee. The form asks for detailed information such as social security numbers, passport information, bank account numbers, and account information. The type of transaction and the volume of cash being transacted by denominations are likewise reported. The financial institution and the individual making the report are identified.

After completing the report the bank sends it to the

IRS Service Centre in Ogden, Utah. There it is placed on computer tapes, supplied to the Treasury Financial Law Enforcement Centre (TFLEC), and made available for law enforcement. With the development of a national computerized currency transaction system, the Act becomes extremely valuable as a criminal research tool. Interesting analytical studies of problem financial institutions, problem depositors, and problem geographic areas can be identified and reviewed. For example, you could ask the computer to list all of the high volume cash deposit within a geographic area or within a certain bank. You could ask for the names of all cash depositors over $ 100,000 within a given time frame. You could match known cash deposits with CTR filed with the International Revenue Service over a given time period to check to see if the CTRs are being properly reported. You also ask for the original CTRs and develop investigative leads from the information given. All of these possibilities, and many more, enable the investigator to develop a picture of how money is being moved and to focus resources to those areas where criminal activity may be taking place.

Report on Foreign Financial Transactions

The Bank Secrecy Act requires two types of foreign transactions to be reported. These include (1) reports on currency entering and leaving the United States, and (2) reports on United States citizens who maintain foreign accounts.

Reports on Currency Entering and Leaving the United States

To have a complete picture of large cash transactions, it is necessary to know what currencies and amounts of those currencies are entering and leaving the United States. The Bank Secrecy Act addresses this issue in Section 5316 of Title 31. The section requires that anyone leaving or entering the country must report currency in excess of $ 5,000. The Treasury Department has implemented this law with the development of a currency reporting form called the Report of International Transportation of Currency of Monetary Instruments. This form is commonly called a Customs Form 4790 or CMIR. The CMIR must be filled "at the time of entry into the United States or at the time of departure, mailing or shipping from the United States..."

There are criminal penalties for failing to file a CMIR. There is also an important distinction with the CMIR that is not present with the CTR. Incorporated into the CMIR Statute is a very broad criminal statute, Title 18 USC, 1001, which makes false statements to the government a crime. Thus, a person who wilfully falsifies a CMIR (understates the amount of currency being transported) can be charged with violation of both the Title 31 Bank Secrecy Act offense and Title 18 false statments offense.

There are also other important statutes which strengthen the enforcement of the CMIR rules. Title 31 USC 5317 permits the government to search for seize, and forfeit the currency which has not properly been reported.

Reports of Foreign Financial Agency Transactions

Title 31 USC 5314 and the corresponding regulations at 31 CFR 103.24 require that a person within the United States who has a financial interest in a foreign account must report certain financial information about his account. This statute has not been utilized very much by law enforcement, but this could change. The Treasury Department has recently published proposed regulations which would allow Treasury to require much more information about foreign accounts and the persons holding them. Requiring persons who hold foreign accounts to report charge cash transactions, under penalty of heavy criminal fines, would help control illegal movements of illicit funds abroad.

Criminal Penalties

Title 31 USC 5322 deals with criminal sanctions for violation of Bank Secrecy Act. The Statute deals in two parts with misdemeanor violations [5322 (a)] and felony violation [5322 (b)].

As in any criminal statute, the first requirement which must be met is intent. To be guilty of this statute, a person must wilfully violate it. To show a wilful violation, the government must prove that the defendant actually knew about the reporting statute and then wilfully and intentionally failed to comply with the requirements. Proving knowledge in these cases is often more difficult than proving the actual failure to file the required reports. Many money couriers who fail to file CMIR have defended themselves on the "lack of knowledge" defence that they did not know that they had to file a report. This defense becomes very critical because there is no attempt provision in this statute. A person must actually, knowingly, commit the offence or there is no offence. To counter such a defence the government must show that the traveller had actual notice of the reporting requirements. To that end, the Customs Service has developed many means of notifying travellers of the law. Large signs in many languages are posted at foreign ports and airports. Custom declarations have the requirements written on them. Many times, uninformed Customs agents will ask the traveller if he or she has any currency to declare and explain the statute to the traveller. It is important to remember that the violation is not the illegal transportation of currency into and out of the country but is only the knowing failure to file the required reports.

The Felony Offenses

The felony provisions of Title 31 are very interesting. The 5322 (b) provisions apply to all violations of the

Act, unless specifically excluded. A felony occurs when the defendant violates the Act or regulations "while violating another law of the United States". The penalty for such a felony violation is imprisonment upto five years and a fine of upto $ 500,000.

The felony provisions apply when there is any violation of United States federal law in conjunction with a Bank Secrecy Act violation. Thus, if one is laundering money without properly reporting it, a violation of the Bank Secrecy Act, and the money came from illegal narcotics violations, then the felony provisions would apply.

The felony statute is also violated when there is "a pattern of illegal activity of more than $ 100,000 in a twelve-month period". Here, the central issue is determining what is a "pattern of illegal activity", since the statute does not define the term. To establish a "pattern of illegal activity", the government must prove that the defendant engaged in reported violation of the Act. An interim issue concerning how many offences made up to a pattern presented in the case of *United States v. Katten-Kassin*, 696 F. 2d 893 (11th cir. 1983). Here the defence argued that a number of offences for failing to file CTRs (and taking place over a long period of time) was part of only one "pattern of illegal acivity" and only one felony could be charged. The government argued, and the court agreed that once a pattern of offences over $ 100,000 within a 12 month period was shown, then each subsequent offence constituted a separate offence. Large criminal fines and term of imprisonment are available for repeat offenders and in narcotics financing multiple violations are common.

I. Civil Remedies

To complete the review of the Bank Secrecy Act, the civil remedies within it must be mentioned.

(a) *Injunctions* : An important but under-utilized feature of the statute is the power of injunction. The Secretary of the Treasury has the power to bring civil action to enjoin a violation of the Act or to enforce compliance with the Act or regulation. There is no prohibition against bringing an injunction in conjunction with a criminal prosecution to enforce future compliance with the Act as well as punish past violations, particularly where a corporate financial institution may be the defendant.

(b) *Civil Fines* : Civil fines are covered by the Bank Secrecy Act. These fines are available for a multitude of infractions and are designed to keep those persons making and keeping records responsible for doing so and civilly liable if they fail in that responsibility.

(c) *Exemptions* : Another area which often touches on criminal prosecutions is the exemption exception. This section provides for certain exemptions from compliance with the Bank Secrecy Act. Most of these exemptions involve domestic financial institutions who have high volume cash customers. For example, a large supermarket chain may have many deposits for which CTRs would normally be required. It would place a large administrative burden on the banks to file each one, therefore, an exemption list is created so that regular high volume cash customer does not have to file CTRs. These exemption lists must be approved by the Treaseery Department. Through the office of compliance. Transactions between financial institutions are also exempt. The exemption issue arises in a criminal case when a domestic financial institution that is a defendant claims that it should not be prosecuted because the person they were dealing with was on the exemption list and no CTR was required to be filed by the bank.

Dissemination of Financial Information

The last major part of Bank Secrecy Act deals with the dissemination of financial information. Congress has imposed certain restrictions on the use of such information. Generally, the information can be disseminated to other agencies (outside of Treasury) to use in criminal, tax, or regulatory investigations or proceedings. Foreign government requests have to be handled through Treasury and be only for criminal investigations. There is, likewise, a requirement of confidentiality as to the records received. They can be disclosed only to persons involved in using the information for criminal, tax or regulatory investigations.

Weakness Within the Act and Pending Congressional Legislation

Congress has now legislation pending that would greatly strengthen the Bank Secrecy Act. House Bill 6031, which has passed the Senate, covers most of the defects in the present lagislation. The Bill is designed to make the enforcement of the Act more effective. These proposed provisions include:

1. Attempt Provision : The Bank Secrecy Act does not currently have an attempt provision. There must be an actual violation ongoing or completed before it can be considered as a possible criminal offence. Thus, a person who plans to violate the law and takes some steps toward that violation may not violate the Act. A common example of such an occurrence would take place at an international airport in the United States where a money launderer would transport a suitcase full of cash to the airport. He would then check the bag. If he did not feel comfortable about the situation (he may have noticed surveillance agents watching him), he would simply leave the airport and have someone on the other end of the flight pick up the bag. If the bag was seized, he still could not be prosecuted because he had not transported money out of the United States. H. Bill 6031 adds the "about to be transported" provision.

2. Reward Authority : There has been limited authority for Federal Agencies to pay rewards for information. In financial cases involving huge sums of money, it is very difficult to develop information concerning criminal activity without the ability to pay rewards. The criminal element's disincentive of death and injury for those who might consider aiding law enforcement usually prevails.

In H. Bill 6031 for the payment of rewards to private citizen of upto 25 per cent of the amount seized or $. 150,000 whichever is less, for information in money laundering cases. This should provide a powerful inducement for information in these types of cases which may be very helpful in developing money laundering investigations.

3. Increased Penalties : H. Bill 6031 substantially increases and changes the criminal penalties section of the Act. Congress realised that the current level of penalties was simply too low in face of the huge amounts of money that was being moved in the illegal market. The proposed changes would make any violation of the Act a felony with a potential term of imprisonment of five years and a potential fine of up to $ 250,000.

There also is a proposed change in the civil penalty for failing to maintain records in Title 31 U.S.C. 5321 (a) (1). The $ 1,000 penalty has been raised to $ 10,000 to encourage financial institutions to take greater care in filling of CTRs properly and to increase compliance.

4. Customs Search Authority : The current statute, Title 31 USC 5317, is confusing. The statute states that Treasury agents may apply for search warrants when they reasonably believe that a monetary instrument is being transported and a report of that instrument has not been filed. The problem with this statute is that it created confusion in Customs Service in dealing with border searches. Customs has broad border search authority to search for contraband without a search warrant. This statute created the impression that, in searches for monetary instruments, a warrant may be required.

H. Bill 6031 clarifies the authority for currency searches to allow for warrantless searches. The proposed statute, however, imposes a "reasonable cause" standard to believe that a monetary instrument is being transported. It appears that if this change is enacted into law, prior to any warrantless search for monetary instruments "reasonable cause" would have to be present and the border search authority would give way to the statute in these types of searches.

5. Other Significant Changes : One major problem in the investigation of major narcotics money laundering cases has been the lack of wire interception authority. In cases where there was ample evidence of illegal money laundering, but no actual drug connections, it was impossible to get permission to intercept communications because Title 18 US C 2516 (1) did not include money laundering as an offence that you could request an interception. H. Bill 6031 removed the language which would allow for wire interceptions in Bank Secrecy Act cases, but the language is in the passed Senate bill and will have to be worked out in Conference.

6. Casino Gambling Regulation: The Treasury Department has proposed new regulations to require gambling casinos to report certain currency transactions. The registered casinos would be required to file a new form similar to a CTR on large currency transactions. The casino would also have a records keeping obligation that would be similar to a bank under the Bank Secrecy Act. This regulation would close a large loophole in the law which allows criminal figures to launder large volumes of money in casinos without reporting requirements attaching to those transactions.

III. Legal Theories Used in Money Laundering Cases

As the case comes closer to completion, the attorney and the agents must begin a careful review of the evidence to determine whether there is a prosecutable case and what the correct charges are. Here, an advanced review of the evidence is important. The evidence must fit the offence and there may be more than one offences that can be charged. The following represents a discussion of current legal theories that have been successfully used in money laundering cases.

A. Conspiracy

The most commonly used weapon in money laundering cases will be a conspiracy offence using Title 18 U. S. C. 371. The offence takes place when two or more persons conspire to commit an offence against or to defraud, the United States. Both parts of the conspiracy section are used in these types of cases. To prove a conspiracy, it is necessary to show that (1) two or more persons conspired or agreed either to commit an offence against the United States or to defraud the United States and that (2) at least one of the conspirators committed an overt act to effect the object of the conspiracy. A conspiracy requires that an agreement to accomplish an illegal objective coupled with one or more overt acts. The agreement to commit the crime and taking a step toward the completion of the agreement is what is prosecuted. Even if the substantive money laundering offence would not be successfully completed, the conspiracy could be a provable offence. The Court in *United States v. Dixon* explained conspiracy best stating that a conspiracy does not require 'mission accomplished', only 'mission attempted'."

In a money laundering conspiracy case then there must be an agreement between two or more persons to violate Title 31 by failing to file CTRs. The normal fact pattern would be where the drug trafficker and his banker agree that the bank will not file CTRs and that the trafficker will pay the banker to see that the forms are not filed.

Another common use of conspiracy law in these types of cases is the so-called Klein conspiracy. In this type of conspiracy charge, the offence is directed at fraud against the government. The theory is that when two or more persons engage in a scheme to launder money and attempt to illegally circumvent the Bank Secrecy Act by failing to file the required forms or by falsely filing the forms, these individuals prevent the Internal Revenue Service or the United States Customs Service from knowing of currency movements, exchanges, or deposits in order to disrupt the agencies obligation to collect that information. In short, the government is the victim of a scheme to impair the function of IRS or Customs Service ability to keep track of currency transactions and, as such, constitutes a conspiracy to defraud.

A scheme to defraud theory can be utilized under the conspiracy law where there is any attempt to interfere with or obstruct a lawful governmental function by deceit, crafts, tricks or at least by means that are dishonest. In money laundering cases, the function being disrupted is the collection of information from currency transactions for use in IRS and Customs investigations.

B. Aiding and Abetting

The legal theory of aiding and abetting can be used in a money laundering prosecution. The theory is that the defendant, by his illegal money laundering or narcotics proceeds, is aiding and abetting the narcotics organisations importation and distribution scheme. Here, the government must show that the defendant provided the money laundering services, that he knew that some of the money was derived from the distribution of controlled substances; and that he acted with the intent to further the crime of distributing or possessing with intent to distribute controlled substances, in other words, that he sought by his actions to make crime succeed.

C. Other Statutory Violations that are Utilized in Money Laundering

The one offense that is used very effectively in money laundering cases has been section 1001 of Title 18. Section 1001 can be used in instances involving the filing of false CTRs and CMIRs in connection with an actual scheme to avoid filing of the forms. Both Title 31 and Title 18 offences can be jointly used as each offence requires different elements of proof. Where a person engaged in a pattern of financial transactions with the intent to evade the reporting requirements of the Bank Secrecy Act there may be violations of both Title 31 and Title 18 USC 1001.

There are three areas of proof that one must deal with when using section 1001 of Title 18 in the prosecution of money laundering offences. These include whether the Title 31 reporting forms are a matter within the jurisdiction of an agency of the United States. Secondly, whether the failure to file or the false filing of CTRs and CMIRs is material. Thirdly, courts have required that the making of false statements to investigative agencies may cause more of a "perversion' of an authorized function than a false statement about pecuniary claims.

This use of a fraudulent statement to conceal is no less material than the use of fraudulent statement to influence a governmental decision in the first instance.

Title 18 USC 1001 may be used even where there is no false statement. The second part of the statute deals with a "concealment or covering up by trick or device a material fact".

It may be possible to charge aiding and abetting as well as Title 18 USC 1001 where false representations are necessary for the success of a money laundering scheme. Proof of all steps in the overall scheme would be proper.

Other "fraud offences" which can be used in money laundering cases effectively are mail fraud and wire fraud. Normally, there will be some mailing or wire transferring in a money laundering case in furtherence of the illegal act of currency movements. Arguably, one could also charge a fraud against the government by use of mail or wire transfering.

Narcotics offences : offences will normally present themselves clearly in these types of cases as money is always an essential part of any "deal". Title 21 drug offenses cases can be brought as "other offences against the government" in money laundering cases. Arguably, one should be able to introduce evidence of narcotics trafficking in financial cases because the drug offence would actually be an element of the Title 31 offence.

D. Conclusion

All large scale narcotic groups will have to deal with currency in order to conduct and remain in business. The larger traffickers will be closer to their money than they will be to the narcotics. Thus it would seem logical that governments should analyze their currency systems to determine if there are any pressure points which can be brought to bear against narcotics financing. The approach the United States has taken is that by requiring the identification and reporting of certain large currency transactions, a "paper trail" can be developed which will link the trafficker with the particular transaction. Criminal groups can be identified and destroyed by analyzing financial information and using the legal penalties outlined herein. If we can develop methods which can identify illegal transactions before they disappear in the stream of legitimate business transactions then we will be successful in this war against drug financing.

After the trafficking group has been identified and prosecuted, it is important to remember that these traffickers can be dealt a very serious blow when they are deprived of their illegally obtained and often hidden assets. It is not enough to arrest someone for narcotics violation and ignore the profits made from that activity. Criminal organizations are more effectively immobilized when the illegal profits or proceeds, the reason for most criminal conduct, are targeted during the investigation and prosecution of the organization's underlying criminal activity. Strong forfeiture laws are as equally important to effective law enforcement as are strong sentences.[1]

The Methodology of Money Laundering

There are as many methods of laundering funds as the mind can conceive. However, they all have basic common characteristics and the various methods are nothing more than variations on a single theme. Some of these methods are given below.

Article 6 of the Council of Europe's convention on laundering, search, seizure and confiscation of the proceeds from crime gives a definition of laundering which is very close to the definition in the 1988 United Nations convention against the illicit traffic in narcotic drugs and psychotropic substances.

In the most simple terms, money laundering is a process by which assets, primarily cash assets, which are derived from illegal activities, are manipulated so as to make them look as if they were derived from legitimate sources.

In the wholesale and street sale of narcotics, large amounts of cash are accumulated. Narcotic transactions are a cash business. Credit cards, bank cheques, letters of credit and other such monetary instruments are not used, primarily because they identify parties to the transaction. The signatures required on bearer negotiable instruments, as well as the account numbers, identify the parties to the transaction and thereby leave what is known as a 'Paper or audit trail" for the investigator. With a cash transaction it is only the cash that must be disposed of, but this is difficult, since narcotics transactions involve millions of dollars. To simply deposit these monies in a bank account raises questions for the tax authorities concerning reported income. For the investigator who has access to a suspect's bank account the same question is asked. How does an individual who earns an alleged fixed income of $ 2,000 a year account for the fact that his bank account contains in excess of $ 1 million? Having dispersed with the trite explanations of winning the lottery, an inheritance from a distant relative in Nepal and other such explanations, the criminal is at a loss to explain his reported income. Rather than provide these explanations it is a relatively simple matter to hide the source of the income through a series of financial transactions which have become known as laundering or covering the source of the income.

Currency Smuggling

The physical transportation of cash is often the beginning of the money laundering process. It is very simple in concepts: the money courier files suitcases with cash, hides it in cargo, or sends it in an international express package. The money is physically transported to a foreign country that has no currency controls and preferably has bank secrecy laws. In the foreign country the cash will be deposited in a bank or other financial institution and from there it can be moved at will. The money is now indistinguishable from the legitimate funds that are routinely transferred throughout the world's financial systems.

This method is largely used to overcome an obstacle erected by the United States. Under the Currency and Foreign Transactions Reporting Act, transporting of more than $ 10.000 in monetary insturments in or out of the United States is required to be reported. Failure to do so can result in civil and criminal penalties and may lead to the forfeiture of the currency.

Throughout the years, hundreds of millions of US dollars were then seized. It is a common procedure that a currency courier, asked by customs if a specific suitcase which is suspected of containing cash belongs to him, simply denies ownership and abandons the one or two million dollars. These are the costs of doing business. A variation of that method consists in travelling with two indentical suitcases, one full or cash, the other filled with personal belonging. Once stopped by the customs officer, the courier would be absolutely astonished about the contents of "his" suitcase and claim that he had confused the suitcases with each other as they looked identical.

In November 1991, US customs special agents arrested a resident alien from Ghana, who was attmepting to smuggle approximately $ 53,000 in US currency out of the United States, $ 6,000 of which she had swallowed in 12 balloons.

Customs inspectors at JFK Airport stopped the passenger as she was about to board a KLM flight for Ghana. When they asked her how much money she was carrying out of the United States, she stated that she had $ 9,000 with her. However, a customs search of her hand-carried bag revealed $ 24,000 concealed in packages of bed sheets: further examination produced two bottles of shampoo filled with 45 balloons, each stuffed with $100 bills for an approximate total of $22,400. She then told the inspectors that she had swallowed a quantity of money-stuffed balloons. A subsequent X-ray confirmed that she had ingested numerous balloons. She was also found to be carrying balloons vaginally. Customs special agents noted that was the first time an individual had been caught using a smuggling method typical of drug traffickers—"swallowing"—to smuggle money out of the country.

Smurfing

Another method used to avoid the paper trail is smurfing.

Incidentally, the term "smurfing" comes from a US investigation in Florida. The investigators discovered that typically the criminal organization would employ a large number of individuals to take cash to many different banks or branches of the same bank.

The smurf would deposit the cash in amounts under US $ 10,000 or purchase cashier's cheques/bank drafts, money orders or other negotiable instruments in order to avoid the CTR (Cash Transaction Report) which financial institutions have to fill in and transmit to the IRS (Internal Revenue Service) for each cash transaction above US 10,000. As the financial institutions are obliged to store all records concerning over US 3,000 for a period of five years, the smurfs very often deposit amounts under this threshold.

Another species of smurfs was uncovered recently in Paris. Dozens of Asian students or Asians without a job queued up at the entrances of luxury leather goods boutiques. They handed over a shopping list to the vendor and paid the bill with brand new 500 franc notes. Subsequently, they loaded their merchandise into cars waiting nearby; 300 of these shopping smurfs were identified. Some boutiques organized sales by receiving the shopping smurfs in groups of ten. The leather goods—suitcases, attache cases, handbags, etc.—were sent to about 20 fake companies in Japan and sold cheaply. It was a sale at a loss, but the goal of the operation was not to make a profit out of the sale of the leather goods but to launder drugs money. Five members of the organization shuttled from Paris to Luxembourg every week in order to fetch one to two million francs per trip.

Structuring

Again in order to avoid the reporting and/or storage of transactions, structuring is used: a huge financial operation is broken up into many single operations of small amounts. As an example, there was a case where a total of US $ 20 million was launderd by transferring it in small amounts to Ecuador with an average of US $ 600 per operation—which amounted to over 40,000 single operations. One could be tempted to think that after making such an effort, they almost deserved the money.

Tax Havens or Financial Havens

Tax havens or financial havens are countries with several characteristics, the principal one being that they have no tax or low tax on most or all types of income. Other important features include (1) bank and commercial secrecy, (2) no currency controls, and (3) ease of buying or establishing companies, banks and/or insurance companies. (Many journals contain advertisements for such companies that cost very little to purchase).

Other important consideration for the money launderers in choosing the tax haven include (1) political and monetary stability, (2) modern communications, (3) ease of access, and (4) geographic location at/or near the path of the business activity.

Much of the business in the haven countries has legitimate purposes and those countries have been utilized for many years by many reputable companies. However, the characteristics of these countries make them ideal for money laundering and illegal activites.

Loan-back Method

By taking cash assets to an offshore or tax haven country that observes bank secrecy, the launderer can deposit them there. He then transfers them to a bank in another country. He will then approach his own bank and request a loan, using as collateral the money on deposit in another country. His bank loans him the money which he is now able to invest in property, business ventures or some other similar activity. When asked by the police or tax authorities to explain his sudden wealth, he points out that he owes the money to the bank. To add a little flavour to his explanation, he may also have created a number of offshore companies or businesses which allegedly own the assests. In fact, the company or business is nothing more thana nominee in an offshore country which represents the launderer.

Use of Front or Shell Companies

Typically, once the money is in the financial haven country it is placed in an account in the name of a front or shell company. The terms "front company" and "shell company" can be used interchangeably—however, a shell company usually has no function other than to act as a conduit for money hiding the true source or nature of the funds. As previously mentioned, these companies can be purchased off-the-shelf and consist of thing more than the establishment papers and usually a local person, normally an attorney, who acts as the trustee or cmpany executive.

A front company may actually conduct some legal business activity to disguise the illegal activity of the money launderer. Most financial haven countries require the involvement of a citizen of that country. These front shell companies can also be "banks" or "insurance companies". Many launderers have more than one front or shell company. They then funnel the funds between these companies to further disguise the source of the money.

In a case in Spain, a criminal organisation built up a small empire of about 500 companies whose only purpose was creating paperwork in order to prove commercial activity and conceal tax evasion and money laundering. In most cases the money launderer does not want to leave the money permanently. The normal method of returning the funds is by wire transfer. This is quick, safe and is difficult for investigators to verify source of the funds.

Other schemes used include claims that the money represents profits from the front or shell company or from a property transaction.

Also used are false claims that the money was from an insurance company settlement. The front company can also purchase property for the hidden benefit of the original source of the funds. Some of these methods do not create a tax liability on the returned funds, whereas some methods do involve a tax liabity. In many cases the money launderer is more than willing to pay taxes on "clean" moeny.

Cash Business

It is not uncommon for the money launderer to have a cash business, such as a restaurant, a coin-operated business, car wash or similar establishment. It is difficult to audit these types of business to actually determine how many pizzas were sold during the day, week, month or year, or how many cars were washed. By taking the money derived from narcotics trafficking and mixing it with money derived from legitimate sources, the launderer can conceal the source of his funds.

There are three different stages in the laundering process: placement, layering and integration. The weakest link in that chain is the moment when the launderers want to get rid of their cash. Here, the professional skills and the integrity of the personnel of financial establishments play a key role. Their knowledge and experience enable them to gain a first suspicion that a transaction might be part of a money laundering scheme. Quite often, fancy explanations are given about the source of the money, and with some commonsense the story sounds too shaky—it "smells". Example : two individuals approached a Scandinavian bank introducing themselves as lawyers representing a colony of Indians in South Africa who wish to rsettle in India. For the purpose, they intend to construct an entirely new city. The lawyers said they were empowered to organize the financial side of the project. According to the plans, the financing of the resettlement of the enterprise was supposed to take place through that Scandinavian bank. Therefore, in the months to come, several hundred million dollars in small denominations (stemming from these Indians) would be deposited in the bank by lawyers.

Another example: three individuals wanted to deposit some Can $ 1,00,000 in cash in a bank in France. When asked about the origin of the money, they explained that they were taxi drivers in Canada and the money represented their savings.

Gambling and Casino

This method is used by money launderers when large amounts of casino chips are purchased for cash or when cash is deposited with the casino for alleged future gambling activity. The money launderer spends some time at the casino but gambles little or not at all. The launderer then redeems his chips or closes his account and requests a cheque either in his name or that of a third party. The cheque is then deposited in the money launderer's account. If there are any questions about the source of the deposit, it is explained that it represents gambling winnings.

However, this method works on a large scale for the movement of substantial funds only with the complicity of the casino management. This explains why—amongst other reasons—organized crime is so keen to get involved in the casino business. In that connection the involvement of the Camoora in the South of France can be guessed.

A variation of this method is to claim winnings from gambling at the horse or dog tracks or through a bookmaker. For a fee or commission the bookmaker can make false entries in the gambling records and testify as to the phoney winnings. The money launderer can frequent the track but not bet or bet only small amounts, yet claim big winnings. This is more difficult in some countries than others, bcause some countries require persons who win large amounts on bets to identify themselves for tax purposes.

Underground Banking System

In some countries there are underground or parallel banking systems. These go by different names such as the Chop, Hundi or Hawala system. In practice they all operate in the same way. They are based on turst and/or family connections. The money launderer will deposit funds with an underground banker. The letter will notify an associate in the receiving country that the funds have been delivered and/or will give the depositor a special mark or chop. The depositor will present this receipt in the foreign country and receive the funds in the local currency less a fee or commission.

These underground banking systems have historically been used to circumvent currency controls established by some countries in order to get profits or savings out of the country or to pay for goods in foreign countries. This is called "flight capital". These systems only work well if both countries have a trading history so that the funds balance out in the books over time.

Use of False or Inflated Invoices

To use this method the money launderer buys or

creates a business in the country where the money is being generated and also the country where in money is to be placed. The launderer then orders goods or service from the company to which the money is to be sent. Two options are now available: either inflate the value of the goods or service invoiced, and the excess amount is the amount of the laundered money; or to send totally false invoices. The advantage in using the first option is that it is much harder to detect. The amount of the expense represented by the false or inflated invoices has the appearance of an expense that can be deducted by the paying company.

For developing countries, a major method of moving Western currencies out of the country is through the over-evalutation of goods purchased. By getting autholrsation from the central bank to not only receive the benefit of his imported goods but is able to convert local currencies at the central bank and have them moved into Western banking centres.

Currency Exchange

By their very nature, currency exchange offices engage in large-scale cash transactions. In many cases they are poorly regulated and are not required to maintain detailed record of their cash transactions. The money launderer normally purchases bank drafts in a foreign currency and these bank drafts are then deposited in a foreign account or used to pay for goods and services in a foreign country. The currency exchange office also wire-transfer the funds to any destinaton desired by the money launderer.

Brokerage Houses

Brokerage houses can be used to launder cash very easily. With a co-operating broker the cash can be brought in and used to purchase stocks or bonds. These securities can be registered in any name or in a front company name. In the case of bearer bonds, no registration is necessary. Each country regulates the securities market and brokers in a different manner. In some countries it is easier than in others to manipulate money laundering through brokers. However, even in the countries with tough regulations, an unscrupulous broker can facilitate money laundering.

Another quite sophisticated method is to manipulate stock in order to launder money. This requires at least two persons or entities working together. The money launderer can hire individuals or set up front companies. These conspirators then take opposite positions in the same company's security. Stock is held in the country where the money is physically located and an equivalent amount of stock is held in the country where the launderer wants to send the money. If the company is small enough or the money to be laundered is large enough, the price of the security can be raised or lowered by money launderers. To get the money out of the country the stock must go down in price. The holder of the stock then "loses" money on the stock when he sells, but the other side of the transaction "gains" or makes a profit when the stock is sold. The amount of the profit is the amount of the money laundered.

Non-transferable Currency

Russian "Mafiosi" and drug dealers used to trade their black money for non-transferable roubles, Purchases include antiques, enterprises, real estate and raw materials. The later can be exported and converted into (white) money.

Another scheme is for foreign companies to buy roubles at black market rates to pay local workers on joint ventures with the government. The higher the rate of exchange, the higher the profits.

Analysis of Some Comparable Cases

As examples, we give below a summary of two of the 14 cases submitted by the French Interpol National Central Bureau. The cases were investigated in France about two to three years ago and the money laundered was derived from drug trafficking.

Fourteen Money Laundering Cases Reported by the French NCB

1. Money Launder Method or Use/Concealment of Assets/Profits*

Investment in—

Gold/paintings/art/cars/planes	8 cases
Personal purchases/foreign tours	7 cases
Foreign real estate	6 cases
Real estate (various)	6 cases
Transfer funds offshore	6 cases
Foreign bank accounts	5 cases
Restaurant/car/travel agency	4 cases
Foreign investment	4 cases
Open local or foreign front company	3 cases
Local bank or savings account	2 cases
Purchase of foreign currency	2 cases
Foreign cheques/Postal money orders	2 cases
Goods for resale or export	1 case

2. Difficulties Encountered in the Investigation

Bank secrecy or lack of co-operation	4 cases
Problem with rogatory commission	4 cases
Evaluation of real estate	2 cases
Delays or inexperience of agent or prosecutor	2 cases

Complexity of international case	2 cases
Tracking cash payment or smurfing	2 cases
Problem with foreign interview	2 cases
Inability to block foreign accounts	1 case

3. Solution Proposed by the Investigators

Better international co-operation	5 cases
Simultaneous investigations of both drugs and money laundering	2 cases
Conduct foreign investigation to locate assets	2 cases
Better co-operation (national)	1 case
International harmonization of laws	1 case
Create special investigative unit	1 case
Block/freeze foreign accounts	1 case
Require banks to obtain custom identification	1 case

* *Note*—Multiple methods were used in a single case

Case No. 1

Origin of the Case

Information from a former drug trafficker, who claimed that several members of the Marseilles underworld had played an active role in the installation of a heroin production laboratory in the United States and were investing their substantial profits in Europe.

Brief Summary of the Case

300 kgs. of morphine base taken from Turkey to the United States by private yacht and then conversion of the morphine into 148 kgs. of heroin in the United States by traffickers from Marseilles. They apparently made a profit of US $ 1,480,000 which they invested in France.

Methods of Laundering at National Level

(1) Cash payments, under cover of suppliers' credits, to firms doing substantial work on converting premises to a discotheque.

(2) Purchase of a large country house by an intermediary with apparently legal bank financing but with partial cash payments from company current accounts by show-business personalities and then cash payments for the conversion of the property into a hotel/restaurant, again under cover of suppliers' credits and with undeclared labour.

(3) Purchase of apartments.

(4) Increase by cash payments of the capital of a company managed by intermediaries acting for the traffickers.

(5) Establishment of companies to acquire and manage property and of other limited liability companies to run the bars, discotheques, snack-bars and restaurants set-up on the premises.

(6) Acquisition of consumer goods and payment for travel around the world.

International Level : Transfers Prior to Laundering

(1) Acquisition of 90 hectares of real estate in France in the name of a company registered in Ireland whose shareholders are Swiss companies. The Irish company is managed by a Swiss *Societe Fiduciaire* (trust company), which gave a French manager power of attorney to manage the property Purchase of the buildings: FFR 1,875,000 (inclusive of tax), purchase of land: FFR 1,25,000 (88 hectares).

(2) Payment for extremely costly work on the property by the Irish company : FFR 1,185,722.

(3) Purchase of a yacht used by the drug traffickers by another company registered in Ireland.

Estimate of the Illegal Profits Involved

— Estimated value of the real estate purchase : FFR 4,500,000.

— Estimated value of one of the nightclubs (the biggest in Europe) : FFR 3,000,000.

— Value of other assets acquired not estimated.

Results of the Investigation

(Penalty : accounts frozen, confiscation etc.) : FFR 828,050 in cash seized when traffickers first questioned. No measures taken in respect of real estate financed by drug traffic profits.

Case No. 2

Origin of the Case

Several drug law enforcement services in the Paris area noticed that a number of seemingly unconnected "dealers" all had accounts with the same Paris branch of a bank.

Brief Summary of the Case

A major Paris bank specializing in financial dealings between France and certain African countries had opened a branch dealing solely with transfers of money by immigrant workers. The latter could open clearing accounts at that branch. A large number of Africans who were suspected of engaging in drug traffic used that banking facility to send the profits they made by trafficking to their own countries.

Methods of Laundering

The traffickers opened one or more bank accounts, giving false names and addresses and transferred their profits claiming that the sums concerned were what they had saved from their wages. This was a method of laundering funds by a succession of small-scale operations.

Investigative Techniques

Within the context of a judicial investigation the people who transferred suspect funds were identified as a result of examining over 12,000 bank slips relating to transfers of sums exceeding FFR 10,000, by almost 710 customers.

Estimate of the Illegal Profits Involved

Over FFR 7,500,000 distributed among 80 suspects.

Money Couriers

It is acknowledged that the weakest point in the money laundering process is the initial stage when the criminals get rid of their bulk cash.

For that reason, in some legislations financial institutions are obliged, or have the right, to inform a competent authority about large, suspicious cash transactions. However, the police and customs can also play an important international role on the tactical level. As we have seen physical transport of cash to individuals abroad is one of the methods used in the placement stage of a money laundering scheme.

Efficient law enforcement work is not only a question of appropriate laws. Sometimes the tools are there and it is only a question of using them or not. However, that decision has to be made individually by the competent authorities in each country.

Money Laundering and Counter Measures

The term money laundering is used to refer to a variety of different processes of putting illegally acquired funds into a financial system where they are converted into banking instrument or other assets in a form that appears to be free from original taint and is clean. Since black money is liable to legal action from enforcement quarters the owner is in urgent need and makes all out efforts to make such money legitimate so that it could be deployed to further continue and expand the/illicit trade to earn more and more murky profits. The mechanism through which this could be accomplished is called "Money Laundering Process".

The channels to which such money transists are companies, corporations, trusts, development houses, banks, etc., or into real estate, the launderer keeping himself at reasonably safe distance from the long arm of law enforcement agencies. Further, as legislations relating to taxation investment and banking vary between countries, launderers and their accomplices can find loopholes in national laws and procedures to suit laundering schemes for concealing the ill-gotten gains.

How Drug Money is Laundered ?

Before discussing the money laundering mechanism, we may have a look at the three different ways of laundering drug money. Such laundering can be done through conversion, i.e., conversion of cash (huge and almost unbelievable amounts of cash are generated in the drug trade because all the drug transactions are made in cash and the money involved runs into billions of dollars) into another form which can be another form or currency or cashier's cheques or money drafts from banks or gold or negotiable instruments or bank deposits or other. Money laundering in this manner flows into two directions, i.e., in furtherance of the launderer's criminal activities and towards the intended method of concealment. Laundering can be done through concealment, i.e., by hiding the criminal proceeds (the larger the amount the more difficult it is) or by purchasing of tangible assets (generally done by less sophisticated criminals) or by secreting the funds off-shore to further remove proceeds from source and also to ensure that the funds are reasonably safe from detection by law enforcement agencies. Laundering can also be done by way of creation of a perception of legitimacy. This is the most sophisticated way to launder money whereby the laundered funds re-emerge as legally earned money to acquire legal companies, corporations, banks, brokerage houses, travelling agencies or other movable and immovable properties. Creation of the perception of legitimacy ensures greater degree of usability in respect of the assets earned through the illegal proceeds.

Methods of Money Laundering

1. *Bank Deposit*: As already stated, incredible amounts of cash are generated in the course of drug transactions. Huge amounts of black money are difficult to handle. The drug trafficker, for reasons of safety and security and also for distancing himself from law enforcement, will definitely not like to store it in some cupboard or safe deposit box or in any secret place at his own home. He feels the immediate necessity, just after the transaction is made, to deposit such money into bank accounts or into other forms of deposit facilities. Huge amounts of cash may be deposited on a single day or in next few days in various branches of the same bank or in various branches of various banks located at one place or at various adjoining places within one area or even at distant areas in a large number of fictitious names to hide the real owner/launderer of the money. In the United States of America this process is called "Smurfing". Under the US Bank Secrecy Act, 1970 each financial institution must file a report of each deposit, withdrawal, exchange of currency or other payment or transfer by, through, or to such financial institutions which involves a transactions in currency of 10,000 US $ or more. So, the Smurfer-type money launderers now tend to deposit amounts of cash below 10,000 US $ to avoid reporting.

2. Loan Back Technique: Narcotics proceeds kept hidden in secret or numbered accounts in a foreign bank can be used as collateral for legitimate loan which may result in a like amount of clean money. The trafficker has, in effect, borrowed his money, which can be used in any manner and its origins can be explained. This is done by the money launderer by depositing his ill-gotten money in a foreign bank first and then by using his front company in his own country to ask for a loan from foreign corporation (which also may be his) by directing it to use his foreign bank account as a collateral, i.e., as a guarantee for the re-payment of money. This is the way he loans money to himself in a manner in which the loan appears to be a legitimate loan and the money gets laundered.

3. Gambling Fronts : This may not be applicable to many of the third world countries which do not have a flourishing casino business. But in the advanced and affluent countries of the west and some other like countries, this mechanism is very much resorted to by the money launderers. In this process, the launderers may go to a gambling casino and deposit money into a gambling account. He may not gamble at all or may not do it heavily. At the time he leaves the casino, he has his money in his account put into cheques that may be payable to third party. Now such cheques can be deposited in some securities firm or other financial institution as legitimate money. This process also gives him immunity from banking regulation.

4. Double Invoicing : Trafficker, using two or more front companies, can conceal the flow of narcotic drug money under th guise of legitimate business transactions. Depending on which direction the money is to flow, he can sell goods or services at artificially high or low prices. For example, a foreign corporation related to him can place an order for puchasing of certain goods worth 20,000 dollars from him and he can secretly arrange with that foreign corporation to get a payment of two lakh dollars in two or more invoices for the sale of the same goods.

5. Foreign Business/Stocks, Real Estate : In this process, the drug money launderer can claim narcotics proceeds as profits from successful foreign business/ stocks or real estate transactions and that way it becomes his legitimate income on which the appropriate tax has been paid. Here the shell corporations in offshore tax havens will be of great use to him. A shell corporation can be a company which was viable at one time but now extinct except for possible stock holders or a corporation created only on paper where stock has been issued and firm registered but with no useful purpose other than to hide its ownership. Shell corporation facilities are rampantly available in offshore tax havens.

6. Direct Investments : Usually, under the guise of a shell corporation or front company, a trafficker can invest drug money in, or buy various types of legitimate businesses, properties etc. If the type of business purchased, or invested in, is one that enables the trafficker to fake an increase in business because of its high cash flow, it can be used to launder money. The trafficker can even pay himself a salary and launder more of his money.

7. Travel Bureau/Money Exchange Houses : One method of concealing the flow of drug money popualr with the traffickers operating between South America and the United States is to conceal the flow within the normal operations of legitimate travel bureaus or currency exchange houses which also may be owned by some traffickers. It is quite common in South America for travel businesses to operate in conjunction with their own currency exchange businesses. If the drug trafficker is connected to such money exchange houses he can get his enormous dirty cash exchanged in another currency in a foreign country through one such money exchange house having its branches in this own country also and then that branch in his own country can be advised to transfer such currency money to his bank account or fictitious bank account in that country. This will look as they are in the habit of dispatching high volumes of money anywhere in the world and laundered money can easily be concealed within such money.

8. Correspondent Account : This may not apply to many countries. However, this mechanism has been found to be rampant in the USA. Foreign banks can have their accounts in a number of US Banks. So, the account will be in the foreign bank's name. Naturally, the foreign bank will also have its representative posted in the USA and their clients will be able to use such account to conduct business in the USA. If a foreigner wants to do some business in the USA, he can have money sent from foreign bank in his country to the same foreign bank's correspondent account in US bank. In this process, the money launderer takes money to the US representative of the foreign bank and gives him the money. The representative deposits the money in the foreign bank's correspondent account in the USA and money is claimed to be owned by the foreign bank. That way the origin of the money and also the true ownership of the money will remain secret.

9. Credit Card Fraud : Credit Card Frauds are normally not found very much in third world countries. However, such frauds are rampant in advanced countries. For an example, a US citizen may get credit card issued by a foreign bank in a tax haven country. The US citizen can buy goods and services on the basis of this credit card and incur a heavy expenditure in US dollars. Now, he can cleverly transfer his drug money to the foreign bank which has issued him the credit card to ostensibly pay for charges and, in the process, can launder a part of his drug money. It will be legal since the credit card has offered him US dollars credit against a foreign account.

10. Currency Smuggling : The method is one of the most popular with drug traffickers because it avoids leaving paper trail. It is often over-looked because of its basic nature. The usual method is to convert cash into cashier's cheques or other types of cheques, although large amounts of cash are also smuggled. All these things are normally smuggled out to the tax havens and introduced into their international financial systems. Once the money is introduced into the financial system, it can be moved at will and it will become virtually indistinguishable from legitimate funds working in the system.

11. Tax Havens : Tax havens have several common features which are attractive to the drug tafficker. They are as follows :

(a) No or low tax on foreign earnings of foreign owned corporations;

(b) Commercial and/or bank secrecy;

(c) Tax incorporation laws;

(d) Goods transportation and communications facilities;

(e) Location near established drug trafficking routes;

(f) No or minimal restrictions on foreign currency exchange or foreign corporations or individuals;

(g) Relatively stable political climate; and

(h) Availability of professional and/or business advice and assistance.

Narcotics traffickers do not use tax havens facilities chiefly because they are tax havens (although most assuredly tax liabilities and benefits are a consideration), but rather, they simply take advantage of the benefits afforded by the international financial system that provides a means to move large amounts of money with some degree of anonymity. Vast amounts of money move through this international financial system on a daily basis and no one knows for certain how much of it represents the proceeds of legitimate business and how much of it is derived from the proceeds of narcotics trafficking and other illicit activities.

Measures to Deter Money Laundering

The measures to deter money laundering are summed up briefly below :—

1.1 Statutory Provision

The statutory obligations of tax payers have been presumably lightened. They include compulsory filing of tax returns by everyone with a taxable income compulsory maintenance of accounts by businessmen and professionals and their compulsory auditing if income or turnover exceeds a specified limit, compulsory canalising of transactions involving payment exceedings Rs. 10,000 through banks, tightening of the benami provisions of law whereby the benamidar has come to be recognised as the owner of the benami property in place of the real owner who had transferred this property benami, and so on.

1.2 Surveys

To bring new assessees into the tax net and strengthen the information base for the detection of tax evasion by existing assessee, a general survey is conducted by the income tax department. This includes a door-to-door survey as well as information gathering from outside agencies.

1.3 Tax Raids and Seizures

Incriminating documents and unaccounted assets seized in raids are utilised for prompt assessment and the swift imposition of penalty, followed by prosecution whereever called for.

1.4 Penalties and Prosecutions

To discourage tax evasion, tax laws also provide for monetary penalties and for the prosecution (and imprisonment) of tax-evaders.

2.1 Measures to Unearth Laundered Money

From time to time in the past, the Government has tried various measures, at least one of them repeatedly, to unearth accumulated amounts of black money in the economy. I shall discuss them briefly.

2.2 Voluntary Disclosure Schemes

So far, five voluntary disclosure schemes (VDS) have been tried out—one in 1951, two in 1965, and another in 1985. Basically, these schemes had one thing in common: they tried to induce tax-evaders to declare their concealed income and wealth in return for taxation at concessional rates and immunity from penalties and prosecution.

2.3 Special Bearer Bonds (SSB) Scheme

In January 1981, through an ordinance, the Government announced a new temporary scheme of SSB. Under this scheme, SSBs of Rs. 10,000 each were issued at par with a maturity period of 10 years. The holders of these bonds would be paid Rs. 12,000/- per bond on maturity in 1991, thus implying a annual compoundrate interest of 1.8 per cent.

2.4 Demonetisation

The Government has resorted to demonetising high-denomination currency on only two occasions, once in 1946 and again in 1978. (In 1978, the currency notes demonetised were of the face value of Rs. 1,000, Rs. 5,000 and Rs. 10,000).

3. New Measures for Dealing with Laundering Money

Various new fiscal measures for dealing with black money were taken by the Government during the year 1991-92. These will now be discussed.

3.1 Special Levy on Deposits in the National Housing Bank

Under the scheme, from the deposits made in this Bank by 31st January, 1992, a special levy of 40 per cent of such deposits was deducted to form the corpus of a fund in the Bank. This fund will be utilized to finance slum clearance and low-cost housing for the poor.

3.2 Waiver of Penalty and Interest

The Income Tax Act contains a provision under which tax payers can avail of the facility of waiver of penalty and interest on the amount disclosed once in a lifetime. For those who have already availed of this facility, it was proposed to give just one more opportunity to disclose their unaccounted income through suitable amendments in the Income Tax Act.

3.3 Tax Deductions

To attract funds from the public for investment in low and middle-income group housing, highway, roads and bridges, non-conventional energy, school buildings and the supply of drinking water, it was proposed to provide for deduction, in commuting the taxable income of a tax-payer, the entire amount paid for financing projects or schemes prompting social and economic welfare.

3.4 Foreign Remittance Scheme

To attract larger inflows of foreign exchange to urgently meet the exceedingly difficult BOP situation, the Government announced two special schemes on 24th July, 1991. Both the schemes were open till 31st January, 1992.

3.5 India Development Bonds

Under the second scheme, the State Bank of India was asked to issue India Development Bond denominated in US dollars. The bonds were available for purchase by non-resident Indians and their overseas corporate bodies. There was no ceiling for investment in these bonds, which have a maturity period of five years.

3.6 Gold Bond Scheme 1992

In the Central Government Budget for 1992-93, besides the gold import scheme 'a gold bond scheme' was also proposed.

Money Laundering and Enforcement Efforts[3]

The organized crime, specially one like drug trafficking generates substantial amount of cash as profit which is further pumped into underworld economy to finance the operations and purchase men and material for the upkeep and betterment of the organization. Such organization, collect a large amount of unaccounted money and enlist service of armed men, who have scant respect for law, become threat to social order and government to ransom and assume invincibility. Since drug trafficking is essentially a trans-national crime, the sharing of profit is also done at international level. The poppy or coca growers get little share in the tainted money while the manufacturers and the distributors reap the maximum benefit. The maximum profit one gets at the consumer's end which is mostly in the wealthy countries, who necessarily become more wealthy to the economic crime money laundering.

The problem of underground money transaction may be more serious in rich countries when considered in absolute terms (in USA the estimates run in the region of $ 50-75 billion) but their economy is also capable of bearing such drains. The conditions in developing countries are different and such transactions destabilize the economy and deplete the resources required for the development efforts to improve the quality of life for their teeming millions. The economic balance is also so delicate that slightest stress on the economy can damage the economic fabric and threaten the integrity of the country which indicates pressing need for introducing effective legal and fiscal measures to check accumulation of unaccounted money or its conversion into white money by dodging the enforcement agencies.

Inadequate Laws and Poor Justice System

In the past our criminal justice system was mostly concerned with arresting and prosecuting the criminals responsible for certain offences. The economic offences were not recognized as crime till very recently and those who collected money by illicit sources were considered successful entrepreneurs. Later on with the increase in financial power of some of the syndicates even the government felt threatened and then the need for introducing accountability for assets so collected, was felt and most of the governments imposed certain restrictions on the commercial transactions. With the growth of the concept of welfare state, certain fiscal measures were introduced in body politic to raise financial resources to finance the Welfare Schemes. The international trade between the countries at different stages of development also attracted corrective measures in the form of customs duty to safeguard the national interest. Both at national and international levels the economic activities which hitherto enjoyed complete freedom, were brought under discipline in consonance with the political philosopohy of the State. With the increased tax liability, the entrepreneur felt robbed of his assets and engaged best brains to outwit the government machinery by bypassing the legal provision by using the loopholes provided in law to check such robbery. In this fight against the State to save one's assets the criminal also joined the race and introduced criminality into the commercial system.

Process of Money Laundering

The concept of money laundering as the subject for serious study is a recent phenomenon and a direct result of the policy of certain governments to prove the dictum that 'Crime does not pay'. The dealer in the drug with good international link-up and backing of political and bureaucratic heavy weight in some of the countries, and fired with the allurement of huge profits, enrolled services of best of the brains to ensure that the assets created from the profits of illegal activities escape the dragnet of enforcement agencies and remain with the criminals to be used for the expansion of the trade and protection of the criminals by corruption and elimination if needed. The process of money laundering aims at concealing the source of ill-gotten money and also the illegal application of the income and then disguise that income to make it appear from legitimate sources. It simply means that 'tainted money' is converted into 'white money', thus making the criminal to appear as successful entrepreneur. The means to convert this 'dirty money' into 'clean money' will differ with the sources of income and the legal and fiscal provisions of the country the trader is operating in. Since there are numerous sources of generation of 'dirty money' so are the means to change its colour depending on the creative imagination and expertise of the criminal entrepreneur.

The illicit money is generated at every level that handles the drug and in every leg of journey to its final destination. The money laundering problems for a large syndicate engaged in the trafficking of drug is more important than the distribution of the drug while to the lone operators at the lowest level distribution problems outweigh the laundering problems. The drug trafficking not only generates illegitimate capital in the hands of the criminals but it also showers liberally the part of the money on the corrupt system of enforcement and political leaders. In certain countries both of them have tried to buttress their position in the cloak of special protection granted in law by way of special provisions of sanction before prosecution.

In the context of drug trafficking there is general awareness about the nefarious activities of the traders and its impact on economy of the nation which necessitated stringent privisions in law to arm police and other enforcement agencies to unearth such transactions but still many areas of the world have a legal and fiscal climate that cater for the growth and prosperity of money laundering industry. The places like Gayman Islands, the Bahamas, the Netherland Antilles, Panama and Switzerland are regarded as safe haven for money laundering. In most of these countries the 'Bank Secrecy Laws' are quite strict and any wilful disclosure may expose the banks and their employees to criminal liability. Such closely guarded bank information system is most favourable for the launderer and most difficult for the investigating officer. Such countries obtain significant competitive advantage in attracting secret money but at the cost of others. This tendency has encouraged others also to impose similar restrictions on banks. Pakistan also introduced criminal friendly foreign exchange regime in 1991 which will make it safe haven for the launderers. Sri Lanka has already assumed notoriety in this connection. A report of the U.S. State Department referred in The Times of India, April 30, 1993 indicated appearance of Sri Lanka and Pakistan as new safe havens for drug money.

There are as many ways of conversion of tainted money with white money, as there are people doing it but the main problem remains the same, i.e., taking out the money from the countries of generation and its receipt in the country of operation in the cloak of strictly legal transaction. As compared to global laundering, domestic laundering is simple, specially, in the countries having complicated fiscal problems and pressure groups claiming protection by way of exemption in one form or other. In investigation, firstly it is very important to identify the money flow and approach to the real source of the flow. The money is visible in the form of assets including deposits in banks, shares in companies and investment in real estate, etc., but the source in the present context, drug trade, remains underground. The other problem, equally difficult is to prove it to the satisfaction of the court that the assets were generated by the conduct of illegal activities and third party interests are not affected in indiscreet way. The evidence to be collected to prove this in court of law may fall in different jurisdictions of the judgement regarding seizure and forfeiture in foreign lands. The money launderers thrive on these weaknesses and devise means to overcome such difficulties in a very intelligent manner.

The money laundering may be conducted through domestic financial transactions. The lone low level operators normally convert cash payments into capital goods such as jewellery, cars, air-conditioners, VCRs, video cameras and other gadgets. Those who are more enlightened and serious in the business create assets and purchase land and houses in urban areas. In former cases, the movable property thus accumulated is totally concealed or it is undervalued on paper. In the second case, real estate is created in fake names or as Benami transaction to conceal the source and to cover the track leading to the real owner. Since our Income Tax laws have not touched agricultural and therefore, exorbitantly inflated returns from it to cover income from dubious sources are shown.

In the wake of proliferation of economic activities, a new class of juristic persons have come into being as Corporations. Through shares, a large amount is collected which defies all investigations if papers are properly maintained with the help of experts. These fake Corporations normally show huge profits to cover inflow of tainted money. Once this tainted money comes to mainstream, it changes its colour. Since the shares are in the names of different share holders, at the time of court proceedings, the interest of third party will come in the way of successful seizure or forfeiture.

The lotteries and other games like, horse race, etc. are other instruments in the hands of maninpulators to conceal the source of income. The winning tickets of lotteries are purchased by small underworld traders and much black money is converted into white money by a simple operation. The actual winner may be a small time farmer who has immense capacity to conceal dirty money, hence, his interests lies only in money. The launderer re-purchases the tickets making payment from black money and gets white money in lieu thereof. The agents of such launderers develop intimacy with the staff of Directorate of Lotteries and on announcement of prize immediately contact the holder of the winning ticket. This racket is going on and corrupt people placed at high places having access to and influence upon the employees connected with Directorate of Lotteries, are getting prize money again and again through dubious means, thus converting their ill-gotten money into white money to ward off any difficulty at the hands of enforcement agencies and Income Tax Department.

Electoral System used for Money Laundering

During elections, the political leaders collect election funds presumably from electorates. In some cases it was reported that amount reported by the political parties was inflated and actually it was not given by the Electorates but by the underground dealers. Similarly the birthday gifts collected by some of the political practitioners is their own money generated from illicit sources and after such public demonstration it becomes white without any trace of its origin. Sometimes, the black money is liberally spent on elections without any mention in the election expenditure statements escaping action under People's Representation Act. The means to launder money are many and exhibit novelty and originality depending on the resourcefulness of the launderer.

The money generated out of trafficking when involving a complicity with other countries makes distribution of assets and further laundering problems assume entirely different dimensions. The countries like USA, UK and Japan have strong currencies and are likely to be accepted by the criminals as it is but in case of other currencies, exchange of currency is involved besides customs laws of the countries concerned. As mentioned earlier, large share of the illicit money is generated in the consumer countries which are incidentally prosperous and their currencies are quite strong in the international fiscal market.

In India the profits of illicit drug trade are reportedly brought in cash in aeroplanes without declarations and then deposited in a bank after exchanging it by making false statement at the point of exchange. Illegal courier service is also rampant in some parts of the country. The money to be remitted to India is given to a courier service in foreign country who telephone or telegram instruct their agents to make payment to the payee in India, the equivalent amount in rupees. The hard currency is deposited in the country, in original itself or in a country having bank secrecy law. Since our country is facing foreign exchange crunch and every effort is made to boost export, therefore, the traders contact these exporters and send their money by over-invoicing and the equivalent amount can be collected in the country. Now with liberalisation and encouragement to N.R.Is it has become very easy to bring foreign money in the country in the shape of loans, etc.

Money Laundering in USA

In countries like the USA the problem has entirely different dimensions. Bank laws are quite strict and the Government is pretty serious about the flow of the currency outside the country. In USA the money laundrers commonly use following methods for transfer of cash outside the country:

(1) Illegally obtained cash may be taken from the United States to a foreign country physically without filing any statement and then deposited in a bank in the country having Bank Secrecy Law.

(2) Cash is deposited in a bank in USA and then transferred by wire or through cashier's cheque to a foreign bank located in countries having Bank Secrecy Laws.

(3) Illicit money is deposited into a fake account (in the name of non-existing person) in USA and then transferred by wire or cashier cheque to a foreign bank.

(4) The money is given to an attorney or an accountant as a cash transfer without filing any statement required in law and then deposited into the trust account of the attorney or accountant and then transferrd by wire or mail and taken physically.

Once the tainted money is taken out of the country, then it may be deposited into banks in fake individual or corporate accounts. Once an account in countries like Switzerland, Bahamas, Panama is opened, it can be taken back into the country after filing regular declaration in the shape of assets acquired in foreign countries as proceeds of a loan granted to a citizen of United States, fee paid to a citizen of U.S. service or by way of

How Drug Money is Laundered and becomes Legal

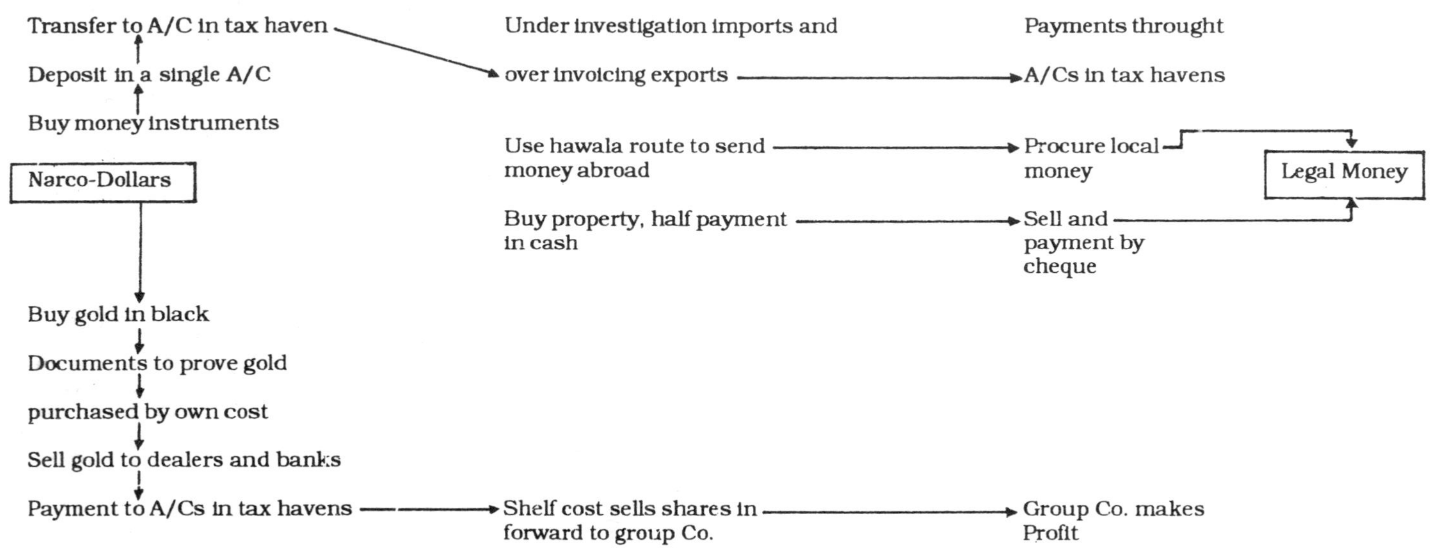

From The Sunday Times of India—April 11, 1993

over-invoicing the goods exported from the country. The net result of this process is that cash generated in the country out of illicit trade in narcotics will be hidden from Law Enforcement Agencies to avoid tax, seizure, forefeiture and the illicit trader will escape criminal liability.

Once cash is physically taken out of the country and deposited in one of these havens, there is no problem in financing illicit activities such as terrorism and anti-establishment activities in the country. The phenomenon may be new but the companies like Lockheed Aircraft Corpn., Gulf Oil Corpn., McDonnel Douglas Corpn., and many others generated tained money which was used for corrupting foreign officials where the latter would again launder the ill-gotten money often with the assistance of the former. From the above, it would appear that criminal activities no more remain a national concern but begin to assume global dimensions and the parties involved are so interdependent that they try to shield each other and defy all efforts of investigating agencies.

The Legislative Action

The impact of drug trade on the most vulnerable section of society and generation of large amount of tainted money which has corrupted the warf and woof of civilized society and has enormous potentialities of being used against democratically elected government has forced the government to enact the NDPS Act, 1985 prevention on September 16, 1985. The enormous potentialities of the black money can be gauged from the Bombay blasts. The premable of NDPS Act *inter alia* mentions that the Act has been enacted" to provide for forfeiture of property derived from or used in illicit traffic in narcotic drugs and psychotropic substances. "NDPS Act in section 27A provides for punishment for financing illicit traffic and harbouring offenders.The quantum of punishment (not less than 10 years but which may extend to 20 years and shall also be liable to fine which shall not be less than one lakh rupees but which may extend to two lakh rupees) also reflects anxiety and determination of Parliament.

The International body, the League of Nations held a convention for the preventation of illicit drug in 1936 followed by U. N. conventions held in 1961 and 1971. These conventions mostly concentrated on production, distribution and trafficking of Narcotic Drugs but the convention of 1988 introduced new areas of co-operation for preventing illicit traffic by sea, intensification in control of trade zone and preventing trafficking through mails. It has further incorporated in Article 5 such measures as may be necessary to enable confiscation of proceeds derived from offences established in accordance with Article 3 or property value of which corresponds to that of such proceeds. In the same Article, the convention enjoined upon the signatories to adopt such measures as may be necessary to enable its competent authority to identify trace and freeze or seize proceeds, property, instrumentalities or any other thing for the purpose of eventual confiscation. The provisions in convention against illicit traffic in Narcotic Drugs and Psychotropic substances Act 1988, are so broad based that if implemented, the traffickers cannot escape the dragnete Law Enforcement Agencies. Government of India ratified this convention in 1988. Being the signatory to convention 1988, Government of India have appended Chapter V-A to NDPS Act 1985, which provides for forfeiture of property derived from and used in the illicit traffic. Again, the provisions are broad-based and it includes not only the sale proceeds but the assets created from them not only in the hands of criminal but also in the hands of associates, relatives and even in the hands of third parties. The peculiar feature of the legal provision in Chapter V-A is that it raises certain presumptions in favour of prosecution and the most important being that the burden of proving that the property specified in the notice served u/s 67 H is not illegally acquired property lies on the person who has received the notice. To avoid fruther delay in adjudications in courts of law, the Act provides for the constitution of Appellate Tribunal u/s 68 N of the NDPS Act. Government of India in exercise of powers conferred by NDPS Act notified the Appellate Tribunal for Foreited Property (procedure) Rule 1989, on 22nd January 1990.

From the above it would appear that Government of India, in consonance with United Nations conventions have made legal provisions to counteract the activities of money launderers. Now, the enforcement agencies are fully equipped to deal with traffickers and their associates, the money launderers.

Collection of Evidence

In international field the difficulty lies in collection of evidence to prove the connection of the assets with the drug trafficking. The Bank Secrecy Law and the principle of the duel criminality come in the way. Since relations between the two States depend on the bilateral basis. United Nations, in its convention 1988 have taken a very pragmatic stand and have provided that if two nations decide to take measures on the basis of bilateral treaty, then the parties shall consider the convention of 1988 as the necessary and sufficient treaty basis. Article 7 envisages that signatories to the convention shall afford one another the widest major or mutual legal assistance in the course of prosecution in the judicial proceedings in relation to the criminal offences related to narcotic drugs. The mutual legal assistance will include taking evidence or recording of statements of persons, service of judicial documents, execution of searches and seizures, examination of objects and sites providing information and evidentiary item identifying or tracing proceeds, property and other things for evidentiary purposes. Not only this, in Article 7 also it envisages that the parties shall facilitate or encourage to the extent possible in their domestic law the presence of availability of persons who assisted in investigation and shall not decline to render mutual legal assistance under Article 7 on the ground of Bank Secrecy.

The net result of these provisions of convention of 1988 is that it surpasses the obstacles of dual criminality and absence of mutual treaty in case of member nations who signed the convention. The removal of difficulties will definitely arm investigating agencies with the additional instruments and will also foster a new era of co-operation between signatories.

References

1. By Charles W. Blau, Deputy Associate Attorney General, United States Department of Justice, USA.

2. List on next page.

   FOPAC Bulletins and Brochure written by A. Freeland, T. Winker and J. Laughton, ICPO-Interpol General Secretariat. ICPR January-February 1993, pp. 2-8.

3. Based on Shri R. C. Dikshit's article Money Laundering and Enforcement Efforts with Special Regard to Drug Trafficking published in CBI Bulletin, August 1993.

49

NARCOTICS AND TERRORISM

Terrorism is commonly defined as the use and threats of violence as a political weapon to achieve control to influence Government policy and to de-stabilise and even over-throw Governments. When we speak of terrorism, we generally refer to groups such as the Red Brigade of Italy, the Red Army Faction in Germany, the American Terrorist Organisation and others.

Consequently, the legal definition of terrorism cannot be used to describe narcotics traffickers and their organisation although the tactics they use at times are at par with those of terrorist organisations. However, while most of the groups involved in narcotics trafficking or production do not meet the legal or traditional definitions of insurgents or terrorists, the fact remains that the use of planned, sophisticated, or threat of violence to achieve goals, in the absence of a political agenda, is terrorism. This is the kind of terrorism that we find in Peru, Colombia and Mexico.

The increasing use of violence against international drug control effort as well as dealing in drugs for arms and the financing of terrorists and other political insurgency through illicit narcotics activities pose severe threats to the national security of the drug producing countries as also to the prospects for successful international narcotics control.

The nexus between Narcotics smuggling and terrorism has become so pronounced that law enforcement agencies have coined the word 'narco-terrorism' to describe the phenomenon. Ethnic groups, operating in India, who fund their militant cause with drug money include not only Punjab extremists but also Sri Lankan Tamils, Pakistani Muhajirs, Afghani Mujahedeens, Nepalese, Nigerian and Iranians. The most active group is of Nigerian terrorists who have enlisted the support of Ghananians, Gambians, Kenyans, Ivorians and Tanzanians as couriers, and, no-where it is more prevalent than in Delhi which has emerged as the major transit point for heroin originating in the Golden Crescent destined for the West.

Drug Related Violence

Drug related violence is on the rise. An international smuggling syndicate sent a professional killer to India in 1980 in order to eliminate a key witness in a case in which substantial pure heroin was recovered from an American and his Canadian girl friend at Palam Airport. The two were couriers for an international drug syndicate. In November, 1984 the US Federal Bureau of Investigation discovered a plot to assassinate the democratically elected President of Honduras. The group of dissidents including a Honduras General, planned first to sell large amounts of cocaine and then use the proceeds to attempt to overthrow the Government. In February 1985, ring leaders of the Mexican narcotic trafficking network conspired and killed the DEA agent Enrique Camarena.

The above examples will prove that in many narcotics producing countries, organisations which describe themselves as political insurgents frequently use terrorist tactics and have become involved directly or indirectly, with narcotics production and trafficking. There are links between drug trafficking, arms smuggling and terrorist groups. Examples of State sponsored narcotics trafficking have also come to notice. However, it is an established fact that narcotics trafficking in Latin America, in Asia, in the Middle East and in Europe is by and large dominated by narcotics traffickers who are governed by their greed and whose ideology is the pursuit of profit. Most of these groups cannot be called terrorists or even political insurgents.

International Lawlessness

The report of the International Narcotics Control Board was correct in saying that "the very security of some States is threatened". The sheer financial power of these trafficking organisations has threatened the political status-quo as traffickers use their millions to influence political decisions, even to elect representatives of trafficker-interests to national leaderships, to buy newspapers and radio stations and to launch high power public relations initiatives.

A few years ago, the Deputy Prime Minister of Malaysia had stated that most severe national security issue that his country faced was narcotics. In September 1984, the US Secretary of State George Schultz stated that he believes that the narcotic network was part of a drive towards international lawlessness which had been increasing dramatically during the past two decades. Comparing the traffickers' blatant disregard for international principles of law, Schultz said that "the modern versions of piracy are narcotics trafficking, terrorism and similar kinds of outlaw behaviour. Money from drug traffickers—and what may be most disturbing is the mounting evidence that some Govenments are involved for their own diverse reasons."

We have come to understand that narcotics trafficking is a key element in global lawlessness which includes terrorism, insurgency, violence and simple disregard for human values. New links between narcotics traffickers and other kinds of organisations are continually confirmed. It is, therefore, clear that ever-intensifying efforts are being challenged with equal intensity by well-armed, well-financed narcotic trafficking organisations. There is no question about their willingness to use murder and assassination as tools for opposition. There is also no denying the fact that lucrative narcotics trade is attracting political insurgents or that terrorists are attempting to exploit the unrest caused by Government suppression of narcotics farming and production. Colossal amounts of profits made by drug traffickers can buy elections, finance supply of arms for insurgency and destabilise legitimate Governments and subordinate democratic processes.

The fact is that narcotic trade predominantly originates in friendly or allied countries and is predominantly controlled by groups and individuals whose primary motivation is financial and not political. Elimination through arrest and prosecution of the narcotic gangs would very definitely have an impact upon drug production in these countries.

The threat of terrorist-style attacks upon narcotics control officials and workers is probably the highest in Colombia and Peru. The world is also concerned about the increased possibility of such attacks in Myanmar (Burma), Bolivia, Mexico and the Golden Crescent countries. Narcotics trafficking most clearly involves political insurgents in Myanmar (Burma) and Colombia. The involvement of the traditional terrorist groups in the narcotics trade remains most pronounced in Europe and the Middle East. We can expect these threats and even attacks as international narcotics control programmes continue to expand and improve.

In the world of narco-terrorism, there are three main protagonists:

1. The insurgent groups such as IRA, the Tamils and the Sikhs.
2. The intelligence services and officials of Government bent on exporting revolution, as is happening to India from across its Western border.
3. Drug gangsters who use political violence and classic terror tactics to further their aims.

The three types of narco-terrorists invariably trade in weapons, launder money, share intelligence, trade false passports, provide each other with safe heavens and render market assistance in a loose global alliance. We are determined to ensure that this unholy alliance of these groups will not diminish our determination to bring this problem under control.

Narco-terrorism in Myanmar (Burma)

Let us now take a look at the problem nearer home, in the South-East Asia. The relationship between insurgency and narcotics trafficking is particularly close in Myanmar (Burma) where most of the principal insurgent groups rely on heroin smuggling to finance their activities. These groups control or influence the main areas of opium production in North-Eastern Myanmar (Burma) where the difficult terrain and shortage of modern military equipment have prevented the Myanmar Government from establishing its authority. The insurgents range from ideological revolutionaries like the Burmese Communist Party (BCP), the ethnic separatists like the Kachin Independence Organisation (KIO) and opium warlords such as the Shan United Army (SUA) which is headed by the legendary Khun Sa.

Whatever their professed objectives, these three groups are heavily involved in the production, transport and sale of heroin. Other groups such as the Karens, are not involved in producing or transporting heroin but tax the trade in their areas. The BCP controls the areas producing the bulk of Myanmar opium crop from which it receives a substantial share in the form of taxes. It has also recently begun to establish refineries to convert opium into heroin and to engage in direct sales of refined opiates of middlemen. The SUA arranges shipment of opium to the Thai-Myanmar (Burma) border area where SUA-controlled and independent refineries process most of the heroin produced in Myanmar (Burma). Competition between this and other Board and the Directorate of Revenue Intelligence (DRI) in 1987 succeeded in arresting 44 Nigerians in 33 seizures of 44 kgs. heroin, 8 kgs. of morphine, 10 kgs. of cocaine and 10 kgs. of amphetamines, the majority of which were seized at Delhi and Bombay airports.

Nigerian Involvement

Latest Interpol reports suggest that in view of the voluminous cargo now exported to West and Europe from both air and sea ports on the Indian sub-continent, these groups having their own communities, have formed a potential for smuggling. Nigerian drug couriers in India, first surfaced in 1984 when several students from that country after obtaining admission in Indian Universities fell prey to Nigerian drug syndicates based in Lagos and became couriers back to Africa. In 1986, 500 Nigerians were arrested in 22 countries with seizures totalling 150 kgs. of heroin, 90 per cent of which had originated from India. Nigerians who made up 65 per cent of the couriers are known as "Stuffers" because of their personal preference of smuggling heroin by stuffing it in their body cavities. In fact, the Nigerian couriers have introduced a new phenomenon to India's Cocaine. Interpol has warned about the spurt in the amount of cocaine seized from one kg. for both 1984 and 1985, to 14 kgs. in 1986.

Narco-terrorism from Golden Triangle and Golden Crescent Countries

Arms and drugs have given rise to organised crime which is now becoming almost impossible to control without a global effort. Even Interpol agrees that the drug traffickers have now acquired the role of the major supplier of funds to terrorist organisations all over the world. The United Nations will have to actively intervene and initiate new strategies. In August, the United Nations Fund for Drug Abuse Control sanctioned $ 20 million to India for its campaign against drug abuse. But this is a drop in the ocean. In the developing countries the drug traffic is the result of complex politico-economic causes which requires a multi-pronged drive involving government and private organisations.

The drug-insurgency linkage dates back to early 50s when, with the communists emerging as a major power in China, a faction of Kuomintang tribes retreated into the Golden Triangle Areas on the borders of Burma, China, and Laos and maintained their subsistence and supply of arms by growing opium and pushing them into the world markets. This legacy continued in the hands of the Burmese communist party, Shan, Kachin, Karen and other insurgent groups (recent intelligence reports indicate that the Burmese Communist party controls 80,000 to 1,20,000 hectares of opium cultivation with a potential yield of 800 to 1000 tones of opium). During Vietnam War, the Hmang tribes controlled the production and transport of the drugs to western countries. In the last few years since the Afghan insurgency came into prominence, Afghanistan emerged as a major producer in the Golden Crescent area controlling a sizeable chunk of the world drug trade. In fact, at present, the biggest flow of drugs into India is from Pakistan and Afghanistan.

Close to the North-east, the Burmese communist party control the illicit cultivation of opium in the Shan State and a number of insurgent groups have a number of refineries to convert illicit opium into heroin. They are, located inside Burma close to the Thai-Burma border, and a few inside Laos, close to the Burmese border. The neighbouring countries like Malaysia and Singapore protect themselves from the onslaught of drugs by imposing capital punishment. So drug-traffickers are pushing bulk quantities of drugs into India. On the other side, large quantities of drugs already transit through India from the Golden Crescent region. Border areas along these countries which are prominent for opium cultivation are Badakhshan and Takhan provinces in North-East Afghanistan, Kunar province east of Kabul, North Waziristan in Pakistan, etc. In early 1980s as the Mujahideen organised their holy war from Pakistan, they looked for possible source of funding. The new Governments that emerged in Afghanistan and Iran after 1980 had drastically reduced the flow of Heroin being smuggled through Kabul and Teheran from the producing areas located along the Pak-Afghan border. The Golden Crescent is now responsible for at least 75% of all Heroin supply to Western Europe and 50 per cent to USA. The following table depicts the dillema India is facing these days because of inflow of heroin from the Golden Crescent. Figures in the brackets indicate this as a percentage of the total quantity of drugs seized in India :

Table A

Years	*Heroin seized (kgs.)*	*Hashish seized (kgs).*
1984	118 (58%)	2316 (44%)
1985	553 (73%)	3988 (38%)
1986	2296 (87%)	10107 (53%)
1987	2301 (84%)	6385 (76%)
1988 (till July)	1536 (86%)	11896 (86%)

The Afghan Affairs

With the Soviet withdrawal and subsequent events in Afghanistan and the Afghan Government's proposed plan to create a non-military zone along the border of Pakistan and less aid available to Afghanistan, the Mujahideen rebels naturally resorted to stepping up production and marketing of drugs. Khalistan terrorists reportedly obtained from Pakistan, a huge quantity of arms, mostly left overs of the Afghan Mujahideens (the quoted price for an AK 47 is as cheap as Rs. 500).

Unconfirmed reports also indicate that ganja, opium and other hard drugs transit through Udalgiri where the contact for such transactions are established with dealers before their despatch to Delhi, Bombay, Bihar and UP Udalgiri is in the Darrang District of Assam and shares a long border with Bhutan. Parties come from Punjab, Bihar and UP and it has also been suspected that arms and ammunitions are also sold to the tribal insurgents by some parties who come from Punjab. In the above context, it is very necessary to tighten control on the India-Burma border, particularly, in view of the political unrest and instability in Burma which has created the necessary climate for increased insurgent activity.

Terrorism and Narcotics

The link between insurgency and narcotics is much weaker in Thailand. Until 1982, Burmese groups, such as SUA operated in Thai territory with relative impunity. However, in January 1982, the Thai government drove the SUA from their headquarters in Northern Thailand and has since initiated an ongoing series of military operations which have severely disrupted the activities of SUA, the Chinese Irregular Force and other Burmese traffickers on the Thai side of the border.

On the home front, the Sikh terrorist rebellion provides another variation on the narco-terror theme. In 1984, after the Blue Star Operation, local newspapers had reported that the soldiers had found quantities of heroin and hashish destined to be sold to finance arms purchases-much like the Tamils in neighbouring Sri Lanka. The direct involvement of Sikh terrorist groups in drug trafficking has not clearly come to notice, presumably for the reason that most of the Sikh terrorist group members are marked persons and cannot openly carry out this trade as they are placed on the alert notices of various intelligence and enforcement agencies. However the indirect involvement of these groups by providing protection or safe passage to drug traffickers on the Indo-Pak border is obvious. So far as the North-Eastern States are concerned, intelligence from reliable sources indicates that the Naga Federation Group (NFG) is involved in trafficking of drug and precious stones since 1981. NFG do this to build up their funds which they require for fresh recruitment and creating activities abroad. NFG headquarters is the main supply centre for drugs. The People's Liberation Army of Manipur (PLA) also receives consignments of drugs. The PLA is in 'eague with the NFG and KIOCC. Limited role in the trafficking of drugs is also reported of the National Social Council of Nagaland (NSCN). NSCN has established contacts with local Muslim leaders of Thoubal and Churachandpur to patronise and protect them. These are Bangla Muslims who are big time traffickers of drugs. NSCN is patronising them with an eye on finding direct routes to Bangladesh from where they wish to export drugs to the western countries.

International Strategy

To contain the threat posed to humanity by the evil of narco-terrorism, it is imperative that international strategies give top priority to crop control. A total ban on cultivation and production of drugs should be established further to be enforced by eradication. An effective international strategy should offer financial and technical assistance where needed for narcotics control projects. Bilateral and multi-lateral assistance should be provided which should be linked to crop control agreements to ensure success in reducing production. Governments of illicit drug producing nations must demonstrate a political will to undertake effective crop control and interdiction programmes. Corruption, which has undermined the control efforts in many source countries must be stemmed out by strong and determined administrations. There is a need for the international community to make common cause in a more vigorous, more widespread and more united effort to control international narcotics production and trafficking.

Narcocracy in Pakistan

Allah was kind to Gen. Zia when the Soviets invaded Afghanistan, making Islamabad a capital ally of Washington and thus giving its ruling *junta* a new lease of life. The US became a fountain of military and economic aid to Pakistan. Along with the money, came an influx of weaponry and drugs which ultimately added to Pakistan's social problems. These have now assumed nightmarish proportions. About 2.5 million drug addicts, more than 100 heroin laboratories, billions of dollars worth of drug business which caters to 70 per cent of the needs of the US heroin market. What has of late come to be known as narco-terrorism sprouted in Pakistan long ago in the form of opium trade.

Paradoxically, the promulgation of the *Hudad* (prohibition) ordinance in 1979 banning the growth and use of opium triggered what is now a flourishing illegal business and industry in Pakistan. The prohibition order had put the opium growers in a quandary over surplus stock. But on being driven underground the growers realised that the right connections offered them more money, provided their opium could be converted into heroin. And they could get more money with less opium.

Drug Money and Arms

The lure of easy money and the need for illegal arms to fight the Kabul regime complemented each other. The narco-trade brought easy money in plenty, so the Mujahideen took to the drug business. The social health of Pakistan was thus sacrificed at the altar of global political strategy. In 1979 (the year of prohibition), ironically there was a bumper crop of poppy yielding 800 tonnes, an all-time record. At the same time, the prohibition order hiked the price of opium from Rs. 1200 to about Rs. 9000 a kg. in Punjab.

Today, one kilogram of heroin with 90 per cent purity fetches about $ 2,600 in Pakistan. The price shoots upto $ 5200 as it crosses the Indian border. When the contraband drug reaches New Delhi, its value goes up further to $ 7000 to $ 7,200. The value of narcotic jumps by 35 to 40 times when it finds itself in a European port, notably Frankfurt. And finally, after crossing the Atlantic to arrive in the promised land, the United States, one kilogram of heroin fetches $ 60,000, according to estimates made by various drug enforcement agencies including the estimate done by Mr. Ikramul Haq, a Pakistani journalist.

Tracing the evolution of narcocracy, a term used for the first time in the context of Gracia Meza in Bolivia, author Ikramul Haq underscores the connivance of the Zia government with the drug racket by using the drug laws with discretion and not really pursuing the anti-drug campaign in spite of western pressure. It also reveals the American dilemma which on the one hand wants to promote the so-called Mujahideen and at the same time could not provide an adequate level of arms supply. Hence, the Mujahideen were forced to go in for narco-trade to supplement their requirements of money for the purchase of military hardware from the open market.

Recycling Income

How do the Pakistani drug syndicate recycle their "income"? A study conducted by Haq over many years reveals that members of these syndicates used the accounts of many of their friends and relatives living abroad to launder the cash they got from the sale of drug in the US and other western countries to remit the money back to Pakistan. These syndicates "could not be questioned about this money in view of the economic coordination committee of the cabinet direction that origin of funds remitted from abroad through banking channel cannot be questioned. The income tax department was also instructed not to question the origin of such remittance vide circular No. (28) ITI/82 dated September 1, 1982 issued by the Central Board of Revenue". A powerful "drug mafia" of Peshawar frequently deposited money in various banks of Europe (different branches of BCCI) and "then requested the bank to wire funds to their accounts in Pakistan", in order to avoid moving cash in bulk, the study found. This money transferred from abroad "has been legalised" as profits earned from business with dummy corporations run by their own persons.

The Victim of Narco-terrorism

The dangerous dimensions of narcotic trade can be gauged from the murder in the sky of Pak President Gen. Zia on August 17, 1988. AC-130 Hercules transport plane took off from Bhawalpur airfield in Pakistan. On board were the top brass of the Pakistan Army, the US ambassador and the highest ranking American military official in Pakistan and the President of Pakistan, General Zia. Five minutes after take-off, the plane crashed in the midst of a desert. All 31 passengers were instantly killed. It was clearly a case of sabotage.

USA Asks Pak to Destroy Heroin Labs

The USA has asked Pakistan to take steps to destroy some 100 to 150 heroin production laboratories in the North-West Frontier Province (NWFP) and arrest their operators as a key measure to stop continued cultivation of opium in Pakistan. A US Embassy spokesman told newsmen that as per a 1988 survey there were 1.2 million heroin addicts in Pakistan while another one million were using other addictive substances. "About 55 tonnes of heroin is required annually to satisfy the heroin addicts in Pakistan", he said adding that Pakistan was producing 165 tonnes of opium per year from which 16 tonnes of heroin could be made. The remaining local requirement was being met by Afghanistan where poppy cultivation was over 600 tonnes.

A recent report by the US administration which, while lauding the efforts of Pakistan government to combat narcotics business had said, "that the international community remained concerned by the presence of other reputed traffickers in the provincial and national assemblies".

Pakistan Behind Extremist-Smuggling Activities

Pakistan has become the "dominating concern" of the law enforcement agencies and Border Security Force because of its continued support to the 'Narco-terrorists'.

Acting on information, BSF Naka party in collaboration with Civil Police laid a joint Naka near a village Mashiwala in the rear of village Tahiwal which falls 3 kms. inside the International Border. On the night of 1/2-8-1987, the Naka party observed movements of two smugglers and on their coming close, they were challenged. The smugglers opened fire upon which the Naka party also fired one round of 9 mm, one VIP cartridge and one round .38 bore revolver in self-defence. The smugglers finally threw their weapons and they were nabbed. They were identified as (1) Phuman Singh S/o Bishan Singh and (2) Smt. Durgo Bai W/o Kundan Singh, both residents of village Tahiwal, Distt, Firozpur. Subsequent searches resulted in the seizure of two polythene bags containing 20 kgs. of heroin and one foreign made revolver of .455 bore along with one fired and 9 live cartridges.

On 27.8.1987, the Barmer Police seized 24 kgs. of charas from truck bearing No. 3949 near village Ganga, P.S. Shiv and arrested three persons (1) Isshya Khan Abada, (2) Virdha Khan and (3) Cirdhan Khan, the owner of the truck. Besides the charas, the police also seized one SIR (foreign make), 800 rounds of ammunition of SLR, 4 pistols with 470 rounds of ammunition alongwith one empty magazine. Acting on information, 61 Bn of BSF Naka party laid in Naka in the area of BOP Bahadukha at BP No. 216/82 about 200 yards from the border on the night of 18.9.1987. At about 0455 hrs. on 19.9.1987 the Naka party observed two miscreants coming from Pakistan to Indian side. When challenged, the miscreants fired on the Naka party and the Naka party returned fire in self-defence. In this armed encounter, one of the miscreants was shot dead on the spot while the other managed to escape towards the Pakistan territory. Search of the area resulted in the seizure of 13 kgs. of opium and two 12 bore live cartridges among other things.

On October 25, 1987, on specific information, the BSF intercepted and seized 9 assault rifles after a brief armed encounter with smugglers in the area of BOP Kulwant of Khem Karan Sector. The weapons were sent to India for Sikh terrorists by one Pakistan smuggler of village Nagar Amin, Distt. Kasur, acting as go between the smugglers and the Indian extremists. Mohd. Akram was a notorious smuggler of narcotic and gold. BSF seized 97 kgs. of charas sent by Mohd. Akram in the area of BOP Udhar Dhariwal on 6-7-1987 after a brief encounter in which one Pakistan carrier was injured and the other died. The injured Pakistan carrier had disclosed that the charas was sent by Mohd. Akram alias Bhella. The injured Pakistan national later died. Of late, a gold consignment was seized by DRI in New Delhi. Mohd. Din, a trusted carrier of Mohd. Akram was arrested alongwith the gold.

On the night intervening 27/28-12-1987, BSF Naka party laid a Naka near BP No. 237/1 to 237/2 ex BOP Chak Khewa and at about 0320 hrs. observed the movements of some smugglers coming from Pakistan to Indian side. When challenged the smugglers opened fire and in the ensuing armed encounter the smugglers managed to escape across the Pakistan border taking advantage of darkness and dense mist. Thorough search of the area resulted in the seizure of 137 kgs. of heroin, one country-made revolver and fired cartridges. A tracker dog was also pressed into service but no suspect could be apprehended. Acting on information, the Customs Preventive Staff, Firozpur assisted by BSF staff search on 13-1-1988, the house of Darshan Singh S/o. Phuman Singh, S/o Gazni Wala, PS Mamdot, Distt. Firozpur. The search resulted in the seizure of five slates of charas weighing 16.5 kgs. On personal search of one of the three occupants, Mohinder Singh S/o. Tulsi, S/o Pojoke Mamdot, Distt. Firozpur, one pistol alongwith 6 live cartridges, was recovered and seized. Mohinder Singh and two others, Gurdeep Singh *alias* Khan, S/o Hoshiar Singh, R/o Pojoke, P.S. Mamdot and Gurmej Singh, S/o Thakar Singh, R/o Gazniwala, P.S. Mamdot were arrested. On interrogation the arrested persons confessed to having brought the narcotics and the arms and ammunition from Pakistan and stated that they crossed over to Pakistan on 9.1.1988 and returned with charas and pistol alongwith cartridges on 13-1-1988.

The Crime Branch, Delhi Police received secret information about some Afghan nationals engaged in illicit trade of narcotics through Pakistan to India and further to European countries. Acting on this, they finally intercepted a truck DEL 3795 in Delhi while it was waiting for the loading of charas from Maruti Van No. DDC 4895 driven by one Neeraj Samadhi *alias* Ali Ahmed, resident of 373, Gagan Vihar. Search of the two vehicles resulted in the seizure of 420 kgs. of hashish some of which was hidden in secret compartments made in the truck. On interrogation, Neeraj Samadhi admitted to having further stocks of hashish at his residence which was latter searched and a further quantity of 237 kgs. of hashish alongwith a French automatic pistol were seized. Tarsem Singh S/o Isar Singh, driver of the truck, Lakhveer Singh S/o Gurmej Singh both R/o village Chauvinda P.S. Kathu Nangal, Distt. Amritsar, Neeraj Samadhi and Jagjit Singh S/o Ajaive Singh of the same village were arrested.

On 22/23-2-88, BSF Naka party at Lakha Singhwala laid Naka in the area BP No. 205/7-8 approximately 200 yards from the international border. At about 0230 hrs., the Naka party observed two miscreants coming from Pakistan, when they were about 25-30 yards away, the Naka party challenged them upon which an armed encounter ensued. Finally, miscreants managed to escape towards Pakistani side under the cover of darkness and standing crops. The area in the vicinity of the incident was thoroughly searched as a result of which approximately 7.880 kgs. of heroin and 8 live cartridges of 12 bore gun were seized. On 14-4-88, Crime Branch, Delhi Police busted a drug smuggling gang with terrorist links in an operation spread over a couple of weeks. During the operation, 17 kgs. of heroin and a lot of arms and ammunition including AK 47 rifles, mauser, 32 bore pistols, etc. were seized.

The police developed an information that Santok Singh of Ludhiana had created a network of trafficking of narcotic drugs obtained through Pakistan between Punjab and Delhi. A trap was laid for him in the Civil Lines area of Delhi and Santok Singh was arrested with two kgs. of heroin which was concealed in the cavities of window panes of his car. Following disclosures made by him, police started tracking the movements of the king-pin Kartar Singh of Gurdaspur, the supplier of the drug and his associates. Later two of his associates Satnam Singh of Gurdaspur and Narinder Singh of Faridpur were arrested with a consignment of 12 kgs. of heroin on 19-4-88 at a South Delhi locality. The arrested men during interrogation revealed that they were involved in the drug trafficking across the Indo-Pakistan border and that proceeds of narcotics smuggled from Pakistan went into buying arms for terrorists.

Patrol Party BOF Maharana ex-64 Btn BSF Jaisalmer detected *pug* marks of two camels having crossed the border between BOP Maharana and Mithbinda (Jaisalmer Sector) on 15-4-88. BOPs were alerted and patrol parties chasing the pug marks seized six bags of charas weighing 385 kgs. three kms. short of village Kula. During further chasing, patrol party also seized one country made single barrel gun. Subsequently, with the help of locals, BSF party apprehended one Indian smuggler namely Ramtala S/o Baloch, R/o Chandru, District Jaisalmer along with two camels. His associate Baryam S/o Bhodu, R/o Mithavin, District Jaisalmer managed to escape. On information provided by Ramtala, 5 kgs. of charas was also seized which was buried underground in the area of Mithavinda.

On 15-4-88, on specific information, BSF engaged armed extremists in the area of BOP old Sundergarh Ajnala, District Amritsar. After the encounter, 5 magazines of Assault Rifles and 1016 rounds were recovered from the area of incident. Subsequently, it was found that the, weapons and the ammunition were sent by two smugglers namely, Aslam Dogar and Buta Dogar both r/o Bathanwala, PS Narang, Distt. Sheikhupura. It will not be out of place to mention that the Customs had seized 175 kgs. of heroin on 6-1-88 from truck No. PJA 8677 at Ludhiana. Smuggling of heroin was arranged by Aslam Dogar from Pakistan.

An armed encounter took place between smugglers and BSF Naka party BOP Rankot (Amritsar) ex Bn 77 BSF near BP No. 95/15 on 19-4-88 at 0200 hrs. One smuggler was killed and another Pakistani national named Mohd. Rafiq Jahanger S/o Sardar Mohd., R/o Bhasin P.S. Batapur District Lahore (Pakistan) was apprehended. Thorough search of the area resulted in the seizure of 72 packets of charas weighing 148 kgs, one country made 12 bore gun and chit specifying tasks allotted by Pakistan Intelligence Agency.

On 19-4-88, BSF recovered 16 missiles from the farm house of Buta Singh R/o Village Kalanpur, Distt. Gurdaspur. During the interrogation it was revealed that the missiles were delivered by Sarwar Khan, R/o Jala now residing in Iklaspur, Tehsil Shakargarh (Pakistan). Sarwar Khan has also been indulging in the smuggling of narcotics.

On 23-4-88, Patiala police intercepted a truck near Jakhwala village on Sirhind Road and seized 104 bags of heroin. The truck had fake registration No., Amar Singh of village Vasarkhi and Gian Singh, occupants of the truck were arrested. Subsequently, the police seized 4 AK 47 Chinese assault rifles, one Thompson machine-gun, one automatic rifle, four .38 revolvers, 3 pistols and 6 magazines of AK 47 rifles. In another incident the CRPF on 26-4-88 recovered and seized one assault rifle and 15 grenades from village Dhelari, P.S. Khalra, Distt. Amritsar. Interrogation of the Indians connected in the smuggling of heroin and arms indicated that these were brought by a Pakistan smuggler namely Kamaruddin, R/o Katloi, District Kasur.

On 11-6-88, 7.390 kgs. of heroin alongwith one AK 47 rifle, one magazine, 800 cartridges and 9 HE 36 grenades were seized in a joint operation by police and customs at village Pella, District Firozpur Punjab after an armed encounter with six smugglers. It led to the arrest of four persons one of the persons carried the drug, the others carried lethal weapons and hand grenades.

Pakistan Uses Drug Money to Back Indian Ultras

According to an American Central Intelligence Agency (CIA) study reports, Pakistan's military intelligence used narcotic profits to fund separatist movements in India. The study found that the heroin trade had penetrated the "highest political circles" of Pakistan and "is becoming the lifeblood" of the economy and government in that country. The intelligence documents asserts that heroin traffickers financed the ruling party, bought enough votes to win seats in the National Assembly and gained access to Prime Minister Nawaz Sharif and President Ghulam Ishaq Khan. The study also alleges that Pakistan's military intelligence used the drug money to finance the war in Afghanistan.

The study says that the then ruling party, the conservative Islamic Democratic Alliance, had 'routinely' tapped narcotics traffickers to bankroll campaigns and so did the ruling parties of the previous administrations.

Sharif's Links

The CIA study, according to Newsday, alleges that one of the top suspected heroin traffickers was related to Prime Minister Nawaz Sharif. "As the Sharif family moved into mainstream of power in Punjab under the patronage of the army, this relative used his influence and the 'untouchability' that his associations ensured to further his alleged heroin enterprise", the report said. The paper quoted the study as saying that another relative of Mr. Sharif functioned as the suspected trafficker's "ambassador" to other drug mafias. Another suspected trafficker is regarded as a close political adviser to the Prime Minister. "These men are all far more than casual or marginal members of the Sharif entourage. Whenever the Prime Minister visits his home in Lahore, he and his brother meet with these men both for business and recreational purposes", NEWSDAY quotes the study as saying.

Drug Baron

The CIA document, according to the paper, says that Pakistan's leading heroin baron "is deeply interconnected in all the key institutions of power in Pakistan, including military intelligence, the presidency and ruling coalition". It says, he offered upto $ 2,000 a vote to win his seat in the national legislature.

The study says that the drug lord, known as the

"king of Khyber heroin", attends the National Assembly without the slightest fear of arrest and has access to the most powerful of the land. He enjoys direct access to President Ghulam Ishaq Khan, the paper quotes the study as saying. It notes that after the previous government of Prime Minister Benazir Bhutto issued warrants for his arrest in 1988, he eluded capture for two years. When her government fell in 1990, the new government promptly quashed the warrants of his arrests, the document says.

The paper quoted Pakistan Embassy Press Attache Zahoor Malik as saying, "If you don't mind, this is a stupid study...totally baseless. The President of Pakistan is one of those few people in the third world who is totally clean". Mr. Malik blamed the heroin problems in his country on Pakistan's role, largely at the behest of the US, in assisting Afghan rebels against Soviet "occupiers".

Like a Condom

"America has used us like a...condom" and turned its back by cutting off military aid and sharply reducing economic assistance when the Soviet troops left Afghanistan, the paper quoted Mr. Malik as saying. The CIA study absolved of involvement in the drug trade, the late dictator Zia-ul Haq, who died in a plane crash in 1988. "Even today, few question his personal honesty, but he did have men in his entourage who used their position to promote criminal interests, including narcotics", the study says. "Two of the pilots used Presidential aircraft to smuggle heroin—one to the United States during a state visit", the report says, but gives no further details.

The study cites, "numerous reports" that Pakistan's powerful military intelligence agency, the Inter-Service Intelligence Directorate, used drug money to fund groups it supports in the neighbouring states. ISI developed funding arrangements with Sikh militants and Kashmiri insurgents in India. Asked whether the top Pakistan officials at the very least condone the drug trade, Mr. Tom O'Grady, the Drug Enforcement Administration's acting chief, said "Precisely". The report, titled 'Heroin in Pakistan: Sowing the Wind', paints a more bleak picture of Pakistan's narcotic problem than that presented by the Bush Administration, whose State Department told Congress that Pakistan was taking "positive steps" to combat heroin.

The study findings, could complicate the Clinton Administration's task of developing its Pakistan policy and cause political explosion. The study portrays Pakistan as beset by the "collapse of ethical standards through (its) society and institutions".[1]

Narco-Terrorism in Indian Context

The dreaded terrorist Jinda, is not a religious fanatic as widely believed. He was initially a professional drug-peddler on the Punjab border. Jinda wanted social recognition and found it as a terrorist with the media turning him into a celebrity. Like Jinda, there are others who have switched over from drug trafficking to terrorism. In fact, the close nexus between drug trafficking on the one hand and terrorism and insurgency on the other is a worldwide phenomenon. At individual level, drug abuse inevitably led to crime—whether it is stealing or prostitution or murder for producing money for the next-dose. Likewise at an organisational level drug trafficking leads to organised, violent crimes of all kinds. Manifestations of this organised violence finds expression in terrorist and insurgent groups and quite naturally, along border areas so that when the going gets rough, these elements can find asylum across the border. In a typical reversible cause-effect relationship, these groups control activities that generate drugs and convert them to hard cash which is needed to buy arms to sustain their fight and to pay their followers.

This position also provides the necessary environment where hostile neighbouring country can freely interact with the traitor and advise and influence the latter's action to its own advantage. The massive narcotic seizures along the land frontiers of India, particularly on Western borders, combined with seizures of arms and ammunition by the enforcement agencies prove these observations.

Insurgency and Drug Peddling in North-East India

While most of the ethnic groups are involved in trafficking of heroin from Golden Crescent, the north-east insurgents continue to bring in small but increasing amounts of contraband drugs from the Golden Triangle, hence, Narcotic Control Bureau sources term the north-east sector equally "vulnerable". Heroin in particular is crossing the Western Myanmar (Burma) frontier into Assam from where instead of coming to Delhi or Bombay, the drug is first carried to Chittagong in Bangladesh. At the same time acetic anhydride is being illicitly transported from India to the clandestine laboratories in the Golden Triangle.

The Punjab Extremists

Taking a cue from the north-east insurgents and groups, the Punjab extremists took full advantage of the emergence of the Golden Crescent as the major source of heroin and India as its major transit channel. It became mutually convenient for both the suppliers and the terrorists. As for the former, they found ready couriers on the labyrinthine Indo-Pak border among the terrorists for whom smuggling became an easy source of funding their militant activities. When the Delhi Police caught the dreaded terrorist Harjinder Singh alias 'Jinda' they did not expect that he would turn out to be a regular smuggler of heroin. Yet they found high grade heroin from him and CBI investigation later led to the discovery that other terrorist gangs including those of

Bairagi and Mathura Singh under his (Jinda's) control dominated drug smuggling along the Indo-Pak border in the Punjab Sector till the middle of 1987.

Interpol reports say that 130 Indians largely Punjab extremists, were arrested with 180 kgs. of heroin in 1986; 166 were arrested with 351 kgs. in 1985 and in 1984, 171 were arrested with 204 kgs. of heroin. The enforcement agencies say the volume has definitely increased. According to the police, catching the young couriers in Punjab has become difficult since most of them had started taking shelter in the Golden Temple. A truck driver arrested in January 1988 along with 50 kgs. of heroin and 400 kgs. of hashish in Ludhiana by the Customs Preventive Authority confessed that the source of supplies was Golden Temple.

India Takes the Lead in Curbing Narco-Terrorism

USA publicly acknowledged India's leading role in this region to curb the narcotic traffic. Mr. Ronald Reagan's praise in March 1989 was followed by the State Department's report lauding this country for arresting narcotic traffickers on the basis of information gathered in the USA. This commendation is significant as some other countries in this region have come in for criticism for their lax approach in tackling this complex problem. The US Administration seems to be slowly veering round to the view that a full scale war against the drug menace can be effective only if its allies in this effort are stable democracies. This is implicit in the report which notes that the overall acreage under opium has probably increased in Pakistan in recent years. It states with dismay the limited use made of the US supplied aerial spraying capability embodied in a thrush aircraft and a helicopter wing. Also India's large seizures of heroin near the Pakistan border in the State of Punjab and Rajasthan are alarming. Similarly, the report notes with concern that the narcotics situation in Nepal has not changed appreciably over the past year. The U.S.A. is reconciled to the fact that a new regime in Kabul once established and secure, may be willing to address itself to the issue but will be faced with so many other pressing problems that drug control may not get a high priority. The lure of a quick and profitable cash crop in the form of opium poppies and heroin must prove extremely tempting for the returning refugees. Thus the U.S.A. is pinning hope on India, the transit point, rather than on its traditional ally Pakistan and its neighbouring Afghanistan, the largest producer of opium in South-West Asia.[2]

If narcotic supplies from this region are a direct threat to the US Society, the massive and continuing arming of Pakistan (even in the changing geo-political situation) remains a serious threat to India's involved reciprocal obligations. In fact, the issues are directly related and it will be futile to work out any arrangement which ignored one of these. In the Western countries, drugs are paralysing their societies while in the poverty-stricken developing countries the arms supplied by the rich nations are having a debilitating effect on the economy.

Deterring the Drug Traffickers

The promulgations of an Ordinance providing for the preventive detention of drug traffickers has not come a day too soon. It is precisely this kind of legal back-up (which enforcement officials had found wanting in the existing Narcotic Drugs and Psychotropic Substance Act) that will help at this juncture to contain the nefarious but highly organised operations of drug smugglers. The Ordinance which covers a wide area of operation in illicit narcotic trafficking will not only strengthen the hands of the enforcement authorities but will also communicate in no uncertain terms that the Government really means business and intends to tackle the flow as well as the source of narcotics supply in the country. For purpose of preventive detention, the Ordinance covers the cultivation of coca, opium poppy, cannabis and the production, warehousing, concealment as well as the consumption of narcotic drugs and psychotropic substances. Its scope extends to the direct or indirect financing of these activities and specific preventive detention for one year in ordinary cases and for two years in areas declared 'highly vulnerable'. The latter category would cover parts of the country where illicit cultivation takes place and also all the chief transhipment points including the border areas. The gravity of narcotic problem in India indicated by the dramatic increase in seizure in the last two years has made it apparent that the country is not just a "transit point" for international narcotics flow any more. Interpol statistics have it that out of the 7.7 tonnes of heroin seized by them last year, 2.7 tonnes came from India. This means that the level of abuse among the local population continues to grow at an alarming rate. The Narcotics Control Bureau has stressed repeatedly that all types of drugs have become "freely available" and that the problem of drug addiction is growing.

While it is clear that there can be no overnight solution to the narcotic menace and few nations caught in the vicious trap of 'narco-terrorism' have found many easy ways out, preventive detention would certainly curb some of the more blatant drug operatiors who have escaped because of sheer lack of legal evidence to pin them down. The Government's determination to strike at the supply side has also been apparent in the recent strikes of the Narcotics Control Bureau against illegal poppy cultivation especially in the inaccessible jungle terrain in Uttar Pradesh, Rajasthan, Madhya Pradesh and in the Tamil Nadu-Kerala border. So far the enforcement officials have had to satisfy themselves with burning the crops but the new Ordinance which has given them the power to arrest should serve a deterrent to those involved in cultivation. It is also essential that the Government address itself to the actual problem of addiction. Preventive detention is necessary for hardened criminals involved in the drug mafia as well as for pushers and financiers, but what of the vulnerable addict who is very often a victim of distressing circumstances and is likely to need much more than the arm of the law to be rehabilitated? These are questions that law-makers should tackle simultaneously—strengthened legal system has to be matched by an enlightened educational and psychological support structure. To provide such a framework for the unfortunate who fall victim to this pernicious habit, the Government must seek the aid of experts—doctors, teachers and social workers all of whom are seriously concerned over the growing menace of drug addiction.[3]

References

1. *Indian Express*, February 25, 1993, p. 9.
2. NEWS FROM NARCONTROL, September 1988, pp. 17-21.
3. *The Hindu*, July 8, 1988.

Note : Burma is now officially known as Myanmar, though the USA and several other countries continue to call it Burma.

50

CHALLENGES OF INTERNATIONAL DRUG TRAFFICKING AND STRATEGY TO CONTROL

Transnationality in Crime

The trend in the sphere of crime has given rise to well-organised syndicates operating in utter disregard of international boundaries. These organised groups indulge in crimes like drug-trafficking, gambling, prostitution and fire-arm-running. For the sake of convenience these transnational crimes can be put in the following categories:

(1) International drug trafficking.

(2) International organised crime enterprises dealing in crimes of traditional nature.

(3) Industrial crimes.

(4) International fraud.

(5) International terrorist activities.

(6) Offences involving works of art and cultural property.

(7) Offences related to environmental pollution.

(8) Miscellaneous.

International drug trafficking and organized crime syndicate are our immediate concern. The trade in drugs is quite a profitable proposition and it attracts the ace underworld criminals. It requires ability to organise a well-knit unit to manage regular supply, efficient delivery system, quick and safe transport, imaginative distribution system and innovative collection and re-investment. The illicit trade has to procure supplies from the region having best suited climate and terrain, bad government, ill-informed social institutions, ill-equipped law and order machinery and unawakened people. Once supply points have been selected, then organized syndicate start proliferating its activities by intimidation, extortion and corruption ultimately resulting into growth of a monster—like organization which can pressurise even the government enjoying sovereign powers. For transport, the syndicate has a well-knit as well as well-insulated network of couriers. Legal and financial wizards can be engaged for laundering of money. At times the syndicates have successfully penetrated legitimate international business with the aim of operating the enterprise on a continuing basis and driving away all healthy competition by predatory methods (intimidation, extortion and corruption coupled with sophisticated modern gadgets and approaches). Once established, these syndicates develop potentiality of destabilizing governments and spread an aura of omnipotence which is quite helpful to them in covering their evil tracks. With such syndicates on the scene, conventional tools in the hands of police and other law enforcing agencies become inadequate and the need for concerted efforts and international will to crush the crime and its beneficiaries. Lack of international co-operation facilitates removal of evidence both oral and material. The dearth of well-equipped laboratories and trained personnel further delay and defeat the process of law. Without international co-operation it is difficult to unearth black money and to trace beneficiaries.

Judicial Assistance

Sine qua non of all international legislation is undisputed recognition of the principle of absolute sovereignty of states and from this follows the usage that no state shall be allowed to perform sovereign functions in the territory of other side state without its explicit concurrence by way of treaty, specific declaration or the provision in any of the several statutes. All functions performed in the sphere of criminal investigation, prosecution, enforcement of judgement, prevention and correction of offenders are covered in the sovereign functions of the state. The search for offenders, collection of evidence for prosecution or defence cannot be collected without co-operation of the states if it involves foreign investigation. Further, some of the material witnesses may be the citizens of other states who cannot be apprehended, persuaded or coerced to appear before the courts of other countries without the active co-operation of the requested state. In modern times no state can afford to ignore this aspect without detriment to their own administration of justice. Though international co-operation in the field of judicial assistance in criminal matters has been recognized as imperative need of our times, yet consensus could not be arrived at so far. The progress of legislation in the sphere of judicial assistance is also far from being satisfactory. Ad hoc arrangements have been made between the states on the principles of reciprocity and double criminality. This proliferation of international treaties and mutual agreements and differences in the system of legal administration have made the situation confusing and may be regarded as a stumbling block in the path of international jurisprudence. In this connection, United Nations should take a lead and a consensus may be arrived at to have uniform procedure in the matters of judicial assistance. Futuristic steps at opportune time can save us from further complications.

Judicial assistance can be grouped in the following categories:

(i) Assistance in Collecting Information at Pre-investigation Stage

The information referred to above can be about places, persons, events or laws that may lead to a formal investigation. Normally such information can be collected by ordinary visitors to the country, but it can also be sought by the requesting state through diplomatic channels. Depending upon the state of relations between the states such information can be allowed to be given to investigating authorities also with routine useful investigative suggestions by their counterparts in the requested state.

(ii) Assistance in Investigation

It includes interrogation, delivery of property, local inspection, service of documents, search and identification, etc. The assistance in investigation can be both informative and formal. The informal assistance can be such as locating a person, interview of a victim or likely witness or suspect, checking of hotel register, documents of companies to check weapons, sale documents to ascertain criminal history of a man and other such allied matters which can be attended to on an informal exchange basis. Such informal assistance on voluntary basis and mutual co-operation between investigating agencies is an exceptionally important ingredient of international co-operation in the field of criminal justice administration.

Formal assistance in criminal matters can be through diplomatic channels:

(a) *Letter Rogatory*—a traditional mechanism for securing evidence from a foreign jurisdiction, "Letter Rogatory" retains the quaint formalities of requesting Seriatim certifications of the validity of the various signatures appended in the course of their transmission and are sent and received through diplomatic channels with numerous ministerial and judicial talismatic niceties to be observed along the way. Once in the hands of receiving judicial authority, the summons issued often produce more lawyers than witnesses" (R.I. Gainer—"International Co-operation, Criminal Justice Administration—A selective over-view"). 'This procedure is slow, irrational, wasteful, expensive and totally unsuited to our times, but at the same time it can be safely regarded as by far the most reliable and time tested mechanism'. Applications are made to the court of requesting country who issues such letters to the court of requested country. The court tribunal of the requested country has discretion to allow or reject the request. Once they allow such requests then they use their own procedure of execution. The effectiveness of such letters depend on the attitude of the courts and relations between the countries.

(b) *Executive agreement regarding judicial assistance*—two or more states can reach the agreement (different from treaties of mutual assistance) to render help in gathering information for investigation and prosecution. This is a recent innovation and more expeditious than other modes. It is based on the principle of reciprocity.

(c) *Treaties*—the states are now increasingly entering into such treaties as give them all necessary help in the field of prevention, investigation, prosecution, enforcement of judgement, transfer of prisoners and correction.

(iii) Assistance in Securing Presence of an Accused for Judicial Trials

The assistance can be rendered in formal as well as informal way. The informal mode is irregular but not illegal. In such cases the criminals are allowed to be abducted by the requesting state and produced in the court. Such methods can be adopted by the officers at the local level but cannot be regarded legal assistance. Other informal methods can be deportation or threatened deportation which also amount to arm-twisting and cannot obtain judicial approval.

Formal modes of securing attendance of the accused can be by way of extradition. Extradition is the age-old and tried mechanism for securing attendance of defendant or accused persons for the purpose of trial. This process is entirely based on the principle of reciprocity and most of the states facing such problems have entered into such treaties.

(iv) Assistance in Enforcing Criminal Judgements

This deals with assistance in the supervision of conditionally released prisoners and also with exchange of prisoners. This form of co-operation has more potentialities as it can render immense help in recovery of fine and forfeiture of property acquired through illicit gains. It would be an encouraging demonstration of foresight if the international community could pave the way for effective execution of such judgements. In this connection future planning under the auspices of United Nations can be of much help.

(v) Transfer of Prisoners

The transfer of prisoners from one country to another is also a new field and is solely based on humane considerations. This type of assistance is rendered reciprocally only.

(vi) Training

International co-operation in the field of training is of great importance. Training provides an insight into the working techniques of the law enforcing arms of other countries and also gives an idea about the play of the mind of criminals in other parts of the world which arms the officer to make future planning in their fight against crime in general and drug trafficking in particular.

International Co-operation in Anti-narcotic Measures

Since the drug menace has not spared any country big or small, rich or poor, white or black. Convention for the Suppression of the Illicit Trafficking in Dangerous Drugs in 1936, followed by Conventions were signed in 1961 and 1971. The concern shown by U.N.O. is evident by the organizations that have come up on international stage recently under the auspices of United Nations which include: (1) Commission on Narcotic Drugs, (2) The Division of Narcotic Drugs, (3) The International Narcotics Central Board, and (4) United Nations Fund of Drug Abuse Control.

In addition to the above organizations, the U.N. also organised Congresses on prevention of crime and treatment of offenders and constituted committees on crime prevention and control.

The World Health Organisation concentrates on the study of medical aspects of drug abuse. W.H.O. also provides expert advice on classification of drugs and treatment of detoxification of addicts. The International Criminal Police Organisation (Interpol) is the topmost international organisation which helps law enforcement agencies of various states in collection, collation and dissemination of intelligence and also in the process of investigation, including analysis of drugs and psychotropic substances.

The ICPO (also known as Interpol) was founded in 1923 and its general Secretariat, located in St. Cloud, France in 1946. In each country, the State Government designated National Central Bureau to have working liaison with ICPO Interpol has drugs sub-divisions in the Police Division of Interpol General Secretariat. One additional Regional Drugs Liaison Office is located in Bangkok, Thailand. The Drugs, Sub-division of Interpol General Secretariat consists of specialised officers drawn from different member states. The Liaison Officer Scheme of the Drugs Sub-division aims at establishing close contacts with the officers of the countries within their jurisdiction to foster international co-operation. The officers or the staff of Interpol do not act as International Police and to that extent the operational action remains in the hands of the police of the countries concerned.

Interpol works in close co-operation with other international organisations such as the Customs Co-operation Council, the International Narcotics Control Board, U.N. Division of Narcotics Drugs and Crime Prevention and Criminal Justice Branch of U.N.O. The "Water edge" operations are one of the oldest measures that have been taken against transnational peddlers of drugs and here comes in picture the Customs Co-operation Council. The widening gap between cost of manufacture of drugs and its sale price have attracted talented people to under-world trade who have brought into play the latest techniques in inventing most sophisticated methods of deceiving the man at 'water's edge', i.e., the Customs officers. When two customs worked in unison the results have been quite amazing. The Customs Co-operation Council has gone a long way to foster such co-operation and it also brought about consensus in certain methods of detection (for instance the controlled delivery technique) in respect of trans-national crime.

However, all the member states do not adhere to the same political philosophy. This results in certain substantive and procedural impediments like the consideration of 'double criminality' (the act should be considered offence in both the countries). 'Political offence' (the act may be considered high treason in one country while in other country it may be treated as political activity). The procedure laid down regarding recovery of property and handing over of incriminating physical and material evidence to court of law and finally the rules regarding evaluation of evidence recorded in a foreign land by persons other than the investigating officer who is available in court for the purposes of examination and cross-examination, also differ form country to country.

The over growing awareness to the damages caused to society by drug abuse, has evoked universal response which is voiced from international forums and in the regional conferences wherein member countries have shown determination to curb trafficking despite differences in political and economic fields. India also signed the final Act of the New U.N. Convention Against Illicit Traffic in Narcotic Drugs and Psychotropic Substances in December 1988 (Shri M.U.N. Rao, Chairman, Central Board of Excise and Custom signed the Act at Hofburg Palace, Vienna). This New Convention introduces new areas of co-operation for preventing illicit traffic by sea, intensification of control in free trade zone and preventing trafficking through mails. At the instance of the Indian delegation a separate article was incorporated in the New Convention which casts an "obligation on the industrialised countries consuming illicit drugs, to support and assist the transit states through a programme of technical co-operation and in case of need provide financial assistance for creation of infrastructure in transit states for drugs interdiction and related matters."

UN Decade Against Drug Abuse and Trafficking

Till very recently the developed countries were pressurising the countries of origin of narcotic drugs to exercise control over production and pretty little was done for controlling the demand of the drugs. In certain countries even the abuse of drugs was not declared an offence but now the situation has changed considerably for the better. It has been realised that trafficking in drugs cannot be stopped by merely controlling production or the supply, its demand also. Mrs. M. Thatcher, British Premier while addressing 600 delegates and ministers from 112 nations on April 10, 1990, voice the same sentiments when she said, "Our efforts to reduce production of drugs and to prevent traffic in them can never succeed while demand for drugs is still there...without customers the drug-traffickers would soon be out of business". The growing realisation of the above facts is bound to foster more amity in international relations and it is sure to open up new areas of co-operation in our fight against criminality in general and drug-trafficking and abuse in particular. It is heartening to note that more ferocious the drug cartels become, the response and the determination to meet the challenge grows stronger as expressed by UN Secretary General, De Cuellar, "By joining forces (forces fighting anti-drug war) we can put an end to the terrible slavery of drug addiction and stop the violence which it causes". U.N. Secretary General also announced (London, April 11, 1990) that next 10 years be designated as the "UN decade against Drug Abuse and Trafficking".

51

THE METHODOLOGY OF INVESTIGATION

The police officer in-charge of a district or the specialist narcotics officer, faced with a problem of drug trafficking or addiction in his area, will have obtained, as his first resource, all the information he can from the uniformed patrol or beatmen normally working in his district. They are the members of the police force most constantly in touch with the people of the neighbourhood and with the best general knowledge of what is happening there. Many other sources of information are, however, available through which the police officer can develop the detailed knowledge he will need to carry out a thorough-going campaign against trafficking or addiction. Depending upon the size and type of the problem and the amount of time and manpower he can devote to it, he may obtain the additional information from any or all of the following sources :—

All police units and other similar control organisations tend naturally to limit themselves to their own particular responsibilities and problems. They also tend to keep their information to themselves, rarely passing it to other organizations. But if asked, they may be willing to pass on information to other police units. It is, therefore, worth knowing where any particular type of information may most easily be obtained, and being on sufficiently friendly terms with the staff of these organizations to be able to get it without difficulty. A good means of developing such co-operation is for officers themselves first to offer any information they may have, which they think might be useful to some other organization or unit, in the hope of later getting some in return.

Sources which may have useful information about drug traffickers include passport controls, foreign exchange controls, frontier and tourist police, custom officials, vehicle registration offices, traffic and port police, ships' officers and officials of shipping companies, airline security officials, inspectors of hotel registers and any licensed premises, travel bureaus, and examiners of any registers kept by pharmacies and clinics dispensing drugs.

Regular "Ground Level" Contacts

People whose work requires them to be in constant touch with members of the public and on good terms with the police, can always be used to supply simple information about the movements and contacts of anyone in whom the police are interested. Examples of such sources of information are the owners, managers and employees of cafes, restaurants, bars, tea and coffee shops, hotels and cabarets; hawkers and street traders; bus conductors and taxi drivers; workers at garages and petrol stations; car cleaners and car park attendants; hotel and shop doormen, night-watchmen, and the like.

(a) Regular Informers

Large number of people, for many different reasons, living in towns or villages, may be willing to give information about drug takers or traffickers, either regularly or casually. Village elders, retired policemen, other government or public servants, anyone wishing to obtain a favour or fearing the ill-favour of the police, can be encouraged to find answers to particular questions. Private motives such as greed, jealousy, hatred, vanity or ambition, can also be exploited, as can a genuine antipathy to drug abuse, friendship with individual police officers, or a sense of social responsibility.

(b) Drug Abusers

A drug user arrested in unlawful possession of a drug is a source of information at least about the types and quality and whereabouts of the drug in his possession, and perhaps also about the identity of the supplier and the price paid for the drug. It may be possible by questioning him closely, either at once or later, to extract more information out of him, and even to develop him into a useful regular informer.

But certainly every police officer should try to estimate as accurately as possible the number of drug addicts in his area of responsibility, the type and quantity of drugs they use, and the prices they pay. He will then have a fair idea of the total amount of drugs being used illicitly in his area, and therefore, of what counter-measures it is necessary or appropriate to use, and of how successful he is in containing the local abuse of drugs.

(c) Seizures and Arrests

Seizures and arrests provide, of course, a primary source of information; they can only be fully exploited, however, if the right information is reported by the officers making the arrests or seizures.

Questions to be answered about *the drug* include its type and weight, what it was wrapped in, how it was hidden, what it was being carried in, and whether there was anything special or unusual about its condition or appearance.

Questions to be answered about *those arrested* include their names and address (and aliases, if any), their descriptions, photographs and fingerprints, where they were when they were arrested, what they were doing, where they were coming from and going to, and anything else that may be known of them locally, including whether they are known to be drug addicts themselves or not.

Information should be supplied at the same time about anyone who may have escaped, or who may be known or suspected to be involved in any way in the offence, even if there is little hope of getting enough evidence to make an arrest. The time, date, and place of the arrest or seizure must also be given, together with the names of the police concerned and the designation of the unit to which they belong.

Attention is invited to paragraph 1 of Part V of this guide, where use of a standarized form of report is suggested to ensure the receipt of correct and full information about every case, and to Appendix "A", where an appropriate form is illustrated together with instructions for its proper completion by reporting units.

The Interrogation of those Arrested

It is always important to try and trace the chain linking the source of supply of the drug with its intended final destination, and to attempt to estimate the size of the illicit traffic in the area. Every person arrested for a narcotic offence should, therefore, be questioned in an effort to know, where he had been coming from and going to when he was arrested, where and from whom and when the drug had been obtained, where and when and to whom it was to have been delivered, what the contact arrangements were, how much he had paid for the drug or was to be paid himself, and for how long he had been operating and on what scale.

It is true that those arrested may be most unwilling to talk to their captors, and that even if they do, they may not tell the truth. Experienced and well-informed police officers can, however, often learn something of value even from lies, and the chance of talking to arrested traffickers should never be missed. On occasions it may prove possible to use an arrested man against other people more important.

It is very often fruitful to treat an arrested person in a friendly way when questioning him, especially if he has been expecting a different approach. Similarly, it may be worthwhile paying occasional visits to men sentenced to terms of imprisonment; they may welcome the opportunity to talk about their previous experience with someone they know.

Searches

Searches of premises or persons can often be productive, and are essential at some stage in every case. An open search, however, inevitably bares the fact that the police have information or suspicions, and it therefore may obstruct future operations; it should not be undertaken without a good expectation of success. Random street searches or routine searches of travellers may tell traffickers nothing of great importance, but their value must be balanced against the delay and nuisance they cause to lawful travellers, and thereby their tendency to alienate respectable members of the public, who otherwise might be prepared to assist the police.

Those taking part in a search of premises, should have a clear understanding of what they are looking for, how they hope to find it, and what will be achieved by its success. The technique of conducting searches is discussed in paragraphs 10-19 of Part VII of this guide.

Under-cover Agents

Police officers or other trustworthy men (or women), specially employed on a regular basis for the purpose, can be sent to a village or town to live there in suitable disguise for periods long enough to win the trust of traffickers and gradually to acquire information about the organization and methods of their business. But it is vital for the safety and continued usefulness of such officers that their connections with the police be thoroughly concealed. They should, therefore, never be used on overt operations or appear in police company, unless they are never again to be employed under cover.

High-grade Information

Although a great deal of information can be cheaply obtained in the various ways already suggested, it usually becomes necessary when faced with organized trafficking to supplement them by a systematic purchase of information not readily obtainable by other means. This is necessary because the offences connected with trafficking are almost all committed in secret or without independent witnesses; the police are, therefore, almost compelled to develop their sources of information by covert means over a period of time, and in the meantime may have to lull offenders into a false sense of security. Since, however, people regularly trafficking in drugs are virtually forced to trust other people, it is always possible that one of them may prove willing to work for the police, and it is this which often provides the police with their best openings. Great care, however, must be taken to ensure that what is being bought is, in fact, good information, and not merely drugs smuggled simply for sale to the police. The police must always be careful not to create their own market.

The organisation of a net-work of high-grade agents is discussed in Part VI of this guide. Here it is only necessary to say that a reliable agent working on instructions is a valuable possession; his identity as a police agent must be scrupulously protected, and he himself should be handled firmly, dealt with fairly, and paid to a definite and understood plan. Any convictions obtained in a court of law as an ultimate result of his information are of no direct interest to him, and his payments should not depend upon them. He should be paid cash, if that is the arrangement on delivery of information. The sanctions available to the police, if he wilfully gives false information, are to stop employing him or to expose him as a police informer. The payment of salaries, bonuses, or casual sums of money in exchange for information should be referred to as "payments to informers", and not as "rewards".

Other Sources

Information may also be obtained by means of the specialised techniques discussed in Part VII of this guide. Their proper use, however, requires the employment of specially trained staff.

THE CENTRALIZATION OF INFORMATION

It is to be expected that police officers infrequently engaged in narcotics cases will seldom know what information it is desirable to record in connection with an arrest or seizure. Useful details may accordingly be lost, either because questions are not asked or because information is not recorded. This may be avoided by the use of a standardized proforma on which to report all arrests and seizures connected with the illicit traffic in narcotic drugs. A model which can be adapted to local requirements is reproduced at Appendix A of this guide. It will also serve as the basis for completing a report to the International Criminal Police Organization (Interpol) and to the United Nations' Division of Narcotic Drugs (see Part VIII below). It can be used to report the loss or theft from lawful custody of drugs liable to abuse.

It should be a rule, however, that cases which at first sight look important or complicated should be handled wherever possible by specialist narcotics officers. Similarly, any worth-while information about the manufacture or trafficking of narcotic drugs should be reported at once to a specialist officer, and not be acted upon by a non-specialist officer without instructions. This is especially important in regard, for instance, to heroin laboratories, where premature or unskilful action will almost certainly lead to the loss of useful evidence, and possibly to total failure. In other cases the local police may make an arrest or carry out a search which was justified on the information available to them but which damages a long under-cover investigation being made by specialist officers into a more important case.

The Recording of Information

It has already been started in Part II of this guide, that it is desirable to centralize at the headquarters of a police force all the information available about the illicit traffic in narcotic drugs. It is also desirable to do this on a national basis where a country has more than one law enforcement agency dealing with narcotic drugs. Certainly in any state with more than a very minor problem of narcotics trafficking or addiction, it is necessary to collect and collate all information on the subject into one office for measuring and assessing, the problem in a rational and economical way, and to give the government accurate reports on which to base its policy.

The mass of information necessarily accumulated in such an office may, in itself, present difficulties in handling. To arrange it systematically, to be sure that all relevant information has been placed on record, and to be able to find any particular piece of information again, requires efficient systems of filing, indexing, and retrieving.

Filing is the ordinary office procedure whereby papers are placed in files or folders, so that they are not lost and are grouped appropriately according to their subject matter. A full discussion of the proper management of an office registry and its filing system cannot be undertaken here, but certain basic rules may be mentioned, for if they are not rigorously enforced, a filing system very soon falls into chaos. They are as follows :—

(i) all incoming papers must first go to the registry to be registered, placed in the right file and serially numbered, before being passed to the appropriate officer for action;

(ii) no un-registered papers or personal copies of papers may be permitted as working papers in the office;

(iii) the opening of new files must be solely the fixed responsibility of named persons, e.g. the head of the registry in agreement with the deputy head of the office;

(iv) the titles of files should be carefully considered so that they define exactly what papers must go into each. They should be neither so wide that their contents become almost miscellaneous nor so narrow that they contain almost nothing. A file title may be as long as is necessary to define its contents precisely. A title such as "Smuggling" is valueless;

(v) files should be grouped and numbered in accordance with a definite system which allows related subjects to be grouped together in neighbouring files and for any subject to be expanded or sub-divided as becomes necessary, while at the same time the file numbers show their relationship to one another. In a police narcotics office the main groups of files might consist of:—

(a) policy files, containing material relating to the implementation of the objectives of the office, whether it has only regional or national responsibilities. They would include records of discussions, recommendations and decisions regarding government, police, or office action; standing working instructions; or the plans of actual or proposed operations;

(b) subject files, containing material dealing with smuggling, illicit trafficking in or manufacture of narcotic drugs, etc.; reports about particular organizations or situations; and other impersonal subjects;

(c) personal files, containing all available information about an individual, who is of sufficient interest to the office to warrant a separate file;

(d) investigation files, which contain the reports and statements of witnesses regarding a particular offence for which someone is to be prosecuted (and which, for that reason, may be allowed to leave the office); and

(e) files containing reports of arrests and seizures obtained from outside offices.

As an example, a part of the group of subject files might be as follows :

SF 3	Smuggling of narcotic drugs into the country (SF stands for Subject File);
SF 3/1	Smuggling of drugs into the country by land;
SF 3/2	Smuggling of drugs into the country by sea; and
SF 3/3	Smuggling of drugs into the country by air.

SF 3 itself would only contain periodical summaries and appreciations of the information contained in the subsidiary files in the group. Each of the other files may be sub-divided in turn when it gets too cumbersome to use conveniently. Its contents could be divided, for instance, into smuggling over a particular named frontier, or of particular named drugs, or by particular named nationalities. Whichever method is chosen, it should be applied consistently to all groups.

Indexing

Indexing is a method, by which any identified person or object, and further information regarding it and its connections, may be found easily and rapidly among the records of an office. It normally involves the making of a card or other simple record which :

(i) identifies the subject indexed;

(ii) records as much information as conveniently possible about the subject; and

(iii) indicates where further information can be found.

Indexing is a continuous process. It is essential that all information should be indexed as soon as possible after its arrival in the office; otherwise it may be overlooked or lost. Care, and if necessary further and special inquiries, must be taken to keep the information accurate and up-to-date.

The maintenance of four indices should be fully adequate to record the details of the illicit traffic in drugs. Of primary importance is a *General Nominal Index* of persons known, or suspected to be engaged in the illicit drugs traffic. Details of all convictions of those known to use drugs should be included, as well as those for narcotics offences. Dates of release of those in prison should be noted. The names of known drug addicts, even though they may never have actually trafficked, may also be included, as may the names of persons engaged in the legitimate trade in drugs. This index does not, of course, replace in any way the records of convicted persons maintained in the main criminal record office; it should, indeed, include cross-references to these dossiers.

The order in which index cards are kept, must be the subject of well thought-out and comprehensive instructions. Generally, it will be found most convenient to maintain a single index, arranged in alphabetical order of surnames, since there is a greater variety of surnames than of first names, but a decision must be made in the light of national practice. It may be thought invidious to group the names of persons engaged in legitimate trades or professions with those of addicts and criminals, but a nominal index of this type is not an official public record, and it will be found simpler to trace a previous mention of a name if the cards are all kept together and arranged in a single order; the identifying details on the card will, at once, show the reason for its existence. Distinctions of occupation or nationality may, however, be shown quite simply by the use of different coloured cards, although it should be appreciated that people may change their occupation or have more than one nationality.

(i) *A Premises Index* should contain details of all premises, addresses, etc., which have come to notice in connection with the illicit traffic. To this may be added, if desired, the addresses of all premises, where drugs under any form of control are lawfully manufactured, processed, stored, or dispensed. It will be a matter for decision how these cards are arranged, but usually divisions into provinces, districts, towns, and then streets should prove satisfactory. The address itself, e.g., 14, Republican Road, should be indexed rather than the name of the building or the company using the premises. Again, coloured cards may be used to differentiate between the uses to which the establishment is put.

(ii) *A Vehicles Index* may record details of all identifiable vehicles used or suspected of having been used by land, sea, or air in the illicit drug traffic, or by anyone known or suspected to be engaged in the illicit traffic. It might be most convenient to arrange these cards simply by reference to the licence numbers of the vehicles, but it is also possible, and perhaps preferable to divide the index into sections, firstly according to type of vehicle, secondly by make, and only thirdly by registration number. Registration of licence numbers can be changed more easily than the type or make of a vehicle.

(iii) *A Subject Index* may be added if desired. This may make it easier to retrieve information about a particular activity or locality, e.g., suspected heroin-making by a particular group of persons, or in a particular locality, or smuggling by a particular airline or route. Usually, however, a good filing system, coupled with proper extracting (see below), should make a subject index unnecessary.

(iv) *Extracting* is a process, wherby information recorded in a report placed in one file, is copied into another file to ensure that the latter file contains all the information available in the office about its own subject matter. For instance, a seizure report may mention a man on whom there is a personal file. It will then be necessary to copy, or to "extract" this piece of information from the original report to the personal file. In such cases, it is important that the copy indicates that it is a copy, where the original material has been filed, and the authority for its contents. This can be done by endorsing on every extract "Copied from report No............................ dated..........written by.....................Original report filed at...........". Similarly, the original report should bear a note to show that the whole or a part of it, has been extracted to the other file.

Registers

It will, however, often be very desirable to maintain certain other information in an easily accessible form. A *register of stolen drugs* will show conveniently the incidence of thefts and the methods used, indicate where preventive action is necessary, and possibly help to identify drugs seized in the illicit traffic.

A *register of arrested persons* and a *register of seized drugs* may be used to facilitate the keeping of statistics regarding the law enforcement effort. Their entries can be simply cross-referenced to the files containing the reports of arrests and seizures. Alternatively, this information may be readily available as a part of the normal criminal statistics of the country.

Passport-sized photographs of an individual, taken full-face and side-view, can be kept in a *register of photographs*, although they should be duplicated in his personal file. This can be a simple album. Each photograph should only be identified on its face by its register serial number; on its back should appear the subject's name (to correspond with that on his Nominal Index Card) and the date of the photograph.

It may be thought desirable to maintain a *"Black List"*. This can be used in several ways, but its basic purpose is to instruct officers to be on the watch for the persons named in it, and to take certain specified action when they are met. For instance, the Customs might subject a person on the list to thorough search and the Immigration report his movement when he passes an international frontier.

A file of *newspaper cuttings* of articles and reports relevant to drug abuse and trafficking is useful. It will show the extent of public interest in and knowledge of the subject, the attitude of the public towards drug abuse and the law enforcement authorities, and perhaps be a source of information which can be used to assist an investigation.

Registry Duties

A method by which the duties of a registry may be arranged is as follows:

Head of Registry	:	Organization of Registry, Control of Registry Staff, and **Control of all aspects of policy regarding files and indexing.**

Incoming Mail Clerk	:	Opening of all letters, Checking of contents, and signing of receipts where used marking of letters to show subject matter and appropriate file maintenance of incoming mail register.
Index Clerk	:	Marking of incoming letters to record names already on index , Indexing of fresh names , Amendment and up-dating of indices.
Files Clerk	:	Placing of incoming letters on file and attaching of connected files, opening of new files on instructions, extracting recording of internal movements of files.
Despatch Clerk	:	Preparing of letters for despatch Register of out-going mail.
Examination Clerk	:	Inspection of all files prior to return to cabinets to ensure that all necessary action has been taken. Periodic examination of files to ensure that those complete or out-of-date are either put away in a store or destroyed.
Secret Files Clerk	:	Custody of files recording identities of informers and agents, and details of secret operations.

The subject of the management of records has been dealt with at some length because the entire law enforcement effort against the illicit traffic in a country as a whole, and indeed, as will be argued later, in a whole region, depends upon the quality of the work done in the registry of a central narcotics office. The purpose of bringing together into one office information about the illicit traffic gathered from all sources throughout a country is defeated if within that office itself the information is not put together correctly and made readily available for use.

52

THE CONTROL OF LAWFUL SUPPLIES OF NARCOTIC DRUGS

As has been stated already, narcotic drugs and psychotropic substances enter the illicit traffic in one of two ways; either they are produced unlawfully or they are diverted at some stage from a lawful channel. To prevent the illegal cultivation of crops or the diversion of part of a lawful harvest is a complex and difficult task; the methods used in the production area range from elaborate licensing systems of lawful cultivation and careful surveillance of all those engaged in harvesting, to the establishment of intelligence-gathering networks in and around the areas concerned, and the detection—sometimes by aerial survey—and destruction of illicit and wild growth. Once natural products enter an officially recognized warehouse, they are, like any other manufactured drug. The problems of securing proper control over the physical safety of both natural and synthetic drugs are similar. To secure adequate standards is a responsibility of the management and a matter of industrial organization, supervised and assisted perhaps by Government experts.

It is, a valid interest of the police to ensure that the processes of manufacture and distribution minimize the risks of diversions into the illicit traffic, as prevent increase in drug addiction or abuse. They will, therefore, be concerned with the physical security of drugs kept in factories, warehouses, pharmacies, hospitals, clinics and retail shops, and with the honesty of those who handle drugs, either in the course of their work or as patients. Forged prescriptions, fictitious requisitions, false books of account, unlawful sales, may all be used to evade measures of control; carelessness in stock or record keeping may facilitate losses or misappropriations; thefts may be committed by people from within or without the premises in which drugs are kept.

The supervision and regulation of the trade in narcotic and other dangerous drugs is generally a function of the public health authorities of a state, but in some countries the police may be required to assist in maintaining the system of inspection. While they cannot, take the place of trained pharmacists, the application of their own expertise can help prevent losses, and their work in this field gives them knowledge which is useful when they come to investigate actual offences against the law.

When they are responsible for doing this work, they should inspect stocks of drugs subject to control at frequent but irregular intervals, and certainly not less than twice a year. These inspections should be directed firstly to ensure that the drugs are kept in physically secure conditions so as to lessen the chances of theft, secondly to ascertaining that the stocks tally with the records, so as to detect any misappropriation of supplies, and thirdly to checking that the records of receipts and issues agree with each other so as to discover any falsification of the books. A further objective at the retail level is to detect any drugs regularly supplied to someone without proper reason; a physician, for instance, in countries where it is lawful to do so, may be deceived into prescribing or over-prescribing a regular supply of a drug to an addict, and the police by a separate investigation may find that the supply is, in fact, being abused.

Acetic Anhydride and Acetyl Chloride

It is appropriate to mention here the contribution which can be made to the suppression of illicit heroin making by the control of the two chemicals, acetic anhydride and acetyl chloride. Either of these is a necessary ingredient for making heroin. If, therefore, they can be entirely prevented from getting into the wrong hands, making heroin will be made impossible. Acetic anhydride is almost always used (about two and a half parts by weight needed for each part by weight of crude morphine base). as acetyl chloride is more unpleasant to handle, three or four times as expensive, and produces less heroin from a given quantity of morphine base.

Acetic anhydride is legitimately used extensively in the production of rayon, photographic film, synthetic glass, aircraft varnishes, colouring matter, pharmaceutical products, and plastics, as also in chemical laboratories for research and analysis. Acetyl chloride does not have the same wide industrial uses, but is much utilized in the preparation of medicines. In a country with a large manufacture or consumption of these chemicals, it is, therefore, impracticable to control their distribution and use by a system of licensing and stock records, especially as the amount diverted to the illicit narcotics traffic is a very small part of the whole. Governments, manufacturers and the police should, however, be aware of the danger of their use in the illicit production of heroin, and the police should make it known that they will pay for information leading to the seizure of diverted supplies.

A Guide for Law Enforcement Officers

The guide, one of a series produced by the United Nations Division of Narcotic Drugs on the abuse of narcotic drugs and psychotropic substances introduced the general duties of police officers, problems involved in combating illicit traffic in narcotic drugs and suggests, possible solutions. It assumes that the reader will already have a sound knowledge of basic police practice and organization.

For the sake of simplicity, the term "police officer" is used throughout for any law enforcement officer engaged in combating illicit traffic in narcotic drugs, whether he is designated police, gendarmerie, customs, coastguard, narcotics control bureau official, or other government officer or agent.

Characteristics of Drug Trafficking

Narcotic drugs and psychotropic substances can only enter the illicit traffic by illicit production or manufacture or by diversion at some stage from legitimate channels. Their subsequent movements in the illicit traffic may be international or limited to one country.

The "*natural*" drugs, because in the main they are produced only in certain parts of the world, are those which hitherto have been commonly subject to international smuggling. The plants from which they are produced are often grown clandestinely in remote mountain areas, in forest clearings, or are interplanted with other crops. The drugs are carried on foot or by animal transport through jungles and across deserts and mountain ranges; in motor vehicles, ships, launches and boats; by commercial air services and in private aircraft; through the post and among commercial consignments of produce and manufactured goods; and hidden in the luggage or on the persons of travellers.

Their derivatives, especially heroin and cocaine, produced almost entirely in illegal, clandestine laboratories, are also major commodities in the international illicit traffic. Both they and the natural drugs themselves are smuggled by amateurs, by people working on their own account, and by professional gangs.

There is an exceedingly large production of "synthetic" drugs in factories and laboratories not only for lawful domestic use but also for international commerce. Their manufacture is less difficult to control than the production of natural drugs. In practice, supplies are illicitly obtained by clandestine production, or by diversion from lawful trade channels through theft, fraud, the misuse of medical prescriptions and excessive prescribing of drugs by physicians, or simply for lack of controls over distribution. Offences are often on a large scale and committed to obtain drugs for re-sale.

In recent years there has been a great increase in the *smuggling* of drugs by amateurs, many of them young people. They are mainly concerned with cannabis (both the leaves and the resin) and hallucinogens, which they smuggle in comparatively small quantities at a time, principally for their own use or that of their friends, or to pay for their holidays abroad. In many urban areas the petty trafficking in drugs by young people constitutes a considerable problem in itself.

Much of the international illicit traffic, however, and especially that concerned with opium, heroin, cocaine, and parts of that concerned with cannabis, is highly organized and carried on in secrecy as commercial enterprises by international gangs specialising in this work. The various stages of the trade are carefully planned and executed so as to give the traffickers their best chance of success, and their profits are so high that all can be well paid. The greatest rewards, of course, are obtained by the organizers and financiers of the trade; the actual carriers of the drugs, usually recruited for individual tasks, may be paid only at modest sums.

The *criminal offences* committed by those engaged in illicit trafficking in narcotic drugs have their own peculiar characteristics. There is seldom a complainant (in the police sense of the word) to give notice that an offence has been committed; very often an examination of the scene of an offence will produce little or no useful evidence. The police or customs officer, therefore, usually has to catch an offender in physical possession of narcotic drugs, in circumstances which show their possession to be unlawful, if he is to obtain evidence to secure a conviction. The problem with which he is faced is to identify and to obtain evidence against those who organize and garner most profit by the illicit traffic. For only by eliminating them can we hope to reduce or end the traffic as a whole.

Finally, experience in many countries has shown that a major factor in the abuse of any drug is simply its availability. A natural development is that if any one drug becomes scarce or too expensive, both dealers and abusers will readily change to another. Sources of supply and the channels through which drugs pass also change in response to demand and the comparative availability of drugs. In general, traffic flows from an area or country where drugs are available and subject to comparatively lax controls to another where there is a demand. Supply and demand thus react upon each other, and both are affected and may be changed by the impact of law enforcement. It is, therefore, necessary for law enforcement authorities to be informed about possible developments regarding the sources of supply, the channels and methods of the illicit traffic, and the abuse of drugs; they also need to be alert to detect any actual changes at the earliest possible moment.

Drug Abusers and the Police

Sympathetic and knowledgeable cooperation among the police, physicians, social workers, the staff of judicial courts and addicts can do much to keep young people away from drugs, or to persuade them to abandon the habit of drug taking. In countries where the police are given some discretion, it will often be more constructive for them to exercise tolerance and understanding and seek to rescue a young addict from his predicament, rather than to enforce the full rigour of the law. This is more likely to be the case if the police officer is the first person in authority whom the young man or woman has met since becoming an addict. When this happens it may be essential that the police officer should be recognised by the addict as a person who understands drugs and addiction and is willing to help him. Such an attitude on the part of the police officer can be justified not only by its particular merits but also by the obligation placed upon him in these countries first to prevent, and only secondly to detect, criminal offences. In countries which have the Parquet system, discretion cannot be exercised by a police officer, but only by a judicial authority; nevertheless, the same considerations apply.

Society and the Police

Law enforcement will always remain vital to any campaign to reduce or eliminate drug addiction. But laws which seek to regulate the personal habits of private individuals are notoriously difficult to enforce, and their successful application often depends in large measure upon their general acceptance by the community. The police should give information and helpful advice to the public, attempt to win understanding and support for their action in the society in which they work. However, the difficulties often experienced in obtaining satisfactory evidence in drug cases, and the impossibility of applying effectively any system of routine checks and examinations will tempt the police into conduct, which may be regarded by the community at large as undue interference with the liberty of the citizen. It is, therefore, desirable for law enforcement in this field to be as discriminating and selective as possible.

In practical terms the job of the police is to reduce to a minimum the amounts of narcotic and other dangerous drugs illicitly available to members of the public. They cannot hope to eliminate completely the supply, but their aim should be to make drugs too scarce and expensive for common illegal use, and trafficking too dangerous to be a widespread occupation. In some countries they may also be able, through direct contact with addicts, to help them seek treatment; in this way they can help to reduce the demand for drugs, and so lesson their own problem.

Other elements necessary in a thorough-going effort to eliminate drug abuse, are the treatment and social rehabilitation of addicts, education of the public, or the prevention of addiction in the first place; success as a whole can only be achieved by concurrent success in each of the parts of the problem. The police should accordingly be familiar with the whole problem and with what is being done in each of its parts to solve it.

53

THE ORGANIZATION OF LAW ENFORCEMENT

Police forces are of several kinds. A force may have authority throughout its national territory or only in one part of it; it may take cognizance of all criminal offences or only of some particular kinds; the way in which it is organized will depend upon its own responsibilities and those of any other police force at work in the state. Custom services also vary considerably in the duties they are called upon to perform. These differences often result less from the ideas of what is the most efficient form of organization than from political or constitutional reasons. Nationally or locally organized police forces work according to the provisions of Law.

The problems facing police forces generally result from modern communications of all sorts, mobility of the criminals, international character of many crimes, and the increasing inter-dependence of states and their economies. Therefore, police and customs authorities should develop the specialized skills, to exchange information, and to act in concert with one another. This is especially true of the suppression of illicit trafficking in narcotic drugs. Much trafficking takes place over very long distances and across international frontiers, and law enforcement officers must be able to work effectively together over the same areas.

Law Enforcement Objectives

In many parts of the world the illicit supply of narcotic and other dangerous drugs has become a highly organized business, and there are plently of men and women (and sometimes children) prepared to act as couriers and carriers of drugs, or to peddle drugs on the streets of a city; the arrest and imprisonment of these numerous but minor offenders will do little by itself to lessen the illicit traffic, for they are often regarded by the persons for whom they work as expendable and easily replaceable. The objective of the law enforcement authorities should, therefore, be to seek to eliminate from the illicit traffic, the men who organize them and supply them with drugs. Sometimes the most effective means of doing this is to use the paddlers and minor offenders, who are more easily found and arrested, to obtain information about the dealers and wholesalers. Through them, in turn, it may be possible to identify and obtain evidence against the organizers (who are often themselves dealers). This is usually very difficult, but simply because these major traffickers are criminals working through associates whom they are compelled to trust, they are vulnerable to pressures which can be exercised by police forces acting systematically over a period of time and using professionally organized counter measures.

On the other hand, much of the illicit traffic in controlled drugs is comparable with minor crime. It consists of breakings into premises, petty thefts, frauds, small-scale smuggling and the sale of small quantities of drugs on street corners or public premises. Often the prevention of these offences is a matter of normal preventive work and routine checks, and their investigation—a matter of normal police inquiries in the locality. Their significance in the context of drug abuse may not become apparent, however, until they are examined as a whole. It may then be found that there is much more trafficking and abuse than had been imagined, or there are particular features in the problem which had not hitherto been appreciated.

Law Enforcement Requirements

Faced with a problem of drug trafficking or drug addiction, police officers will find it advantageous, to include action against drug traffickers as a part of normal policing. Others may find it necessary to engage wholly or partly on specialist drugs duties. Whatever the size of the problem, police force should centralize in its headquarters all information about it, including details of its own activities, such as the making of arrests and the seizures of drugs. It should also possess a command structure which provides a method of deciding and putting into action its policy towards drug addiction and trafficking, and a means of communicating and co-operating with other law enforcement organizations within the state.

Within any one country, however, there are almost certainly at least two authorities, viz., Police and Customs, concerned with law enforcement against illicit trafficking, and there may be several more. If they are to work together effectively, it is necessary for them all to centralize and share their information and to provide a means of co-ordinating their action at one single point.

In countries with a national police force it is usually more convenient for its headquarters to provide the necessary services, since it probably has the greater resources for information-gathering and action. One organization or another should, however, be made clearly responsible for this work. When this would create undue difficulty or promote undesired rivalries, it may be necessary to establish an independent, specialized government office or a joint bureau. A separate organization will nevertheless be handicapped if it is separated from the national criminal investigation department. Whatever the organization the national criminal records must be within reach of enforcing officers. The function of the central office should be, at the minimum, to collect full information of the problem of narcotic and other controlled drugs to co-ordinate their action against illicit trafficking, and to co-operate directly on law enforcement matters with similar organizations at the national headquarters of other states.

It is a matter for discussion whether additional powers should be given to a central narcotics office. Where there is a problem of any size it is probable that its staff will find it difficult to perform its duties to its own satisfaction merely by recording and analysing information supplied to it by other organizations. It will probably feel the need to develop its own sources of information, to pursue its own investigations, and to bring to fruition its own cases. It will, therefore, seek to obtain its own executive powers.

Need for a Specialist Department

If a central office is to be anything more than an office of record and a means of liaison between different authorities, its staff will need to become specialists in order to make full use of the information they receive, to undertake their own investigations and to direct others in law enforcement, and to develop the use of various techniques for their own purposes. The bigger and more complex the problems of drug abuse and drug trafficking, the greater is the need for a specialist office of this type.

In countries with a centralized national police force exercising general police powers it is normal for this specialized ability to be developed at its headquarters, usually as a part of the criminal investigation department. There will then be no problem in regard to executive functions, since its officers will have the normal powers of a police officer. Elsewhere it may be necessary to make special arrangements to confer the necessary powers upon officers.

The methods of work and techniques which can usefully be adapted by a specialist department are set out elsewhere in this body.

Co-ordination of Government Action Against Drug Abuse

A central office doing the work described above will satisfactorily co-ordinate the efforts against the illicit traffic of all enforcement authorities in the state. Law enforcement by itself, however, cannot prevent drug abuse. Other measures are equally necessary. The addicts need to be treated and rehabilitated; the public, and especially young people, have to be educated to the dangers of drug abuse; the lawful supply of drugs has to be limited to the quantities required for medical and scientific purposes; the characteristics of drug abuse have to be analysed and understood.

Government policies in these fields require to be co-ordinated to become effective. An inter-ministerial or inter-departmental committee will often provide a suitable means. The specialist narcotics law enforcement office will provide. The information required by the committee. In turn, this has an advantage for the police. They will gain a fuller understanding of the whole problem of drug abuse, and become familiar with workers dealing with its various aspects. Physicians, psychiatrists, pharmacologists, magistrates, law-makers, social workers, prison officers, and possibly men with other qualifications as well as law enforcement officers, may all be involved in the formation and execution of a complete policy, and they will work better for understanding each others' attitudes.

Qualities Required of Specialist Officers

The need to centralize information and to co-ordinate action among law enforcement authorities implies the need for them to trust each other. A police organization which is itself honest and competent, and requires help from another one, must give it information, or request action by it, on the assumption that it too is honest and competent. Often the mere fact that the assumption is clearly implied will be enough to ensure that it is justified. In any case, to refuse to supply information for fear that it will be misused is to fail before starting.

Where there is a problem of any complexity, the police officer must be able not only skilfully to investigate cases but also patiently study the criminals themselves, their organizations, and their methods of work. This requires detailed, specialist knowledge, systematic collection and analysis of information, and the application of professionally organized skill; the more expert and highly organized the criminals are themselves, the more necessary it is for the police and other law enforcement authorities to be themselves professional specialists. Ideally, the narcotics specialists should be of first-class quality, have had a good basic police training, and possess a thorough knowledge of, and experience in, detective work. Only after he has acquired this background and to developed additional skills and knowledge will it be possible for him to combat smuggling and trafficking of narcotic drugs. Furthermore, he needs to be properly organized, adequately equipped, politically supported, and sure of a quick and reliable judicial process.

Police Action in a Rural District

In a rural district without serious problems of trafficking or addiction the local police unit will be able to deal satisfactorily with the situation. Where there is organized trafficking in drugs, it may be necessary to appoint narcotics officers with specialist knowledge who can devote time and use techniques to develop information, and co-ordinate action in their own and neighbouring districts.

The problems of illicit trafficking which many police units in rural districts may face vary enormously; it is, therefore, dangerous to attempt to indicate the general lines upon which they should act. In the following paragraphs, however, a sketch is given of what it might be appropriate to do in a rural district where there is smuggling of narcotic and other dangerous drugs on a scale large enough to warrant action. It is assumed that the commander of the local district police unit is faced with one of two kinds of trafficking, or both :—

(i) local trafficking, intended to supply consumers living within his own area of responsibility; or

(ii) transit trafficking, which is merely passing through his area on its way to somewhere else.

He should be equally concerned to combat both types.

In order to do this he and all his subordinate commanders, each in respect of his own area, should be thoroughly familiar with all the suspected traffickers, living or working in the district, their known methods of work and means of transport, the routes they use and the places they frequent, and their friends and associates. A system of patrol routes, check-points, ambush-places etc., should be arranged to keep under observation all the places, where drug offences may be committed. While specifying all routes and places, it should be appreciated that the ones actually in use would be liable to be changed frequently.

Whatever other duties any police patrol may have, it is also empowered to seek information concerning the smuggling of drugs. Subordinate policemen should, therefore, be carefully and clearly instructed before leaving on a patrol, on how to conduct themselves, what information to try and obtain, what to look for and where, and what questions to ask and of whom. They should also be trained to be suspicious of any movements or events which appear to have no legitimate purpose. They should already know their lawful powers to use in given circumstances. On their return from patrol they should be thoroughly be briefed. If this produces nothing useful, fresh attempts should be made to obtain information in different ways or with different men.

Officers dealing with travellers should, in particular, be on the watch for strange patterns of behaviour, such as a peculiar time of the day or night to be at a frontier post, an unusual season of the year to be there, an unlikely destination, unduly bulky clothing, the presence of smells or plastic wrappings (to camouflage or suppress the smell of a natural drug), or the making of a surprising number of journeys. They should also look out for particular types of persons known to engage in drug smuggling, such as local tribesmen, or young people (not only those dressed as "hippies") and foreign workers travelling from countries where drugs are known to be comparatively easy to obtain.

These particular methods should be supplemented by normal routine checks and observations, and by the collection of the information described later.

District training programmes should, from time to time be used to provide information to subordinates about the illicit traffic and to give general instructions on how to collect and report information. It may be of advantage to hold regular monthly meetings of representatives of all departments in the district concerned with illicit trafficking, when information can be exchanged and reviewed (e.g., the prices of drugs on the black market, operating methods of traffickers, routes used, seizures and arrests made and the information thereby obtained, convictions recorded, information required, effects achieved in the past month, etc.), and when action in the coming month can be decided. The general aim will always be to get to understand how the illicit traffic operates in the district, and then to use appropriate methods against it, rather than to apply routine stereotyped measures.

Operational Records

In a police district headquarters it would be useful to maintain special records relating to the illicit traffic in drugs, as aids to preventive action and summaries of operations performed. They should be brought up to date daily with the latest relevant information obtained not only from within the district, but also from neighbouring units and from the headquarters of the police force. A simple record can be kept by using the following means :—

(i) **A route map.** Topographical map of the district,

covered with a sheet of talc or acetate paper, on which can be marked the locations of all known or suspected smugglers and traffickers, the routes they are known or suspected to use or which they might use, the places where they meet or rest or eat or sleep, and any other information about them it may be useful to see at a glance. Another colour pencil may be used to mark the police patrol routes, check-points, etc.

(ii) **An incident map.** This is another copy of the same map covered in the same way on which can be marked, using different colours for different types of offences, the places where incidents have occured. Each one should be marked with its date and a serial number. Depending upon how soon on an average these maps become congested with markings, the talc should be wiped after it has been in use for a stipulated number of months. If it is desired to preserve for a longer period the information recorded in this manner, a second pair of maps may be brought into use. When their period of use has been compeleted, it is possible to wipe and use again the first set.

(iii) **An incident diary.** This is simply a list of incidents, each with a serial number to correspond with that shown on the Incident Map, arranged in chronological order and recording brief details to supplement the information on the map.

(iv) **A suspect's file.** This should contain a page or a section for each known or suspected smuggler, trafficker, dealer or peddler who is or has been in the district, recording in diary from every piece of information which can be obtained about him, and giving for each entry, the source of the information and an assessment of its reliability.

Police Action in a Town

What has already been said about the scale of effort and size of organization required to combat illicit trafficking in a rural district applies equally to an urban district. The problem may be small enough to be dealt with adequately by general duties police. When drug trafficking has started in a town, however, it will probably become necessary at a comparatively early stage for officers to receive special training in the prevention and detection of drug offences, even if they do not become full time specialists. Later on, it may become necessary to appoint one or more officers to the sole task of combating trafficking and peddling, and even to give all members of the force in the town some training in the subject.

The problems may range from large-scale trafficking as part of some international operation to the distribution of "pep-pills" (amphetamine tablets) at a young people's party. The main work of the police in a town, however, will often be concerned with the retail end of the trade in illicit drugs and with diversions of pharmaceutical drugs from their lawful channels of supply. Such places as tea-houses, coffee-bars, discotheques, dance-halls and other places where addicts may congregate or drugs may be traded will repay attention, while the physical safety of drugs stored in pharmacies, dispensaries and doctor's clinics or surgeries ought also to be of concern to the police, Together information about the movements of the suspected traffickers and their customers, a network, similar to that described for a rural district, should be established. Uniformed men on beat duty, detectives on other duties, traffic and other specialist units, may all be briefed to find the answers to particular questions of a simple nature.

Where the amount of information obtained justifies the work, maps and files similar to those already described, may be maintained in urban police stations and district headquarters. It may be helpful to maintain in addition a file of premises where drugs have been discovered, so that relevant information is easily available when it becomes desirable to take action against owners or tenants.

Police Action in Areas Where Drug Plants are Grown

The conditions in which narcotic plants are grown vary tremendously from one country to another. The problem of control may be to detect and destroy illicit growth or to prevent diversion of the lawful harvest into illicit channels. Appropriate police action will depend upon local circumstances and the government's methods of controlling lawful production; it is not proposed to discuss the subject further here, except to say that police officers stationed in the growing areas should be familiar with local methods of control and the local population to ensure that there is no abuse of the system. Again, a local intelligence network will enable police resources to be used to the greatest effect.

General Customs Action

It is normal practice for a Customs Administration to have a preventive or investigation branch concerned with the detection and prevention of smuggling, but it is a fact that the speedy clearance through customs of a large flow of passengers, luggage and goods is not compatible with intensive routine checks and elaborate examinations. Since there is a world-wide tendency to liberalize frontier controls, smuggling can only be countered effectively by the application of known information to selected targets. It is possible, for instance, to examine with some care passengers, luggage and freight arriving from regions known to be sources of narcotics smuggling; or to pay special attention to goods arriving from suspect consigners, or addressed to suspect consignees; or to explore more fully the circumstances of a shipment which appears unusual.

Training

Enough has already been said to make it clear that all law enforcement officers likely to come into contact with drug abusers or traffickers should be given basic training on the subject of narcotic drugs. As a minimum, information should be provided to enable them to:

(i) recognize the commoner drugs of abuse by inspection;

(ii) carry out simple chemical tests to establish preliminary identification where physical inspection is inadequate;

(iii) know the more usual methods of administering drugs of abuse and their more usual effects;

(iv) recognize the common symptoms of a person under the influence of a drug; and

(v) be familiar with the general characteristics of the illicit traffic in drugs in their area of jurisdiction.

Information and training of a more detailed kind can be included in courses for detective and more senior officers, and appropriate lectures inserted into all refresher training. Police bulletins or other publications should include descriptions of interesting or educational cases, summaries and analyses of work against the illicit traffic, and notices regarding new drugs or methods of supply for which police should become alert. It may be particularly valuable to bring to notice methods known to have been used (not only nationally but also in other parts of the world) to hide drugs in commercial goods or vehicles and new routes which may be developing. Any conclusions reached by research or intelligence sections after study and analysis of reports received both locally and from abroad should be disseminated not only to the executive branches concerned but also to the training establishments of all authorities concerned to combat the illicit traffic. It cannot be too strongly emphasized that although different authorities in a country may have different executive functions in regard to the illicit traffic in narcotic and other dangerous drugs, they all have the same need for information to help train their staffs and perform their duties.

54

ORGANIZATION OF INTELLIGENCE GATHERING

A country with a specific drug problem needs a specialized department which acts as a law enforcer, has specialist techniques, and gathers information.

The centralized co-ordination of information will check improper use of powers.

Intelligence gathering may be defined as deliberately setting out to collect information from a variety of sources, both open and covert, so that it is as systematic and complete as possible; of putting it together so that it forms a picture; of interpreting the meaning of that picture; and finally of putting it to use, so that the executive authorities have a clear understanding of their problems, and can plan and conduct their operations with knowledge and, indeed, intelligence. Intelligence, is a collection of digested information necessary or desirable to have before deciding a course of action.

A Central Narcotics Intelligence Bureau

For the sake of simplicity it is convenient to refer to an office doing this work as a Central Narcotics Intelligence Bureau (CNIB). Its purpose should be :

(i) to provide a central pool of information about illicit trafficking on which the government can base its policy;

(ii) to collect itself, and to help other enforcement units to collect, good quality intelligence which can then be used against the illicit traffic, both at national and international level;

(iii) to act as a clearing house for information both between the various law enforcement agencies within the country and between them and the appropriate foreign and international authorities; and

(iv) possibly to carry out investigation of criminal cases and collect the evidence necessary to prepare criminal prosecutions.

Subjects for Intelligence

The subjects to be covered by intelligence on illicit trafficking should at least be on a countrywide scale, and should also include those movements in the international trade, both licit and illicit, which may affect the country concerned. They can be grouped under :

(i) the sources, both persons and places, from which illicit supplies originate;

(ii) the routes along which these drugs travel, and the people who handle them;

(iii) the places where drugs are stored, bulk supplies are divided into smaller quantities, e.g., heroin; are manufactured, and the people who organize and carry out this part of the trade;

(iv) the identification of consumers and addicts and the means and places used for the retail distribution of illicit drugs;

(v) any means of communication, transport or manufacture which has been or will probably be used for illicit drugs;

(vi) the protection of the licit trade from theft and abuse; and

(vii) the assessment of the effectiveness of the measures taken by the government as a whole against illicit trafficking.

The first step in gathering intelligence should be to receive all information about the seizure of drugs and the arrest of drug offenders. This can most easily be achieved by the use of the standardized reporting form already described in Part V of this guide, one copy of which in each case should be sent to the CNIB. Secondly, all information of more than minor and purely local significance obtained by the processes described in Part IV above should be reported to the CNIB. Further possible sources of information, obtainable by more specialised means, are described in Part VII of this guide.

The collection and distribution of good quality intelligence and its exploitation can be achieved if a CNIB :

(i) distributes the funds for other or subordinate enforcement agencies to establish regular and systematic intelligence networks;

(ii) provides the channel through which they exchange operational intelligence and so co-ordinate their under-cover investigations;

(iii) creates its own high-class intelligence network, aimed at the major and difficult targets of the illicit traffic;

(iv) possibly establishes its own branch offices in provinces where the illicit traffic is heavy in order to facilitate intelligence gathering and the co-ordination of action;

(v) has available specialist officers ready to help when desirable in the actual investigation of difficult or lengthy cases; and

(vi) distributes useful information to other or subordinate enforcement agencies. This may be done by weekly, monthly or quarterly summaries of cases reported and information received; by regular appreciation of the state of the illicit traffic; by newsletters to promote interest and expertise; or by selected information supplied to help a particular investigation.

Necessary Sections in an Intelligence Bureau

How precisely the above functions are carried out must depend upon the organization of the government's administrative machinery. In order to carry them out, however, such a central office would require :

(i) a registry, where the storage, indexing, and retrieval of information can be quickly and competently done;

(ii) a production unit, concerned with the active seeking of fresh intelligence by specialist means described later, the control of the specialist staff required for this purpose, and the custody of the working files of agents and informers;

(iii) desk officers to study and assess the information coming in, each specializing in a particular aspect of the work, to provide detailed facts and guidance for the benefit of senior staff, to prepare statistics, summaries and appreciations as required, and to record details of cases, prosecutions and the results of trials; and

(iv) investigation officers, available to carry out investigations and overt operations in the field, and to liaise with other police units, so that the latter are kept fully informed of the latest developments, are trained in acceptable techniques, and are warned of anything particular to be on the watch for.

The Organization of Intelligence Agents

In this guide the term "agent" is used not to denote a type of law enforcement official, but a person who regularly supplies information in accordance with instructions given to him. It is useful to distinguish between such a person and an "informer", who volunteers information on a more or less casual and infrequent basis. In a country of any size, an intelligence network will probably be more conveniently arranged in two tiers. In the lower (but not less important) one, there would be agents whose knowledge of the illicit traffic is essentially local. They would be controlled at local level, for example, by district commanders of police as set out in Part II of this guide, and their information would, in the main, be used to combat local trafficking. They should, indeed, be confined strictly to local affairs, and as soon as they begin to obtain information of wider import or appear likely to be able to do so, the CNIB should be informed and asked for instructions. In any case, the CNIB would need to know and keep a record of their identities and targets, how much they were being paid and how useful they were proving, but probably nothing more.

On the higher level there would be agents whose duty it would be to provide information about the more important and professional organizers and financiers of the illicit traffic, and they and their information require long-term development, skilful handling, and central direction and control.

Central control is necessary in order to :—

(i) decide on which aspects or parts of the illicit traffic it is necessary to seek information by secret means;

(ii) decide what organizations or persons or places can be "penetrated" by an agent in order to obtain this information;

(iii) decide where there are the best opportunities to obtain this "penetration";

(iv) recruit and control the agents intended to gain access to the expected information so as to avoid duplication or mis-direction of effort, buying the same information twice, and errors in assessing the worth or accuracy of information; and

(v) ensure that agents are handled in secure conditions, and their information used in such a way that their identities and future safety and usefulness are protected.

Careful attention to these procedures will guard against a dissipation of effort, and lead to a concentration upon what is really necessary and to the safety of secret operations. The more care that is taken with these matters, the easier it becomes to recruit, handle and retain useful agents; conversely, nothing ruins an intelligence agency more quickly than a reputation for carelessness in preserving the secrecy of its covert sources of information.

An officer operating at a local level cannot see these requirements as a whole, and, therefore, cannot know what is needed to meet them. If decisions are to be made intelligently and applied over a countrywide area in a coherent manner, there must be one authority which can state the requirements, see that they are met, and avoid both duplication and gaps.

The Handling of Informers and Agents

It has already been stated in Part IV of this guide that many different persons may be willing to assist the police with information, for many different reasons. The person who is most likely to possess valuable information about criminals, or to be able to get into a position where he can obtain it, is, however, usually a confidant of criminals, and probably one himself; his motives for becoming a police agent or informer may be simple but are probably complex. If full and correct use is to be made of him, it will be essential to find out what they are, for as a criminal himself he will be basically opposed to the forces of the law. Among those more commonly encountered are :

(i) fear of the law : he may hope to receive lenient treatment for his own offence in exchange for information;

(ii) fear of other criminals : he may have offended his associates or rivals and seek protection from them;

(iii) revenge : he may think himself unfairly treated by his associates and try to make trouble for them;

(iv) animosity : he may wish to eliminate a rival, for personal or business reasons;

(v) egotism : he may wish to be well-thought-of or to feel self-important;

(vi) financial gain : he may just want money; and

(vii) repentance : he may simply want to reform and lead an honest life.

Whatever his original motives, a well-handled informer may come to develop a genuine sense of loyalty to his handler, and although he will naturally expect to be paid for his work, he will do what he is told and genuinely seek the information he has been asked for. An officer handling an agent should, therefore, seek to develop this mutual confidence. An observance of the following principles will help this process :

(i) The officer should arrange at an early stage for a thorough background check to be made of the informer, so that he can assess why he has given information and what more he may be able to obtain.

(ii) He should always be seen to be the person in charge. He must say what he wants and see that he gets it; he must not allow the informer to control the relationship.

(iii) He should be scrupulously fair and polite, but not over-friendly. He must not adopt a patronising or contemptuous attitude, or address his informer in derogatory terms. He should never make a promise unless he knows he can keep it, and having made one he must keep it. He must be very careful to protect his identity as a police informer and must be punctilious in seeing that the informer, too maintains proper security precautions. Arrangements for meetings must be properly worked out and punctually observed. The basis for paying for information must be agreed and then adhered to.

(iv) He must de-brief the informer thoroughly at the first and all later meetings. While a fully-trained agent under good control may supply without prompting all the information the handler wishes to obtain, usually it is only systematic probing by the handler which will produce all that is available. Sometimes, indeed, the informer may not realize what is important in his information, and normally the handler will not wish to give away the extent of his own information.

(v) As soon as possible the handling officer should check his own information with other available facts, and then report it fully and accurately to his superiors, so that it remains on record. It will thus be available for future use in assessing the value of the agent in developing fresh cases.

Payments for Information

There are two main methods of paying for information. The first is that of payment by results: a man says he knows where some drugs are to be found, then or later; as a result they are found by the police, and the informer gets paid for his information. This is a very common method and is often perfectly acceptable. If, however, a man makes a practice of selling information to the police, the suspicion must arise that he may be stimulating the illicit traffic in order to make an income from informing, since the more cases he can "discover", the greater his rewards. When, therefore, an informer has proved his continuing usefulness, it may be better to employ the second method of paying for information.

The casual informer then becomes a regular agent, and is paid a regular monthly salary linked with a system of bonuses for especially valuable information. In this way the temptation to the agent to instigate offences is lessened, while the police can ensure that the agent keeps working satisfactorily by varying his bonuses or threatening to stop paying him altogether.

It may sometimes be desirable to test the truth of a statement before paying for it, but the general rule should be that information is paid for promptly and well. It is, however, necessary to realize that while it is permissible to pay for information leading to a seizure of drugs, this is less valuable than information leading to arrests and convictions. As in the case of rewards to police officers, payments for information should not be assessed solely in relation to the quantity of drugs seized or the number of arrests made, but to the importance of the case and the value of the information. In any case, whether payments are made only in exchange for definite information or on the basis of a salary and bonuses, they should certainly not be more than the local black-market price of the drugs seized; otherwise, an agent may think it beneficial to organise supplies of drugs simply for the purpose of, in effect, selling them to the police.

Agents' Files

It is a matter of opinion what record should be maintained of isolated pieces of information contributed by casual informers, although ideally the identity of every supplier of information should be recorded together with details of his information, and each such report should be linked to the record of any further investigation and/or court proceedings. Certainly, however, once an informer is working regularly, the production unit in the CNIB (*see* paragraph 8 (ii) above of this Part) should maintain for him two work files, one containing details of his identity, handling arrangements and payments, and the other copies of all his reports. Both should undoubtedly be protected carefully from improper disclosure.

"Traps" and "Agents Provocateurs"

It is perhaps advisable to end this Part with a warning. When using informers, agents, or under-cover officers, the police must be very careful that they do not themselves commit, or aid or abet the commission of, an offence. It is, therefore, necessary for them to recognise clearly a legitimate "trap". A legitimate "trap" is when the police have good reason to believe that the person whom it is intended to trap is going to commit the offence anyway, first-hand evidence can be obtained in no other way. An agent provocateur, on the other hand, is a person who provokes or persuades someone else to commit an offence which he would not have done but for the persuasion.

The dividing line between what is lawful and unlawful in this respect differs from country to country, mainly in accordance with the case law of each national jurisprudence. Police officers engaged in drugs duties, therefore, need to have a clear understanding of their own law on this subject.

Surveillance

Surveillance is indispensable for combating secret, organized crime, and an almost essential preliminary step for the successful search of premises used for drug trafficking. Without it, a search is no more than a chance with it, success can be made almost certain.

Surveillance may be defined as the secret and continuous watching of places, persons or objects, in order to gain information about the activities or identities of individuals. It may be static, as when a building is kept under observation, or moving, as when a person is followed on foot or in a vehicle; and may range from one man keeping simple watch by himself for a few hours to an elaborate operation equipped with its own premises, cameras, vehicles, and radio transmitter/receivers.

The objective of surveillance is:

(i) to ascertain the habits of a suspected person;

(ii) to establish a suspect's connection with certain premises;

(iii) to identify a suspect, or his associates, or visitors to certain premises;

(iv) to check on deliveries of goods or materials to certain places or persons;

(v) to confirm information already obtained by other means;

(vi) to make a search or arrest at a time when an offence is actually being committed; or

(vii) to obtain evidence for use in a court of law.

To be productive, surveillance requires planning and preparation. Among the matters to which attention should be directed are the following:

(i) the object of the operation;

(ii) the collection of all information already available which may be useful;

(iii) a thorough reconnaissance of the area and premises, whether the locality is not already known;

(iv) the appearance of the watchers;

(v) their understanding of what they are to do in any probable circumstances, the tactics they are to adopt, the signals they are to use, and the communications they are to make;

(vi) their equipment with necessary items such as note-books and pencils; photographs of suspects they do not know; money and small change for meals, drinks, telephone calls, taxis, etc.; cigarettes, matches, and a newspaper or magazine to appear to engross themselves with when they may have to wait around; means of identification if they are questioned too closely by local police; and possibly a concealed camera;

(vii) arrangements for their relief and possible reinforcement;

(viii) the creation of a satisfactory cover story; and

(ix) details of command.

Since the suspects under surveillance are presumably engaged in criminal activities and, therefore, more conscious of their surroundings than the average law-abiding citizen, they will be difficult to follow and observe; full use must, therefore, be made of what may be limited opportunities. If surveillance is constantly used, the officers engaged on this work should not be employed on other duties as well, or their identities as police officers may become known.

Listening Devices

Listening devices are normally adaptations of commonly used electronic machines designed to eavesdrop without the subject being aware that he is being overheard. Their use requires considerable technical knowledge and in many countries is subject to legal restrictions, often based upon constitutional considerations.

Checks on Communications

It will be clear that the more highly organized professional traffickers become, the more they must depend upon using the normal methods of communication. Even though they may use all possible precautions to preserve the secrecy or conceal the real meaning of their meetings and messages, they inevitably become vulnerable to skilled and delicate investigations in this area. It, therefore, becomes profitable to trace the methods they use, and to search for such things as contacts, meeting places, letters, telegrams, telephone calls, hotel bookings, hand luggage, transfers of money, and notes in pocket books or on scraps of paper. If these things can be exploited without the knowledge of the traffickers or under cover of routine action by other agencies, e.g., a customs search at a port, they are likely to be much more useful. For some of these operations, it is often necessary to obtain a judicial warrant.

It should be noted that any of these specialised techniques may be used to confirm and strengthen information obtained by another one. For example, information from an agent that two suspects are in contact with each other may be confirmed by surveillance or a check on communications. Surveillance itself may be supplemented by information given by e.g. clerks at the reception desk of a hotel or waiters in a restaurant.

Searches

The successful search of premises is often the culmination of a long and difficult confidential investigation. On the other hand, it is not unknown for months of work to be ruined by a badly prepared or casually conducted search. It should, therefore, be considered as a deliberate operation, and only undertaken at the particular stage of an investigation, which is thought likely to produce the best results. Often this will be at the final stage, when all the various links in the organisation under investigation have been satisfactorily identified, but much will always depend upon circumstances and events outside the control of the police.

Speed, surprise, and completeness are the three essentials in a search of both premises and persons. It should be arranged to give the police the advantage of favourable circumstances and to obtain the maximum possible amount of evidence and information. In considering whether to make a search, the following questions should be debated:

(i) the objective of the search;

(ii) whether the information already available is sufficient to justify overt action;

(iii) whether the objective can be attained by any other means;

(iv) whether surveillance can be mounted without giving warning;

(v) whether there is time to plan the search;

(vi) whether the information already available is sufficient to justify applying for any necessary legal authority (this question is separate from that at (ii) above); and

(vii) what would be the likely consequences of a failure.

If it is decided to make a search, a plan should be carefully worked out and attention should be paid to:

(i) making a quiet and unseen approach to the premises from different directions;

(ii) making the search at a time when as many as possible of the gang, including its more important members, are both present and off their guard;

(iii) ensuring that the best ways of approaching and entering the premises are known and used, and that any dangers and obstacles in the way are understood;

(iv) making as sure as possible that the occupants have no opportunities to raise an alarm, escape, or hide or destroy evidence;

(v) the possibility of meeting armed resistance;

(vi) the need to make a thorough and complete search, including an examination for finger-prints, even though some evidence may be found at an early stage;

(vii) the possible need to carry out searches on other premises at the same time; and

(viii) making sure that all those taking part in the search know both the whole plan and their own parts in it.

The searchers themselves should know what they are looking for and be able to recognize as evidence what may not at first sight appear incriminating; they should be familiar with likely hiding places, both on persons and in premises. It is particularly necessary that specialist officers should search premises, known or suspected to contain a heroin-making laboratory, for what is clear evidence to them may mean little or nothing to other officers. All persons found on the scene should be searched at once to prevent their attempting to hide or destroy any evidence. Householders or persons renting individual rooms should be present to witness a search of their own property; persons already searched should be kept separate and out of touch with those still to be searched. Precautions should be taken to avoid any later accusations that drugs were "planted", and if possible, procedures devised to enable evidence to be produced if necessary to refute them. More than one officer should be able to give evidence of having seen a piece of evidence before it is touched or moved. It goes without saying that during a peaceful search no persons should be injured or property damaged.

When complicated structures have to be searched, it is usually desirable to obtain help from specialists with a particular knowledge of that type of structure. This has often proved useful in such comparatively simple cases as motor vehicles or modern blocks of flats; it is essential when dealing with ships or aircraft.

When searching a person it should be remembered that an unlawful drug may be a solid, powder or liquid and will probably take up little room. In deciding how thorough a search to make, an officer should consider how much time the suspect has probably had in which to conceal a drug and what his purpose was in doing so. A few cannabis cigarettes, for instance, are unlikely to be well hidden if the carrier is surprised by a raid on a private house; on the other hand a quantity of ISD being carried through a customs check may be almost impossible to detect.

(a) *Clothing* should be examined. Apart from the pockets, attention should be given to hat bands and ribbons, coat linings, shoulder pads, coat lapels, belts, trouser turn-ups, and the insides of socks and shoes. Wallets, purses, books, personal papers, handkerchiefs and so on should all be opened. Drugs may be concealed in cigarette lighters, packets of cigarettes, boxes of matches, lipstick tubes, fountain pens and torches. Bottles and other containers appearing to contain legitimate liquids or tablets may, in fact, contain unlawful drugs. Blotting paper or a piece of cloth may be soaked in LSD.

(b) *Luggage* is often used to convey drugs in false bottoms or sides, or in the lining. Toilet articles and cosmetic containers may contain drugs. Clothing and other things in luggage should also be examined.

(c) *Motor vehicles* used to smuggle substantial quantities of drugs often have a secret compartment built into them. Other hiding places could be the petrol tank, the spare tyre, behind the headlights, the seats, under the floor, behind the dashboard, in the doors, and in the roof.

(d) *Documents* found during a full-scale search should be closely examined, for they may provide suggestions for further investigation. Letters, envelopes, address-books, telephone numbers, and so on may all produce useful leads.

Technical Devices

There are many occasions when searches have to be made without elaborate preparation. The Customs authorities, for example, because of the enormous volume of international traffic and the general tendency of governments to liberalise frontier formalities, cannot fully apply a policy of rigorous checks and controls to all goods and passengers entering or leaving a country, while the detection of narcotic drugs is difficult because they take up little room and so can be easily concealed. Devices are, therefore, being developed to assist the authorities to pick out possible suspected persons, packages and vehicles.

The use of scenting dogs to smell out both opium and cannabis, infrared sensors allied to cameras to determine whether cavities which ought to be empty, in fact, contain nothing. Closed circuit TV cameras can be exploited in premises such as airport lounges to keep watch on peoples' movements, and so possibly help to identify furtive or other behaviour which may suggest drug smuggling, or now common place.

Physical Evidence

All physical evidence found during a search should be collected, whether or not it seems important at the time. It is necessary for the efficient prosecution of offenders in any subsequent trials that all pieces of evidence should be clearly labelled so that they can be identified again and not confused with other material, packed separately so that they cannot become contaminated, and fastened securely so that they cannot be tampered with or stolen. The report on the search should carefully list and describe what was found, including any packings and markings, and state exactly where and by whom each item was found; this should be supported if possible by plans, sketches or photographs. When a building has been searched, it should also include a list of all the rooms searched and of the people found on the premises. Care should be taken to preserve and record the chain of evidence so that witnesses can be made available to state what happened to any piece of physical evidence from the time it was found until the time of its production in court.

The identification of a narcotic drug or psychotropic substance for legal purposes should always be done by a technically qualified person in a chemical laboratory. An initial identification of a suspect substance, sufficient to justify further investigation, can often be made, however, by a law enforcement officer himself. Mere inspection of the substance may sometimes be enough, but usually a better result can be obtained by the use of various chemical re-agents. Their use depends upon the fact that a small mixture of a particular drug and a particular re-agent will always result in the development of a known colour; a series of tests will thus enable an officer to identify a substance. At present tests are available for the identification of opium, morphine, heroin and codeine; cannabis leaves and resin; cocaine; and barbiturates. It is possible to make up a supply of re-agents with the necessary simple equipment into portable boxes, while in some countries test kits are commercially available.

It is, however, desirable to give two warnings. Firstly, some of the re-agents are strong acids or are inflammable, and must, therefore, be handled with care and with due regard to the dangers involved in their use. Secondly, some practice is necessary to satisfactorily identify the colours produced by the mixtures of drug and re-agent. For these reasons, identification kits should only be issued to law enforcement officers who have been fully trained in their use.

Laboratory Analysis

The police can immensely assist themselves and the laboratory staff by asking the right questions when sending samples of suspected drugs for laboratory analysis and identification. To ask "what is this substance?" is far too wide an inquiry and might require much unnecessary work to answer. To ask "is this heroin?" might be too narrow a question, since one might be given the answer "no" without qualification, when in fact it was morphine. It is, therefore, good practice to say what the police think or suspect the substance to be and to ask whether or not it is a narcotic drug or psychotropic substance, and if so which one.

In addition to establishing the identity of a drug, laboratory analysis of seized drugs can be put to further uses. Two are of particular interest to police forces. As soon as possible after it comes into police hands, an analysis to determine the purity of a drug is a safeguard against its adulteration while in store; a later analysis, before destruction will clearly show whether it has, in fact, been adulterated since the first analysis. The same analysis is also a valuable guide to the availability of supplies, particularly of heroin, on the illicit market. The police will know the price of heroin in the streets of a town; chemical analysis will reveal to them its quality; these two factors regularly put together in terms of place and time will provide a reliable index of the effectiveness of the law enforcement effort. Changes in the index can very easily be shown by a graph indicating these two factors.

The other use to which laboratory analysis can be put, concerns identification of the origin of a drug. For instance, provided that a comprehensive range of authenticated samples (i.e. samples certified to have been grown in a particular named area) is available for reference, it is possible to analyse opium in a laboratory to determine broadly its area of origin. This possibility is perhaps of limited use to the police, since it may only confirm what is already known or presumed, but a more direct use of laboratory analysis can be made with manufactured drugs. A lawfully manufactured tablet or capsule, for example, can usually be identified through a series of comparison tests with known samples of manufacturers' drugs, while drugs illicitly made can often be compared with each other to determine whether or not they came from the same source of manufacture or whether they were adulterated in the same batch. Before asking a laboratory to undertake such work, the value to the police of the expected results should be carefully weighed against the work involved, which may be very high.

Police Diaries

Many police forces require all their officers to keep an official diary in a note-book carried on their persons. The pages are serially numbered and alterations or erasures may not be made. Notes are written at the time, events take place, and thus provide a contemporary record. This has two advantages: it provides what is likely to be a more accurate account of an officer's actions than a diary written later, and it may provide a good defence against false accusations against the officer's honesty.

An Incentive System

It is perhaps not wholly out of place to note in this Part that law enforcement officers, like anyone else, usually work better if their good work is noticed and rewarded. Conditions of work and types of staff vary considerably from country to country, so incentive schemes must also vary. They should be deliberately framed, however, so that they are, and remain, effective, and lead to the results required.

Low salaries and poor conditions of service and work are in themselves a disincentive, since a police officer will feel he has no motive to work hard or for long hours or to risk his health or life; on the contrary, he may feel justified in a measure of dishonesty and in receiving a part of his income from the traffickers themselves. The first incentive to honest effort is, therefore, an adequate salary regularly and promptly paid.

The second necessary incentive is the recognition that to do his work properly a police officer engaged in combating the illicit drug traffic has to incur certain expenses in the normal course of his work. These include not only money he necessarily spends on himself, such as the costs of transport, lodgings and meals when on duty and away from home, but also minor sums of money paid for casual information or other services. These should be promptly repaid, or he should receive a small monthly allowance to cover their average cost.

If these requirements are met, and a police officer feels himself a member of a respected, socially-accepted, well-trained and well-paid organization, his primary satisfaction will be obtained by doing his job well, and a system of making additional payments will be less essential as a means of encouraging zeal. Nevertheless, whatever the conditions in a particular country, some system of recognizing outstanding work will always be necessary, and it is worth considering the various forms available.

It can be said at once that a system of paying a fixed sum for the seizure of a definite quantity of a drug is more likely to increase than reduce the illicit traffic. Such a system:

(i) creates a constant and unlimited demand for the drug at a known price;

(ii) provides a level below which the price of the drug cannot fall;

(iii) encourages the police officer to concentrate upon the seizure of the actual drug and to ignore everything else;

(iv) encourages the police officer to keep information to himself or to a few favoured colleagues, so as to share the reward among as few people as possible;

(v) strongly discourages co-operation and the exchange of information among different officers and agencies, and even countries;

(vi) may easily lead to jealousy and ill-feeling among colleagues who ought to be on terms of the greatest confidence and goodwill;

(vii) may reward sheer good luck or chance rather than zeal and hard work; and

(viii) may actually discourage serious attempts to destroy the illicit traffic and gangs of traffickers, since part of the incomes of a sizable number of police officers may depend upon the continued existence of trafficking, upon which, in effect, they merely levy a toll.

The Objective

The purpose of the enforcement authority is the destruction of the gangs engaged in the illicit traffic, and not merely the seizure of what is inevitably no more than a part of the drugs they trade in. A reward system should recognize this purpose, and should accordingly attempt to assess not merely the quantity of drugs seized, but also the importance of a particular case judged in relation to the whole of the illicit narcotics traffic in the country, the difficulty and danger of the investigation, its quality and completeness, and the contributions to its success made by individual officers.

It can, therefore, be laid down as a general principle that a reward system should not be based upon a fixed tariff. Nor should it be based upon regular additions to salaries, by whatever name they are called, for these soon come to be regarded by their recipients as a part of their salaries to which they are entitled by right, and so equally soon they lose their point as an incentive. They may, however, as already noted, often be paid as an appropriate allowance in lieu of making claims for minor expenses or for regular overtime work.

Among the means which can correctly be used to discriminate among those deserving of reward, and to commend outstandingly good work, are the following:—

(i) an informal oral recognition of good work given by a superior officer. The more senior the officer in relation to the recipient, the greater is the praise;

(ii) an official letter which specifically mentions the good work of particular officers. Again, the praise is the greater if the officer is more senior.

(iii) an official letter of praise to a particular officer which is placed in his personal record of service;

(iv) an official commendation of an officer which is given publicity among his colleagues by being included in e.g. the weekly orders of the police force concerned, and becomes part of his record of service;

(v) in police forces where promotion is partly or wholly based upon length of service in a rank, credit of a number of months towards the next promotion;

(vi) an immediate promotion to a higher rank;

(vii) the award of an appropriate medal or decoration

by the head of state; and

(viii) the payment of a sum of money, either by itself or in combination with one of the other rewards already mentioned. In such cases the amount should be decided not only in accordance with individual merit, but also in proportion to the status of the recipient.

It is necessary to distinguish between a system of this sort for recognizing and rewarding the zeal and devotion to duty of police officers, and the payment of money to informers for information supplied. The former should not be used to finance the latter. Confusion can be avoided if payments to informers are referred to in that phrase, and the term "reward" is confined to payments to police officers.

There is no reason why rewards should not ordinarily be paid from funds openly provided for the purpose. On occasions, however, it may be desirable to conceal some particular activity or the connection with a case of some particular officer, when it would be appropriate to pay a reward from secret funds. Each organization should have its own accounting.

A Police Museum

Most police forces maintain museums where exhibits and other articles connected with criminal cases or police work generally are kept. It is also useful for them to have a library of complete investigation files of cases of outstanding technical interest. A well-stocked museum is not only a place of interest, but can provide valuable instruction. Exhibits relating to narcotic drugs should include specimens of the various drugs likely to be found in the country concerned, together with wrappings and markings taken from actual seizures; apparatus and substances used in the illicit manufacture of drugs; and a selection of hiding places in which drugs have been found.

55

INSTRUCTION CONCERNING ANTI-NARCOTIC KITS

The purpose of this kit is to provide law enforcement officers with rapid and simple tests for preliminary field identification of the drugs most commonly encountered in illicit traffic. Because of the preliminary nature of these tests, it is imperative that in all cases where presumptive positive results have been obtained, the suspected material should be submitted to an authorised laboratory for further confirmatory analysis.

— Always clean and dry the equipment (test-tube spot-plate and spatula) immediately after use.

— Never allow transfer pipettes or tips of reagent flasks to come into contact with the suspected material, as contamination of the reagents will yield false results.

Warning

This kit contains corrosive concentrated acids. Therefore:

— Handle the reagent flasks so labelled with care, avoiding contact with any part of the body.

— If the contents come into contact with the body, flush immediately with water, apply the acid neutralizer solution N provided in the Kit and seek medical attention.

— In the event of spillage, apply the acid neutralizer solution N to the area of the spill and flush with water.

This kit also contains some volatile and inflammable reagents. Therefore :

— Do not allow smoking when the kit is in use.

— Always replace and tighten caps on reagent flasks immediately after use.

— Conduct the tests in the open or in an adequately ventilated area and protect the reagents against lengthy exposure to sun and heat.

Do not sniff or taste suspected material or reagent
Testing methods (see also Charts I and II)

TEST A

Opium

1. Place a small amount of the suspected material (the size of a match-head) on the spot-plate provided in the kit.
2. Add two or three drops of water with the glass rod or spatula provided in the kit, smear the sample against the spot-plate for one or two minutes.
3. Transfer a drop of the liquid to another part of the spot-plate.
4. Add one drop of reagent A_1.
5. Add three drops of reagent A_2.

   Colour * * * indicates a positive result.

Morphine, Codeine, Heroin

1. Place a small amount of the suspected material (the size of a match-head) on the spot-plate provided in the kit.
2. Add one drop of reagent A_1.
3. Add three drops of reagent A_2.

   Colours in the range * * * to * * * are positive for morphine.

   Colours in the range * * * to * * * are positive for codeine.

   Colours in the range * * * to * * * are positive for heroin.

Amphetamines, Mescaline

1. Place a small amount of the suspected material

(the size of a match-head) on the spot-plate provided in the kit.

2. Add one drop of reagent A_1.
3. Add three drops of reagent A_2.

Colours in the range * * * to * * * are positive for amphetamines.

Colours in the range * * * to * * * are positive for mescaline.

TEST B

Marijuana, Hashish, Hashish Oil

1. Place a small amount of the suspected material (the size of a match-head) in one of the test-tubes provided in the kit.
2. Add small amount (the size of a match-head) of reagent B_1.
3. Add 25 drops of reagent B_2 and shake for a minute.
4. Add 25 drops of reagent B_3 and shake for two minutes.
5. Allow the test-tube to stand for two minutes.

Colours in the range * * * to * * * in the lower liquid layer indicate a positive result. Ignore colour of the upper layer.

TEST C

Methamphetamine, Methylphenidate

1. Place a small amount of the suspected material (the size of a match-head) on the spot-plate provided in the kit.
2. Add one drop of reagent C_1.
3. Add three drops of reagent C_2.

Colours in the range * * * to * * * are positive for methamphetamine.

Colours in the range * * * to * * * are positive for methylphenidate.

TEST D

Barbiturates

1. Place a small amount of the suspected material (the size of a match-head) on the spot-plate provided in the kit.
2. Add three drops of reagent D_1.
3. Add three drops of reagent D_2.

Colour * * * indicates a positive result.

TEST E

Cocaine, Methaqualone

1. Grind the suspected material to a fine powder if it is in tablet form, and place a small amount of the suspected material (the size of a match-head) in one of the test-tubes provided in the kit.
2. Add one drop of reagent E_1 and shake for 10 seconds.
3. Add three drops of reagent E_2 and shake for 10 seconds.

Colour * * * indicates a positive result.

* * * for exact colour ranges please refer to the charts supplied by the UNO in their Drug Testing Kit.

FLOW CHART I

NATURE OF SUSPECTED MATERIAL

Herb
Resin
Oil

Resin
Powder

Cigarettes
Smoked cigarettes
and pipes and other
suspicious material

? Marijuana
? Hashish
? Hashish oil

? Opiate

Try TEST A for:

Opium
Morphine
Codeine
Heroin

grounds for suspicion

Negative result

Positive result

Try TEST B for:

Marijuana
Hashish
Hashish oil

Negative Result

Positive Result

Stop testing

Only if grounds for suspicion

SUBMIT TO LABORATORY

FLOW CHART II

NATURE OF SUSPECTED MATERIAL

Impregnated paper
or
Gelatine film

Tablets
Capsules
Powders
Liquids

? Amphetamine | **? Opiate** | **? Barbiturate**

Try TEST A for:
Amphetamines
Mescaline

Try TEST A for:
Opium
Morphine
Codeine
Heroin

Try TEST D for:
Barbiturates

Positive Result | **Negative Result** | **Positive Result** | **Negative Result** | **Positive Result** | **Negative Result**

Try TEST E for:
Cocaine
Methaqualone

Try TEST A for:
Opium
Morphine
Codeine
Heroin

Positive Result | **Negative Result**
Try TEST D for:
Barbiturates

Amphetamines
Mescaline
Try TEST C for:
Methamphetamine
Methylphenidate

Positive Result | **Negative Result** | **Positive Result** | **Negative Result**

Stop testing

Try TEST E for:
Cocaine
Methaqualone

only if grounds for suspicion

Positive Result | **Negative Result**
Stop Testing

Only if grounds for suspicion

SUBMIT TO LABORATORY

56

THE ROLE OF CRIMINAL JUSTICE SYSTEM[1]

The detection by the police, the public prosecutor, the court, of drug misuse makes it possible for the misuser to get the necessary medical help (in particular during the first days of detention in which withdrawal usually takes place and also in case of drug associated diseases, like *hepatitis* and *Aids*, etc.) Legal identification is, therefore, a pre-requisite for the contribution that the criminal justice system can and should make to the social rehabilitation of drug misusers.

Nevertheless, to achieve better identification of drug misusers, attention should be given to the negative consequences namely, the stigmatisation and marginalisation of drug misusers, especiall of those who only, occasionally take drugs. Although adequate information on the offenders misusing drugs but not identified is lacking, there is a general feeling that while few drug addicts are not identified, a large number of non-addicted drug misusers go undetected by the criminal justice system.

Whether a drug misuser is detected or not by the criminal justice system depends on a wide range of circumstances. First of all, the nature of the offence committed has a direct bearing on the likelihood of the detection of drug misuser. Thus, while for drug related offences (drug consumption, detention, trafficking, etc.) as well as certain "typical" ordinary offences (i.e. burglary in a pharmacy) the identification of an offender as a drug misuser is either automatic or the result of systematic attention to this factor by the criminal justice system. The same is however not true for other offences committed by drug misusers. It is precisely these drug misusers accused or convicted of ordinary offences who are in increasing number coming into contact with the criminal justice system.

Secondly, detection, of persons fully addicted to drugs, is easy because of their withdrawal symptoms and need for urgent medical help and medical examination carried out systematically in prisons or in police stations. Inversely, the majority of those "common" offenders taking drugs who are simply given a warning or a fine, seem to pass undetected by the police or the court.

An important aspect on which the identification sometimes depends is the interests of the offender in the framework of the criminal proceedings taken against him. Thus, the nature of the charges against the misuser, the consideration of drug addiction as aggravating or extenuating circumstance, or as an additional offence to be added to the common offence which originally provoked the arrest, the greater or less likelihood of release pending trial, a probation sentence, etc. if the fact of drug misuse is known by the court, are *inter alia* factors which influence the perception of his interests by the misuser, or his defence counsel, often determining whether the drug misuse will be disclosed or concealed during the proceedings. Moreover, the interest of the misuser as regards measures which can be taken while in detention, like placement in the prison surgery or in a re-inforced security quarter, treatment with methadone, greater availability of sports activities, etc. will also influence his willingness to make known his drug misuse.

Other factors that identity drug misusers are their typical life style, often revealed by the police enquiry, habits or acquaintances, and also the findings of drug addicts' search which often follow the arrest of common offenders, in particular, thieves.

Drug Misuse and Criminal Proceedings

There is general agreement on the importance of identifying all offenders, by the criminal justice system, who misuse drugs. The two main reasons for this are the following :

(a) The court's knowledge of an accused's drug misuse permits better assessment of criminal responsibility and fairer application of penalties, for offences committed under the influence of drug or in a state of withdrawal or in order to obtain the dose needed. In this connection the problem of the reverse situation, namely, that offenders (particularly drug traffickers) falsely claiming to be drug addicts in order to obtain a more lenient sentence or better conditions in detention, anticipated release, etc. must be noted. Proper identification of real drug abusers will prevent professional drug traffickers from taking advantage of provisions not intended for them.

(b) The detention by the police, the public prosecutor, the court, the prison staff, etc. of drug misuse makes it possible to give the misuser the necessary medical help.

The Customs and Central Excise Collectorates in addition to their basic responsibilities have taken effective steps to combat drug trafficking and eradication of illicit cultivation of Cannabis in different parts of the country. In 1989-90 the efforts were stepped up against smuggling operations to safeguard the economic health of the country.[2]

The importance of a drug misuser's sincerity when undertaking to follow treatment is important addicted offenders into, therapeutic communities, had failed in the past, as many of the former addicts would make promises which would facilitate their release and then discontinue the treatment and abscond is viewed seriosly afterwards.

The courts are often reluctant to release accused persons pending trial if they are known to misuse drugs and this was mainly due to the fact that relapse into drug abuse and the subsequent criminal recidivism is viewed seriously. Certain judges consider a period of detention, in some cases, be the best way of protecting the misuser from himself and provide the court with an effective base to orientate the misuser towards treatment.

Release on Parole

A wide range of measures and devices are available to the criminal justice system in the different national legislations to avoid the detention of offenders, misusing drugs. The measures could be taken at varying stages of the proceedings, from the criminal enquiry (waiving of prosecution, release pending trial) to the trial (suspension of the judgement, conditional discharge, probation sentence, postponement of the execution of the sentence) or the serving of a prison sentence (anticipated release). Some measures are of general application to all offenders, whereas others are restricted to minor offenders, consumers of "soft" drugs, first offenders, juvenile offenders, etc. Prosecuting agencies, courts, prison authorities, etc. should try to encourage drug misusers to opt for treatment.

Although there is agreement on the importance of early detection of drug misusers, systematic screening of all offenders coming into contact with the criminal justice system, even when no drug factor or connection exists in the offence involved, is impracticable. More attention is necessary during the police inquiry to the possible part played by drugs.[3]

Different Means of Avoiding Imprisonment

The general question of whether the imprisonment of drug misusers having committed an offence should be avoided as far as possible has been discussed frequently. While the situation of detention very often annulls all motivation for the misuser's rehabilitation the prospect of full freedom could operate as a useful lever to orientate the offender toward appropriate treatment. In certain cases there is no better alternative to imprisonment in order to sever the offender's links with his habitual criminal environment, avoid the immediate recurrence of new offences and ensure his withdrawal from drugs.

The importance of proper selection of drug misusers to be released, whose sincerity and suitability for treatment should be ascertained, as well as the useful role that the social services could play in this connection needs to be stressed.

Lastly, the gravity of the offence committed and of the sentence received are, essential factors to be considered when deciding the granting of release. Drug misusers should not be treated favourably compared to offenders, not taking drugs, both as a requirement of justice and as a deterrant.

Compulsory Treatment

A basic distinction was made in this field between physical withdrawal and psychological therapy. Physical withdrawal is in many countries an unavoidable consequence of detention, as the misuser cannot, in principle, obtain drugs once in prison; in other countries, this withdrawal does not always take place, as the prison doctor may decide to supply the addict, at least for some time, with drugs on prescription. The fact that the misuser cannot have access to drugs while in detention cannot be regarded as compulsory treatment and that the supervision and assistance of the prison medical services, responsible for the health of all detainees, should be ensured in all cases.

The general principle that any medical treatment requires the consent of the patient. This is particularly true for psychotherapy provided to drug misusers not only for reasons of professional ethics but also for effectiveness, since the free co-operation of the patient is of the highest importance for any psychological withdrawal treatment.

On account of his health needs, and in the interest

of his rehabilitation, a drug addict remanded to custody or serving a prison sentence may be placed in a special centre, hospital or prison surgery without first seeking his consent. This does not constitute an instance of compulsory treatment.

Lastly, the condition under which physical detoxification takes place often have an influence on the effectiveness of any further treatment. In effect, drastic withdrawal with no medical assistance may have psychological costs which could hinder any subsequent psychotherapy and also provoke an attitude of reserve or even rejection in the patient which will be an obstacle for the success of the treatment.

Release Pending Trial

The courts are often reluctant to release accused persons pending trial if they are known to misuse drugs and this is mainly due to the fact that relapse into drug abuse and criminal recidivism is regarded as likely for this type of offenders.

It is desirable that, in the framework of the legal provisions in different countries governing the release of accused persons pending trial, discrimination against drug misusers in comparison to other offenders should be avoided as far as possible. Private institutions, associations, etc. could play a useful part in assisting and providing help to released drug misusers, thus overcoming the court's apprehensions and making their release easier.

The Role of the Social Services in Criminal Proceedings

The discussion of this item covers both court services (probation services, social enquiry services, forensic surgeons, etc.) and services independent from the criminal justice system (local health units, hospitals, drug counselling centres, etc.). The different social services play useful roles in providing information and expertise to the court and medical aid and social assistance to the accused drug misuser. Also their intervention in the criminal proceedings (through expert reports, social enquiries, etc.) could be instrumental in ensuring better contact between the criminal justice system as a whole and the social and health services. These contacts become difficult in the case of social services not integrated into the law enforcement system.

The court services and probation services sometimes lack specific training regarding the problems of drug misusers, as well as sufficient staff to provide assistance and follow up for an ever-increasing number.

Fines

Fines appear to be particularly appropriate for drug trafficking and illegal import offences. The concept of a double jeopardy for the same crime exists in several countries which punished illicit drug imports twice: by the customs authorities and by the criminal courts.

As regards drug misusers who commit a given offence (drug peddling, theft, etc.) in order to finance their habit, the use of fines raises additional problems. Another difficulty with the fining of drug misusers having committed an offence is that it does not give the court proper leverage to orientate the misuser toward appropriate treatment. If the fine is paid, no other measure can be taken to encourage treatment, whereas if the fine is not paid, it is often replaced by a prison sentence. Experts emphasize the need for measures alternative to imprisonment when a fine is not paid, for example community work, which would probably be more suitable for offenders with limited resources and in particular for those among them who misuse drugs.

Possibility of Treatment in Prison

Some specialists consider it "impossible and undesirable" to treat addicts in prison, whereas others feel that, because the prison removes the addict from his former environment, it may constitute a favourable setting to begin treatment. As regards physical withdrawal, in a majority of countries, this takes place automatically in prison.

As regards the possibility of psychotherapeutic treatment in prison, some experts doubt that such a treatment could be effective in prison because they feel there was an inherent conflict between constraint and therapy and also because keeping the drug misusers among other prisoners, which was necessary if their marginalisation was to be avoided, made it impossible to provide community therapy. Other experts thought that psychological therapy provided in prison could achieve adequate results provided it was based on the sincere consent of the prisoner and supplemented upon release by the appropriate assistance and follow-up.

Contacts with the Outside World

A general policy is pursued in many states to open the prison to society ensuring the highest possible level of contacts between prisoners and the outside world. As prison is an intrinsically exceptional and abnormal life environment, contacts with the free world can play a useful part in preparing the inmate's re-integration into the normal society once he is released. This general principle seems to be particularly true for detained drug misusers, as their motivation toward rehabilitation can be greatly encouraged by measures allowing, for example, the maintenance of family contacts. A balance should be struck between the importance of allowing contacts with the free world and the risks of these contacts for the continuance of physical withdrawal which is an essential contribution of custody to the misuser's rehabilitation. Particular control measures like the requirement that all prisoners released on short term leave undergo urine tests before being released and after being reintegrated into the prison, need to be study.

Drug-free Sections

Netherlands, Norway and Sweden have set-up what are called "drug-free sections or units" or "contract units." These are separate departments or special prisons where placement is based on the prisoner's undertaking to abstain from drugs and to accept frequent control measures, in particular urine tests. These units are intended for strongly motivated addicts and provide them with particular support. The system permits a relaxation of restrictions on contacts with the outside world (prison leave, semi-open regime, etc.) and also the organisation of group therapy in prison.

Criminal justice Norms, Standards and Policies

By "norms" and "standards" is meant here the United Nations position with respect to prevention and control of crime and deviance, including some theoretical aspects that may be built into such standards and norms.

There exists a very serious and growing concern of Member States of the United Nations about drug abuse and its ramifications through the contemporary world. Member States are calling for new measures that will help more effectively to fight this phenomenon and for revaluation of the old ones. At the Fifth United Nations Congress on the Prevention of Crime and the Treatment of Offenders (Geneva, 1-12 September, 1975) it was observed in this connection that "penal measures, if they were not to be excluded, should in no way prevent the application of treatment and social re-integration...sic Governments should follow the 1972 Protocol Amending the Single Convention on Psychotropic Substances of 1971. They should also provide, either as an alternative or in addition to correction or punishment that drug addicts and drug abusers, where appropriate, undergo measures of treatment. It was generally accepted, however, that as yet very little evidence of the success of the treatment was available".[4]

It is evident from above that existing international drug control instruments provide a basic, relatively broad framework for combating drug crimes, whereby there is enough flexibility and space for making adjustments to particular concerns of Member States and for drafting of new provisions with a view to regulating fight against the drug phenomenon. Within this broad framework it is also clearly recognized that, as far as possible, different measures should be applied to drug users and to drug traffickers. The measures against the latter group, as viewed by the Fifth United Nations Congress, should, *inter alia*, take into account:

(i) need for an international convention on judicial assistance and extradition procedures, whether the drug offenders should be recognized as extraditable offences;

(ii) that, in conformity with national legislation, the drug offenders who had been convicted in one country but had escaped, served the sentence in the country in which they had taken refuge or were found, if extradition was not feasible;

(iii) that illegal traffick in drugs should be considered as a trans-national crime; and

(iv) that all knowledge in the field of law enforcement should be stored, and relevant information, in particular, concerning new methods and routes used by illegal traffickers, should be constantly exchanged.

Owing to various factors outside and inside the United Nations system, including, for instance, changes in the position of United Nations Congress on the prevention of crime and the treatment of offenders and of the Committee on Crime Prevention and Control *vis-a-vis* Member States, consideration and follow up of the above recommendations serves here as a departure point for substantive presentation but not as an analysis of the implementation of these recommendations after almost ten years since their adoption.

With this qualification in mind, we shall focus now on two kinds of measures, concerning drug traffickers and drug users by commenting on them from the present standpoint of the United Nations criminal justice and crime prevention programme.

Measures against Illicit Drug Traffickers

(i) *International Judicial Assistance*—The idea of drafting a special international convention on judicial assistance with regard to illegal traffick in drugs was not taken in the United Nations system but only recently within the Organization of American States. Leaving this matter aside, for the time being, it should be noted here that the United Nations Secretariat initiated its work on a broader item involving judicial assistance in all criminal justice matters among the nations, which implicitly would apply to illicit drug traffickers by the very nature of their activity. This broader item has been, since its inception at the Sixth United Nations Congress (Caracas, 25 August—6 September, 1980)[5], thoroughly considered by the Crime Prevention and Criminal Justice Branch and subsequently been elaborated by non-governmental organizations in consultative status with the Economic and Social Council. In effect, the Seventh United Nations Congress (Milan, Italy, 26 August—6 September, 1986) will be presented with a draft "Model Agreement on the Transfer of Foreign Prisoners", which is *inter alia*, based on the following principles :

Firstly, it recognizes that the transfer of a prisoner finally and definitely convicted of crime abroad to his country of origin or residence should be effected at the earlier possible stage of serving the sentence.

Secondly, such a transfer should be conducted on the basis of mutual respect for national sovereignty and jurisdiction.

Thirdly, this transfer may take place only when the offence giving rise to the conviction is punishable by deprivation of liberty by the judicial authorities, of both sentencing State and administering State, according to their national laws.

Fourthly, a transfer should be based on the agreement of both of these States and the informed consent of the prisoner, that is only with his/her expressed free will.

This draft model agreement is intended to serve as a specimen for interested Governments which in negotiating specific bilateral or multilateral agreements may wish to follow principles of the model. By its nature, the model agreement does not include the question of extradition, referred to earlier, nor does it include items that could be pertinent to its aims but which, probably were not considered to be of a primary importance for the execution of the sentence, as for example, an international register of convictions. Such instruments of co-operation may anyway be established. However, exchange of information on convictions and the granting of excerpts from national crime registers on request already fulfils, to a large extent, the functions of an international conviction register[6]. What matters here is the most practical means for providing such information, especially in the case mentioned already under sub-paragraph (ii) of this chapter. This important concern has, to some extent, been taken into account in another draft document which is briefly presented below.

At the initiative of Venezuela, the Organization of American States elaborated a "Draft Convention on the Traffic in Narcotics and Psychotropic Drugs and Related Activities", which is now lying for consideration before the General Assembly. For the purpose of that convention, it is proposed therein that "illicit activities" denote: "the traffick in, distribution, supply, manufacture, processing, refinement, transportation, extraction, preparation, production, cultivation, preservation, storage, management, organization, financing and facilitation of the traffic in any of the substances or their raw materials to which this convention refers" (Art. 3)[7].

The international co-operation on penal matters is dealt with in Article 9. This article proposes, *inter alia*, that the contracting parties shall assist each other to "identify, detain and try" the alleged drug traffickers, and confiscate and destroy the trafficked substance.

The draft convention further provides that the contracting parties shall co-operate with one another and with the competent international organizations by compiling information and documents pertinent to investigation of an alleged illegal trafficking so as to make it easier to bring those involved in it to trial. Such files should be exchanged for the purpose of instituting legal proceedings in a country where the offender is being tried and should, presumably, include information on previous convictions abroad for purposes of establishing recidivism. Finally, "the states shall not either adopt laws or take measures of any other kind that could infringe upon the international obligations that they have incurred to identify, arrest, extradite and punish those found guilty of the offences described in this Convention". Consequently, these offenders should be extraditable whenever extradition is called for, and if the offender has not been already tried and sentenced for them in another country.

The two drafts described herein differ considerably between themselves in the degree of specificity. The draft convention provides detailed provisions on the exchange of information relating to pending criminal investigation and also, quite understandably, formulates conditions for extradition of the illegal drug trafficker. From the standpoint of the draft model agreement such provisions strengthen international drug control measures, whereas the draft model agreement itself may assist the government in the post-conviction stage of the criminal justice process. Absence of any such provisions in the draft convention suggests that both drafts may be complementary to one another, thus contributing to a more co-ordinated approach to the control of drug crimes.

(ii) *International Condemnation of Illegal Trafficking in Drugs*—The idea that illegal trafficking in drugs should be internationally condemned as "a transnational crime" could not make further progress, neither at the Fifth United Nations Congress nor for a long time after. However, only recently this idea was dealt with in even more explicit and rigid terms in the so-called "Quito Declaration against Traffick in Narcotic Drugs", signed by the Governments of Bolivia, Colombia, Equador, Nicaragua, Panama, Peru and Venezuela. This declaration, transmitted to the 39th session of the General Assembly, emphasized that in view of the volume, magnitude and extent of the traffick in narcotic drugs, this phenomenon represented a threat to the world and was harmful to the very existence and further development of the human being. Recognizing these detrimental consequences, the declaration proposed "for the consideration of the competent international organizations" that traffic in narcotic drugs "should be considered a crime against humanity, with all the legal consequences implicit therein"[8], that is by placing this crime under international jurisdiction alongwith *apartheid*, genocide and other crimes. A similar formulation, as a matter of fact, is contained in the draft convention on the traffic in narcotic and psychotropic drugs, whereby in the preamble and in Article 2 the contracting parties define such illegal traffick as "a serious international offence against mankind".

It must be specially emphasized here that behind such formulation dramatic picture of illegal drug trafficking and its multiplicity of consequences has to be constructed. It is estimated that about two-fifth of the population of Latin America consumes cocaine, and that there is an ample evidence that bribery, corruption, violence and other crimes are corollary to trafficking and all these undermine the social fabric to an extent previously unknown. Reported allegations of involvement of high-ranking government officials in illegal traffick, not only in narcotic drugs, may pinpoint the detrimental role of drugs as being at the heart of some countries' structures of power.

(iii) *International co-operation in Law Enforcement against Illegal Trafficking in Drugs*—This matter lies clearly within the competence of the United Nations Division of Narcotic Drugs and the Commission on Narcotic Drugs with its Sub-commission on Illicit Drug Traffic and Related Matters in the Near and Middle East[9]. For this reason it shall be only pointed out that along with a self-evident need for co-operation of the law enforcement officials in fighting illegal trafficking in drugs, there is also a clear-cut need for specialization in detection of drug crimes. The results of the Second United Nations Survey of Crime Trends, Operations, Criminal Justice Systems and Crime Prevention Strategies (1975-80) show that, indeed, several countries follow the path of specialization by setting up police units or departments with the primary aim to control drug crimes. Another form of specialization was reported from the Philippines, the Federal Republic of Germany, Ireland and the Bahamas where specific in-service training is provided for police officers dealing with drug crimes. In this respect the solution of Sweden is worth noting: also prosecutors received specialized training in dealing with drug crimes.

(iv) *Capital punishment as a sanction against Illegal Trafficking in Drugs*—Both the Single Convention on Narcotic Drugs and the Convention on Psychotropic Substances call for "adequate" penal measures which should be applied against drug traffickers. Such measures, if the crime is "serious", may particularly involve "imprisonment or other penalty of deprivation of liberty". It may follow from this that the contracting parties are in a position to apply quite a variety of measures available in the national criminal legislation which, in other words, may or may not involve imprisonment and could be spread on the continuum starting from mild fines to capital punishment.

Such a broad criminal policy framework would encourage to analyze current developments in penal sanctions throughout the world and, at the same time, incline to find out what is the place of such penal sanctions in a wider context of criminal justice norms and standards formulated in the United Nations. However, such a goal would be practically unattainable, for there is not enough comparable data available on the dynamics and patterns of penal sanctions against illegal drug traffickers. It would have been also politically unfeasible as it could be considered to imply that national criminal policies are contrasted with international norms and standards without due consideration to the socio-economic, cultural and also political background against which national policies are built.

For the first reason, the analysis which will follow here restricts itself to the areas on which relatively more information is available, that is, to the area of capital punishment. Secondly, this analysis is not meant to contrast the practice of imposing capital punishment with the United Nations position on this question; in the situation of "compartmentalization" of the United Nations activities in the field of crime prevention and control, such an analysis seeks simply to bring to attention the substance of both national and various United Nations developments.

According to the report of the Secretary General on the question of capital punishment containing information on it for the years 1974-1978[10], out of 74 countries which replied to the survey, on which the Secretary General's report was based, only four countries (Algeria, Indonesia, the Philippines and Syria) reported provision of capital punishment for drug crimes. Two of them reported, that this penalty may be applied for "recidivist trafficking in drugs" (Syria) or the "manufacture of drugs, the sale, administration, delivery, distribution and transportation of prohibited drugs resulting in death" (the Philippines). From other source[11], it is known that Burma, Egypt, Iran and Malaysia provide for a possibility of improving the death penalty for illegal drug trafficking, but as in the previous cases, there exists a variety of specific situations when such punishment could be meted out. In Iran, the severity of the penalty of such offences as the import, export, manufacture, but also possession varies according to the amount seized. Thus, if any of the prescribed activities involve opium in quantities of 50 kg to 2 kg, the offender is liable to imprisonment at hard labour from 3 years to 15 years in addition to being fined. If other narcotic drugs are involved in quantities of one to 10 g, the penalty is the same. However, if the acts are committed with knowledge and intent, the penalty can be increased to capital punishment when more than 10 g are involved.

In Egypt, the penalties seem to be even more stringent. Import/export without authorization, production, extraction, separation or manufacture for sale, for possession, purchase, sale, transportation and a number of other activities may result in death penalty and, in addition, fines from 30,000 to 10,000 pounds.

In Burma, the minimum penalty for importing, exporting or dealing in narcotic drugs with countries outside Burma is 10 years imprisonment. The maximum term is life imprisonment or death.

Finally, till 1983 in Malaysia trafficking could result in a life term and whipping or the death penalty[12], and since that time only a mandatory death sentence is imposed.

A survey of drug trafficking penalties conducted in 1979 in 125 countries[13], revealed that, in sum, in ten countries, including the Republic of Korea and Singapore, the death penalty could be meted out for trafficking in opium, or its derivatives, in six for trafficking in cocaine, and in five, in cannabis. However, it is not clear whether these data inform of separate instances of existing capital offences in such an altogether sizable number of countries, or, what is also possible, that in these countries there are at the same time several capital offences involving different drugs so that, in effect the total number of countries providing for a possible imposition of the death sentence is smaller.

Ambiguity in this matter does not blur, however, the fact that capital punishment is quite a popular means envisaged for suppressing illegal traffic in drugs. It exists at least in seven countries of vast Asian region and in other regions. Consequently this trend cannot be explained only by social, cultural and other factors of each country because this popularity transcends borders of geopolitical subregions and regions. It rather could be explained by the often held belief that the death penalty is the most traditional means of formal social control without which the nations may not see fit to effectively counteract the existence and possible spread of the drug phenomenon.

However, such policies seem not to account for the desire to abolish capital punishment, which is expressed in Article 6, para 6 of the International Covenant on Civil and Political Rights ("nothing in this article shall be invoked to delay or prevent the abolition of capital punishment..."), and also for para 2 of this article which states, *inter alia*, that "the sentence of death may be only imposed for the *most* serious crimes" (emphasis added)[14]. Such a phrase was recently qualified by the Economic and Social Council which in its resolution 1984/50 (Annex) stated that "most serious crimes" should be understood only as "intentional crimes, with lethal or other extremely grave consequences"[15]. This formulation basically precludes the interpretation that illegal trafficking in drugs belongs to this category of crimes because it lacks *dolo directi*.[16]

Secondly, there is no convincing evidence that the death penalty may be an "adequate", i.e., effective measure in deterring illegal trafficking; at least the evidence corroborated by a thorough research.[17] Moreover, sophisticated research on the relationships between the capital punishment and criminal homicide, certainly a primary capital offence, points out the irrelevancy of one element to another, even if a retributivist look at it shows to the contrary.[18]

It is self-evident that illegal drug trafficking lacks such a straight retributive relationship. Furthermore, available empirical data do not shed much light on effectiveness of capital punishment to the incidence of illegal drug trafficking, that is on the second reason for its use. Accounts on the respective deterrent effect of death penalty are here very conflicting and, additionally, so emotional and value laiden, that it is difficult to formulate an objective position on the issue.

On one side of this debate stand, among others, decision-makers and criminal justice practitioners in Asian countries which resorted to capital punishment. According to one source, use of death penalty in Burma, Iran, Malaysia and Singapore was believed by some of them to have a positive effect on decrease of illegal drug trafficking and subsequent abuse of heroin.[19] The same source, however, quotes one criminal justice practitioner from Singapore who suggested that "on the contrary, heroin seems to be more widely used than ever before" and adds that according to press publication of 1981 from another country (Malaysia) despite imposing harsh measures the authorities reported a widespread use of illegal drugs.[20]

On the other side of capital punishment debate stand, among others, the researches. Quoted here A.S. Trebach critically reviews the above opinions, suggests that the death penalty has no deterrent effect and finally emphasizes that resort to it, desirable as it may seem, "rarely solves human problems. It is necessary, therefore, to work out the painful details of a new world order for the control of dangerous drugs, which must begin on the middle ground between the extremes, where few matters are black and white, where most are grey and where a patient sense of human balance is crucial. In other words, ingenius compromise, cautious experimentation and readjustment in the face of new experience must become the tools of creating that new order. Those who believe in extreme solutions or in rigid and hoary ideologies will find such work inhospitable".[21]

The above conflict of conclusions, probably, cannot be solved without finding out whether empirical evidence for or against application of capital punishment in cases of illegal drug trafficking is reliable and valid or not. However, both conclusions fall short of sufficient evidence as it will be demonstrated below :

Firstly, one would need to know clearly whether the deterrent effect of death penalty is to influence cases of illegal drug trafficking (including those undetected) or is to reduce indirectly consumption of specified drugs, or both. In case of death penalty applied to criminal homicide the relation between "punishment" and "crime" is clear. In the situation discussed, presumably, both dependent variables, that is "illegal drug trafficking" and "drug consumption" plus an intermediate variable that is "illegal possession" would need to be considered for correlation with an independent variable, e.g. "capital punishment."

Secondly, any correlation would need to account for a specific time lag between the death penalty and subsequent drug-related behaviours (one or two weeks lag will cause other results there one year lag).

Thirdly, since it is assumed that capital sentences may have a real deterrent effect only when they are executed, the total number of imposed sentences may be irrelevant to statistical analysis.

Fourthly, and in consequence of the previous point, only executions which are publicized by mass-media should be accounted for, since otherwise the message that "you may be executed for such and such drug crime" may not reach the potential offenders.

Fifthly, setting the question of extradition apart, as in illegal drug trafficking are frequently involved foreign offenders, one would need to consider whether a possible deterrent effect of executing a foreign drug trafficker will have the same bearing on the domestic offender.

Sixthly, any measure of the deterrent effect should be based not only on the results of relationship between the death penalty and drug crimes, but concurrently, between this penalty, that is imprisonment. In other words, only the "additional" deterrent effect of a publicized capital punishment execution *over* imprisonment may be counted. Here, however, we may run into serious difficulties as in some countries of the region not only imprisonment, but also recently established "preventive detention" is applied to illegal drug traffickers.

Lastly, several other factors should be carefully studied with a view to include them and to make the test methodologically flawless.[22]

In sum, then, only after thorough observance of very elaborate rules of criminological enquiry one may arrive at substantiated conclusions on the deterrent effect of capital punishment on the rate of illegal drug trafficking. But even without such an enquiry and only accepting indisputable evidence that on balance the death penalty does not have a perceptible influence on homicide rate, one may be sceptical whether such an influence will be visible in case of illegal drug trafficking.

The sixth point takes us directly back to the question of norms and standards, since the institution of "preventive detention" may be read in the context of Article 9 of the International Covenant on Civil and Political Rights[23] and, probably, considered as independent of its provisions inasmuch as this institution provides other legal arrangements, i.e. whose primary aim, in the strict judicial sense, is not "punishment" or "penalization" but incapacitation.

None of the above cursory observations on international norms and standards and national policies should be interpreted as suggesting existence of any clear-cut disparities between these two elements. Setting "preventive detention" apart[24], both the Single Convention and the Convention on Psychotropic Substances clearly make it possible that the contracting parties may resort to any "adequate punishment", even if particular preference is given to imprisonment or "other penalties of deprivation of liberty". First of all, these observations rather point out at incompatibility provisions among international norms and standards themselves, which seems, however, to be quite understandable. Provisions of the covenant restricting use of the death penalty to "most serious crimes" and those relating to detention were adopted after eleven years of work in 1966, that is five years later than the Single Convention. Naturally, both the Single Convention and the Convention on Psychotropic Substances could not account for further development, (i.e., ECOSOC resolution), although the latter 1971 Convention could have related itself to the provisions of the covenant.

In line with the above reasoning would be the final conclusion that in the future national criminal legislation against illegal drug trafficking may need to relate itself also to norms and standards which developed on the top of international drug control instruments. Especially, the concept of "serious offences" would need to be reviewed and uptodated in the context of the provision of the covenant and the recent standard provided under ECOSOC resolution.

Measures Concerning Drug Users

(i) *Punishment for Use of Drugs*—For quite a long time it has been understood that measures applied to drug users should, in principle, differ from those imposed on drug traffickers. Desirable as this position is, in reality it is frequently not followed. One of the surveys revealed that while trafficking penalties are generally somewhat more severe than possession penalties, 55 countries (out of 125) do not distinguish in any manner between possession or trafficking; *all* illegal activities involving drugs carry the same maximum penalty.[25]

A less comprehensive but more indepth study by the United Nations Division on Narcotic Drugs found that many of the legal systems penalize the users or possessers (for personal use) very severely. "In the absence of statistical studies on the incidence of such penalties on illicit traffick, it is difficult to reach hard-and-fast conclusions; but we accept the crux of the narcotic abuse problem is not the addict but the addict's source of supply, it would seem that severe penalties directed against the user are often a sign of society's frustration at not being able to, or not desiring to, strike more effectively the real root of the evil".[26]

(ii) *"Nothing Works" Argument*—In view of this often non-existence distinction between measures applied to drug traffickers and drug users, what follows starts with the issue of punishing drug takers.

> Consider the statement which justifies punishment: "Doing something leads to bad effects or even to disease-eating unhealthy food, smoking, continuously taking drugs...is not itself a disease.... The person who exposes himself to pneumonia may catch it. But exposing oneself is not a disease. Both the cigarette addict who gets cancer and the heroin addict who as a result of his initial voluntary exposure contracts an overpowering craving have exposed themselves to what they eventually contracted, but the exposure is no more a disease than sensual intercourse is when it exposes syphillis.... One does volunteer for addiction—few persons become addicted unless they volunteer.... If we threaten to punish why one who develops cancer or catches the flu, no fewer persons will contract either. Threats do not prevent people from contracting diseases. But if we punish *becoming* addicted, at last becoming addicted is not a disease. It is not beyond the control of the person who becomes addicted, even if being addicted were a disease".[27]

This "get tough" approach has its substantiation in one American survey, according to which the treatment model does not work.[28] "Nothing works" argument extended over the entire prison rehabilitation efforts in the United States and has considerably influenced changes in criminal justice policy in that country since the second half of 1970's. This important survey covered 231 evaluation studies conducted in the USA and elsewhere between 1945 and 1967. A major finding was: "with few isolated exceptions the rehabilitative efforts that have been reported had no appreciable effect on recidivism".

(iii) *What works under which circumstances?*—A reaction to these findings "elsewhere" was certainly less unequivocal than in the United States. The Inter-regional Preparatory Meeting of Experts for the Seventh United Nations Congress on its provisional agenda item "Criminal justice processes and perspectives in a changing world" considering reported failure of the treatment model noted *inter alia* that "a certain degree of scepticism on the part of administrators and policy-makers *vis-a-vis* such theoretical fashions was a healthy attitude, in that it avoided wide policy swings that later came to be regretted."[29] At the first glance the case of rehabilitation of imprisoned drug offenders seem, however, to fall exactly under the category of "nothing works". A recent survey conducted by the United Nations Division of Narcotic Drugs to which replied 67 countries, indicated that in the opinion of the respondents "treatment in poisons was often reported to have low success rate."[30] But here, no prior rigid evaluation of such programmes was made, hence caution must be exercised as to conclusiveness of this statement. At this training course it was suggested that such a relatively successful programme is implemented in Singapore. Having the benefit of a detailed description of it by L. Yew[31] it needs only to be emphasized here that the programme consists of two principal parts involving drug treatment and rehabilitation and a day release scheme. The basic goal of the first mandatory phase is treatment of committed addicts (including those convicted for consumption offences) and of the second voluntary phase, i.e., aimed at re-integration of the cured addicts on a "half-way" house basis into the mainstream of society. It was reported that after two years' follow-up those who completed two phases avoided relapse into addiction than those who underwent the first phase only, even if in both situations relapse rates increased over the period under review, that is from February 1980 to August 1982. This increase, however, was far more considerable for the first, in a sense, "control group" of addicts (32%) than for those who received "full treatment" (13%). In another paper from Thailand, it was suggested that also there reintegration of drug abusers was successful, about 70% of probationers convicted for consumption offences completed probation without violations of its conditions. Available literature suggests that successful treatment and re-integration programmes also exist in the United States, where "nothing works" argument was most popular. For instance, "California Civil Addict Programme" evaluated in 1977 involved treatment of voluntarily committed addicts convicted of felonies and misdemeanours, and relied on such measures as therapy, schooling and occupational training, which took first seven months, after the addicts were transferred to outpatient status under close parole supervision. Interviews made with 750 men five to twelve years after admission revealed that those who remained in the programme improved on measures of drug use, employment and criminality more so than a comparably matched group who were discharged from the programme on a technicality shortly after admission. At least during the commitment period and less so during the post-treatment period, the group remaining in the programme significantly reduced their substance abuse and related criminal behaviour.

These three examples are not being re-evaluated here, but will rather serve as a departure point for making some general observations on "what works under which circumstances" in direct and indirect relation to treatment and rehabilitation programmes. Firstly, the more comprehensive are evaluation methods, the more certain one may be of the results of the programme. The interviews conducted with the addicts are more reliable than official statistical records, likewise self-reported criminality surveys provide a fuller picture of the incidence of crime than the data on police recorded crimes only. In this context, one may suggest that the more control staff is involved in checking for symptoms of addiction, the more probable determination of their real presence, hence also more chances for an "increased" detection rate. Although available literature does not corroborate this point, abundant material on "crime waves" created by pooling more resources into law enforcement may indirectly substantiate the above suggestion.

Secondly, the more rigid are conditions for treatment and rehabilitation, the bigger chances may exist for the success of the programme, but also the more probable become infringement of various rules lined up for determination of "recidivism". In other words, and generally, the more relaxed conditions of the programme, the bigger margin of ambiguity in evaluation of the programme's outcome.

Thirdly, specified methods of determining someone's "addiction" may also have an impact on the success rate of the programme. For example, if a test administered to the alleged recidivist addict produces results contested by two specialists and there is no policy that benefit of the doubt precludes a positive finding, then this policy directly increases a relapse record.

Fourthly, the length of a follow-up period, whether one, two, five or twelve years, may have an impact on how truly someone became psychologically independent of using drugs. The longer the period, the more reliable data in this regard.

Fifthly, and this especially refers to larger countries where exist many rehabilitation programmes, only a centralized system of information on patients may assist in assuring relatively reliable data on relapse. In other words, without such system and provided that interviews are not a day-to-day survey instrument, one should not say that "our programme is successful" because the former patient may show up in another place seeking treatment again, without acknowledging past addiction.

Sixthly, any changes in a relapse may have nothing to do with the terms of reference of the programme, but with general changes e.g. in demographic patterns. Assuming, for the sake of this paper, that treatment and rehabilitation can be most effectively administered to males between 16-20 years of age, one should make a comparative longitudinal analysis of this group's representation in the context of overall population trends, then among all known drug addicts and next among those treated and rehabilitated. In other words, conclusion that the programme "works" better or worse than before should be proceeded by an analysis of demographic changes within and outside a reviewed group. Such an analysis may show that as a result of ageing of society, the addicts also get older and, therefore, became less amenable to behaviour modification than earlier.

Finally, there exists quite a variety of other factors which should be accounted for, starting from different forms of behaviour modification (individual versus group), meaningful employment opportunities, personal life attitudes, etc., all of which may be relevant to the outcome of treatment and rehabilitation, hence to revision rate. As these were already exhaustively dealt with in UNAFEI before, they shall not be mentioned here.

Conclusions

In view of a perceived threat posed by illegal drug trafficking and related crimes to national security, stability and resilience it may be understandable why the national policies seem to justify resort to extreme measures. By resorting to them both the law-makers and society, at large, hope to curb the drug phenomenon. Yet when the postulate that "there ought to be the law" is satisfied, stark realities may frustrate previous expectations: the law may not deliver what was intended to. Further, it may raise the question whether it offers fair and humane treatment to those against whom it applies. Experience shows that laws should be formulated and operated with a reasonable degree of flexibility and should account for a variety of economic, social, cultural and other factors, if promise embodied in it, is to be fulfilled. The failure of criminal sanctions in general and the death penalty in particular to function as effective means of formal social control may reflect the fact that they are sometimes not articulated and imposed in terms of their direct implications in the potential offenders, immediate social environment. The chances to "get away" with luck from the criminal justice system for illegal trafficking in drugs are high. It is not suggested here that these chances will decrease, if the penalties will become harsher and harsher, nowhere some drug-related behaviour will be criminalized, for even the extreme penalty of death in the wake of socio-economic developments is a doubtful means to achieve a desired end. As a matter of fact, the society itself encourages to risk death essentially by instructing people to be irrational about death even if it's close by. Indeed, risking life may be a virtue, as it shows bravery, courage, patriotism and loyalty. Rewards for engaging in illegal activity are high: not only psychological gratification by the immediate environment but also a demonstrable financial gain. It is a trivial statement, but perhaps it needs to be reiterated: the core of the drug problem lies in the socio-economic realities of the present world. Criminal law only can react to these realities by discerning the areas where it should be and could be successfully applied.

Finally, one needs to emphasize that prison rehabilitation of drug abuse offenders is probably the last means of doing something with the addicts who go to prison because of drug taking is punished or because of commission of other punishable act. By its nature, prison is the worst kind of setting where any success in reformation comes hardest. Any other measures should be pursued by frustrated legislators before they decide to send a criminal, or for that matter, a drug addict there. Alternative means e.g. (suspension of imprisonment pending effects of commission to an out-patient treatment programme) should be first considered and tried. Perhaps only those offenders who show that their drug involvement demonstratably and significantly increases the chances of self-harm should be treated under strict regimes, including prison setting and all others as much as possible outside of it.

Even if something works in crime prevention and control field, those are the last measures to be helpful to such a case.

References

1. Adopted from *International Criminal Police Review*—March-April, 1988.
2. *The Times of India*, New Delhi, Saturday, Jan. 28, 1989.
3. In this connection, it could be mentioned that in 1981, 58000 automobile accidents in the state of New York were due to drug use. (NEWS FROM NARCONTROL, Oct., 1988).
4. Report of the Fifth United Nations Congress on the Prevention of Crime and the Treatment of Offenders (Geneva, 1-12 September, 1975), United Nations, New York 1976 (A/CONF 56/10, paras 73-75).
5. Specifically, the Congress requested that the Committee on Crime Prevention and Control should "give priority to the development of a model agreement for the transfer of prisoners with a view to presenting it to the General Assembly for consideration as soon as possible.
6. *See* further, General Report for the XIIIth Congress of the International Association of Penal Law, prepared by T-Vogler, *Revenue Internationale de Droit* Penal, Vol. 55/1984, pp. 17-20.
7. Definition of "narcotic drug and psychotropic substances" which are included under the Single Convention on Narcotic Drugs and Protocol amending it and the Convention on Psychotropic Substances, but also includes "any others that the World Health Organization may have found" and which generally speak after normal functioning of the central nervous system or the use of which may produce results similar to those caused by any of the substances indicated in the three Acts mentioned above (Document OEA/Ser.G. CP/doc. 1497/84, 13 September, 1984).
8. General Assembly, Report of the Economic and Social Council International Campaign Against Traffic in Drugs, doc. A/39/407.
9. This submission has a working relationship with a number of international organizations and entities, e.g. with the tenth meeting of operational heads of National Narcotics Law Enforcement Agencies (HNNLEA) which operated in middle Asia.
10. Economic and Social Council, Capital Punishment, Report of the Secretary-General, doc. E/1980/9, Add. 1-3 Corr. 1-2.
11. United Nations Division of Narcotic Drugs, Memorandum on Penal Sanctions for Narcotic Offences. A comparative study of selected countries, November 1982 (mimeographed).
12. *See* further, Memorandum, *op. cit.*, paras, 39-55.
13. G.F. Uelmen, Punishing Drug Offenders. An international and comparative perspective, *Journal of Drug Issues*, Fall 1981, pp. 373-77.
14. Both the Single Convention (Art. 36, para 1) and the Convention on Psychotropic Substances (Art. 22, para 1, sub-para (a)) refer only to "serious offences".
15. Economic and Social Council, Report of the Committee on Crime Prevention.
16. This kind of criminal intent extends only to the stage of entering the drug into the market through the chain of actions starting from cultivation to offering for sale. *Exceptionally*, however, there may be a direct intent of causing someone's death. e.g. the Philippine Dangerous Drugs Act of 1972 (Article III, Section IV) provides for the death penalty for traffickers when a particular drug was "the proximate cause of the death of a victim thereof", that is presumably, when it was offered by someone who was clearly aware of a direct poisonous effect of the substance (for instance, adulterated or counterfeit heroin), but not in the case of overdoses of a drug of a pure quality.
17. At the International Preparatory Meeting of Experts for the Seventh United Nations Congress on the Prevention of Crime and the Treatment of Offenders on Congress provisional item "Formulation and Application of United Nations Standards and Norms in Criminal Justice" (Varenna, Italy, 24-28 September, 1984) the following position was formulated: "many participants said that research on the issue of deterrence should be intensified. Indeed, the burden of proving that the death penalty had a deterrent effect was on those who favoured its retention. Some participants said that the claim of deterrence was a myth that should be exploded by sustained research and writing.... Participants felt that it was necessary to inform the public as well as decision-makers that the application of the death penalty at a time of rising crime rates did not, in fact, reduce the incidence of crime, as shown by abundant empirical evidence on the dynamics and patterns of crime" (Report, paras 69 and 67 respectively).

18. *See* further: Sixth United Nations Congress on the Prevention of Crime and the Treatment of Offenders, Capital Punishment. Working paper prepared by the Secretariat, doc. A/CONF. 87/9.
19. A.S. Trebach, The lesson of Ayatollah Khalkhali, *Journal of Drug Issues*, Fall 1981, pp. 384-86.
20. *Ibid.*
21. *Ibid.*, pp. 393-94.
22. Th. Sellin, The death penalty retribution or deterrence, UNAFEI *Resource Material Series* No. 13/1977, pp. 41-52; S. Redo, The question of capital punishment in the United Nations: Recent developments, Cahiers de Defesa Sociale, *Bulletin of the International Society of Social Defence*, No. 1/1984 (in print) and the entire volume of the *Journal of Criminal Law and Criminology* No. 3/1983.
23. "...No one shall be subjected to arbitrary...detention... except on such groups and in accordance with such procedure, as are established by law. Anyone...detained on a criminal charge shall be brought promptly before a judge or other officer authorized by law to exercise power and shall be entitled to trial within a reasonable time or to release.... Anyone who is deprived of his liberty by...detention should be entitled to take proceedings before a court, in order that court may decide without delay on the lawfulness of his detention and order his release if the detention is not lawful.... Anyone who has been the victim of unlawful...detention shall have an enforceable right to compensation."
24. *See* further: H.F. Rawlings, *habeas corpus* and preventive detention in Singapore and Malaysia, *Malaya Law Review* No. 2/1983, pp. 324-50, and report of the 67th UNAFEI International Training Course for a legal description of the Institution.
25. G.F. Uelmen, *op. cit.*, p. 374.
26. Memorandum, *op. cit.*, para 65.
27. E. van den Haag, *Publishing Criminals, Concerning a Very Old and Painful Question*, Basic Books, New York 1975, p. 127.
28. R. Martinson, What Works?—question and answers about prison reform, The Public Interest 35/1974, pp. 22-53; D. Lipton, R. Martinson, J. Wilks, *The Effectiveness of Correctional Treatment*, Praeger Publishers, New York, 1975.
29. Report of the International Repraly Meeting for the Seventh United Nations Congress on the Prevention of Crime and the Treatment of Offenders (Budapest, 4-8 June 1984), para 65.
30. Preventive and treatment measures to reduce drug abuse. Summary of responses to a survey of national programme, *Bulletin on Narcotics*, Vol. XXXV, No. 3/1983, p. 38.
31. L. Yew, Singapore's drug treatment and rehabilitation programme. Lecture for the UNAFEI's 67th International Training Course on the Integrated Approach to Drug Problems.

57

PRISONS AND DRUG CONTROL

When once the drug has taken hold on these people, they will do anything to acquire supply. The charges against dope fiends are usually petty crimes, they steal just enough to prepare them to obtain a supply of the drug. There are many ways to smuggle in the drug. On one occasion, a case of condensed milk was sent to a "fiend", did not show signs of having been tampered with. On prying off the lid, the searcher found morphine in the milk, and a note to the effect that more was coming. On another occasion a package of cigarette was sent to a prisoner, the government seal was unbroken. When opened, the cigarettes were found wrapped in silver leaf and upon examination proved to be loaded with morphine tablets. On another occasion two books were sent to an habitue, the searcher noticed a white powder on his desk. He opened the book, but could not find any trace of a drug. He slit open the binding and found two packages, one containing morphine and the other heroin. On still another occasion a magazine was sent in which, upon examination by the warden disclosed a picture pasted on one of the pages, under which a large amount of morphine was secreted. The ways in which drug smuggling is attempted are too numerous to mention in this book.[1]

Social Service Institutions

Outside institutions and associations can help drug misusers within the framework of the policy to open the prison as far as possible to outside society. Although the organisations of therapeutic communities and group therapy in prison are regarded by most experts as impractical, it is agreed that the intervention in prison of public or private institutions, drug counselling centres, therapeutic communities, etc. which could even detach some of their staff to work in prison on a permanent basis as 'Liaison Units (anti-users)' could be extremely useful to reinforce the assistance given to drug misusers by the prison services, encourage their motivation for treatment and prepare and organise participation upon release in a suitable therapy scheme, for example, in a therapeutic community.

After Prison Follow-up and Care

Follow-up programmes for reintegration into normal life devised in co-operation with the social services (in particular private institutions, voluntary associations or therapeutic communities), the prison authorities and the detained misuser himself, would be an extremely useful way of preparing his release and ensuring appropriate after-prison care.

In this connection though, practical problems like the insufficient staff of probation services or the inadequate communication between, on the one hand, medical practitioners and the staff of therapeutic communities and, on the other, judicial and prison authorities raises difficulties for the control and follow-up by the criminal justice system.

In view of the above, following definite conclusions are derived :

(i) The penal justice system is necessary and useful with regard to the treatment of offenders who are drug addicts, although its most important aspect is its contribution to the fight against drug trafficking.

(ii) It can constitute a starting point for encouraging drug addict offenders to undergo treatment.

(iii) It is a useful additional aid to their treatment.

It is, therefore, essential that governments should give the criminal justice system, the resources it needs to find solution to the problems of drug misusers.

Cracking Down on Drugs in Prison

Recent studies by the Correctional Service of Canada (CSC) and Queen's University have shown that as much as 70 per cent of federal penitentiary inmates have substance abuse problems. This comes as no surprise to those who have worked inside prisons. But there is growing recognition that a concentrated effort must be made to address this problem before real rehabilitation among any inmate population can be hoped for. This was the subject of a conference on the substance abusing offender, held in Kingston and co-sponsored by CSC, Queen's University and the Alcoholism and Drug Addiction Research Foundation.

A description of the problem inside Canadian penitentiaries would shock the uninitiated. While 70 per cent of inmates have substance abuse problems, nearly 80 per cent admit to using drugs, including alcohol, on the day they committed their offences. Fifteen per cent of federal inmates have a history of conviction for drug possession or trafficking or both.

These problems appear to blossom within the walls of our institutions. Solicitor General James Kelleher noted that nearly 60 per cent of the 200 major incidents of violence recorded by CSC in the last three years involved drugs. "And when I say violence," he continued, "I am referring to the worst kind : murder, serious assautls on staff and inmates, major disturbances, hostage takings and suicide". Cst. Rod Carscallen, a member of the Ontario Provincial Police Penitentiary Squad, paints an even more frightening picture. The squad, formed in 1984, investigates crimes both directly and indirectly associated with federal prisons in Ontario, particularly in the Kingston area. Carscallen and the other members of the squad have investigated 14 institutional murders, and he says that those were related in some way to substance abuse itself or drug trafficking.

Ken Allen and Karl Niemann are Institutional Preventive Security Officers at Joyceville Institution, just outside Kingston. They told conference delegates that alcohol is by far the most common and cheapest drug to be found inside institutions. Any food item containing sugar can be used to make it, including jam and ketchup. Alcohol is often combined with other drugs such as cannabis or valium which are smuggled in from the outside.

Such trafficking within and into institutions is not unlike organized crime activity outside, says Dave Boyd who is a security officer at Kingston Penitentiary, a primarily protective custody facility. The drugs and the money generated are not handled by those who make the ideals but rather by willing or coerced inmates lower down in the prison hierarchy. Money comes in and out of the institution through inmates on temporary absence programmes, institutional visits by everyone from friends to grandmothers.

Vehicles for smuggling money or drugs are as varied as the people who do it. Allen and Niemann described musical greeting cards filled with three or four grammes of marijuana, shoes with the heel hollowed out, frisbees and baseballs thrown over the institution's fence to eager hands—even inmates who have ingested money, syringe and containers filled with pills and other assorted drugs, hoping to pass them unnoticed later. Eighty per cent of offenders use drugs on the day of their offence.

What drives inmates to go to such lengths to obtain and abuse intoxicating substances? What causes them to ingest everything from home brew, paint thinner, shoe polish and windshield washer fluid to valium and cocaine and what can be done to help them? To date, no one seems to have much of an answer to these questions. Dr. Edward Gottheil, a professor of psychiatry at Thomas Jefferson University in Philadelphia, told conference delegates that researchers are still unclear about what links substance abuse with criminal behaviour. Gottheil said studies have been shown that some substances like cocaine and alcohol are related to increases in aggressiveness, and that stimulation of specific parts of the brain result in various forms of violent behaviour.

These results indicate a biological factor in violence related to substance abuse but Gottheil said that psychological factors also have an effect, noting that shame, fear and family background also contribute to criminal acts by substance abusers. He urged that more direct experimental studies on humans be undertaken in this area to draw together and pursue what is already known about the link between substance abuse and violence, and to arrive at some definite and useful answers.

Treatment of Drug Addicted Prison Inmates

The treatment of substance-abusing criminals is even more encumbered by uncertainty than the cause of their ailments. Dr. William Miller, professor of psychology and psychiatry at the University of New Mexico, enumerated for delegates the range of existing strategies for dealing with substance abuse. They fall into two basic categories; those that deal with the substance abuse itself, and those that are aimed at other areas—physical or behavioral—which have a bearing on the substance abuse, and include medication, aversion therapy, psychotherapy, group therapy, family counselling, and even acupuncture and surgery.

Proponents of these programmes have claimed varying degrees of overall success, with such strategies as behavioural self-control therapy, social skills training, community reinforcement and marital counselling appearing to be the most helpful. But Miller points out that the overall success rate is not an overly useful fact to know, since some treatments are more effective with certain kinds of people than others. Level of dependence, age, social stability—these are all variables in what Miller calls a very heterogenous inmate population, and the process of matching the individual to a programme is crucial. It is also a process that is often underemphasized in penitentiaries due to lack of resources, facilities or even ignorance, He should not accept the idea that all approaches to the treatment of substance abuse have equal merit for everyone.

Despite the lack of specific knowledge on the nature and treatment of inmate substance abuse, it is not a problem that can be ignored until answers are found. The cost of inmates, their families, institutional staff and society itself is too high. Solicitor General James Kelleher outlined steps being taken in federal prisons to address the problem. Part of the $ 210 million National Drug Strategy launched last year, these initiatives cover everything from treatment to research to strenghtened enforcement. For example, CSC has implemented a new training course for staff. By the end of March, several thousand staff members had been trained in such areas as drug recognition and effective intervention techniques. More recently, a collaborative programme has been developed between CSC and the RCMP which will see both agencies undertake a reciprocal training programme to orient new and current personnel to substance abuse and trafficking problems unique to the penitentiary setting.

But the inmates themselves must also be informed, and CSC has launched an information programme to make new inmates aware of the dangers of drug abuse as well as to advise them to treament programmes available in the institution.

There are now almost 100 programmes available to federal inmates with substance abuse problems and in 1986-87, CSC spent more than $ 1.3 million on community services and other programmes to fight substance use. One of the most prolific and universally recognized of these is Alcoholics Anonymous, which is available, to all federal inmates. A similar programme, Narcotics Anonymous, exists in 13 federal institutions.

But there is certainly no general consensus that A.A. and its spin-offs, no matter how popular, are the answer for all drug-abusing offenders. Consequently, many institutions have developed their own programmes. Their strategies range from one-to-one counselling and education to confrontation. One of the most successful is the Bentwood Recovery Programmes at Collins Bay Institution in Kingston which boasts a 75 per cent completion rate among its clientele.

A model program at Joyceville Institution is aiming its efforts specifically at inmates about to be released. Designed to help inmates cope with their problems and to identify community-based services that will support them following release, the programme is currently being field-tested and will be implemented at other institutions, if it proves effective. Urinalysis has also been established in three federal Institutions in Ontario. Widely used in the United States, the procedure remains controversial in Canada. But the Correctional Service believes that it can play an important part in reducing the demand for drugs inside institutions because treatment and the granting of privileges can be tied to test results. Expansion of the urinalysis programme has been postponed for now by a Quebec Superior Court ruling that the procedure could not be used randomly but only when an inmate was suspected of being under the influence of drugs.

Finally, additional resources will be devoted to stemming drug trafficking into and with penitentiaries. The RCMP will be assigning drug enforcement officers to work directly with CSC institutions to reduce the availability of drugs. and the service is considering the use of RCMP dogs to aid in general searches. CSC is also pursuing improved relationships with local police departments, the OPP and the **Surete du Quebec** in the detection and apprehension of people smuggling drugs into institutions.

Recent press stories have suggested that drug use may be commonplace even in provincial holding facilities. A former inmate at the Ottawa-Carleton Regional Detention Centre told the Ottawa Citizen that cocaine, hashish and other drugs are readily available. Clearly, the problem is becoming increasingly wide-spread at all levels, both in Canada and in the United States, where federal officials recently classified illegal drug use by inmates as a top level offence along with murder and hostage-taking. Staff and most inmates themselves, who are often the victims of such crime, are hopeful that the initiatives taken by the Canadian federal authorities will go a long way in solving the problem of inmates substance abuse and trafficking in our penitentiaries[2].

Prison and Drug Treatment

The treatment of a drug addict begins in prison immediately after his admittance. He is thoroughly searched by the keeper for hidden drugs. It is necessary that the keeper be an experienced searcher, so as not to overlook any drug, syringe or needle. The prisoner is stripped and every bit of body and clothing is searched. Particular attention is paid to the seams and pleates of the clothing. Quite frequently packages of drugs are found rivetted in the seams of trousers or pleates of shirts. Extra-hidden pockets are often found, like a detachable heel or concealing drug below the arm-pits in a pouch or in paste form.

Women inmates require a very thorough search. Each female suspect should be examined vaginally for it is not rare to find the morphine in a condom secreted in the vagina.

In treating drug habituals the following must be taken into consideration :

1. Physical condition of patients.
2. Period of addition.

3. Amount taken daily.
4. Reason, if any, for taking the drug.
5. Kind of drug taken.

Battling Drugs Behind Bars

Drug abuse has become so rampant in the Westlinge prison, HEERHUGOWAARD, Netherlands, that authorities are opening a drug-free wing. The move reflects an increasingly pragmatic approach in many European countries towards battling drugs behind bars. The restricted access prison block for 40 prisoners has been added according to the Warden Jan Vanden after a "wave of complaints about disgusting behaviour" among prison addicts. Westlinge's drug problem received heavy publicity when an inmate serving a five-month sentence for fraud and selling stolen goods said in an interview with 'De Telegraff', the nation's largest newspaper, that he had failed to return from a weekend furlough because he was frightened by the drug abuse in the prison.

Fact of Life

The Westlinge wing and other similar drug-free penal facilities around western Europe, are an admission by corrections authorities that drug abuse is a fact of life in prison. They also indicate a growing attempt to safeguard non-drug users in prison from the temptations of illicit drugs and the often violent behaviour that surrounds trafficking. European nations have come up with various plans to separate users and non-users of illicit drugs, ranging from drug-free wings like the one planned at the Westlinge facility to entire prisons in Sweden which are out-of-bounds to the drug-dependent. Addicted inmates make up a significant slice of the western European prison population, ranging from 6.2 per cent of prisoner in Britain to 40 per cent in Sweden according to figures obtained by Associated Press. Jacques Van Vliet, a Dutch Justice Ministry spokesman, estimated that 34 per cent of the Dutch prison population of about 5,400 prisoners are addicted to illicit drugs. The drug problem in Dutch prisons can be traced to both the liberal penal system and the relatively easy availability of illicit drugs in this nation where abuse is treated as a social ill rather than a criminal offence.

The Westlinge prison in this town 32 km north of Amsterdam is a so-called "half-open" facility, where inmates serving upto six months sentences enjoy limited freedom of movement including regular weekend leave. Prisoners live in four-person units, sharing two bedrooms and a living room within five 40-man residential blocks.

Lax Security

"Given the character of Westlinge, secutiry is on the lax side....people return from weekend leave and take drugs with them", warden Mr. Van Den Berg said adding that he lacks the staff to do the body searches on returning inmates. Dutch law currently bans mandatory urine tests on prisoners, but parliament is expected to pass a bill enabling legislation within the next few months. "The only way to cope with the problem is to increase repression within institution, and we have got neither the money nor the inclination to do so", he said.

Sweden has several drug-free prisons open only to inmates who are not drug dependent according to the Swedish Correctional Board spokesman, Mr. Ville Karlstrom, who conceded that "there is a (drug) problem which is hard to cope with" in the general prison population. In the neighbouring Norway, large prisons feature so-called drug-free" Contract wings", where prisoners are required to sign a contract allowing random urine tests and other checks. The system, which the Norwegian Justice Ministry spokesman, Mr. Hege Tveit, said is working successfully, is also foreseen for the planned drug-free wing at Westlinge. Although the number of nations giving preferential treatment to non-addicted prisoners is growing, other countries such as Belgium and Britain reject such compromise steps in favour of trying to keep all illicit drugs outside their prison walls, according to authorities in those nations.

In Britain, body searches on prisoners are compulsory after visits by friends and relatives and most prisons have sniffer dogs to facilitate drug detection. Physical contact between visitors and prisoners, such as kissing and embracing has been restricted to stem smuggling.

The Belgian Government also addressed the prison drug problem by trying to keep its prison system completely drug-free, according to a Belgian Corrections Department spokesman.

Althogh the legal provisions in force in most countries with regard to the resources available to actually implement these provisions. It is, therefore, essential that governments should give the criminal justice system the resources it needs to find a solution to the problem of drug misuse.

References

1. New York *Medical Journal*, Vol. 100 November 14, 1914, Dr. Perry M. Lichtenstein.

2. Liaison Vol. 14, No. 4—April 1988.

58

CRIMINALITY IN CONTEMPORARY WORLD*

UNAFEI, has published "Criminal Justice in Asia: the Quest for an Integrated Approach"[1] which reflects the current trends in prevention and control of crime.

At a global level the need for co-ordination of criminal justice operations was very strongly emphasized by the Sixth United Nations Congress on the Prevention of Crime and Treatment of Offenders (Caracas, 25 August—5 September, 1980). The Caracas Declaration adopted unanimously by this Congress underlined the importance of comprehensive policy on formulation and planning in criminal justice, and the Congress in separate resolutions 1 and 2 stressed also the need for better information on all aspects of crime prevention and control, recognizing the necessity for the entire machinery of criminal justice to be responsive to changing economic and social conditions in the society in which it operates[2].

This, general position of the international committee on co-ordination of activities of criminal justice agencies, covers a variety of aspects and control of criminality relating to drug abuse is one such aspect. This chapter focuses on a number of matters concerning this specific aspect.

The phenomenon of drug abuse has assumed such serious proportions that concerted steps are needed at the global, regional and national levels to curtail its spread. In the course of regional preparatory meetings for the Seventh United Nations Congress on the Prevention of Crime and the Treatment of Offenders held in Milan, Italy, 26 August—6 September, 1986, these proportions were perceived and reported by Member States. At the Asia and Pacific Regional Preparatory Meeting it was, *inter alia*, observed that "Crime across national boundaries was increasing significantly, especially illegal trafficking...of drugs....(Also) drugs abuse...seemed to grow with increased urbanization and unemployment. The relationship between drug abuse and drug related crimes, and between such crimes and the expansion of tourism and other aspects of economic development was particularly mentioned.[3]" At the Latin American Regional Preparatory Meeting, Member States reported that there was "alarming increase of criminality and delinquency in the region...(in) offences related to drug trafficking and consumption.[4]" Similarly, at the African Regional Preparatory Meeting it was noted that "the...main category of crime that had recently increased swiftly in most of the region was the use of and trafficking in drugs.... That was seen to be problem requiring inter-governmental collaborations.[5]" The Asian and Pacific Regional Preparatory Meeting added to the already expressed concern about dimensions of drug problem in that region that "drug related criminality was particularly a complex phenomenon because of the two quite different types of activities involved...(the) price difference between the legal and illegal markets (1 to 1,000) caused illicit trade to flourish.... In turn, illicit traffic led to organized smuggling, often involving violence against or corruption of authorities"[6].

These statements show a contemporary concern about the extent of drug problem and is supported by the following statistical account (Table 1):

Table-1. Totals of Drugs Reported Seized in the World

Drugs	*Year*		
	1975	*1980*	*1982*
	(in kg.)	(in kg.)	(in kg.)
Opium	50,000	51,786	45,753
Morphine	694	1,495	2,202
Heroin	1,708	2,510	6,153
Cocaine	2,405	11,820	12,092

* by Slawomir M. Redo—Social Affairs Officer, Crime Prevention and Criminal Justice Branch, United Nations. The views expressed here are those of the author and do not necessarily reflect the view on the United Nations Secretariat. Responsibility for these views also rests with the author alone.

Source: United Nations Division of Narcotic Drugs, Review of Trend in Drug Abuse and Illicit Traffic, doc. MNAR (1) 1984 (rounded up figures); Commission on Narcotic Drugs, Review of the Illicit Traffic in Narcotic Drugs and Psychotropic Substances during 1977, doc. E/CN. 7/630 (rounded up figures).

In Asia, the Far East Region and in Oceania, the reported seizures were shown in Table 2 :

Table-2. Totals of Drugs Reported Seized in Asia, the Far East Region, and in Oceania and as a Per Cent of World Seizures

Drugs	Year					
	1975 (in kg.)	%	1980 (in kg.)	%	1982 (in kg.)	%
Opium	13,418	26	7,462	14	11,626	25
Morphine	169	24	64	4	514	23
Heroin	429	25	385	15	1,266	20
Cocaine	7.5	0.03	7	0.05	9.5	0.07

Source : Ibidem

Asia, Far East, oceania, thus, occupy crucial place in narcotic drug trafficking. However these statistics do not reveal the whole truth because :

(i) the above data include only selected narcotic drugs and not seizures of raw material for such drugs (i.e. coca leaf) or seizures of cannabis and its resin. In 1982, the seizures of cannabis and its resin amounted in the region to 116,713 kgs. and 4,116 kgs. respectively (increase over 540% in seized cannabis and 235% in seized cannabis resin, both in comparison with 1975);

(ii) no information was provided on reported seizures of synthetic narcotics, stimulants, depressants, LSD and other hallucinogens;

(iii) although the precise estimates are not available, rough assessments made in 1970s by INTERPOL suggested that seizures reached usually no more than 5-10% of a total volume of illegal trafficking[7]. Improved law enforcement, increased rate of recorded illegal drug trafficking, but it would be impossible to specify their degree for the year 1982.

(iv) No comparable information is available on the extent of drug taking in the world or in specific regions, but according to some older estimates in mid 1970s there were about 2 million people dependent on heroin who consumed between 20 and 30 tons of illegally produced heroin every year. Other 0.5 to 1 million persons were dependent on opium, 0.5 million of persons who consistently used cocaine and about 40 to 50 million used *cannabis sativa*. No reliable world-wide estimate of number of persons dependent on synthetic drugs or psychotropic substances exist is available[8].

(v) Similarly, no in-depth information exists on the collaterals of drug problems: bribery, corruption, illegal sales in arms, forgeries, burglaries, etc. although it is clear, that such forms of drug related criminality exists and are on the increase.

In view of the above observations, it might be concluded that the drug problem, whether assessed in the statements, like those made in the course of preparations for the Seventh United Nations Congress or in statistical terms for the Commission on Narcotic Drugs and Other Relevant Bodies, dramatically falls short of showing all intricacies and real extent of this very dangerous phenomenon.

Drug Criminality and Co-ordinated Reaction of Criminal Justice System

In view of a reported increase of drug related criminality, a coherent reaction of the criminal justice system against this phenomenon may be crucial in its control and prevention. Not one criminal justice agency, but their system should counteract incidence of drug criminality. The old approach to crime proved itself ineffective, and, therefore, liable to criticism:

> "It immediately becomes obvious that every step of the process is dependent upon all the others if the system is to work. Those who think the solution is as simple and plain as "honest judges" may be little naive. Raising penalties...is useless if the police do not catch the guilty, or, worse, if arrests an innocent. The most honest and efficient police system will be pointless if its prosecutors are inept or, worse, corrupt. Dedicated prosecutors can't make a case out of botched police work or work through inefficient and graft-ridden courts. The best judiciary is dependent upon wise laws, honest cops and good prosecutors. But even with all these, if we sent out criminals to a place where we could simply forget about them—our jails, our system would certainly be a failure. If we neglect our jails, they would be cesspools of crime as finishing schools for criminals[9]."

At the International Preparatory Meeting of Experts on provisional agenda item of the Seventh United Nations Congress which is entitled "Criminal Justice Processes and Perspectives in a Changing World" held in Budapest from 4-8 June, 1984, the above concern on inefficiency of the criminal justice system supplemented by a clearcut example, by no means unique, which particularly refers to suppression of illegal drug trafficking in one developed country. This country is actually facing a large-scale illegal importation of cocaine and marijuana from Latin America. With the establishment of a task force concentrating on suppression of illegal drug trade, there was a considerable increase in the number of arrests to such an extent that the judiciary and correctional institutions were incapable of handling the case load. All but the most serious smuggling cases resulted in dismissals or token sanctions. Clearly the criminal justice system was not in a position to administer justice in an appropriate manner.

By pointing at this example, I would like to suggest that "co-ordination" of criminal justice operations is a prerequisite to deliver speedy and fair justice. This term seems to be more suitable than "integration" which implies a great deal of consensus on the goals and objective(s) of all criminal justice agencies. In line with this view, desirable as such consensus might be theoretically and politically, in practice both, consensus and conflict play role in criminal justice. It is frequently emphasized that existence of conflict may be a healthy factor, which prevents excessive influence of one control agency over another, and allows for a due process of law. However, "co-ordination" of criminal justice system operations also remain in the sphere of *desiderate*, so, in a sense, this terminological consideration may only help to clarify the subject under discussion.

The preliminary results of the Second United Nations Survey of Crime Trends, Operations of Criminal Justice System and Crime Prevention Strategies both clarify and amplify the drug crime. This survey was prepared by the United Nations Secretary General, and its results are now elaborated by the Crime Prevention and Criminal Justice Branch of the United Nations Centre for Social Development and Humanitarian Affairs for the Seventh Congress. The Questionnaire for the survey, which covered 1975-1980 period, contained the "drug crimes" as separate offence category whose incidence was measured at the stage of recording them by police or other law enforcement agency, at the stage of conviction, both relating to the "case" and the "offender", and at the stage of imprisonment as a "drug offender".

Recorded Drug Crimes

The "drug crimes" referred to in the questionnaire, following a technical explanation of terms provided therein denoted "international acts that may involve cultivation, production, manufacture, extraction, preparation, possession, offering for sale, distribution, purchase, sale, delivery on any terms whatsoever, brokerage, despatch, despatch in transit, importation and exportation of drugs or psychotropic substances (in accordance with)...the Single Convention on Narcotic Drugs of 1961 and other regulations adopted in pursuance of the provisions under the Convention on Psychotropic Substances of 1971". Table 3 presents a cross-regional breakdown of recorded drug crimes (1980) and as a per cent of change-over 1975-1980 period.

Table-3. Recorded Drug Crimes per 100,000 Population by Regions and as a per cent of Change 1975-1980

Region	*Regional Average*	*Totals for 1980*	*% of Change 1975-1980 (Second column)*
I	171	1,027	+ 67
II	271	542	+ 3.66
III	35	106	– 1.91
IV	1.66	3.32	+ 71
V	86	1,290	– 12
VI	7.85	15	+ 1.32
VII	16	97	+ 10

The following countries were included in the regional analyses:

I Bahamas, Belize, Chile, Colombia, Panama, Trinidad and Tobago.

II Canada, USA.

III Madagascar, Senegal, Seychelles,

IV Poland, Yugoslavia.

V Austria, Denmark, Finland, France, Germany, Federal Republic of Germany, Greece, Ireland, Israel, Italy, Netherlands, Norway, Sweden, United Kingdom (Northern Ireland, Scotland).

VI Kuwait, Qatar.

VII Bangladesh, Fiji, Japan, Korea, Republic of Pakistan, Thailand.

The Table shows that in the four out of seven "regions", the drug crime rate seems to rise. Of a special concern should be here both American (Region I and II). In Western Europe (V) the reported decrease may be due to Finland whose rate of drug crimes dropped from 748 in 1975 to only 32 in 1980 per 1,00,000 inhabitants. In Asia and Pacific Region percentage of recorded drug crimes rose by 10 in the six-year period.

In general, the above regional overview has only an illustrative value, for in case of each region different countries contributed to comparatively "higher" or "lower" than elsewhere "totals". Also such an amalgamation of data from seven countries of the region into one regional total does not help analyse intra-regional differences.

Table-4. Recorded Drug Crime Rate in Countries of the Asia and Pacific Region per 100,000 Population in 1975 and 1980 and as a per cent of Total Recorded Crime plus as a per cent of Change in Recorded Drug Crimes in 1975-1980.

Country	1975 Rate	1975 % in Structure	1980 Rate	1980 % in Structure	% of Change 1975-1980
Bangladesh	0.013	0.007	0.012	0.008	- 5
Fiji	2.21	0.07	1.11	0.05	- 49
Japan[1]	14.06	1.27	30.2	2.60	+ 114
Korea, Rep. of	1.45	0.08	2.76	0.13	+ 19
New Zealand[2]	93	1.25	276	2.52	+ 183
Pakistan[3]	26	12	12	6.38	- 55
Philippines[4]	1.83	1.84	2.60	5	+ 14
Singapore[5]	80	9.7	65	5.27	- 19
Sri Lanka[6]	28	4.51	42	10.1	+ 48
Thailand	44	5.02	52	11.2	+ 17

1. Including drug crimes under control law on stimulant drugs.
2. Change in statistical reporting system (1978).
3. Change in law in 1979.
4. Rate for 1977 instead of 1975; no data on a total number of crimes available.
5. Cases prosecuted.
6. Rate for 1977 instead of 1975.

A cursory analysis of these data shows that in six out of ten countries the recorded crime rate increased (most significantly in Sri Lanka and New Zealand), and in four of these decreased (most significantly in Fiji and Pakistan). In case of Pakistan these decreases may be related to introduction of Islamic Law in 1979, but no further information was forwarded as to the possible substantive or other factors which might have a bearing on reported drug crime rate. A change in statistical reporting system had no doubt an effect on the incidence of reported drug crime in New Zealand. In 1975-77 and 1978-80 period and, therefore, data for the first period were disregarded. Singapore reported on the number of cases prosecuted, hence these data cannot be relevant for intra-regional comparisons, but are pertinent only to Singapore itself. In case of Bangladesh the recorded drug crimes rate per 100,000 inhabitants was extremely low. In absolute numbers only 10 crimes were reported in 1975 and 11 in 1980. So a decrease of 5% due to an increase in population may say more about creation of statistics than of any real extent of drug crimes.

It would be interesting, however, to see whether the result could be, determined by using two sets of data: recorded drug crimes and reported drug seizures. For instance, a considerable increase in recorded drug crimes in Sri Lanka over 1977-80 period seems to be confirmed by a parallel increase in reported decrease of drugs in that country. While in 1977, it was 21,463,386 kg. of cannabis, its resin and plants plus 20 kg. of raw opium were reported seized, in 1980 about 67 per cent more of these drugs were seized. This seems to be, however, more a coincidence of trends than their inter-relationship, since in Bangladesh in 1980 about 65 per cent more drugs (opium, cannabis) were reported seized than in 1975, but at the same time there was decrease in a reported drug crime rate. In addition, without knowing more about the way in which reported crimes were recorded (one continuous activity may be registered as a sequence of separate crimes, or one organized large activity may be registered as one crime only) it is very difficult to arrive at any further clear observation. Finally, even if it is evident what kind and amount of drugs were reported seized, nothing can be known from the survey, if, thc activity concerned with the seized drug was recorded in national criminal justice statistics was without reference to the same drug.

In view of these very specific methodological difficulties, it would be more opportune to approach the question of criminal justice system response to drug phenomenon from another angle. To determine the degree of such response, let us find out what was the percentage of drug crime in the overall structure of recorded criminality. Table 4 shows that in three (Pakistan, Singapore and Thailand) out of 10 countries of the region the proportion of drug crimes in all recorded crimes seemed to be quite sizable, ranging from 5% to 12% in 1975 and from 5% to 11% in 1980. But because of factors mentioned above, caution must be exercised in assessing reliability of such data. In six out of seven other countries, presence of these crimes in the overall structure of criminality was almost visible both in 1975 and 1980. This could be interpreted either as an indication of dearth of drug criminality or, what could be more probable in some countries, as almost total lack of reaction of law enforcement agencies to drug criminality. Recalling estimates made by INTERPOL it is suggested that the second interpretation is more plausible and, in addition, concurs with a current view of criminal justice statistics as a result of an interaction process between the actual amount of law-breaking and the response of the authorities and the society to these acts. In other words, such statistics does not measure crime "by itself", but rather reflects the police workload, their decisions within the framework of all these activities, movements in investigations, changes in criminal justice policies and willingness to report a crime. To conclude, the above data may be accepted as confirmation of a very limited involvement of law enforcement agencies in combating the drug phenomenon.

Convictions for Drug Crimes

The data sent by countries responding to the questionnaire on the Second Survey allow also for analysing the incidence of drug crimes and drug offenders at the conviction stage. Convictions were defined very broadly and referred to "pronouncements of guilt in respect of penal code offences or equivalent by a criminal court or some legal body duly authorised to do so under national law whether later on upheld or not". The questionnaire included in this item also adjudications of juveniles, so these comprised data categories could be comparable with other general category of "recorded crimes" which implies that the age of the perpetrator is unknown. Table 5 presents available data along with the data on rate of convicted drug offenders. Both main items contain also information on percentage of such convictions of drug offenders within a total number of decisions made in the case or on the perpetrator. These tables seem to reinforce the observations based on analysis of data from Table 4 that in Singapore and Thailand convictions for drug crimes are relatively as large in rate and proportion as at the stage of recording of drug crimes. However, the data on convicted offenders provided in Table 5 should be treated with great caution, as in some replies they exactly match the data on convictions (e.g. Japan and New Zealand). In some other countries (e.g. Thailand) in the number of persons convicted for drug crimes were included persons convicted for "alcoholism offences", hence all the data had to be disregarded. Only the data from the Republic of Korea seemed to be reliable, suggesting that the rate of convicted drug offenders (0.67 per 100,000 population) was two and a half times higher than the rate of convictions apparently due to possible involvement of two or more perpetrators in one organized crime.

Table-5. Rate of Convictions including Adjudications for Drug Crimes and of Convicted Drug Offenders (Adult/Juvenile) per 100,000 Population and as a per cent of Total Number in 1980

Country	*Convictions* Rate	%	*Convicts* Rate	%
Bangladesh	0.0011	0.010	—	—
Indonesia	0.21	0.44	—	—
Fiji	—	—	—	0.17
Japan	12	29	—	—
Korea, Rep. of	0.26	0.004	0.67	0.10
New Zealand	147	1.52	—	—
Singapore	33	14	—	—
Thailand	51	10	0.90	—

Imprisonment for Drug Crimes

The Second Survey provided also information on prisoners sentenced for drug crimes. Information in this regard was supplied by only five countries of the region and is shown in Table 6. Two countries (the Philippines and Singapore) indicated that the data supplied were not entirely reliable for the reasons specified in the table.

Table-6. Rate of Imprisoned Drug Offenders (Adult/Juvenile) per 100,000 Population and as a per cent of Total Number of Imprisoned Offenders

Country and Year	*1 Year* Rate	%	*2 Years* Rate	%
India (1980)	—	—	0.11	1.32
New Zealand (1978, 80)	5.45	7.7	7.11	8.51
Philippines[1] (1975, 80)	0.086	0.17	0.026	0.07
Singapore[2] (1975, 79)	21	21	37	34
Sri Lanka (1975, 79)	11	27	16	47

1. Adults only.
2. Figures included persons imprisoned for consuming controlled drugs.

Because of the latter country no continuum is possible with data for Table 5, but within the purview of the qualification made for Singapore on the contents of data provided for Table 6 and Table 4 it may be observed that the criminal justice system of this country quite consistently processes the drug crimes and offenders. This is also in the case of Sri Lanka (Tables 4 and 6). However, a more comprehensive and thorough view of criminal justice system performance requires a separate insight into this matter, which is provided in Table 7. This table was constructed to illustrate a flow of drug crimes and/or offenders on the assumption that a number of drug crimes recorded by the police was equivalent to 100 per cent and that a recorded crime is next processed by subsequent criminal justice agencies, i.e. when the alleged perpetrator is apprehended, charged, sentenced and imprisoned. The entire process may happen with a different time lag. The first variant assumes that an offender is imprisoned for his drug crime at the year of its commission (e.g. 1980). Second variant assumes that between commission of an offence and imprisonment elapsed two years time (1979-80), and according to the third variant, that this took three years (1978-80). For instance, in pursuance of the first variant for Singapore (1980) the relative percentage of cases/offenders processed would be 51% of convictions and 57% of prisoners serving sentence of a drug crime. According to the second variant the respective figures would be 66% and 74%, and according to the third variant, 46% and 52%. In case of Japan the first variant, for example as of 1978, shows that convictions for drug crimes were only 38% of the total of crimes recorded and there was only 16% of prisoners serving sentences for drug crimes out of 100% of drug crimes recorded. In case of New Zealand, if an offender was convicted in 1979 for a crime recorded in 1978 and sent to prison in 1980, then the relative percentages purporting to illustrate the dynamics of criminal justice process would be 44% and 4.77%. In other words, out of 100% of drug crimes recorded in 1978, 44% of these were acted upon by the court, as of 1979, and only in 4.77% of cases the offender went to prison as of 1980.

Table-7. Flow of Drug Crimes/Offenders through Criminal Justice System (Stage of Criminal Justice Process)

Country & Year	*Crimes Recorded (%)*	*Convictions (%)*			*Convicts (%)*			*Prisoners (%)*		
		*I**	*II**	*III**	*I**	*II**	*III**	*I**	*II**	*III**
Bangladesh										
1978	100	5.80	—	—	—	—	—	—	—	—
1979	100	6.25	—	—	—	—	—	—	—	—
1980	100	9.09	6.25	5.88	—	—	—	—	—	—
Fiji										
1978	100	—	—	—	81	—	—	—	—	—
1979	100	—	—	—	57	—	—	—	—	—
1980	100	—	—	—	71	57	31	—	—	—
Japan (I)										
1978	100	38	—	—	—	—	—	16	—	—
1979	100	39	—	—	—	—	—	19	22	23
1980	100	41	43	46	—	—	—	21	—	—
Korea, Rep. of										
1978	100	52	—	—	—	—	—	—	—	—
1979	100	49	—	—	—	—	—	—	—	—
1980	100	9	19	17	—	—	—	—	—	—
New Zealand										
1978	100	57	—	—	—	—	—	3.62	—	—
1979	100	44	—	—	—	—	—	3.31	—	—
1980	100	53	66	99	—	—	—	2.57	3.21	4.77
Singapore										
1978	100	36	—	—	—	—	—	68	—	—
1979	100	40	—	—	—	—	—	98	—	—
1980	100	51	66	46	—	—	—	57	74	52
Sri Lanka										
1978	100	—	—	—	—	—	—	—	—	—
1979	100	—	—	—	—	—	—	46	—	—
1980	100	—	—	—	—	—	—	37	63	52
Thailand										
1978	100	52	—	—	—	—	—	—	—	—
1979	100	72	—	—	—	—	—	—	—	—
1980	100	98	84	75	—	—	—	—	—	—

* First variant—as if crime is processed within one year (1980), second variant as within two years 1979-80, third variant as within three years (1978-80).

1. However, if data on number of drug crimes under the law on stimulant drugs were excluded, relative percentages of convictions (I variant) would be as follows; 724, 856 and 753; II variant—948; III variant—861. Percentages of drug prisoners would be as follows: I variant—310, 425, 385, II variant—485, III variant—440.

The above mentioned matrix served here only as a theoretical demonstration for a simple evaluation of the flow of cases/persons through criminal justice system. Such matrixes would have been more adequate and hence, useful if more were known on, e.g.—

(i) the average actual length of criminal justice process for "crimes recorded" involving drug crimes (i.e., from recording stage through upheld conviction of effective incapacitation);

(ii) relative frequency, with which the courts resort to imprisonment in comparison with alternative sanctions (the more often and longer imprisonment, the less frequent releases and the more inmates in prison, hence, the stronger pressure on the courts not to resort to imprisonment);

(iii) about the drug offender at the law enforcement stage (one unit of court).

Prevention of Drug Crimes

The data obtained through the Second Survey allow also for making some general observations on what is regarded as a "tail-end" of criminal justice system; more, however, because of a general underestimation of its role than a true value of crime prevention.

In brief, the Second Survey requested information on a number of possible activities aimed at prevention of crime which could be principally divided in two groups: those intended to improve quality of life and thus indirectly influencing the level of crime, and those directly designed to prevent crime. The first group included, *inter alia*, provision of leisure time upgrading community life by subsidies, providing welfare and financial assistance. The second group dealt, among others, with preventing policing, use of special architectural techniques, citizen crime prevention associations, publicity campaign, visits to correctional institutions for psychological deterrence.

As it has been expected, received replies were very impressionistic and general, and thus contrasted with quite detailed answers on the questions relating to operations of other components of criminal justice system. Such a way of responding was very understandable as the reported activities could not be quantified in terms of "hard" indicators of performance (like e.g. the recorded crime rate). Moreover, as a general rule, no central crime prevention agency exists which could have been in a position to collate such a dispersed information for presentation in the reply to the Second Survey. In addition, information provided to questions on indirect crime prevention strategies may have an entirely different meaning; for instance, improvement in quality of life may for one country be equal with food supply or liquidation of slums, but for other countries these may imply further refinements in social security scheme, including better pensions.

However, it being understood that so far not much has been done to update knowledge on existing crime prevention strategies in the framework of co-ordinated reaction to criminal justice system against criminality, such replies should be regarded but a first step which contributes to a more comprehensive approach to drug crimes, especially, since no clear indications were given whether they were applied to prevention of illegal trafficking or illegal use of drugs. Such a differentiation is very critical if we wish to know what we are talking about. Within the purview of these qualifications, information on prevention of drug crimes could be summarised as follows:

(i) several countries mentioned their efforts to combat drug criminality through *legislation*. The responses do not always indicate the contents of such legislation but it might be reasonably assumed that these efforts involve criminalization of certain acts or imposing of new penalties for drug crimes (e.g., the Philippines);

(ii) some countries report that *publicity, anti-drug campaigns* and consultation on drug issues take place (the Philippines) and a number of others refer to such activities in general terms without pointing at the prevention of drug phenomenon (Bangladesh, Fiji, New Zealand, Japan and Thailand);

(iii) some countries emphasized that through *increased effectiveness of law enforcement* agencies it was intended to obtain better results in prevention of drug crimes (in the Philippines police received special in-service training);

(iv) a number of other replies suggested that there existed an array of other crime prevention activities, however, no particular reference was made to their applicability for prevention of drug crimes.

The very scanty information on prevention of drug crimes should not suggest that such activity does not exist, but rather confirm, the difficulties in their recording and reporting and uncertainty over what one really can expect of prevention in comparison with traditional *post delicto* repressive way of controlling these crimes.

The existence of this dilemma could be demonstrated on the basis of a reply of one country of the region which, on one side, reported a plethora of general crime prevention activities, but, on the other side, observed that: "the main problem (in implementing and evaluating crime prevention activities) is an apparent apathy in all levels of the community. In the police, as in other countries, reactive police policies take precedence over proactive policies. How can you measure the worth of crime prevention activities?"

Towards Evaluation of Crime Prevention Activities

To answer this very legitimate question, let us observe that in criminology exist a vast number of theories of crime and deviance and quite a large number of techniques to evaluate effectiveness of operations of criminal justice system. While some of these theories may have specific relation to the drug problem, others account for it through a broader spectrum of controlling all undesirable behaviours and through searching of common denominators of these behaviours. Likewise, whereas some techniques could be applicable to assess the performance of most components of criminal justice system (e.g. matrix of criminal justice process), some other techniques may prove suitable mostly for a very selected area of intervention.

The replies of the Second Survey indicate that these quite theoretical concerns relating to methods of evaluation of crime prevention activities are confronted with stark realities. Most countries reported obstacles in making such evaluations resulting from shortage of financial resources because of organizational shortcomings and lack of modern technologies[14]. In effect, and this is very relevant to most of the countries of Asia and Pacific region, the only, if any, method pursued in this regard relies on data from traditional criminal justice statistics and thus, invites inherent to it methodological difficulties.

The above observation was not meant here to discourage from resorting to such a traditional technique, but rather to emphasize its limitations, which, if recognised, may still be of a great help as better techniques are lacking. For instance, this method was reported to be in use in Singapore for measurement of effectiveness of an experimental Neighbourhood Watch Scheme. In the same country, other more complex techniques were applied with regard to other projects. No doubt, surveys on self-reported crime and delinquency and victimization surveys come immediately to mind as "other complex techniques". These may be used prior to introduction of a specific drug crime prevention activity and after its cessation. On this issue exists quite a vast criminological literature[15] that may be of assistance should conditions for application of these techniques materialize. So long this cannot happen, regional exchange of experiences on prevention of drug crimes may not be viable.

Conclusion

This chapter seeks to overview the operations of criminal justice system in countries of Asia and Pacific Region with regard to drug crimes on the basis of the results of the Second United Nations Survey of Crime Trends. Operations of Criminal Justice System and Crime Prevention Strategies (1975-80). Specifically, attempt was made to look at these operations from the angle of co-ordination of activities of criminal justice agencies and consideration of possible "bottlenecks" in these agencies' approach to the drug problem.

The proceeding analysis suggests the following tentative conclusions:

(i) in view of the reported increase in seizures of illegal trafficked drugs and due to other inter-related factors not considered here, reaction of criminal justice systems ranged from an increased involvement in processing of drug crimes and offenders to a lack of such an involvement;

(ii) available data may be quite illustrative in a general demonstration of the flow of drug crimes/offenders through criminal justice system, provided that attention will be paid to reliability of these data. At the same time, dearth of many of the requested data may indicate that more detailed evaluation of performance of criminal justice agencies is not yet possible. These data, if available, may be very useful for self-evaluation of such performance by criminal justice practitioners;

(iii) more attention should be paid to provision of specific drug crime prevention activities and their subsequent evaluation. Availability of such information may be important for a more meaningful regional exchange of experiences.

References

1. B.J. George (Ed.) *Criminal Justice in Asia*: the Quest for an Integrated Approach, UNAFEI, Tokyo, 1982.
2. Sixth United Nations Congress on the Prevention of Crime and the Treatment of Offenders, report prepared by the Secretariat, United Nations, New York 1981, doc. A/Conf. 87/14, pp. 3-6.
3. Report of the Asia and Pacific Regional Preparatory Meeting for the Seventh United Nations Congress on the Prevention of Crime and the Treatment of Offenders, doc. A/CONF. 121/RPM/2 para. 13.
4. Report of the Latin American Regional Report Meeting, doc. A/CONF. 121/RPM/3, para. 7.
5. Report of the Regional Preparatory Meeting, doc. A/CONF. 121/RPM/4, para 13.
6. Doc. A/CONF. 121/2/RPM. *op. cit.*, para 13.
7. Fifth United Nations Congress on the Prevention of Crime and the Treatment of Offenders, Changes in Forms and Dimensions of Criminality—International and National, working paper prepared by the Secretariat, doc. A/CONF. 56/3, para 40.
8. S. Redo, Narkomania. *Aspekty Prawnokarnei Kryminologiczne* (Drug Abuse, Legal and Criminological Aspects), Nicolas Copernicus University Press, Torun 1979, p. 106.
9. Jerry Barrican, quoted in: C.C. Campos. Development in Criminal justice planning and administrations in the Asia and Pacific region, with particular reference to the conflict of the objectives between the different agencies and recent changes of priorities and practices observable in the region. Consultant's Paper on Topic 2 of the provisional agenda item of the Seventh United Nations Congress, p. 38 (mimeographed).
10. G.O.W. Mueller, criminal justice processes and perspectives in a changing world, experts paper for the Inter-regional Meeting of Experts on Topic 2 of the provisional agenda item of the Seventh United Nations Congress, p. 13 (mimeographed).
11. One country (Sri Lanka), with data for 1977, should be analysed separately.
12. See, in this connection, F. Bruno, *Combating Drug Abuse and Related Crime*. Comparative research on the effectiveness of socio-legal preventive and control measures in different countries on the interaction between criminal behaviour and drug abuse, United Nations Social Defence Research Institute, Publication No. 21, funded by the United Nations Fund for Drug Abuse Control, Fratelli Plombi Edition, Rome, 198, pp. 58-63.
13. For a more sophisticated analysis see: G.P. Whitaker, Ch. D. Phillips, (Ed.), *Evaluating Performance of Criminal Justice Agencies*. Sage Publications, Beverly Hills-London-New Delhi 1983.
14. See further: Preventive and Treatment Measure to Reduce Drug Abuse: Summary of Responses to a Survey of National

Programme, *Bulletin on Narcotics*, Vol. XXV, No. 3, July-Sept. 1983, pp. 33-40.

15. See, for example, U. Buikhuisen, L.J.M. d'Anjou, ***Analysing Evaluative Research.*** Research and Documentation Center, Ministry of Justice, Netherlands 1975. The authors distinguish five designs: (i) the pre-test—post-test (control group) designs; (ii) the non-equivalent (control group) design; (iii) the time series experiment; (iv) the ex-post correlational design; and (v) the one group (pre-test—post-test) design. These measures differ greatly as to their sophistication and methodological rigidity. Those mentioned under (iii) and (v) seem to be the simplest ones, but all should probably be carried out by outside teams with a strong research and statistics background. However, alternative and simple methods exist which allow for an internal evaluation of crime prevention activity.

See further: M.J. Seng, Criminal Justice Evaluation: Exploring Some Myths, *Journal of Police Science and Administration* No. 2/1983, pp. 206-10.

59

DRUG DEPENDENCE : PREVENTION AND TREATMENT

Despite ever increasing awareness of the drug problem, searching studies into the causes of drug addiction, the responses made at the international level and encouraging feed back received from correctional centres, the drug abuse situation continues to remain grave and in most parts of the world shows signs of deterioration. A new pattern is emerging in the use of drugs and new regions are being opened up to drugs, both natural and synthetic.

The struggle against drug abuse is first and foremost a fight against the degradation and destruction of the human being. The cost of drug abuse is immense, particularly when we take into account the crime and violence that emerge in its wake and the ensuing impact on moral values leaving alone generation of black money that further corrodes moral fibre and degenerates humanity. The drug abuse and its debasing influence block national development programme in many countries and cause a seroius drain on badly needed developmental resources. Thus drug control is an international challenge and failure on this front will threaten the very existence of human society. The international community has accepted this challenge, which is reflected in various conventions and international and regional meetings of the countries threatened by this menace.

The drugs specially those of organic origin, being used from early days have assumed now international dimensions in the wake of new strides made in the field of transport and communication. The wide gap in prices of these drugs in the place of origin and in the place of supply has attracted sharpest brains to the illicit trade.

Suppliers and Users

The countries have been divided into users and suppliers categories. India is situated in between 'poppy bowl' of S.E. Asia, known as 'golden triangle' covering Laos, Myanmar and Thailand and areas of 'Golden Crescent' covering Afghanistan, Iran and Pakistan. By virtue of its strategic geographic position and proximity to poppy producing regions, India has become a transit zone for international trafficking. Some of the smaller countries of South East Asia have introduced a very stringent system of punishment to deal with drug addiction and trafficking which also diverts supply from South East Asia to countries of South Asia for illicit international trade. Recent seizures more than evidence this trend. However, in recent years a comparison of seizure with respect to illegal import and attempted export does indicate that a sizable amount of drugs is being consumed within the country itself. This disturbing trend called for immediate measures to control production, manufacture of illicit drugs transport and international trafficking with heavy hands.

From time immemorial certain plants or their parts have been used by people for relieving pain, and obtaining pleasure. During last two centuries with the development of science active ingredients of these plants were found and were made available to societies after extraction or chemical synthesis. Some of these drugs originally used for medicinal purposes developed dependence among the users who ended up as drug addicts. The dependence or addiction could be psychic dependence or physical dependence. The psychic dependence relates to a situation where "there is a feeling of satisfaction and a psychic drive that requires periodic or continuous administration of the drugs to produce pleasure or to avoid discomforts. Indeed, "this mental state is the most powerful of all the factors involved in cronic intoxication with psychotropic drugs and with certain types of drugs it may be the only factor involved."

Physical dependence can be described as "an adaptive state that manifests itself by intense physical disturbances when the administration of the drug is suspended or when its action is affected by the administration of a specific antagonist. These disturbances i.e., the withdrawal or abstinence syndromes, are made up of specific arrays of symptoms and signs of psychic and physical nature that are characteristic for each drug type. These conditions are relieved by readministration of the same drug or of another drug of similar pharmacological action within the same generic type. No overt manifestation of physical dependence is evident if an adequate dosage is maintained. Physical dependence is a powerful factor in reinforcing the influence of psychic dependence upon continuing drug use or relaps to drug use after attempted withdrawal."

These drugs when used for non-medical purposes cause concern to society as their abuse can be harmful to the user, to the society and at times to both. A durg has a capacity to interact with a living organism to produce a state of psychic or physical dependence or both. When taken for non-medical purposes it is the subject matter of grave concern to the society and it has been generally accepted that their production sale and consumption should not remain uncontrolled by the Government with a view of restricting their use to legal, medical and scientific requirements.

Types of Drugs

The drugs are usually referred as narcotic drugs and psychotropic substances. These narcotic drugs and psychotropic substances can be divided on the basis of their origin in the following two groups :

(i) Natural drugs.

(ii) Synthetic drugs.

(i) Natural Drugs— Natural drugs are mostly derived from plants (i) Hemp poppy, (ii) the coca bush and (iii) the hemp cannabis. These drugs have been placed under the National Control by the Single convention on Narcotic Drugs of 1961.

(ii) Synthetic Drugs—These drugs include nearly 100 substances which have been included in the list of the drugs control under 1971 convention of psychotropic substances.

The narcotic drugs and psychotropic substances comprise a range of materials which can be classified according to their effect upon a person's mental activity and psychic state as follows:-

(i) Depressants—These drugs are useful for medical purposes : for relieving pain, 'Analgesic'; for inducing or maintaining sleep; 'Hypnotics'; for suppressing nervous excitement 'Sedatives'; or for relieving anxiety 'Tranquillizers'.

(ii) Stimulants—These drugs affect mental activity and induce state of excitement and can be useful for mental concentration and activity. They also suppress appetite.

(iii) Hallucinogenic Substances—Produce hallucination or illusion. The most widely used hallucinogenic substance (LSD) is the best non-psychodynamic drug. It produces disturbances in perception, thinking mood or behaviour.

(iv) Cannabis—The products derived from Cannabis Sativa are known under hundreds of diffenent names in different parts of the world.

The drugs as defined in international convention are being used in many parts of the world for many centuries but their use was confined to medicinal purposes. The use for non-medical purposes was limited to those who were living in periphery of a society and abuse of drugs was never taken as a civilised way of living in enlightened society. This type of restricted use did not affect health of the society, therefore, did not attract social reformers in the earlier period.

There has been an overall revolution in contacts among different people which led to opening of drugs to various regions, certain new drugs were also produced which spread drug abuse. The multiple drug use as well as sequential substitution of one drug for another was common but simultaneous use of more than one type of drug is a recent phenomenon which is the direct result of availability of various types of drugs and now the drug abuse is not confined to oral administration or smoking but it has acquired new dimensions as certain salts can be used for injection without fear of blocking the circulatory system. Certain inhalants have also come in the market.

Depressants—Opium, Opium Alkaloids and Synthetic Narcotic

(i) Opium—Opium is the coagulated juice of the poppy plant botanically known as Papaver, Somniferum. It was medically administered for cough and diarrhoea as well as to reduce pain and induce sleep but now it has been largely replaced in therapy by its isolated Alkaloids i.e. Morphine and Codeine and also Synthetic Narcotic such as Pethidine and Methodone. In due course of time raw opium which was a useful therapeutic agent has become an important raw material for the manufacture of Morphine and Codeine.

Opium as a drug of pleasure was originally eaten, or drunk as an infusion. Later on smoking of opium also found its way along with tobacco in the countries of South-West, South and South-East Asia for smoking purposes. The problem of opium addiction is almost restricted to some countries in Asia and Far East.

The opium consists of the partly dried latex obtained from unripe capsules of opium poppy. Incisions are made in the capsules of the poppy which produce a mulberry like substance which hardens and turns to a dull reddish brown gum. This gum is raw opium which is put in boiling water and subjected to filtration followed by purification roasting. This prepared opium is used for smoking purposes.

Opium is dark brown also black in colour. It has a pungent slightly ammoniacal smell which may be compared to molasses, very sticky to touch. The drug is usually found in following forms:

—Dark brown sticky resin.

—Hard dark brown wafers

—Small dark brown sticks and granules.

—Large light brown lumps.

—A dark brown spiritous liquid.

(Opium cocktail).

—Very dark strong smelling, charred lumps.

—"Opium drugs".

(ii) Morphine—Morphine was widely used for short term acute pain resulting from surgery, fracture, burns etc. and in the stage of terminal illness such as cancer. With the advent of Narcotic drugs produced synthetically the extent of use and the therapeutic importance of Morphine has been considerably diminished but it serves as the standard of pain relief by which other Narcotic drugs are measured. The Morphine dependents have become addicts during long term treatment with morphine, they are called as "Therapeutic addicts". At times morphine is also used by Heroin addicts as a substitute.

The general term 'morphine' applies to both morphine base as well as its salt. Morphine is extracted from all parts of the poppy plant excepting the seeds. The capsule contains the largest proportion of morphine. The morphine is found in the following form:

—As powder

—In aqueous solution

—Tablets

—Salts with various acids such as Acitate;

—Hydrochloride, Sulphate, Tartrate.

The morphine is generally adiminstered by abusers by intravenous injection commonly referred as "shooting'. At times morphine solution available in ampoules is also taken by addicts without 'straining' with the help of codeine wool or filter. The pills of morphine are usually crushed into powder and prepared for injection. Normally the injection is put into the veins inside the arms. A belt, tie or similar item is placed around the body of the arm to tighten the vein which is known as "bringing up" the vein. The needle point is then struck into veins and the plunger of the syringe is pulled back so that blood oozes out. This process is known as Spiking and is intended to remove air bubbles. The euphoric condition is obtained within seconds.

For medical purposes the applied morphine is 15 mg. In case of acute pain 30 mg. may be taken yet an abuser who has become tolerant to morphine may use as much as 60 mg. of morphine a day.

(iii) Heroin—Heroin is 5 to 10 times more potent than morphine and its produce is most addictive of all drugs known. The intense euphoria produced by the drug has made heroin the most popular morphine derivative among the addicts. The addicts enjoy the fruits immediately after an injection. The victim develops tolerance very rapidly so that the larger quantity must be injected to gain the same effect.

The Heroin can be administered as follows:—

(a) Injection.

(b) Inhalation.

(a) Injection—This form of abuse is mainly prevalent in Western countries. The first experience can be by way of sniffing and second stage can be subcutaneous injection skin-popping and in the third stage the addict progresses with intravenous injection known as 'mainlining'. Adulteration of Heroin 'cutting' with milk, sugar, quinine, etc. makes it less potent with the result the addict does not get more than 3 to 5 per cent heroin in the white stuff available in the streets, on the long road from the smuggler to the street addict. Pure heroin is diluted to the extent that it loses much of its potency. If some one offers higher percentage of heroin in the street then the result can be disastrous for the addict who may even die from over-dosage. Further medical complications might be caused by substances used for adulteration. Unsterilised needle can also carry disease such as AIDS. These complications are additional to physical deterioration caused by heroin. Multiple drug usage is quite common among heroin abusers and in certain countries mainlining heroin cocaine mixture ("Speed balls") is quite common. Some of the addicts have started even mixing Amphetemines to counter-act the depressing effect caused by heroin and to intensify flash.

(b) Heroin Inhalation—A common mode of administration of Heroin by addicts in the East and Far East is by inhalation. The techniques adopted are as follows:—

—Chasing the dragon

—Ack-Ack or firing the anti-aircraft gun.

Chasing the dragon from substances in which heroin is mixed with several parts of adulterating powder on a piece of tin foil which has been folded longitudinally. The mixture is heated with lighted match with powder and the molten mass is rolled back and forth on the tin foil while addict inhales the fumes which evolve. The fumes appear to take a shape of undulating tail of a dragon hence 'chasing the dragon'. If the fumes are inhaled through a rectangular match box cover this is known as playing the 'mouth organ'.

Firing Anti-air Craft Gun—The light end of a cigarette is dipped into heroin. The cigarette is smoked with the head slightly tilted backwards so that the lighted end of the cigarette points towards sky like the anti-air craft gun hence the name Ack-Ack. A variant of Ack-Ack is smoking of cigarette containing heroin.

Under the influence of Heroin the abuser is generally lethargic and indifferent to his environment and personal situation. Chronic use may lead to both physical and psychological dependence as the quantity of the drug increases the addict's activities become increasingly drug centred. When drug supplies are cut off characteristic withdrawal symptoms develop.

(iv) Codeine and Other Morphine Derivatives—Codeine occurs in Opium in low concentration. Therefore, it is less addictive than Morphine. Codeine is an effective cough suppressant. Normally codeine is prepared by conversion of morphine. The codeine addicts are relatively rare.

The addiction may develop after prolonged use of codeine. Other morphine derivatives are not very important as far as dependence is concerned.

(v) Synthetic Narcotic—The term synthetic narcotics is generally applied to the drugs which have come in the market as Analgesics since the discovery of Pethidine some 40-45 years ago. These synthetic drugs also have addiction potential. No doubt addiction is slower to develop and it is less vicious than morphine or heroin. Some of the drugs of this category are Pethidine and Methodone. There are a number of Therapeutic addicts who have become dependent on this drug during long term medical treatment. The incident of abuse of synthetics is relatively high among the professional people like nurses and physician and students of Medical Colleges.

Depressants of the Category of Hypnotics, Sedatives and Tranquillizers

(vi) Barbiturates—These drugs have been widely used for Therapy for nearly 60 years. They are being used for medical purposes and these may be considered as safe medicines in small doses or even in large doses under proper medical supervision. In certain countries these drugs are not subject to medical control, therefore, their availability is not restricted by medical prescriptions. This lack of control gives rise to opportunities for their abuse. After regular and prolonged use of barbiturates, there is great danger of dependence. The uncontrolled use of these drugs may lead to multiple abuse i.e., barbiturates can be used along with other intoxicating substances. This phenomenon has greatly contributed to the growth of an illicit market which has assumed alarming proportions in a number of countries.

The barbiturates intoxication closely resembles alcoholic intoxication. Chronic abuse of barbiturates develops tolerance; both physical and psychological dependence. In a barbiturate addict, abrupt withdrawal is extremely dangerous. Convulsion, which may follow withdrawal can be fatal. There can be a period of mental confusion and delirium and hallucinations may occur.

Barbiturates used in Association with Heroin

(a) Chasing the dragon in Asia and the Far East is the use of a mixture of heroin and barbiturates.

(b) Barbiturates are used for adulteration in heroin.

Barbiturates Used in Association with Alcohol

This type of abuse is very dangerous because of the increased toxicity of barbiturates in the presence of alcohol in man.

Barbiturates Used in Association with Stimulants

There are some pharmaceutical preparations in the market which contain such ingredients.

(vii) Tranquillizers

Large quantities of tranquillizers are used in therapy. In many countries production, trade and distribution of these substances are not subject to effective control.

(viii) Inhalation of Solvents ("Glue Sniffing")

All these products first stimulate and produce euphoria but they are all central nervous system depressants and these include model aeroplane glue, lacquers, diluents, cleaning fluids. The cases of inhalation of petrol have also been noted.

The adolescents pour glue in a paper or plastic bag, place it in front of the face and inhale the fumes. Another method is to soak a handkerchief with glue and apply it to the mouth. Still another method is to pour glue into a container and then the container is heated. Petrol fumes are usually inhaled directly.

It is well known that many solvents can damage the kidneys, liver, heart, blood and nervous system. The practice easily induces adolescents to begin experimenting with Narcotic drugs or other psychotropic substances.

Abuse of Stimulants

(i) Coca Leaf

Coca leaves are leaves of the evergreen shrub *Erythroxylon coca* which grows in Western South America. They have been chewed for centuries in some parts of South America by highland peoples. They are also the raw material for the manufacture of cocaine.

The habit of coca chewing is conditioned by a

number of unfavourable social and economic factors. Coca, by diminishing the sensation of hunger, has an unfavourable effect on the nutritional state of the habitual chewer in contradiction to the erroneous claims that coca-leaf chewing in some way acts as a substitute for food. Malnutrition, and conditions, in which nutritional deficiencies are important disease determinants, occur more frequently in coca chewers than among non-chewers. Coca chewing leads to a state of relative indifference. The work performance of coca chewers is, contrary to some erroneous beliefs, adversely affected by chewing.

(ii) Cocaine

Cocaine, a powerful stimulant drug, is the principal alkaloid of coca leaf. Cocaine can be extracted from coca leaves or manufactured by chemical synthesis. It was first used as a local anaesthetic a century ago, but its importance for therapy disappeared with the introduction of other anaesthetic drugs.

Cocaine is not very commonly used in India though it is quite popular with labour class in America.

The stimulant effect of cocaine results in excitability, talkativeness and a diminished feeling of fatigue. Cocaine may produce a sense of euphoria and increased muscular strength; it may also produce anxiety, fear and hallucinations. Stimulation is followed by a period of depression. In overdosage, cocaine may so depress respiration and heart function that results in deaths.

(iii) Amphetamine-type

The term "amphetamine" includes all derivatives of the chemical group amphetamine. The therapeutic value of such medicines is rather limited, being confined, in practice, to two types of cases, in which they are used (i) as stimulants, and (ii) for the treatment of obesity. The stimulating effect of these substances is thought to be responsible for loss of appetite.

The trends in the abuse of amphetamines and the forms which it takes are very varied and complex, and what follows serves only to give a general idea of the seriousness and the consequences of the different forms of such abuse.

(a) *Abuse of Stimulants to Increase Concentration or to Eliminate Fatigue and Sleepiness* : It is amphetamines which are used most frequently for doping athletes during competition. Since, under the effect of amphetamines, the athlete no longer feels fatigue he is able to go beyond the limit of his physical strength, and this, combined with over-exhaustion and certain atmospheric conditions (heat, etc.), may result in death.

The abuse of amphetamines by drivers of motor vehicles, particularly lorry drivers, has assumed substantial proportions in sereval countries. Under the action of the stimulant the driver does not feel sleepy, but sleepiness comes suddenly when the effect wears off; this explains the fact that among those responsible for serious accidents there is a considerable number of drivers who fall asleep while driving after having taken stimulants.

The danger of the abuse of amphetamines by young people is not so direct and immediate, but the development of dependence resulting from the taking of amphetamines by students to help concentration in the course of their studies and examinations and by young factory workers to do away with the need for sleep during the week-end, constitutes a potential danger of addiction.

(b) Abuse of stimulants to produce a change in mood: It is chiefly middle aged women, including a large number of obese persons who start taking anorexigenic medicines regularly and repeatedly on a doctor's prescription. The stimulating effect of anoretic amphetamines is the principal reason why a substantial proportion of such persons increase the dosage in the course of time, and dependence develops.

(c) Intravenous injection of stimulants : The taking of excessive doses of amphetamines intravenously is spreading among young people in certain countries. Pharmaceutical preparations generally contain 10 to 20 milligrammes of active constituent in an ampoule or tablet, while drug addicts take several grammes (i.e. one hundred times as much) intravenously (there are descriptions in medical literature of cases in which "travellers" have taken 1 gramme every two hours, that is to say 12 grammes per day for a week). Intravenous administration is the most dangerous form of all the types of abuse of amphetamines. Psychosis often results from such abuse.

Persons with a dependence on heroin sometimes try to replace it by amphetamines or take amphetamines in combination with heroin.

Abuse of Hallucinogens

Hallucinogens constitute a heterogenous class of substances not under international control, belonging to a wide variety of chemical and pharmacological groups, abuse of which for the purpose of causing hallucinations and illusions has been noted.

The principal problems created by the use and abuse of hallucinogens, which are very different from those posed by the abuse of depressants and stimulants, are as follows:—

(a) the known hallucinogens are not medicines;

(b) since, except for scientific experiments, there is no "lawful consumption" of hallucinogenic substances, no distinction can be made between their use and abuse, for the present, their consumption is an abuse;

(c) the production of trade in and "consumption" of hallucinogens being, in the vast majority of cases, illegal acts, no one really knows either the quantities produced or the number of consumers.

Among the large number of substances used at present, the main one is certainly LSD.

"LSD" (=LSD 25) is an abbreviation of the German expression for lysergic acid diethylamide (Lysergsaure diathylamid). LSD is a synthetic or more precisely a semi-synthetic compound, since it is not found in any natural substance; but lysergic acid, the raw material required for its production, is a natural product found in ergot of rye (Secale cornutum).

For a "psychedelic trip", 100-250 microgrammes LSD is used in the form of a diluted solution or in capsule form. Drops of the solution are usually taken on a lump of sugar.

Some people use STP or a mixture of LSD and STP. The letters "STP" (Serenity, Tranquillity, Peace) designate a synthetic compound produced exclusively in "underground" laboratories.

Other hallucinogens the abuse of which is wide spread are derivatives of tryptamine in particular dimethyltryptamine (DMT) and diethyltryptamine (DET), but according to "local sources" other substances are used.

In addition to these "classic" substances "New" toxic substances are often also taken, and these experiments frequently end in the hospital for those trying them.

Abuse of Cannabis

In small amounts, cannabis acts as a mild euphoriant and sedative; in larger doses, its effects resemble more closely to those of the hallucinogens than any other pharmacological cateogry; tetrahydrocannabinol, its active principle, is a very potent hallucinogenic drug.

The cannabis plant or the crude drug derived from it and folk preparation of cannabis are known under hundreds of different names. The intoxicating constituents of the cannabis resin (known, *inter alia* as hashish) are found in the flowering tops, particularly of the female plant. The resin contains most, if not all, of the active principal tetrahydrocannabinol (THC).

Tetrahydrocannabinol (THC) is a potent hallucinogen, its pharmacological effects show several similarities to those of LSD, i.e. changes in mood, alterations in sense of time, distortion in visual and auditory perception, depersonalization, de-realizsation, auditory and optical hallucination; but there are indications that the two drugs probably act by different biochemical mechanisms, and their mental effects are somewhat different.

There are also indications that the effects of cannabis depend on the amount of THC in the "cannabis" used, because most of the phenomena experienced with LSD, such as depersonalization, marked visual and temporal distortion, and hallucinations, have been observed with sufficiently large amounts of marijuana and especially with hashish.

The great differences in potency of the cannabis in different preparations could be explained on the basis of differences in the THC concentration.

In common parlance the use of term "cannabis" differs from this definition, and the expresssions "cannabis", "marijuana" or "hash" embrace a range of different forms of "cannabis", such as the plant *(cannabis sativa)*, the leaves or flowering tops of this plant, the pure resin and different preparations of which the parts of the plant or its resin are the base. It is impossible to compare the principle active components of the different types of "cannabis". (Let us imagine a situation in which the term "opium" included the poppy plant, poppy straw, raw and prepared opium, tincture of opium, opium extract, pantopon, morphine, codeine, heroin, etc.) in the majority of cases it is impossible to establish from the available statistics of case of abuse, whether the abuse involves the utilisation of the dried leaves of the plant, the flowering tops of the female plant, the resin preparations, or mixture of cannabis and other plants, such as tobacco.

The impression is gained, however, that among those in favour of the legalization of marijuana, many have never used cannabis in its concentrated form and only have experience of mixtures of cannabis and tobacco or other plants, or have only smoked dried leaves of cannabis.

In certain countries, the consumption of cannabis has been a traditional custom for centuries. This is particularly the case in Islamic countries where the consumption of alcohol is prohibited. Similarly in India, Hindus also use dried leaves on certain festivals while some become dependant on it as a result of continuous use. One might imagine that with the development of these countries and the change in the ancient social structure, the traditional consumption of cannabis would perhaps disappear.

In certain countries in Africa, Asia and the Far East, cannabis has been commonly used by the lower socio-economic groups for centuries; in North America and in certain European countries, cannabis smoking is a new phenomenon. During the past two decades the use of cannabis has spread among the younger generation in some of the large metropolitan cities, and, although reliable data are not available, it is clear that the spectacular increase is taking place among students and other middle and upper class groups in urban centres, specially in the United States. The phenomenon is closely connected with the abuse of other hallucinogenic drugs by the same section of youth, and with the other problems arising from the use of "mind-expanding drugs".

The prevention and control of drugs abuse and the rehabilitation of drug addiction is interwoven in the concept of Welfare State. The directive principles of State Policy in our Constitution set forth the goal under Article 47 which provides that the State shall regard the raising of the level of the nutrition and the standard of living of its people and the improvement of public health as among its primary duties. Particularly, the State shall endeavour to bring prohibition of the consumption except for medicinal purposes to intoxicating drugs and of drugs which are injurious to health.

To achieve the goal set forth in the form of constitutional provisions, the Government have initiated an integral policy to reduce supply of drugs by regulating production, distribution, possession, sale and export, to punish the traffickers and treat the victims who have developed drug dependence.

Attention is being paid to improve social, cultural and economic conditions, which were found responsible for drug abuse. Rehabilitation of de-addicted persons is also attracting social attention, both under the patronage of government agencies and voluntary organisations.

Legislative Action

In consonance with the tenets of modern jurisprudence, the victim under the influence of intoxicant, escapes even the criminal liability in certain conditions.

The drug addict does not attract any penal provisions or disqualification under any law except criminal provision controlling social behaviour of an individual. Despite the laxity in the legislative provision the government even before Independence was sensitive to this health hazard. The anxiety of the government is reflected in promulgation of various Acts for proper control of Narcotics as well as other psychotropic substances.

As early as 1893 Royal Indian Hemp Commission was constituted by the then government of India, which came to the conclusion that occasional and moderate use of hemp might be beneficial for health, though excessive use can be injurious. Since the Commission was constituted to give opinion on taxation laws in pre-independent India, therefore, the conclusion arrived at may not be of any use in the present context. In the year 1904 Ewens carried out a sample survey, but that was confined to narrow field. During the year 1930 Dhungi Bhoy also carried out a survey but its result was also not of much consequence. During the year 1939 Chopra and Chopra reported on long-term cannabis use, based on data on 1238 cannabis users. Chopra and Chopra also arrived at the conclusion that moderate use of cannabis, either for medicinal or habitual use had no appreciable harmful physical effect, while heavy use by smoking cause a bronchitis or chest infection and it was also not associated with any moral degradation.

As far as Opium is concerned, as early as 1857 first Opium Act was promulgated. Then again in 1878 another Opium Act was promulgated to control production transport, manufacture and sale of opium and opium products. During the first half of 20th Century various other acts were promulgated, such as The Dangerous Drug Act 1930, The Drugs and Cosmetics Act 1940. After Independence, problem received some pointed attention and Drug and Magic remedy (Objectionable Material) Act 1954, The Drug Control Act 1950 came into being.

After Single Convention of Narcotics Drug 1961 and Convention Psychotropic Substances 1971 this problem received further attention. The problem of illicit trafficking was also studied in a proper perspective and legislation to control growth of black economy was brought about in the form of Foreign Exchange Regulation Act 1973, Conservation of Foreign Exchange and Prevention of Smuggling Activities Act 1974, The Smugglers and Foreign Exchange Manipulators Act 1976 to provide powerful weapon against the smugglers and drug traffickers.

With the passage of Narcotic Drugs and Psychotropic Substances Act of 1985, an attempt has been made to "consolidate and amend the law relating to Narcotics Drugs to make stringent provisions for the control and regulation of operations relating to Narcotic Drugs and Psychotropic Substances and for matters connected therewith". The important points of Narcotic Drugs and Psychotropic Substances are as under:—

(i) Provision has been made to have Narcotic Drugs and Psychotropic Substances Consultative Committee and also have Sub-Committee/ Committees for the consideration of particular matter.

(ii) The Central Govt. has also appointed Narcotic Commissioner to exercise all powers and perform all functions relating to superintendence of cultivation of opium poppy and production of opium.

(iii) Stringent punishment has been provided for contravention of provisions of Act and Rules made under it.

(iv) For the first time the provision has been made to release the offender with his consent from undergoing treatment and also to release him on execution of a bond.

(v) Provision has also been made to obtain security for abstaining from commission of offence.

(vi) Certain presumptions have also been raised in this Act, such as presumption of culpable mental state, presumption of possession of illicit articles, presumption of possession of certain documents in certain cases.

(vii) Bar of jurisdiction.

(viii) The schedule appended with the Act gives list of Narcotic Drugs and Psychotropic Substances.

(ix) Power of Government to establish centres for identification, treatment, education, after-care, rehabilitation, social re-integration and supply of drugs of addicts registered with the Government.

The administration of law related to Narcotic Drugs and Psychotropic Substances is the responsibility of following agencies:

(i) Department of Narcotics.

(ii) Department of Central Excise & Customs.

(iii) Department of Revenue Intelligence.

(iv) Department of Central Intelligence Bureau.

(v) Border Security Force.

(vi) Central Bureau of Investigation.

(vii) State Police and State Excise.

Criminal Behaviour and Drug Addiction

The jurists, criminalogist, psychologist, social reformers and policemen agree to the fact that criminality is accentuated by drug addiction.

The addiction and trafficking in drugs makes the subject delinquent *ipso facto*. Further the compulsion dictates a particular mode of behaviour on the part of drug addicts.

Strictly speaking drug addiction is a public order crime in which drugs are administered voluntarily in violation of law and as such it is a crime against the perpetrator himself and the society.

It has been proved by medical evidence that drug addicts who have become dependent on depressants become so indifferent to their environment and depressed that the commission of crimes involving violence, such as murders, rape become beyond their capability. However, the application of stimulants make the users more excitable and more violent. Normally athletes and players dope themselves to give better performance on the field. At times criminals take these stimulants with or without alcohol to prepare themselves for the commission of ghastly crimes. The financial compulsion also leads to commission of crimes of particular nature, such as theft, robbery, burglary, forgery, embezzlement and fraud, shop lifting and counterfeiting etc. The majority of drug users derive the supply from gang of hardened criminals, who are traffickers in drugs. By constant association with such hardened criminals the addicts are also initiated in the underworld of crime.

Measures for Treatment, Rehabilitation and Social Reintegration

As far as police is concerned, the Narcotic Drugs & Psychotropic Substances Act, 1985 has for the first time made provisions for the drug addicts. Section 27 of the said Act provides for punishment for those who are found in possession of small quantities of drug for self use. Further Section 39 provides for medical treatment as an alternative to imprisonment. The provision for furnishing bonds and release of the offender after due admonishing are the new provisions on which will depend the future treatment and reintegration system. Section 71 also provides that the Government may in its discretion establish centres for identification, treatment, education after careful rehabilitation and social reintegration of addicts. In the same Section provision has been made for the supply of narcotic drugs and psychotropic substances to the addict on medical advice which gives an indication for the line of treatment to be adopted in various centres likely to be established.

The treatment, rehabilitation and social integration of drug dependent persons is a continuous process to achieve drug free existence of such persons and to re-establish them as useful members of the society with proper attitude towards life. All the measures for the treatment, rehabilitation, social re-integration and to prevent relapse should be so dove-tailed that once a man is taken to rehabilitation centre he comes out as a useful member of the society in general without any stigma, whatsoever. Psychological strength is very important and without it the subject will not be able to cope with the problems of leading a balanced life and avail himself of the opportunities provided in the community.

Objectives of Treatment and Rehabilitation

(a) To achieve a state of abstinence and a more respectable way of life.

(b) To stabilise and supervise a drug addict in order to facilitate rehabilitation and social re-integration.

(c) To achieve a general reduction in drug abuse and trafficking.

Treatment Measures

The treatment measures will normally include the following :

(i) Management of the effects of overdose.

(ii) Management of emergencies associated with the withdrawal from drugs.

(iii) Medical-psychiatric emergencies arising from drug abuse.

(iv) Detoxification of individuals.

(v) Maintenance of addicted persons on an appropriate diagnosis.

(vi) Administration of narcotic antagonist which blocks the effects of opiates, making their administration futile.

(vii) Helping addicted persons to achieve a drug free existence on an out-patient basis.

Re-integration Measures

(i) Programmes to upgrade the educational qualifications and skills of the addict.

(ii) Vocational training.

(iii) The finding of suitable families willing to accept younger addicts who lack a suitable family setting.

(iv) The operation of half-way houses to enable addicts to move gradually from an institutional setting to an independent life.

(v) Provision for follow-up by way of periodical medical examination and monitoring about the job contents of the detoxified client.

(vi) Psychological approach to sustain the motivation of the addict for detoxification.'

Measures to reduce the demand for drugs must also be fully supported by the measures to reduce the illicit supply. Efforts should be made to strengthen or maintain control of pharmaceutical products which are liable to abuse, and to strengthen law enforcement agencies to suppress illicit cultivation, manufacture and distribution.

The evaluation to determine the effectiveness of the various measures to reduce illicit demand is also very important. The evaluation should be a continuous process which will determine revision, rejection or further strengthening of programme for treatment, rehabilitation and re-integration measures.

The problem of drug abuse can be fought only with the help of society in general. Any governmental or voluntary agency cannot fight this menace alone. To fight this menace one has to create atmosphere which is not conducive for drug use or its continuance and for this, participation of broader section of society is needed. Society's permissive attitude has to be changed and it is to be prepared to bear extra cost for the controls and efforts on discouraging use of drugs and also for providing attractive alternatives and correctional services.

Treatment and Re-habilitation Strategies

Treatment and rehabilitation can be categorised as intervention designed

(a) to contain, control or eliminate drug misuse;

(b) to provide changes in life-style (behaviour);

(c) to treat medical complications associated with drug misuse.

This intervention category includes: short-term de-toxification; long-term de-toxification and substitute therapy or maintenance; psychiatric treatment, such as psychotherapy and psychoanalysis.

Short-term Detoxification

Current research suggests that there is no optimum method for de-toxification applicable to all patients. Therefore, it is appropriate to individualize the de-toxification procedure. Decisions concerning the rate of detoxification, the use of methadone or other de-toxification methods, and whether de-toxification should be done on an in-patient basis, whether dosing is blind or open, self-regulated or physician-regulated, cannot be generalized and have to be based on a clinical assessment of the patient's needs, and on the experience and resources of the treatment staff.

The objective of the de-toxification procedure is to provide symptomatic relief from the abstinence syndrome while physical dependence is being eliminated. As there is cross-tolerance and cross-dependence between all opiates, in cases of opiate dependence (e.g. heroin or morphine), any of them can be used to eliminate withdrawal symptoms and to detoxify the addicted person gradually. Methadone has been widely used in the de-toxification process, particularly in the 1970s. With the discovery of narcotic antagonists, such as naloxone and naltrexone, de-toxification treatments have achieved greater success. Other pharmaceutical products, such as clonidine, have also been used by some therapists. Acupuncture has been used for de-toxification instead of chemical substances, but its effectiveness and potential role in de-toxification need to be studied further.

The "cold turkey" method of de-toxification does not involve the use of any substance to minimise the discomforts of withdrawal symptoms. This method is used in Singapore; its duration being about one to three weeks. Counselling and other rehabilitative services may begin during the de-toxification peroid. The de-toxification period may last from 21 to 45 days. The majority of short-term de-toxification programmes are conducted on an in-patient basis.

Long-term De-toxification and Substitute Therapy or Maintenance

The substitution of illicit opiates by a licit drug, such as methadone, is used in a number of countries. As in short-term de-toxification, decisions about appropriate treatment needs are based on an assessment of the individual client and her/his particular circumstances. The goals of treatment are to assist drug misusers control their chaotic drug misuse and to provide them with tools to function adequately within society. Substitute drug therapy (maintenance) seeks to redude illicit drug use and its associated criminal and anti-social behaviour and to eliminate the medical complications associated with unsupervised self-administration of illicit drugs. The provision of substitute drugs permits some people to achieve stability and to concentrate their attention on resolving personal problems. For example, a method of therapy has been developed in out-patient programmes in the Netherlands and the United Kingdom. It lasts for six months or more, during which time methadone in an oral form is prescribed and a range of supportive therapies are offered. The goal is complete de-toxification by the end of the programme.

Prescribing substitute drugs without a time-limit, usually referred to as maintenance, is a controversial modality. Before the rapid increase in drug misuse among young people in the United Kingdom in the early 1960s, it was the accepted treatment in that country. Initial approaches to drug treatment between 1963 and the early 1970s involved the long-term prescription of heroin and cocaine, both as a means of attracting and retaining drug misusers in treatment and as a means of alleviating some of the pathological effects of drug misuse without necessarily achieving full recovery. Also in the 1960s work began with the use of methadone as a substitute for opiates. In a maintenance programme, methadone had some advantage over heroin, in that it was a longer-acting drug requiring less frequent administration. The results suggested that those persons who were admitted into methadone maintenance had considerably less criminal involvement and were more able to function effectively within the community.

The whole concept of maintenace has been undermined because while many other programmes have been evaluated as successful even though they have shown relatively small number of abstinents (drug free), maintenance has been severely criticized as unethical and ineffective because it does not aim at or abstinence. Programmes for methadone maintenance have rarely been evaluated on the basis of their success in reducing reliance on illicit drugs, and criminal and anti-social activity or in improving the social functioning of the maintained individual.

The methadone maintenance remains controversial. During the past decade several countries, in addition to the United States, have introduced methadone maintenance programmes either on a pilot basis or as a permanent service (for example, the Netherlands, Sweden and Switzerland). The most outstanding example in the argument for methadone maintenance is found in Hong Kong, where statistics strongly suggest that clients who choose to be on maintenance can function normally in their everyday life. Furthermore, there is evidence that drug-related crime in Hong Kong has substantially decreased since the establishment of maintenance programmes.

It must be emphasized that, in all instances where substitute drugs are used, a range of support services are also needed and are often provided. Thus, substitute therapy can be seen as having two components: long-term de-toxification, commonly on an out-patient basis; and maintenance on a substitute drug for an unlimited period of time. The distinction between the two processes, maintenance and long-term de-toxification, is often misunderstood.

Psychiatric Treatment

Psychiatric treatment is provided in both in-patient and out-patient services, sometimes in conjunction with substitute therapy. It offers a large variety of treatment programmes for drug addicts, including milieu therapy, the application of behaviour modification, family therapy, logotherapy and individual and group psychotherapy.

The psychological approach emphasizes the importance of individual motives and attitudes in the occurrence of drug misuse. The relationship between internal motives and beliefs and external behaviour, gives rise to two important corollaries in approaching the treatment of drug-dependence. The first is the assumption that all behaviour is a result of certain motives and, therefore, all drug-dependent behaviour is motivated. The second is that a change in attitude may result in a change in behaviour.

The treatment of drug-dependent persons should take into account two aspects, described below. The first is to find an alternative to drug use. If it is accepted that drug-dependent behaviour is motivated, then the motives can be isolated and, with a little imagination, alternatives can be provided. These alternatives are not the exclusive domain of any one profession. They require co-operation between physician, sociologist and psychologist and the logistic support of administrators for their implementation.

The second aspect relates to attitudinal change and how this change can be brought about in drug-dependent persons. By far the most important tools by which to change attitudes are education and information. Just as it is impossible to consider approaches to treatment without recourse to the underlying aetiological theory, it is impossible to conceive an approach to treatment that can succeed without due attention being given to the role of, and necessity for education.

Intervention to Achieve Changes in Life-style

The intervention modalities that are aimed at achieving changes in life-style include therapeutic communities, traditional medicine, acupuncture, yoga and meditation, spiritual or religious approaches (often in combination with other methods), and self-help groups.

(a) Therapeutic Community: Therapeutic communities (TCs) have been established in many countries and were initially modelled after the Maxwell Jones communities for psychiatric patients. The model essentially insists on a high degree of participation by patients in decision-making and programme organization.

It has been found necessary to restructure the democratically oriented TC models in response to changes in the residents' demographic, personality structure and pattern of drug dependence. Tchtenhagen indicates that the democratic TC model is more adequate, at least in Europe, for abusers of hallucinogenic substances who are looking for a "new identity", whereas persons who are dependent on amphetamines and heroin, and have severe social and personality problems and whose addiction is motivated by an attempt to avoid depression and feelings of annihilation, are less prepared to cope with the unstructured, democratically oriented TCs. The majority of TCs, however, are structured in a hierarchical way and resemble the type of concept adopted by Phoenix House or Daytop, New York. This type of TC is more acceptable as a treatment modality for intravenous drug users.

There are also elements within TCs that have been less successful. For example, drop-out rates within the first four weeks of admission is relatively high (approximately 50 per cent). It is important for TCs to examine their programmes to identify the factors that lead to this high drop-out rate and to develop new approaches that can assist in the retention of residents. It may be helpful to develop a greater participation of residents in planning not only the operation but also the programme of the TC. The rules of many TCs are oriented towards an ideal life style, which is difficult to maintain outside the artificial life of the community. For instance, complete honesty or direct confrontation are not concepts readily used within the work place or the family. If residents are not assisted to cope with the transition from the TC into the wider community, they may return to drug misuse as a means of coping with the personal and social crises.

(b) Traditional Medicine : The success and content of traditional medicine approaches depend very heavily on the cultural and social environment where they are practised. Reports indicate that certain methods are being used in India, based on the Ayurveda medicines.

In Malaysia, traditional healing is based on archaic animistic beliefs, i.e. a person is vulnerable to the influence of the supernatural as well as to physical causes. Drug addiction is seen as a result of incorrect behaviour (predisposing cause) and the presence of a supranatural element, which has negative effects on physical well-being. In view of the cultural or religious background of the population, traditional healers (bomoh) can provide an important treatment modality for some drug addicts. The treatment consists of de-toxification, with the help of herbal medicine, physical exercise and balanced nutrition to build up physical health. In other parts of the world, in particular in Africa and Latin America, similar treatment approaches exist.

(c) Acupuncture : Acupuncture is used in detoxification process in Hong Kong, the United States, Vietnam and elsewhere. Ear acupuncture (without electrical stimulation) can be applied and taught effectively in clinical or non-clinical settings at low cost. It is reported that initial ear treatment relieves 90 per cent of the withdrawal symptoms related to heroin or morphine addiction. Ear acupuncture has been shown to have a significant preventive effect in the reduction of physical craving patterns. Acupuncture is used in both medical and non-medical settings.

(d) Yoga and Meditation: In certain settings, yoga and meditation are used as an important part of behavioural treatment. They include breathing techniques, body postures, and meditation and relaxation techniques as means of providing clients with the tools to improve their personal strength to free themselves of their habits. They have shown certain success in improving inter-personal skills after repeated sessions.

(e) Religious Approach: In certain countries, religion plays an important role in the treatment and rehabilitation process. The use of religious teaching as a psychological reinforcement tool has proved to be successful in some countries, e.g. Egypt and Thailand. The role of mosque, the church or the temple can be significant in the process of counselling, care and social reintegration.

(f) Self-help Groups: The therapeutic community may be considered an in-patient self-group, in which drug misusers work together to achieve abstinence; out-patient self-help groups have also been developed in a number of countries. These self-help groups, such as Narcotics Anonymous, provide a meeting point where former drug misusers and those who have not yet achieved abstinence (but wish to achieve it) can support and encourage others to become and remain drug free. They, therefore, have provided both the means of gaining motivation to seek treatment and support to the reintegration of former drug misusers.

Another form of self-help group has been an activity in which drug misusers join together to press for adequate treatment and rehabilitation services and to act as a voice for the needs of drug misusers. An example of this kind of self-help is the Junkies Union in the Netherlands.

Self-help groups have been fromed by the parents and relatives of drug misusers. These groups have recognized that they need support themselves and by sharing experience and learning from each other, they may be able to assist drug misusers more effectively.

Intervention to Treat Medical Complications Associated with Drug Misuse

Drug use and misuse can result in medical complications arising from the effects of the drug, the adulterations and impurities in street drugs, certain modes of drug use and inadequate personal and health care. Treatment must address these issues in order to help drug users to restore their health. For instance, outreach workers, by contacting drug misusers not yet ready to seek help, can act as community health workers and educate misusers about safer modes of drug administration, the dangers of using street drugs, basic hygiene and improvements in the diet.

Other treatment modalities also have an important role in treating medical complications or improving the health care of drug misusers. General medical services are often involved in the health care of drug misusers. For instance, hospital emergency rooms may have to deal with drug overdoses and accidents occurring as a result of drug intoxification. In addition to providing immediate treatment, they can act as agents in referring drug misusers to specialized drug services. A particular important role in the health care and treatment of medical complications of drug misuse is played by general practitioners and nurses, in the detection of drug misuse and the referral of patients. The early detection of drug misuse before the development of serious medical or social problems allows effective treatment and rehabilitation within the community without resort to more expensive specialized drug services.

It can be seen that, throughout the process of treatment and rehabilitation, there is considerable need for interlinkages between different professions and modalities. No single approach in itself can readily offer all resources. The sharing of work between non-specialized and specialized services offers the most efficient and effective strategy. Specialized services should be developed only where existing community resources cannot meet the established needs of the community.

Treatment in Prisons

Many drug misusers are arrested and placed in a prison, either on remand before trial or to serve a prison sentence after trial. While in prison, the drug misuser will normally be drug-free. The opportunity is, therefore, given to provide treatment and rehabilitation to reduce the likelihood of a return to drug misuse on release from prison. The range of treatment and rehabilitation interventions are described here subject to modification according to individual needs and environment.

Social Reintegration Strategies

Social reintegration is the process whereby the benefits gained from treatment and rehabilitation are sustained and the drug misuser is assisted to adapt to living within the wider community without recourse to indiscriminate or, if possible, any drug misuse. It may involve: a return to the family; the provision of new housing; the completion of basic education or further education; skills training; unpaid, part-time or full-time employment; an introduction into social and cultural activities and, through this, the development of new friendship networks; an introduction into a self-help group to provide mutual support; residence in a halfway house; or the provision of the names and telephone numbers of people to turn to at times of crisis so that the individual does not feel the need to turn again to drug misuse to resolve personal problems.

Drug misusers often have a poor education, having failed to complete their formal education, and an unsatisfactory work history, with frequent periods of sickness, temporary employment or an extended period of unemployment. When employers can afford to be highly selective in choosing employees, drug misusers are at a considerable disadvantage because they have little evidence to show that they are competent and efficient employees.

Treatment and rehabilitation agencies began to recognize this problem in 1970s. They observed that individuals who had successfully completed a particular programme were unable to adapt to life in the community. In many parts of the world, therefore, specific strategies were developed to assist in and sustain the transition into the community. In some countries, halfway houses are provided where individuals live as they begin employment or resume their education. In these houses, the residents have responsibility for their own lives, preparing food, cleaning and managing their own money. In the houses, other former drug misusers and a small number of therapeutic staff provide a means of support and assistance to residents in coping with the stress of learning to live independently.

Changes have also been made in the length of the therapeutic programme so that greater time can be devoted to the reintegration programme, with many therapeutic communities devoting 6 months out of a 12 or 15 months programme to reintegration. Some countries have developed specific work training and employment opportunities as a means of assisting the former addict to learn or relearn work habits, which can help to convince employers of the suitability of the former addict as an employee. Reintegration can be achieved through effective assistance to employees who develop drug problems while at work. Schemes have been developed that seek to identify employees with drug problems at an early point in their drug use. By offering treatment at this stage and by continuing to employ individuals at the work place during treatment, drug users can be successfully reintegrated into the community before they have damaged themselves or their employers' businesses and before more expensive and longer treatment and rehabilitation is necessary.

Relapse

Drug misuse may be cyclical over a period of years. Many people who engage in drug misuse grow out of their dependence after a number of years. Even in the course of drug misuse, there are often short periods of abstinence. It is not uncommon for misusers to drop out of treatment or rehabilitation before completing a programme. Therefore, programmes must be prepared to readmit those who have dropped out in order that they might have a further opportunity to achieve control over their drug misuse and, in the long-term, abstinence from drugs.

There is increasing evidence that, although many drug misusers may experience a temporary relapse at the end of treatment and rehabilitation, after a period of few weeks or months they achieve long-term stability and abstinence. This strongly suggests that the opportunities offered by treatment, rehabilitation and social reintegration can provide an important means of reducing the demand for drugs and containing drug problems at an earlier stage than if no such help were provided.

Concluding Comment

All treatment, rehabilitation and social reintegration approaches that have been described in this chapter are not applicable in every country, and each one must be adapted to the prevailing situations, including the social, cultural and religious circumstances of a given country.

Awareness against Drug Abuse

"A person can lose control of his senses. In several cases drugs and death are linked together". Dr. G.C. Munjal, Professor and Head, G.B. Pant Hospital, said to a social workers and counsellors meeting jointly organised by the Society for Promotion of Area Resource Centres (SPARC), Delhi School of Social Work and the National Institute of Social Defence, Delhi University. Listing various types of drug abuses, Dr. Munjal said, "Brown sugar is rapidly becoming the common drug abuse in major metropolitan cities as well as in some towns". Dr. Dalbir Singh, Deputy Director General, Narcotics Control Bureau, dwelt on Drug Trafficking and Psychotropic Substances Act. He said "Under this Act carrying of beyond a certain quantity is non-bailable offence".

Problems

Thanks to modern techniques of data gathering and processing, it is now possible in many countries to measure precisely the extent of drug use for medicinal purposes as prescribed by doctors. Drug abuse, however, defies precise measurement, owing to its complexity and to its diffusion throughout a neighbourhood, city, region or nation.

'Under international drug control treaties, the use and possession of specific drugs and psychotropic substances are limited exclusively to medical and scientific purposes, misuse and abuse are, therefore, defined as illicit and are distinguished from the strictly medical use (i.e., medical administration or dispensing) by the legal status of the drug in question. Data contained in records of law enforcement agencies and of the judiciary will, therefore, only relate to unlawful use, whereas those from hospital emergency rooms will also include incidents caused by the use of substances that are not (or not yet) under control. The World Health Organization is developing a standardized nosology for drug abuse. The classification system will be valuable in clarifying the use of drug related terms, fostering communication allowing for collaborative approach.

Surveys of illicit drug use can be unreliable for a number of reasons, including utilization of inappropriate methodology, under reporting because of fear of exposure and the inability to sample certain user groups by survey techniques. In addition, owing to differences in statistical capabilities and approaches between countries, methods appropriate in one country may not be applicable in another. Rersearch into the drug problem should be carried out by means of statistical studies and epidemiological surveys, in accordance with the circumstances.

Suggested Courses of Action

The responsible authorities, in cooperation with all other ministries and services concerned or the national co-ordinating body where it exists, and with due regard for the national legislation safeguarding individual freedoms and rights to privacy could:

(a) review current methodology for epidemiological studies of drug abuse and, where necessary, sponsor research leading to the development of more valid and reliable methodologies and instruments for assessing the extent of drug abuse;

(b) review present methods of data collection by statistical, medical, legal and demographic services;

(c) provide for the systematic collection of data on sub-populations of the drug abusers from records of the police registers of deaths, courts (including coroners' courts), hospital emergency rooms, drug treatment centres, prisons, mental hospitals, psychiatric clinics, correctional institutions, social security and welfare organizations, schools and universities, the armed forces, employers, trade unions, as well as community agencies, institutions and associations;

(d) establish training programmes for personnel involved in the collection and analysis of data;

(e) establish central records for storing, analysing and evaluating data;

(f) determine drug abusing patterns in special groups and environments, such as boarding schools, army bases, work places, prisons, hospitals; and

(g) determine, in respect of each type of drug abuse:

   (i) the prevalence of active use (e. g., number of current users within a given period; say, the last year or the last month);

   (ii) incidence of first use (e. g., number of new cases per year);

   (iii) age at onset of use and the demography of users;

   (vi) drug of abuse initially used and/or other drugs currently used;

   (v) the source of supply of currently abused drugs;

   (vi) prevalance of use by sub-groups (e. g., males and females);

   (vii) extent of use of particular drugs which cause diseases or affect special health conditions (e. g. pregnancy).

(h) establish machinery for monitoring trends of abuse and evaluating the effectiveness of instruments, preventive policies, facilitate training, and the planning of prevention strategies; and test periodically the validity of the methodology used.

A country lacking a system for estimating the extent of drug abuse could develop such a system by stages, possibly with the assistance of WHO or any other international organization. The appropriate authorities could undertake or sponsor studies to determine the frequency or gravity of drug addiction (or addiction to particular drugs) in specific population groups and the connection (if any) that exists between drug addiction and a variety of social, economic and cultural factors. These studies could include both particular and general factors related to the individual, the family, and society. It is recommended that the results of the studies be brought to the attention of the agency or authority responsible for the social services that need to be mobilized in order to lessen the illicit demand for drugs.

Universities, research institutes and other academic institutions should conduct studies to determine the implications of present patterns of drug abuse in terms of predicted demographic developments. If the appropriate authority has reason to suspect, in the light of statistical or other data that a change has taken place or is about to take place in the pattern of consumption of addictive drugs (e. g., shift towards a newly "fashionable" drug among drug users, or a marked rise in illicit demand in a particular locality), it should instruct the appropriate agencies to investigate the circumstances and possible causes of the suspected phenomenon promptly and to report back with recommendations for dealing with the situation. In countries where "an early warning" system of this nature does not already exist, the authority concerned may consider the establishment of such machinary if in its opinion the situation warrants it.

At the Regional and International Level

Regional organizations should make comparative studies of drug abuse patterns in countries in the region and inquire into reasons for variations in drug abuse (patterns, methods, etc.). Formal agreements should be established for international collaboration, using comparable methodology and instrumentation, so that resulting data can provide measures of international drug abuse patterns. Field testing of the common methodology and instrumentation is essential.

National focal points of States in the region concerned might share information of drug abuse and on methodologies for determining its incidence and prevalence, as well as exchanging experiences. Ongoing information collection, analysis and dissemination activities within the United Nations system, including in particular the regional commissions, should include data on the full range of drug abuse issues in their areas of responsibility in order to permit an integrated approach to activities touching on drug abuse (e. g. activities concerning social development policy and planning, integrated rural development, popular participation, women and youth). Focal points could be invited to collaborate closely with the regional commissions in this effort.

Regional organizations should establish training programmes for personnel engaged or about to be engaged in epidemiological surveys; expert knowledge available in the first instance for this purpose.

The United Nations Secretariat (Division of Narcotic Drugs), in association with WHO and regional and international organizations as well as non-governmental organization, should disseminate information on the experiences of States in developing an appropriate methodology for assessing drug abuse. The Division of Narcotic Drugs, in co-operation with WHO, should prepare and publish handbooks suggesting methodologies for collecting and analysing data on drug abuse. The United Nations Fund for Drug Abuse Control (UNFFDAC), in co-operation with WHO and other international organizations should include among its priorities providing of assistance to States, to support their standardized systems of case registration surveys of the causes, extent and patterns of drug abuse, and for training specialized personnel to collect, analyse and evaluate data. Technical assistance in the planning and carrying out of epidemiological surveys should be provided to States by international organizations, such as WHO, and by countries with the necessary expertise.

Organization of Comprehensive Systems for the Collection and Evaluation of Data

Crude indicators of future trends in drug use and abuse are available from many sources, including the records of medical practitioners, hospitals, social insurance agencies, school administrators, police, customs authorities and courts. Medical associations world-wide should also be urged to co-operate in assembling this information in collaboration with WHO. This information should be taken into account, subject to respect for confidentiality, in the formulation of national policies intended to prevent abuse and reduce demand for the drugs in question.

To be useful to policymakers, this information, which generally exists only in the records of the agency or person concerned, needs to be collected, collated and analysed, preferably by a system or machinery designed for this specific purpose, with particular emphasis on the consistency and comparability of national and regional data.

Suggested Courses of Action

The appropriate national authority could, as a beginning appoint a working group to devise simple, easy-to-use, reliable standardized instruments that may:

(a) Facilitate the gathering of common information about drug abuse from agencies with different aspects of the abuse;(b) Be used by all entities concerned for collecting basic descriptive data; and (c) Be used nationally as a foundation for epidemiological surveys.

At the local and national levels, a simple framework could be created for the collaboration of data sheets. The data could be made freely available, without disclosure of names or personal particulars to agencies and individuals involved in research and or planning. Mechanism could be established for sharing the data among appropriate policy-makers and drug abuse professionals. Sharing encourages common data collection techniques and makes it possible to give early warning of new drug use patterns.

Field trials of the reliability and validity of any instruments should be conducted and independently analysed by appropriate researchers. In addition to sponsoring measures and activities designed to reduce the illicit demand for drugs, the ministry concerned (health, education, social welfare, as the case may be) might inquire into the multiple possible causes of drug abuse and, as appropriate, recommend action for eliminating the causes, or at least the most serious of them. Surveys of the social and family circumstances of drug addicts or groups of addicts or persons at risk to drug abuse, their housing, employment, level of education and any other characteristic that may appear as a contributory cause of drug abuse. The effectiveness of the programme and techniques utilized by States to reduce drug abuse ought always be evaluated.

In view of the constant evolution of the pharmaceutical industry, intensive research into the properties of newly discovered drugs and their effects on the human organism and the risk of abuse, the appropriate authority could make regulations or issue orders (insofar as these do not already exist) directing local health authorities, medical practioners, pharmacists, directors of educational establishments and research institutes to report to it promptly any case that has come to their notice of persons showing symptoms of incipient addiction to a newly or recently discovered drug or substance or to a pharmaceutical product, previously considered innocuous, which is found to possess narcotic or hallucinogenic properties. The authority concerned might consider whether the product or substance in question should be placed under control and to what extent.

At the Regional and International Levels

Standardized instruments developed by appropriate national authorities should be field tested and, if found useful, made available for the collection of comparable data. National authorities should determine what kinds of data are needed to address key issues, e.g., incidence, prevalence, risk factors and aetiology. Governments might collaborate with each other and with appropriate international organizations, e.g., WHO and the Division of Narcotic Drugs, to identify common risk factors and to determine if these factors are predictive.

Provisional data sheets for pilot studies should be initially drawn up in a small number of States; if found useful, such studies might be carried out in a larger number of countries. Technical assistance might be made available for the purpose of standardized record keeping, making due allowance for the social and cultural setting in which instruments will be used.

The committees established in New York and at Vienna by international non-governmental organizations having consultative status with the Economic and Social Council could act as clearing-houses of relevant information for dissemination to national non-governmental organizations concerned with drug abuse.

In regions where the dimensions of illicit drug trafficking and drug abuse are very imperfectly known and where the laws and the practices of the agencies concerned with the suppression of these illegal activities vary greatly from country to country a unit at the headquarters of the regional commission might co-operate with existing specialized bodies of the United Nations system and other international organizations in identifying the needs of their member countries for strengthening drug control. Where there is a need for improved law enforcement to combat illicit drug trafficking, such a unit could usefully co-operate with the International Criminal Police Organization (ICPC/Interpol) and the Customs Co-operation Council (CCC) in their joint regional programme involving, *inter alia* the holding of regular working meetings of senior police and customs officials, may be given grants of support from sources within the United Nations System.

Whenever illicit use and traffic in drugs give rise to common action, governments could establish a regional centre for research and education for fighting the drug abuse as envisaged in article 38 of 1961 convention.[1]

Global Drug Abuse Worse Than Before

Drug abuse is worse than before and the world's top narcotics body is complaining that some governments are amiss in their reporting and obligations. At the same time, the International Narcotics Control Board (INCB) says that legitimate medical needs for opiates are not being fully met. INCB, a United Nations Agency reports that various drug control measures are not working or are simply ineffective because many governments "repeatedly fail to comply with their reporting" obligations under international conventions on narcotic drugs and psychotropic substances.

In the annual report, INCB Secretary Abdel-Aziz Bahi cites instances of high government officials working in close cooperation with illicit drug traffickers. But he refuses to confirm or deny that many governments are infiltrated by drug syndicates. Although the annual report named Cuba, Mr. Bahi would not identify the countries whose officials are involved in drug dealing. The report says the number of drug seizures, drug-related deaths and drug patients admitted to hospitals indicated the situation was " worse than ever before". It complains that over 50 States provided only partial or incorrect data to the INCB. The INCB, taking an unusual step, named the Central African Republic, Kampuchea, Gambia, Gibraltar, Israel, Mongolia, Nepal, St. Lucia, Sierra Leone and Vietnam as the countries having supplied no information at all to the Board. The 1989 annual report is accompanied by a special section prepared in cooperation with the World Health Organisation (WHO). It notes: "Despite the existence of excessive stocks information available to the Board and to the WHO suggests that the need for opiates for legitimate medical purposes is not being fully met. Thus, patients suffering from conditions which might be treated with opiates often cannot obtain them. This situation exists in both developed and developing countries. But the situation was more acute in the latter.

The special section notes that countries with inadequate funds for health care were not in a position to provide essential drugs including codeine, morphine and pethidine. In 1987, 71 per cent of morphine meant for medical purposes was used in just five countries with a total population of 620 million or 12 per cent of world population. In 1988, an average 200 tons of morphine equivalent were consumed. Of these 42.4 per cent went to Europe, 33.1 per cent to the Americas, 16.6 per cent to Asia, 4.6 per cent to Africa and 3.2 per cent to Oceania.

The INCB is a quasi-judicial international body of independent experts elected by the UN Economic and Social Council to endeavour under various drug control treaties "to limit the cultivation, porduction, manufacture and utilization of narcotic drugs and psychotropic substances to the amounts necessary for medical and scientific purposes". It is also mandated "to ensure that the quantities of these substances necessary for legitimate purposes are available". It has to guide governments in preventing the illicit cultivation, production, manufacture and trafficking in and use of these substances. The Board was also recently entrusted to monitor the international trade and use of chemicals used in making hard drugs.

Although the INCB did not identify directly the countries violating the convention on psychotropic substances, the report hinted that many culprits are in Western Europe. The Board says large quantities of psychotropic substances were exported to small developing countries on the basis of bogus import licences. Besides, free ports and free zones like Hong Kong receive huge amounts of the substances for re-export by broker compaines. Explaining the trend, Mr. Bahi says that 95 per cent of heroin abused was dangerously adulterated. Most of these originated in Myanmar (Burma) or Afghanistan. Consumption of adulterated heroin was particularly more dangerous.

While the programme for eradication of illicit cultivation of poppy plants was not progressing well enough successful efforts to spread the illicit poppy cultivation in other continents were continuing. Mr. Bahi cites the eradication efforts of new illegal poppy plantation in Mexico and the discovery of plants in four areas of Columbia where two heroin laboratories and two million opium poppy plants were destroyed. Similarly, 1300 hectares, in Guatemala were found to be under illicit poppy cultivation, described as "a matter of serious concern".

The Board gives some interesting insight into the drug situation in Asia. One of this is that during its mission to Nepal last year, the government expressed its helplessness since under Hindu religion it was obliged to supply huge quantities of marijuana to priests "during the festivals of Maha Shivratri". The Board itself found "the distribution of a maximum quantity of 40 kilograms of marijuana once a year to priests should not raise practical problems".

In Thailand the Board found there was an alarming increase of AIDS among heroin addicts, during the last four years. For the reporting period 40 per cent of the addicts were infected by AIDS. The situation in Laos had worsened because of the bumper illicit crop of poppy caused by favourable weather conditions. In Malaysian schools, there was an alarming increase in the abuse of marijuana, heroin, hallucinogens and volatile solvents. After introducing drug tests for students, the Malaysian Government is now planning to introduce compulsory drug test for all public servants.

In the Near and Middle East, the Board hopes to send missions to Afghanistan, Iran and Pakistan. "Spreading heroin abuse and the lack of progress in reducing illicit poppy cultivation" in Pakistan was a cause for concern to the Board.

In Western Europe, seizure of cocaine, heroin and other drugs had gone up six-fold but still illicit drugs were getting inexpensive indicating large supplies. The trend of increased drug-related deaths and crime was continuing. With the abolition of frontiers within the European Community in 1992 the narcotic law enforcement in the territory may have to be reorganised. The EC countries are planning such reorganisation.

In the USA, the appearance of cocaine-based smokable "crack" has led to higher incidence of violent crimes. It has also led to a 28-fold increase since 1984 in the hospitalisation of "crack" victims.

Classification, Effect and Identification of Drugs

Drug abuse, i.e. the use of drugs except for medical and scientific purposes, is a centuries-old phenomenon. Myriad variety of drugs have been known to exist and abused for different effects on human mind. Even the experts and the knowledgeable people would not be able to identify or classify them or claim full knowledge of their effects. "Narcotic Drugs"embrace a number of substances whose abuse has created serious problems to physical and psychic health of the individual and society. The 'Single Convention' on Narcotic Drugs, 1961 covered these drugs in a single instrument. The complexity of the problem increased several fold when another class of drugs, called "psychotropic substances" covered under another comprehensive instrument 1971-convention, came to be abused. In India, the new enactment 'Narcotic Drugs and Psychotropic Substances Act, 1985,' covers in a single instrument most of the drugs/substances commonly being abused. Narcotic drugs and psychotropic substances are natural or synthetic products which fall in the following categories:

Classification of Drugs of Abuse

Narcotic Drugs and Psychotropic Substances may be classified as under:

Effect on Mental Activity

Following are their effect upon a person's mental activity and psychic state:

(i) *depressants* of the central nervous system, which slow down mental activity:

(ii) *stimulants* of the central nervous system, which cause states of excitement;

(iii) *hallucinogen* which produce hallucinations or illusions, i.e. distortions of the sense; and

(iv) *Cannabis* which in small quantity acts as a depressant, but in larger doses has effects similar to those of a hallucinogen.

(Ref: United States: 1973 a guide for law enforcement officers).

Natural and Synthetic Drugs

The natural and synthetic drugs may be classified as under:

(a) **Depressants:** Dextrombromide (e.g. Plaflum); Methadone (e.g., Dolphin, Physeptone); Normethadone (e.g. Ticarde); Pethidine (e.g., Demerol, Meperidine, Dolantin).

(b) **Barbiturates-Tranquillizers:** Other sedatives such as Glutethimide (e.g., Doriden) Methaquolone (e.g., Mandrax).

(c) **Stimulants**: Amphetamines such as Amphetamine (e.g., Benzedrine, Dexedrine, Durophet).

Combinations of a barbiturate and amphetamine (e.g. Drinamyl) are also popularly abused.

(d) **Hallucinogens:** Such as, Peyoti (Mescaline) 'Morning glory' seeds, DMT, DET, Lysergic Acid Diethylamide (LSD).

While classifying the drugs, some of the pharmaceutical terms may be understood. They are:

(i) Tablets (compressed on a punch machine);

(ii) Capsules (powder or drug kept in shells);

(iii) Pills (doses to be swallowed);

(iv) Tinctures (extracts of plants material in alcoholic solution);

(v) Syrups (solutions in sugar water). Then there are proprietory names of drugs (given by the manufacturers), chemical names (according to chemical composition) and internationally known proprietory names (INN). For example: INN:

methequolone.

Chemical name: 2-methy 1-3-0-toly 1-4 (3H): Quinezolinone.

Proprietory names: About thirty names of which Mandrax (the best known), Dityl, Haladorn, Sedatyl, Toquilone, are examples.

Recognition, Abuse and Effects of Drugs

1. Depressants

Opium, opium alkaloids and synthetic narcotics: Opium is coagulated juice of the poppy plant, scientifically known as *Papaver somniferum.* Opium poppy is not a narcotic drug. It is the source of the opiates, i.e. opium morphine, heroin and codeine. Opium Poppy is an annual plant, 2 to 4 feet in height, with 4-petal flowers in white, pink, red, violet and purple colours. The head or capsule of poppy in oval shape, size of a nut, contains plant seeds among other material.

Raw opium is taken out from latex which oozes from the lanced capsules. It is kept in plastic wrappers to keep it moist.

Prepared opium is suitable for smoking or eating, following the process of cooking and fermentation kept in slabs or sticks.

Burnt opium is the remnant on the pipe used for smoking, contains higher morphine concentration. May be used again, mixed with fresh opium.

Opium dross is burnt opium mixed with dross, boiled and filtered water may contain little morphine, mostly charcoal, considered ineffective.

Opium was medically a very important drug, administered as pain reliever, sleep inducer and for cough and diarrhoea. Its therapeutic role was taken over by isolated alkaloids (morphine and codeine) and synthetic narcotics (pothidine and methadone).

Opium as a drug of process: Earlier eaten, drunk or taken as an infusion, smoking is a comparative recent phenomenon. It is also possible to inject opium after dissolving in water a method prevalent in Western Europe.

For use, opium pill is held in a pair of tongs or on needle point over a small fire and fumes inhaled through a pipe.

Morphine: Manufactured from opium or extracted directly from poppy straw. Widely used for acute pain, resulting from surgery, fractures, burn, etc and terminal illnesses. Therapeutic use taken over by synthetic narcotics and analgesic drugs. Abusers are mostly the former patients or medicalmen. Morphine is still a measure of pain relief.

Morphine Base: Morphine (7 to 14 per cent in opium) extracted from opium through hydrate lime, water, ammonium chloride, source of heat, and filtering apparatus—compressed blocks with brand names, such as "999", or "O.K.". In white or pale yellow to brown colour, smells like acid.

Morphine salts: Morphine, for medical use, available in salt form, odourless, white crystalline powder.

Heroin: Derived from morphine by acetilation (diacetyl- morphine), 10 times more powerful than morphine, most additive of all the known drugs. Produces intense euphoria and the addict drifts into a dream-like trance where reality seems far away. Tolerance to the effect develops rapidly requiring larger intake. May be administered through (a) intravenous injection, (b) inhalation. First experience by sniffing (snorting) followed by subcutaneous (skin-popping), intravenous injections. Adulteration (cutting) of heroin with milk sugar, quinine, etc to increase profit, also reduces the effect leading to more intake.

Effects: Abusers feel lethargic and indifferent to environment and to personal situations. Chronic abuser is always in poor health due to indigestion, blockage, physical and psychological dependence.

Dull senses: Depressed central nervous system. Sense of well-being. Lack of co-ordination. Troubles of respiration and circulation, stupor and coma.

Codeine and other Morphine Derivatives

Codeine (methylomorphine), a conversion of morphine, an effective cough suppressant. Addiction is relatively rare, larger quantities to continue for a long period for dependence. (It is also true for opium and morphine.)

Synthetic narcotics: The term used for addiction producing substances since the discovery of pethidine, 30 years ago. When marked, claimed to be devoid of addiction potential. Proved otherwise. Pethidine, methodone and non-methodone are useful therapeutic agents but they create addicts amongst physicians and nurses.

Recognition of opiate abusers: Through discovery of heating apparatus (lamp or charcoal fire) needle, matches, opium dross or soot (cooking spoon). Addicts may be recognised by needle marks, black or blue scabs and scars. Under the influence of opiates, abuser appears drowsy, apathetic, uncommunicative and while in urge suffers withdrawal symptoms, watery eyes, running nose, aches, itching, yawning, goose-flesh dilation of the pupils. Addict will be irritable, restless, nervous and sleepless, sweating and will complain pain and nausea, suffer diarrhoea. He may wear dark glasses to conceal pupils and long sleeved shirts to conceal needle marks. Effects may be different from drug-to-drug and person-to-person.

1. **Other Depressants**

Barbiturates or sleeping pills depressed central nervous system. Made synthetically as capsules and tablets. Variety of drugs prescribed by physicians. Extent of abuse vary from place-to-place and time-to-time, lack of control due to therapeutic use, also leads to dependence. Barbiturates may be used with heroin ("chasing dragon" in Asia and the Far East), alcohol and stimulants.

Known barbiturates/sedatives: Such as methaqualone and glute thimide, being widely abused and also banned.

Recognition of barbiturates abusers: Appear humoured, slow in reaction, slurring of speech, staggering, tendency to fall or drag things, readiness to laugh or weap, heavy sleep. Symptoms similar to alcoholics. Addiction after prolonged use. Similar symptoms found among tranquillizers abusers.

2. **Stimulants**

Coca shrub/leaves grown in South America (Peru and Bolivia), chewed for centuries. Plants 5" in height, grown in damp climate, leaves smooth and oval shape. Coca chewing to diminish hunger, caused indifference and ill health. Cocaine the principal alkaloid, white crystalline powder, odourless acid and fluffy treated with hydrochloric acid to form soluble salt. Called "snow", when mixed with heroin called "speedball". Maybe sniffed with nostrils or injected. Abusers having red nose, broken veins, needle marks, rapid loss of weight, vomiting and excitability.

Amphetamines: Stimulants of central nervous system, used medically for narcolepsy. White powder in tablets and capsules in different colours, also in ampules, makes tired, abuser feels alert for a while, but gets depressed and eager soon. Popular names, "pep-pills". US slang "eye-openers", "wake-ups", "uppers" "truck drivers", and "co-pilots", etc. Common brand name—Benzedrine, dextroamphetamine. Recognition: abuser pretends fully well and confident to overcome sleepiness. Becomes talkative, excitable, elated and restless. The abuser has trembling hands, enlarged pupils, perspiration and insomnia, nose and mouth give bad breath, licking lips and scratching of the nose. Used by students, long distant drivers, athletes and strenuous living persons. Mixture of stimulants and depressants possible; slang name "speedball".

3. **Hallucinogens**

Constitutes groups of substances, wide variety of chemicals and pharmacologicals. Problems created by use/abuse are often different from depressants and stimulants: (i) Hallucinogens are not medicines. (ii) No lawful consumption i.e. no distinction in use and abuse.

Production for trade and consumption mostly illegal. LSD (Lysergic Acid Diethylamide) main substance, a semi-synthetic compound manufactured with lysergic acid, and ergot of rye for a "psychedelic trip", 100-250 microgrammes of LSD used as diluted solution or in capsule. Traditional abuse of hallucinogens among Indian tribes of South America with fungi and other psychotoxic plants. On a "psychedelic trip" a feeling of being at one with the universe, as a part of group activity is exercised. Myth of "perfecting the personality" and "broaden consciousness". Unlike opiates, giving solitary experience and causing dependence, hallucinogens are less dependence-causing. However, extreme harmful reactions may result in suicides, murders and attempted murders during the trips. Complex reaction like "depersonalization" in mental images of body and a feeling of detachment, and of unreality made prolonged.

4. **Cannabis**

It means "the flowering and fruiting tops of cannabis plant" (excluding seeds and leaves unaccompanied by tops). Grown in Europe, Asia and Africa for thousands of years, first used medically in ancient China. Used as an intoxicant for hundreds of years in India. Chemical cultivation in Northern America, Central and South America. Small groups of intellectuals and abusers: Hashish Users' Club in Paris in 1885s wrote accounts of their experiences. Mild euphoric and sedative. In larger doses, resembles hallucinogens. Causes change in mood, alters sense of time, starts visual and auditory perception, depersonalization, optical and auditory hallucinations. Cannabis marijuana or hashish only different forms. Cannabis plant converts into hashish through different methods. "Ganja" is young leaves and flowering tops of cultivated female plants, its resin formed into cakes. "Bhang" older or matured leaves of plant often used as ingredients in sweet meat, etc or to make syrup. Cannabis' effect is quickly obtained by smoking. A social activity done in groups. Effects felt in 15 to 30 minutes. Abuser is recognised by stimulated behaviour, laughter, drowsy, redening of the eyes, dry mouth, hunger, perspiration, pale face, stained fingers and smell of cannabis smoke in breath and throats.

Volatile Chemicals: Group of substances, neither drugs nor under international control. Fumes, however, cause intoxication, may be dangerous, even cause deaths. Effects appear rapidly and last about half-an-hour. Produce sensation, dizziness, blurring of vision, ringing in the ears, and slurred speech. The abuser may appear dazed and drunk, may even vomit, feel sleepy and even lose consciousness. There are dangers of fire and explosion if the user mishandles the substance.

(Prepared on the basis of Authentic Reports).

References

1. *Courtesy* : CBI Bulletin, R.C. Dikshit, ADG.
2. *Bulletin on Narcotics*, Vol. XL, No. 1, 1988—Treatment options in responding to drug misuse problems.
3. *News from Narcontrol*, Oct. '88.

60

SOCIO-MEDICAL STRATEGIES AND THERAPEUTIC PROGRAMMES

The supply and demand for illicit drugs has become a major theme in the efforts to control drug problems. The international Conference on Drug Abuse and Illicit Trafficking, which was held at Vienna in June 1987, adopted a comrehensive multi-disciplinary outline and a declaration in which the essential balance between the two elements was emphasized as follows:

(i) Supply reduction is the foremost concern.

(ii) Demand reduction is to be achieved through multi-reformative measures.

(iii) Sustained efforts to control emergence of substitute drugs is necessary in the light of Single Convention on Narcotic Drugs, as amended by the 1972 protocol;

(iv) or Convention of psychotropic Substances 1971;

(v) is indicative of the rapid emergence of alternatives to controlled drugs as drugs of misuse. "Demand reduction programmes may inevitably respond to drug problems as they appear. For instance, the control of six amphetamines-like substances was considered by the Commission on Narcotic Drugs every year from 1955 to 1967; only in 1969, after consistent pressure from the Government of Sweden, supported by the Government of the United Kingdom of Great Britain and Northern Ireland and the United States of America, and because of the health and social problems caused by these drugs was the adoption of a resolution called for effective international controls on those substances (4-6)".[1]

Efforts to reduce the harm caused by drug misuse and to reduce prevalence of drug problems may be placed in three main categories:

(a) making and maintaining contact with drug misusers;

(b) treatment and rehabilitation strategies; and

(c) social re-integration strategies.

Intervention Strategies

Modalities of treatment may differ in different parts of the world depending on the socio-cultural environment, the type of drug used and other factors, such as age, sex, religion and deployment status.

It is essential for therapeutic activities to be well planned, with both long range and short term objectives, long range therapeutic goals might be:

(a) to reduce the extent and seriousness of problems associated with the non-medical use of dependence-producing drugs;

(b) to reduce the prevalence and incidence of drug dependence;

(c) to improve the personal and social functioning of drug dependent persons; and

(d) to develop an ongoing system for the continued planning evaluation and modification of programme activities, as indicated by changing needs and the results of appraisals of the effectiveness of different policies, approaches and methods in use.

Traditional Medicine

The success and content of traditional medicine approaches depend very heavily on the cultural and social environment where they are practised. Reports indicate that certain methods are being used in India, based on the Ayurveda and other forms of traditional medicine and practices.

In Malaysia, traditional healing is based on archaic animistic beliefs, i.e., a person is vulnerable to the influence of the supernatural. Drug addiction is seen as a result of incorrect behaviour and the presence of a supernatural element. In view of the cultural or religious background of the population, traditional healers (bomoh) can provide an important treatment modality for some drug addicts. The treatment consists of detoxification, with the help of herbal medicine, physical exercise and balanced nutrition to build up physical health.

Acupuncture

Acupuncture is used in the detoxification process in Hong Kong, the United States, Vietnam and elsewhere. Ear acupuncture (without electrical stimulation) can be applied and taught effectively in clinical settings at low cost. Ear acupuncture is reported to have a significant preventive effect in the reduction of physical craving pattern. Acupuncture relieves withdrawal symptoms, prevents the craving for drugs and increases the rate of participation of patients in long term treatment programmes. The best results have been obtained by treating patients in an open group setting, using acupuncture points in the external ear with needles used for electrical stimulation.

Yoga and Meditation

In certain settings Yoga and Meditation are used as an important part of behavioural treatment. They include breathing techniques, body postures, meditation and relaxation techniques, as means of providing clients with the tools to improve their personal strength to free themselves of this habit. They have shown definite success in improving inter-personal skills after repeated sessions.

In certain countries religion plays an important role in the treatment and rehabilitation process. The use of religious teachings as a psychological re-inforcement tool has proved to be successful in some countries, e.g. Egypt and Thailand. The role of the mosque, the church or the temple can be significant in the process of counselling, care and social re-integration.

Self-help Group

The therapeutic community may be considered an in-patient self-help group in which drug misusers work together to achieve abstinence; out-patient self-help groups have also been developed in a number of countries. These self-help groups, such as Narcotic Anonymous, provide a meeting point where former drug misusers and those who have not yet achieved abstinence but wish to achieve it can support and encourage others to remain drug-free. Another form of self-help group has been an activity in which drug misusers join together to press for adequate treatment and rehabilitation services and to act as a voice for the needs of drug misusers. An example of this kind of self-help is the Junkies Union in the Netherlands.

Withdrawal of Narcotic Drugs

The patient's medical history, taken at the time of admission, includes a detailed account of his use of drugs. The patient is given a thorough physical examination and, if there is no contradiction to the withdrawal of drugs, is sent to the withdrawal unit of the hospital. The patient is observed in the ward for withdrawal symptoms—sweating, slavination, rhinorrhea, yawning, dilation of pupils, muscular fishing, quorexia, diarrhoea, insomnia and an increase in blood pressure and temperature.

Following the withdrawal procedure, the patient begins to regain his appetite, and his weight and strength. Physical symptoms of irritability, restlessness and difficulty in getting proper rest may last for several weeks.

Preparation for Discharge

For admission until discharge every effort is made to help the patient become drug free and to learn to live without having to resort to the use of drugs. Social workers also conduct individual and group psychotherapy and, in addition, function as the liaison between the patients' family and the community, and assist in post discharge planning and follow-up, giving due regard to family situation. All physically able patients are given a vocational assignment, and every effort is made to help them learn good, habits, work and live with others.

Another activity important to patient's morale is the re-creational programme which offers not only entertainment but also an opportunity to become interested in sports.

Intravenous Drug Abuse

Intravenous drug abusers constitute 25 per cent of the cases of the "acquired immuno deficiency syndrome (AIDS)" in adults in the United States of America and 21 per cent such cases in Europe. The potential for the rapid spread of the human immunodeficiency virus (HIV) among intravenous drug abusers exists because such drug abusers commonly share drug injection equipment. Thus the primary AIDS prevention strategy must be to help addicts stop using drugs and to encourage more intravenous drug abusers to enter treatment. Aids risk reduction counselling must also be provided to intravenous drug abusers and their sexual partners to prevent the sexual spread of HIV (Human immunodeficiency virus).

Effects

Although individual variations are frequent, the following sequence is fairly well representative of a typical heroin experience:

The rush: Moments after the heroin is maintained, the user will experience a "rush" or "flash" as the nervous system reacts violently to the drug's presence. Addicts describe the rush as an extraordinarily pleasurable sensation—one that is similar in many ways to a sexual orgasm, only much more intense involving the entire body.

The nods: Next, the user experiences a lingering state of euphoric bliss. Fatigue, tension and anxiety fade away. Feelings of inadequacy and inferiority are replaced by a relaxed contentment. The user becomes dissociated from his or her surroundings—no matter how depressing they may be in reality. During this period, the user will frequently give the appearance of being in a languid stupor—hence the slang phrase, "on the nods".

In addition to the mood altering changes for which it is used, heroin also produces a wide-range of other effects. For example, it constricts the pupils, causes constipation, increases the respiration rate and depth. In addition, heroin addicts of both sexes frequently experience a depressed sexual drive, in the case of female users, menstrual periods may be intermittent.

Dependence

Heroin is one of the most seductive drugs known. Its ability to enslave those who use it is legendary. The body quickly builds a system to respond to many of its pleasurable effects and psychic and physical dependence develop rapidly. In the beginning heroin produces intensely pleasurable sensation, however, with the passage of time, the quantum of pleasure is less. The rush is muted, the experience dulled to lethargy.

Symptoms of Withdrawal

The severity and nature of the withdrawal symptoms that a sudden stop in heroin use brings vary considerably according to the make-up of the addict. The following account is fairly typical.

"I started to get sick, quick, real quick. At first I just reacted a lot, then I started to vomit. I had such bad stomach cramps that I couldn't stop vomiting. Soon I didn't have anything left to vomit, dry heaves kept on. By that time I was all cramped up and I had the cold sweats. I cried and screamed a lot, but that didn't do any good...My eyes were running, so was my nose. I had awful gas and then I got terrible diarrhoea. I cropped all over myself all the time, but I couldn't help that. All of a sudden I'd get real hot, then real cold. My body was covered with goose lumps... I would have given anything in the world for some I would have given you myself for a nickel by".[2] [sic]

Heroin addicts frequently use other drugs as well. Barbiturates and other sedatives are used to enhance the effect of heroin that has been overly 'cut' (adulterated) and as a substitute when it is not available.

Recognising the Abuser

Short of laboratory tests, there is no positive way of telling whether or not an individual is using heroin or any other narcotic. However, there are a number of signs that may indicate heroin abuse:

(i) Constricted or 'pinned' pupils of the eyes—that fail to respond when a light is shown into them.

(ii) Scars (tracks) or discoloration over the veins caused by repeated intravenous injections.

(iii) Abscesses caused by infected needles.

(iv) Stuporous, lethargic behaviour

(v) A generally rundown physical appearance. Heroin abusers frequently ignore their personal hygiene and nutritional needs.

(vi) Cigarette-burned fingers, abscesses are frequently in semi-curation condition during which they suffer from a diminished perception of pain.

(vii) Early signs of withrawal such as a running nose, sweating and anxiety.

(viii) The presence of narcotics paraphernalia such as needles, syringes, cookers, strings or cords that are used as torniquets and capsules, envelopes or paper or foil packages with traces of white powder.

The Role of Medical Profession

Dr. Davinder Mohan, Head of the Department of Psychiatry at the All-India Institute of Medical Sciences, Delhi and Sister Jessie Abraham, social worker and counsellor at the Shanti Dham Drug Detoxification and Rehabilitation Centre at Dahisar, Bombay, have been jointly awarded the Indian Merchants' Chamber Platinum Jubilee Endowment Trust award for outstanding work in the field of drug addiction. The winners were selected from among 27 nominations by a panel of judges comprising Mr. Justice C.S. Dharmadhikari of the Bombay High Court, Dr. M.D. Bangalee, Vice-Chancellor of Bombay University, Dr. M.S. Gore, Adviser to the Tata Institute of Social Sciences and Mr. M.V. Kamath, Journalist.

Outstanding Contribution

The award for "outstanding contribution towards tackling the problem of drug addiction by creating awareness, providing treatment, rehabilitating addicts and strengthening enforcement measures", carries a cash prize of Rs. One lakh which will be shared equally by Sister Jessie and Dr. Mohan.

The announcement of awards was made by Dr. Nalin K. Vissanji, President of the DMC and Mr. Pravinchandra V. Gandhi, Chairman of the Trust. They said in addition to the winners, the trustees had decided to give certificates of appreciation and Rs. 5,000 each to five other persons working in the field of drug addiction. They are Mr. Russel Pinto, a former addict who is now Director of Seva Dham Drug Rehabilitation Centre, Bombay; Dr. Anand Nadkarni, Consultant Psychiatrist at KEM Hospital; Mr. Nitin M. Khambati, Chairman of the Lions Drug Awareness Committee and Dr. Anil Awachat of the Muktangan De-addiction Centre attached to the Central Institute of Mental Health and Research, Pune.

Profiles

Dr. Mohan has been working towards establishing profiles of drug users and studying the extent of drug abuse in the country through uninterrupted research since 1976. "He is, perhaps, the only person in India who has delineated the problem of opium use in Punjab and Rajasthan rural communities", according to the IMC. A member of the global Expert Advisory Committee on Drug Dependence of the World Health Organisation (WHO), he has initiated or carried out almost all the research in India since 1976. He has also been involved with formulating Central Government policies on the drug problem.

Sister Jessie has been working in the field of alcohol addiction since 1976 after her training at the TISS. She gradually moved into the field of drug addiction when the problem began to manifest itself in Bombay after 1980, and now works both with Alcohol and Rehabilitation Centre at the Shanti Dham Drug Detoxification and Rehabilitation Centre at Dahisar. "Sister Jessie has been relentlessly pursuing the cause of fighting drug addiction by creating awareness among the vulnerable section of society by arranging talks at educational institutions, besides training social workers and teachers. She has also undertaken studies and surveys of drug addicts, including one of drug peddlers who were arrested by the Bombay Police during the past five years".

Prevention through Education

Comprehensive and effective education programmes are a necessary part of measures for counteracting drug abuse worldwide. In many countries drug abuse has already spread to the various age and population groups. It is essential that all individuals in the formal and general education system, as well as their families, should be enlightened about the risks of drug abuse. Prevention through education should be considered as a continuous process and a lengthy exacting action, aiming to seek and improve understanding of the long-term and immediate causes of recourse to drugs, and help young people and adults to find solutions to their difficulties without resorting to drugs. Curricula and programmes for enhancing the community's awareness, developed as part of an agreed national strategy, should be so constructed as to strengthen the population's motivation to avoid drug abuse.

Indications are that the impact of preventive education is the greatest when it:

(a) takes place in its appropriate social economic and cultural setting;

(b) is integrated into the overall framework of academic social and cultural learning;

(c) promotes a healthy drug-free life-style as a primary goal, as opposed to placing emphasis on abstinence from drugs and on the negative effects of drug abuse.

(d) reaches individuals before they are exposed to the drug sub-culture and other influences that contribute to initial drug use;

(e) does not involve elements that evoke curiosity or desire to experiment with narcotic drugs (detailed "positive" descriptions of euphoria etc), but clearly indicates the negative, harmful consequences of drug abuse and emphasizes the positive effects of alternative activities and life-style free from narcotic drugs and psychotropic substances; and

(f) does not contain details which might make access to illicit drug easy, such as detailed descriptions of methods and routes of illicit traffic in narcotic drugs, places of origin of illicit production, non-medical uses of narcotic drugs, etc.

Care should be taken that preventive education in school and out-of-school does not become yet another addition to curricula and that additional costs are avoided. Where advisable, attention should be paid to integrating progressively into school curricula and out-of-school activities elements concerning the prevention of drug abuse.

Action at the National Level

The appropriate authority could establish a multi-disciplinary unit with trained teaching personnel. The unit would perform such functions as:

(a) Prescribing or recommending for all levels of educational institutions the development of drug abuse prevention curricula and instructional materials, taking into account cultural values and traditions and emphasizing the benefits of a healthy drug-free life-style;

(b) Preparing training and educational materials for the unemployed and undereducated, especially the young people, to assist them in developing vocational and self-employment skills;

(c) Within the school system, teacher education and motivation are essential to ensure the effective education of children regarding drugs and healthy lifestyles. Therefore, it is necessary to prepare training materials and to conduct training programmes for teachers and counsellers that will enable them to instruct their students in the advantages of a drug-free life.

(d) Preparing basic information publications on drug

abuse and the different models of prevention, Particular attention should be given to ensuring that preventive instruction at school is not instituted as an isolated, one-time measure, but rather that it is designated as a long-term programme accompanied by preventive measures involving students, parents, workers, the clergy, doctors and pharmacists. Only a comprehensive approach at the regional and local levels will have a chance of success;

(e) Undertaking periodic reviews to the curricula characteristics and publicity material to determine their effectiveness, and recommending necessary adjustments.

(f) Preparing programmes and information relative to promoting drug-free sport, cultural and leisure time facilities and activities.

At the level of pre-service training, all educational personnel should be familiarized with problems of drug abuse and prevention techniques. In-service courses should be orgainsed in specialized areas for certain categories of personnel.

Seminars and training courses should be organised to help parents recognize the symptoms of drug abuse at the early stages and to enable them to educate their children to recognize the damaging physical and mental effects of drug abuse.

School boards and authorities responsible for private and denominational schools should similarly formulate school policies aimed at preventing drug abuse, consonant with their curricula, and develop training and instructional material for classroom and related community use. Interested members of the training staff and school and guidance counsellors should be given training to advise students and parents about the dangers of drug abuse.

An important milieu for the education of the young is the family. Programmes should be developed to educate parents about the harmful effects of drug use and about methods to develop the competence skills of their own children.

In view of the vulnerability of young persons to the temptations offered by certain narcotic and psychotropic substances and to the solicitations of "drug pushers", publicity and educational programmes to alert young people and society in general, may be launched.

The national co-ordination agency, where it exists, or the appropriate authority could consider favourably the idea of establishing a small body composed of experts, representatives of agencies and other persons familiar, with the situation regarding drug abuse in the country (e.g. teachers, social workers, members of the police force, representatives of national non-governmental organizations and others) who would be given the mandate to design a programme to reduce and prevent the illicit demand for drugs.

At the Regional and International Levels

Regional bodies should consider establishing regional training, resource and information centres for individuals involved in designinig curricula, in order to evaluate and develop teaching techniques consonant with the cultural patterns of the region and to disseminate these techniques by means of regional courses and exchange of personnel.

The United Nations Educational, Scientific and Cultural Organization (UNESCO) and other international and non-governmental organizations could encourage the creation of co-operative networks among institutions and provide data on strategies, methods and experience in prevention through exchange of educational personnel. Specific initiatives should be encouraged to promote exchanges of experience between countries having similar problems.

In cases where projects with an educational content are envisaged for the purpose of preventing and reducing the illicit demand for drugs, the Government or regional or international bodies or non-governmental organizations sponsoring or initiating the projects, may, if they lack funds, apply for support to the United Nations Fund for Drug Abuse Control which should give priority to such projects.

Programmes and projects of United Nations Organizations, especially the regional commissions, dealing with vulnerable population groups should include information about the scope and nature of drug abuse problems, their causes, and preventive and remedial actions as they relate to these groups.

Where large groups of refugees are receiving care (including health and education care) from an intergovernmental organization and where cases of drug abuse have been noted among them, the organization concerned should investigate such cases and ascertain their frequency and gravity and, in the light of the information, issue warnings of the dangers of drug abuse (especially among the young) within the refugee population and, if possible, undertake an educational, cultural and therapeutic programme for refugee groups at risk. Any relevant activity should be carried out in co-operation and in accord with the concerned State.

Prevention in the Workplace

The multiple and harmful consequences associated with the drug abuse in the workplace and in occupational activities include deterioration of productivity and performance, defective quality of products, loss of qualified employee, accidents, loss of income and professional disqualification—all of which are costly to the individual, the employer, and the community. Where alcohol abuse prevention programmes exist, consideration should be given to applying the experience gained in these programmes to the prevention and reduction of the abuse of drugs in the workplace.

Drug abuse by individuals in a wide-range of sensitive occupations can result in disaster. Measures for dealing with drug abuse in the workplace and in occupational activities must have a significant prevention component.

Suggested Courses of Action

Following courses of action are suggested for drug abuse prevention:

(i) At the national level

With a view to preventing or reducing drug abuse in the workplace, the authority concerned could: (a) publicize information in the workplace warning of the risks of drug abuse; and (b) set up national training workshops for supervisors, programme developers and others; issue guidelines for preventing or reducing drug abuse in the workplace, and for treating and rehabilitating the drug addicts.

Special emphasis should be placed on the urgent need to prevent abuse of drugs among persons regarded as role models. Professional and occupational associations should accordingly develop drug abuse prevention programmes for their members. Employers' and workers' organizations should develop joint action programmes.

Medical authorities and medical laboratories should design reliable testing methods for determining the presence of drugs in the human organism, identifying the drug user and ensuring safety of the public.

(ii) At the regional and international levels

The Governments of the countries in a region could, through the ministries concerned or through regional bodies or non-governmental organizations, identify regional resources for training and programme development. They should organize regional seminars, for the exchange of experiences and dissemination of relevant information. The International Labour Organisation (ILO) could invited to initiate and participate actively in the intensification of the exchange of experiences at the regional and international levels.

The ILO, in cooperation with UNESCO could co-ordinate international efforts to prevent and reduce drug abuse in the workplace, by distributing multi-media resources kits and to rehabilitate former drug abusers. It could also monitor the use of these kits and results thereof.

The Division of Narcotic Drugs, in collaboration with WHO and ILO, should promote and harmonize national efforts by developing internationally acceptable guidelines, criteria and methodologies for national testing programmes; a central source of reference standards of major drug metabolites should be established to serve national laboratories.

Treatment Options in Responding to Drug Misuse Problems

Reducing both the supply of and demand for illicit drugs has become a major theme in the efforts to control drug problems. The International Conference on Drug Abuse and Illicit Trafficking, held at Vienna in June 1987, adopted a Comprehensive Multidisciplinary outline and a Declaration in which the essential balance between the two elements was emphasized. At the national and international levels, over the years, supply reduction has consistently been of the foremost concern; it will re-emerge as the predominant theme as the links between illicit drug trafficking and terrorism, international financial dealings and arms supply become obvious. It is important, however, to recognize the demand-reduction efforts as essential to supply-reduction efforts to achieve realizable goals.

Reviewing patterns of drug misuse over the last 25 years it is clear that, as control is achieved over one drug, new drugs emerge as drugs of misuse. The regular addition of new drugs controlled under either the Single Convention on Narcotic Drugs, as amended by the 1972 Protocal or Convention on Psychotropic Substances 1971 is indicative of the rapid emergence of alternatives to controlled drugs as drugs of misuse. Demand-reduction programmes must inevitably respond to drug problems as they appear.

If the efforts to reduce the supply of the demand for drugs are a response to the emerging and identified problems, it is inevitable that the initial goal of any intervention must be the limitation of harm to the individual, society, and the international community. The elimination of drug problems, while properly the ideal, is a goal that has been and will continue to be frustrated by the emergence of new drugs. Demand-reduction, therefore, has a major role to play as means of limiting the growth of drug problems and reducing the harm caused by drug misuse.

Within the efforts to reduce the demand, there are different intervention strategies. These may be aimed at reducing: (a) the number of new drug users; (b) the harm caused by drug use; and (c) the number of drug users within the community.

Each of these strategies has a vital role to play, and they are interdependent.

Efforts to reduce the harm caused by drug misuse and to reduce prevalence of drug problems may be placed in three main categories: (a) Making and maintaining contact with drug misusers; (b) Treatment and rehabilitation strategies; and (c) Social reintegration strategies.

In the development of demand reduction strategies, it is important to make an assessment of the nature and level of drug problems within a given community. This may be done through surveys and research or through a review of available indicators such as drug-related overdoses, drug users that come to the attention of the authorities, and arrests of drug offenders. Knowledge of patterns of drug misuse provides a basis for targeting intervention strategies, reducing the likelihood of limited resources being misdirected. For instance, the occasional use of psychotropic substances may require concentration on preventive education, accurate information about limiting harm to themselves and others avoiding harmful practices such as intravenous injection.

The concepts of treatment, rehabilitation and social reintegration have come to imply particular intervention approaches. Treatment is commonly viewed as a medical intervention through the use of substitute drugs, detoxification or the use of other medical approaches. If the goal of treatment is to assist the individual to control over her or his drug use problem, a range of non-medical approaches are equally applicable. For some drug misusers rehabilitation and social integration imply a process of restoration to the state prior to the onset of drug misuse. It is important to recognize that, for many persons, it will be a process of rehabilitation and integration, that is, creating new circumstances and ways of living that the individual has never experienced before. It is perhaps here that one of the major differences between drug and alcohol interventions can be most readily seen. Drug misuse most commonly affects young people with little experience of life. Alcohol problems more commonly affect adults, who have already substantial experience of life, including that relating to employment and personal relations. Thus, for adults, rehabilitation and social reintegration are a process of restoration, but, for drug misusers, there is a need to identify means of supporting new ways of living and of dealing with new experiences without resort to drug use.

Given the relative immaturity of drug misusers they should be taken into treatment compulsorily. It may be argued that, for the protection of the community, drug misusers should be removed so that they do not draw others into drug misuse. There have been different experiences with compulsory treatment, which may in part be because of the different goals pursued. For instance, the arrest and imprisonment of drug misusers commonly involves the compulsory detoxification. Experience has shown, however, that if no other intervention strategy is employed, the individual rapidly returns to drug misuse on release from prison. Where the goal of the intervention strategy is abstinence from drugs, treatment and rehabilitation is most effective when the misuser wishes to achieve abstinence. Without this desire or motivation to change, temporary benefits may be gained by the individual and the community, but relapse into drug misuse is likely to follow rapidly after discharge. Thus, intervention strategies must not only offer the techniques of treatment rehabilitation and reintegration but also stimulate the desire of the drug misuser to change.

Return to Drugs

A common experience in services dealing with drug misusers is that the individual returns to drug misuse after the completion of treatment. It is at times argued that this represents a failure of the services to reduce the level of drug misuse within a community. The corollary is that emphasis should be placed on supply reduction because of the ineffectiveness of demand-reduction approaches. Such arguments are false. It would be as valid to argue that supply-reduction approaches should be abandoned because, despite the resources devoted to this area, the supply of drugs has continued to grow. Drug misuse is an activity that can dominate the interests and concerns of the individual to the exclusion of all other considerations. In such circumstances, it is not surprising that a number of attempts to achieve control over drug misuse may be necessary before the individual has sufficient personal and emotional strength to regain independence from drugs. For some, this may require intervention strategies that are in the first instance aimed at containing drug misuse and reducing the involvement of the misuser in criminal and anti-social behaviour. For others, it may involve temporary removal from the community to provide a safe setting, away from the temptation of drugs, where living without drugs can be experienced. A growing body of evidence on relapse and relapse prevention shows that, while a drug misuser may return to drug misuse after treatment and rehabilitation, this is for many a temporary phase. If relapse is seen as any return to illicit drug use, then treatment and rehabilitation may be regarded as a failure. However, after a period of six months or more of contained drug use or abstinence, relapse is greatly reduced and return to chaotic drug misuse is minimized.

Current treatment approaches and methods include treating individuals who display alterations in mood, perception and thinking processess as well as modifying the client's environment. Often therapeutic activities are designed to influence the client's family, her or his home and social situation and/or the required vocational or educationl skills. As a result of these approaches, certain facilities, such as hostels and halfway houses, have been utilized in some countries, and various disciplines, such as psychiatry, general medicine, psychology, sociology and social work, education, industrial rehabilitation and vocational guidance, have been brought together in a holistic concept of treatment. Recently, it has been realized that it is difficult if not impossible to separate the concepts of treatment and those of rehabilitation.

Holistic Concept

Several approaches to the treatment of drug-dependent persons exist within the holistic concept. Each of the professional disciplines concerned with drug dependence has its unique approach to treatment. Therefore, the most effective approach to treatment is an inter-disciplinary one. Given that intervention strategies require a range of approaches and involve interaction between the drug misuser, family, culture and community, it is inevitable that no one approach, agency or profession can or should have a monopoly on expertise or services. A range of disciplines and expertise is required from medicine to education and training for employment. In this, the family, community groups, religious organizations and self-help organizations have played an important role through the support they have been able to offer both to intervention strategies and to the individual drug misuser. Where community action has been mobilized to reduce the demand for drugs, prevention, treatment, rehabilitation and social reintegration have all benefited. Because drug misusers inevitably come into contact with a range of services, voluntarily or involuntarily, it is important that these services are able to offer appropriate interventions. In some cases, The interventions are provided by governmental services. For instance, within the national health care system or in prisons; in others, they are provided by non-governmental or private organizations. If demand-reduction strategies aimed at reducing the harm to the individual and society are to be effective, it is essential for the services to be available to the individual whenever she or he needs them. These services should be provided at an easily accessible place and should not be restricted by the ability of the individual to pay for them.

The general concept of treatment could be defined as: (a) a process of crisis solving intervention; and (b) a process to introduce changes in life-style. In the absence of such objectives, the success of therapy may be seriously endangered.

Long-range therapeutic goals are: (a) to reduce the extent and seriousness of problems associated with the non-medical use of dependence producing drugs; (b) to reduce the prevalence and incidence of drug dependence; (c) to improve the personal and social functioning of drug-dependent persons; and (d) to develop an ongoing system for the continued planning, evaluation and modification of programme activities, as indicated by changing needs and the results of appraisals of the effectiveness of different policies, approaches and methods in use.

Immediate therapeutic objectives are: (a) to bring drug-dependent persons into continuing contact with "helping" personnel; (b) to encourage those with whom contact is made to accept one or more types of help to improve their life situations; (c) to improve the personal and social functioning of drug dependent persons; (d) to reduce the number of medical complications associated with the unsupervised self-administration of illicit drugs; (e) to reduce the unlawful or other socially unacceptable behaviour associated with obtaining and using drugs; (f) to help reduce the diversion of licitly produced drugs to illicit channels for non-medical use; and (g) to help prevent the development or to reduce the extent of a black market in illicit drugs by reducing the demand for such drugs.

The broad functions to be performed in the treatment and rehabilitation of drug dependent persons are : case finding; diagnosis or case evaluation; the establishment of immediate and long-term objective with respect to the person involved; the formulation of a management or treatment strategy, preferably with the participation of the client; the management of medical emergencies and complications associated with the basic problem; behaviour modification; the provision of rehabilitation services (medical, vocational or social); the provision of social services; the provision of continuing follow-up and after-care services; and the periodic evaluation of the effectiveness of therapeutic activities, including rehabilitation.

Overlapping Strategies

It is clear that there is an overlap and inter-relationship between intervention strategies. Because treatment, rehabilitation and social reintegration strategies are all concerned with improving the personal and social functioning of the drug-dependent individual. The particular approach adopted within a given strategy is dependent upon the personnel available and the stage of the drug-using career of the individual involved. Ideally, it is built on the traditions of youth and community work and has now become an effective means of contacting disadvantaged groups. It was adopted for use with drug misusers who were seen as particularly disadvantaged and isolated from the rest of the community. In recent years, outreach programmes have been directed towards other groups of misusers such as women, the elderly and minorities.

Obviously, reaching drug misusers is the most difficult part of intervention strategies. Recently, treatment programmes have begun to experiment with aggressive outreach approaches to contact and draw misusers into treatment before they lose social and economic supports and become a serious burden to the community.

For outreach work to be effective, trust must be established between the misusers and the outreach worker. Without this, it is possible to make contact but difficult not to retain it. Outreach work also requires the trust and co-operation of other agencies concerned with drug misuse in a community, such as the police and social and health services, because there will be little if any immediate return from such intervention strategies until trust is established with drug misusers. For instance, it may take as long as six months to establish effective contact with drug misusers, and, therefore, no referrals to treatment or rehabilitation may be made for some time. The outreach worker, however, may be able to advise and assist drug misusers to limit their drug use and to engage in less dangerous drug misusing behaviour druring this period.

The sequence of stages of intervention should be followed. It is possible, for the drug misuser to make contact with the service that she or he regards as most appropriate for his or her needs at a particular point in time. For instance, an individual may, establish contact with an outreach worker, for entering a rehabilitation hostel. This may prove unsuccessful on the first occasion. Hemay then enter a treatment programme to control her or his drug use and to detoxify before returning to rehabilitation.

Making and Maintaining Contact with Drug Misusers

Drug misuse is a covert activity that only comes to the notice of the family and community when it has become both a personal and a public problem. The goal of "case finding", that is making and retaining contact with drug misusers, should be defined at the outset. It might be:

(a) to identify and make contact with drug users at an early stage to avoid or limit the likelihood of social or medical complications ;

(b) to make contact with drug misusers who are unaware of the treatment and rehabilitation services available to them and to provide them with a link to these services; and

(c) to estalish contact with drug misusers who are not yet prepared to make contact with helping services and to provide them with counselling, advice and support in order to reduce the harm which may arise from continued drug misuse.

Outreach work was first developed as a means of reaching "unattached" young people in the United States of America.

Outreach work is most commonly carried out by youth workers or community nurses who have received training in providing help to people away from established centres. Such workers are commonly found in a number of European countries as well as in north America. Recently, former and current drug misusers have been recruited as outreach workers because of their knowledge and experience. Within particular communities, religious leaders and significant figures have also acted as contacts with drug misusers, identifying particular needs and assisting the misusers to achieve effective contact with specialized helping services.

This form of intervention has led to the identification of new and emerging needs and community proposals for drug misuse services. Among these have been telephone counselling and advice lines, self-help groups for the parents and families of drug misusers and self-help groups for drug misusers themselves. If outreach were a more common modality, it is likely that a much higher proportion of drug misusers would seek help. One of the most important limitations in extending outreach programmes, however, is the general lack of treatment and rehabilitation services available in most countries. It is unfortunate that relatively little has been written on outreach work among drug misusers, though it appears to be a valuable means of promoting an effective demand-reduction intervention strategy.

Acupuncture Treatment for Drug Addiction

Acupuncture has become a well-established therapy for various health disorders. It has an ancient tradition in Asian medicine and has been increasingly used by Western medicine. The World Health Organization (WHO) has published a list of health disorders for which acupuncture has been found to be helpful. The safety and convenience of acupuncture also makes it an attractive method of treatment for drug-addicts.

The magnitude and diversity of drug abuse problems have over-weighed current health care resources. Various methods are currently used for the treatment of drug-dependent persons, but the acceptability and reported usefulness of these methods vary from one society to another. It is desirable to use a treatment method that is effective and low-costing. In order to meet this requirement and to help attain the objective of the WHO to provide health care to all by the year 2000, different methods of treatment need to be explored. Acupuncture offers an option for drug-dependent persons.

Hospital Programme

Over the past 13 years, Lincoln Hospital in America has used acupuncture as the primary method for treating drug-addicted persons. It is a city-owned hospital located in the South Bronx, which is mainly populated by lower-income people. The acupuncture programme for the drug addicts received on an out-patient basis 200 addicted persons daily for detoxification. At Lincoln Hospital, 40 per cent of new patients dependent on crack give a series of clean urine tests several weeks after treatment. Acupuncture greatly increases participation of drug addicts in long-term treatment programmes.

The results obtained in research carried out at the University of Minnesota and the Downstate Medical Center in the United States have confirmed that acupuncture substantially reduces withdrawal symptoms and the craving for drugs. The study of acupuncture for alcoholics conducted during 1984 and 1985 by Hennepin County Hospital in Minnesota was particularly impressive with its clear-cut positive results. Positive results have been obtained using acupuncture detoxification of patients dependent on cocaine, opiates, phencyclidine or alcohol.

Drug addicts are impressed by the fact that daily acupuncture can relieve withdrawal symptoms reliably. They also discover that acupuncture can prevent drug craving even if none of their personal problems are resolved. Nevertheless, it is very important to integrate the acupuncture treatment into counselling and other psycho-social services provided for drug-dependent persons.

Acupuncture detoxification is a very convenient treatment modality for a patient. The actual process of acupuncture takes several minutes per patient. It has been learned that for each patient the same set of acupuncture points should be used at each visit, regardless of the type of drug that has been abused. The best acupuncture results are obtained where patients received treatment in an open-group setting, in which they are able to relax easily. Some women hold their small babies in their laps during treatment. Children usually remain calm in this type of setting, and no one is distracted. New patients are able to observe the treatment process so that they do not fear the pain of needle insertion.

Addicted persons are attracted by the opportunity to reduce their inner turmoil. It has been described elsewhere that violent and paranoid drug abusers have been substantially helped by the Lincoln Hospital acupuncture programme. Acupuncture detoxification is a non-judgemental nurturing process. It blends well with most crisis-counselling and long-term rehabilitation programmes. The method of acupuncture used for the treatment of drug-dependent persons is also appropriate for the treatment of persons suffering from stress.

Mechanism of Action

The action mechanism of acupuncture has not been fully understood in physiological terms. It probably involves the homeostatis actions of the automatic nervous system, various neurotransmitters and elements in the pituitary-subcortical axis. It is important to focus on the following functions of acupuncture :

(a) Relief of withdrawal symptoms and craving;

(b) General relaxation and homeostatis; and

(c) Apparent enhancement of mental and physical functioning by activating the reserve capabilities of the body.

In traditional Chinese medicine, that lack of a calm inner tone in a person is described as a condition of "empty fire" (xu huo), because the heat of aggressiveness burns out of control when the calm inner tone is lost. It is easy to be confused by the *xu huo* that many addicts exhibit and to conclude that the main goal should be the sedation of excess fire. In fact, addicts themselves take this approach in the extreme by using sedative drugs. The *xu huo* condition represents the illusion of power, and illusion that leads to more desperate chemical abuse and senseless violence. Acupuncture helps patients with this condition to restore their inner control.

It has been observed that the use of acupuncture points located in the external part of the ear gives the best result in the treatment of drug-dependent persons. The following points are used: "sympathetic", "shen men", "kidney", "liver" and "lung". At Lincoln Hospital, no electrical apparatus is used. In the 1970s, several acupuncture programmes applied the electro-acupuncture method for the treatment of drug-dependent persons, but it was observed that this method was short-acting and had to be used in an in-patient setting.

Limitations of Acupuncture Detoxification

There is a risk of infection if the acupuncture needles are not cleaned and sterilized properly. At present, this risk is compounded by the spread of the human immunodeficiency virus (HIV), which causes acquired immunodeficiency syndrome (AIDS). At Lincol Hospital, a double autoclave procedure for sterilization is used so that the technician preparing the needles is protected from an accidental puncture. As an extra-precaution, needles from patients with diagnosis of AIDS are destroyed after use.

Another difficulty in this field is an unfamiliarity with the use of acupuncture in the treatment of drug dependence. The procedure is nearly painless and causes the rapid onset of a gratifying sense of relaxation, but, on first exposure, most patient express fear of the pain of the needle insertion and are confused by the idea that little needles can cope with their big problems. This fear is easily solved by letting prospective patients observe the actual process of treatment.

After acupuncture detoxification patients should be provided with counselling, social support and measures for social reintegration.

Current Programme of the National Acupuncture Detoxification Association

The National Acupuncture Detoxification Association (NADA) was formed in 1985 to support the use of acupuncture in the treatment of drug-dependent persons in various clinical and socio-cultural settings. Successful acupuncture components have been included in conventional programmes for the reduction of drug abuse in a number of countries. NADA has helped the Los Angeles Sheriff's Department to set up a programme as a potential alternative to imprisonment. The programme is designed to use acupuncture as the primary means of detoxification in the county jail and to treat upto 500 drug-addicted persons daily in this large facility.

NADA has conducted successful training programmes in social settings with different cultural and economic backgrounds. It has trained physicians and related staff in using the technique and philosophy of acupuncture. Senior NADA consultants are licensed in traditional Chinese acupuncture. They also have recognized professional degrees for substance abuse treatment. Several have considerable international experience and multi-lingual skills necessary for cross-cultural co-operation in technical assistance programmes. The promotion of cross-cultural expertise in this field requires learning traditional Chinese medicine and translating it into a concrete drug abuse treatment programme. NADA has accomplished this goal in a clinic, which has received little financial support, particularly in terms of research funding. Training is conducted either in a community or hospital setting.

The training programme of NADA is of shorter duration than most other acupuncture training programmes, which usually have long traditional curricula. While such curricula are important for the treatment of various health disorders, they have little relevance to the use of acupuncture for treatment of drug-addicted persons. Numerous training programmes for physicians provide short courses in the use of electro-acupuncture. The use of electro-acupuncture for the treatment of drug-dependent persons is not as effective as the traditional use of needles and point locations.

Training Provided in Local Settings

NADA has set up acupuncture components in two native American programmes on alcoholism. Each of these programmes has been able to provide 20 acupuncture treatments daily. Ear acupuncture is used as a technique. In spite of the high rate of alcoholism among native Americans, these programmes had previously applied methods of detoxification, which were expensive and culturally less acceptable. Although native Americans have a coherent set of cultural values that can form the basis of long-term rehabilitation, the two programmes lacked a method of reaching members of the tribe who were actively involved in drinking. Acupuncture helped them overcome this problem and in one week the number of patients in the programmes showed a 10-fold increase. Many more patients were able to achieve advanced levels of rehabilitations, because the use of acupuncture had laid a more solid foundation for long-term recovery. The chairperson of NADA gave a workshop in October 1985 at Szeged, Hungary, at the large alcoholism hospital, Nagyfa, which is administered by the Hungarian police. The physicians there had only a minimum understanding of acupuncture prior to the workshop; after intensive training, they were able to establish an acupuncture programme which subsequently reported beneficial results in 60 per cent of alcoholics treated with acupuncture.

In December 1986, a NADA specialist conducted a two-week acupuncture training programme for staff of the local drug abuse prevention and treatment programme by St. Xavier's Service Centre operated by the Jesuit Ader at Kathmandu, Nepal. From 6 to 10 drug-dependent persons were treated each during the training programme, and, during the second week of training, the NADA trainer was able to observe that the trainers performed treatments correctly and used appropriate sterile precautions. A videotape and slides were used in the training sessions. Following the acupuncture training, the local drug abuse programme developed an extensive network among nearby Nepalese families, while previously it had not been able to use any method of detoxification of drug addicts that was acceptable to patients and suitable to their indigenous cultural setting. Of the first 60 patients treated at St. Xavier's Service Centre, 40 per cent have shown clear improvement, according to the Director of the Centre.

Drug-demand Reduction

The illicit demand for drugs, which is at present an immense problem has become even more critical with the recent increasing spread of HIV causing AIDS. Encouraged by the Health Department of San Francisco, NADA has begun to operate a pilot programme. The goal of this programme is to treat both active and intermittent abusers, as well as people who are under high level of stress, which may lead to drug abuse. It is assumed that acupuncture in conjunction with other drug-demand reduction programmes, by reaching this entire range of patients, can make a significant impact on the illicit demand for drugs. The programme is projected to treat approximately 1,000 people per day. The basic ear acupuncture protocol developed by NADA has been shown to be effective for a diverse range of patients. NADA has carried out stress-relief programme for the staff of a large city agency, as well as for labour organizations. More than 50 persons were treated per hour and all sessions were well-accepted. Using the example of Lincoln Hospital, it is possible to operate programmes that can treat 500 persons daily.

Relevance to Developing Countries

The advantages of acupuncture detoxification have particular significance for developing countries. The acupuncture protocol developed by NADA is simple to apply at low cost. It has been well-accepted by Latin American immigrants, rural native Americans and urban black people in the United States. None of these groups had any prior experience with acupuncture. In addition, it is well-known that acupuncture is traditionally popular in China, East Asia and parts of South-East Asia.

Acupuncture does not require special equipment or special testing procedures. Acupuncture needles can be easily sterilized in a steam autoclave or by other procedures. An acupuncture detoxification clinic can be set up in a matter of minutes in any location where people can sit comfortably. Training programmes have been conducted succesfully in a variety of simple treatment settings.

Further Development of Training Programmes

NADA is prepared to offer training programmes in any part of the world. The acupuncture training has been most effectively conducted at the site of an on-going programme for the prevention and reduction of drug abuse. In such a setting, classroom discussions lead directly to the mastery of skills on a practical, clinical basis. Furthermore, it is essential to combine acupuncture with social support measures, rehabilitation and social reintegration. This too can be done most effectively in a setting of an on-going programme, although an acupuncture detoxification training programme should not be limited only to staff of an on-going programme.

Medical Abuse and Therapeutic Measures

Drug abuse arises from a basic inability to fulfil needs and desires satisfactorily. The probability of taking drugs increases when the holistic growth of life is blocked and the person's creative intelligence cannot express itself fully. The individual experiences a vicious spiral of frustration, tension and diminished creativity. The individual cannot use his own creativity and intelligence to escape from this vicious circle. This is why so many rehabilitation programmes fail when the external support of counsellor, doctor, alternative drug, etc is removed. The inner resources of the individual have not been developed. Restoration of creative intelligence is thus the one ingredient necessary to bring success to any rehabilitation programme.

The Transcendental Meditation Programme: Its Contribution to Rehabilitation

Transcendental Meditation (TM) as taught by Maharishi Mahesh Yogi is a simple, natural, effortless and easily learned technique. It is practised for 15-20 minutes twice daily sitting comfortably with the eyes closed. TM requires neither a special environmental setting nor special equipment, and since it is non-religious, it requires no belief or faith. During the practice the mind settles down, to a silent yet fully awake or state of awareness or transcendental consciousness. This state of consciousness is fundamentally different from waking, dreaming and sleeping states and is accompanied by a distinct set of psychophysiological correlates showing unique deep rest, relaxation and high orderliness in brain functioning as measured by EEG studies. This unique state of mind and body has been described as a "ground state" of the system sharing the holistic properties of ground states of physical systems. In particular, a close relationship has been seen between the characteristics of this state and those of the unified field of all the laws of nature described by recent theories in quantum physics. This close connection has provided an understanding of the reason for the holistic improvements resulting from TM.

To date, over 250 research studies have been conducted in research institutions in over 25 countries. Many parameters of mental and physical health show distinct improvements, and there is a holistic development of all aspects of the personality as a result of the practice. Specifically, a number of studies have provided evidence of benefits in the field of rehabilitation. The reduction of drug abuse is a natural side-effect of the overall development of the person irrespective of the reasons for abusing drugs and irrespective of the particular drugs being abused. Therefore, the TM programme leads also to reductions in requirements for prescribed tranquillizers and prescriptions in general and non-prescribed drugs. The programme involves no attempt at prohibiting drug use once the TM technique has been learned. It does not interfere with any of the existing rehabilitation programmes and can be implemented at low cost. The practice of TM has also been found useful for the therapists themselves.

Research Studies

One of the first research programme in 1969 found that of 143 subjects who were regular users of marijuana, hallucinogens or hard drugs before beginning TM, 119 discontinued all drug use and 22 reduced drug use by 50 per cent or more, whereas only two continued regular use. Similar results were seen in a more comprehensive study on 1,862 subjects. In the six-month period before they started TM, about 80 per cent of the subjects had used cannabis. After six months of practice, 37 per cent continued to use cannabis. After 21 months, only 12 per cent continued to do so. Comparable changes occurred in the use of hallucinogens, barbiturates, amphetamines and narcotics. There results aroused great interest and many more studies followed and confirmed those findings.

Research published in 1974 found a marked decline in the use of cannabis in meditators, while little change occurred among matched control subjects. The degree of decline was correlated with the number of months the persons had been practising TM. In a prospective study, 10 experimental and 10 control subjects, matched for past drug use, were monitored for drug consumption over a three months period. Subjects in the experimental group who were instructed in the meditation technique showed a marked decrease in drug usage, whereas the control subjects maintained a high usage level. Psychological tests of both groups indicated improvements in mental health in the experimental group in comparison to the non-meditating control group.

Reduction in anxiety together with a reduction in the use of drugs, alcohol and tobacco was seen in several studies.

The preventive effect of TM was reported in a study where the use of cannabis by two groups of students, matched for age and sex, was monitored over a six-month period. None of the individuals in either group used cannabis prior to the study. It was found that whereas the number of people in the control group using marijuana rose from 0 to 30 per cent during the months of study, in the TM programme group after four months of meditation only 6.5 per cent were using cannabis.

Institutes in which a majority of the members practise TM are unlikely to be confronted with the problem of drug usage. A large body of research has shown that when one per cent of the population is practising TM, or the square root of one per cent is practising the TM Sidhi programme together in one place the result shows crime, sickness and accident rates of that area are markedly reduced.

A prospective one-year study assessed that with the utility of TM in a drug rehabilitation centre with 76 known drug abusers the total drug consumption was reduced by 57.6 per cent. The use of cannabis products was reduced by 58.2 per cent, amphetamines and barbiturates by 96.3 per cent, opiates by 78.1 per cent and hallucinogens by 90.8 per cent. In a detailed study of results, the authors found a trend of four phases in the decrease of drug abuse. Due to high motivation, there was a remarkable decrease in drug usage in phase I. When individuals realistically confronted their situation, former conflicts and problems led to a return to limited drug usage in phase II. In phases III and IV drug usage was again reduced due to growing self-confidence and acquired skills in managing problems.

That results of a study investigating the TM programme as an approach to secondary drug prevention showed after an average of 19 months' practising TM, usage level of all substances (including prescribed tranquillizers and prescription in general) declined significantly among TM participants, while those of controls did not. The length of time of meditating and regularity of meditation were positively correlated with decreased drug usage. This was also found in another study where subjects who regularly practised TM showed lower drug usage levels than those who were irregular in their practice, even though both groups had shown similarly high usage levels prior to learning the technique.

Conclusions

In over 15 research studies the Transcendental Meditation programme was associated with significant decrease in the use of a variety of substances. Research has also found that TM develops a more orderly style of neurophysiological functioning that enables meditators to respond more effectively to stressful situations. These results support the hypothesis that the TM programme creates a balanced state of physiological and psychological functioning, in which potentially detrimental behaviour is naturally avoided. This offers new hope in a field in which the prevailing view is that "nothing works".

The unique and systematic benefits of TM deserve the immediate and serious attention of all those involved in drug rehabilitation and prevention programmes. Organisation and trained teachers of TM exist in over 100 countries, and special instruction programmes have been designed and utilized in the field of drug abuse. The potential benefit to any country is enormous in terms of the recovery of previously wasted human resources. All organisations working in this area ought to make use of this opportunity.

Family Therapy for Addiction

In order to understand addiction, it is necessary not only to focus on its medical symptoms but also to examine the constellation of behavioural interactions, reflective of psycho-social development, in the context of the family, the community and wider socio-economic networks in which the family functions and to which the family relates. All these influence, stimulate and regulate individual behaviour. Thus, an important task of a therapist in the process of treating drug-addicted persons is to gain insight into the subtle, overt and complex dynamics of social interactions and not merely into the psycho-dynamics of individuals.

The theory and concepts of family therapy, which came to be recognized as an independent discipline in the 1950s help therapists to understand these interactions. Whereas classical medicine focuses on the treatment of individuals, family therapy regards the network of relationship in which the individual interacts as its unit of treatment. The family is regarded as a psychological unit, the internal processes of which help establish and maintain emotional balance.

Until the middle of the 1970s, many of those trying to cope with the syndrome of addiction largely relied on autodidatic methods that frequently had strong moral overtones. The International Council on Alcohol and Addictions (ICAA), which was founded in 1907 and is one of the oldest organizations dealing with addiction, set up a special section on the family in 1979 to apply the results of research in family therapy to the field of addiction. This underlines the short history of the union between family therapy and addiction.

Changing Social Attitudes and Family Patterns

In the late 1960s, in certain areas of the world, drug taking was identified with protest against authority, with a search for identity or as a mystical experience. The use of alcohol was viewed as an attempted demonstration of adulthood and as an identification with the peer group. The drug abuser of that time might be today someone who casually takes medically prescribed benzodiazephine. In general, there is now a tendency towards shifting consumption from illegal to legal drugs, combining drugs and alcohol; and increasing drug consumption at home.

There are many indications of change and variations in family patterns, especially in countries of North America and Western Europe. The number of legally defined marriages is decreasing. The tendency for couples to live together without a legal bond is increasing to such an extent that, for certain age groups, unmarried couples are more numerous than married couples. In Sweden and the United States of America, one out of every two or three marriages presently results in divorce. In Denmark, France, the former Union of Soviet Socialist Republics and in the United Kingdom of Great Britain and Northern Ireland, this figure was one in three in 1985. In France, about 20 per cent of all children born in 1986 were out of wedlock. In Denmark and Sweden, the figure for births out of wedlock is presently over 40 per cent, appearing to warrant a re-definition of the term "legitimacy".

Addiction and Abuse in the Family

Research and studies have pointed to the inter-linkages between abuse (corporal, emotional, sexual) and addiction in family situations. Members of families affected by drug dependence are often abused. A substantial number of children of such families are victims of various forms of child abuse. The duration of child abuse tends to last longer with the presence of drug dependency in the family environment.

Crime and victimization data in a number of countries indicate a relatively high incidence of homicide and assault among family members. Females are most frequently the victims. In fact, a family member is the most important single victim category as far as violence is concerned, and this is a considerably under-reported phenomenon. In attempting to understand the development and consequences of the types of behavioural characteristics and patterns in some families affected by addiction, it is useful to examine inter-relationships and interactions in the family, which some authors refer to as sub-systems. These are: (a) between partners; (b) between partners and children; and (c) between children.

Drug abuse is an indication of dysfunction within family sub-systems. In this respect, it is more beneficial to administer treatment within the framework of the network of relationships in which the individual interacts (e.g. the family) than to treat the individual in isolation. In this connection, the following questions emerge: Are the changing patterns of family life resulting in an isolation of the family from other social reference points? To what extent can the family be regarded as the patient? Is a dysfunctional family a symptom carrier of a "breakdown" taking place within the larger social system?

Characteristics of Families Affected by Drug Dependence

A family-therapy approach has made a definitive contribution in the prevention of drug abuse. Families affected by addiction characterize trends, moderately extant in all families, to a pathological extreme. The repetition of alcohol-related problems over a number of generations in the same family is a well-documented fact. Partners of addicted persons who remarry unconsciously seek similar personality traits in prospective partners. This is found to repeat itself over a number of remarriages.

Children of addicted persons who are not "problematic" (e.g., in terms of school achievement) in early childhood can develop difficulties in their relationships in marriage. Children who have become surrogate parents in affected families tend to develop a sensitivity for the feelings and needs of others yet do not manage to attain sufficient control or understanding of their own feelings. This hinders the process of successfully building up marital relationships. Some of the children of drug-dependent and/or abusive parents marry at a very early age. A child is often anxious to move out of this type of family as quickly as possible, often establishing a relationship with the first possible partner. A family environment in which there is the presence of addiction and/or abuse does not lend itself to stable relationships outside the family, and such relationships are often marked by failure.

Research has found that the so-called "social atom" of addicted persons displays considerable deformation, a devastated core of inter-personal relationships. The central zone, which refers to meaningful relationships with friends and colleagues, has practically no cohesion; the peripheral zone, which refers to relationships with all other persons in everyday life, is relatively crowded. This is an indication of the degree of the lack of personal identity, in the case of drug dependence.[4]

Family Therapy

The classical relationship between the medical doctor and the patient reinforces passivity by fulfilling the expectation of the patient that she or he can be cured by an outside agent. This model of a treatment relationship is to be avoided in the field of drug dependence. A more suitable approach would be to integrate "holistic" medicine and the concept of family therapy, with due regard to qualified practice in the field of treatment of drug dependence. Holistic medicine activates resources within the individual in order to cope with personal disturbances, and family therapy views a patient in the context of mental and physical stabilizing factors in an interactive environment.

The goal of family therapy, in general, is to avoid focusing primary attention on the affected individual. It is highly desirable to take a holistic approach and, thus, make a contribution to the prevention and treatment of familial abuse and drug addiction in the context of a network of relationships within the family and between the family and society. Experience shows that the treatment of drug-dependent persons in a family setting is most beneficial when it is provided on a multi-disciplinary basis. Such treatment helps the affected person to utilize all his or her potential resources to overcome the problem.

Multiple family therapy (MFT) is a form of treatment that can be of great assistance in taking families out of isolation. It enables families to develop social contacts and relationships outside the family and to relativize fixations and dependencies, helping to place family interaction into perspective. The therapy group, which can include up to 10 families, offers an opportunity to rebuild the core and zones of the social atom or network. Feelings of emotional rejection, isolation and helplessness, disturbance and other difficulties can manifest themselves more easily in a MFT setting.

The Role of the Family Physician

Medical practitioners and specialists are frequently confronted with both abuse and drug addiction in the course of their daily practice. But sensitivity to the early warning signs of familial abuse and addiction is critical. When a chronic condition has already set in, there tends to be resistance to change, and therapy is difficult. Family physicians are often trusted by families over a number of generations, and many cases of abuse and drug addiction come to their attention. The medical training of general practitioners, however, may not be adequate to enable them to recognize these problems at an early stage. Continuing education of family physicians is beneficial in dealing with such problems. Providing family physicians with various opportunities to exchange experiences and information and to acquire the essential knowledge and skills would enable them to deal more expeditiously and efficiently with problems related to drug dependence, at various stages, in the family context.

Guidelines for Parents

Youthful drug abuse is a wily, unpredictable beast. It may be snarling right outside their doors, yet many parents choose to ignore its ominous presence. In one US mid-western school district, for example, only one in five parents of seventh to twelfth-standard students thought his or her child smoked marijuana. In reality, over half the students had tried it, and nearly one-third admitted smoking pot regularly. Drug abusers are not just deprived kids. They can be anyone's children.

Parents can steer their children away from drug and alcohol-abuse, if they know what to do. Here are seven steps may serve as a guideline for parents:

(1) Educate at Home: Some parents believe, wrongly, that if they discuss drugs, or drinking, they somehow condone them. Other parents leave drug education to the schools. But children are not getting all the facts from school drug curricula. Something as crucial to your child's welfare as drug addiction should never be entrusted entirely to outsiders. Who has more influence on or a greater interest in your child than you? Before you can begin to educate your kids, however, you must educate yourself using accurate, up-to-date material. There is no surer way of alienating a child and tainting your credibility than to spout misinformation, particularly when talking to a teenager who may already have some drug experience. Up-to-date data are essential because new drug studies can contradict outdated but well-established beliefs. For instance, many in the medical field once believed that marijuana was relatively safe, but the American Medical Association now states: "Marijuana is a dangerous drug. Regular use can constitute a hazard, especially for adolescents." Furthermore, today's growers are producing marijuana that is up to 20 times stronger than that of a generation ago. Material on marijuana published before the mid-1980s might as well be discussing a different drug.

(2) Show and Tell: In teaching your children, your most eloquent statement is what you do. Anyone who thinks he can indulge in own dependencies and rear drug-free kids is fooling himself. Tobacco is the first dependency inducing substance tried by most youth. I had been an occasional smoker for years but stopped the day I found my son then just two, sucking on a cigarette he had pulled out of a pack. For me there was no greater incentive for breaking the habit.

The US National Council on Alcoholism states, "The best predictor of the drinking habits of adolescents is the attitude and behaviour of their parents." After an exasperating day at work do you say, "I need a drink"? This can teach your child a bad way to handle stress are inclined to use alcohol and drugs.

If your child asks why you drink, you might respond, "I'm an adult, and I take an occasional drink because I enjoy the taste." Eliminate "I need" from your vocabulary. And if he counters, "Then why cannot I drink?" you say, "Because you are not an adult, it is against the law, and it can be physically harmful. When you are of legal age, the choice will be yours."

(3) Start when they are young: Anti-drug education should begin as early as possible—at five years, even four. Just as you tell your child that touching a hot stove will burn his fingers, remined him repeatedly that drugs are harmful and can kill him. By the time he is offered his first joint, he'll associate it with danger.

Pre-schoolers are able to grasp basic concepts about drugs. If you wait until your children reach adolescence to discuss drugs, it may be too late.

(4) Prohibit any Use: Your posture should be unequivocal: your youngster is to touch no drugs or alcohol. In a California survey of eleventh-standard students, 21 per cent claimed parents either allowed or were neutral to adolescent drinking, and seven per cent declared parents held similarly casual attitudes about pot smoking. Some parents rationalized "I'd rather my kid drank or smoked pot at home, supervised, than out on the street". Admittedly if we are speaking about legal-age young people and alcohol, this has some merit. But giving tacit approval to illegal drugs in the home is wrong. Parents who do so are condoning criminal behaviour, destroying any basis for discipline, and inviting their kids to reel out of control.

(5) Do not Lecture: When talking to your children on substance abuse, broach the subject casually, perhaps while watching TV together or driving in the car. Don't be a moralizing preacher. Let the facts, the most persuasive evidence, work for you. Talking about cocaine, I said to then eight-year-old Adam, "Some people use drugs to avoid pain—not physical pain but the kind you feel inside when you are unhappy or upset. Unfortunately, we cannot always avoid feeling pain; sometimes it is even good for us. When people take a drug to get rid of their unhappiness, the drug wears off and the pain comes back. Only now it feels worse because the person has not solved the problem. If he continues to use cocaine, it gnaws away at his body like mice eating cheese. The drug no longer masks his pain, but by then he cannot stop using it".

(6) Promote Alternatives: Offer your children something better than drugs to make them feel good and to fill spare time. Try sports, arts, pets, volunteer work, gardening, reading—the list is endless.

Kids should be exposed to as many interests as possible. Young children are naturally inquisitive and generally receptive to activities parents suggest. Physical exercise yields a multitude of benefits. Because it stimulates the brain's production of endorphinsa, morphine-like substance that alleviates pain and fear, producing a natural high of sorts—kids discover early that there are alternatives to chemically induced sensations. They also learn important lessons about the gratification derived from self-discipline, from setting and achieving a goal.

(7) Provide Ways to Say No: Because the offers generally come from their friends, saying no can brand a kid an outcast. Explain to your child that you understand the difficulties of the situation. Say, for example, "I cannot pretend that refusing to do what your friends are doing is simple; it is not. But it is worth it. You won't always have to live with those other kids, but you will have to live with yourself." When "No thanks" seems hard to say, one statement other kids will usually accept is: "My parents would ground me for the rest of my life." Better yet, your child might turn around the peer pressure and say, "Drugs are boring. I cannot believe you need to take that stuff."

Help your child by role-playing. You—or an older sibling—might portray someone cajoling him to drink or take drugs. Stimulate a confrontation and repeat it until his responses are natural and self-assured.

Despite the pervasiveness of youthful drug and alcohol use, parents do have an advantage: because children are impressionable, the chances of moulding their values about drugs and alcohol are excellent. But move quickly, before pro-drug peers influence your kids. Take pre-action and you may never need to react to a drug crisis.

Drug Addiction Amongst Students

Among the college students, where a drug sub-culture always existed, the major reason for being booked into smack is, as psychiarist Alan d'Souza of Bombay's J.J. Hospital puts it: "intense peer group pressure". While in the 70s it was ganja, hashish and LSD that were the rage among the students, in the 80s with the prices of heroin dropping to about that of a bottle of beer smack is in-thing. "It has become fashionable for students to smoke heroin", laments Moonis Raza, Vice-Chancellor of Delhi University, the heroin addiction is acquiring serious proportions. In the more prestigious colleges of Delhi, the evil is speading its fangs. Bombay University found the drug addiction among students had become so acute that the authorities have formed a drug addiction cell to chalk out programmes to educate students. In Madras, student counsellors estimate that at least 15 per cent of the students had tried smack at some time or other and at least seven per cent have become regulars. Madras based psychiatrist S. Vijoykumar says, "Initially they smoke in groups but later it becomes a totally individual addiction. Then their grades in college come plumming down."

The Heroin Plague

Not everyone is convinced that the heroin plague is caused mainly by availability and low prices, especially the poor. Dr. L.P. Shah's colleague Dr. Nadkarni of KEM. Hospital is of the view that most of the slum dwellers' children are addicted to heroin.

The epidemic is not immediately apparent because it has no stereotype addicts—like the jeans-clad, long-haired hippies who symbolised the drug sub-culture in the West. In India, heroin with its deadly addictive qualities has become the great leveller, enslaving students, auto-rickshaw pullers, taxi-drivers and petty vendors. It has not spared factory workers, young professionals either.

According to an INDIA TODAY survey, "In Delhi and Bombay heroin addiction has reached serious levels. In Calcutta, it is the lower income groups like petty-shop-owners, taxi-drivers and vendors who have been badly affected, while students are now getting hooked on the smack bug too. But in Madras, it is mostly students who come for treatment. What is alarming is that a majority of the addicts coming for treatment are males in the age group of 16 to 25 in the prime of youth. Trapped in heroin acid fumes, they have been reduced to the level of gombies willing to resort to anything, even crime, to pay for their daily fix."[5]

Addiction Leads to Petty Crime and Drug-Peddling

So powerful is the craving for heroin that nothing stops an addict to get his daily dose. Many of the addicts turn peddlers too. These children took to tobacco, alcohol and finally drugs like heroin. Dr. Nadkarni says, "This is the price we have to pay for urbanisation."

Heroin Addiction—Wide-spread Phenomenon

An epidemic of heroin addiction is sweeping through the country and hitting new high. The plague has spread with amazing rapidity and hit almost every section of society—students, young professionals, taxi-drivers, petty vendors and lower government officials. The problem has acquired more dangerous dimensions with the large scale influx of heroin into India.

The estimated figures often put up are only tip of the iceberg because researchers estimate. Police intelligence puts the number of addicts in the country at a conservative one lakh, which is increasing at the rate of 15,000 every year and psychiatrists are now talking of a heroin plague sweeping the country. At the All India Institute of Medical Sciences (AIIMS) which treated a record 1044 patients last year, psychiatrist Devendra Mohan said, "What we are seeing is an epidemic. It is only when they run out of money, start borrowing and stealing or their health fails that most addicts seek medical help." According to Ved Marwah, former Police Commissioner of Delhi. "There seems to be a growing nexus between narcotics and organised crime and terrorism." His fear about smack financing terrorism was confirmed when police arrested Jagdish Singh Narialwala.

Some Successful Programmes

Barbara Zugor, director of the "Do Drugs, Do Time" programme, explains that once someone goes through the initial arrest proceedings, he or she can opt for entrance into the one-year TASC rehabilitation programme, where he or she will undergo daily treatment in lieu of going through the court system. Participants must pay a fee for the programme which is based on a sliding scale. (The programme offers aid to those who cannot pay). They then must successfully complete the full year to avoid the traditional criminal system. Zugor says that 96 per cent of these offenders have never been treated before. TASC gives them the opportunity to get help, while freeing up court time and jail space. She says that 68 to 70 per cent of TASC participants successfully complete the programme. "Do Drugs, Do Time", according to Zugor, "is a real viable option for some".

Teen Court

Gary's "Teen Court" also is proving to be a viable option for some—some teens. Teen Court "uses the power of peer pressure to positively affect the teenagers and help them learn responsible ways to act," says Shirley Caylor, Associate Director of the centre. A first time offender can choose the programme over the traditional court route. An offender appearing in Teen Court will find a jury, prosecuting attorney, bailiffs and court reporters who are all teenagers. The judge is an adult attorney who volunteers his time. The players are trained and guided through the process by additional volunteer attorneys.

The jury asks questions throughout the trial and then pronounces the sentence which always includes future service on Teen Court. "Over 400 kids have volunteered (to serve in Teen Court) from every school, junior high and senior high", says Caylor. "They are usually creative and very appropriate in their sentencing". Many sentences include individual or family counselling community service and restitution. "Over 50 per cent of the kids that come through have substance abuse problems, and many of the others have family members with problems", Caylor says.

Teen Courteers may leave the process at any time, even after sentencing, and instead go through juvenile proceedings. If they choose to accept their sentences, and successfully serve them, their records are wiped clean. Caylor says that since Teen Court began in 1990 only four participants have become repeat offenders, "We have a 96 per cent success rate", she says.

For the success, the brewing industry's Beer Institute, which helps to support local drug and alcohol prevention programmes, chose Teen Court as one of its grant recipients. In 1989 the Institute developed its Community Assistance Fund. "We didn't want to reinvent the wheel", says Director Jeff Becker, "so we decided to give grants to about 50 programmes in 26 states". In addition to "Teen Court two other programmes that have received grants from the institute are the Midnight Basketball League, a programme targeting young men who are school drop-puts and/or jobless and AARDY's Army, a programme aimed at elementary school children.

Midnight Basketball

Midnight Basketball originated in 1986 in Prince George's County, Maryland, because of one man's intolerance of the escalating crime and drug-related activity. Van Standifer, former manager of Glenarden Township, was concerned about the group of young men, aged 17 to 25, because "they are levelled high-risk and written off".

Recently having become the National Midnight Basketball League, the programme now exists in Washington, New York City, Atlantas, Chicago, Neptune, New Jersey, Daytona Beach, Florida, Austin, Texas, Topeka, Kansas and Charles County, Maryland. And that is not to mention the 20 or so communities, Standifer says, are planning to join up.

Standifer explains that cities or counties that join the national league are asked to sign a letter of agreement that outlines the set of parameters for running their local leagues. No game may begin before 10 p.m. (The games traditionally run from 10 p.m. to 2 a.m. prime time for drug and (gang-related activity). Two uniformed officers must be assigned to each game. Ballplayers must be between the ages of 17 to 25, and if the group has the entire range of ages, it must be separated into two playing groups. 17 to 21 and 22 to 25. And uniforms or jerseys must be purchased from league-authorised manufacturers.

Once a community signs the letter, it becomes a member of the National Midnight Basketball League Board of Directors. "The idea is to promote the concept of midnight basketball because we believe it is a positive force against the drug crime", says Standifer. And statistics prove the late-night games are doing just that. Since the programme began, there has been a 60 per cent drop in reported crime in Prince George's County. Of course, the main issue is funding. The only way the programme can survive, says Standifer, is through public-private partnership. "There will never be enough tax dollars to keep it going".

The same is true for the East Hampstead, Connecticut—based AARDY's Army—the only educational drug programme ever to be endorsed by the California School Boards Association, the largest such association in the USA. The non-prcfit AARDY stands for Alcohol-drug Abuse Really Destroys You.

The programme developed through founder Bob Alexander's 11-year old speaking circuit, which includes drug and alcohol educational talks to students of all ages and schools throughout the country. Alexander, professional cartoonist, was the victim of a drunk driving accident and knows first-hand what drug abuse can do. He developed a programme specific to younger children simply because, he says, "that's the time to reach them".

When schools or school systems adopt the programme they receive a kit for each child and a teacher's manual. Each kit cost $ 5 and contains a membership card, a certificate of membership, some stickers and the programme's information newsletters. AARDY and various other cartoon friends teach the children about peer pressure and positive self-esteem. As city and county officials face the dilemma of cleaning up the drug scenes in their home towns, they might remember that successful drug efforts have one common factor: coordination. When agencies, community groups, police departments and school systems work together funding for enforcement, education, treatment and research comes more easily.

Drugs and Voluntary Agencies

Differing viewpoints on the role of non-governmental organisations in tackling the drug menace brought into focus simmering discontent among the doctors of Government-aided and Government-run hospitals across the country. The Government policy of seeking the assistance of NGOs to fight drug abuse came in for criticism at a round-table in New Delhi.

The Role of Voluntary Organisations

Dr. D. Mohan, Professor and Head of the Psychiatry Department of the All India Institute of Medical Sciences, came down sharply on the whole idea which he said only went to create parallel organisations. The fault lay in the planning, Dr. Mohan said referring to the mushrooming drug de-addiction centres, mostly run by NGOs. When a complete set of infrastructure is in the form of Government hospitals, CGHS dispensaries and primary health care centres across the country existed, it was wrong to create additional infrastructure through these NGOs, he said.

When drug addiction and drug abuse assume menacing proportions, doctors come into the picture. Ultimately, it had to be the doctors who would cure the patients. So when so much infrastructure already existed, it would be an unwise decision to use up more of the scarce resources for setting up similar facilities. Also de-adiction centres will not be able to deal with people who take to injecting heroin.

It is an unfortunate situation when the health sector and the NGOs compete with each other for the Government funds. NGOs should an fact be in extended arm of the health sector and not function as a parallel arm. In fact, if a NGO has to be paid to involve the community and work among it, then it is nothing but another form of Government employment," [Sic] Dr. Mohan said.

Addressing the round-table, organised by the Indian Committee of Non-Governmental Organisations for the United Nations to observe the international day against drug abuse and illicit trafficking, Dr. Mohan said some of the measures being contemplated in the country to fight the menace had already been proved ineffective. "We should draw lessons from history", he said. Citing an example, he said the concept of these drug de-addiction centres had totally failed in Thailand way back in the 60s. The slogan should be "better health first and de-addiction centres later", he said.

Dr. Mohan's strong views on the Government perception of the problem and its means to fight them evoked an equally strong reaction from the representatives of the NGOs present at the round-table whose discussion theme was "Role of NGOs in drug abuse control in India—a retrospective". Ms Manjari Saxena outlined the positive steps being taken by several NGOs in the fight against drug abuse. She said that despite the magnitude of the problem these voluntary organisations were doing yeomans service for those caught in the drug net.

Spills in Transit: Mr. M.M. Bhatnagar, Director General, Narcotics Control Bureau, also appreciated the role NGOs were playing in the fight against drug abuse. He is of the view that though India served as a transit point for several drugs on their way to the Western nations, some quantity did leak out. The spillover effect, though not serious at the moment, could assume alarming proportions if effective steps were not taken. Even after the legislation was strengthened and punishment for drug offences made more stringent, these were inadequate to meet the challenge.

Mr. Bhatnagar explained the measures the NCB, the nodal agency which coordinates the law enforcement agencies in the fight against drug traffickers, was taking. These included stricter enforcement of rules and identifying and destroying illicit poppy cultivation.

Ms. Najma Heptullah, Deputy Chairman of the Rajya Sabha, inaugurating the round-table discussion, suggested that members of Parliament be involved in the fight against the social menace. Ms. Asha Das, Joint Secretary, Ministry of Welfare, admitted that till date no survey or study had been carried out by the Government to quantify the extent of the menace. However, some limited studies undertaken in selected towns showed that drug abuse was prevalent among the weaker sections of the society. The Government had a limited budget but was still funding various NGOs, she said.

"Anti-drug Bodies not doing Enough"

While most people in the four metropolitan cities are aware of the magnitude of the threat from drug abuse, nearly half of them do not think the enforcement agencies are doing enough to fight the drug menace. About one-fourth of the respondents in a survey, conducted recently by the Operations Research Group (ORG) are not aware of the corrective facilities which are now available in these cities for de-addiction. The salient findings of the ORG survey, carried out in 1,150 households in Bombay, Calcutta, Delhi and Madras, revealed that 46 per cent of adults in the four metros felt that the various enforcement agencies like the police, customs and excise, were not taking adequate steps to curb the drug menace. This belief was strongest in Bombay with 66 per cent of the respondents thinking that enforcement agencies were not doing enough. Thirty-five per cent of Delhi and Madras respondents and 45 per cent from Calcutta shared this view. The ORG survey found that in Bombay city, about 10 per cent of the respondents felt that drinking alcohol was more dangerous than taking drugs. There were no variations in the response from different age-groups or between the sexes. With regard to curing users, 94 per cent of adults in Delhi were aware that drug addiction could be cured compared to 86 per cent in Bombay and Calcutta. However, 27 per cent of the respondents were unaware that corrective facilities for drug addicts were available as against 14 per cent in Delhi and 21 per cent in Madras.

The ORG survey found that awareness of co-addiction facilities was generally low among the youth, with a little over one-third of the under 24 age-bracket not aware of the corrective facilities. Half the adults in the four cities were not aware of the severe punishment that could be imposed on those caught possessing drugs. About 56 per cent of Bombay respondents said they did not know about the kind of penalisation that could be imposed on persons in possession of narcotics. Such an awareness about punishment provision is even lower among younger age groups, as many of 56 per cent of those under 24 are not aware of the severity of the punishment as against 46 per cent among adults of over 35 years.

According to the ORG president, Dr. N. Bhaskara Rao, Doordarshan could play a more active role in promoting information about corrective treatment facilities rather than limiting itself to mere reporting on drug trafficking. Another ORG functionary, Mr. N. K. Singhal, said the survey was commissioned by the Ministry of Welfare to study the effect and inadequacies of the Narcotic Drugs and Psychotropic Substances Act of 1985. Mr. Singhal further said it was not enough to seize drugs and arrest the drug peddlers. The objective should be to put the kingpins of the trade out of commission.

Another problem that had arisen out of the scheme of offering rewards for drug seizures was that enforcement agencies were unwilling to share information. The fear of having to share the reward had led to a degree of lack of coordination among different agencies fighting the drug menace.

Synanon : A Movement for Drug Control

Synanon got its name from an addict who was trying to say 'Seminor'. The term 'Synanon' is used to refer to the entire-organisation, but when it is spelled with a lower-core it refers only to the meetings occurring in the evenings among small groups of six to ten members. Each evening, all members are assigned to such groups, and membership in the groups rotated so that one does not regularly interact with the same six or ten persons. The aim of these meetings is to 'trigger feelings' and to allow what some members refer to as a "Catharsis".

"Synanon", an organisation of former drug addicts, was founded in May 1958, by a member of Alcoholics Annonymous with the assistance of an alcoholic and a drug addict. In December 1958, Volkman (a new addict) heard about the two dozen ex-addicts living together in an abandoned store, and she obtained permission of the Synanon Board of Directors to visit the group daily and to live in during the weekends. In July 1959, she moved into the girls' dormitory of the groups now, larger quarters and continued to reside at Synanon House until June 1960. Cressey visited the house on Volkman's invitation in the Spring of 1960 for one year beginning in July 1960.

Admission

Not every addict who knocks on the door of Synanon is given admission. However, the only admission criteria is 'exposed willingness' to submit one's self to a group that hates drug addiction. Synanon adopted Cressey's following principle—

> "If criminals are to be changed, they must be assimilated into groups which emphasize values conducive to law-abiding behaviour and, concurrently, alienated from groups emphasizing values conducive to criminality. Since our experience has been that the majority of criminals experience great difficulty in securing intimate contacts in ordinary groups, special groups whose major common goal is the reformation of criminals must be created".[6]

Although it is impossible to say whether Synanon selective admission process inadvertantly admits those addicts who are most amenable to change, no addict has been refused admission on the ground that his case is "hopeless" or "difficult" or that is "unreliable".

Indoctrination

In the admission process, and throughout his residence the addict discovers over and over again that the group to which he is submitting is anti-drug, anti-crime and anti-alcohol. This emphasis is an unwitting implementation of Cressey's second principle:

> "The more relevant the common purpose of the group to the reformation of criminals, the greater will be its influence on the criminal members' attitudes and values".

Indoctrination makes clear that Synanon exists in order to trap addicts of drugs, not for purpose of recreation, vocational education, etc. The indoctrination techniques envisage 'example is better than precepts'.

Group Cohesion

The daily programme on Synanon is consistent with Cressey's third principle, and appears to be an unwilling attempt to implement that principle:

> "The more cohesive the group, the greater the members readiness to influence others and the more relevant the problem of conformity in the group norms. The criminals who are to be reformed and the persons expected to effect the change must, then, have a strong sense of belonging to the group: between 'we' feeling. The reformers consequently, should not be identifiable as correction workers, probation or parole officers or social workers".

The daily programme has been deliberately designed to throw members into continuous mutual activity. In addition, to the free, unrestricted interaction in small groups called 'Synanons', the members meet as a group at least twice a day. There are weekend recreational activities, and holidays, wedding anniversaries, and birthdays are celebrated. Each member is urged "Be yourself", "Speak the truth", "Be honest" and this kind of action in an atmosphere that is informal and open, quickly gives participants a strong sense of "belonging". Since many of the members have been homeless drifters, it is not surprising to hear frequent repetition of some comment to the effect that "this is the first 'have' I ever had".

Holding addicts in a house once they have been allowed to enter is a strong appeal to ideas such as

> "Its O.K. boy. We've all been through it before."
>
> "For once you are with people like us you've got everything to gain here and nothing to lose."
>
> "You think you're tough. Listen, we've got guys in here who could run circles around you, so

quiet yours."

"You are one of us now, so keep your eyes open, your mouth shut and try to listen for a while."

"May be you'll learn a few things."

"Hay tough baby. We won't let you."

Status Ascription

Cressey's fourth principle is:

> "Both reformers and those to be reformed must achieve status within the group by exhibition of 'pro-reform' or anti-criminal values behaviour patterns. As a novitiate...he is a therapeutic parasite and not actually a member until he accepts the group's own system for assigning status."

The house has an explicit programme for distributing status to members in return for staying off the drugs, and later for actually displaying anti-drug attitudes. Every member is expected to go through 'three stages of growth', indicated by Roman numerals below:

Stage I—A

Sick kicking addicts

New Comers | Non-workers

Stage I—A-B

Automobile crew
Barber
Electricity
House cleaning
Kitchen help

Laundry
Library
Maintenance crew
Plumbing
Service crew

Stage I—B

crew
Junior co-ordinators
Kitchen crew chief

Nursery heads
Office workers
Service crew chief

Stage—II

Assistant Directors
Assistant School
Business Manager
Department Chief

Office Manager
Directors
Senior Co-ordinators
Work

Stage—III

Workers

Board of Directors
Policy makers

Addict Rehabilitation

In 1955 Cressey listed five principles for applying Edwin Sutherlands theory of differential association to the rehabilitation of criminals.[7] This article is now frequently cited in the sociological literature dealing with group therapy, "therapeutic communities" and total institutions. According to the chief of the United States Narcotics Division "The Community should restore the former addict to his proper place in society and help him avoid association that would influence him to return to the use of drugs."[8]

Cressey gives five rules for implementing this directive to "restore", "help" and "influence the addict". These rules derived from the sociological and social psychological literature on social change, crime prevention, group therapy, communications, personality change, and social change were designed to show that sociology has distinctive, non-psychiatric theory that can be used effectively by practitioners seeking to prevent crime and change criminals.

Rehabilitation of Drug Addicted Persons in India

Before 1985, there was a serious lack of facilities for the treatment and rehabilitation of addicted persons. In response to this situation, the Nav Chetna Drug De-addiction and Rehabilitation Centre was established in Varanasi, in December 1985, sponsored initially by the Kashi Club. In the early days of its operations in 1965, the centre provided counselling services to occasional and moderate drug-dependent persons, and promoted drug awareness education for the general public.

In 1987, the Nav Chetna Centre began to offer clinical treatment on an out-patient basis. In April 1987, the centre provided full-fledged residential rehabilitation, supported by a grant from the Ministry of Welfare, Government of India, making all requisite services available in one institution, including de-toxification, medical care, counselling, educational and vocational guidance, yoga therapy and after care.

Yoga Therapy

When addicts stop using drugs, their rapidly improving physical health should not be regarded as the main criterion to judge their real condition. A craving for drugs does not automatically stop after detoxification. The post detoxification period requires constant adjustment and the removal of distractions in order to achieve co-ordination of mind, body and senses. Yoga as a system for rejuvenation, becomes a natural ally, offering a systematic method of achieving this goal in a relatively short period of tenure.

Yoga therapy satisfies the aim of drug rehabilitation by bringing about a holistic development in mental and physical health and in social behaviour. The metabolic and breath rates and cardiac output decrease with yoga therapy. In fact, the principal aim of the Yoga is to help a person overcome physical discomfort and emotional tensions by training her or him to forget unpleasant experiences and to face stressful situation boldly. Thus yoga offers new avenues for the promotion of positive health and protection against stressful events.

The package of tailored programmes of yoga therapeutic exercises being provided in the centre is listed below:

Asans: (Body postures)

Shavasana: It is a relaxation exercise. It helps bring progressive muscle relaxation, balance nervous system overactivity and reduce restlessness.

Asishmatsyendrasana and matsyendrasana: These asanas help alleviate indigestion and loss of appetite(s) and indirectly help offset abdominal problems and the craving for drugs.

Sarvangasana: This asana has a very beneficial effect on the endocrine glands, toning up the whole nervous system and increasing considerably the strength of the brain and nerves. It is rejuvenation exercise.

Sirshasana: It is a yoga 'head stand'. It sends a rich glow of refreshing blood to the brain, relaxing the mind and body and improving concentration, body metabolics and brain functioning.

Pranayam (Breathing): Pranayam, derived from Prana (the life breath) and ayam (pare), constitutes the yoga science of breath control.

Nadi Shodhana: It helps concentration by increasing amino-acid levels in the brains. It gives stability to otherwise restless ex-addicts and after prolonged regular use, results in an increase in alpha wave activity.

Bhramari: The bhramari exercise helps to relieve the tension of mind and strengthen veins and tissues of the brain.

Bhartrika: A large amount of fresh air is supplied to the body by the bhartrika pranayam and at the same time, carbon dioxide is driven out.

When practised regularly, this exercise results in an increase in the oxygen/carbon dioxide ratio and a decrease in the presence of non-adrenaline metabolites in the urine, the body is relaxed and the mind alert.

Kriyas (purification exercises)

Purification exercises (kriyas) help solve symptomatic condition (e.g., watering from eyes and nose, insomnia) that ex-addicts have to live with for quite some time and help to restore the normal sleep cycle. Some of the kriyas are:

Neti: In Sutra Neti, a cotton thread is passed through a nostril and brought out through mouth; then it is repeated with the other nostril. In Jal neti, lukewarm salted water is passed through one nostril and brought out through the other one, this is repeated with both nostrils.

Dhauti: Dhauti is the process of gradually swallowing a muslin surgical cloth soaked in salt water and then taking it out slowly.

Krunjal: Krunjal involves drinking and then vomiting salted lukewarm water. This helps in cases of dyspepsia by regulating the normal periteslic movements of the intestines and removing excessive gastric acid.

Nauli: Nauli removes the impurities in the lining of the intestines, it also alleviates constipation, obesity and dyspepsia. The liver, spleen and glands are influenced positively and the stomach remains light and the appetite increases, promoting physical and mental alertness. Apart from relieving some of the physical problems of an ex-addict, nauli imparts a source of cleanliness.

Treatment Approach

Motivation to abandon drug addiction is perhaps the single most important factor for successful rehabilitation. Law enforcement and educational and clinical treatment alone and a variety of programmes involving these approaches have not been sufficient to stop drug abuse.

Most drug addicted persons who visit the centre are required to undergo a detoxification process for about six days. Then individual and group therapy and counselling sessions begin, followed by bio-energetic exercise classes. Individuals are handled carefully and monitored round-the-clock by the persons who are well-acquainted with the characteristics of them requiring drug abuse treatment.

The approach of the Nav Chetna Drug De-addiction and Rehabilitation Centre is to build up individual motivation and initiative, to promote personal development and to bring about improvements in maturation attitudes and behaviour through a therapeutic oriented programme. The way to this approach lies in creating physical and psychological conditions that optimize the natural tendency of the individuals system to self-actualize and eventually stabilize.

(a) In individual therapy, the drug dependent person is encouraged and helped to modify her or his attitudes and behaviour and to build and strengthen initiation and confidence to overcome the drug addiction problems.

(b) In group therapy, various subjects are discussed; new ideas are generated; morale is boosted; individuals are encouraged to talk freely about drug dependency; and yoga exercises are organised. In this way, by providing effective medical and therapeutic aversion programmes, the centre can tackle many of the problems that prolong drug dependency.

References

1. Bulletin on Narcotics, Vol. XL No. 1. 1988—Treatment option in responding to drug misuse problems—E. TOWGUE AND TUKNEA, p. 4.
2. The Drugs Abuse: Paul Fuqeves, p. 44.
3. UN Bulletin on Narcotics, Vol. XL, No. 1, 1988—Special issue on treatment, rehabilitation and social reintegration of drug dependent persons; pp. 35—41.
4. UN Bulletin on Narcotics, Vol. XL, No. 1, 1988—Special issue on treatment, rehabilitation and social reintegration of drug dependent persons, pp. 57—60.
5. *India Today*, March 15, 1986, p. 119.
6. Narcotic Addiction—John A. O'Donnell and John C. Ball; Differential Association and the Rehabilitation of Drug Addicts—Rita Volkmann, p. 214.
7. Donald R. Cressey "Changing Criminals: The Application of the Theory of Differential Association", American Journal of Sociology— Sept. 1985.
8. Harry J. "Drug Addiction"—Encyclopaedia Britanica-VII, 1960, pp. 677-79.

61

UNITED STATES WAR ON DRUGS

In a joint television address, Ronald Reagan and Nancy hoped to escalate Washington's widening war on drugs to the level of a national crusade: "Those who are killing America and terrorizing it with slow but sure chemical destruction will see that they are up against the mightiest force for good. They will have no dark alleyways to hide in". The House, with 392 to 16 votes, approved a sweeping omnibus bill that called for immediate deployment of enough U.S. Military force to halt the flow of narcotics in America within 45 days. The legislation included the death penalty for drug racketeers who commit murder, mandatory life sentences for those convicted of selling drugs to children for the second time and a controversial provision to ease the rules of evidence in all Federal criminal prosecutions. It also increased Federal spending on virtually every aspect of the drug war—from stopped enforcement activities to prison construction to expanded treatment and drug-abuse education programme.

The broad outlines of the administration programme include a joint effort with the private sector to reduce drug abuse on the job as well as intensified government campaign against drug-trafficking. The government employees would be subject to mandatory drug testing and second offenders would be liable to discipline or dismissal. Thus, the deluge of smuggled cocaine is now the top-priority target. America's war against drug abuse really begins at home.

Good Conduct Certificate

Each year, Pakistan, like other countries linked to the drug trade, must obtain a certificate of good conduct required by U.S. Law. Countries that fail certification, risk losing loans and aid from the United States and international organisations controlled by Washington. However, uncertainty prevailed regarding the co-ordination of anti-narcotic activities between U.S. and Pakistan following the death of General Zia-ul-Haq in mid-August 1988.

While the U.S. government claims to have made it clear to the government in Pakistan that narcotics control remains a high priority issue in the US-Pakistan relations, a report of the Bureau of International Narcotics Matters (BINM) of the U.S. government says, "It is impossible to predict at this moment what the future may hold." The government of Pakistan's eradication efforts produced rather an unsatisfying result. The bureau's international narcotics control report stated that although the Pak government centered its eradication efforts on preventive enforcement activities, but, it worked in only a few areas like Gadoon and Malakand. The programme was never applied to the major opium growing areas. As a result, majority of the farmers continued to grow opium and the overall acreage cultivated probably increased. "In addition, limited use was made of the US-supplied aerial spraying capability embodied in a Thrush aircraft and a helicopter wing".

There was disagreement between the US and Pakistani officials on the effectiveness of their joint programmes in 1987. While the Pakistan Narcotics Control Board (PNCB) claimed reduction in the opium production, from 130 to 80 tonnes, agents of the Drug Enforcement Administration of the US in Pakistan pointed out that Pak officials "concentrate on surveying traditional poppy growing areas and areas where the poppy ban has been extended and does much less in areas not covered by development agreements, largely, tribal areas on the western border where most cultivation has shifted".

Giving his view on the narcotics situation in Pakistan and the role of his government, Mr. George P. Shultz, Secretary of State, in an interview told that his government was working very hard with Pakistan on this problem and he thinks very "strongly, co-operatively with the government of Pakistan. President Zia, before his death, and I am sure his successors will carry on in the effort to eradicate crops and to get at the laboratories where these drugs are processed and the movement of drug to India for transhipment purposes". [Sic] He continued to say that the shipment "is aimed at the U.S. and once again that brings into play our own interdiction along our own borders where we are working increasingly hard and effectively, but also doing something about the big market here".

Even as Mr. Shultz's claim that his government was working "collaboratively" with Pakistan on narcotics problem, most of enforcement agencies in his government express dissatisfaction over the approach of the Pak officials in implementing the existing laws or following suggestion made by the US government as part of their joint anti-narcotic programmes.

Positive Efforts

Positive efforts at eradication have been, thus, counter-balanced by increases in poppy cultivation in areas where the ban on cultivation is not in place. The DEA reported that "It is believed that these areas, including a portion of Dir, and tribal areas like Bajaur and Mohmad, and small areas of Hazara, now account for 60 per cent of the poppy production in Pakistan".

The administrator of the DEA, Mr. John C. Lawn, said while referring to a report from the US Department of Justice that the overall opium production estimates are "complicated by uncertainty about the number of Pakistani addicts as well as the volume of raw opium exported from Pakistan and Afghanistan to Iran for its addict population". He said: "A report survey of drug abuse in Pakistan, funded and supervised by the Federal Republic of Germany, indicates that heroin and opium abusers require 340 MT of opium or its equivalent. Even allowing a generous margin for error, it is clear that Pakistan is not an importer of opium".

But Pakistan, in the eyes of Federal agents is a major centre for money laundering. The structure of the Pakistani financial system, the existence of the 'hundi' system, and the large number of overseas Pakistanis who regularly transmit funds in a variety of forms make the Pakistani financial picture "extremely complicated". For example, interest rate differentials between Pakistan and the rest of the world are responsible for large movements of funds. While drug money is undoubtedly involved, the situation is not driven by drugs. In fact, several reports to analyse drug related funds have concluded that most Pakistanis "particularly drug traders prefer to keep large foreign exchange holdings overseas".

DEA agents also point out that law enforcement personnel in Pakistan are poorly paid and are "susceptible to bribes from drug traffickers. Law enforcement agencies, have not, except, in a handful of cases, arrested major drug traffickers, financiers and organisers despite repeated concern from the US side about the need to target enforcement efforts against major narcotics traffickers in Pakistan".

The office of the Associate Attorney-General, Mr. Francis Keating and officials from the DEA jointly reported to the concerned department in White House about the adequacy and inadequacy of legal and law enforcement measures in Pakistan. The report says that the Pakistani legal system is over-burdened with delays in prosecution of criminal cases, including narcotics cases. The effectiveness of the asset seizure law enacted in 1987 has yet to be tested in Pakistan courts. Though evidence collected by wire-tapping or electronic eaves-dropping is admissible in Pakistani courts, there are no legal guidelines for its acquisition in narcotics cases. Thus, in the absence of legal authority, enforcement agencies in Pakistan feel constrained by the civil liberties statutes, even though evidence gained through such methods can be key to providing conspiracy.

While the Federal officials express their dissatisfaction over the Pakistani approach in controlling the drug problem their views on the issue in Afghanistan is not apparently very clear. Afghanistan is not only a part of Golden Crescent (Pakistan-Afghanistan-Iran) but happens to be the largest poppy grower and biggest producer of heroin.

The involvement of Afghan rebels—the Mujahideen—in drug trafficking from that region and "US covert aid" to the Mujahideens over this decade is no more a secret. With all the money and arms flowing through Karachi the Mujahideens based in Pakistan are now reportedly planning to turn to other plans also to raise funds for their jihads (holy war). It has been extensively reported in the media that they sell their weapons in the open arms market in Peshawar and they smuggle opium and manufacture heroin destined for Europe and the US.

"The Afghan government is taking steps to develop prevention, treatment and rehabilitation programmes. Seizures of opium and cannabis within Afghanistan have doubled while the heroin seizure increased ten-fold".[1]

Strengthened activities are thereof evident as the government has also enacted legislation providing severe penalties for traffickers.

The Drug Crisis

George Bush perhaps never knew what happened in Brooklyn, N.Y. USA, Lebanon and Pakistan, but it would have helped him to fight America's war on drugs. The news from Brooklyn was brutal: Maria Hernandez, 34-year-old—mother of three—was murdered by an unknown gunman who fired five shots through her bedroom window at 4.38 in the morning. Mrs. Hernandez and her husband, Carlos, were well-known crusaders against drug trafficking in Brooklyn's Bushwick section. Police investigators who had no leads in the case, said her husband was the intended target. In Lebanon, meanwhile, police were stunned to discover that a 7-year-old boy and a 9-year-old girl had been doing make-believe drug deals in a local playground. In a depressing variation on the old lemonade-stand game, the kids apparently sold lawn clippings as pot, sugar and cocaine; they even kept records of their sales in an old fishing-tackle box. "I don't care what Bush says, this is bad;" Police Capt. Bernie Reillt said, "We just lost the war on drugs".

Two small episodes in America's long struggle against narcotics: A Colombian-style assassination on the night-time streets of Brooklyn and a couple of small-town kids who think drug trafficking is fun. Neither event will define the strategy that George Bush had for the drug war to be won, but both underscore its urgency. That strategy was being shaped by Bush's drug czar, William Bennet; Attorney-General, Dick Thornburgh, Secretary of State James Baker and other key officials. Months in preparation, the Bennett plan was making the rounds as a 235-page policy paper was notable for its cautious outlook. Indeed, its primary goal—a 50 per cent decrease in US drug consumption over the next decade—seemed almost embarrassingly modest in the light of a new Federal study showing that drug use declined by 37 per cent since 1985. Perhaps as a consequence of that encouraging trend, the word from the White House was that Bush, despite the staunch rhetoric of his Inaugural (this scourge will stop), soft-pedalled the drug war after all.

Bush and Bennet had a firm congressional deadline for their strategy. As a result, there was no question that the Bennet plan, modified by inter-departmental wrangling and presidential review in Kennebunkport, could be announced on the day after Labour Day. The White House seemed divided on the question of how the plan was to be unveiled. Previously-leaked predictions that Bush would make a televised address to the nation—the first of his presidency—were being challenged. According to White House sources, those rumours could be traced to Bennett's Office of National Drug Control Policy, and they were regarded as an attempt to pressure the boss to a no-no in the team-oriented Bush Administration. "Bennett does have a tendency to announce things and then check in", one White House official said, "Not having a prime-time speech might help curb that habit".

But far larger political calculations were involved as well. Bennett, for all his caution, clearly believed the drug crisis was a matter of fundamental importance to the nation. He also felt, and rightly, that Bush's vigorous participation would send the right message to two key audiences: the cocaine producing nations of Andean Latin America and the turf-conscious chieftains of the Federal drug bureacracy, some of whom clearly resented Bennett's attempts to intrude on their domains. Against all that, from the White House's view, was the blunt fact that Bennett himself was predicting a long and costly war with no quick solution and no crowd-pleasing Federal initiatives. If that was true, Bush's political advisers reasonned, it would be better to protect the President from a potentially controversial issue. In fact, some advisers suggested that Bush should announce the drug strategy in a decidedly low-key way. After all, one aide said, the Bennett plan was "just an interim report".

But it wsn't. The new strategy was a pivotal change in the U.S. war on drugs and the first public test of Bush's willingness to lead. Bennett's paper, though remarkable for its measured judgement of the nation's drug epidemic, contained no bold promises and none of the drama that some in Congress would like to see. Bennett was arguably correct in asserting that crack cocaine was by far the most critical drug problem facing the nation, and he was likewise correct that tougher law enforcement was the necessary first response. It is broadly encouraging, as he said, that large number of Americans were turning away from drugs. But it was also true, as he argued, that America was "now fighting two drug wars, not just one. The first and easiest was against (the) 'casual' use of drugs by many Americans, and we were winning it. The other, much more difficult war was against addiction to cocaine", especially crack. "On this second front", Bennett wrote, "we are losing badly."

Main Choices

His strategy was based on an elaborate overview of policy options that included education, treatment, law enforcement, the interdiction of drug smuggling and U.S. foreign policy in Latin America. It rejected legalizing drugs as "a national disaster", and it likewise rebuted the view that drug abuse could be fought only by attacking the social causes of crime and addiction in low-income neighbourhoods. His main choices were unglamorous: Bennett supported more Federal aid for local law enforcement, which was losing the battle against crack-related crime, and the expansion of State and Federal prison systems to house the rising tide of drug offenders. He wanted a bit more federal money for education and treatment, and he favoured tougher legal sanctions against casual drug users: the criminal-justice system, he argued, could deter demand and coerce uses into treatment if wisely used.

One section of the Bennett plan was still secret : the all-important discussion of U.S. policy toward the cocaine-producing and trafficking nations of Latin America. It was clear from the available portions of Bennett's report, however, that his basic approach was to intensify the Federal government's fumbling and uncertain campaign against the Colombian drug cartels, the vast criminal syndicate that had flooded America with cheap cocaine. Enriched by revenues estimated in the billions—and protecting their operations with a systematic campaign of murder and intimidation—the cartels had thwarted the Colombian government's attempts to control the drug trade for years. They were also corrupting government officials in countries from Bolivia to Mexico and all across the Caribbean basin.

There was little question that interdiction, the Reagan administration's primary strategy against Latin drug trafficking, had failed. Although the U.S. Customs Services and the Coast Guard were intercepting ever-increasing amounts of smuggled cocaine, the price and availability of the drug in cities all across the nation demonstrated that interdiction alone was not enough. As a result, hard-line drug warriors like former Customs Commissioner William von Raab argued the United States must sharply escalate diplomatic pressure on Latin governments, and some hawks argued that US military forces should be sent to fight the cartels on their own turf. The military option was a non-starter. Both the Pentagon and the host countries were strongly opposed to an active role for US forces, and Bennett had publicly rejected the idea of unilateral US involvement. He was also inclined to reject the State Department's preferred tactic, which was to try eradicate the coca-plant fields in Bolivia, Colombia and Peru. Instead, this strategy argued for substantial improvement in U.S. intelligence-gathering actively targeted at cocaine trafficking, coupled with increased US law-enforcement and economic aid to Latin America. Though Bennett's report diplomatically worded, he and his staff were plainly dismayed by the turf rivalries among key agencies like the Drug Enforcement Administration, the Customs Service and the State Department and they were critical of DEA's failure to computerize and share its information.

What happened now was up to George Bush and Congress. Bennett, according to his aides, had won most of the key policy battles so far and the whole plan, including its foreign-policy component, was ready for adoption by the President. Its total cost would be something like $ 7 billion in the first year which was about $ 1 billion more than the administration wanted to spend. But its effectiveness would depend on Congress's ability to resist micromanaging and grandstanding—and on Bush's willingness to confront the worst international crime crisis of this century.

The Federal Express Connection

What does an enterprising drug dealer do these days when he has to get a shipment of cocaine, or even anabolic steroids, to a customer overnight? Sometimes he does exactly what many legitimate American businessmen do to send their cargo. He calls Federal Express.

The Justice Department conducted an in-depth business analysis of how drug trafficking works in the United States. What did they find out? For one thing, some drug gangs apparently have decided that, macho codes notwithstanding, it's not always worth the risk to make a midnight run across the border themselves by car, boat or even airplane. Instead, this generation of drug dealers has developed its own variety of methods for transporting its contraband goods. Some dealers prefer a corporate approach—such as hiring Federal Express or other speedy delivery services. One gang in northern Alabama badly relied on the U.S. Postal Service, often using Express Mail. Its typical package contained one kilo of cocaine, though once the dealers sent 242 grams of crack through the mail.

Other dealers now challenge authorities to sophisticated games of hide-and-seek that make the car-door hiding place in "The French Connection" seem primitive by comparison. Last year authorities in Boston found 190 pounds of Asian heroin stowed away in a bean-sprout washing machine shipped from Hong Kong. Cops in Philadelphia fished more than 2,400 pounds of cocaine out of large containers marked, ANCHOVIES IN SALT WATER. That shipment, says U.S. Attorney Michael Baylson, "was virtually indistinguishable from the surrounding cans which actually contained anchovies." Caribbean dealers orchestrated a scheme for smuggling drugs among records and tape cassettes bound for music stores in the United States.

Federal Express, responding to the Justice Department's report, pointed out that it had turned over 6,000 suspicious shipments to authorities since 1979. "We are trying extremely hard to put a clamp wherever we can", says Federal Express spokesman Armand Schneider, who said that the company had trained its handlers in "methodologies" for detecing drug shipments. Still, a company whose business has suffered since the dawn of the fax revolution could take some cold comfort in the report. "It is unfortunate", says Schneider, "but we are probably the most reliable delivery service, so we probably do get singled out."[2]

US Keen to Maintain Opium Market Stable

A joint Indo-US working group proposed to discuss alleged diversions of India's illicit opium crop as well as the illegal cultivation of the narcotics to prevent misuse by heroin traffickers, according to Assistant Secretary of State in the Bureau of International Narcotics Matters Melvyn Levitsky. India and Turkey produce nearly 80 per cent of all opiates worldwide.

Mr. Levitsky told a group of foreign journalists attending a three-week long "News and current affairs drug workshop" that the U.S. Government was keen to maintain the stability of the opium market and help the 200,000-odd farmers in India lawfully growing opium, mainly for pharmaceuticals. However, Turkey processed opium by using straw instead of lancing the opium pod. India, he felt, could adopt the same process to maintain stricter control over production as it had barely 200-300 inspectors to oversee the processing of opium.

Mr. Levitsky said the drug issue confronted the whole

world and increasingly, there was an international consensus that it was a threat to the stability of our systems and democracy. He said about 65 per cent of all illegal drugs produced in the world was consumed in the U.S. "so we recognise our responsibility for the problem." Mr. Levitsky said that President George Bush's drug strategy acknowledged that the U.S. demand was the major element of the problem, but that did not mean other countries were absolved from their responsibilities. Whether it was their own demand problems or the fact that they were in an area where drugs were grown or processed.

Mr. Levitsky said casual use of drugs in America was going down, but regular and addictive use of hard drugs, especially "crack and cocaine" had expanded significantly along with the money and the violence involved. He said demand was an important aspect of the problem but supply was crucial as well. "Basically, we are working with other countries in the foreign policy area to decrease the supply of drugs flowing not only to the United States but worldwide, and to help other countries work on their own demand problems within their societies", he said. Mr. Levitsky said that there was a diplomatic responsibility to presuade and encourage other governments to fight against illegal drugs by working together in international bodies like the United Nations. This was reflected in last year's United Nations convention on illegal drug trafficking", he said.

Mr. Levitsky said the second aspect was a problematic one. "We have programme around the world to work with other governments by providing assistance to the local police, communication equipment, vehicles and loan assets like helicopters and other aircraft. We also have a large training programme. The budget for this activity was 130 million dollars", he said. Mr. Levitsky regretted that heroin production had exploded in the last year in Asia, specially in Burma and Laos, where opium crops were larger than before there was a high degree of trafficking to Europe from where it came to the United States. The U.S. had co-operative programmes in Thailand and Pakistan, where an agreement was reached with Prime Minister Benazir Bhutto's government to help establish an elite anti-narcotics force.

In Burma, Laos, part of Afghanistan and Iran where opium was grown in large quantities and where the US government had little access, it was influencing its allies to urge these countries to solve their drug problems. In Africa, some countries provided air facilities for transit of drugs. Mr. Levitsky gave Nigeria as an example where local drug barons in league with Asian traffickers use the country as a transit point for illegal heroin going to Europe and the United States. "We have picked up a number of Nigerian citizens, but they tend to be small fish", he conceded.[3]

Anti-drug Package

The Senate voted overwhelmingly for a bipartisan $ 9.4 billion anti-drug package with leaders of both parties saying their two weeks of contentious bargaining had produced an effective and much needed plan. "This is the first step", said Democratic Senator, Robert Byrd, his party's chief negotiator during the talks. "Weak shots, across the bow will not cripple the garbage scow of drug violence and drug dependency."

Agreement of the proposal, nailed down and ended a fight between the two parties that had blocked Senate passage of spending bills for the fiscal year. Its eye on the calendar, the Senate passed two spending measures for the fiscal year 1990. A $ 14.4 billion foreign aid package and a $ 157.4 billion measures for labour, health and education programmes.

The Senate action came after Republicans lifted the road-block they threw in front of spending measures while they and Democrats searched for a compromise on a broad package of anti-drug programmes. Senate leaders said that when added to the $ 7.9 billion Mr. Bush had already proposed, the new anti-drug package would bring previous year's spending on the drug problem to $ 9.4 billion, some $ 600 million of the total, however, had previously been proposed by Congress. Of the remaining $ 900 million senators agreed to seek, $ 100 million would be for law enforcement and the rest for treatment, education and rehabilitation. Overall, the package would tack nearly $ 3.2 billion to the government's $ 5.7 billion anti-drug effort this year.

Bush's War on Drugs

The President, Mr. George Bush, had announced, a nation-wide televised address, an eight billion dollar war on drugs and offered the resources of the US armed forces to any nation that wants them to fight the drug lords. He planed an "anti-drug summit" with the presidents of the three nations which were the source of the problem in Latin America—Colombia, Bolivia and Peru. The President's strategy included intensifying efforts against drug smugglers on the high seas, in international airspace and on the US borders. Other steps included educating school children, providing treatment for more people on drugs, punishing casual users and strengthening law enforcement.

Mr. Bush called cocaine and its derivative crack the gravest domestic problem the US had faced in decades. Drugs, he said, were sapping the strength of America as a nation. Mr. Bush emphasised that it was war that could only be won if the entire nation was involved. The Democrats announced their support for the President's goals but wondered whether he would succeed any more than other Democratic and Republican Presidents who too had declared similar wars on drugs. Senator Joseph Bidden, giving the official Democratic reply on television immediately after Mr. Bush's address, said that the President's funding did not match his rhetoric. He pointed out that the Pentagon was spending over $ 300 billion for a war which the country hoped it would not have to fight.

Limited War

According to Senator Joseph Bidden the $ 8 billion provided for the war on drugs, a 'limited war' would not yield results. The violence associated with drugs was spewing out all over America. "This day alone", said Mr. Bidden, 'violent' drug offenders will commit more than 1,00,000 crimes. The anti-drug message reached only 50 per cent of the school children targeted by the drug pushers. The money provided by the President would reach only another five per cent. Mr. Bidden called for an "international strike force" to flush out the "narco terrorists".

Mr. Bush said in his address that there was both good news and bad news on the drug front. In 1985, a government survey estimated that 23 million Americans were using drugs on a current basis. i.e., they had used it at least once the preceding month. The year before that number fell to 14.5 million. However, while casual drug use had declined, frequent use had doubled. The 22-minute speech was a blend of emotion and tough talk with the President issuing a warning to drug deal ers that penalties would be hard. The focus on drug abuse had become an even bigger issue in the US since the crackdown in Colombia following the killing of a presidential candidate by drug traffickers.

President Bush's Initiative

President George Bush and his national security team worked on a new "Andean stratagy" in stepped-up effort to tackle the drug problem at its source. The three Andean nations—Bolivia, Colombia and Peru—produce most of the world's cocaine. The President spelled out details of the plan in a nationally televised speech on September 5, 1989. National Security Adviser Mr. Brent Scowcroft described the Medellin drug cartel's declaration of war on the Colombian Government as a "new ingredient" in the administration's drug strategy. "Colombia is not just a small embattled country that happens to be threatened by drug cartels. It is the pipeline where most of the drugs are refined and it is the pipeline through which drugs come to the United States", Mr. Scowcroft said.[4]

Protection to Colombian Anti-drug Machines

Meanwhile in Washington, Colombian Justice Minister Monica de Greiff asked for $ 19 million in new aid to protect Colombian officials from the cartel. The money would be used to purchase security devices including armoured cars, bullet proof vests and metal detectors. "We will try to accommodate the request", State Department spokesman Richard Bouchers stated. "We all have a major stake in going after the drug cartels", Mr. Scowcroft added, "and the Colombians are bravely standing up now to take them on, and we want to do everything we can to help". Mr. Scowcroft also revealed that about $ 365 million will be committed to the Andean Strategy, including the $ 65 million announced last week of August 1989.

An official said the new strategy would include measurable goals as raising drug prices, restricting availability, making it convenient to nut illicit and dangerous to sell them, easing access to treatment and improving interdiction. Meanwhile, Senator Richard Lugar, Indiana Republican and ranking member of the Senate Foreign Relations Committee, said that the exorbitant prices paid to farmers to grow crops, combined with the abjectly poor Latin American economies, make eradication of the drug trade painfully difficult.

The Senator, who had returned from a 14-day tour of nine south and central American countries, said US programmes aimed at eradicating the drug supply abroad had barely begun to address the crisis. The flow of drugs into the US could not be halted unless American demand for drugs plummeted, he said. According to the Senator, an oft-repeated, "solution for eradicating the drug supply substituting coca cultivation with other cash crops—is unrealistic because Latin farmers can earn 4 to 10 times more by raising coca than they could by growing any other crop".[5]

Bush Plan to Kill Drugs at Source Fails

The US President, Mr. George Bush's idea to combat cocaine at the South American source was all but dead, just months after its inception, according to officials who cited poor planning and a lack of coordination as the reasons for the failure. The plan to eradicate and intercept cocaine in Peru and Bolivia—the world's main coca plant producers—also suffered serious setbacks because the two Andean nations were rife with corruption, and political problems. In Peru, the largest coca leaf grower, an ambitious US programme got halted because increasing guerilla activity threatened American personnel.

South America was to be the battleground for stopping the cocaine flow into the United States. Mr. Bush, in his presidential campaign, called for an "international strike force to hit back, quick and hard and often at foreign drug operations" to stop narcotics production at the source. Most of the world's coca leaves, the source of cocaine, grow in Peru's Upper Huallage Valley, a lush region at the foothills of the Andes, known as the "Cocaine Cradle". The sprawling region is a stronghold of Leftist guerillas known as Sendero Luminos, or "Shining Path". US efforts to operate in the region and set-up Mr. Bush's strike force were halted because of the security threat, said a spokeswoman for the State Department's Bureau of International Narcotics Matters. Ground-level eradication, already inefficient because it was done manually slowed further. Aerial spraying, which would effectively stop the plant's four harvests a year, was banned by Bolivian law and repeatedly put off in Peru.

Attempts by about a dozen agents of the Drug Enforcement Administration (DEA) to locate and destroy makeshift cocaine laboratories and clandestine airstrips had been curtailed by the guerilla activity, forcing agents to move out of the region and ferry in by helicopter for daytrips. With the strike force approach stalled, DEA agents were falling back on long standing programmes of advising local authorities. They continued to train Peruvian Police, while work went on in the Upper Huallaga Valley to shore up a helicopter base to offer enough security against the guerillas for American personnel to operate full time in the region. The work, according to a State Department official, included "everything from lighting and wires to clearing the area outside the fence so you can see who's coming in". At the same time, DEA agents were undergoing a course tailored for them by the US army. Mr. Charles J. Gutensohn, chief of cocaine investigations for the DEA, said in an interview that the training was primarily to teach the agents to deal with Peru's jungle environment and the guerillas operating there.

US Seeks Japanese Help on Drug Menace

Winding up a six-day visit to Japan, the U.S. Vice-President, Mr. Dan Quayle, said he expected cooperation from Tokyo in the war against drugs and in promoting democracy. Mr. Quayle said he discussed with the Prime Minister, Mr. Toshiki Kaifu, a letter from the U.S. President, Mr. George Bush, requesting more Japanese support in the war against illicit drugs, particularly in Latin America.[6] Japan could help Colombia, Peru and Bolivia liberate themselves from the drug trade by providing assistance in crop substitution to enable farmers to earn adequate incomes from legal crops and in general economic assistance.

Drug Shipments to US Dry Up

The flow of drugs into the United States dried up since Colombia's latest round of violence began, raising the price of cocaine and shrinking the size of *crock rocks*. The Miami, Florida-based US Coast Guard group saw import activity suddenly grind to a halt, said Lt. Cmdr. Jeff Karonis. The group has accounted for 97 per cent of the more than 4540 kg. of cocaine intercepted by the entire services in 1989.

Mr. Pat O'Brien, outgoing head of the US customs service in Miami, said agents from this district, by far the busiest in the nation, had been tracking three to five air drops a week prior to the Colombian crackdown. That came to a standstill", he said. "There was only one right after the violence started, and we weren't even liable to confirm that".

The only major cocaine haul—363 kg. on a freighter in Tampa, Florida, had been loaded and shipped before the violence began. Undercover agents trying to make cocaine buys had been put off for two to three weeks, and even though that the period was about up, there were still no shipments ready, he said. The summer months had traditionally been the traffickers' busiest. In addition, money laundering activity tracked by customs in sputtering traffickers were still moving wire transfers for old cocaine shipments, but there were few new payments.

The US Drug Enforcement Administration (DEA) also spotted the slowdown in drug activity, but agents there were sure if it marked a true disruption. "But we don't know if that's because of what's happening in Colombia, or if entrepreneurs are trying to take advantage of what's happening to raise the price", said the Miami DEA spokesman, Mr. John Fernandes. Mr. Fernandes said the price of large cocaine shipments in the two weeks had increased from $ 11,000 to $ 15,000 a kg. to $ 18,000 to $ 19,000 with individual kilogram sales reaching as high as $ 23,000. The metro-Dade police Cmndr. Steve Rothlein, head of the country's narcotics bureau, said "street prices for a kilogram of cocaine have shot up about $ 7,000. Crack, on the other hand, continues to sell for $ 10 a rock—but some of the rocks are getting smaller".

Mr. O'Brien attributed the disruption in part to the confiscation of planes used to carry cocaine from processing plants in southern Colombia to coastal staging areas. The heads of the cartels were also having difficulties trying to control their operation while on the run. A reputed money launderer for the Medellin cocaine cartel arrived in the United States, the first person extradited from Colombia during the crackdown on drugs traffickers.

Hours after the extradition began two men on a motor cycle tossed a dynamite bomb at the Banco Central Hiptecario in Medellin, the cocaine centre of Colombia, said a Medellin police official. The blast caused $ 45,000 in damage and threats of violence by drug traffickers opposed to any extraditions to the United States. The Government said that the President, Mr. Virgilio Barco would use his state of siege powers to name military mayors of violence plagued cities. Alvaro Garcia, spokesman for the Government's office of Boyaca State, said the mayor, Mr. Marino Estrada of Puerto Boyaca, 95-150 km. north of Bogota, would be replaced by a military officer, Army Maj. Alvaro Matallana.

Puerto Boyaca had been a stronghold of right-wing death squads that the government said were controlled by Colombia's notorious cartels. It is in a region rife with cocaine processing laboratories. The decree was the latest in a series of measures Mr. Barco had taken to quell violence by drug traffickers. The opposition Conservative Party issued a communique declaring unconstitutional Mr. Barco's decision to impose military rule on some towns. And the Medellin Mayor, Mr. Juan Gomez yesterday called the military substitution decree "a violation of the constitution". But Mr. Barco's Minister of Government, Mr. Orlando Vasquez, asked the Congress yesterday to make the anti-drug emergency measures permanent. He said such action "is compelled by tragic circumstances". Colombia's Supreme Court debated on whether the special measures were legal. Eduardo Martinez Omero became the first drug figure extradited to the United States under the special emergency powers decreed by Mr. Barco.

He arrived in Atlanta under heavy guard and later appeared with a lawyer for preliminary federal court hearing. He told the judge through an interpreter that he spoke only Spanish and did not fully understand the charges against him. Mr. Feldman agreed to postpone the hearing until Monday and ordered Mr. Martinez held without bail. Prosecutors said they would ask that he be denied bail. Mr. Martinez, 35, was charged in Atlanta in connection with the laundering of $ 15 million worth of illicit cocaine cash.

The war of attrition going on between the U.S. Administration and General Manual Autonic Noriega of Panama City reached its culmination on December 20, 1989 when a full-fledged military invasion of Panama was launched to "protect American lives, restore the democratic process, preserve the integrity of Panama Canal treaties and apprehend General Manual Antonio Noriega" the Panamanian strongmen and virtual ruler, who was now under indictment in the US on drug trafficking charges.

General Noriega had taken refuge in the Papal Nuncio in Panama City just before the commencement of operations against him. He had come out of the Papal Mission to surrender before the American forces to face charges of drug trafficking in the Miami court. Right from the day his hide-out was detected to the final day of his surrender, the Papal Embassy in Panama City was virtually put under siege—thousands of Panamanians and US forces had encircled it and a constant battery of cat-calls, booing, jeering under flood lights, coupled with rising crescendo of slogan-shouting was mounted to unnerve Mr. Noriega and 17 of his comrades who had taken shelter along with him. When on December 29 two of his trusted generals came out to surrender, it was only a matter of time before Noriega would have given up. All escape routes for Mr. Noriega had been sealed. There was no possibility of taking advantage of darkness as the entire area of Papal Nancio compound was under floodlights. Noriega's wife had taken shelter in the Cuban Embassy. Mr. Noriega had himself set on the collusion course with the USA, by making the Panamanian Assembly, a body of hand picked stooges, declare a state of war. Noriega while making belligerent postures and constantly making insulting remarks against the U.S. administration had never anticipated that things could come to such a sorry pass.

Noriega got advance information of American invasion and took to flight. Noriega's surrender or capture raised the fundamental question, whether any political fugitive, having sought asylum, could be given the facility to migrate to the country, which had granted him the asylum or could be arrested from outside the foreign mission. During Noriega's sojourn, confusion prevailed over the interpretation of international law on the score. There was urgent need for granting diplomatic immunity to such political fugitives, for asylum meant nothing in the absence of impunity against arrest.

Mr. Noriega was the usurper of Panamanian Presidency. He deliberately annulled the May elections. Mr. Guillermo Endara who had won in the presidential race, was threatened and beaten up. Mr. Endara had taken refuge in the Papal Nuncio. It was a queer chance that the Papal Nuncios' complex, in which Mr. Noriega had come to surrender, had given shelter earlier to Mr. Endara. Noriega, like the late Nicolai Ceausescu of Romania, had built a monstrous security system. PDF (Panama Defence Forces) whose 15000 men were fiercely loyal to him, gave a tough fight to the Americans. Had not Noriega asked his men in the western region to surrender, there would have been much blood bath. General Maxwell Turman, commander-in chief of the US Southern Command, acknowledged that US forces were fighting a 'real war' as they struggled to suppress at least 2000 centrally controlled and well armed supporters of deposed Panamanian leader, Manual Antonio Noriega.

The PDF was now in total disarray. The police duties in Panama City were being handled by the Americans. The Americans would act as occupation force. The world opinion was hostile to them. The sooner Mr. Endara was able to raise a credible security apparatus the better it was for him. However, President Endara was having tough time because he had no other alternative but to rely on the old PDF members who were once loyal to Noriega.

It was suspected that Gen. Noriega was working in close co-operation with Columbian drug cartel. There was no documentary evidence yet on the degree of his collusion with Columbian drug traffickers. He was instrumental in providing navigational facilities to drug traffickers through Panama Canal. A cache of 5 kg. of cocaine found from one of his houses, together with his vast financial assets spoke of his active involvement in drug trafficking.

US had asked friendly European governments to freeze the assets of Noriega believed to be 10 million dollars in Britain, France, Switzerland and Luxembourg. The British Government which took a sympathetic view of American invasion as Mrs. Margaret Thatcher put it "all those who had criticised Gen. Noriega must clearly support the Americans in their actions", was the first to obtain British High Courts order to force disclosures of the amount Gen. Noriega had in the banks in U.K. The orders were sought following US Government's request. French authorities blocked all currency accounts of Noriega, who was reported to have transferred some 23 million francs into France. It was yet premature to say how much assets he had in other countries, besides in Europe and America. Only a full investigation would reveal the real dimensions of his financial empire.

Mr. Noriega had refused to stand trial in the US He had expressed his apprehensions about the trial in spite of Mr. Bush's firm assurance of justice to Noriega. Mr. Noriega was aware of severe punishments that awaited him, if he was found guilty. The sentence could land him in perpetual imprisonment in the U.S. His apprehensions were sharpened by his inability to manipulate evidence that could get him out of the legal stranglehold. However, there was another side to the story too. There were a few sceptics who thought Mr. Bush would not have gone that far but for the fact that he had some personal scores to settle with Noriega. It was rumoured that Mr. Noriega was acting as frogman for CIA at the time of Mr. Bush's tenure in the CIA. However, since Mr. Noriega was in American custody, it would be interesting to watch what damaging revelation Noriega hurled at Mr. Bush.

Noriega was unpopular with the Panamanians. The way Panamanians behaved at his surrender testify to his unpopularity. Within a month two dictators disappeared from the world scene. However unpopular Mr. Noriega might have been in Panama City, however much he might have invited the American wrath, how oppressive his rule might have been and no matter, what his financial bunglings and involvements in drug trafficking, the central issue was whether any member-nation of UNO could invoke the charter of the right of self-defence to launch a naked aggression against an unpopular ruler.[7]

United States Package to South American States

The US President, Mr. George Bush's administration was willing to send military advisers to Peru and Bolivia to help in the war against narcotics, a drug official Mr. William Bennet, said. Mr. Bennet, however, denied published reports that the administration had a secret plan to send U.S. special forces on drug fighting missions in Combay zones in the two Andean mountain neighbours of Colombia. "There was no plan for any special forces to accompany troops in Peru and Bolivia into combat missions", Mr. Bennet said on the U.S. television network ABC's— 'This Week with David Brinkley' programme. "That is not in the plan....there is no such plan", Mr. Bennet said. "We see now in Colombia the presence of American trainers working with the Colombians, giving them advice, training them on equipment. This is the kind of thing we would anticipate if Peru and Bolivia take the step", Mr. Bennet said. Mr. Bennet was responding to a report in the Sunday editions of the "Washington Post" that Mr. Bush had a secret national security decision directive including new "rules for engagement" for US special forces in the three Andean countries. He did not specifically deny that Mr. Bush had signed the directive, but insisted the administration did not intend to send troops into combat in Latin America. About two dozen members of the U.S. special forces based in Panama had been rotating into Colombia, training military forces there, for about two years. The directive, the "Post" said, Mr. Bush had signed would apparently expand the role. Details of the plan would be worked out over the next two weeks, the newspaper reported.

The report quoted a senior administration official saying that "several hundred" U.S. military advisers could be sent to the three nations under the "Andean initiative" announced by Mr. Bush in his nationally televised address. A secret section of the plan would authorise the US special forces to accompany military patrols into so-called "safe areas" in the three nations, including Peru's Upper Huallaga valley. The region is a coca growing jungle that US officials said was dominated by Leftist "Shining Path" guerillas. "As we have said about the troops on the ground in Colombia, it is a dangerous environment", Mr. Bennet said, "they are not going to seek out combat or conflict. And we just all hope that nothing befalls them". The proposed step drew mixed reviews from Democrats in the Congress. "It could be a good idea", said Senator Joseph Bidden who had emerged as the Democratic spokesman on Mr. Bush's plan for the war on drugs. "I can say that I do not dismiss the notion of using U.S. troops in certain circumstances in the Andean effort".

Mr. Bidden said the administration had not yet given him a report on role of the military in the drug war.[8]

US Crackdown on Peru Drug Gangs

Peruvian government forces launched a major offensive against drug gangs and their allies—Maoist guerillas who control led much of the lush Upper Huallaga river valley on the eastern slopes of the Andes. Armed US narcotics agents trained in tropical combat were preparing to move into a heavily fortified police garrison built with funds from the United States, in the heart of Peru's coca-rich jungle. The United States campaign in Peru was the centerpiece of its drive against cocaine trafficking in Andean nations, which supplied the coca leaf from which the drug was refined.

US officials say the American drug agents were not concerned with the guerillas' politics, only with their drug activity. But the U.S. drug-fighting team's presence in the valley had led to speculation that the United

States could be drawn into overall war against the "Shining Path" guerilla group. "Sixty to 70 per cent of the world's coca production comes out of here. Peru, of all the countries, offers the best opportunity to do significant damage to the coca trade", said Mr. Craign Chretien, Chief of the US Drug Enforcement Amdinistration's (DEA) operations in Peru. The anti-drug base was at Santa Lucia (100 kms) north-west of Tingo Maria. After it was completed in October, as many as 50 Americans operated directly from the Upper Huallaga region, providing training and air support to some 300 police rangers assigned to the garrison.

The Andean Commission of Jurists, a human-rights organisation, noted that the base was intended for the drug fight, "observers are concerned that in the medium term it can be converted into a spring-board for the internationalisation of the counter-insurgency fight in Peru". "It's probably the most hostile situation the DEA works anywhere", Mr. Chretien said in Lima.[9]

US Aid of Little Help in Drug War : Colombia

US and Colombian officials said most of the equipment in a $ 65 million emergency package from the United States was not the type that is most needed to fight the drug war and was unlikely to have a major effect against the traffickers. The officials said the aid served mainly as a symbolic show of American support. "Our operation against the narcotic traffickers will be essentially the same as before", a senior police officer said. Major General Miguel Gomez Padilla, the chief of the 70,000-member national police force, said, "The total package is more suitable for conventional warfare than the kind of struggle we are waging here against narcotics traffickers".

The emergency aid was announced by the President, Mr. Geroge Bush, in the face of a rapidly escalating war between the Colombian government and drug-trafficking interests. The national police had been conducting 85 per cent to 90 per cent of the anti-narcotics operations in Colombia, but American and Colombian officials said that about 85 per cent of the $ 65 million in emergence aid was being spent on equipment for the army and air force. One reason, the officials said, was that the equipment was coming from American military stockpiles and was more compatible with the army and air force than with the police. "The package did not have a whole lot what the police would like to have had" and American official said, "but it was welcome as a symbolic gesture". Since 1989 Septermber 3 the Colombians had received eight A-37 jets, seven helicopters, 20 jeeps, three ambulances, some field medical kits, radios, machine guns, and a pair of C-130s, alongwith crates of spare parts for the transpsort planes. Most of the equipment had not yet been put into operation. But over the weekend, the C-130s were used to ferry troops on raid against traffickers in the Magdalena River Valley north of Bogota in which more than 40 ranches owned by narcotics traffickers were seized. "We are grateful for any assistance", a senior police officer said, "but we could have done much more with the proper equipment".

The United States had not said how the $ 261 million in aid for Colombia, Peru and Bolivia, which President Bush announced on Septermber 5 in a speech on drug policy would be distributed and what it would contain. The police said their biggest disappointment was that they received none of the sophisticated equipment they wanted for their intelligence bureaus.[10]

Drug Kingpin to be Extradited to US

The US Attorney General, Mr. Dick Thornburgh, said he expected the first alleged Colombian drug figure to be extradited to the US. Mr. Thornburgh also said that if reports were true that the Colombian justice minister, Mrs. Monica De Greiff, had resigned because of the death threat against her, he would not be surprised if she sought political asylum in the US. "If that is true, of course she would be welcome as other refugees from terrorism and the kind of violence that affects Colombia have always been", he said in an interview to the U.S. TV network CBS' programme—"Face The Nation".

Mrs. De Greiff was in the US according to an administration official who spoke on condition of anonymity. He refused to say whether she had resigned. Senior officials at the State Department, Justice Department, the Drug Enforcemnt Administration, the Immigration and Naturalisation Service and the Office of the National Drug Control Policy said that they were unaware of Mrs. De Greiff's status.

Mr. Thornburgh said efforts to extradite Eduardo Martinez Romero, the most prominent alleged Colombian figure arrested in recent government sweep against the cartel, were proceeding and "we expect to have him delivered to U.S. authorities". Money laundering a term for various methods of circumventing Federal currency laws by transferring illicitly raised cash, such as profits from drug trafficking, into bank accounts in such a way that its origin is hidden.

Romero reputedly was involved in the Colombian cartel's money laundering operation, but was not one of a dozen Colombian drug traffickers included in a U.S. "most wanted" list sent to Colombia. The three men at the top of the US list were Pablo Escobar, Jorge Ochoa and Jose Gonzalo Rodriguez Gacha. All evaded the government crackdown and were believed to be in hiding.

Mr. Thronburgh also said unconfirmed sightings of some of Colombia's drug cartel kingpins had been reported, but he suggested there was little to indicate any of them would captured soon. "As in most fugitive situations, there are sightings of these figures, but they are very clever, they are well financed", Mr. Thornburgh said. The US President, Mr. George Bush, announced plans for a $ 64 million military package to Colombia that included the use of US military advisers and equipment.

Poll Findings

Meanwhile, a poll had found that one-half of the US public favoured deploying US troops against drug dealers overseas, although most believe drug traffickers would try to retaliate violently. Fifty-three per cent of the 600 adults in the Gallup poll for "Newsweek" magazine said they would favour sending troops to fight or arrest drug traffickers in Colombia or other major drug exporting centres. Thirty-four per cent opposed such action and 13 per cent said they were not sure. At the same time, 64 per cent said they believed the drug cartels would retaliate with "increased violence against Americans" if more foreign drug traffickers were brought to this country for prosecution. Nineteen per cent did not expect retaliation and 77 per cent were unsure.

The poll found 85 per cent support penals against casual users of illegal drugs, such as the temporary suspension of a driver's licence.[11]

US Offers to Extradite Drug Dealers

The US President, Mr. George Bush, said that the United States was prepared to co-ordinate the extradition of suspected Colombian drug dealers after the assassination of a leading Colombian presidential candidate. "We in the United States should not forget that others are paying very high costs for the unchecked rampage of the international criminals trafficking in cocaine and undermining the lives of law-abiding citizens",

The Colombian President, Mr. Virgillio Barco, said he was reinstating under state-of-emergency powers, an extradition treaty between the United States and his country that was held void on a technicality in April 1988 by the Colombian supreme court. Mr. Barco announced the measure after Senator Luis Carlos Galan, a leading candidate to succeed Mr. Barco as Colombia's President, was shot just before he was scheduled to give a campaign speech in Soacha, about 32 kilometres south of Bogota. At least 10 others were wounded in the attack, which police blamed on drug lords. Hours earlier, a provincial police chief, who had led a campaign against drug lords was shot by five gunmen at a short distance from his home. In Washington, the State Department issued a statement saying that the US was appalled the "latest round of vicious attacks by the Colombian drug cartels against innocent Colombians".

Mr. Bush called Mr. Barco's reinstatement of the extradition treaty a "courageous move" and said he welcomed it in the fight against the narco-traffickers.

Meanwhile Colombia awoke in shock and mourning over the murder of Mr. Luis Carlos Galan, the country's favoured presidential candidate. Hundreds of thousands of people flocked to his coffin, to protest the government's inability to win its vicious war with Colombia's powerful drug cartels. Galan, 46, who was struck by sub-machinegun fire, was the presidential candidate of the Liberal Party. He was leading by wide margins in all opinion polls for the elections in May 1990. As news of Mr. Galan's death spread through the country, groups of people came out on the streets shouting "justice, justice", and clamouring for firm action against the drug cartels that claimed responsibility for Mr. Galan's death. Mourners shouted at Mr. Barco to launch an all-out military offensive against the drug cartels, responsible for most of the cocaine exported to Europe and the US and thousands of dead in drug-related violence.

Mr. Barco had consistently refused to negotiate with the druglords who had never made their demands known. He cautioned Colombians to be prepared "to experience more pain and sufferings" in the "war against the nation" unleashed by drug rings. Drug traffickers often co-operated with leftist guerillas, giving them arms and ammunition in return for information and proteciton. The druglords, on the other hand, stood up to Mr. Barco and his war on drugs, declaring they would keep killing government officials and judges.

"Now the fight is with blood", Medellin Cartel declared in a brief communique. The threat by the world's most powerful drug organisation was broadcast by RCN radio network to "La Prensa", one of the capital's major dailies. "We do not want peace. We have screamed for it, but will not beg for it", said the document signed by "the extraditables".[12]

G-7 Join Hands to Track Down Drugs Money

Officials from the world's richest nations agreed to join forces in tracking billions of dollars of illegal drugs money being laundered through international banks, a U.S. diplomatic source said. The meeting at the French Finance Ministry, broadened a task force commissioned by the US and its main enonomic partners at the summit of a group of seven (G-7) in Paris. G-7 officials from the U.S., France, Japan, Britain, West Germany, Italy and Canada, were joined for the inaugural session of the task force by Switzerland, Sweden, Austria, Belgium, Luxembourg, Spain, Australia, and the European Commission. "The experts had a first exchange of views on measures to fight money laundering, in particular from the drug trade", the U.S. source said. They discussed the scale of the problem and how to increase judicial and administrative co-operation. "From this work should emerge realistic and concrete proposals for action to be submitted to the heads of the 15 participating countries", said a brief concluding summary "This first meeting has allowed us to note the unanimous willingness of all the participants to work and achieve quickly effective measures".

The US embarking on anti-drugs campaign directed primarily at Latin America's cocaine traffickers, stressed that it did not plan to bully foreign banks or clog up the financial system in order to track down laundered drug profits. The treasury official told reporters before the meeting that there was no desire to pillory any country or to impose tight U.S. style rules for reporting large cash movements. Strict U.S. laws require banks to report cash transactions of over $ 10,000 to help identify profits from crime. Once the money is smuggled into the world financial system, bank secrecy can make it almost impossible to trace. Illegal drugs rake in an estimated $ 100 billion a year in the U.S. and upto $ 300 billion globally.

US Banks Take the Lead in Drug War

If the US is fighting a war on drugs, then it is on the money laundering front that the anti-drug generals see their greatest chance of success. The frontline troops in this war, however, are not law enforcement agencies, but banks and financial institutions are. The impact of this way of doing business was a major topic of two-day symposium on money laundering sponsored by the University of Miami's Graduate School of International Studies and a coalition of anti-drug and banking groups. What has turned money laundering—the art or science of converting illicitly-earned funds into "legitimate" money into an important issue in the US is the near-obsession with illegal drugs which is gripping the country. Since previous efforts have failed to curb the lucrative drug business, politicians and law enforcement officials now hope that cutting profitability of the business may succeed where past efforts failed. Mr. Dick Thornburgh, US Attorney General, told the symposium that by attacking the money laundering the Government hoped to "break the drug dealers by impoverishing them".

The principal legal tools the Government is using are : the misleadingly named Bank Secrecy Act of 1986, the Money Laundering Control Act of 1986 and the Anti-drug Abuse Act of 1988. The new laws require financial institutions to report details of many financial transactions, they are being pressed to act as agents of the Government, reporting to the authorities "suspicious" transactions of their clients.

$ 250 m US Aid to Tackle Drug Cartels

Mr. Bush met with his top advisers in the drug war at his vacation home in Kennebunkport, Maine, to discuss the strategy. But sources said that the focus was Colombia, but the strategy and that no specific decisions were made on changing plans for aid to that country. Mr. Bush announced that the United States would give Colombia $ 65 million in military equipment such as helicopters in addition to some $ 2.5 million for protection of the Colombian judiciary that had been promised previously. The Congress last year set aside $ 5 million for such portection. The Colombian justice minister, Ms. Monica de Greiff asked for $ 19 million in aid for judicial protection.

Mrs. De Greiff met Mr. William Bennett, Director of the Office of the National Drug Control Policy who developed the drug strategy. The drug strategy will urge states to impose tough laws against drug users, but won't threaten to withhold highway funds from the states that fail to do so. In addition, a proposed $ 25 million national drug intelligence centre, to be run jointly by the Drug Enforcement Administration and the FBI was eliminated from the strategy due to a last minute bid by the treasury department to develop a money laundering intelligence centre and combine all the elements. However, of the elements included in the draft strategy the emphasis was on targetting drug abusers who fuel the drug traffic, the hiking of funds for drug treatment by 49 per cent and the need for massively increased aid to the Andean countries—remain the same as proposed originally by Bennet. Copies of the draft strategy were leaked to the press shortly after it was produced in late July. Mr. Bush had said he would present the drug plan in a nationally televised speech on Septermber 5, 1989, which he did.

References

1. *The Times of India*, New Delhi, October 27, 1988.
2. *Newsweek*, August 21, 1989.
3. *The Pioneer*, Lucknow, September 29, 1989, p. 5.
4. *The Times of India*, New Delhi, September 1, 1989, p. 1
5. *The Hindustan Times*, New Delhi, August 31, 1989, p. 13.
6. *Indian Express*, New Delhi, Wednesday, September 27, 1989, p. 7.
7. *The Pioneer*, Lucknow, January 10, 1990, p. 5.
8. *The Times of India*, New Delhi, September 12, 1989
9. *The Times of India*, New Delhi, September 11, 1989.
10. *The Times of India*, New Delhi, September 15, 1989—Section 2.
11. *The Times of India*, New Delhi, Tuesday, August 29, 1989.
12. *The Times of India*, New Delhi, August 21, 1989—Section 2, p. I.

62

REGIONAL AND INTERNATIONAL CO-OPERATION: CONVENTIONS AND OBLIGATIONS

Drug crimes transcend national boundaries. Recognizing the importance of international co-operation in combating the drug menace nation-states have entered into a number of international conventions since 1912.

The International Conventions in force at present are :

(i) The Single Convention on Narcotic Drugs, 1961 as amended by the 1972 protocol;

(ii) The Convention on Psychotropic Substances, 1971 and

(iii) The UN Convention against Illicit Traffick in Narcotic Drugs and Psychotropic Substances, 1988.

Besides the above the SAARC Countries have signed a regional convention on drugs, namely, the SAARC Convention on Narcotic Drugs and Psychotropic Substances. This Convention has been in force with effect from 15th September, 1993. This Convention resembles the UN Convention Against Illicit Traffic in Narcotic Drugs and Psychotropic Substances, 1988 in many respects.

While the Single Convention on Narcotic Drugs 1981 and the Convention on Psychotropic Substances 1971 mainly emphasise the regulation of the use and trade of Narcotic Drugs and Psychotropic Substances respectively, the thrust of the UN convention Against Illicit Traffick in Narcotic Drugs and Psychotropic Substances is strenghtening international co-operation to combat illicit traffick in drugs. Article (2) of the Convention which outlines the scope of the Convention states in unambiguous terms that the purpose of the Convention is to promote co-operation among the parties so that they may address more effectively the various aspects of illicit traffick in narcotic drugs and psychotropic substances having international dimensions.

Means of international co-operation could be established in the areas of search for wanted criminals, communication of information about new trends of illicit traffick, sources from which drugs are obtained, methods employed etc., jointly effecting controlled delivery, a technique adopted to identify the organisers of illicit shipment of drugs, extradition, confiscation of proceeds, properties, instrumentalities, etc., connected with drug crimes, extending mutual legal assistance in investigation, prosecutions etc. of drug offences, monitoring import/export of Illicit drugs and precursors and essential chemicals so as to prevent their diversion, identification and elimination of illicit cultivation of drug-bearing plants. etc.

Co-operation under the Single Convention on Narcotic Drugs, 1961

The Convention *vide* article 4 requires each member state, *inter alia*, to co-operate with other states in the execution of the provisions of the Convention. Under Article 35, the member countries are required, *inter alia* to—

(i) assist each other in the campaign against the illicit traffick in narcotic drugs.

(ii) co-operate closely with each other and competent international organisations of which they are members with a view to maintain a co-ordinated campaign against the illicit traffick.

(iii) ensure that where legal papers are transmitted internationally for the purpose of prosecution the transmittal be effected in an expeditious manner to the bodies designated by the parties; this requirement shall be without prejudice to the right of a party to require that legal papers be sent to it through the diplomatic channel.

Article 36 states that it is desirable to include drug offences as extraditable crimes in any extradition treaty which in future will be concluded between any of the parties, and, as between any of the parties which do not make extradition conditional on the existence of a treaty or on reciprocity, be recognised as extradition crimes; provided that extradition shall be granted in conformity with the law of the party to which application is made, and that the party shall have the right to refuse to effect the arrest or grant extradition in cases where the competent authority considers that the offence is not sufficiently serious.

The Convention on Psychotropic Substances, 1971 has identical provision. Under the 1961 and 1971 Conventions the parties have very broad discretion to decide whether a certain act is 'sufficiently serious' to warrant extradition. A strict obligation to extradite drug offenders is not deductable from these conventions.

It is generally accepted theory that extradition may be based either on a binding international agreement (of multilateral or bilateral character) or, in the absence of a treaty obligation, on domestic extradition law. Most common-law countries are not prepared to grant extradition in the absence of a valid treaty, whereas, the civil law countries, whose legal system is based on Roman law, are to a large extent both legally in a position and certainly to do so, and therefore, do not insist on an extradition agreement. It is the prevailing theory that extradition, whether or not based on a treaty, is of a contractual character. That is to say, extradition is never a unilateral legal act, but rather a transfer and acceptance of rights and obligations by the states involved which invariably require a certain standard of agreement. This means that no obligation to extradite is instituted by domestic legislation. There is no source in general international law for the assumption of an obligation to extradite solely on the basis of domestic law; it is rather for the requested state to decide in which way to react to any request for extradition. The general situation in extradition relations is, however, modified for drug offences to the extent that a basic obligation to extradite for those parties which do not make extradition conditional on the existence of treaty can be inferred from those Conventions.

Co-operation under the UN Convention Against Illicit Traffick in Narcotic Drugs and Psychotropic Substances, 1988

The Single Convention on Narcotic Drugs, 1961 and the Convention on Psychotropic Substances, 1971 have not provided sound basis for any meaningful co-operation between the nations to combat illicit traffick in drugs. Recognising the need to have a comprehensive international instrument to promote and strengthen international co-operation in the field the UN Convention Against Ilicit Traffic in Narcotic Drugs and Psychotropic Substances, 1988 was brought into effect.

The Convention incorporates elaborate provisions for mutual co-operation in the areas of extradition, mutual legal assistance, transfer of proceedings, training etc. controlled delivery, eradication of illicit cultivation of drug-bearing plants, confiscation of proceeds, property instrumentalities etc. derived from illicit trafficking of drugs.

Article 6 of the Convention which deals with mutual co-operation in the matter of extradition requires that the offences, established in accordance with the Convention, shall be included as an extraditable offence in any extradition treaty existing between the parties. The parties undertake to include such offences as extraditable offences in every extradition treaty to be concluded between them. The Article further states that, a party in whose territory an offender is found shall, if it does not extradite him, submit the case to its competent authorities for the purpose of prosecution. If extradition, sought for the purpose of enforcing a sentence, is refused because the person sought is a national of the requested party, the requested party shall, if its law so permits and in conformity with the requirements of such law upon application of the requesting party, consider the enforcement of the sentence which has been imposed under the law of the requesting party or the remainder thereof.

Extraditon is the only effective way by which offenders present in another's territory could be prosecuted. To effect extradition, the state in whose territory the offender is found or arrested agrees to surrender that person to a third state which is competent and willing to prosecute the offender and seeks his extradition (extradition for prosecution). If the offender has already been convicted in the state requesting extradition. The surrender enables that state to implement the sentence (extradition for execution). Thus extradition necessarily involves two states. Sometimes a third-state would also be involved for instance if an offender whose extradition has been granted must be transmitted through the territory of a third state (extradition in transit), the consent of such third state would also be necessary.

Controlled Delivery

"Controlled Delivery" is an advanced technique which if implemented with utmost cooperation between nations would achieve excellent results. In applying the technique, the transit state provisionally waives its right to seize the consignment of drugs and arrest the courier if there is possibility of identifying the entire syndicate in operation including the financiers. Usually, the transit state reserves its right to request extradition of the courier at a later stage, once the shipment has reached its ultimate destination. The 1988 UN Convention requires the parties to take the necessary measures, within their possibilities, to allow for the appropriate use of controlled delivery at the international level, on the basis of agreements or arrangements mutually consented to. Decision to use controlled delivery shall be made on a case-by-case basis and may, when necessary, take into consideration financial arrangements and understandings with respect to the exercise of jurisdiction by the parties. The convention further states that illicit consignments whose "Controlled Delivery" is agreed to, may, with the consent of the parties concerned, be intercepted and allowed to continue the narcotic drugs or psychotropic substances intact or removed or replaced in whole or in part.

Article 5 of the Convention provides for execution of a request of one country to another country in whose territory proceeds, property, instrumentalities etc. are found for confiscation of such property.

The Convention incorporates intensive provisions for providing mutual legal assistance by member countries to each other. Such mutual assistance is to be afforded for taking evidence or statement from persons, effecting service of judicial documents, executing searches and seizures, examining objects and sites, providing information and evidentiary items, providing originals or certified copies of relevant documents and records including bank, financial, corporate or business records and identifying or tracing proceeds, property, instrumentalities or other things for evidentiary purposes. Upon request the parties shall facilitate or encourage, to the extent consistent with their domestic law and practice, the presence or availability of persons, including persons in custody who consent to assist in investigations or participate in proceedings. Bank secrecy cannot be a reason to decline to render mutual legal assistance. The Convention also required each member country to designate an authority or authorities to execute requests for mutual legal assistance or to transmit such request to competent authorities for execution. Transmission of requests for mutual legal assistance or any communication related thereto shall be effected between the authorities designated for the purpose; this requirement shall be without prejudice to the right of a party to require that such requests and communications be addressed to it through the diplomatic channel and, in urgent circumstances, where the parties agree, through channels of the International criminal police organisation, if possible. For refusal of mutual legal assistance, reasons shall be given.

The Convention envisages certain safeguards to witnesses, experts or other persons who consent to give evidence in a proceeding or to assist in an investigation, prosecution or judicial proceeding in the territory of the requesting party. Such person shall not be prosecuted, detained, punished or subjected to any other restriction of his personal liberty in that territory in respect of acts, emission or conviction prior to his departure from the territory of the requested party.

The ordinary cost of executing a request shall be borne by the requested party, unless otherwise agreed by the parties concerned. If expenses of a substantial or extra-ordinary nature are required, the parties shall consult to determine the terms and conditions under which the request will be executed as well as the manner in which the cost shall be borne. The convention also requires that the parties shall consider, as may be necessary, the possibility of concluding bilateral or multilateral agreement or agreements that would serve the purposes of, give practical effect to, or enhance the provision of the Convention (Article 7), the Convention *vide* article 8 requires the parties that consideration shall be given to the possibility of transferring to one another proceedings for criminal prosecution of offences, in cases where such transfer is considered to be in the interest of the proper administration of justice.

Establishing Communication Channels

Recognising the importance of establishment of channels of communication etc., the Convention *vide* article 9 requires that the parties shall on the basis of bilateral or multilateral agreements or arrangements, *inter alia*—(1) establish and maintain channels of communication between their competent agencies to facilitate the secure and rapid exchange of information concerning all aspects of drug offences, (2) cooperate one another in conducting enquiries, with respect to offences established in accordance with the Convention having an international character, concerning the identity, whereabouts, and activities of persons suspected of being involved in drug offences, the movement of proceeds derived from the commission of such offences, narcotic drugs, psychotropic substances and precursors and essential chemicals given in Table I & II of the Convention etc., (3) in appropriate cases establish joint teams to carry out the provisions, (4) provide, when appropriate, necessary quantities for analyatical or investigative purpose, (5) facilitate effective coordination between their competent agencies and services and promote the exchange of personnel and other experts, including the posting of liaison officers. The Convention also requires that the parties shall assist one another to plan and implement, research and training programmes designated to share expertise in certain areas. *Vide* article 10 the Convention requires, that the parties shall cooperate directly or through competent international or regional organisations, to assist and support transit states and in particular, developing countries in need of such assistance and support, to the extent possible through programmes of technical cooperation on interdiction and other related activities.

Article 14 of the Convention, *inter alia*, requires cooperation between the parties to increase the effectiveness of eradication of drug-bearing plants. Such cooperation may, *inter alia*, include support, when appropriate, for integrated rural development leading to economically viable alternatives to illicit cultivation. Factors such as access to markets, the availability of resources and prevailing socio-economic conditions should be taken into account before such rural development programmes are implemented. The Convention requires (*vide* article 17) cooperation to the fullest extent possible to suppress illicit traffic by sea, in conformity with the international law of the sea.

Co-operation Under the SAARC Convention on Narcotic Drugs and Psychotropic Substances

The SAARC Convention, which has been drafted on the lines of the 1988 UN Conventon has identical provision as that of the latter in respect of prosecution extradition, mutual legal assistance, eradication of illicit cultivation of drug-bearing plants and exchange of information. The Convention also emphasises the necessity for cooperation *vide* article 2 under the heading 'Scope of the Convention' that the purpose of the convention is to promote cooperation among member states, so that they may address more effectively the various aspects of prevention and control of drug abuse and the suppression of illicit traffic in narcotic drugs and psychotropic substances which are specific to the SAARC region.

It need not be emphasised that unless specific legislations are enacted by each country to give effect to the obligations arising under the conventions it would not be feasible to give effect to the requests when received. Taking this into consideration, the SAARC Convention *vide* article 2 (2) calls the member counrties to take necessary measures including legislative and administrative meausres to carry out their obligations under the Convention, in conformity with the fundamental provisions of their respective domestic legislative systems.

Indian laws provide for extensive framework to provide mutual legal assistance to other countries as envisaged under the 1988 UN Convention Against Illicit Traffic in Narcotic Drugs and Psychotropic Substances and the SAARC Convention.

Balanced Approach for Demand Reduction

Under the chairmanship of Mr. Fulani Kuajafa of Nigeria, the 36th session of the United Nations Commission on Narcotic Drugs met in Vienna from 29th March to 7th April, 1993. In his opening statement Mr. Giorgio Giacomelli Executive Director of the United Nations International Drug Control Programme (UNDCP) and Director General of the United Nations office at Vienna, said that the International community was facing a time of uncertainty, since the world after growing accustomed to a bipolar system, had entered a new and undefined era. One factor that would have an effect on the final outcome of that ongoing process was the world's struggle against illicit drug production, trafficking and consumption.

With regard to demand reduction activities he said that within a framework of a balanced approach, the programme had pursued preventive and rehabilitative efforts, particularly in cooperation with other United Nations bodies and non-governmental organisations. Also, it had increasingly fulfilled its role in coordinating drug control activities, in cooperation with the United Nations Children's Fund (UNICEF), the International Fund for Agricultural Development (IFAD) and the United Nations Population Fund (UNFPA).

The success of the programme's partnership with other international organisations depended on a clear delineation of their respective roles and on whether it could achieve a better balance between its demand reduction and illicit crop reduction activities. He asked the commission to make recommendations as to whether the UNDCP should accelerate the process and focus its resources more assertively in the demand reduction sector. Further, the UNDCP had to devote more resources to various activities, including the promotion of research, the collection and sharing of information and the development of specialized expertise.

The commission this time indeed gave much more attention to the question of demand reduction than formerly. In this regard, the representative of Jamaica, Dr. Winston Davidson presented a proposal for holding regional meetings of authorities dealing with demand reduction activities. Taking part in the commission's debate on demand reduction efforts, he said that self-sustainable strategies should be developed with the assistance of the international community, taking into account the characteristics and needs of each region. Most speakers called for greater emphasis on demand reduction, since law enforcement activities could have only a limited effect on controlling the supply of narcotic drugs.

Two popular initiatives concerning the depenalization of drug consumption and the legalization of hashish were under way in Switzerland, the representative of that country told the Commission. He said that his government now sought to alter the current drug policy in the country. The representative of the Netherlands told the Commission that the existence of controls concerning substances used in the manufacture of illicit drugs should not substitute for examining the question of whether drug abuse policies served their intended purpose. Addressing the Commssion as it concluded its discussion of the 1992 report of the International Narcotics Control Board (INCB) he said his country's experience with a liberal approach showed that putting emphasis on demand reduction, while it had not eradicated drug abuse, had produced a "generally acceptable level of containment of drug abuse and related crimes".

Most speakers, however, agreed with the INCB's view that liberalisation would be counterproductive and would produce harmful effects on health and social well being. The Minister of the Interior of Bolivia urged the commission to take steps to lift restrictions on the traditional use of coca leaves. He said that his country had succeeded in reducing the production of coca paste by 50 per cent and in the last four years, a larger quantity of drugs had been seized than in the last 20 years.

Drug trafficking had decreased but still continued, since it followed the basic economic law of supply and demand. Bolivia produced coca leaf, as part of a tradition pursued for thousands of years for religious, cultural and medical purposes. Coca leaf was not cocaine, the production of which had been imposed from outside; the restrictions imposed on coca leaf by international conventions should be lifted, since it was not harmful to the Bolivia people. However, the Government was aware that it was necessary to eliminate excessive production of coca leaf, needed for the manufacture of cocaine. The Bolivian peasant was cultivating coca leaf for lack of valid alternatives. The Government reiterated its commitment to the international community to eliminate the remaining 25,000 hectares of coca leaf, although with enormous sacrifices, in the next five years. There was need for private capital to generate alternative developement.

Other matters before the commission included the role of rural development in combating illicit drug cultivation, proposals to reduce penalties on drugs, organised drug crime, the implementation of international controls on precursors and the impact of drug trafficking in transit countries.

Appendix 1 to Chapter 62

The Single Convention on Narcotic Drugs, 1961

(As amended by the Protocol of 1972)[1]

The Conference,

Recalling that the Preamble to the Single Convention on Narcotic Drugs, 1961, states that the Parties to the Convention are "concerned with the health and welfare of mankind" and are "conscious of their duty to prevent and combat" the evil of drug addiction.

Considering that the discussions at the Conference have given evidence of the desire to take effective steps to prevent drug addiction.

Considering that, while drug addiction leads to personal degradation and social disruption, it happens very often that the deplorable social and economic conditions in which certain individuals and certain groups are living predispose them to drug addiction.

Recognizing that social factors have a certain and sometimes preponderant influence on the behaviour of individuals and groups.

Recommends that the Parties:

1. Should bear in mind that drug addiction is often the result of an unwholesome social atmosphere in which those who are most exposed to the danger of drug abuse live;
2. Should do everything in their power to combat the spread of the illicit use of drugs;
3. Should develop leisure and other activities conducive to the sound physical and psychological health of young people.

Preamble

The Parties.

Concerned with the health and welfare of mankind.

Recognizing that the medical use of narcotic drugs continues to be indispensable for the relief of pain and suffering and that adequate provision must be made to ensure the availability of narcotic drugs for such purposes.

Recognizing that addiction to narcotic drugs constitutes a serious evil for the individual and is fraught with social and economic danger to mankind.

Conscious of their duty to prevent and combat this evil.

Considering that effective measures against abuse of narcotic drugs require co-ordinated and universal action.

Understanding that such universal action calls for international co-operation guided by the same principles and aimed at common objectives.

Acknowledging the competence of the United Nations in the field of narcotics control and desirous that the international organs concerned should be within the framework of that Organization.

Desiring to conclude a generally acceptable international convention replacing existing treaties on narcotic drugs, limiting such drugs to medical and scientific use, and providing for continuous international co-operation and control for the achievement of such aims and objectives.

Hereby agree as follows....

Article 21

Limitation of Manufacture and Importation

1. The total of the quantities of each drug manufactured and imported by any country or territory in any one year shall not exceed the sum of the following:

(a) The quantity consumed, within the limit of the relevant estimate, for medical and scientific purposes;

(b) The quantity used, within the limit of the relevant estimate, for the manufacture of other drugs, of preparations in Schedule III, and of substances not covered by this Convention;

(c) The quantity exported;

(d) The quantity added to the stock for the purpose of bringing that stock up to the level specified in the relevant estimate; and

(e) The quantity acquired within the limit of the relevant estimate for special purposes.

2. From the sum of the quantities specified in paragraph 1 there shall be deducted any quantity that has been seized and released for licit use, as well as any quantity taken from special stocks for the requirements of the civilian population.

3. If the Board finds that the quantity manufactured and imported in any one year exceeds the sum of the quantities specified in paragraph 1, less any deductions required under paragraph 2 of this article, any excess so established and remaining at the end of the year shall, in the following year, be deducted from the quantity to be manufactured or imported and from the total of the estimates as defined in paragraph 2 of Article 19.

4. (a) If it appears from the statistical returns on imports or exports (Article 20) that the quantity exported to any country or territory exceeds the total of the estimates for that country or territory, as defined in paragraph 2 of Article 19, with the addition of the amounts shown to have been exported, and after deduction of any excess as established in paragraph 3 of this article, the Board may notify this fact to States which, in the opinion of the Board, should be so informed;

(b) On receipt of such a notification, Parties shall not during the year in question authorise any further exports of the drug concerned to that country or territory, except

(i) In the event of a supplementary estimate being furnished for that country or territory in respect both of any quantity over-imposed and of the additional quantity required, or

(ii) In exceptional cases where the export, in the opinion of the Government of the exporting country, is essential for the treatment of the sick.

Article 21 bis

Limitation of Production of Opium

1. The production of opium by any country or territory shall be organized and controlled in such manner as to ensure that, as far as possible, the quantity produced in any one year shall not exceed the estimate of opium to be produced as established under paragraph 1 (f) of Article 19.

2. If the Board finds on the basis of information at its disposal in accordance with the provisions of this Convention that a Party which has submitted an estimate under paragraph 1 (f) of Article 19 has not limited opium produced within its borders to licit purposes in accordance with relevant estimates and that a significant amount of opium produced, whether licitly or illicitly, within the borders of such a Party, has been introduced into the illicit traffic it may; after studying the explanations of the Party concerned, which shall be submitted to it within one month after notification of the finding in question, decide to deduct all, or a portion, of such an amount from the quantity to be produced and from the total of the estimates as defined in paragraph 2 (b) of Article 19 for the next year in which such a deduction can be technically accomplished, taking into account the season of the year and contractual commitments to export opium. This decision shall take effect ninety days after the Party concerned is notified thereof.

3. After notifying the Party concerned of the decision it has taken under paragraph 2 above with regard to a deduction, the Board shall consult with that Party in order to resolve the situation satisfactorily.

4. If the situation is not satisfactorily resolved, the Board may utilize the provisions of Article 14 where appropriate.

5. In taking its decision with regard to a deduction under paragraph 2 above, the Board shall take into account not only all relevant circumstances including those giving rise to the illicit traffic problem referred to in paragraph 2 above, but also any relevant new control measures which may have been adopted by the Party.

Article 22

Special Provision Applicable to Cultivation

1. Whenever the prevailing conditions in the country or a territory of a Party render the prohibition of the cultivation of the opium poppy, the coca bush or the cannabis plant the most suitable measure, in its opinion, for protecting the public health and welfare and preventing the diversion of the drugs into the illicit traffick, the Party concerned shall prohibit cultivation.

2. A Party prohibiting cultivation of the opium poppy or the cannabis plant shall take appropriate measures to seize any plants illicitly cultivated and to destroy them, except for small quantities required by the Party for scientific or research purposes.

Article 23

National Opium Agencies

1. A Party that permits the cultivation of the opium poppy for the production of opium shall establish, if it has not already done so, and maintain, one or more government agencies (hereafter in this article referred to as the Agency) to carry out the functions required under this article.

2. Each such Party shall apply the following provisions to the cultivation of the opium poppy for the production of opium and to opium :

(a) The Agency shall designate the areas in which, and the plots of land on which, cultivation of the opium poppy for the purpose of producing opium shall be permitted.

(b) Only cultivators licensed by the Agency shall be authorized to engage in such cultivation.

(c) Each licence shall specify the extent of the land on which the cultivation is permitted.

(d) All cultivators of the opium poppy shall be required to deliver their total crops of opium to the Agency. The Agency shall purchase and take physical possession of such crops as soon as possible; but not later than four months after the end of the harvest.

(e) The Agency shall, in respect of opium, have the exclusive right of importing exporting, wholesale trading and maintaining stocks other than those held by manufactures of opium alkaloids, medicinal opium or opium preparations. Parties need not extend this exclusive right to medicinal opium and opium preparations.

3. The government functions referred to in paragraph 2 shall be discharged by a single government agency if the constitution of the Party concerned permits it.

Article 24

Limitation on Production of Opium for International Trade

1. (a) If any Party intends to initiate the production of opium or to increase existing production, it shall take account of the prevailing world need for opium in accordance with the estimates thereof published by the Board so that the production of opium by such Party does not result in overproduction of opium in the world.

(b) A Party shall not permit the production of opium or increase the existing production thereof if in its opinion such production or increased production in its territory may result in illicit traffic of opium.

2. (a) Subject to paragraph 1, where a Party which as of 1 January, 1961 was not producing opium for export desires to export opium which it produces, in amounts not exceeding five tons annually, it shall notify the Board, furnishing with such notification information regarding:

(i) The controls in force as required by this Convention respecting the opium to be produced and exported, and

(ii) The name of the country or countries to which it expects to export such opium;

and the Board may either approve such notification or may recommend to the Party that it not engage in the production of opium for export.

(b) Where a Party other than a Party referred to in paragraph 3 desires to produce opium for export in amounts exceeding five tons annually, it shall notify the Council, furnishing with such notification relevant information including:

(i) The estimated amounts to be produced for export;

(ii) The controls existing or proposed respecting the opium to be produced;

(iii) The name of the country or countries to which it expects to export such opium;

and the Council shall either approve the notification or may recommend to the Party that it not engage in the production of opium for export.

3. Notwithstanding the provisions of subparagraphs (a) and (b) of paragraph 2, a Party that during ten years immediately prior to 1 January, 1961 exported opium which such country produced may continue to export opium which it produces.

4. (a) A Party shall not import opium from any country or territory except opium produced in the territory of:

(i) A Party referred to in paragraph 3;

(ii) A Party that has notified the Board as provided in subparagraph (a) of paragraph 2; or

(iii) A Party that has received the approval of the Council as provided in subparagraph (b) of paragraph 2.

(b) Notwithstanding subparagraph (a) of this paragraph, a Party may import opium produced by any country which produced and exported opium during the ten years prior to 1 January, 1961 if such country has established and maintains a national control organ or agency for the purposes set out in Article 23 and has in force an effective means of ensuring that the opium it produces is not diverted into the illicit traffick.

5. The provisions of this article do not prevent a Party:

(a) From producing opium sufficient for its own requirements; or

(b) From exporting opium seized in the illicit traffick, to another Party in accordance with the requirements of this Convention.

Article 25

Control of Poppy Straw

1. A Party that permits the cultivation of the opium poppy for purposes other than the production of opium shall take all measures necessary to ensure:

(a) That opium is not produced from such opium poppies; and

(b) That the manufacture of drugs from poppy straw is adequately controlled.

2. The Parties shall apply to poppy straw the system of import certificates and export authorisation as provided in Article 31, paragraphs 4 to 15.

3. The Parties shall furnish statistical information on the import and export of poppy straw as required for drugs under Article 20, paragraphs 1 (d) and 2 (b).

Article 26

The Coca Bush and Coca Leaves

1. If a Party permits the cultivation of the coca bush, it shall apply thereto and to coca leaves the system of controls as provided in Article 23 respecting the control of the opium poppy, but as regards paragraph 2 (d) of that article, the requirements imposed on the Agency therein referred to shall be only to take physical possession of the crops as soon as possible after the end of the harvest.

2. The Parties shall so far as possible enforce the uprooting of all coca bushes which grow wild. They shall destroy the coca bushes if illegally cultivated.

Article 27

Additional Provisions Relating to Coca Leaves

1. The Parties may permit the use of coca leaves for the preparation of a flavouring agent, which shall not contain any alkaloids, and, to the extent necessary for such use, may permit the production, import, export, trade in and possession of such leaves.

2. The Parties shall furnish separately estimates (Article 19) and statistical information (Article 20) in respect of coca leaves for preparation of the flavouring agent, except to the extent that the same coca leaves are used for the extraction of alkaloids and the flavouring agent, and so explained in the estimates and statistical information.

Article 28

Control of Cannabis

1. If a Party permits the cultivation of the cannabis plant for the production of cannabis or cannabis resin, it shall apply thereto the system of controls as provided in Article 23 respecting the control of the opium poppy.

2. This Convention shall not apply to the cultivation of the cannabis plant exclusively for industrial purposes (fibre and seed) or horticultural purposes.

3. The Parties shall adopt such measures as may be necessary to prevent the misuse of, and illicit traffick in the leaves of the cannabis plant.

Article 29

Manufacture

1. The Parties shall require that the manufacture of drugs be under licence except where such manufacture is carried out by a State enterprise or State enterprises.

2. The Parties shall:

(a) Control all persons and enterprises carrying on or engaged in the manufacture of drugs;

(b) Control under licence the establishments and premises in which such manufacture may take place; and

(c) Require that licensed manufacturers of drugs obtain periodical permits specifying the kinds and amounts of drugs which they shall be entitled to manufacture. A periodical permit, however, need not be required for preparations.

3. The Parties shall prevent the accumulation, in the possession of drug manufacturers, of quantities of drugs and poppy straw in excess of those required for the normal conduct of business, having regard to the prevailing market conditions.

Article 30

Trade and Distribution

1. (a) The Parties shall require that the trade in and distribution of drugs be under licence except where such trade or distribution is carried out by a State enterprise or State enterprises.

(b) The Parties shall:

(i) Control all persons and enterprises carrying on or engaged in the trade in or distribution of drugs;

(ii) Control under licence the establishments and premises in which such trade or distribution may take place. The requirement of licensing need not apply to preparations.

(c) The provisions of subparagraphs (a) and (b) relating to licensing need not apply to persons duly authorized to perform and while performing therapeutic or scientific functions.

2. The Parties shall also:

(a) Prevent the accumulation in the possession of traders, distributors, State enterprises or duly authorized persons referred to above, or quantities of drugs and poppy straw in excess of those required for the normal conduct of business, having regard to the prevailing market conditions; and

(b) (i) Require medical prescriptions for the supply or dispensation of drugs to individuals. This requirement need not apply to such drugs as individuals may lawfully obtain, use, dispense or administer in connection with their duly authorised therapeutic functions; and

(ii) If the Parties deem these measures necessary or desirable, require that prescriptions for drugs in Schedule I should be written on official forms to be issued in the form of counterfoil books by the competent governmental authorities or by authorized professional associations.

3. It is desirable that Parties require that written or printed offers of drugs, advertisements of every kind or descriptive literature relating to drugs and used for commercial purposes, interior wrappings of packages containing drugs, and labels under which drugs are offered for sale indicate the international non-proprietary name communicated by the World Health Organization.

4. If a Party considers such measures necessary or desirable, it shall require that the inner package containing a drug or wrapping thereof shall bear a clearly visible double red band. The exterior wrapping of the package in which such drug is contained shall not bear a double red band.

5. A Party shall require that the label under which a drug is offered for sale show the exact drug content by weight or percentage. This requirement of label information need not apply to a drug dispensed to an individual on medical prescription.

6. The provisions of paragraph 2 and 5 need not apply to the retail trade in or retail distribution of drugs in Schedule II.

Article 31

Special Provisions Relating to International Trade

1. The Parties shall not knowingly permit the export of drugs to any country or territory except:

(a) In accordance with the laws and regulations of that country or territory; and

(b) Within the limits of the total of the estimates for that country or territory, as defined in paragraph 2 of Article 19, with the addition of the amounts intended to be re-exported.

2. The Parties shall exercise in free ports and zones the same supervision and control as in other parts of their territories, provided, however, that they may apply more drastic measures.

3. The Parties shall:

(a) Control under licence the import and export of drugs except where such import or export is carried out by a State enterprise or enterprises;

(b) Control all persons and enterprises carrying on or engaged in such import or export.

4. (a) Every Party permitting the import or export of drugs shall require a separate import or export authorisation to be obtained for each such import or export whether it consists of one or more drugs.

(b) Such authorisation shall state the name of the drug, the international non-proprietary namc if any, the quantity to be imported or exported, and the name and address of the importer and exporter, and shall specify the period within which the importation or exportation must be effected.

(c) The export authorisation shall also state the number and date of the import certificate (paragraph 5) and the authority by whom it has been issued.

(d) The import authorisation may allow an importation in more than one consignment.

5. Before issuing an export authorisation the Parties shall require an import certificate, issued by the competent authorities of the importing country or territory and certifying that the importation of the drug or drugs referred to therein, is approved and such certificate shall be produced by the person or establishment applying for the export authorisation. The Parties shall follow as closely as may be practicable the form of import certificate approved by the Commission.

6. A copy of the export authorisation shall accompany each consignment, and the Government issuing the export authorisation shall send a copy to the Government of the importing country or territory.

7. (a) The Government of the importing country or territory, when the importation has been effected or when the period fixed for the importation has expired, shall return the export authorisation, with an endorsement

to that effect, to the Government of the exporting country or territory.

(b) The endorsement shall specify the amount actually imported.

(c) If a lesser quantity than that specified in the export authorisation is actually exported, the quantity actually exported shall be stated by the competent authorities on the export authorisation and on any official copy thereof.

8. Exports of consignments to a post office box, or to a bank to the account of a Party other than the Party named in the export authorisation, shall be prohibited.

9. Exports of consignments to a bonded warehouse are prohibited unless the Government of the importing country certifies on the import certificate, produced by the person or establishment applying for the export authorisation, that it has approved the importation for the purpose of being placed in a bonded warehouse. In such case the export authorisation shall specify that the consignment is exported for such purpose. Each withdrawal from the bonded warehouse shall require a permit from the authorities having jurisdiction over the warehouse and, in the case of a foreign destination shall be treated as if it were a new export within the meaning of this Convention.

10. Consignment of drugs entering or leaving the territory of a Party not accompanied by an export authorisation shall be detained by the competent authorities.

11. A Party shall not permit any drugs consigned to another country to pass through its territory, whether or not the consignment is removed from the conveyance in which it is carried, unless a copy of the export authorisation for such consignment is produced to the competent authorities of such Party.

12. The competent authorities of any country or territory through which a consignment of drugs is permitted to pass shall take all due measures to prevent the diversion of the consignment to a destination other than that named in the accompanying copy of the export authorisation unless the Government of that country or territory through which the consignment is passing authorizes the diversion. The Government of the country or territory or transit shall treat any requested diversion as if the diversion were an export from the country or territory of transit to the country or territory of new destination. If the diversion is authorised, the provisions of paragraph 7 (a) and (b) shall also apply between the country or territory of transit and the country or territory which originally exported the consignment.

13. No consignment of drugs while in transit, or whilst being stored in a bonded warehouse, may be subjected to any process which would change the nature of the drugs in question. The packing may not be altered without the permission of the competent authorities.

14. The provisions of paragraphs 11 to 13 relating to the passage of drugs through the territory of a Party do not apply where the consignment in question is transported by aircraft which does not land in the country or territory of transit. If the aircraft lands in any such country or territory, those provisions shall be applied so far as circumstances require.

15. The provisions of this article are without prejudice to the provisions of any international agreements which limit the control which may be exercised by any of the Parties over drugs in transit.

16. Nothing in this article other than paragraphs 1 (a) and 2 need apply in the case of preparations in Schedule III.

Article 32

Special Provisions Concerning the Carriage of Drugs in First-aid Kits of Ships or Aircraft Engaged in International Traffic

1. The international carriage by ships or aircraft of such limited amounts of drugs as may be needed during their journey or voyage for first-aid purposes or emergency cases shall not be considered to be import, export or passage through a country within the meaning of this Convention.

2. Appropriate safeguards shall be taken by the country of registry to prevent the improper use of the drugs referred to in paragraph 1 or their diversion for illicit purposes. The Commission, in consultation with the appropriate international organizations, shall recommend such safeguards.

3. Drugs carried by ships or aircraft in accordance with paragraph 1 shall be subject to the laws, regulations, permits and licences of the country of registry, without prejudice to any rights of the competent local authorities to carry out checks, inspections and other control measures on board ships or aircraft. The administration of such drugs in the case of emergency shall not be considered a violation of the requirements of Article 30, paragraph 2 (b).

Article 33

Possession of Drugs

The Parties shall not permit the possession of drugs except under legal authority.

Article 34

Measures of Supervision and Inspection

The Parties shall require:

(a) That all persons who obtain licences as provided in accordance with this Convention, or who have managerial or supervisory positions in a State enterprise established in accordance with this Convention, shall have adequate qualifications for the effective and faithful execution of the provisions of such laws and regulations as are enacted in pursuance thereof; and

(b) That government authorities, manufacturers, traders, scientists, scientific institutions and hospitals keep such records as will show the quantities of each drug manufactured and of each individual acquisition and disposal of drugs. Such records shall respectively be preserved for a period of not less than two years. Where counterfoil books [Article 30, paragraph 2 (b)] of official prescriptions are used, such books including the counterfoils shall also be kept for a period of not less than two years.

Article 35

Action Against the Illicit Traffic

Having due regard to their constitutional, legal and administrative systems, the Parties shall:

(a) Make arrangements at the national level for co-ordination of preventive and repressive action against the illicit traffick; to this end they may usefully designate an appropriate agency responsible for such co-ordination.

(b) Assist each other in the campaign against the illicit traffick in narcotic drugs;

(c) Co-operate closely with each other and with the competent international organisations of which they are members with a view to maintaining a co-ordinated campaign against the illicit traffick;

(d) Ensure that international co-operation between the appropriate agencies be conducted in an expeditious manner; and

(e) Ensure that where legal papers are transmitted internationally for the purposes of a prosecution, the transmittal be effected in an expeditious manner to the bodies designated by the Parties; this requirement shall be without prejudice to the right of a Party to require that legal papers be sent to it through the diplomatic channel;

(f) Furnish, if they deem it appropriate to the Board and the Commission through the Secretary-General, in addition to information required by Article 18, information relating to illicit drug activity within their borders, including information on illicit cultivation, production, manufacture and use of, and on illicit trafficking in drugs; and

(g) Furnish the information referred to in the preceding paragraph as far as possible in such manner and by such dates as the Board may request; if requested by a Party, the Board may offer its advice to it in furnishing the information and in endeavouring to reduce the illicit drug activity within the borders of that Party.

Article 36

Penal Provisions

1. (a) Subject to its constitutional limitations, each Party shall adopt such measures as will ensure that cultivation, production, manufacture, extraction, preparation, possession, offering, offering for sale, distribution, purchase, sale, delivery on any terms whatsoever, brokerage, dispatch, dispatch in transit, transport, importation and exportation of drugs contrary to the provisions of this Convention, and any other action which in the opinion of such Party may be country to the provisions of this Convention, shall be punishable offences when committed intentionally, and that serious offences shall be liable to adequate punishment particularly by imprisonment or other penalties of deprivation of liberty.

(b) Notwithstanding the preceding subparagraph, when abusers of drugs have committed such offences, the Parties may provide, either as an alternative to conviction or punishment or in addition to conviction or punishment, that such abusers shall undergo measures of treatment, education, after-care, rehabilitation and social reintegration in conformity with paragraph 1 of Article 38.

2. Subject to the constitutional limitations of a Party, its legal system and domestic law.

(a) (i) Each of the offences enumerated in paragraph 1, if committed in different countries, shall be considered as a distinct offence;

(ii) Intentional participation in conspiracy to commit and attempts to commit, any of such offences, and preparatory acts and financial operations in connection with the offences referred to in this article, shall be punishable offences as provided in paragraph 1;

(iii) Foreign convictions for such offences shall be taken into account for the purpose of establishing recidivism; and

(iv) Serious offences heretofore referred to committed either by nationals or by foreigners shall be prosecuted by the Party in whose terrtory the offence was committed, or by the Party in whose territory the offender is found if extradition is not acceptable in conformity with the law of the Party to which application is made, and if such offender has not already been prosecuted and judgment given.

(b) (i) Each of the offences enumerated in paragraphs 1 and 2 (a) (ii) of this article shall be deemed to be included as an extraditable offence in any extradition treaty existing between Parties. Parties undertake to include such offences as extraditable offences in every extradition treaty to be concluded between them.

(ii) If a Party which makes extradition conditional on the existence of a treaty receives a request for extradition from another Party with which it has no extradition treaty, it may at its option consider this Convention as the legal basis for extradition in respect of the offences enumerated in paragraphs 1 and 2 (a) (ii) of this article. Extradition shall be subject to the other conditions provided by the law of the requested Party.

(iii) Parties which do not make extradition conditional on the existence of a treaty shall recognize the offences enumerated in paragraph 1 and 2 (a) (ii) of this article as extraditable offences between themselves, subject to the conditions provided by the law of the requested Party.

(iv) Extradition shall be granted in conformity with the law of the Party to which application is made, and, notwithstanding subparagraphs (b) (i), (ii) and (iii) of this paragraph, the Party shall have the right to refuse to grant the extradition in cases where the competent authorities consider that the offence is not sufficiently serious.

3. The provisions of this article shall be subject to the provisions of the criminal law of the Party concerned on questions of jurisdiction.

4. Nothing contained in this article shall affect the principle that the offences to which it refers shall be defined, prosecuted and punished in conformity with the domestic law of a Party.

Article 37

Seizure and Confiscation

Any drugs, substances and equipment, used in or intended for the commission of any of the offences, referred to in Article 36, shall be liable to seizure and confiscation.

Article 38

Measures Against the Abuse of Drugs

1. The Parties shall give special attention to and take all practicable measures for the prevention of abuse of drugs and for the early identification, treatment, education, after-care, rehabilitation and social reintegration of the persons involved and shall co-ordinate their efforts to these ends.

2. The Parties shall as far as possible promote the training of personnel in the treatment, after-care, rehabilitation and social reintegration of abusers of drugs.

3. The Parties shall take all practicable measures to assist persons whose work so requires to gain an understanding of the problems of abuse of drugs and of its prevention, and shall also promote such understanding among the general public if there is a risk that abuse of drugs will become widespread.

Article 38 bis

Agreements on Regional Centres

If a Party considers it desirable as part of its action against the illicit traffic in drugs, having due regard to its constitutional, legal and administrative systems and, if it so desires, with the technical advice of the Board or the specialized agencies, it shall promote the establishment, in consultation with other interested Parties in the region, of agreements which contemplate the development of regional centres for scientific research and education to combat the problems resulting from the illicit use of and traffic in drugs.

Article 39

Application of Stricter National Control Measures than those Required by this Convention

Notwithstanding anything contained in this Convention, a Party shall not be, or be deemed to be, precluded from adopting measures of control more strict or severe than those provided by this Convention and in particular from requiring that Preparations in Schedule III or drugs in Schedule II be subject to all or such of the measures of control applicable to drugs in Schedule I as in its opinion is necessary or desirable for the protection of the public health or welfare.

Schedules[2]

List of Drugs Included in Schedule I

ACETORPHINE	(3-*O*-acetyltetrahydro-7)(1-hydroxy-1-methylbutyl)-6, 14-endoethencoripavine)
ACETYLMETHADOL	(3-acetoxy-6-dimethylamino-4, 4-diphenylheptane)
ALLYLPRODINE	(3-allyl-1-methyl-4-phenyl-4-

	propionoxypiperidine)
ALPHACETYLMETHADOL	(alpha-3-acetoxy- 6- dimethylamino-4, 4-diphenylheptane)
ALPHAMEPRODINE	(alpha-3-ethyl-1-methyl-4-phenyl-4 -propionoxypiperidine)
ALPHAMETHADOL	(alpha-6-dimethylamino-4, 4-diphenyl-3-heptanol)
ALPHAPRODINE	(alpha-1, 3-dimethyl-4-phenyl-4-propionoxypiperidine)
ANILERIDINE	(1-*para*-aminophenetyl-4-phenylpiperidine-4-carboxylic acid ethyl ester)
BENZETHIDINE	(1-(2-benzyloxyethyl)-4-phenylpiperidine-1-caroxylic acid ethyl ester)
BENZYLMORPHINE	(3-benzylmorphine)
BETACETYLMETHADOL	(beta-3-acetoxy-6-dimethylamino-4,4-diphenylheptane)
BETAMETHADOL	(beta-6-dimethylamino-4, 4-propionxypiperidine)
BETAPRODINE	(beta-1,3-dimethyl-4-phenyl-4-propionoxypiperidine)
BEZITRAMIDE	(1-(3cyano-3,3-diphenylpropyl)-4-(2-oxo-3-propionyl-1-benzimidazolinyl)-piperidine)
CANNABIS and CANNABIS	RESIN and EXTRACTS and TINCTURES OF CANNABIS
CLONITAZENE	(2-*para*-chlorbenzyl-1-diethylaminoethyl-5-nitrobenzimidazole)
COCA LEAF	
COCAINE	(methyl ester of benzoylecgonine)
CODOXIME	(dihydrocodeinone-6 -carboxymethyloxime)
CONCENTRATE OF POPPY STRAW	(the material arising when poppy straw has entered into a process

	for the concentration of its alkaloids when such material is made available in trade)
DESOMORPHINE	(dihydrodeoxymorphine)
DEXTROMORAMIDE	((+)-4-[2-methyl-4-oxo-3, 3-diphenyl-4-(1-pyrrolidinyl) butyl] morpholine)
DIAMPROMIDE	(N-[2-methylphenethylamino propyl] propionanilide)
DIETHYLTHIAMBUTENE	(3-diethylamino-1, 1-di-(2'-thienyl)-1-butene)
DIFENOXIN	(1-(3-cyano-3, 3 -diphenylpropyl)- 4-phenylisonipecotic acid)
DIHYDROMORPHINE	
DIMENOXADOL	(2-dimethylaminoethyl-1-ethoxy-1, 1-diphenylacetate)
DIMEPHEPTANOL	(6-dimethylamino-4, 4-diphenyl-3-heptanol)
DIMETHYLTHIAMBUTENE	(3-dimethylamino-1, 1-di (2'-thienyl)-1-butene)
DIOXAPHETYL BUTYRATE	(ethyl-4-morpholino-2, 2-diphenylbutyrate)
DIPHENOXYLATE	(1-(3-cyano-3, 3-diphenylpropyl)- 4-phenylpiperidine-4-carboxylic acid ethyl ester)
DIPIPANONE	(4, 4-diphenyl-6-piperidine-3-heptanone)
DROTEBANOL	(3, 4-dimethoxy-17 -methylmorphinan-68, 14-diol)
ECGONINE,	its esters and derivatives which are convertible to ecgonine and cocaine
ETHYLMETHYL-	(3-ethylmethylamino-1, 1-di
THIAMBUTENE	(2' thienyl)-1-butene)
ETONITAZENE	(1-diethylaminoethyl-2-*para*-ethoxybenzyl-5 -mitorbenzimidazole)
ETORPHINE	(tetrahydro-7-(1-hydroxy-1-methylbutyl)-6, 14-en-

	doetheno-oripavine)
ETOXERIDINE	(1-[2-(2-hydroxyethoxy) -ethyl]-4-phenylpiperidine-4-carboxylic acid ethyl ester)
FENTANYL	(1-phenethyl-4-N -propionylanilinopiperidine)
FURETHIDINE	(1-(2-tetrahydrofurfuryloxyethyl)-4-phenylpiperidine-4-carboxylic acid ethyl ester)
HEROIN	(diacetylmorphine)
HYDROCODONE	(dihydrocodeinone)
HYDROMORPHINOL	(14-hydroxydihydromorphine)
HYDROMORPHONE	(dihydromorphinone)
HYDROXYPETHIDINE	(4-*meta*-hydroxyphenyl-1-methylpiperidine-4-carboxylic acid ethyl ester)
ISOMETHADONE	(6-dimethylamino-5-methyl-4, 4-diphenyl-3-hexanone)
KETOBEMIDONE	(4-*meta*-hydroxyphenyl-1-methyl-4-propionylipiperidine)
LEVOMETHORPHAN*	((–)-3-methoxy-N -methylmorphinan)
LEVOMORAMIDE	((–)-4-[2-methyl-4-oxo-3, 3-diphenyl-4-(1-pyrrolidinyl) butyl] morpholine)
LEVOPHENACYLMORPHAN	((–)-3-hydroxy-N -phenacylmorphiman)
LEVORPHANOL*	((–)-3-hydroxy-N -methylmorphinan)
METAZOCINE	(2-hydroxy-2, 5, 9-trimethyl-6, 7-benzomorphan)
METHADONE	(6-dimethylamino-4, 4 -diphenyl-3-heptanone)
METHADONE	(4-cyano-3 -dimethylamino-4,
INTERMEDIATE	4 -diphenylbutane)
METHYLDESORPHINE	(6-methyl-delta-6 -deoxymorphine)
METHYLDIHYDROMORPHINE	(6 -methyldihydromorphine)
METOPON	(5-methyldihydromorphinone)
MORAMIDE INTERMEDIATE	(2-methyl-3-morpholino-1, 1 -diphenylpropane carboxylic acid)
MORPHERIDINE	(1-(2-morphelinoethyl)-4-phenylpiperidine-4-carboxylic acid ethyl ester)
MORPHINE	
MORPHINE METHOBROMIDE	and other pentavalent nitrogen morphine derivatives
MORPHINE-N-OXIDE	
MYROPHINE	(myristylbenzylmorphine)
NICOMORPHINE	(3, 6-dinicotinylmorphine)
NORACYMETHADOL	((±)-alpha-3-acetoxy-6 -methylamino-4, 4 -diphenylheptane)
NORLEVORPHANOL	((–)-3-hydroxymorphinan)
NORMETHADONE	(6-dimethylamino-4, 4-diphenyl-3-hexanone)
NORMORPHINE	(demethylmorphine)
NORPIPANONE	(4, 4-diphenyl-6-piperidino-3-hexanone)
OPIUM	
OXYCODONE	(14-hydroxydihydrocodeinone)
OXYMORPHONE	(14-hydroxydihydromorphinone)
PETHIDINE	(1-methyl-4-phenylpiperidine-4-carboxylic acid ethyl ester)
PETHIDINE INTERMEDIATE A	(4-cyano-1-methyl-4-phenylpiperidine)
PETHIDINE INTERMEDIATE B	(4-phenylpiperidine-4-carboxylic acid ethyl ester)
PETHIDINE INTERMEDIATE C	(1-methyl-4 -phenylpiperidine-4 -carboxylic acid)

1. Including all amendments up to the end of 1976.

PHENADOXONE	(6-morpholino-4, 4-diphenyl-3-heptanone)
PHENAMPROMIDE	(N(1-methyl-2-piperidinoethyl) propionanilide)
PHENAZOCINE	(2'-hydroxy-5, 9-dimethyl-2-phenethyl-6, 7-benzomor- phan)
PHENOMORPHAN	(3-hydroxy-N -phenethyl-morphinan)
PHENOPERIDINE	(1-(3-hydroxy-3 -phenylpropyl)-4 -phenylpiperidine-4-carboxylic acid ethyl ester)
PIMINODINE	(4-phenyl-1-(3 -phenylaminopropyl) piperidine-4-carboxylic acid ethyl ester)
PIRITRAMIDE	(1-(3-cyano-3, 3 -diphenylpropyl)-4-(1 -piperidino)-piperidine-4 -carboxylic acid amide)
PROHEPTAZINE	(1,3-dimethyl-4-phenyl-4 -propionoxyazacycloheptane)
PROPERIDINE	(1-methyl-4-phenylpiperidine-4-carboxylic acid isopropyl ester)
RACEMETHORPHAN	((±)-3-methoxy-N -methyl-morphinan)
RACEMORAMIDE	((±)-4-[2-methyl-4-oxo-3, 3 -diphenyl-(1-pyrrolidinyl) butyl] morpholine)
RACEMORPHAN	((±)-3-hydroxyl-N -methyl-morphinan)
THEBACON	(acetyldihydrocodeinone)
THEBAINE	
TRIMEPERIDINE	(1, 2, 5-trimethyl-4-phenyl-4-propionoxypiperidine); and

The isomers, unless specifically expected, of the drugs in this Schedule whenever the existence of such isomers is possible within the specific chemical designation;

The esters and ethers, unless appearing in another Schedule, of the drugs in this Schedule whenever the existence of such esters or ethers is possible;

The salts of the drugs listed in this Schedule, including the salts of esters, ethers and isomers as provided above whenever the existence of such salts is possible.

*Dextromethorphan ((±)-3-methoxy-N-methylmorphan) and dextrorphan ((±)-3-hydroxy-N-methylmorphinan) are specifically excluded from this Schedule.

List of Drugs included in Schedule II

ACETYLDIHYDROCODEINE	
CODLINE (3-methylmorphine)	
DIHYDROCODEINE	
ETHYLMORPHINE	(3-ethylmorphine)
NICOCODINE	(6-nicotinylcodeine)
NICODICODINE	(6-nicotinyldihydrocodeine)
NORCODEINE	(N-demethylcodeine)
PHOLCODINE	(morpholinylethylmorphine)
PROPIRAM	(N-(1-methyl-2-piperidinoethyl)-N-2-pyridylpropionamide); and

The isomers, unless specifically expected, of the drugs in this Schedule whenever the existence of such isomers is possible within the specific chemical designation;

The salts of the drugs listed in this Schedule, including the salts of the isomers as provided above whenever the existence of such salt is possible.

List of Preparations Included in Schedule III

1. Preparations of Acetyldihydrocodeine, Codeine, Dihydrocodeine, Ethylmorphine, Nicodicodine, Norcodeine, and Pholcodine

When compounded with one or more other ingredients and containing not more than 100 milligrams of the drug per dosage unit and with a concentration of not more than 2.5 per cent in undivided preparations. [Sic]

2. Preparations of propiram containing not more than 100 milligrams of propiram per dosage unit and compounded with at least the same amount of methylcellulose.

3. Preparations of cocaine containing not more than 0.1 per cent of cocaine calculated as cocaine base and preparations of opium or morphine containing not more than 0.2 per cent of morphine calculated as anhydrous morphine base and compounded with one or more other ingredients and in such a way that the drug cannot be

recovered by readily applicable means or in a yield which would constitute a risk to public health.

4. Preparations of difenoxin containing, per dosage unit, not more than 0.5 milligram of difenoxin and a quantity of atropine sulphate equivalent to at least 5 per cent of dose of difenoxin.

5. Preparations of diphenoxylate containing, per dosage unit, not more than 2.5 milligrams of diphenoxylate calculated as base and a quantity of atropine sulphate equivalent to at least one per cent of the dose of diphenoxylate.

6. Pulvis ipecacuanhae et opii compositus

10 per cent opium in powder

10 per cent Ipecacuanha root, in powder well mixed with

80 per cent of any other powdered ingredient containing no drug.

7. Preparations conforming to any of the formulae listed in this Schedule and mixtures of such preparations with any material which contains no drug.

List of Drugs Included in Schedule IV

ACETORPHINE(3-*O*-acetyltetrahydro-7)-(1-hydroxy-1-methylbutyl)-6, 14-endoetheno -oripavine)

CANNABIS and CANNABIS RESIN

DESOMORPHINE (dihydrodeoxymorphine)

ETORPHINE (tetrahydro-7)-(1-hydroxyl-methylbutyl)-6, 14-endoetheno-oripavine)

HEROIN (diacetylmorphine)

KETOBEMIDONE (4-*meta*-hydroxyphenyl-1-methyl-4-propionylpiperidine); and

The salts of the drugs listed in the Schedule whenever the formation of such salts is possible.

References

1. United Nations Conference to consider Amendments to the Single Convention on Narcotic Drugs 1961, held at Geneva, 6-24 March, 1972.
2. Including all amendments up to the end of 1976.

Appendix II to Chapter 62

The Convention against Illicit Traffic in Narcotic Drugs and Psychotropic Substances, 1988

The Parties to this Convention,

Deeply concerned by the magnitude of and rising trend in the illicit production of, demand for and traffic in Narcotic Drugs and Psychotropic Substances, was pose a serious threat to the health and welfare of human beings and adverse affect the economic, cultural and political foundations of society,

Deeply concerned also by the steadily increasing inroads into various soul groups made by illicit traffic in Narcotic Drugs and Psychotropic Substances and particularly by the fact that children are used in many parts of the world an illicit drug consumers market and for purposes of illicit production, distribution and trade in Narcotic Drugs and Psychotropic Substances, which entails a danger of incalculable gravity,

Recognising the links between illicit traffic and other related organised criminal activities which undermine the legitimate economies and threaten stability, security and sovereignty of States,

Recognising also that illicit traffic is an international criminal activity was suppression of which demands urgent attention and the highest priority,

Aware that illicit traffic generates large financial profits and wealth earnings transnational criminal organisations to penetrate, contaminate and corrupt structures of government, legitimate commercial and financial business and society at all its levels,

Determined to deprive persons engaged in illicit traffic of the process of their criminal activities and thereby eliminate their main incentive for doing,

Desiring to eliminate the root causes of the problem of abuse of Narcotic Drugs and Psychotropic Substances, including the illicit demand for such drugs and substances and the enormous profits derived from illicit traffic,

Considering that measures are necessary to monitor certain substances, including precursors, chemicals and solvents, which are used in the manufacture of Narcotic Drugs and Psychotropic Substances, the ready availability of which has led to an increase in the clandestine manufacture of such drugs and substances.

Determined tc improve international co-operation in the suppression of illicit traffic by sea,

Recognising that eradication of illicit traffic is a collective responsibility of all States and that, to that end, co-ordinated action within the frameworks of international co-operation is necessary,

Acknowledging the competence of the United Nations in the field of control of Narcotic Drugs and Psychotropic Substances and desirous that the international organs concerned with such control should be within the framework of that Organisation,

Reaffirming the guiding principles of existing treaties in the fields of Narcotic Drugs and Psychotropic Substances and the system of control which they embody,

Recognising the need to reinforce and supplement the measures provided in the Single Convention on Narcotic Drugs, 1961, that Convention as amended by the 1972 Protocol Amending the Single Convention on Narcotic Drugs, 1961, and the 1971 Convention on Psychotropic Substances, in order to counter the magnitude and extent of illicit traffic and its grave consequences,

Recognising also the importance of strengthening and enhancing effective legal means for international co-operation in criminal matters for suppressing the international criminal activities of illicit traffic,

Desiring to conclude a comprehensive, effective and operative international convention that is directed specifically against illicit traffic and that considers the various aspects of the problem as a whole, in particular those aspects not envisaged in the existing treaties in the field of Narcotic Drugs and Psychotropic Substances,

Hereby agree as follows :

Article 1

Definitions

Except where otherwise expressly indicated or where the context otherwise requires, the following definitions shall apply throughout this Convention :

(a) "Board" means the International Narcotics Control Board established by the Single Convention on Narcotic Drugs, 1961, and that Convention as amended by the 1972 Protocol Amending the Single Convention on Narcotic Drugs, 1961;

(b) "Cannabis Plant" means any plant of the genus Cannabis;

(c) "Coca bush" means the plant of any species of the genus Erythroxylon;

(d) "Commercial carrier" means any person or any public, private or other entity engaged in transporting persons, goods or mails for remuneration, hire or any other benefit;

(e) "Commission" means the Commission on Narcotic Drugs of the Economic and Social Council of the United Nations;

(f) "Confiscation", which includes forfeiture where applicable, means the permanent deprivation of property by order of a court or other competent authority;

(g) "Controlled delivery" means the technique of allowing illicit or suspect consignments of narcotic drugs, psychotropic substances, substances in Table I and Table II annexed to this Convention, or substances substituted for them, to pass out of, through or into the territory of one or more countries, with the knowledge and under the supervision of their competent authorities, with a view to identifying persons involved in the commission of offences established in accordance with Article 3, paragraph 1 of the Convention;

(h) "1961 Convention" means the Single Convention on Narcotic Drugs, 1961;

(i) "1961 Conventions as amended" means the Single Convention on Narcotic Drugs, 1961, as amended by the 1972 Protocol Amending the Single Convention on Narcotic Drugs, 1961;

(j) "1971 Convention" means the Convention on Psychotropic Substances, 1971;

(k) "Council" means the Economic and Social Council of the United Nations;

(l) "Freezing" or "seizure" means temporarily prohibiting the transfer, conversion, disposition or movement of property or temporarily assuming custody or control of property on the basis of an order issued by a court or a competent authority;

(m) "Illicit traffic" means the offences set forth in Article 3, paragraph 1 and 2, of this Convention;

(n) "Narcotic drug" means any of the substances, natural or synthetic, in Schedule I and II of the Single Convention on Narcotic Drugs, 1961, and that Convention as amended by the 1972 Protocol Amending the Single Convention on Narcotic Drugs, 1961;

(o) "Opium poppy" means the plant of the species *Papaver somniferum* L;

(p) "Proceeds" means any property derived from or obtained, directly or indirectly, through the commission of an offence established in accordance with Article 3, paragraph 1;

(q) "Property" means assets of every kind, whether corporeal or incorporeal, movable or immovable, tangible or intangible, and legal documents or instruments evidencing title to, or interest in, such assets;

(r) "Psychotropic Substance" means any substance, natural or synthetic, or any natural material in Schedules I, II, III and IV of the Convention on Psychotropic Substances, 1971;

(s) "Secretary-General" means the Secretary-General of the United Nations;

(t) "Table I" and "Table II" mean the correspondingly numbered lists of substances annexed to this Convention, as amended from time to time in accordance with Article 12;

(u) "Transit State" means a State through the territory of which illicit narcotic drugs, psychotropic substances and substances in Table I and Table II are being moved, which is neither the place of origin nor the place of ultimate destination thereof.

Article 2

Scope of the Convention

1. The purpose of this Convention is to promote cooperation among the Parties so that they may address more effectively the various aspects of illicit traffic in Narcotic Drugs and Psychotropic Substances having an international dimension. In carrying out their obligations under the Convention, the Parties shall take necessary measures, including legislative and administrative measures, in conformity with the fundamental provisions of their respective domestic legislative systems.

2. The Parties shall carry out their obligations under this Convention in a manner consistent with the principles of sovereign equality and territorial integrity of

States and that of non-intervention in the domestic affairs of other States.

3. A Party shall not undertake in the territory of another Party the exercise of jurisdiction and performance of functions which are exclusively reserved for the authorities of that other Party by its domestic law.

Article 3

Offences and Sanctions

1. Each Party shall adopt such measures as may be necessary to establish as criminal offences under its domestic law, when committed intentionally :

(a) (i) The production, manufacture, extraction, preparation, offering, offering for sale, distribution, sale, delivery on any terms whatsoever, brokerage, despatch, despatch in transit, transport, importation or exportation of any narcotic drug or any psychotropic substance contrary to the provisions of the 1961 Convention, the 1961 Convention as amended or the 1971 Convention;

(ii) The Cultivation of opium poppy, coca bush or cannabis plant for the purpose of the production of narcotic drugs contrary to the provisions of the 1961 Convention and the 1961 Convention as amended;

(iii) The possession or purchase of any Narcotic Drug or Psychotropic Substance for the purpose of any of the activities enumerated in (i) above;

(iv) The manufacture, transport or distribution of equipment, materials or of substances listed in Table I and Table II, knowing that they are to be used in or for the illicit cultivation, production or manufacture of Narcotic Drugs or Psychotropic Substances;

(v) The organisation, management or financing of any of the offences enumerated in (i), (ii), (iii) or (iv) above;

(b) (i) The conversion or transfer of property, knowing that such property is derived from any offence or offences established in accordance with sub-paragraph (a) of this paragraph, or from an act of participation in such offence or offences, for the purpose of concealing or disguising the illicit origin of the property or of assisting any person who is involved in the commission of such an offence or offences to evade the legal consequences of his actions;

(ii) The concealment or disguise of the true nature, source, location, disposition, movement, rights with respect to, or ownership of property knowing that such property is derived from an offence or offences established in accordance with sub-paragraph (a) of this paragraph or from

(c) Subject to its constitutional principles and the basic concepts of its legal system:

(i) The acquisition, possession or use of property, knowing, at the time of receipt, that such property was derived from an offence or offences established in accordance with sub-paragraph (a) of this paragraph or from an act of participation in such offence or offences;

(ii) The possession of equipment or materials or substances listed in Table I and Table II, knowing that they are being or are to be used in or for the illicit cultivation, production or manufacture of Narcotic Drugs or Psychotropic Substances;

(iii) Publicly inciting or inducing others, by any means, to commit any of the offences established in accordance with this article or to use Narcotic Drugs or Psychotropic Substances illicitly;

(iv) Participation in, association or conspiracy to commit, attempts to commit and aiding, abet ting, facilitating and counselling the commission of any of the offences established in accordance with this article.

2. Subject to its constitutional principles and the basic concepts of its legal system, each Party shall adopt such measures as may be necessary to establish as a criminal offence under its domestic law, when committed intentionally, the possession, purchase or cultivation of Narcotic Drugs or Psychotropic Substances for personal consumption contrary to the provisions of the 1961 Convention the 1961 Convention as amended or the 1971 Convention.

3. Knowledge, intent or purpose required as an element of an offence set forth in paragraph 1 of this article may be inferred from objective factual circumstances.

4. (a) Each Party shall make the commission of the offences established in accordance with paragraph 1 of this article liable to sanctions which take into account the grave nature of these offences, such as imprisonment or other forms of deprivation of liberty, pecuniary sanctions and confiscation.

(b) The Parties may provide, in addition to conviction or punishment for an offence established in accordance with paragraph 1 of this article, that the offender shall undergo measures such as treatment, education, aftercare, rehabilitation or social reintegration.

(c) Notwithstanding the preceding sub-paragraphs, in appropriate cases of a minor nature, the Parties may provide, as alternatives to conviction or punishment, measures such as education, rehabilitation or social reintegration, as well as, when the offender is a drug abuser, treatment and aftercare.

(d) The Parties may provide, either as an alternative to conviction or punishment, or in addition to conviction or punishment of an offence established in accordance with paragraph 2 of this article, measures for the treatment, education, aftercare, rehabilitation or social reintegration of the offender.

5. The Parties shall ensure that their courts and other competent authorities having jurisdiction can take into account factual circumstances which make the commission of the offences established in accordance with paragraph 1 of this article particularly serious, such as:

(a) The involvement in the offences of an organised criminal group to which the offender belongs;

(b) The involvement of the offender in other international organised criminal activities;

(c) The involvement of the offender in other illegal activities facilitated by commission of the offence;

(d) The use of violence or arms by the offender;

(e) The fact that the offender holds a public office and that the offence is connected with the office in question;

(f) The victimisation or use of minors;

(g) The fact that the offence is committed in a penal institution or in an educational institution or social service facility or in their immediate vicinity or in other places to which school children and students resort for educational, sports and social activities;

(h) Prior conviction, particularly for similar offences, whether foreign or domestic, to the extent permitted under the domestic law of a Party.

6. The Parties shall endeavour to ensure that any discretionary legal powers under their domestic law relating to the prosecution of persons for offences established in accordance with this article are exercised to maximize the effectiveness of law enforcement measures in respect of those offences and with due regard to the need to deter the commission of such offences.

7. The Parties shall ensure that their courts or other competent authorities bear in mind the serious nature of the offences enumerated in paragraph 1 of this article and the circumstances enumerated in paragraph 5 of this article when considering the eventuality of early release or parole of persons convicted of such offences.

8. Each Party shall, where appropriate, establish under its domestic law long statute of limitations period in which to commence proceedings for any offence established in accordance with paragraph 1 of this article, and a longer period where the alleged offender has evaded the administration of justice.

9. Each Party shall take appropriate measures, consistent with its legal system, to ensure that a person charged with or convicted of an offence established in accordance with paragraph 1 of this article, who is found within its territory, is present at the necessary criminal proceedings.

10. For the purpose of co-operation among the Parties under this Convention, including in particular, co-operation under Articles 5, 6, 7 and 9, offences established in accordance with this article shall not be considered as fiscal offences or as political offences or regarded as politically motivated, without prejudice to the constitutional limitations and the fundamental domestic law of the Parties.

11. Nothing contained in this article shall affect the principle that the description of the offences to which it refers and of legal defences thereto is reserved to the domestic law of a Party and that such offences shall be prosecuted and punished in confirmity with that law.

Article 4
Jurisdiction

1. Each Party:

(a) Shall take such measures as may be necessary to establish its jurisdiction over the offences it has established in accordance with Article 3, paragraph 1, when :

(i) The offence is committed in its territory;

(ii) The offence is committed on board a vessel flying its flag or an aircraft which is registered under its laws at the time the offence is committed:

(b) May take such measures as may be necessary to establish its jurisdiction over the offences it has established in accordance with Article 3, paragraph 1, when:

(i) The offence is committed by one of its nationals or by a person who has his habitual residence in its territory;

(ii) The offence is committed on board a vessel concerning which that Party has been authorised to take appropriate action pursuant to Article 17, provided that such jurisdiction shall be exercised only on the basis of agreements or arrangements referred to in paragraphs 4 and 9 of that article;

(iii) The offence is one of those established in accordance with Article 3, paragraph 1, sub-paragraph (c) (iv), and is committed outside its territory with a view to the commission, within its territory, of an offence established in accordance with Article 3, paragraph 1.

2. Each Party :

(a) Shall also take such measures as may be necessary to establish its jurisdiction over the offences it has established in accordance with Article 3, paragraph 1, when the alleged offender is present in its territory and it does not extradite him to another Party on the ground :

(i) That the offence has been committed in its territory or on board a vessel flying its flag or an aircraft which was registered under its law at the time the offence was committed; or

(ii) That the offence has been committed by one of its nationals;

(b) May also take such measures as may be necessary to establish its jurisdiction over the offences it has established in accordance with Article 3, paragraph 1, when the alleged offender is present in its territory and it does not extradite him to another Party.

3. This Convention does not exclude the exercise of any criminal jurisdiction established by a Party in accordance with its domestic law.

Article 5

Confiscation

1. Each Party shall adopt such measures as may be necessary to enable confiscation of:

(a) Proceeds derived from offences established in accordance with Article 3, paragraph 1, or property the value of which corresponds to that of such proceeds;

(b) Narcotic Drugs and Psychotropic Substances, materials and equipment or other instrumentalities used in or intended for use in any manner in offences established in accordance with Article 3, paragraph 1.

2. Each Party shall also adopt such measures as may be necessary to enable its competent authorities to identify, trace, and freeze or seize proceeds, property, intrumentalities or any other things referred to in paragraph 1 of this article, for the purpose of eventual confiscation.

3. In order to carry out the measures referred to in this article, each Party shall empower its courts or other competent authorities to order that bank, financial or commercial records be made available or be seized. A Party shall not decline to act under the provisions of this paragraph on the ground of bank secrecy.

4. (a) Following a request made pursuant to this article by another Party having jurisdiction over an offence established in accordance with Article 3, paragraph 1, the Party in whose territory proceeds, property, instrumentalities or any other things referred to in paragraph 1 of this article are situated shall :

(i) Submit the request to its competent authorities for the purpose of obtaining an order of confiscation and, if such order is granted, give effect to it; or

(ii) Submit to its competent authorities, with a view to giving effect to it to the extent requested, an order of confiscation issued by the requesting Party in accordance with paragraph 1 of this article, in so far as it relates to proceeds, property, instrumentalities or any other things referred to in paragraph 1 situated in the territory of the requested Party.

(b) Following a request made pursuant to this article by another Party having jurisdiction over an offence established in accordance with Article 3, paragraph 1, the requested Party shall take measures to identify, trace, and freeze or seize proceeds, property, instrumentalities or any other things referred to in paragraph 1 of this article for the purpose of eventual confiscation to be ordered either by the requesting Party or, pursuant to a request under subparagraph (a) of this paragraph, by the requested Party.

(c) The decisions or actions provided for in subparagraphs (a) and (b) of this paragraph shall be taken by the requested Party, in accordance with and subject to the provisions of its domestic law and its procedural rules or any bilateral or multilateral treaty, agreement or arrangement to which it may be bound in relation to the requesting Party.

(d) The provisions of Article 7, paragraphs 6 to 19 are applicable *mutatis mutandis*. In addition to the information specified in Article 7, paragraph 10, requests made pursuant to this article shall contain the following:

(i) In the case of a request pertaining to subparagraph (a) (i) of this paragraph, a description of the property to be confiscated and a statement of the facts relied upon by the requesting Party sufficient to enable the requested Party to seek the order under its domestic law;

(ii) In the case of a request pertaining to subparagraph (a) (ii), a legally admissible copy of an order of confiscation issued by the requesting Party upon which the request is based, a statement of the facts and information as to the extent to which the execution of the order is requested;

(e) Each Party shall furnish to the Secretary-General the text of the any of its laws and regulations which give effect to this paragraph and the text of any subsequent changes to such laws and regulations.

(f) If a Party elects to make the taking of the measures referred to in subparagraphs (a) and (b) of this paragraph conditional on the existence of a relevant treaty, that Party shall consider this Convention as the necessary and sufficient treaty basis.

(g) The Parties shall seek to conclude bilateral and multilateral treaties, agreements or arrangements to enhance the effectiveness of international co-operation pursuant to this article.

5. (a) Proceeds or property confiscated by a Party pursuant to paragraph 1 or paragraph 4 of this article shall be disposed of by that Party according to its domestic law and administrative procedures.

(b) When acting on the request of another Party in accordance with this article, a Party may give special consideration to concluding agreements on :

(i) Contributing the value of such proceeds and property, or funds derived from the sale of such proceeds or property, or a substantial part thereof, to inter-governmental bodies specialising in the fight against illicit traffic in and abuse of Narcotic Drugs and Psychotropic Substances;

(ii) Sharing with other Parties, on a regular or case-by-case basis, such proceeds or property, or funds derived from the sale of such proceeds or property, in accordance with its domestic law, administrative procedures or bilateral or multilateral agreements entered into for this purpose.

6. (a) If proceeds have been transformed or converted into other property, such property shall be liable to the measures referred to in this article instead of the proceeds.

(b) If proceeds have been intermingled with property acquired from legitimate sources, such property shall, without prejudice to any powers relating to seizure or freezing, be liable to confiscation up to the assessed value of the intermingled proceeds.

(c) Income or other benefits derived from :

(i) Proceeds;

(ii) Property into which proceeds have been transformed or converted; or

(iii) Property with which proceeds have been intermingled

shall also be liable to the measures referred to in this article, in the same manner and to the same extent as proceeds.

7. Each Party may consider ensuring that the onus of proof be reversed regarding the lawful origin of alleged proceeds or other property liable to confiscation, to the extent that such action is consistent with the principles of its domestic law and with the nature of the judicial and other proceedings.

8. The provisions of this article shall not be construed as prejudicing the rights of *bona fide* third parties.

9. Nothing contained in this article shall affect the principle that the measures to which it refers shall be defined and implemented in accordance with and subject to the provisions of the domestic law of a Party.

Article 6

Extradition

1. This article shall apply to the offences established by the Parties in accordance with article 3, paragraph 1.

2. Each of the offences to which this article applies shall be deemed to be included as an extraditable offence in any extradition treaty existing between Parties. The Parties undertake to include such offences as extraditable offences in every extradition treaty to be concluded between them.

3. If a Party which makes extradition conditional on the existence of a treaty receives a request for extradition from another Party with which it has no extradition treaty, it may consider this Convention as the legal basis for extradition in respect of any offence to which this article applies. The Parties which require detailed legislation in order to use this Convention as a legal basis for extradition shall consider enacting such legislation as may be necessary.

4. The Parties which do not make extradition conditional on the existence of a treaty shall recognise offences to which this article applies as extraditable offences between themselves.

5. Extradition shall be subject to the conditions provided for by the law of the requested Party or by applicable extradition treaties, including the grounds upon which the requested Party may refuse extradition.

6. In considering requests received pursuant to this article, the requested State may refuse to comply with such requests where there are substantial grounds leading its judicial or other competent authorities to believe that compliance would facilitate the prosecution or punishment of any person on account of his race, religion, nationality or political opinions, or would cause prejudice for any of those reasons to any person affected by

the request.

7. The Parties shall endeavour to expedite extradition procedures and to simplify evidentiary requirements relating thereto in respect of any offence to which this article applies.

8. Subject to the provisions of its domestic law and its extradition treaties, the requested Party may, upon being satisfied that the circumstances so warrant and are urgent and at the request of the requesting Party, take a person whose extradition is sought and who is present in its territory into custody or take other appropriate measures to ensure his presence at extradition proceedings.

9. Without prejudice to the exercise of any criminal jurisdiction established in accordance with its domestic law, a Party in whose territory an alleged offender is found shall :

(a) If it does not extradite him in respect of an offence established in accordance with article 3, paragraph 1, on the grounds set forth in article 4, paragraph 2, subparagraph (a), submit the case to its competent authorities for the purpose of prosecution, unless otherwise agreed with the requesting Party;

(b) If it does not extradite him in respect of such an offence and has established its jurisdiction in relation to that offence in accordance with article 4, paragraph 2, subparagraph (b), submit the case to its competent authorities for the purpose of prosecution, unless otherwise requested by the requesting Party for the purposes of preserving its legitimate jurisdiction.

10. If extradition, sought for purposes of enforcing a sentence, is refused because the person sought is a national of the requested Party, the requested Party shall, if its law so permits and in conformity with the requirements of such law, upon application of the requesting Party, consider the enforcement of the sentence which has been imposed under the law of the requesting Party, or the remainder thereof.

11. The Parties shall seek to conclude bilateral and multilateral agreements to carry out or to enhance the effectiveness of extradition.

12. The Parties may consider entering into bilateral or multilateral agreements, whether *ad hoc* or general, on the transfer to their country of persons sentenced to imprisonment and other forms of deprivation of liberty for offences to which this article applies, in order that they may complete their sentences there.

Article 7

Mutual Legal Assistance

1. The Parties shall afford one another, pursuant to this article, the widest measure of mutual legal assistance in investigations, prosecutions and judicial proceedings in relation to criminal offences established in accordance with article 3, paragraph 1.

2. Mutual legal assistance to be afforded in accordance with this article may be requested for any of the following purposes :

(a) Taking evidence or statements from persons;

(b) Effecting service of judicial documents;

(c) Executing searches and seizures;

(d) Examining objects and sites;

(e) Providing information and evidentially items;

(f) Providing originals or certified copies of relevant documents and records, including bank, financial, corporate or business records;

(g) Identifying or tracing proceeds, property, instrumentalities or other things for evidentiary purposes.

3. The Parties may afford one another any other forms of mutual legal assistance allowed by the domestic law of the requested Party.

4. Upon request, the Parties shall facilitate or encourage, to the extent consistent with their domestic law and practice, the presence or availability of persons, including persons in custody, who assist in investigations or participate in proceedings.

5. A Party shall not decline to render mutual legal assistance under this article on the ground of bank secrecy.

6. The provisions of this article shall not affect the obligations under any other treaty, bilateral or multilateral, which governs or will govern, in whole or in part, mutual legal assistance in criminal matters.

7. Paragraphs 8 to 19 of this article shall apply to requests made pursuant to this article if the Parties in question are not bound by a treaty of mutual legal assistance. If these Parties are bound by such a treaty, the corresponding provisions of that treaty shall apply unless the Parties agree to apply paragraph 8 to 19 of this article in lieu thereof.

8. Parties shall designate an authority, or when necessary authorities, which shall have the responsibility and power to execute requests for mutual legal assistance or to transmit them to the competent authorities for execution. The authority or the authorities designated for this purpose shall be notified to the Secretary-General. Transmission of requests for mutual legal assistance and any communication related thereto shall be effected between the authorities designated by the Parties; this requirement shall be without prejudice to the right of a Party to require that such requests and

communications be addressed to it through the diplomatic channel and, in urgent circumstances, where the Parties agree, through channels of the International Criminal Police Organisation, if possible.

9. Requests shall be made in writing in a language acceptable to the requested Party. The language or languages acceptable to each Party shall be notified to the Secretary-General. In urgent circumstances, and where agreed by the Parties, requests may be made orally, but shall be confirmed in writing forthwith.

10. A request for mutual legal assistance shall contain:

(a) The identity of the authority making the request;

(b) The subject matter and nature of the investigation, prosecution or proceeding to which the request relates, and the name and the functions of the authority conducting such investigation, prosecution or proceeding;

(c) A summary of the relevant facts, except in respect of requests for the purpose of service of judicial documents;

(d) A description of the assistance sought and details of any particular procedure the requesting Party wishes to be followed;

(e) Where possible, the identity, location and nationality of any person concerned;

(f) The purpose for which the evidence, information or action is sought.

11. The requested Party may request additional information when it appears necessary for the execution of the request in accordance with its domestic law or when it can facilitate such execution.

12. A request shall be executed in accordance with the domestic law of the requested Party and, to the extent not contrary to the domestic law of the requested Party and where possible, in accordance with the procedures specified in the request.

13. The requesting Party shall not transmit nor use information or evidence furnished by the requested Party for investigations, prosecutions or proceedings other than those stated in the request without the prior consent of the requested Party.

14. The requesting Party may require that the requested Party keep confidential the fact and substance of the request, except to the extent necessary to execute the request. If the requested Party cannot comply with the requirement of confidentiality, it shall promptly inform the requesting Party.

15. Mutual legal assistance may be refused :

(a) If the request is not made in conformity with the provisions of this article;

(b) If the requested Party considers that execution of the request is likely to prejudice its sovereignty, security, *ordre public* or other essential interests;

(c) If the authorities of the requested Party would be prohibited by its domestic law from carrying out the action requested with regard to any similar offence, had it been subject to investigation, prosecution or proceedings under their own jurisdiction;

(d) If it would be contrary to the legal system of the requested Party relating to mutual legal assistance for the request to be granted.

16. Reasons shall be given for any refusal of mutual legal assistance.

17. Mutual legal assistance may be postponed by the requested Party on the ground that it interferes with an ongoing investigation, prosecution or proceeding. In such a case, the requested Party shall consult with the requesting Party to determine if the assistance can still be given subject to such terms and conditions as the requested Party deems necessary.

18. A witness, expert or other person who consents to give evidence in a proceeding or to assist in an investigation, prosecution or judicial proceeding in the territory of the requesting Party, shall not be prosecuted, detained, punished or subjected to any other restriction of his personal liberty in that territory in respect of acts, omissions or convictions prior to his departure from the territory of the requested Party. Such safeconduct shall cease when the witness, expertor other person having had, for a period of fifteen consecutive days, or for any period agreed upon by the Parties, from the date on which he has been officially informed that his presence is no longer required by the judicial authorities, an opportunity of leaving, has nevertheless remained voluntarily in the territory or, having left it, has returned of his own free will.

19. The ordinary costs of executing a request shall be borne by the requested Party, unless otherwise agreed by the Parties concerned. If expenses of a substantial or extraordinary nature are or will be required to fulfil the request, the Parties shall consult to determine the terms and conditions under which the request will be executed as well as the manner in which the costs shall be borne.

20. The Parties shall consider, as may be necessary, the possibility of concluding bilateral or multilateral agreements or arrangements that would serve the purposes of, give practical effect to, or enhance the provisions of this article.

Article 8

Transfer of Proceedings

The Parties shall give consideration to the possibility of transferring to one another proceedings for criminal prosecution of offences established in accordance with Article 3, paragraph 1, in case where such transfer is considered to be in the interests of a proper administration of justice.

Article 9

Other Forms of Co-operation and Training

1. The Parties shall co-operate closely with one another, consistent with their respective domestic legal and administrative systems, with a view to enhancing the effectiveness of law enforcement action to suppress the commission of offences established in accordance with Article 3, paragraph 1. They shall, in particular, on the basis of bilateral or multilateral agreements or arrangements:

(a) Establish and maintain channels of communication between their competent agencies and services to facilitate the secure and rapid exchange of information concerning all aspects of offences established in accordance with Article 3, paragraph 1, including, if the Parties concerned deem it appropriate, links with other criminal activities;

(b) Co-operate with one another in conducting enquiries, with respect to offences established in accordance with Article 3, paragraph 1, having an international character, concerning :

(i) The identity, whereabouts and activities of persons suspected of being involved in offences established in accordance with Article 3, paragraph 1;

(ii) The movement of proceeds or property derived from the commission of such offences;

(iii) The movement of narcotic drugs, psychotropic substances, substances in Table I and Table II of this Convention and instrumentalities used or intended for use in the commission of such offences;

(c) In appropriate cases and if not contrary to domestic law, establish joint teams, taking into account the need to protect the security of persons and of operations, to carry out the provisions of this paragraph. Officials of any Party taking part in such teams shall act as authorised by the appropriate authorities of the Party in whose territory the operation is to take place; in all such cases, the Parties involved shall ensure that the sovereignty of the Party on whose territory the operaton is to take place is fully respected;

(d) Provide, when appropriate, necessary quantities of substances for analytical or investigative purposes;

(e) Facilitate effective co-ordination between their competent agencies and services and promote the exchange of personnel and other experts, including the posting of liaison officers.

2. Each Party shall, to the extent necessary initiate, develop or improve specific training programmes for its law enforcement and other personnel, including customs, charged with the suppression of offences established in accordance with Article 3, paragraph 1. Such programmes shall deal, in particular, with the following :

(a) Methods used in the detection and suppression of offences established in accordance with Article 3, paragraph 1;

(b) Routes and techniques used by persons suspected of being involved in offences established in accordance with Article 3, paragraph 1, particularly in transit States, and appropriate conter measures;

(c) Monitoring of the import and export of narcotic drugs, psychotropic substances and substances in Table I and Table II;

(d) Detection and monitoring of the movement of proceeds and property derived from, and narcotic drugs, psychotropic substances and substances in Table I and Table II, and instrumentalities used or intended for use in, the commission of offences established in accordance with Article 3, paragraph 1;

(e) Methods used for the transfer, concealment or disguise of such proceeds, property and instrumentalities;

(f) Collection of evidence;

(g) Control techniques in free trade zones and free ports;

(h) Modern law enforcement techniques.

3. The Parties shall assist one another to plan and implement research and training programmes designed to share expertise in the areas referred to in paragraph 2 of this article and, to this end, shall also, when appropriate, use regional and international conferences and seminars to promote co-operation and stimulate discussion on problems of mutual concern, including the special problems and needs of transit States.

Article 10

International Co-operation and Assistance for Transit States

1. The Parties shall co-operate, directly or through competent international or regional organisations, to assist and support transit States and, in particular, devel-

oping countries in need of such assistance and support, to the extent possible, through programmes of technical co-operation on interdiction and other related activities.

2. The Parties may undertake, directly or through competent international or regional organisations, to provide financial assistance to such transit States for the purpose of augmenting and strengthening the infrastructure needed for effective control and prevention of illicit traffic.

3. The Parties may conclude bilateral or multilateral agreements or arrangements to enhance the effectiveness of international co-operation pursuant to this article and may take into consideration financial arrangements in this regard.

Article 11

Controlled Delivery

1. If permitted by the basic principles of their respective domestic legal systems, the Parties shall take the necessary measures, within their possibilities, to allow for the appropriate use of controlled delivery at the international level, on the basis of agreements or arrangements mutually consented to, with a view to identifying persons involved in offences established in accordance with Article 3, paragraph 1, and to taking legal action against them.

2. Decisions to use controlled delivery shall be made on a case-by-case basis and may, when necessary, take into consideration financial arrangements and understandings with respect to the exercise of jurisdiction by the Parties concerned.

3. Illicit consignments whose controlled delivery is agreed to may, with the consent of the Parties concerned, be intercepted and allowed to continue with the Narcotic Drugs or Psychotropic Substances intact or removed or replaced in whole or in part.

Article 12

Substances Frequently Used in the Illicit Manufacture of Narcotic Drugs or Psychotropic Substances

1. The Parties shall take the measures they deem appropriate to prevent diversion of substances, in Table I and Table II, used for the purpose of illicit manufacture of Narcotic Drugs or Psychotropic Substances, and shall co-operate with one another to this end.

2. If a Party or the Board has information which in its opinion may require the inclusion of a substance in Table I or Table II, it shall notify the Secretary-General and furnish him with the information in support of that notification. The procedure described in paragraphs 2 to 7 of this article shall also apply when a Party or the Board has information justifying the deletion of a substance from Table I or Table II, or the transfer of a substance from one Table to the other.

3. The Secretary-General shall transmit such notification, and any information which he considers relevant, to the Parties, to the Commission, and, where notification is made a Party, to the Board. The Parties shall communicate their comments concerning the notification to the Secretary-General, together with all supplementary information which may assist the Board in establishing an assessment and the Commission in reaching a decision.

4. If the Board, taking into account the extent, importance and diversity of the licit use of the substance, and the possibility and ease of using alternate substances both for licit purposes and for the illicit manufacture of Narcotic Drugs or Psychotropic Substances, finds :

(a) That the substance is frequently used in the illicit manufacture of a Narcotic Drug or Psychotropic Substance;

(b) That the volume and extent of the illicit manufacture of a Narcotic Drug or Psychotropic Substance creates serious public health or social problems, so as to warrant international action, it shall communicate to the Commission an assessment of the substance, including the likely effect of adding the substance to either Table I or Table II on both licit use and illicit manufacture, together with recommendations of monitoring measures, if any, that would be appropriate in the light of its assessment.

5. The Commission, taking into account the comments submitted by the Parties and the comments and recommendations of the Board, whose assessment shall be determinative as to scientific matters, and also taking into due consideration any other relevant factors, may decide by a two-thirds majority of its members to place a substance in Table I or Table II.

6. Any decision of the Commission taken pursuant to this article shall be communicated by the Secretary-General to all States and other entities which are, or which are entitled to become, Parties to this Convention, and to the Board. Such decision shall become fully effective with respect to each Party one hundred and eighty days after the date of such communication.

7. (a) The decisions of the Commission taken under this article shall be subject to review by the Council upon the request of any Party filed within one hundred and eighty days after the date of notification of the decision. The request for review shall be sent to the Secretary-General, together with all relevant information upon which the request for review is based.

(b) The Secretary-General shall transmit copies of the request for review and the relevant information to the Commission, to the Board and to all the Parties, inviting them to submit their comments within ninety days. All comments received shall be submitted to the Council for consideration.

(c) The Council may confirm or reverse the decision of the Commission. Notification of the Council's decision shall be transmitted to all States and other entities which are, or which are entitled to become, Parties to this Convention, to the Commission and to the Board.

8. (a) Without prejudice to the generality of the provisions contained in paragraph 1 of this article and the provisions of the 1961 Convention, the 1961 Convention as amended and the 1971 Convention, the Parties shall take the measures they deem appropriate to monitor the manufacture and distribution of substances in Table I and Table II which are carried out within their territory.

(b) To this end, the Parties may :

(i) Control all persons and enterprises engaged in the manufacture and distribution of such substances;

(ii) Control under licence the establishment and premises in which such manufacture or distribution may take place;

(iii) Require that licensees obtain a permit for conducting the aforesaid operations;

(iv) Prevent the accumulation of such substances in the possession of manufacturers and distributors, in excess of the quantities required for the normal conduct of business and the prevailing market conditions.

9. Each Party shall, with respect to substances in Table I and Table II, take the following measures :

(a) Establish and maintain a system to monitor international trade in substances in Table I and Table II in order to facilitate the identification of suspicious transactions. Such monitoring systems shall be applied in close co-operation with manufacturers, importers, exporters, wholesalers and retailers, who shall inform the competent authorities of suspicious orders and transactions.

(b) Provide for the seizure of any substance in Table I or Table II if there is sufficient evidence that it is for use in the illicit manufacture of a Narcotic Drug or Psychotropic Substance.

(c) Notify, as soon as possible, the competent authorities and services of the Parties concerned if there is reason to believe that the import, export or transit of a substance in Table I or Table II is destined for the illicit manufacture of Narcotic Drugs or Psychotropic Substances, including in particular information about the means of payment and any other essential elements which led to that belief.

(d) Require that imports and exports be properly labelled and documented. Commercial documents such as invoices, cargo manifests, customs, transport and other shipping documents shall include the names, as stated in Table I or Table II, of the substances being imported or exported, the quantity being imported or exported, and the name and address of the exporter, the importer and, when available, the consignee.

(e) Ensure that documents referred to in sub-paragraph (d) of this paragraph are maintained for a period of not less than two years and may be made available for inspection by the competent authorities.

10. (a) In addition to the provisions of paragraph 9, and upon request to the Secretary-General by the interested Party, each Party from whose territory a substance in Table I is to be exported shall ensure that, prior to such export, the following information is supplied by its competent authorities to the competent authorities of the importing country:

(i) Name and address of the exporter and importer and, when available, the consignee;

(ii) Name of the substance in Table I;

(iii) Quantity of the substance to be exported;

(iv) Expected point of entry and expected date of despatch;

(v) Any other information which is mutually agreed upon by the Parties;

(b) A Party may adopt more strict or severe measures of control than those provided by this paragraph if, in its opinion, such measures are desirable or necessary.

11. Where a Party furnishes informations to another Party in accordance with paragraphs 9 and 10 of this article, the Party furnishing such information may require that the Party receiving it keep confidential any trade, business, commercial or professional secret or trade process.

12. Each Party shall furnish annually to the Board, in the form and manner provided for by it and on forms made available by it, information on:

(a) The amounts seized of substances in Table I and Table II and, when known, their origin;

(b) Any substance not included in Table I or Table II which is identified as having been used in illicit manufacture of Narcotic Drugs and Psychotropic Substances, and which is deemed by the Party to be sufficiently significant to be brought to the attention of the Board;

(c) Methods of diversion and illicit manufacture.

13. The Board shall report annually to the Commis-

sion on the implementation of this article and the Commission shall periodically review the adequacy and propriety of Table I and Table II.

14. The provisions of this article shall not apply to pharmaceutical preparations, nor to other preparations containing substances in Table I or Table II that are compounded in such a way that such substances cannot be easily used or recovered by readily applicable means.

Article 13

Materials and Equipment

The Parties shall take such measures as they deem appropriate to prevent trade in and the diversion of materials and equipment for illicit production or manufacture of Narcotic Drugs and Psychotropic Substances and shall co-operate to this end.

Article 14

Measures to Eradicate Illicit Cultivation of Narcotic Plants and to Eliminate Illicit Demand for Narcotic Drugs and Psychotropic Substances

1. Any measures taken pursuant to this Convention by Parties shall not be less stringent than the provisions applicable to the eradication of illicit cultivation of plants containing narcotic and psychotropic substances and to the elimination of illicit demand for Narcotic Drugs and Psychotropic Substances under the provisions of the 1961 Convention, the 1961 Convention as amended and the 1971 Convention.

2. Each Party shall take appropriate measures to prevent illicit cultivation of and to eradicate plants containing narcotic or psychotropic substances, such as opium poppy, coca bush and cannabis plants, cultivated illicitly in its territory. The measures adopted shall respect fundamental human rights and shall take due account of traditional licit uses, where there is historic evidence of such use, as well as the protection of the environment.

3. (a) The Parties may co-operate to increase the effectiveness of eradication efforts. Such co-operation may, *inter alia*, include support, when appropriate, for integrated rural development leading to economically viable alternatives to illicit cultivation. Factors such as access to markets, the availability of resources and prevailing socio-economic conditions should be taken into account before such rural development programmes are implemented. The Parties may agree on any other appropriate measures of co-operation.

(b) The Parties shall also facilitate the exchange of scientific and technical information and the conduct of research concerning eradication.

(c) Whenever they have common frontiers, the Parties shall seek to co-operate in eradication programmes in their respective areas along those frontiers.

4. The Parties shall adopt appropriate measures aimed at eliminating or reducing illicit demand for Narcotic Drugs and Psychotropic Substances, with a view to reducing human suffering and eliminating financial incentives for illicit traffic. These measures may be based, *inter alia*, on the recommendations of the United Nations, specialised agencies of the United Nations such as the World Health Organisations, and other competent international organisations, and on the Comprehensive Multidisciplinary Outline adopted by the International Conference on Drug Abuse and Illicit Trafficking, held in 1987, as it pertains to governmental and non-governmental agencies and private efforts in the fields of prevention, treatment and rehabilitation. The Parties may enter into bilateral or multilateral agreements or arrangements aimed at eliminating or reducing illicit demand for Narcotic Drugs and Psychotropic Substances.

5. The Parties may also take necessary measures for early destruction or lawful disposal of the narcotic drugs, psychotropic substances and substances in Table I and Table II which have been seized or confiscated and for the admissibility as evidence of duly certified necessary quantities of such substances.

Article 15

Commerical Carriers

1. The Parties shall take appropriate measures to ensure that means of transport operated by commercial carriers are not used in the commission of offences established in accordance with Article 3, paragraph 1; such measures may include special arrangements with commercial carriers.

2. Each Party shall require commercial carriers to take reasonable precautions to prevent the use of their means of transport for the commission of offences established in accordance with Article 3, paragraph 1. Such precautions may include:

(a) If the principal place of business of a commercial carrier is within the territory of the Party:

(i) Training of personnel to identify suspicious consignments or persons;

(ii) Promotion of integrity of personnel;

(b) If a commercial carrier is operating within the territory of the Party:

(i) Submission of cargo manifests in advance, whenever possible;

(ii) Use of tamper-resistant, individually verifiable seals on containers;

(iii) Reporting to the appropriate authorities at the earliest opportunity all suspicious circumstances that may be related to the commission of offences established in accordance with Article 3, paragraph 1.

3. Each Party shall seek to ensure that commercial carriers and the appropriate authorities at points of entry and exit and other customs control areas co-operate, with a view to preventing unauthorised access to means of transport and cargo and to implementing appropriate security measures.

Article 16

Commercial Documents and Labelling of Exports

1. Each Party shall require that lawful exports of Narcotic Drugs and Psychotropic Substances be properly documented. In addition to the requirements for documentation under Article 31 of the 1961 Convention, Article 31 of the 1961 Convention as amended and Article 12 of the 1971 Convention, commercial documents such as invoices, cargo manifests, customs, transport and other shipping documents shall include the names of the Narcotic Drugs and Psychotropic Substances being exported as set out in the respective Schedules of the 1961 Convention, the 1961 Convention as amended and the 1971 Convention, the quantity being exported, and the name and address of the exporter, the importer and, when available, the consignee.

2. Each Party shall require that consignments of Narcotic Drugs and Psychotropic Substances being exported be not mislabelled.

Article 17

Illicit Traffic by Sea

1. The Parties shall co-operate to the fullest extent possible to suppress illicit traffic by sea, in confirmity with the international law of the sea.

2. A Party which has reasonable grounds to suspect that a vessel flying its flag or marks of registry is engaged in illicit traffic may request the assistance of other Parties in suppressing its use for that purpose. The Parties so requested shall render such assistance within the means available to them.

3. A Party which has reasonable grounds to suspect that a vessel exercising freedom of navigation in accordance with international law and flying the flag or displaying marks of registry of another Party is engaged in illicit traffic may so notify the flag State, request confirmation of registry and, if confirmed, request authorisation from the flag State to take appropriate measures in regard to that vessel.

4. In accordance with paragraph 3 or in accordance with treaties in force between them or in accordance with any agreement or arrangement otherwise reached between those Parties, the flag State may authorize the requesting State to, *inter alia:*

(a) Board the vessel;

(b) Search the vessel;

(c) If evidence of involvement in illicit traffic is found, take appropriate action with respect to the vessel, persons and cargo on board.

5. Where action is taken pursuant to this article, the Parties concerned shall take due account of the need not to endanger the safety of life at sea, the security of the vessel and the cargo or to prejudice the commercial and legal interests of the flag State or any other interested State.

6. The flag State may, consistent with its obligations in paragraph 1 of this article, subject its authorisation to conditions to be mutually agreed between it and the requesting Party, including conditions relating to responsibility.

7. For the purposes of paragraphs 3 and 4 of this article, a Party shall respond expeditiously to a request from another Party to determine whether a vessel that is flying its flag is entitled to do so, and to requests for authorisation made pursuant to paragraph 3. At the time of becoming a Party to this Convention, each Party shall designate an authority or, when necessary, authorities to receive and respond to such requests. Such designation shall be notified through the Secretary-General to all other Parties within one month of the designation.

8. A Party which has taken any action in accordance with this article shall inform the flag State concerned of the results of that action.

9. The Parties shall consider entering into bilateral or regional agreements or arrangements to carry out, or to enhance the effectiveness of, the provisions of this article.

10. Action pursuant to paragraph 4 of this article shall be carried out only by warships or military aircraft, or other ships or aircraft clearly marked and identifiable as being on government service and authorised to that effect.

11. Any action taken in accordance with this article shall take due account of the need not to interfere with or affect the rights and obligations and the exercise of jurisdiction of coastal States in accordance with the international law of the sea.

Article 18

Free Trade Zones and Free Ports

1. The Parties shall apply measures to suppress illicit traffic in narcotic drugs, psychotropic substances and substances in Table I and Table II in free trade

zones and in free ports that are no less stringent than those applied in other parts of their territories.

2. The Parties shall endeavour:

(a) To monitor the movement of goods and persons in free trade zones and free ports, and, to that end, shall empower the competent authorities to search cargoes and incoming and outgoing vessels, including pleasure craft and fishing vessels, as well as aircraft and vehicles and, when appropriate, to search crew members, passengers and their baggage;

(b) To establish and maintain a system to detect consignments suspected of containing narcotic drugs, psychotropic substances and substances in Table I and Table II passing into or out of free trade zones and free ports;

(c) To establish and maintain surveillance systems in harbour and dock areas and at airports and border control points in free zones and free ports.

Article 19

The Use of the Mails

1. In conformity with their obligations under the Conventions of the Universal Postal Union, and in accordance with the basic principles of their domestic legal systems, the Parties shall adopt measures to suppress the use of the mails for illicit traffic and shall co-operate with one another to that end.

2. The measures referred to in paragraph 1 of this article shall include, in particular:

(a) Co-ordinated action for the prevention and repression of the use of the mails for illicit traffic;

(b) Introduction and maintenance by authorised law enforcement personnel of investigative and control techniques designated to detect illicit consignments of narcotic drugs, psychotropic substances and substances in Table I and Table II in the mails;

(c) Legislative measures to enable the use of appropriate means to secure evidence required for judicial proceedings.

Article 20

Information to be Furnished by the Parties

1. The Parties shall furnish, through the Secretary-General, information to the Commission on the working of this Convention in their territories and, in particular:

(a) The text of laws and regulations promulgated in order to give effect to the Convention;

(b) Particulars of cases of illicit traffic within their jurisdiction which they consider important because of new trends disclosed, the quantities involved, the sources from which the substances are obtained, or the methods employed by persons so engaged.

2. The Parties shall furnish such information in such a manner and by such dates as the Commission may request.

Article 21

Functions of the Commission

The Commission is authorised to consider all matters pertaining to the aims of this Convention and, in particular:

(a) The Commission shall, on the basis of the information submitted by the Parties in accordance with Article 20, review the operation of this Convention;

(b) The Commission may make suggestions and general recommendations based on the examination of the information received from the Parties;

(c) The Commission may call the attention of the Board to any matters which may be relevant to the functions of the Board;

(d) The Commission shall, on any matter referred to it by the Board under Article 22, paragraph 1 (b), take such action as it deems appropriate;

(e) The Commission may, in conformity with the procedures laid down in Article 12, amend Table I and Table II;

(f) The Commission may draw the attention of non-Parties to decisions and recommendations which it adopts under this Convention, with a view to their considering taking action in accordance therewith.

Article 22

Functions of the Board

1. Without prejudice to the functions of the Commission under Article 21, and without prejudice to the functions of the Board and the Commission under the 1961 Convention, the 1961 Convention as amended and the 1971 Convention:

(a) If, on the basis of its examination of information available to it, to the Secretary-General or to the Commission, or of information communicated by United Nations organs, the Board has reason to believe that the aims of this Convention in matters related to its competence are not being met, the Board may invite a Party or Parties to furnish any relevant information;

(b) With respect to Articles 12, 13 and 16:

(i) After taking action under sub-paragraph (a) of this article, the Board, if satisfied that it is necessary to do so, may call upon the Party concerned to adopt such remedial measures as shall seem under the circumstances to be necessary for the execution of the provisions of articles 12, 13 and 16;

(ii) Prior to taking action under (iii) below, the Board shall treat as confidential its communications with the Party concerned under the preceding sub-paragraphs;

(iii) If the Board finds that the Party concerned has not taken remedial measures which it has been called upon to take under this sub-paragraph, it may call the attention of the Parties, the Council and the Commission to the matter. Any report published by the Board under this sub-paragraph shall also contain the views of the Party concerned if the latter so requests.

2. Any Party shall be invited to be represented at a meeting of the Board at which a question of direct interest to it is to be considered under this article.

3. If in any case a decision of the Board which is adopted under this article is not unanimous, the views of the minority shall be stated.

4. Decisions of the Board under this article shall be taken by a two-thirds majority of the whole number of the Board.

5. In carrying out its functions pursuant to sub-paragraph 1 (a) of this article, the Board shall ensure the confidentiality of all information which may come into its possession.

6. The Board's responsibility under this article shall not apply to the implementation of treaties or agreements entered into between Parties in accordance with the provisions of this Convention.

7. The provisions of this article shall not be applicable to disputes between Parties falling under the provisions of Article 32.

Article 23

Reports of the Board

1. The Board shall prepare an annual report on its work containing an analysis of the information at its disposal and, in appropriate cases, an account of the explanations, if any, given by or required of Parties, together with any observations and recommendations which the Board desires to make. The Board may make such additional reports as it considers necessary. The reports shall be submitted to the Council through the Commission which may take such comments as it sees fit.

2. The reports of the Board shall be communicated to the Parties and subsequently published by the Secretary-General. The Parties shall permit their unrestricted distribution.

Article 24

Application of Stricter Measures than those Required by this Convention

A Party may adopt more strict or severe measures than those provided by this Convention if, in its opinion, such measures are desirable or necessary for the prevention or suppression of illicit traffic.

Article 25

Non-derogation from Earlier Treaty Rights and Obligations

The provisions of this Convention shall not derogate from any rights enjoyed or obligations undertaken by Parties to this Convention under the 1961 Convention, the 1961 Convention as amended and the 1971 Convention.

Article 26

Signature

This Convention shall be open for signature at the United Nations Office at Vienna, from 20 December 1988 to 28 February 1989, and thereafter at the Headquarters of the United Nations at New York, until 20 December 1989, by:

(a) All States;

(b) Namibia, represented by the United Nations Council for Namibia;

(c) Regional economic integration organisations which have competence in respect of the negotiation, conclusion and application of international agreements in matters covered by this Convention, references under the Convention to Parties, States or national services being applicable to these organisations within the limits of their competence.

Article 27

Ratification, Acceptance, Approval or Act of Formal Confirmation

1. The Convention is subject to ratification, acceptance or approval by States and by Namibia, represented by the United Nations Council for Namibia, and to acts of formal confirmation by regional economic integration organisations referred to in Article 26, sub-paragraph (c). The instruments of ratification, accep-

tance or approval and those relating to acts of formal confirmation shall be deposited with the Secretary-General.

2. In their instruments of formal confirmation, regional economic integration organisations shall declare the extent of their competence with respect to the matters governed by this Convention. These organisations shall also inform the Secretary-General of any modification in the extent of their competence with respect to the matters governed by the Convention.

Article 28

Accession

1. This Convention shall remain open for accession by any State, by Namibia, represented by the United Nations Council for Namibia, and by regional economic integration organisations referred to in Article 26, sub-paragraph (c).

Accession shall be effected by the deposit of an instrument of accession with the Secretary-General.

2. In their instruments of accession, regional economic integration organisations shall declare the extent of their competence with respect to the matters governed by this Convention. These organisations shall also inform the Secretary-General of any modification in the extent of their competence with respect to the matters governed by this Convention.

Article 29

Entry into Force

1. This Convention shall enter into force on the ninetieth day after the date of the deposit with the Secretary-General of the twentieth instrument of ratification, acceptance, approval or accession by States or by Namibia, represented by the Council of Namibia.

2. For each State or for Namibia, represented by the Council for Namibia, ratifying, accepting, approving or acceding to this Convention after the deposit of the twentieth instrument of ratification, acceptance, approval or accession, the Convention shall enter into force on the ninetieth day after the date of the deposit of its instruments of ratification, acceptance, approval or accession.

3. For each regional economic integration organisation referred to in Article 26, sub-paragraph (c) depositing an instrument relating to an act of formal confirmation or an instrument of accession, this Convention shall enter into force on the ninetieth day after such deposit, or at the date the Convention enters into force pursuant to paragraph 1 of this article, whichever is later.

Article 30

Denunciation

1. A Party may denounce this Convention at any time by a written notification addressed to the Secretary-General.

2. Such denunciation shall take effect for the Party concerned one year after the date of receipt of the notification by the Secretary-General.

Article 31

Amendments

1. Any Party may propose an amendment to this Convention. The text of any such amendment and the reasons therefor shall be communicated by that Party to the Secretary-General, who shall communicate it to the other Parties and shall ask them whether they accept the proposed amendment. If a proposed amendment so circulated has not been rejected by any Party within twenty-four months after it has been circulated, it shall be deemed to have been accepted and shall enter into force in respect of a Party ninety days after that Party has deposited with the Secretary-General an instrument expressing its consent to be bound by that amendment.

2. If a proposed amendment has been rejected by any Party, the Secretary-General shall consult with the Parties and, if a majority so requests, he shall bring the matter, together with any comments made by the Parties, before the Council which may decide to call a conference in accordance with Article 62, paragraph 4, of the Charter of the United Nations. Any amendment resulting from such a Conference shall be embodied in a Protocol of Amendment. Consent to be bound by such a Protocol shall be required to be expressed specifically to the Secretary-General.

Article 32

Settlement of Disputes

1. If there should arise between two or more Parties, a dispute relating to the interpretation or application of this Convention, the Parties shall consult together with a view to the settlement of the dispute by negotiation, enquiry, mediation, conciliation, arbitration, recourse to regional bodies, judicial process or other peaceful means of their own choice.

2. Any such dispute which cannot be settled in the manner prescribed in paragraph 1 of this article shall be referred, at the request of any one of the States Parties to the dispute, to the International Court of Justice for decision.

3. If a regional economic integration organisation referred to in Article 26, sub-paragraph (c) is a Party to a dispute which cannot be settled in the manner prescribed in paragraph 1 of this article, it may, through a State Member of the United Nations, request the Council to request an advisory opinion of the International Court of Justice in accordance with Article 65 of the Statute of the Court, which opinion shall be regarded as

decisive.

4. Each State, at the time of signature or ratification, acceptance or approval of this Convention or accession thereto, or eash regional economic integration organisation, at the time of signature or deposit of an act of formal confirmation or accession, may declare that it does not consider itself bound by paragraphs 2 and 3 of this article. The other Parties shall not be bound by paragraphs 2 and 3 with respect to any Party having made such a declaration.

5. Any Party having made a declaration in accordance with paragraph 4 of this article may at any time withdraw the declaration by notification to the Secretary-General.

Article 33

Authentic Texts

The Arabic, Chinese, English, French, Russian and Spanish texts of this Convention are equally authentic.

Article 34

Depositary

The Secretary-General shall be the depositary of this Convention.

IN WITNESS WHEREOF the undersigned, being duly authorised thereto, have signed this Convention.

DONE AT VIENNA, in one original, this twentieth day of December one thousand nine hundred and eighty-eight.

Annex

Table I	*Table II*
Ephedrine	Acetic anhydride
Ergometrine	Acetone
Ergotamine	Anthranilic acid
Lysergic acid	Ethyl ether
1-phenyl-2 propanone	Phenylacetic acid
Pseudoephedrine	Piperidine
The salts of the substances listed in this Table whenever the existence of such salts is possible.	The salts of the substances listed in this Table whenever the existence of such salts is possible.

63

GLOBAL CONTROL MECHANISM

The narcotic drugs have had a unique place in medical science to relieve human suffering and ailments. The ambivalent character of these substances being indispensable for relief of pain and suffering but addictive and destructive when misused or abused led the societies from the earliest times to provide for regulatory mechanism for restricting their availability to medical or religious purposes. Since the mid-nineteenth century, drug abuse has been spreading in many a country as a result of many factors including, in particular, the increased availability of the narcotic drugs; the expasnion of communications, the socio-economic changes, migration and rapid urbanization, erosion in moral and religious values and above all, ruthless exploitation of humanity by hardened criminals for their quick personal gains.

In order to deal with the increasing drug problem, the community of nations have, since the early 20th Century, evolved a global control mechanism intended to limit the availability of drugs to medical and scientific needs because it was realised that no single country, howsoever resourceful it might be, could succeed alone in this Herculean endeavour of combating drug menace. In the succeeding paragraphs, the historical developments for global control on illicit traffic is briefly recapitulated.

It may be noted that in the 18th Century, the opium poppy and the coca leaf were recognised as the main commodities for the purpose of trade and revenue in the producing countries. The exportation of opium into China by foreign traders is said to have led to war of 1840 between Great Britain and that country. Despite the opposition, China became the epicentre of attraction for the international smuggling rackets. By the end of the second opium war between China and Great Britain in 1855, not only was the cultivation of opium in China permitted but its import from India was also legalised. The policy of unrestricted production, distribution and sale of opium by the Britishers in India and, its free export to China invited criticism in the international circles. As a result and at the behest of the Chinse administration, in 1906 a ten years' agreement between China and the British Government was reached during which period, the cultivation of opium poppy and its use was to be prohibited in China and its import from India reduced in equal proportion so as to cease completely by the end of 10 years.

Early Efforts to Curb Trading in Narcotics

It was about this time that the international community became conscious of the threat posed by the traffic in drugs to the health and the well-being of the people. Public opinion demanded that sale, distribution and trade in narcotic drugs could, no longer, be viewed as a legitimate commercial activity, free Government control, but should be confined within permissible limits for medicinal and scientific use only under multilateral international agreements.

The first step in the direction of international cooperation in narcotics control was taken in 1909 when, on the initiative of the United States, 13 countries with interests in the Far-East, appointed an Opium Commission which met in Shanghai and adopted a series of resolutions regarding the control of traffic in opium. This conference ultimately, led to the building up of a system of control over narcotic drugs to be carried out by several international agencies entrusted with its policy-making, administrative and semi-judicial functions. The first organ of international control machinery was the "League of Nations Advisory Committee on the Traffic in Opium and other Dangerous Drugs." This Committee was the harbinger for a series of future conventions with a view to develop and tighten the system.

The first International Opium Convention was signed at The Hague on 23rd Jaunary, 1912. It formulated the basic principles of international control which have retained their validity to the present day. During the intervening half-century, an impressive body of international treaty law had gradually developed: between 1912 and 1953 negotiated in quick succession, the following 9 international instruments were adopted, put to the test of implementation and successfully modified in the light of experience gained:-

1912 Convention:	International Opium Convention, signed at The Hague (23.1.1912).
1925 Convention:	International Opium Convention, signed at Geneva (19.2.1925).
1931 Convention:	International Convention for Limiting the Manufacture and Regulating the Distribution of Narcotic Drugs, signed at Geneva (13.7.1931).
1936 Convention:	The Convention of 1936 for the Suppression of the Illicit Traffic in Dangerous Drugs.
1925 Agreement:	Agreement concerning the manufacture of internal trade in and use of prepared opium, signed at Geneva (11.2.1925).
1931 Agreement;	Agreement for the control of opium smoking in the Far East signed at Bangkok (27.11.1931).
1946 Agreement:	Protocol of 1946 Amending the Agreements, Conventions and Protocols on Narcotic Drugs, concluded at The Hague on 23rd January, 1921; at Geneva (11.2.1925) (19.2.1925 and 13.3.1931); at Bangkok (27.11.1931), and at Geneva (26.6.1936).
1946 Protocol:	Protocol signed at Paris (19.11.1948) bringing under international control drugs outside the scope of the convention of 13th July, 1931 for limiting the manufacture and regulating the distribution of narcotic drugs, as amended by Protocol signed at Lake Success (11.12.1946).
1953 Protocol:	Protocol for limiting and regulating the cultivation of the poppy plant, the production of international and wholesale trade in and use of opium signed at New York (23.6.1953).

Since the main drugs of abuse, in the first half of this century were opium, morphine and cocaine, the international treaties adopted during this period were designed to bring about production of these drugs and their derivatives to the level of the world's medical and scientific needs and in meeting this objective to suppress illicit traffic therein. These also provide for review and monitoring of the implementation by the competent international organ, viz. Permanent Central Opium Board, the predecessor of the International Narcotics Control Board.

Early Efforts

One of the first known attempts in Europe to control drugs was made in France in 1845. Later in Britain the 1868 Pharmacy and Poisons Act gave a small measure of control over opium and its preparations. The 1908 Pharmacy and Poisons Act amended the 1868 Act and part I of the new schedule included "opium and all preparations or admixtures containing 2 per cent or more of morphine". Also for the time, sales of cocaine was restricted by the inclusion of 'Coca, any preparations or admixtures of containing 1 per cent or more of coca alkaloids'. Within a decade controls on international level had begun, culmination in 1920 with Britain's first Dangerous Act.

Britain's Role

The Society for the Study of Addiction to Alcohol and Other Drugs was founded in London in 1884 and has been studying scientific problems related to addiction to drugs and alcoholism for the eighty-four years since its foundation. Dr. Normann Kerr was the joint president from 1884 to 1899. The British journal of addiction is the official journal of the Society. Dr. J. T. Kelyanack was appointed as its first editor in 1903. Dr. J. Y. Dent who was well known for his work on the treatment of alcoholism with aphomorphine, was editor of the journal from 1941 to 1961.

The important part Britain played in the system of international control was partly related to her position as a leading world power in 1900 with interests in the Far-East. The British moral conscience had also been stirred by the opium wars with China. Sir Harry Greenfield cites the instance of a British official Howard Becker writing to Palmerston as far back as 1837 deprecating the "Vast prohibited traffic in a vicious luxury". In 1893 the concern in Britain about drug taking in India was reflected by the appointment of a Royal Commissioner to review the position of opium and cannabis smoking on the Indian sub-continent. In 1906 the House of Commons passed a resolution without a division reaffirming its belief that "the opium trade between India and China is morally indefensible."

In 1906 China again banned the cultivation of the opium poppy. Britain in 1908 agreed to reduce exports from India for an experimental period of three years, provided China also reduced imports from other countries. It was in this spirit "to help the Chinese" for which a meeting was held in the Palace Hotel in Shanghai in 1909. The Shanghai conference consisted of the representative of 13 countries including Britain, China, Russia, Japan, France and the U.S.A. They had no power to sign a diplomatic act, but unanimously adopted 9 resolutions, which are summarisd below:—

(i) The use of opium in any form other than for medical purposes was held by almost every participating country to be matter for prohibition.

(ii) Drastic measures should be taken by each government to institute proper control over opium and any derivatives which appeared to produce similar effects and were liable to produce similar abuse.

(iii) It was to be the duty of all countries to adopt reasonable measures to prevent the shipment of opium and its alkaloids to any country which prevented its entry.

(iv) All governments possessing concessions or settlements in China which had not yet taken effective action towards the closing of opium ought to take this step as soon as possible.

The efforts of Shanghai delegates succeeded in arousing strong International Opium Convention held at The Hague in 1912. This differed from the Shanghai Conference as it led to an international treaty rather than being merely a general conference and the aim was "to advance a step further on the road opened by the International Commission of Shanghai in 1909; determined to bring about the gradual suppression of the abuse of opium, morphine and cocaine". It was accepted that the production of raw opium should be controlled and its distribution regulated. The manufacture and use of morphine, cocaine and their respective salts was to be limited to legitimate medical purposes and each party was to use its best endeavours to control or cause to be controlled the manufacture, import and export of these drugs by restricting their use by authorised persons.

In 1914, as part of the general war measures, a restriction was placed on the export of opium. In 1916 it became apparent that cocaine was being peddled to and by prostitutes in London, a number of cases were reported to the police of it being given to Canadian soldiers. In 1916, the Commissioner of London Police wrote that it was necessary 'to stamp out the evil now rapidly assuming huge dimensions'. Eight years later, Britain passed the first Dangerous Drugs Act on the 1st September, 1920. The Act itself which is short, containing only 17 sections, sticks to the convention norms.

Initiative by United States

In the 1920s all the European powers with Far Eastern possessions in which opium smoking was permitted were assailed by protectionist-minded critics in the United States. The European powers retaliated as best as they could by pressing the United States to report on its own prohibition scheme in the Philippines or by blaming China for producing too much. Japan was criticised in the late 1930s for sending large quantities of opium into China which was later referred to by the Chinese government "as a form of genocide". After World War II, when opium smoking was banned in the Far East, Thailand was blamed for not following suit and thereby nullifying the efforts of Thailand's neighbours to solve their own problems. China was upto 1956, regularly accused by the United States and Formosa of deliberately encouraging the production of opium for political ends.

In the United States in 1914, two years after The Hague convention, the Government introduced the Harrison Act. From 1920 to 1970 international control of Narcotics continued to develop. First under the aegis of the League of Nations, and later under the United Nations. The reluctance of the members to ratify. The Hague convention continued even after the World War I (1914-1918), "the experience was in fact so disappointing that at the end of the war, the victorious powers decided vigorously to underwrite the international campaign against narcotics abuse.[1]

International Efforts Through League of Nations

Further, steps were taken, Firstly, a provision was inserted into the peace treaties which automatically brought the convention into force between the signatories of the Treaty of Versalles, Article 295 of the Treaty states "ratification of the peace treaty should be deemed in all respects equivalent to ratification of the (Hague) convention; secondly, a special provision was also included in the covenant of the League of Nations, and Article 23 states that the members of the League of Nations will entrust the league with the general supervision over the execution of agreements with regard to...the traffic of opium and other dangerous drugs".

On 15th December 1920, at the first Assembly of the League of Nations, a resolution by the British Government was passed creating an Advisory Committee with powers to exercise a general supervision over the traffic of opium and the dangerous drugs and to secure full co-operation of the various countries in the field. At the first meeting of the Committee in May 1921, a recommendation was accepted that an enquiry should be made into the world requirement of drugs. As a result it was revealed that the production of raw manufactured drugs far exceeded the requirement. This finding prepared the ground for the first and second opium conferences held in Geneva in 1925 and ultimately led to the 1925 Dangerous Drugs Act in Britain. The Geneva convention came into operation in September 1928. Its major influence was in the extensive machinery it created for international control.

At The Hague convention there had been considerations as to whether Cannabis should be included in international control. In 1923, the South African delegate proposed to the League's Advisory Committee that cannabis should be classified as one of the habit-forming drugs. At the 1925 Geneva convention the Egyptian delegate at an early meeting submitted a proposal that cannabis should be brought within the ambit of the convention, because it was a narcotic and could produce a serious delusion and strong physical agitation, it predisposed to acts of violence and produced a characteristic strident laugh. Chronic Hasteism was a more serious condition, it was said to produce "physical ailments, heart troubles, digestive troubles, etc."

The Geneva convention came into operation in September 1928. Its major influence was in the extensive machinery it created for international control. Britain as a party to the Geneva convention amended the 1920 and 1923 Dangerous Drugs Acts "so far as it is necessary to enable effect to be given to Convention signed at Geneva," and introduced the 1925 Dangerous Drugs Act. This act was timed to coincide with the ratification of the convention and came into operation in 1928.

Smuggling

As late as 1930 neither Turkey nor Persia, both opium producing countries, had ratified either. The Hague or Geneva convention. A factory in the Netherlands in 1927-28 had exported to the Far East 860 kg of morphine, 2711 kg of Heroin and 40 kg cocaine. Advisory Committee in April 1928 reported that hundreds of kilos of morphine were being exported from France ostensibely to Germany. The Health Committee of the League of Nations believed that 250 tons of raw opium and 6 tons of cocaine were enough for the world requirement, but in the first six months Turkey exported 2282 kg of Morphine and 4348 kg of Heroin !

In 1931 Limitation Convention was held which was attended by 54 states. China and U.S.A also attended this time. The 1931 convention differed from the 1925 convention in that all estimates were to be binding, as its main aim was to close the channels throug which drugs escaped in to the illicit traffic. The Drug Supervisory Body was to work with the permanent Central Opium Board, in that the latter performed a semi-judicial function whilest the former had the duty of determining the totals and estimates for each country and territory. The 1931 Convention also added new drugs to the list, but those were confined to the phenanthene alkaloids of opium, and ugomine alkaloids of the coca leaf. As a result of the 1931 convention, Britain introduced the 1932 Dangerous Drugs Act.

Pre-war Efforts

The last major international convention before World War II was in 1936. It was convened as an another attempt to suppress the illicit traffic in dangerous drugs. Its object was "to strengthen the measure intended to penalise offences... (and)...to combat by the methods most effective in the present circumstances the illicit traffic in the drugs and substances covered by the convention". This convention came into operation in1939, one month after the outbreak of the World War II. Britain did not ratify it on the grounds that it interfered with her right to decide her own penalties. By 1939 international control was firmly established and the Headquarters of the Opium Board and Supervisory Body shifted to Washington. The Opium Board reported that during 1939-45 "in general government furnished estimates required, and the Limitation Convention operated to a satisfactory degree".

United Nations

The Control System under the United Nations began on 12th February, 1946. The General Assembly of United Nations adopted a resolution "to take necessary measures to ensure the continued exercise of those functions and powers of a technical and non-political character conferred by certain international instruments on the League of Nations". On 3rd June, 1948, the protocol on Narcotic Drugs signed at Lake Success, New York became operative, and in 1950 Britain passed the Dangerous Drugs (Amendment) Act, which transferred the duties and powers to the United Nations.

The 1948 protocol signed in Paris on 19th November, attempted to plug the gaps in the control system. The protocol does not, of course, limit the production of opium. In 1953 it was seen only an interim measure and it was hoped that opportunities would be available during the discussions on the single convention to introduce more effective control. The final important agreement, the 1961 Single Convention, came into operation on 13th December, 1984. It was a collective work which involved the commission and its secretariat, the United Nations Division of Narcotic Drugs, the Opium Board and the Supervisory Body, together with the World Health Organization and Interpol. The convention was finally adopted by a Plenipotentiary Conference in 1961 in which 71 governments took part together with a host of other organisations such as the International Federation of Women Lawyers and the International Conference of Catholic Charities.

The general feature of the Single Convention is still the system of estmates which were introduced by earlier treaties. However, the Permanent Central Opium Board and the Drugs Supervisory Body were amalgamated and are now called the International Narcotics Control Board. By 1970 about half the members of the United Nations had ratified the Single Convention. For the remainder, the provisions of the old convention are still in force, although in practice the Single Convention is observed by all countries whether they have signed it or not.

International Initiative

The first effective attempt to control drugs of addiction resulted from the convention held at The Hague in 1912. Its purpose was to control international traffic in narcotic. Its articles are rather vaguely worded and do not bind the signatories to anything very definite. It contains phrases such as 'they will use their best endeavours'. However, it initiated some positive moves like :

'The general suppression of the abuse of drugs must be aimed at. International agreement for this purpose is necessary and advantageous.'

'National laws should be enacted to limit the manufacture, sale and use of drugs to medical and legitimate purposes, and illegal possession should be considered a penal offence.'

The Hague Convention and the Aftermath

When war broke out in 1914 only six countries had ratified the articles of The Hague convention. Signature of the peace treaty after the great war was made to include signatures of The Hague countries, both for the victorious and the defeated signatories. In spite of this ratification of The Hague convention, the drug abuse increased tremendously, alarmed by which the First Assembly of the League of Nations in 1920 created an Opium Advisory Committee to "assist and advise the council." In 1925 second convention was held at Geneva. It was considered necessary to regulate the international trade in drugs. As a result of this convention a second body was created. The Permanent Control Board. This consisted of eight members who 'by their technical competence, impartiality and disinterestedness, shall command general confidence.' In 1925 cannabis was also placed on the same basis as opium.

The third international conference was held in 1931 and led to the limitation convention. The object was to limit the manufacture of narcotic drugs, as far as possible, to the world's medical and scientific needs. In continuation fourth convention was held in 1936. It was aimed at declaring 'illicit drug trafficking a crime of international character.' But soon efforts of the League of Nations were eclipsed by the World War II which soon engulfed the entire world.

In 1961 under the auspices of United Nations and its specialised agency the World Health Organisation, a single convention of Narcotic Drugs was signed in New York and came into force in 1964. The signatories to this treaty set up an International Narcotics Control Board which replaced the Permanent Control Board and the Drug supervising body established under previous conventions. The World Health Organisation itself set up an Expert Committee on Dependence Producing Drugs.

Most countries of the world recognised the abuse potential of these drugs and as early as 1912, they attempted to enact laws and restriction which would control these drugs. All of these efforts, however, met with little success.

United States in the Cauldron

The United States bears the tragic distinction of being the leader in the explosion of drug substances—legal and illegal—in the work place and socially. Added to this, U.S. industry is losing vast sums in lost productivity every year because of employees who have an alcohol problem or snort, swallow, inhale or inject illicit substances. The toll takes the forms of slowed productivity, abstensions and irrational decisions. This is like throwing sand in the nation's industrial engine.

One U. S. study has shown that drug users are 31 times likely to be in a plant accident, are five times more likely to file a workman's compensation claim, sever three times the average out of side benefits, fraction at 67 per cent of their work potential, are more likely to steal from this company to support their drug habits.[1]

The substance abuse is not just occurring among the uneducated, experimenting young society, or losers, or drop-outs. It is prevalent among educated middle and upper class adults—the people on whom industrial and advancing societies depend for leadership and creative energy. The abuse of drugs and alcohol socially and also in the work place is a major factor why growth of U. S. productivity has been lagging behind that of other industrial nations, none of which has a drug problem as serious.

In Britain also, drugs are being taken excessively by people in jobs all the way from the executive suite to production lines, says a confederation of British Marketing Booklet—"A man stood incharge of an automobile plant will cause vastly more damage than a man at a single lathe. Some jobs carry such a high risk that even a small excess can be partially lethal", adds the book.

Russian society, too, has been seriously hurt by widespread alcoholism. Alcohol abuse is so undermining the Russian family unit and the industrial economy that Russian leaders have been forced to wage an intensified battle against drinking. Drug abuse is also reported to be a growing problem in some Russian cities and regions.

Asian Plague

Across the Asian continent, drug abuse is no longer a matter of minor concern, it is now an epidemic in many eastern drug producing nations. "It is in every town and every social strata", says Toaha Quereshi, Director of the Pakistan Narcotics Control Board. "It's completely out of mind".

In recent years, the drug chieftains of the Golden Triangle, which includes parts of Thailand, Laos and Burma and the Golden Crescent, which cuts across Pakistan, Afghanistan and Iran, have seen their heroin traffic to the West slowed as drug users turned to South American cocaine. So they have begun to dump top grade heroin at lowered costs on home markets. Addiction, once largely a western affliction, has hit these nations with a vengeance. One report indicates that in Thailand drug related offences now account for more than 50 per cent of crimes in the nation throughout the Asian sub-continent. Corruption of local officials goes hand in hand with the growth of underground drug traffic.

Nose for Narcotics

Nose for narcotics sends a US Customs service dog along a conveyor belt at New York's John F. Kennedy International Airport. Dogs trained to detail heroins, cocaine and marijuana cover major parts of the entry, shifting luggage and planes after passengers disembark. Federal authorities annually spend nearly 1.5 billion dollars on all drug enforcement. Yet they estimate they seize less than a tenth of the heroin entering the U.S. believed to be more than four metric tonnes each year.

The EL PASO Intelligence Centre in Texas was created in 1974 to deter the renewed influx of heroin from Mexico following the severing of the trade that led from Turkey through Marseilles, France to the U. S., the so-called French connection. The centre now collects and disseminates worldwide narcotics information. Coast Guard helps them with the watch staff seven days a week, 24 hours a day under joint supervision of US agencies—Immigration, FBI, LAA, Internal Revenue, Drug Enforcement, Marshals Service and Customs.

Heroin increasingly figures as a medicine of exchange, a sort of illicit currency for shady dealings on a large scale, there's growing evidence of close nexus in many parts of the world between drug trafficking and arms smuggling, subversion and international terrorism. What greases the channels of drug trafficking is official corruption in all countries.

As soon as one trafficking route is put under pressure, a new one takes over, if supply is reduced from one area, it will be replaced from another.

Severe Penalties

In spite of the everyday seizure, the drug habit is on the increase. The menace is silent and invisible on the surface but devastating underneath. The nefarious trade net-work is widespread in the under-world of crime, politicians and the police. The task is stupendous because the penalties are not severe and enforcement of laws not so effective. The cumulative experiences of the U.S., Mexico, Colombia and Asian countries like mainly Thailand and Pakistan, show that the root of it is corruption. India, despite its efforts to curb the traffickers have been accused by the U.K. as the main conduit for trade in heroin. If it is so, what is the attitude or effort of Indian Government and where the fault exists. To begin with, the law enforcement agencies both at the Centre and the States, were unable to respond quickly and effectively in the absence of any Narcotics Act. As a result dangerous criminals got bails for petty amounts and quite a few even escaped it. The convicted brought to trial were awarded a few months imprisonment or had to pay a small fine. In contrast, Singapore and Malaysia, a decade ago, prescribed mandatory death sentence for drug peddlers. Iran and China used firing squads from 1979-80 and even the mild Britain introduced life sentence for heroin smugglers in October 1984. Indian Government, in this direction, brought the Narcotic Drugs and Psychotropic Substances into force on September 18, 1984. Thus, one problem was solved by this legislation. It provided the main equities for better law implementation.

In 19th Century Opium was exported by the British from India and China, John Ruskin's famous jibe at his own countrymen, on how British ships were sailing up and down the China seas and altering, for the benefit of a foreign nation, the common highway man's demand "your life or your money into your life and your money" as true today as the Opium War of 1842. Thus, India's main legislation on narcotics that lasted till September 1984, was the contribution of British Government under the Opium Act of 1878 which introduced a Licence system to grow, possess, transport, import or export or sell opium but mainly it was a source of revenue to the Government. This Act, after the discovery of morphine and heroin, became outdated.

In the 1983-84, the poppy-production area was reduced from 9,000 to 5,700 acres in U.P., Madhya Pradesh and Rajasthan compared to licences issued to about two and a half lakh farmers on 50,000 acres. In spite of it, total production continued to be constant at about 288 tonnes.

India, till recently, maintained herself as a model to other Asian countries when the leakages were non-substantial and the Narcotics Department saw to it that all opium production was sold to the government, but the discovery of other processed drugs promoted organised trading by gangs, who, especially on Thai-Malaysian border used such devices internally stuffing the murdered cleaned-up babies with heroin and as sleeping babies in young mothers arms across. Similarly the coffins of American service-men killed in Vietnam became the buster's paradise in trafficking. These heinous tricks could be detected only after a long time.

The U. S. and India, it seems, woke up to the menace very late. Interpol, after the United Nations' Single Convention of Narcotic Drugs in 1961, transmitted warning signals to its member nations. India signed the International Treaty to enforce it on December 13, 1964. The Opium Act 1878 and the Dangerous Drugs Act were only bows and arrows to fight the heroin menace. The Indian Narcotics department had the staff and expertise to control the opium growth but not heroin-addiction which was soon to sweep the nation like an epidemic. Consequently, a small branch in the CBI was set up to curb the menace and it could only function as a co-ordinating and reporting agency rather than deal with field problems ruthlessly. India, till early '70s kept under the thumb the effective control the acreage of opium growing, compulsory sale to Government and the rigid check on opium factories in Ghazipur (U.P.), Neemuch (M. P.) and other States.

The U.S., first of all, used heroin for medical purposes later to be hit up with its addiction during the Vietnam War. In India, the number of such addicts was very low till the end of Sixties but in early Seventies, it came to a quantum jump when the well known smugglers were detained under MISA but soon, illicit trade resumed as these traders were released. It was, at this juncture, that the amendment in law, a strong enforcement machinery and rigorous punishment could have helped in a premature death of traffic-boom in India which was then most needed. The problem during this period was solved temporarily but the hazards were to leave indelible imprints.This laxity on the part of Indian Government assumed monstrous proportions in the eighties.

Economic Crime

Narcotics smuggling is a pure monetary crime. Success in the direction is possible only when the expense incurred by the big operator exceeds the income of law enforcement unit man. The profits to middleman is thousand times more than the estimated price of seized heroin or hashish. It can be calculated from the basic starting point that opium, sold for a mere $ 10 in South East Asia and Afghanistan, multiplies into worth a million dollars on the American Streets in the form of heroin. Thus seizures or confiscations at early stages do not cause any worthwhile damages. The real loss occurs only when the material moves closer to the U. S. and Western Europe market. The U.N. official figures of heroin seizures show a marked increase per year, 1951 less than 100 kgs., 1977 2337 kgs.

Anti-narcotics drive in India, Hong-Kong, Indonesia, West Asia and Pakistan fails to effectively function because of large scale corruption prevalent in the police and patronage of the corrupt politicians.

The effective and successful implementation of this Act needs maintenance of comprehensive criminal records of the new and previous convicts all over, the neglect of which poses grave threats to internal security who emerge as National Criminals. The National Criminal Record Bureau may too help in this direction.[2]

Preventive Steps

Douglas Hurd, the British Home Secretary, signed the Drug Traffic Offence Act which will help checking drug trade in England and Wales. Under the Act the first offender is the one who helps the illegal trader in the narcotics or is connected, in some way, with the income and property of the trader which can be confiscated; the second offender will be one who will provide cocaine-kit or narcotic drugs. Such offenders will be sentenced to 14 years and 6 months imprisonment respectively. The Act has come into force from 30th of September, 1986.[3]

Indo-U.S. governments realising the threatening dimensions of international drug trafficking, agreed to form a body of Home, Finance, External Affairs and other departments such as Enforcement and Narcotics Bureau of India from both the countries, which will fight out the menace of this trade and function as an ongoing body for intelligence sharing and cooperation pursuing over all objectives of the joint policy. The body will seek help from international organisations like the interpol to detect economic crimes to improve the activities of the body. Mr. Edwin Meese, the U. S. Attorney General, on his visit appreciated the stringent law passed by India as a strong stop in this direction. He also emphasized on other steps such as confiscation of property purchased with illegal funds from drug trafficking and a nation-wide effort to educate young people not to fall a prey to this deadly habit.[4]

Indo-Australian Efforts

The DRI in collaboration with Australian Narcotics Bureau busted international network of narcotic smugglers operating in India through a novel method of cargo switching. Following a tip, the officials swooped upon the contraband hidden in four crates of incense sticks (Agarbatti). The investigations revealed that the two arrested persons and their accomplices had taken these 4 genuine crates and four others containing contraband to Palam Airport. The genuine ones were put up before the officials for examination and later the other four were also taken into airlines warehouse. In collusion with employees of the clearing agents, the custom seal was put on all the eight crates. The four genuine were left at another location in the airport but the culprits did not have enough time to remove the original cargo from the godown. The Australian national Robert Richard supposed to have visited India in November paid Rs. 40,000/- as the first instalment for shipment.

India Signs Anti-drug Measures Document

In an effort to co-ordinate action against drug abuse and smuggling, on 17th August, 1988, two project documents were signed between the United Nations Fund for Drug Abuse Control and the Government of India project No. AD/IND/88/352 that provide for assistance of U. S. $ 7.5 million for strengthening law enforcement measure against illicit drug trafficking. The other project No. AD/IND/88/3/353 provides for U. S. $ 12.5 million for development of drug abuse prevention, treatment, rehabilitation and control measures. This assistance is to be provided over a period of five years.

A team consisting of 8 members headed by Mr. J. Gammelgarrd who had been posted as UNIDAC's Field Advisor to India, visited India from 26th September to 9th October, 1988 for formulation of addiction work plan for the year 1989 i.e. the first year of the five years project. The team visited the sites where the project was to be implemented like the Narcotics Control Bureau's Zonal offices at Bombay, Madras, the Regional Training Institutes of Directorate of Training; The Central Forensic Science Laboratory, Delhi and Central Revenue Control Laboratories at Delhi. Bombay and Madras. Some organisations dealing with de-addiction were also visited by the team at Delhi, Bombay and Calcutta. Discussions were held in detail with the concerned implementation agencies and a detailed plan for the year 1989 to each of the projects was worked out.[5] The work plan for strengthening of law enforcement measures provides for the following items:

(i) Setting up a computer system with centralised data base at NCB Headquarters and Miow computers with printers at the zonal units. The system provides for future expansion and link with Narcotic Commission's computers, customs and police computers. The project includes provision of application Software and tracing for the personnel.

(ii) The project provides for a tele-communication net-work with 100 watts. H. F. Radios numbering 16 to be installed in the NCB Headquarters as well as Zonal Units. 25 watts. VHF, F. M. Radios numbering 72 are to be installed in the NCB offices as well as in the vehicles for quick communication for operational purposes. 2 watts, hand-held radios will be provided for surveillance and intelligence work.

(iii) Surveillance equipment in the form of electronic surveillance kits, image intensifying monocular equipment, night vision binoculars, poloroid and ordinary cameras will be provided to NCB offices. Two jeeps, one pick-up vehicle and 4 motor-cycles will be provided to each zonal unit.

(iv) Twenty fellowships for training abroad in various aspects of narcotics enforcement, training techniques, etc. will be organised for NCB, Customs and Police officers. Equipment in the form of Photocopiers, Overhead-projectors, Slide-projectors, VCRs and T. V. Monitors, I.B.B.T. compatible personal computers is also being provided to Directorate of Training as well as NCB officers for better organisation of Narcotics activities.

Strengthening of Forensic Science and Revenue Control Laboratories for Narcotics testing work is also covered in both the projects. It is proposed to strengthen all the 14 Forensic Science Laboratories besides modernisation of C.F.S.L. and C.R.C.L.

Monitoring of Licit Opium

The control over the production of licit opium has also been envisaged through strengthening of controls by providing equipment to the laboratories attached to Neemuch and Ghazipur government opium factories. The R.S.D. Laboratory at the Neemuch Factory is also being strengthened. It is proposed to use the expertise available in this laboratory for preparing reference standard samples of narcotic drugs and psychotropic substances which would also be supplied to United Nations for supply to other countries. The Narcotics Commissioner's office is being strengthened with modern equipment and communication facilities.

Drug Abuse Prevention Through Education

This plan is to be implemented by the University Grants Commission in the Department of Education. The work plan provided for curricular development activities through seminars and regional workshops.

Drug Abuse Prevention Through Health Programmes

This part of the project is being implemented by Ministry of Health and Family Welfare. The work plan incorporates implementation of a manpower development centre for development of health services in the area of drug dependence. A programme for training 9,600 personnel in this field includes production of training manuals, audio-visual aids, equipment support., etc. A drug abuse monitoring system will also be developed.

Rehabilitation and Social Reintegration of Ex-drug Addicts

The work plan provides for funding of 2 identified projects non-government organisations in the first year. It provides for fellowships and study tours for studying the rehabilitation activities in the Asian as well as European countries. This is to be implemented by the Ministry of Welfare.

SAARC Steps to Curb Drug Menace

The standing committee of Foreign Secretaries of South Asian Association for Regional Cooperation has agreed that member states should take effective steps to stop drug trafficking through bilateral arrangements. The report of the technical committee adopted by it for consideration of the ministerial meeting suggests that drug trafficking be made punishable by adequate measures. It also suggested that drug enforcement agencies of the SAARC countries should maintain dossiers on drug traffickers and suspects and exchange information periodically. Recognising the need for adopting pragmatic and meaningful programmes, the committee has recommended that to check drug trafficking on the borders, there should be frequent meetings.

In order to discuss the problem of drug trafficking and to suggest ways and means to stop drug trafficking through bilateral arrangements, the drug enforcement agencies of the member states should ensure closer co-operation and promptly communicate with relevant inter-governmental, international and regional organisations.

The committee said the member countries who have not done so should comply with the United Nations Convention on Narcotic Drugs of 1961, Protocol of 1972 amending the same convention and the U. N. convention on Psychotropic Substances 1971.

In order to reduce the demand for drugs, the committee felt that it was imperative that comprehensive measures be taken for identification, treatment and rehabilitation of drug addicts.

Preventive Measures

The SAARC Committee suggested that treatment facilities, should be readily available through the existing health delivery system so as to make them most effective. The treatment should be provided in hospitals, other health centres, in penal and correctional institutions and community based institutions which provide voluntary treatment. Specialised national resource centres should be formed to develop appropriate treatment models to send the requirement of people, with drug problems in each socio-economic setting. The expertise achieved in these centres should be exchanged regularly through workshops, seminars and study groups.

Emphasizing the role of preventive education, the committee recomended the implementation of comprehensive programmes at local vocational levels to check the increase in drug alone. A meaningful teachers orientation programme should also be introduced to impart preventive education in schools and colleges.

International Co-operation

There are two international treaties of particular interest to law enforcement authorities engaged in combating the illicit traffic in narcotic drugs. The first is the 1936 Convention for the Suppression of the Illicit Traffic in Dangerous Drugs, and second is the 1961 Single Convention on Narcotic Drugs. Whether a particular government has put either or both of these treaties into force in its own territory is a matter for local inquiry.

The 1936 Convention is directed against the illicit traffic, and had as its purpose the strengthening of the measures already taken by previous international agreements to penalise offenders and to combat in the most effective manner the illicit traffic in drugs. The parties agreed to make laws to punish severely, particularly by imprisonment, various offences connected with illicit trafficking. Those offences were to be included as extradition crimes in treaties between the signatories. In countries which do not permit the extradition of their nationals who have committed abroad offences relating to narcotic drugs are to be tried and punished as if they had committed the offences in their own country. Subject to certain qualifications, foreigners living in one country may be punished for offences committed in another.

The Convention also required its signatories to set up "a central office for the supervision and co-ordination of all operations necessary to prevent" the offences specified in the Convention, and "for ensuring that steps are taken to prosecute persons guilty of such offences". The main tasks of these central offices are to centralise all information of a nature to facilitate the investigation and prevention of offences, and to keep in close contact both with other official bodies dealing with narcotic drugs in the country and with the central offices of other countries. In particular, they are to communicate to each other information of investigations into and operations against illicit trafficking, particulars regarding the identities and descriptions of traffickers so as to facilitate checks on their movements, and details of discoveries of clandestine drug-making factories.

The Convention also laid down three methods by which information and requests could be exchanged by governments:

(i) preferably by direct communication between the competent authorities or through the central offices of each country;

(ii) by direct correspondence between the Ministers of Justice of the countries concerned; or

(iii) through the consular or diplomatic representatives of a country to the designated authority of the other government.

It should be noted, however, that the Convention has not yet been ratified by enough governments, and is therefore not legally in force.

The 1961 Single Convention, which came into force in 1964, had as its main purpose the restatement, codification and amplification of earlier treaties, except for the 1936 Convention. It limits the production, manufacture, trade, import, export, distribution and use of narcotic drugs exclusively to medical and scientific purposes, and seeks to combat drug addiction itself by requiring government to provide all possible facilities for the treatment, care and social rehabilitation of drug addicts. It places narcotic drugs in one of four categories according to the danger posed by its unregulated use and the extent of its abuse; each category is subject o measures of control of varying severity.

Among other provisions, it

(i) continues the system requiring all governments to furnish annual estimates of the quantities of narcotic drugs needed for medical and scientific purposes in the country, so that manufacture may be limited to legitimate world demand;

(ii) extends the agreement whereby all international trade in narcotic substances is regulated by a system of import certificates and export authorisations supervised by the International Narcotics Control Board;

(iii) requires all governments to furnish statistical returns of their receipt, consumption and disposal of all narcotic substances; and

(iv) obliges all governments to maintain a special administration to put into effect the provisions of the Convention.

As regards action against illicit trafficking, it provides that signatories shall:

(i) make arrangements at the national level for the co-ordination of preventive and suppressive action against the illicit traffic, for which they may usefully designate an appropriate agency responsible for this co-ordination;

(ii) assist each other in the campaign against the illicit traffic in narcotic drugs;

(iii) co-operate closely with each other and with the competent international organisations of which they are members, in order to maintain a co-ordinated campaign against the illicit traffic;

(iv) ensure that international co-operation between the appropriate agencies is conducted in an expeditious manner; and

(v) ensure that where legal papers are transmitted internationally for the purposes of a prosecution they are sent promptly.

A third international treaty will become important. This is the Convention on Psychotropic Substances, 1971, which is intended to create obligations on states to being under national and international control substances which are being abused, or may be abused in future, to the detriment of public health. These include depressants of the central nervous system such as sleeping pills of the barbiturate type and many tranquillisers, stimulants such as many amphetamines, and hallucinogens such as LSD and mescaline.

The Convention provides four different classes of control. The most dangerous drugs such as LSD will be practically under direct government control. Drugs in schedule II are to be controlled in much the same way as morphine is now. For drugs in the other two schedules there is to be a simple system whereby governments will notify each other of their export.

Although probably of no direct concern to police forces, the *International Narcotics Control Board*, it should be mentioned, is the authority set up by governments to regulate the international lawful trade in narcotic drugs. It consists of a body of independent experts elected in their personal capacities, and seeks to limit the production and distribution of drugs to the quantities needed for legitimate medical and scientific purposes. It is unnecessary to describe in this chapter its methods of work.

These provisions of international law require to be put into action to become effective. Frontiers by themselves are no bar to drug traffickers; they should not be allowed to hamper the efforts of the world's law enforcement services. The considerations which lead to an acknowledgement of the need to centralise national effort also lead to the acceptance of the need for international co-operation.

This can only be achieved by the law enforcement authorities of the governments concerned making direct contact with each other and working together in close co-operation to pursue common aims. It is necessary to operate at three levels: locally across the frontier between two countries, internationally from one headquarters to another, and thirdly by the exchange of information with international organisations. These means may be supplemented and given support through diplomatic channels.

Local Co-operation Across a National Frontier

When the competent authorities of two governments have agreed that their enforcement services will work together to suppress smuggling, certain steps should be taken to facilitate actual co-operation. They include:

(i) a common understanding and agreement between the local authorities actually working in the frontier areas about where exactly the frontier lies, how it is marked, and what are the regular legal requirements to be complied with by a person wishing to cross it, whether he is a local resident or lives further away;

(ii) an exchange of maps of the areas on both sides of the frontier, showing its actual line, the frontier posts of both countries, the location of towns, villages, roads, tracks, rivers, water points and other facilities, together with agreed names and spellings;

(iii) a statement of what means of communication are available, *e.g.*, radio links, telephones, postal services, etc.; what methods have been authorised for use; and with whom communication may be maintained;

(iv) a recongnition of the areas of jurisdiction of each organization on either side, and a common understanding of the legal powers and administrative authority exercisable at each level of the various forces engaged; and

(v) agreement on the rights, if any, of police and frontier guards to "close pursuit", and of how to go about seeking the extradition, or return by less formal procedures, of fugitive offenders. This might include agreement on how to deal with persons of dual nationality or dual domicile.

Agreement and understanding on these points, while clearly not completely within the sole discretion of the police frontier authorities, can be re-inforced in the field by frequent personal contacts between officers in charge of frontier posts and units, by clear orders regarding what they are allowed to settle among themselves, and by regular meetings of more senior officers to review the working of these arrangements and to make sure that they are producing good results.

When this framework has been established it should be simple to work out means to exchange up-to-date information regarding:

(i) the identities of known or suspected smugglers, and details about them and their movements;

(ii) seizures of smuggled goods, giving details wherever possible of their places of origin;

(iii) arrests and convictions of smugglers;

(iv) details of methods and routes used by smugglers; and possibly

(v) details of each other's counter-measures.

A network of common information extending throughout a zone on both sides of a frontier provides a far better means of preventing and detecting offences than a so-called "line" of frontier posts, particularly if it is backed-up by joint action. Every smuggler knows that a "line" is more gap than obstacle.

Co-operation between National Authorities

Whereas the local co-operation on a frontier which has already been described is intended to prevent as far as possible the actual carrying of smuggled goods across the frontier, particularly in regions where frontiers traverse open, difficult or sparsely inhabited territory, police co-operation at the national level is designed to help in longer-term or more complex cases. For instance, to combat smuggling by commercial air or sea lines almost certainly requires the collation of information and the concerting of police and customs action between two countries at national, rather than local level. Similarly, the delicate investigation of an established international smuggling organization may require months of work, which must be agreed and carried out together, before the authorities are ready to strike.

Co-operation of this kind requires an almost daily flow of detailed information about specific cases and of requests for action. The information to be exchanged should consist of anything and everything that may concern two or more national authorities and help them to do their jobs better. It would include information about arrests and seizures, the identities and movements of suspects, routes and methods used in individual cases, requests to investigate or identify particular methods or persons or routes, as well as summaries and analyses. But above all, it would include the exchange of intelligence, so that traffickers can actually be arrested in the course of their smuggling. It is, therefore, necessary that the quickest possible means be used to transmit information. To facilitate these exchanges of information in an orderly and expeditious manner requires an international means of co-operation, an understanding of what can be achieved by its use and an adherence to its methods of work. (In this connection, the Interpol publication "Methods of Record Keeping in the I.C.P.O. (Interpol) Criminal Information Office" is a useful work of reference.)

Use of International Organizations

The main international organization available to law enforcement authorities is the International Criminal Police Organization (Interpol). It specialises in the international fight against criminals, and has as its aims the improvement of direct contacts and exchanges of information between countries, and the development of co-ordinated police action. Its role is essentially a technical police one, for in the present state of international law and national sovereignties it can have no executive powers of its own. To achieve its aims, Interpol has established at the international level a general secretariat at its headquarters in Paris, and at the national level in many countries an Interpol National Central Bureau (NCB).

A National Central Bureau (NCB) of Interpol ensures liaison with

(i) the other enforcement agencies in the same country;

(ii) National Central Bureaux of other countries; and

(iii) the General Secretariat of Interpol in Paris.

The General Secretariat acts as a depository and clearing-house for information, and provides a means by which the police forces of its member countries can co-operate and communicate with each other. In 1970, for instance, its radio network transmitted 150,000 messages.

In a member country with a national police headquarters of some kind both the Interpol NCB and a Central Narcotics Intelligence Bureau will almost certainly be parts of its headquarters. In countries where the two offices are not parts of the same administration, the NCB normally provides help to a CNIB by acting as its channel of liaison. Where this local co-operation cannot for some reason be established, then in accordance with article 33 of the Interpol Statute and with the agreement of the General Secretariat, the CNIB may establish its own direct link with Interpol.

In many countries police action against the illicit traffic is the responsibility of several authorities which are independent of each other (police, gendarmerie, customs, etc.). Where this is so, Interpol advocates sound co-ordination between them and encourages in each country the establishment of a single point of reference.

This machinery of international co-operation makes it possible to :

(i) ask for investigations to be carried out in other member countries, or for particulars concerning a specific case or person;

(ii) publish notices giving particulars of the identity or description of any "wanted" or fugitive trafficker;

(iii) ask, by means of international warning notices or regional information bulletins, that leading international traffickers be kept under surveillance;

(iv) provide a detailed monthly account, in tabulated form, of particulars of the illicit traffic;

(v) furnish reports on narcotics trafficking in particular regions;

(vi) facilitate direct contacts between officers of the enforcement services in different countries, for instance by providing introductions;

(vii) help member countries to obtain books and articles relating to a wide range of problems by distributing a quarterly list of selected publications; and

(viii) provide an opportunity for officers of the enforcement services to take part in Interpol seminars.

The United Nations Division of Narcotic Drugs, which provides the secretariat of the United Nations Commission on Narcotic Drugs, is also of use in combating the illicit traffic. Its principal uses to law enforcement authorities are:

(i) its annual review of the illicit traffic, prepared in conjunction with Interpol;

(ii) its regular summaries of reports of seizures made throughout the world;

(iii) its occasional papers on particular aspects of trafficking;

(iv) its lists of merchant seamen and members of air-crews convicted of narcotics offences;

(v) its reports listing cases of narcotics smuggling through the mails;

(vi) its very large collection of literature on all aspects of narcotics;

(vii) its laboratory, which can do a great deal to help governments set up their own narcotics laboratories and advise on methods for the identification, quantitative determination and in some cases determination of the origin of drugs; and

(viii) the technical assistance it can render to governments by the sending of training or consultative missions, the holding of seminars and study groups, the grant of fellowships, the lending of films, and the assignment of expert advisers.

Reports of all cases of International Importance or interest are made to Interpol and the United Nations Division of Narcotic Drugs on a prescribed form common to both organisations, which is reproduced at Appendix C. Interpol National Central Bureaux should complete reports as fully and quickly as possible, so that both organizations can fulfil their roles in combating the illicit traffic and foster effective international cooperation.

Government Diplomatic Action

The international collaboration so far described in this Part is essentially technical in character. It may be supported by national diplomatic and political action. In order that a government may come to appreciate the need for any action in this field, decide what is the best thing to do, and carry it out in a well-informed manner, it is necessary for it first to be supplied with accurate and detailed information. This is best done, not by submitting a series of reports on individual cases, but by preparing periodical summaries and analyses of the illicit traffic and the extent of drug abuse. They can be submitted to the government together with recommendations for any action which is thought desirable by the reporting authority. The best organization to prepare these reports, certainly as far as the illicit traffic is concerned, is a Central Narcotics Intelligence Bureau.

Retrospect

Despite the widely-held ambition to bring under effective control the production and distribution of all narcotic drugs and other psychotropic substances, and the hope of eradicating all drug abuse, it is only prudent to assume that the laws against drug trafficking will continue to have to be enforced. It is, therefore, necessary for all law enforcement authorities engaged in this duty to ensure that their measures remain efficacious whatever the changing circumstances, and to keep in view their real objective, the destructions of the organisations and lines of supply which together are responsible for the major part of the illicit traffic. Emphasis should always be placed upon the improvement of the quality of the work done.

Whether or not all the techniques and resources mentioned in this guide are available or necessary to the national authorities of a country, and whatever the administrative responsibilities of a particular law enforcement authority, the means to be employed in attempting to achieve the objective should be of constant concern. It is impossible for a police force to be equally strong at all points, and impossible for it to destroy the illicit traffic by the use of routine methods uniformly and unimaginatively carried out. In order, therefore, to understand how the illicit traffic works in a region, and so how best to use selective action against it, it is necessary to ascertain what are its more important features at any particular time, and then to concentrate adequate attention and effort upon them, even at the cost of the temporary neglect of other work in this field. This can only be done by the collection, analysis, and use of intelligence relating to all aspects of that traffic.

The arrest of an individual in illegal possession of drugs, however deserving in itself, is therefore of no great consequence unless he can be related to some other part of the organisation or chain to which he almost certainly belongs. A major aim in arresting an individual should, therefore, be to obtain information from him, and then to use it in order to identify and take action against his organization or suppliers. This is especially true of investigations into the illicit manufacture of drugs, where the seizure of a laboratory and the exemplary punishment of its operators will do far more to reduce the sale of these drugs than the arrest of many of the pedlars or couriers handling its products.

Finally, it is necessary to assess the effectiveness of the law enforcement effort as a whole, free from considerations of policy-making and prestige at any level, and unembarrassed by prejudice or wishful thinking. This requires a continuous and detailed knowledge of which gangs have been broken up and what remain at work, what channels of supply have been stopped and what potential alternatives there are, what trends are showing in the demand for and the supply of drugs, and an objective and dispassionate judgement of what remains to be done. Ignoring the temporary effect of market "rigging" by a monopoly, the most effective single index of the situation at any particular time is almost certainly to be found in the retail price and quality of illicit drugs actually bought by their users, for it is only at this final stage of the traffic that all the variable factors in a cmplicated commerce are brought together and measured in a free market.

Sample of some prescribed reports charts, registers cards etc., has been given below.

Appendix-A

NARCOTICS INFORMATION REPORT

For CNIB use only

1. Reporting Unit	IR file No.
2. Serial No. of report	Date
3. Date of report	Entered in
First report or follow-up report	registers
previous reference, if any	Entered in
	Indices

Details of the drugs poised

4. Kind and weight of drug	Interpol/UN
5. Time and date of seizure	Report
6. Place of seizure	Laboratory
7. Place where drug was hidden	Report
If in a vehicle, state :	Further
Type of vehicle	Reports
Registration No.	Name of driver
Route being followed	Name of owner
	Place where hidden

8. Number of packages in which

the drug was contained
9. Place of origin of drug
10. Other exhibits seized
11. Exhibits delivered to
on (date)

Details of persons arrested

	Accused No. 1	Accused No. 2	Accused No. 3
12. Name			
Father's name			
Alias			
Sex			
Nationality			
Tribe			
Place of residence			
Lawful occupation			
Date of birth			
Place of birth			
Identity Card No.			
Date of issue			
Place of issue			
Distinguishing marks			
Is he an addict ?			
How was he involved ?			

Details of other persons involved but not arrested

	Suspect No. 1	Suspect No. 2	Suspect No. 3
13. Name			
Father's name			
Alias			
Sex			
Nationality			
Tribe			
Place of residence			
Lawful occupation			
Date of birth			
Place of birth			
Identity Card No.			
Date of issue			
Place of issue			
Distinguishing marks			
Is he an addict ?			
How was he involved ?			
Why has he not been arrested ?			

Judicial Proceedings

	Accused No. 1	Accused No. 2	Accused No. 3
14. Date case referred to Public Prosecutor			
15. Accused charged under Article of Law			
16. Case No. and Name of Court			
17. Date of end of trial			
18. Sentences : Offender No. 1			
Offender No. 2			
Offender No. 3			
19. Results of appeals, if any			

Any further information or comments

(especially any information which suggests the intended destination of the drugs)

20.

Names and ranks of officers and man engaged in the discovery

21.

Signature of reporting officer

Title

Instructions

How to Complete a Narcotics Information Report

Every seizure of narcotic drugs or other psychotropic substances under control, and every arrest of a person in connexion with offences related to these drugs, must be reported :

By the Police to :

By (e.g.) the Customs to :

In addition to the normal reference to the Ministry of Justice or Criminal Court.

2. A report to these authorities is made by completing this form. No other report is necessary. It is only necessary to complete one form in respect of each case. This will be despatched by post at the same time as the

arrested persons are handed over to the Public Prosecutor.

3. It will not always be possible, however, to answer all the questions on the form at that time. When further information becomes available (e.g. more arrests are made, complices become known, the result of a trial is notified, etc.) it should be reported by filling up the relevant parts of a fresh form. This new form will be given its own serial number but the words "first report" will be crossed out and the serial number of the earlier report about the same case will be given at "previous reference". This process may go on for as long as there is anything new to report. If, on the other hand, the report is the first one about a case, than the words "follow-up report" and "previous reference" will be deleted.

4. Copies of the report will be sent to the following:

top copy :

second copy :

third copy :

last copy : for filing in the office of origin.

5. It is emphasised that it is essential that as much information as possible should be given in these reports. Any detail placed on record in the Central Narcotics Intelligence Bureau may become significant in a later investigation.

6. The following notes are intended to help officers fill up the form. The numbers given below refer to the numbered items on it.

1. This refers to the unit making the report.
2. Each unit making reports will serially number its own reports, beginning with No. 1 at the New Year.
4. Describe the kind of drug (e.g. raw opium, heroin, etc.) and give its weight in grams.
6. Describe the place as accurately as possible. Give the name of a village or path, the exact address in a town.
7. Say exactly where the drug was found, e.g. on a donkey, hidden in a roll of cloth, placed in a specially made compartment in a truck, under the sitting-room carpet, etc.
8. Describe the wrappings in which the drug was found, e.g. six paper packets of heroin, three plastic bags of opium, etc.
9. Give not only the country of origin, but if possible the district and even the village or house. Say whether this is known for sure, or whether it is only thought to be true. If the origin is unknown, go back as far as possible.
12. Give as much information as possible about each person arrested. Under "name" give his name and surname. Be as exact as possible about his address, particularly if it is in a town. Under "distinguishing marks" note anything which would help someone who has never seen the person to recognise him, e.g. a pock marked face, a broken nose, a aquint, a scar, etc. Under "how involved" state briefly what the person was doing with the drug, e.g. selling, transporting, keeping it for his own use, driving the vehicle, etc. If there is not enough space here for all the accused, continue on plain paper. Remember that these particulars must be so complete and accurate that from them the person and his address can always be identified again, and his role in the case plainly seen.
13. Give as much similar information as possible about any other person(s) who are known or suspected to be concerned in any way with the case. Say whether it is known for sure, or whether it is only thought to be true. Under "how involved" explain the connexion between this person and those arrested. Remember that it is wanted to know from where and whom the drug had come, and to whom and where it was intended to go; further investigations may have to be made starting from the information given here.

14-19. Every effort must be made to obtain information from the Public Prosecutor about the progress of the trial, and to report its result as soon as possible. Any other information which may be obtained from him which may alter or emplify the information already reported should be made the subject of a follow-up report.

20. Any information not already given should be recorded here, e.g., how the drug got to where it was seized, where it was intended to go, the amount of its adulteration, its selling price, etc.

7. Attach to the report a photograph and set of finger prints of each person arrested. Write on each the name of the person and the report number and office of origin.

Appendix B

Proformas of Index Cards and Registers to be used in a Specialist Drugs Record Offices

Personal Particulars Index Card

Name	Sex
Father's name	Nationality
Alias	Tribe
Residence	Lawful occupation

Date of birth	Identity Card No.
Place of birth	Date of issue
Distinguishing marks	Place of issue
Area in which active	CNIB PF No.
	CNIB SF No.
Associates	Finger Print No.
	Photograph No.

On Reverse

CNIB Ref. No.	Date	Brief details of report

Addicted Person's Index Card

Face of card : the same as a Personal Particulars Index Card

On Reverse

Married, single or divorced	Origin of addiction
No. of wives	Year first addicted
No. of children	Drug (s) used
No. in family addicted	Method of administration
Daily income	Amount taken daily
Education	at cost of
	Dates of treatment

N. B. Where a person is both an addict and a trafficker the two cards can be fastened together.

(face of card) *(back of card)*

Premises Index Card

Address	Information		
Description	CNIB	Date	Date Details of report
Owned by	Ref. No.		
Occupied by			
Lawfully used for			
Illicitly used for			
by			
when			
Connected with			

(face of card) *(back of card)*

Vehicle Index Card

Registration No.	Information		
Type	CNIB	Date	Details of report
Make	Ref. No.		
Registered owner			
Kept at			
Used for			
Illicitly used for			
by			
when			
where			
Vehicle connected with			
Final disposal of vehicle			

Subject Index Card

Subject	Names involved	Report filed at

Appendix C

Sender: Date:

Addresses : Secretariat General
ORGANISATION
INTERNATIONAL E
DE POLICE
CRIMINELLE

1. 26 rue Arnengaud
92 St. Cloud,
France

Reference no.

2. Division of Narcotic
Drugs, United
Nations, Geneva.

INFORMATION ON A CASE OF ILLICIT TRAFFIC IN NARCOTIC DRUGS OR ON A SEIZURE

Report to the United Nations and to the International Criminal Police Organisation/Interpol

Section A— Information on the Drug Seized

1. Kind and weight (in kilogrammes and grammes) of the drug seized :

____

____

2. Place and date of the illicit transaction or of the seizure :

____

____

3. Packing, labelling, trade-mark, etc. of the drug seized

____

____

4. Type of transport used : give name, owner, nationality, registration number of ships, aircrafts and/or other vehicles used :

5. Route followed by the drug (including places of trans-shipment and ports of call) :
6. Place where the drug was to have been sent :

This place is* : unknown
presumed
certain

Section B—Origin

7. Place where the drug was obtained :

This place is* : unknown
presumed
certain

8. Place where the drug was converted :

This place is* : unknown
presumed
certain

9. Place of production either of the natural product, or of its derivative, or of the synthetic product :

This place is* : unkown
presumed
certain

Section C—General Description of the Case of Traffic or the Seizure

10. Summary of the case :

11. Means by which the drug was obtained (purchase, theft, misappropriation, etc.)

12. Means of concealment, in particular hiding places used by the traffickers :

13. Methods and means of investigation used in the case :

14. Additional information : (a) in the case of a clandestine laboratory, indicate apparatus seized :
(b) drug addict paraphernalia seized :

15. Information (name, date and place of birth, nationality, occupation, residence, etc.) about the persons, whether arrested or at large, implicated in the case and about the part they played :

16. Legal or administrative measures (indicate sentence, place and date of trial as well as legal basis for the measures) :

Section D—International Co-operation Following the Seizure

17. Indicate whether the Government (s) of the country (countries) concerned has (have) been consulted about the seizure :

For the I.C.P.O. INTERPOL only :

18. Attach photographs and fingerprints (three copies of each), record of previous convictions, etc. of persons referred to under question 15.

Signature, stamp and designation

Universal Efforts

Drug trafficking is already recognised as a CRIME AGAINST HUMANITY by almost all the countries in the world. The fight against drug trafficking is, therefore, of major importance for all. While in some countries this task ranks fairly high in terms of priorities due to the extent of drug abuse, in others, this problem may still be considered secondary, either because it is thought that drug abuse is not really a problem or because the problem may not have been properly assessed. What-

Register of Arrested Persons

CNIB serial no.	Unit and serial no.	Date of report	Quantity and drug	Other exhibits	Number arrested	Offences	Court and Case no.	Disposal of persons charged	Year Month Date of disposal	Result of Appeal

Register of Seizures

CNIB serial no.	Unit and serial no.	Date of report	Raw Opium	Quantity of Drug Prepared Opium	Morphine Base	Heroin etc. as appropriate	Year Month Heroin Labs.	Illicit cultivation

Register of Stolen Drugs

Date of Report	Office of origin	Type of drug lost/stolen	Quantity	Unit Packing	Where lost/stolen	By/from whom	Short details

ever be the reason, no country can afford to be complacent about this CRIME AGAINST HUMANITY for it recognises no territorial barriers and possessed the potential for establishing national Governments and threatening the security of certain states.

A basic tenet for all persons and agencies engaged in drug law enforcement is that success comes only from co-operation with other national, regional and international agencies. Drug traffickers have many advantages over those trying to disrupt their illicit transactions. Large trafficking organisations have the financial resources required to pay for the man-power, equipment, transport and anything else they need; they have the time and material to prepare for the concealment of drugs. They can pick the time, place and method of introducing drugs into a country and they are prepared to flout all laws and regulations. Enforcement agencies, on the other hand, usually lack financial resources which would provide them with the necessary manpower and equipment. To overcome these disadvantages, they require a free and regular flow of information among one another and a willingness to co-operate and work together. We must all be conscious that the war against drug is not the responsibility of any one country but of all nations.

In order to be successful in our endeavour, regular and timely exchange of information between all the affected countries is extremely essential. This exchange may help in identifying couriers, monitoring their movements and eventually apprehending them. There have been some very positive results in this field as experienced by us and as known to us about the experience of several other countries co-operating with one another either bilaterally or through the international organisations like the Customs Co-operation Council and ICPO/INTERPOL. Unfortunately, this exchange is not always reciprocated and consequently some countries may not favour this form of co-operation. The lack of proper communication facilities could be one obstacle which needs to be overcome. The aim should be to make use of any existing communication system either directly or if necessary through the drug law enforcement agency of another co-operating country for exchange of information between source, transit and recipient countries. Any system which will enable a rapid exchange of timely information on persons, modes of conveyance and transport and *modus operandi*, will pay rich dividends. There should also be feed-back on information received. Apart from the exchange of operational information which needs to be done on immediate basis, there is also a need to update all other drug law enforcement agencies about the new trends in drug trafficking. This is already being done by the ICPO/INTERPOL through their Weekly Drug Messages and by the Customs Co-operation Council through the Enforcement Bulletins every fortnight. There are similar other publications by different countries like the "News from NARCONTROL"—a monthly publication issued by the Indian Narcotics Control Bureau, circulated to enforcement agencies in different countries. These publications have been well received by the agencies all over the world but these can only be as goods as all of us make them by providing information quickly to the issuing agencies so that it can be promptly disseminated.

Vienna Conference 1989

The International Conference on Drug Abuse and Illicit Trafficking was convened at Vienna from 17th June to 26th June, 1987 as an expression of political will of nations to combat the drug menance on a world-wide basis. The conference made a major breakthrough by deciding that a balanced approach was needed to deal with this plague affecting society and that the prevention of drug abuse and treatment and the rehabilitation of drug abusers should be accorded the same importance in policy and action as the reduction of supply and illict traffic.

The Comprehensive Multi-disciplinary Outline of Future Activities in Drug Abuse Control, also adopted unanimously by the conference, is a compendium of practical action for Governments, the United Nations System, inter-governmental and regional organizations, non-governmental organizations, academic institutions and individuals to take in combating drug abuse and illicit trafficking. In its four chapters, 35 action targets are set that are realistically attainable over the next 10-15 years, existing problems are defined and a course of action designed to attain these objectives is suggested.

Vienna Conference 1987

The outline comprises four chapters, covering to main elements involved in the fight against drug abuse and illicit trafficking and the subjects included in the agenda of the Conference : the prevention and reduction of illicit demand; the control of supply; action against illicit trafficking; and treatment and rehabilitation. Each chapter indicates specific targets, particularizing the objective to be attained and the action to be taken at the national level (by governments, professional associations, academic institutions, non-governmental organizations, communities, parents and individuals); at the regional level.

Vienna Convention—1988

While discussing the potential of international co-operation, we must not forget to mention the U.N. Convention against Illicit Traffic in Narcotic Drugs and Psychotropic Substances adopted on 19th December, 1988 at Vienna. The convention has been adopted by the international community in recognition of the fact that eradication of illicit traffic is a collective responsibility of all states and that, to that end, co-ordinated action within the framework of international co-operation is necessary.

It is heartening to note that the need for having better co-operation is being recognised by all of us engaged in drugs control. The last couple of years have shown a marked improvement in the field of intelligence and investigation co-operation. Our effort in India in the last two years, coupled with the co-operation of a number of agencies like the Pakistan Narcotics Control Board, United States Drug Enforcement Administration, British Customs, Royal Canadian Mounted Police, Narcotics Branch of Federal Republic of Germany and such concerned agencies in a number of other countries have produced good results.

It was on account of this co-operation that, we could identify a large number of drug trafficking organisation having tentacles in various countries, and were also able to neutralise them by apprehending the top drug traffickers and seizing multi-kilogram quantities of narcotic drugs. Further, we could generate evidence by causing investigations in other countries which could subsequently be utilized in legal proceedings against these traffickers. We have taken up a number of joint operations, using the latest investigation techniques, with the drug law enforcement agencies of several foreign countries and were able to proceed against all the persons involved. Such operations also build up intelligence data which can be of immense use in future investigations.

There is, however, a lot of scope or improving co-operation in the field of intelligence sharing and investigation. Any intelligence collected by a source country needs to be passed on to the concerned transit or consumer country within the shortest possible time so that it could be acted upon well in time. Similar intelligence is often collected by the transit or the consumer countries which needs to be developed and acted upon by the source country and should, therefore, be shared with the source country expeditiously. To prove this point, I would like to give details of a case worked out jointly by the agencies in Pakistan, India and Japan assisted by the U.S. Drug Enforcement Admintstration and the Drug Liaison Officer of the H.M. Customs and Excise posted in India. A telex was received by the Indian Narcotics Control Bureau in the morning of 22nd July, 1987 from Director, Intelligence and Investigation (Customs and Excise), Islamabad informing that eight suspects had left Lahore Airport for New Delhi in two batches on 20th July and 21st July, 1987. They were suspected to carry drugs concealed in baggage or bodies while leaving for India. We were further informed that they were likely to travel by a particular airlines. Immediate investigations were conducted by the officers of the Indian Narcotics Control Bureau and four of the suspects were located in a particular hotel in New Delhi. It was also gathered by the NCB officers that the other suspects had already left for Tokyo on 22nd July itself by an Alitalia flight. It was also learnt that they had further booking from Tokyo to London. Authorities in Tokyo were immediately alerted by us directly and also through the INTERPOL to ensure action. The four suspects staying at the New Delhi hotel were kept under constant watch and at the appropriate time, their rooms were searched, which resulted in recovery of 1.420 kgs. of heroin packed in 15 ball-shaped packages covered with black insulating tape. These were to be carried in their body cavities by the persons arrested. The Japanese authorities on the other hand, on receipt of information, arrested five persons in Tokyo and seized 1.2 kilograms of heroin from their possession. Investigations made by Japanese authorities revealed that part of this heroin was to be despatched to U.S.A. in a postal package to a person whose particulars were passed on to the U.S. DEA for further necessary action.

The success of this case is due, largely, to the timely exchange of operational intelligence and the effective investigations conducted by all the agencies concerned. The importance of co-operation in the field of intelligence and investigations cannot be overemphasized.

Investigating Narcotics Cases

Success of enforcement action in any operation is measured not in terms of quantity of drugs seized but by the conviction of traffickers arrested in the operation. With punishment becoming increasingly severe, the Courts are also becoming increasingly sensitive in the matter of conviction of the accused. As such, quality of evidence produced before the Courts by the prosecution has to be improved. Drug trafficking being an international and even intercontinental operation, the trail of evidence left behind by the traffickers also gets spread over in various countries. Under these circumstances, international co-operation in collecting evidence from the concerned countries and production of the same before the trial courts is of prime importance. Not only the documentary evidence like hotel records, airlines tickets, telephone toll records, etc. But also the oral evidence tendered by investigating officers who might have investigated the case in their own country is sometimes crucial for successful prosecution of offenders in another country.

In the last couple of years, India has not only received legal assistance from several countries but also reciprocated in rendering such assistance to other countries, This legal assistance has been in the form of taking evidence or statement from persons, effecting service of judicial documents executing searches and seizures, providing information and evidentiary documents and tendering oral evidence in trial courts. Article 7 of the U.N. Convention against Illicit Traffic in Narcotic Drugs and Psychotropic Substances adopted in December 1988 is a welcome development in the field of Mutual Legal Assistance. The said Article provides for the measure of such an assistance in investigation, prosecutions and judicial proceedings in relation to drug offences.

International Agreement

When we talk about international co-operation in curbing the drug menace, we have to consider the aspect of legal provisions available in statutes of individual countries. To cover up the deficiencies in law as well as to codify the extent of co-operation in matters of mutual interest, it is imperative that the countries have bilateral/multilateral treaties. This facilitates not only the availability of pre-identified contact points to pass on the operational information but also removes ambiguities which might create obstacles at crucial moments in absence of well defined agreements. Diplomatic maturity gained by international community in the last several decades has brought the world to a stage where international treaties/agreements signed by one national government are duly honoured by its successors despite ideological differences, except in compelling circumstances.

Over the years India has entered into several such fruitful arrangements, not only with its immediate neighbours like Pakistan, Burma, Nepal and Sri Lanka but also with countries of Europe and North America. One result of such arrangements with consuming countries in Europe and North American continent is posting Drug Liaison Officers in India from USA, UK Canada and Federal Republic of Germany. Availability of DLOs of these countries has greatly enhanced the extent of co-operation between these countries and India resulting not only in seizures of large quantities of narcotic drugs but also liquidation of international syndicates which were, until recently, operating with impunity. The co-operation has also resulted in generation of enormous data on traffickers and trends which will be useful in future.

Extent of co-operation with Pakistan has been codified in the form of bilateral arrangements, two meetings of which have taken place so far. There has been an increasing trend in exchange of information between the two countries, both being very important to each other, given the volume of illicit drug trade between the two. In the first two months of 1989, 94% of heroin seized (721 kgs. out of 767 kgs. was seized on the Western land borders of India, while being attempted to be smuggled into India). As envisaged in the arrangements reached, the exchange is expected to be not only in the form of details of seizures having trans-border ramifications, but also in the form of lists and dossiers of gangs and individual traffickers, operating on either side.

In terms of exchange of information with Nepal, the steps taken by both the countries in their respective territories have brought down the trans-border smuggling to a very large extent. But with Burma also, the two sides have agreed in principle to exchange information. Some exchange has already taken place in the form of list of operators alongwith available details. Similar bilateral arrangements are at various stages of finalisation with Mauritius, Thailand and Nigeria. Recently a high level Indian delegation visited Nigeria and explored the possibility of entering into a comprehensive bilateral agreement. It can be hoped that a treaty between the two countries will make a positive impact on the international drug scene.

Victory in war depends upon victory in individual battles. While the international conventions adopted by world community have shown increasing maturity in terms of international co-operation as evidenced by the U.N. Convention Against Illicit Traffic in Narcotic Drugs and Psychotropic Substances in December 1988, a far greater maturity, coupled with a high degree of boldness will have to be shown by individual member countries in more vulnerable areas of the world in the form of comprehensive bilateral/multilateral treaties. In this context, a reference to Paragraph 20, Article 7 of the U.N. Convention of 1988 is necessary which provides for the parties to consider the possibility of concluding bilateral or multilateral agreements or arrangements in the field of mutual assistance. Under the diverse economic conditions of different countries and the quantum of resources made available by each country to fight the drug war, there is a huge difference in the system of storing intelligence/information and related data on traffickers. While for the advanced and affluent countries of the West it is easy to collect and store detailed information regarding an individual suspect, the same may not be possible for the countries of southern part of the globe. To cite but one example: in a country such as India the practice of keeping record of date of birth, place of birth, etc. is not very common while this information is vital for investigation being made in European and other Western countries. Difficulties arise when an exchange of information has to take place between two countries of diverse social and economic environments.

To counter this situation, a standardized format will have to be devised to store information with data bank of enforcement agencies. This will cut down the time loss in exchange of desired information to a considerable extent. Various inputs for the formats may be finalised taking into account the socio-economic conditions of more vulnerable parts of the world. As far as drug trade is concerned these formats will have to be standardized for individual traffickers, organisation, trends, *modus operandi*, routes, money, laundering centres, etc.

Difficulties also arise in the case of passports issued by various countries to its individuals. These should be a well defined set of information regarding the passport holder which must be available in passports issued by different countries. Having identified the narcotic drugs as a common enemy, and war against drugs as the collective responsibility, let us seek to secure the fullest co-operation of all nations in combating illicit drug traffic. We have no option but to strengthen each other's hands in this war against drugs for this menace recognises no territorial jurisdiction and poses a serious threat to all of us alike.

Multinational Operation

Much has already been said about the worsening drug situation in India and the world. Although India has taken significant steps of augmenting the enforcement activity, instituting stringent legislative authority and educating the public to effectively counter the spread of this menace, the spectre has not been mitigated. Seizures of narcotic drugs are reported as frequently; number of persons involved in drug trafficking has not gone down; the demand (illicit) for narcotic drugs and psychotropic substances has not shown a reversal and above all, alarming reports about the growing abuse among a large number of populace regularly appear.

Since drug trafficking, over the years has become as complicated as the working of a multinational business empire involving billions of dollars/rupees worldwide, it is but natural that this activity gradually acquired organisational skills. Just like any other business or trading activity, the prime objective of drug trafficking is profit. Like any other organisational activity, it requires a lot of planning faithful execution. Careful planning is all the more essential because the commodity involved i.e. Narcotic Drugs is outlawed all over the world. Minutest details must be planned because in this trade, markets from which huge profits can be cornered are thousands of miles away from where these drugs are produced and all the steps right from the time of manufacture and processing through their transit and subsequent sale and abuse are against the law.

Let us now see what laws govern drug trafficking. Not the laws instituted by the Government to curb this evil but those factors leading to and controlling the operational aspects of drug trafficking.

The New Convention

A major piece of international co-operation has been adopted by a ministerial-level conference in Vienna. The New Convention Against Illicit Drugs and Psychotropic Substances will come into force 90 days after 20 countries have ratified it. The prospects for ratification are good, because it has been singed by 43 countries with 55 others putting their name to the final report of the conference.

The convention will make it easier to confiscate money and property traceable to illegal drugs. It also facilitates extradition of those involved at all levels of the business and ensures that they are tried and punished in countries where they have committed a crime. Though adopted by consensus, problems remain. Colombian Ambassador Enrique Parejo Gonzalez, who survived an assassination attempt two years ago because of his stand on extradition, felt that the Article on extradition should have been stronger. "It was possible because of the existing legislation of different countries", he said. "But the most important fact is that there is a political will of all governments to apply the rules of extradition".

US Attorney-General Richard Thornburg was also dissatisfied with the clause, and many delegations expressed reservations about extradition of their nationals. Said Thornburg: "we would have preferred accused drug trafficking rendered up for extradition, irrespective of their nationality, wherever they may be found". But he praised the Convention as an "extraordinary international commitment". The big question is how far nations will be willing to modify domestic laws, particularly those perceived as diminishing sovereignty, to accommodate the Convention.

Raymond Kendall, Secretary-General of the International Criminal Police Organisation (Interpol), estimates that 35-40 per cent of Interpol's activity is related to the illicit drug trade, which on US figures ranks second in value only to arms. What happens when the drug trade or the laundering of money occurs at the highest levels of government? "We know that there are certain countries where for all sorts of reasons the situation is difficult because of the amount of money involved", comments Kendall. "Corruption can influence very high levels of government. But at the end of the day one has to make a practical and positive approach to these matters and not necessarily assume that such a situation, which after all exists in a very small number of countries, cannot be resolved by the international community. What is extremely important in dealing with this kind of international problem is that there should be a clear indication that there is a political will to do something about this. And that is clearly what the Convention will give us the maximum amount of support in what we are doing".

The articles relating to seizure and confiscation of proceeds are regarded as vague. Disposal of seized assets is a key issue that remains for individual countries to interpret. Bolivian Foreign Minister Bedregal Gutierrez, the President of the conference which adopted the convention, said of trafficking : "It is without any kind of respect for sovereignty, without any kind of respect for human and moral value".

Previous global conventions against drugs aimed to stop the diversion of legel drugs into the legal market.

They did not concentrate on illegal trafficking or on the identification and prosecution of traffickers. Thornburg said : "the convention gives us a clear message, that those involved in international drug trafficking have no safe haven in the world today for their illegal activities".

Increasing International Response

Humanity is at the crossroads today and the slightest indiscrition on the part of men, who matter, can result into total self-destruction removing even the traces of most of the organisms inhabiting our planet. The problems, besetting our road to universal welfare, are varied and many but the problem of drug abuse and illicit trafficking is one of those problems which affect every state without any difference as regards to race or stage of development.

The unprecedented growth of means of transport, communication and financial institutions coupled with the technological development of highest order, has improved the quality of human life and has brought all the convenience needed at the door-steps of the man but not always without strings. The scientific and technological revolution in its wake, has demolished the age-old natural barriers which confined the Homosapiens for millenniums in a few pockets specially favoured by Mother Earth in terms of bounties of nature. The Islands of enlightenment, progress and prosperity are now co-existing with the oceans of abysmal depth of ignorance, misery and poverty which give rise to many problems resulting from their interactions. The gains of science and technology could not be concealed for long from the alluring gaze of needy millions similarly the symptoms of ailments are not confined to the afflicted bodies only. One of the ailments which have left scars on the face of humanity, is addiction to drugs. The drug addiction created demand and crime syndicates came with the supply and in the process, the first casualty is health of society and social institutions which are repositories of accumulated knowledge acquired through generations and harbingers of new ideas which facilitated technological advancement.

The progress in the field of science and technology has been at breakneck speed but the supporting philosophy and social institutions failed and failed miserably to cope with the changes brought about by scientific research and material gains that followed. Undue caution and systematic inertia in adaptation has reduced our social institutions including judicial system, to ineffective tools to arrest the deviations and unhealthy non-conformity. These institutions are to be remodelled and changed in view of changing problems and we have to infuse in them greater adaptability to suit the needs of the time. Individual nations have already embarked on the task of judicial construction and repair but in the field of international judicial system these efforts are lamentable. We continue to think in terms of nation, states and international system of adjudication is yet to emerge. We have to provide teeth to the International Court of Justice at The Hague so as to enable it to bark and bite. Jurisdiction of this court can be extended to certain categories of criminal offences also.

Trans-Nationality in Crime

For the sake of convenience, trans-national crimes can be put in the following categories.

(1) International drug trafficking.

(2) International organised crime enterprises dealing in crimes of traditional nature.

(3) Industrial crime.

(4) International fraud.

(5) International terrorist activity.

(6) Offences involving works of Art and Cultural property.

(7) Offences relating to environmental pollution.

(8) Miscellaneous.

International drug trafficking and organized crime syndicates are our immediate concern. The trade in drugs is quite a profitable proposition and it attracts the cream of underworld criminals because it requires ability to organise a well-knit unit to manage responsibility, regular supply, efficient delivery system, quick and safe transport, imaginative distribution system and innovative collection and re-investment. The illicit trade has to procure supplies from the regions having best suited climate and terrain, ill-suited government, ill-informed social institutions, ill-equipped law and order machinery and unawakened people. Once supply points have been selected, the organised syndicate starts proliferating its activities by intimidation, extortion and corruption ultimately resulting into the growth of a monster like organization which can pressurise even the government enjoying sovereign powers. For transport, the syndicates have a well-knitted but at the same time well-insulated network of couriers. After successful delivery, the fool-proof distribution and collection system is another imperative need of the enterprise. Legal and financial wizards can be engaged for laundering of money. At times these syndicates have successfully penetrated legitimate international business with a view to either parasitically 'eating away the substance of the enterprise and leaving an abandoned shell' or with a aim to operating 'the enterprise on continuing basis and driving away all healthy competition by predatory methods' (intimidation, extortion and curruption) coupled with sophisticated modern gadgets and approaches. Once established these syndicates develop potentiality of destabilizing governments and spread an aura of omnipotence which is quite helpful to them in covering their tracks. With such syndicates on the scene, conventional tools in the hands of police and other law enforcing agencies become inadequate and here comes the need for concerted efforts and international will to crush such crime and its beneficiaries. Lack of international co-operation and delay in formal approach to sovereign states facilitates removal of evidence both oral and material. The dirth of equipped laboratories and trained personnel further delays and defeats the process of law. Without international co-operation it is difficult to unearth black money and to reach from Bank to beneficiaries.

Judicial Assistance

Sine qua non of all international legislation is undisputed recognition of the principle of absolute sovereignty of states and from this follows the usage that no state shall be allowed to perform sovereign functions in the territory of other state without its explicit concurrence by way of treaty, specific declaration, or the provision in any of the several statutes. All functions performed in the sphere of criminal investigation, prosecution, enforcement of judgement, prevention and correction of offenders are covered in the sovereign functions of the state. The search for the offenders, collection of evidence for prosecution or defence cannot be collected without co-operation of the states if it involves foreign investigation. Further, some of the material witnesses may be the citizen of other states who cannot be approached, persuaded of coerced to appear before the courts of other countries without active co-operation of the requested state. In modern times, no state can afford to ignore this aspect without detriment to its own administration of justice. Though international co-operation in the field of judicial assistance in criminal matters has been recognised as imperative need of our time, yet consensus could not be arrived at so far. The progress of legislation in the sphere of judicial assistance is also far from being satisfactory. Ad-hoc arrangements have been made between the states on the principles of reciprocity and double criminality. This proliferation of international treaties and mutual agreements and difference in the systems of legal administration has made the situation confusing and may be regarded as a stumbling block in the path of growth of international jurisprudence. In this connection United Nations should take a lead and a consensus may be arrived at to have uniform procedure in the matters of judicial assistance. Futuristic steps at opportune time can save us from future complications.

Judicial assistance can be grouped in following four categories:

(1) Assistance in Collecting Information at Pre-investigation State—

The information referred to above can be regarding places, persons, events or laws that may lead to a formal investigation. Normally such information can be collected by ordinary visitors to the country but can also be sought by the requesting state through diplomatic channels. Depending upon the state of relations between the states such information can be allowed to be given to investigating authorities also with routine useful investigative suggestions by their counterparts in the requested states.

(2) Assistance in Investigation—

It includes interrogation, delivery of property, local inspection, service of documents, search and identification, etc.

The assistance in investigation can be both informal and formal. The informal assistance can be such as locating a person, interview of a victim or likely witness or suspect, checking of hotel register, documents of companies which are openly available to check weapons sale documents to ascertain criminal history of a man and other such allied matters which can be attended to on informal exchange basis. Such informal assistance on voluntary basis and mutual co-operation between investigating agencies is an exceptionally important ingredient of international co-operation in the field of criminal justice administration.

Formal assistance in criminal matters can be through displomatic channels.

(a) Letter Rogatory—A traditional mechanism for securing evidence from a foreign jurisdiction. "Letters rogatory retain the quaint formalities of requiring seriatim certifications of the validity of the various signatories appended in the course of their transmission and are sent and received through diplomatic channels with numerous ministerial and judicial talismatic niceties to be observed along the way. Once the hands of receiving judicial authority, the summons issued often produce more lawyers than witnesses." (R. L. Gainer "International Co-operation, Criminal Justice Administration—A selective overview)". (This process is slow, irrational, wasteful, expensive and totally unsuited to our times, but at the same time it can be safely regarded as by for the most reliable and time-tested mechanism). Such requests are made to the courts of requesting country who issue such letters to the court of requested country. The court/tribunal of the requested country have discretion to allow or reject the request. Once they allow such requests then they use their own procedure of execution. The effectiveness of such letters depend on the attitude of the court and relations between the countries.

(b) Transfer of Prisoners—The transfer of prisoners from one country to other is also a new field and is entirely banned on humane considerations. This type of assistance is entirely based on reciprocity.

(c) Training—International co-operation in the field of training is of immense importance. Training provides an insight into the working techniques of the law enforcing arms of other countries and also gives an idea about the play of the mind of criminal in other parts of the world which arms the officer to make futuristic planning in their fight against crime in general and drug trafficking in particular.

It is really a paradox that the need for co-operation in the judicial field has been repeatedly voiced from international forum and still a large number of states have not accepted in letter and spirit the conventions of United Nations and have not executed treaties for mutual co-operation what to say of futuristic planning in this field which may bring about homogeneity in the law of nations which may regulate judicial interaction. Some countries still subscribe to the old school of thought that "The war on crime is a domestic affair of the state concerned. Co-operation between the states in fighting crime is possible in respect of the problems which the states themselves consider acceptable and to the extents to which they do so". (Lipowscheck—Text book on International Law of the Academy of Science of the U.S.S.R.).

Since the drug menace has not spared any country, big or small, rich or poor, white or black, the response has also come from all directions for signing of Convention for the Supervision of the illicit Traffick in Dangerous Drugs in 1936, followed by Conventions signed in 1961 and 1971. The concern shown by U.N.O. is evident by the organizations that have come up on international stage recently under the auspices of United Nations which include:

(1) Commission on Narcotic Drugs.

(2) The Division of Narcotic Drugs.

(3) The International Narcotic Control Board.

(4) United Nations Fund of Drug Abuse Control.

In addition to above organisations, U.N. also organised Congresses on prevention of crime and treatment of offenders and constituted Committees on crime prevention and control.

The World Health Organisation concentrates on the study of medical aspect of drug abuse. W.H.O. also provides expert advice on classification of drugs and treatment and detoxification of addicts. The International Criminal Police Organization (Interpol) is the most important international organization which helps law enforcement agencies of various states in collection and dissemination of intelligence and also in the process of investigation, including analysis of drugs and psychotropic substances.

The I.C.P.O. (also known as Interpol) was founded in 1923 and its General Secretariat was located in St. Cloud, France in 1946 where it is working even now. In each country, the State Government designated National Centre Bureau to have working liaison with I.C.P.O. Interpol have drugs sub-division in the Police Division of Interpol General Secretariat. One additional Regional Drugs Liaison Office is located in Bangkok, Thailand. The Drugs Sub-Division of Interpol General Secretariat consists of specialist officers drawn from different member states. The Liaison Officer scheme of the Drugs sub-division aims at establishing close contact with the officers of the countries within the jurisdiction to foster international co-operation. The officers or the staff of Interpol do not act as International Police and to the extent the operational action remains in the hands of the police of the countries concerned.

I.C.P.O. (Interpol) works in close co-operation with other International Organizations such as the Customs Co-operation Council, the International Narcotics Control Board, U.N. Division of Narcotics, Drugs and Crime Prevention and Criminal Justice Branch of U.N.O.

The "water edge" operations are one of the oldest measures that have been taken against transnational peddlers of the drugs and have led to formation of the Customs Co-operation Council. The widening gap between cost of manufacture of drugs and its sale price has attracted talented people to under-world trade who have brought into play latest techniques in inventing most sophisticated methods of deceiving the man at the 'water's edge', i.e., the Customs Officer. The international co-operation and quick transmission of information is the crying need of the day to detect the crime and prevent it by futuristic planning. When two customs worked in unison, the results have been quite amazing. The Customs Co-operation Council has gone a long way to foster such co-operation and it also brought about the concensus with respect to certain methods of detection (for instance the controlled delivery techniques) in respect of trans-national crime.

The role of international conventions and co-operation in various fields cannot be over-emphasized but the fact remains that all the member states do not adhere to one and the same political philosophy which results into certain substantive and procedural impediments like the considerations of 'double criminality' (the act should be considered offence in both the countries). 'Political Offence' (the act may be considered high treason in one country while in other country it may be treated as political activity) and procedure laid down regarding recovery of property, procedure regarding handing over of incrimating physical and material evidence to court of law and finally the rules regarding evaluation of evidence recorded in foreign land by persons other than the investigating officers who are available in court for the purpose of examination and cross-examination.

The evergrowing awareness to the damage caused

to society by drug abuse, has evoked universal response which is voiced from international fora and in the regional conferences wherein member countries have shown determination to curb trafficking despite differences in political and economic fields. India also signed the final act of the new U.N. Convention against illicit traffic in Narcotic drugs and Psychotropic substances in December 1988 (Shri M.U.N. Rao, Chairman Central Board of Excise and Cotton signed the Act at Hufburg Place, Vienna). This new convention introduces new areas of co-operation for preventing illicit traffic by sea, intensification of control in free trade zone and preventing trafficking through mails. At the instance of Indian delegation a separate article has been incorporated in the new convention which casts an obligation on the industrialised countries which happen to be consumers of illicit drugs, to support and assist the transit States through a programme of technical co-operation and in case of need, provide financial assistance for creation of infrastructure in transit States for drugs interdiction and related matters.

Till very recently, the developing countries were pressurising the countries of origin of narcotic drugs to exercise control over production and pretty little was done for controlling the demand of the drugs, in certain countries even the abuse of drugs was not declared an offence but now the situation has changed considerably for betterment of human society. It has been realised that trafficking in drugs cannot be stopped by merely controlling production or the supply but it has to take the demand also into consideration. Mrs. M. Thatcher, British Prime Minister, while addressing 600 delegates and ministers from 112 nations on April 10, 1990, voiced the same sentiments when she said, 'but efforts to reduce production of drugs and to prevent traffic in them can never succeed while demand for drugs is still there...without customer the drug traffickers would soon be out of business'. The growing realisation of the above facts is bound to foster more amity in international relations and it is sure to open up new areas of co-operation in our fight against criminality in general and drug trafficking and abuse in particular. It is heartening to note that more ferocious the drug cartels become the response and the determination to meet the challenge is equally strong. "By joining forces fighting anti-drug war, we can put an end to the terrible slavery of drug addiction and stop the violence which it causes. U.N. Secretary General said and announced (London, April 11, 1990) that next 10 years be designated as the 'U.N. decade against Drug Abuse and Trafficking'.

References

1. Ranborg B. D.—International Control of Narcotics—Contemporary Problems, Vol. 22, No. 1.
2. How drugs destroy Nation—Donald Schoweder—Plain Truth—April, 1989.
3. *Times of India*, 29.3.86/31.3.86—Tackling the Heroin Menace—N.S. Saxena.
4. *Dainik Jagran*, October 29, 1986.
5. *Indian Express*—March 26, 1986.

64

UNITED NATIONS' VITAL ROLE IN DRUG CONTROL

All the previous treaties were subsumed into the 1961 Convention known as 'The Single Convention on Narcotic Drugs, 1961' By this Convention, the Governments of the contracting parties, for the first time extended the whole apparatus of international control to all narcotic substances.

1971 Convention

Yet another event took place in 1971 when under the aegis of the United Nations, a Convention providing for international control mechanism in respect of altogether different drugs which were not covered under the 1961 Convention but were equally deleterious, known as "Psychotropic Substances 1971". The psychoactive substances have immense medical values but when abused, leads to addiction causing human misery. This convention also embodies the cardinal principles that the availability of psychotropic substances should be limited to legitimate purposes and effective measures against abuse of such substances require co-ordination and universal action.

In the first decade of the working of the Single Convention on Narcotic Drugs, 1961, certain deficiencies were noticed with regard to the reporting system to the concerned organs, adequate penal sanction, proper and effective monitoring system, etc. With a view to fill up the required gaps, the Single Convention on Narcotic Drugs, 1961 was amended by the Protocol of 1972 making the same a wholesome international instruments covering all aspects of narcotic drugs.

The Single Convention on Narcotic Drugs, 1961, as amended by the 1972 Protocol and the Convention on Psychotropic Substances, 1972 which enshrine similar basic principles in providing adequate control measures on narcotic drugs and psychotropic substances respectively. The salient features of these conventions are as follows:

(i) The objective, extent and applicability of the Conventions in respect of drugs and substances have been indicated in clear terms.

(ii) On the basis of the abusive potentiality, nature and medical uses, the drugs and substances have been categorised in various schedules.

(iii) Elaborate provisions for scheduling, re-scheduling and de-scheduling of drugs and substances on the recommendation of the World Health Organisation have been incorporated.

(iv) The Policy Making Body called, the U.N. Commission on Narcotic Drugs, which is a functional commission of ECOSOC, has been empowered to take final decision in such matter.

(v) For proper implementation and monitoring of the working of these treaties, a control organ with the U.N. System called, "International Narcotics Control Board" was constituted comprising 13 members, duly elected by Economic and Social Council (ECOSOC).

(vi) The contracting parties are required to establish Special Administration (National Centralized Coordination Agencies) to function as nodal points for assisting each other in carrying out the mandates given by the Conventions; exchange of information and submission of Annual Reports on the working of Treaties to the U.N. Secretary General.

(vii) The Convention also makes it mandatory on the part of the Contracting parties to furnish to the Secretary General such information as may be necessary for the performance of its functions and, in particular, an annual report on the working of the Convention, text of all laws and regulations from time to time promulgated to give effect to the provisions of the Convention, information relating to cases of illicit traffic detected in the country of report, etc., as also estimates of drug requirements, area under opium poppy cultivation, imports and exports of drugs and poppy straw, seizures of drugs and disposal thereof, stock of drugs as on 31st December of the year under report, etc. Limitation has also been imposed on the manufacture and importation of the drugs.

Salient Features of the new UN Convention Against Illicit Traffick in Narcotic Drugs and Psychotropic Substances—Adopted by the UN Conference of the Plenipotentiaries held in Vienna in November-December 1988 and Report of the Indian Delegation on its Participation in the Conference :

In this part, the salient features of the new UN Convention against illicit traffic in narcotic drugs and psychotropic substances have been briefly given. Government of India had deputed a high level delegation to attend the Conference. It was a matter of great honour for India that Shri M.V.N. Rao, Chairman, Central Board of Excise & Customs was unanimously elected as Chairman of the Drafting Committee of the Plenipotentiary Conference. The highlights of the contributions made by the Indian delegation and some of its achievements have also been briefly mentioned in the ensuing paragraphs.

Background Leading to the Convention of the Plenipotentiary Conference

The, drug abuse and illicit traffic situation in the world has steadily deteriorated causing serious concern to the international community. Even though the Single Convention on Narcotic Drugs, 1961 as amended by the 1972 Amending Protocol and the UN Convention on Psychotropic Substances, 1971 which are the current treaties forming the corner stones of international drug control system have been reported to be working well, in the context of the deteriorating situation, a need had been felt to strengthen and supplement them in certain areas to effectively deal with all aspects of illicit traffic menace. As a matter of fact, the idea of having a new Convention to specifically deal with the illicit traffic, was inspired by QUITO, NEW YORK and LIMA declarations signed by some of the Latin American countries in 1984 and 1985. These declarations sounded profound alarm at the seriousness of the drug menace. They also reflected the will and determination of the people to combat the same which culminated in holding of international conference at the Ministerial level in June 1987, *inter alia*, adopting a Political Declaration through which the world leaders committed themselves to comprehensive international action against drug abuse and illicit traffic, and also stressed for urgent but careful preparation and finalisation of a new Convention against illicit traffic.

The UN Commission on Narcotic Drugs, in pursuance of the mandate contained in General Assembly Resolution 39/141 of 14th December, 1984 and also in the subsequent Resolutions of ECOSOC, initiated action for preparation of a Draft Convention as a matter of priority at its 31st Session in February 1985 which considered the various aspects of the problem as a whole and, "in particular, those not envisaged in the existing International Conventions." The Governments were invited to identify necessary elements which they considered appropriate to be included in the Draft Convention. Majority of the governments communicated the elements, of which the Commission selected 14 elements for preparation of the initial text of the Draft Convention, *vide* its Decision 1 (S-IX) of 14.2.1986. These elements related to tracing; freezing and forfeiture of proceeds of drug trafficking; strengthening of extradition obligations; control on specific chemicals and precursors used in illicit manufacture; control on commercial carriers; co-operation amongst drug law enforcement agencies; improvement in communication system; training assistance; exchange of expertise; exchange of Drug Liaison Officers; consideration of special problems of transit states; mutual legal and judicial assistance in investigating and prosecutional matters; controlled delivery; adequacy of sanction for drug offences; strengthening of mutual co-operation to check drug traffic on high seas; prevention and control of illicit cultivation; extension of control in free trade zones and free ports; prevention of illicit traffic through mail; and prevention of receipt, possession, etc. of equipment needed for manufacture of drugs. The elements identified and included in the said initial draft convention also took into account some of the elements, communicated by India, which related to tracing, freezing and forfeiture of proceeds of drug crime; streamlining the extradition procedures; technique of controlled delivery; co-operation across frontiers and removal of deficiencies in certain areas in the Single Convention on Narcotic Drugs, 1961 and the convention on Psychotropic Substances, 1971.

The Draft Convention was considered by three Sessions of Open-Ended Inter-Governmental Expert Groups (held in 1987 and one in early 1988). At the February 1988 Session, the UN Commission on Narcotic Drugs, reviewed the text of the Draft Convention, as finalised by the three Expert Groups and decided that certain articles thereof, should be referred to the Pleni-potentiary Conference to be convened to adopt the Convention. Vide its Resolution 1988/9, the ECOSOC decided to convene a Review Group meeting to review the Draft text of certain articles of the Convention and consider necessary organisational matters for holding of the Plenipotentiary Conference from 26th November to 20th December, 1988. Accordingly, the Review Group, in compliance with the mandate of the aforesaid ECOSOC Resolutions considered and finalised the draft Convention and the organisational matters relating to the holding the Plenipotentiary Conference and the draft provisional Rules of Procedure in respect of that Conference. It also finalised draft provisional agenda of the Pleni-potentiary Conference and prepared the Working Document E. COBF. 82/3 which formed the basis of deliberations of the Plenipotentiary Conference.

This Convention is a comprehensive, effective and operative international law directed specificially against illicit traffic; considers all aspects of the illicit traffic menace, in particular, those not envisaged in the existing treaties, namely—the Single Convention on Narcotic Drugs, 1961 and the Convention on Narcotic Drugs, 1961 and the Convention on Psychotropic Substances, 1971. It reinforces and supplements the measures provided in the aforesaid treaties. It contains 19 substantive Articles. The new convention contains directives for demand reduction. It also defines the scope of this Convention as one to permit co-operation among the parties, *inter alia*, underlying the principle of non-interference in the domestic affairs of each other.

This Convention provided for—

(i) mandatory sanction for the punishment of serious drug offences—earlier Convention did not provide for such mandatory sanction;

(ii) identification, tracing, freezing, seizure and forfeiture of drug-related proceeds and properties—earlier Conventions did not contain any such provision;

(iii) mandate for extradition and also streamlining the procedure related thereto—earlier Conventions were not so specific about this;

(iv) lays down a detailed procedure for affording mutual legal assistance in investigation and prosecutional matters and also enlarges co-operation on operational matters—earlier Conventions, except generally providing for co-operation between the parties in suppression of illicit traffic, etc. did not lay down any specific procedure on these matters;

(v) exchange of information on operational matters, training of enforcement officers and other experts, etc.—no such provisions were available in the earlier Conventions;

(vi) controlled delivery techniques of permitting the substances to move under surveillance of enforcement officers for identification and apprehension of persons involved—no such specific provision was available under the existing Conventions;

(vii) control of chemicals and precursors at international and national levels, which are frequently used in illicit manufacture of narcotic drugs and psychotropic substances—there was no such provision in the earlier Conventions;

(viii) introduces new area of co-operation for preventing illicit traffic by sea; intensification of control in free trade zones and preventing illicit traffic through mails—earlier Conventions had no specific provision on these.

To make this Convention effective and strong, no provisions for giving reservation or declaration in the matter of implementation of certain articles have been provided—earlier Conventions contained no specific provisions on this.

Highlights of Indian Achievements

The major achievements of the Indian delegation has been in terms of getting an Article on co-operation and assistance for transit states, particularly, developing countries incorporated in the Convention, which concept had been opposed throughout by the Western Bloc countries. Article 6 bis "international co-operation and assistance for transit states" was incorporated in the Draft Convention at the instance of India at the 10th Special Session of the UN Commission on Narcotic Drugs in February 1988. At the Plenipotentiary Conference, the industrialised countries such as USA, UK, France and Netherlands, with the co-sponsorship of Kenya and Thailand moved an amendment to replace the concept of 'transit states' by introducing the concept of 'transit traffic' in Article 6. This was a tactical move to wipe out not only the concept of 'Transit States' from Article 6 but also to eliminate Article 6 from the Convention. The Indian delegation anticipating this move with its attendant implication, submitted an amendment to Article 6 with the co-sponsorship of 11 transit countries, adding one more paragraph to provide for entering into bilateral or multilateral agreement between the parties for enhancing the effectiveness of this Article. The Indian delegation did extensive lobbying for mustering support of developing transit countries. Netherlands has also submitted an altogether different draft of Article 6 which diluted the entire concept of technical and financial support to be made available to the transit countries as contained in the original draft text submitted by India.

The Indian delegation, while formally introducing the amendment traced the historical development of the concept of 'Transit State', *inter alia*, pointing out that this concept was a well recognised one and adopted over many years now. In this connection, the recommendation of the Inter-regional HONLEA meeting of 1986,

World Conference of 1987 and the Commission's observation on the question of alleviating the problems of transit states and the need for incorporation of suitable provisions in the convention for providing financial support on bilateral and multilateral basis to such states was referred to and strongly reiterated. Most of the Latin American countries, namely—Argentina, Bolivia, Colombia, Mexico, Bahama, etc.; Asian countries namely—China, Nepal, Bangladesh, Sri Lanka, Indonesia, Iran, etc.; African countries namely—Egypt, Nigeria and Mauritania and certain Communist Bloc Countries including USSR supported the stand of the Indian delegation for retention of the concept of 'Transit State'. The conference decided to consider the amendment moved by India on Article 6 and after a great deal of discussions, adopted the same by consensus with certain modifications. In our view, adoption of Article 6 can be reckoned as a remarkable achievement of the Indian delegation. Other notable achievements of the Indian delegation are as follows:

Article 3— **'Confiscation of Proceeds Related to Drug Crimes':**

Indian concepts 'forfeiture', 'Competent Authority' and reversal of 'onus of proof' have been suitably got incorporated.

Article 9— **'Other Forms of Co-operation and Training':**

Despite the best efforts of the Western Bloc Countries to eliminate the concept of 'Transit State' from this Article, the Indian delegation succeeded in getting the support for retention of this concept.

Article 12— **'Substances Frequently Used in Illicit Manufacture of Narcotic Drugs and Psychotropic Substances':**

At the instance of the Indian delegation, a paragraph providing for control and monitoring mechanism at national level for such substances was got incorporated. Through an amendment Bill, necessary provisions for such controls in our national law were already introduced. In this connection, it may be mentioned that the Western Countries which are the major manufacturers of acetic anhydride, etc. were opposed to this idea.

Article 13— **'Materials and Equipments':**

The original text of this Article was very sketchy. Indian delegation took a lead in providing a self-contained draft of this Article.

Article 14— **'Measures to Eradicate Illicit Cultivation of Narcotic Plants and to Eliminate Illicit Demand for Narcotic Drugs and Psychotropic Substances':**

The Indian delegation succeeded in getting a paragraph, on easy destruction/lawful disposal of seized/confiscated drugs after retaining necessary samples of such drugs, incorporated.

Article 17— **'Illicit Traffic by Sea':**

This Article deals with suppression of illicit traffic by sea. The original text of the corresponding Article 12, as drafted by the UN Secretariat contained a specific reference to the protection of rights, obligations and exercise of jurisdiction of the coastal states in different marine zones in terms of the UN Convention on the Law of the Sea, 1982. In the course of review of the original text, however, due to strong opposition from the Western Bloc, and particularly from USA, the reference to the UN Convention on the law of the sea was omitted. India, as a signatory to the aforesaid Convention, had joined Brazil in stressing for retention of the original text prepared by the UN Secretariat. At the Conference, an Informal Working Panel was constituted to go into the draft of this Article and after a prolonged negotiation lasting for more than a week, a compromise draft was formulated which could take care of the interests of different groups, particularly, the countries which are signatories or parties to the UN Convention on the Law of the Sea. Since most of the countries from the Western Bloc, are not signatories or parties to the 1982 Convention, consensus was to omit a specific reference to the UN Convention on the Law of the Sea and substitute the same by the "International Law of the Sea". In the final draft, which has been adopted by the Conference, reference to different marine zones has also been avoided and the interests of coastal states have been sought to be taken care of through specific safeguard clause contained in para II of Article 17. The Indian delegation played a very active role in formulation of the compromise draft as discussed above to ensure that there is no interference of third parties in the contiguous zone (This was the advice of the Ministry of External Affairs also). This Article as adopted finally does not confer unwarranted rights on the third parties in the matter of suppression of illicit traffic in the contiguous zone where the coastal states may take necessary action to suppress Customs and fiscal offences under their national law in terms of Article 33 on the UN Convention of the Law of the Sea, 1982.

Assembly Adopts Global Programme, Proclaims UN Decade Against Drug Abuse

A 30-point Political Declaration and wide-ranging Global Programme of Action were adopted by representatives of some 157 States attending the seventeenth special session of the General Assembly held from 20 to 23 February, 1990 in New York. The Assembly session was devoted to the question of international co-operation against illicit production, supply, demand, trafficking and distribution of narcotic drugs and psychotropic substances.

Of the approximately 110 speakers who addressed the Assmebly in plenary, 50 were ministers.

Referring to the resources available for drug abuse control, in his opening statement, the President of the General Assembly, Joseph N. Garba (Nigeria), said that the international organizations engaged in this field "need and deserve far greater allocations than those prevailing now". He urged Member States to be realistic about the cost when considering what more to ask of the United Nations in drug abuse control.

Resolving to protect mankind from the scourge of drug abuse and illicit trafficking, the Assembly proclaimed the period from 1991 to 2000 as the United Nations Decade Against Drug Abuse. In this context, it calls for effective and sustained national, regional and international actions to promote the implementation of the Global Programme of Action. The Declaration calls for intensified efforts in the fields of health, education and public information, including well-targeted campaigns involving the relevant United Nations specialized agencies and NGOs, as an essential measure to reduce illicit demand. It urges the international community to increase economic and technical co-operation to developing countries and to facilitate trade flows in support of viable alternative income schemes, such as crop substitution programmes.

States are encouraged to ratify or accede to the United Nations drug control treaties and, to the extent they are able to do so, to apply provisionally the terms of the 1988 United Nations Convention Against Illicit Traffic in Narcotic Drugs and Psychotropic Substances. Emphasis is put on the need to strengthen and enhance United Nations structures for drug abuse control and to provide additional financial and human resources for United Nations activities in this field.

The Global Programme stresses the need for collective and simultaneous action in combating drug abuse and illicit trafficking, noting that States are not in a position to deal with this scourge individually.

The Comprehensive Multi-disciplinary Outline adopted at the International Conference on Drug Abuse and Illicit Trafficking is to be used as a basis for developing balanced strategies aimed at combating all aspects of drug abuse and illicit trafficking.

The programme outlines the various aspects of such action, including prevention and reduction of demand for illicit drugs; treatment, rehabilitation and social reintegration of drug addicts; control of supply; suppression of illicit trafficking; measures against the use and laundering of drug-related money; strengthening of judicial and legal systems.

Urging states to give high priority to activities for the prevention and reduction of drug abuse, the Programme emphasizes the role of the United Nations as an advisory centre for collecting, analyzing and disseminating information and experience in demand reduction. States are requested to establish and promote national drug abuse assessment and data collection systems. To this end, they should set up data banks on the basis of the International Drug Abuse Assessment System being developed by the Division of Narcotic Drugs.

The Co-ordinator of all United Nations drug control-related activities, nothing that existing United Nations mandates in drug abuse control were extensive, pointed out to the Committee that to fully implement those mandates and additional ones the Assembly might identify, the Secretary-General had suggested the governments might consider an exceptional increase to the regular budget as well as greatly increased extra-budgetary flows.

If the international community were to resolve to release savings of a mere 1 per cent of the estimated value of the current total global defence expenditure, a massive $ 10 billion per annum would become available to help eradicate drug abuse, poverty and every kind of social deprivation. Speakers before the Assembly's Committee included representatives of UNESCO, WHO, IFAD, UNDP, ILO and the Organization of American States.

The special session of the General Assembly was convened pursuant to resolution 44/16 adopted by the Assembly in response to calls for action made during its 44th regular session by numerous world leaders. The Political Declaration and the Global Programme of Action are found in document A/RES/S-17/2 of 15 March, 1990.

Commission submits Statement to GA

Special attention was given to the enhancement of the role of the United Nations in the fight against drug abuse and illicit trafficking at the eleventh special session of the Commission on Narcotic Drugs, held from 29 January to 2 February, 1990 at Vienna. Some 100 States were represented.

The Commission's views, outlined in a special statement and a chapter of its report, were submitted to the General Assembly at its seventeenth special session on co-operation in matters of international drug abuse control, pursuant to Assembly decision 44/410.

The statement underlined a series of general principles for the development of a global programme of action, including a multisectoral approach to health and social problems associated with drug abuse as well as a commensurate strengthening of the United Nations drug control organs and their secretariats. To permit consideration of the various views and proposals put forward in the limited time available, the Commission established an Ad Hoc Committee of the whole.

Speakers before the Commission included the Colombian Minister of Justice, Gaston Ponce Caballero; the Minister of Home Affairs of Zambia, General G.K. Chinkuli; and Georgine Dufoix, national co-ordinator for drug abuse control activities in France.

In the absence of the outgoing Chairman, the session was opened by the Director-General of the United Nations Office at Vienna and Co-ordinator of all United Nations drug control-related activities, Margaret J. Anstee. Statements were also made by the president of the International Narcotics Control Board, Benjamin Huyghe-Braeckmans, the Executive Director of the United Nations Fund for Drug Abuse Control, Giuseppe die Gennaro, and the Director of the Division of Narcotic Drugs, Francisco Ramos-Galino.

The Commission adopted eight resolutions and 11 decisions and recommended four draft resolutions to the Economic and Social Council for adoption.

Elected Chairman was Edward Babayan (USSR), who had also served as Chairman of the 27th session of the Commission held in 1977.

Other officers were Robert J. Samson (Netherlands), Francisco Cuevas-Cancino (Mexico) and Yahya Mahmassani (Lebanon) Vice-Chairmen; Maurice Randrianame (Madagascar) Rapporteur.

Organized by the Division of Narcotic Drugs, the Commission's special session was also attended by representatives of 3 specialized agencies, 9 intergovernmental organizations and 29 non-governmental organizations in consultative status with the Economic and Social Council.

The 40-member Commission serves as the main policy-making body in the United Nations for matters of international drug control. Catherine Volz (USA), newly appointed Chief of the Treaty Implementation and Commission Secretariat Section, served as Secretary of the Commission.

UNITED NATIONS

DIVISION OF NARCOTIC DRUGS. VIENNA

INSTRUCTIONS FOR USE

SPECIAL INSTRUCTIONS

READ CAREFULLY BEFORE USING KIT

Important

The purpose of this kit is to provide law enforcement officers with rapid and simple tests for preliminary field identification of the drugs most commonly encountered in illicit traffic. Because of the preliminary nature of these tests, it is imperative that in all cases where presumptive positive results have been obtained, the suspected material should be submitted to an authorized laboratory for further confirmatory analyses.

In order to obtain reliable results with the tests included in this kit:

— Always clean and dry the equipment (test-tube, spot-plate and spatula) immediately after use;

— Never allow transfer pipettes or tips of reagent flasks to come into contact with the suspected material, as contamination of the reagents will yield false results.

Warning

This kit contains corrosive concentrated acids. Therefore:

— Handle the reagent flasks so labelled with care, avoiding contact with any part of the body;

— If the contents come into contact with the body, flush immediately with water, apply the acid neutralizer solution N provided in the kit, and seek medical attention;

— In the event of spillage, apply the acid neutralizer solution N to the area of the spill and flush with water.

This kit also contains some volatile and inflammable reagents. Therefore:

— Do not allow smoking when the kit is in use;

— Always replace and tighten caps on reagent flasks immediately after use;

— Conduct the tests in the open or in an adequately ventilated area and protect the reagents against lengthy exposure to sun and heat.

DO NOT SNIFF OR TASTE SUSPECTED MATERIAL OR REAGENTS

TESTING METHODS

(See also flow charts I and II)

TEST A

Opium

1. Place a small amount of the suspected material (the size of a match-head) on the spot-plate provided in the kit.
2. Add two or three drops of water; with the glass rod or spatula provided in the kit, smear the sample against the spot-plate for one or two minutes.
3. Transfer a drop of the liquid to another part of the spot-plate.
4. Add one drop of reagent A_1.
5. Add three drops of reagent A_2.

   Colour * * * indicates a positive result.

Morphine, codeine, heroin

1. Place a small amount of the suspected material

(the size of a match-head) on the spot-plate provided in the kit.

2. Add one drop of reagent A_1.
3. Add three drops of reagent A_2.

Colours in the range * * * to * * * are positive for morphine.

Colours in the range * * * to * * * are positive for codeine.

Colours in the range * * * to * * * are positive for heroin.

Amphetamines, mescaline

1. Place a small amount of the suspected material (the size of a match-head) in on the spot-plate provided in the kit.
2. Add one drop of reagent A_1.
3. Add three drops of reagent A_2.

Colours in the range * * * to * * * are positive for amphetamines.

Colours in the range * * * to * * * are positive for mescaline.

TEST B

Marijuana, hashish, hashish oil

1. Place a small amount of the suspected material (the size of a match-head) in one of the test tube provided in the kit.
2. Add a small amount (the size of a match-head) of reagent B_1.
3. Add 25 drops of reagent B_2 and shake for a minute.
4. Add 25 drops of reagent B_3 and shake for two minutes.
5. Allow the test-tube to stand for two minutes.

Colours in the range * * * to * * * in the lower liquid layer indicate a positive result. Ignore colour of the upper layer.

TEST C

Methamphetamine, methylphenidate

1. Place a small amount of the suspected material (the size of a match-head) on the spot-plate provided in the kit.
2. Add one drop of reagent C_1.
3. Add three drops of reagent C_2.

Colours in the range * * * to * * * are positive for methamphetamine.

Colours in the range * * * to * * * are positive for methylphenidate.

TEST D

Barbiturates

1. Place a small amount of the suspected material (the size of a match-head) on the spot-plate provided in the kit.
2. Add three drops of reagent D_1.
3. Add three drops of reagent D_2.

Colour * * * indicates a positive result.

TEST E

Cocaine, methaqualone

1. Grind the suspected material to a fine powder if it is in tablet form, and place a small amount of the suspected material (the size of a match-head) in one of the test tubes provided in the kit.
2. Add one drop of reagent E_1 and shake for 10 seconds.
3. Add one drop of reagent E_2 and shake for 10 seconds.

Colour * * * indicates a positive result.

* * * For exact colour ranges please refer to the charts supplied by the UNO in their Drug testing kit.

FLOW CHART I

NATURE OF SUSPECTED MATERIAL

HERB
RESIN
OIL

RESIN
POWDER

CIGARETTES, SMOKED CIGARETTES AND PIPES, AND OTHER SUSPICIOUS MATERIAL

? MARIJUANA
? HASHISH
? HASHISH OIL

? OPIATE

Try TEST A for:
Opium
Morphine
Codeine
Heroin

Negative result

Positive result

IF GROUNDS FOR SUSPICION

Try TEST B for:
Marijuana
Hashish
Hashish oil

Negative result | Positive result

STOP TESTING

ONLY IF GROUNDS FOR SUSPICION

SUBMIT TO LABORATORY

FLOW CHART II

NATURE OF SUSPECTED MATERIAL

IMPREGNATED PAPER OR CAPSULES GELATINE FILM

TABLETS

POWDERS LIQUIDS

? AMPHETAMINE

? OPIATE

? BARBITURATE

Try TEST A for:
Amphetamines
Mescaline

Try TEST A for:
Opium
Morphine
Codeine
Heroin

Try TEST D for:
Barbiturates

Try TEST C for:
Methamphetamine
Methylphenidate

Positive result

Negative result

Negative result

Positive result

Positive result

Negative

Positive result

Try TEST E for:
Cocaine
Methaqualone

Try TEST A for:
Opium
Morphine
Codeine
Heroin
Amphetamines
Mescaline

Try TEST C for:
Methamphetamine
Methylphenidate

Positive result

Negative result

Try TEST D for:
Barbiturates

Positive result

Negative result

Try TEST E for:
Cocaine
Methaqualone

Positive result

Negative result

STOP TESTING

Positive result

Negative result

STOP TESTING

ONLY IF GROUNDS FOR SUSPICION

ONLY IF GROUNDS FOR SUSPICION

SUBMIT TO LABORATORY

65

SMOKING

The lure that draws a child to his first cigarette is subtle and complex. There is no easy explanation for it, but it is a giant step on the road to addiction to nicotine and disturbingly also alcohol and possibly drugs. For some, it is an initiation ceremony which may set the tobacco habit for life. Tobacco is the operative word because manufacturers are trying to put a variety of tobacco products in the market, not just cigarettes.

There are specific laws restricting purchase by young people. Cigarettes are not generally accessible to them. If young people see smoking as normal social behaviour, they will see it as acceptable. The influence of parents and media is especially relevant. Youngsters who have been exposed to smoking in the family become familiar with the gestures and habits associated with smoking as well as the kinds of satisfaction associated with it. The image of smoking as a normal adult attribute consequently gets reinforced in their minds.

Despite laws to restrict advertising and access to tobacco, many young people still see tobacco use as socially acceptable and this is often reinforced by parental habits and the media. The implied satisfaction, the hint that experimentation is daring, the attraction of the forbidden fruit, are all backed up by clever advertising which continues to link tobacco habits with fun, maturity and an image of the 'modern' man or woman. Indeed, there is evidence that advertising directly influences the decision to start smoking. In countries where tobacco advertising has been restricted, there has been a reduction in smoking among young people.

The lure to experiment with behaviour associated with adulthood may be heightened by the fact that it is generally prohibited for children to smoke and thus it can be a means of expressing opposition to adult authority and independence. This experimentation often occurs in the company of peers, as this context is where approval is being sought. The motivation to smoke may be related to a wish to try to set oneself apart from friends by being 'more adult' or to gain acceptance of friends through conforming to their attitudes and actions. Peer pressure is used by children to encourage each other to smoke and far outweighs any fear of long-term health risk. This is particularly so as young people generally feel themselves to be healthy and may not yet be able to associate current behaviour with future consequences, something which comes with greater maturity. The environment they live in, is an important clue to young people's perception of what society expects of each member. The clinical moves against tuberculosis were also accompanied by a general system of fines to stop spitting in public places and now, of course, it is unacceptable at work and leisure venues will support the healthy development of young people.

Resist the Temptation

The tobacco sector has a sinister perspective: it is acutely aware that young people represent its commercial survival. To compensate for the large number of adult smokers it loses because they stop smoking or die, the tobacco sector needs continually to recruit new smokers. In the U.S. alone it is estimated that 3000 adolescents take to smoking every day. A large number of them will die prematurely due to cancer and other smoking-related diseases or suffer chronic ill-health.

The school thus becomes a vital instrument in health education and awareness about the dangers of tobacco. It has to take a lead in promoting positive responses to non-smoking to combat the pressure that may encourage youngsters to take to tobacco. In the strategy to address the student, three elements seem foremost—the classroom (represented by the teacher and curriculam), the ethos of the school buildings and facilities and the local environment (including the family). There is no one single, proven best way to promote non-smoking and prevent the use of tobacco but only many inter-related variables that influence the growth, development and lifestyle of children and young people.

Health education on smoking has evolved from the Sixties a didactic top-down and victim-blaming stand that often also relied on fear. 'Smoking can kill you, it is major cause of cancer, therefore, do not smoke'. This was true of areas other than smoking as well. Even the children's posters exhibited at the 5th World Conference on Smoking and Health in Winnipeg in 1983 used the "scare approach" showing an over-abundance of skulls, tombs, bones and similar gory scenes.

What was the result of this didactic stand ? Smoking rates among children and adolescents kept on increasing during the Sixties and Seventies, particularly among girls and young women. Now a more balanced approach is suggested (although fear may still have a role), with emphasis on the positive alternative to smoking—a healthy, happy life, satisfaction by sports activities and the like. Greater care is given to knowing at what stage the students themselves have reached in their knowledge, attitudes, values and skills. This offers them greater scope for participation, a key concern in the recommendations of the World Health Assembly, Geneva in May 1989. Students require the opportunity to contribute actively in their own development, acquiring not only the information and knowledge but also the motivation, skills and opportunity to lead life-styles conducive to health.

School-based Smoking Prevention Programme

Evidence exists that school-based smoking prevention programmes can be successful, at least in delaying the on-set of smoking among adolescents if not in complete prevention. Young people may still experiment with risk-taking behaviour including smoking. But they will be able to make decisions later on based on their experiences. It is generally accepted that the later the smoking habit is picked up, the less are the chances of becoming an established heavy smoker and the easier it will be to quit.

Perhaps the most influential force at school is the pressure exerted by the group. It is a classic feature of the growing up process. Its importance of devising strategies is now more clearly understood. Role-playing, the ability to say 'NO' to a cigarette and social influence programmes are part of the skills that should be included in this approach. Often it is the desire to be accepted, to be one in the group that leads to experimenting with cigarette-smoking. Canada, Finland, Israel and U.S. are just some of the countries that have acknowledged the value of peer-led anti-smoking initiatives.

In the North Karelia Intervention Project for lifestyle-related diseases in Finland, a two-year school-based smoking intervention programme focussed on 13-17 years olds ran from 1978 until 1980 and was evaluated in 1982. Its aim was to increase the percieved social desirability of non-smoking and to teach skills on how to resist the pressure to smoke. It used 14 and 15 years olds of both sexes as peer leaders and as the point of first contact for delivering the anti-smoking message to the classrooms. In some schools, the programme was supervised by specially trained health educators, in others by locally available resources like school teachers, temperance associations and youth groups. A third group of schools did not have any anti-smoking programme and served as controls.

Before the programme started, 2-10 per cent of the children reported smoking at least once or twice a month. As the children grew up, the percentage of smokers also increased as would be expected. However, the increase in smoking among the children of the intervention schools was much lower (20 per cent) than among the children from the control of non-intervention schools (30 per cent). This effect was visible not only at the end of the two-year programme, but also in the long-term, two years after the programme ended and the children were 17 years old adolescents. But how soon should anti-tobacco programmes start in schools. Canada is implementing a programme 'Smoke-free : for a new generation of non-smokers', which comprises an information competent for children aged 3-6; other kindergarten approaches are of interest.

The Social environment

The environment of the home, school and community in which young people grow up and get socialised has a major influence on their health behaviour. The school, to varying degrees, reflects the culture norms and values of the wider society. This is a two-way process. Where non-smoking is the social norm, the task of the school in promoting it, is considerably easier.

Many factors in society contribute to the development of a non-smoking norm. Health education and health promotion help raise the level of awareness and knowledge. They also help provide the motivation for children either to resist starting or to give up smoking and tobacco use. The mass media also have a major role. Other interested parties, apart from the school, such as cancer associations and the health professionals combine to create a supportive environment for the introduction of legislation. This covers control of tobacco sales and promotion, ad-taxation and the establishment of smoke-free environments in public transport, hotels, restaurants and public places. All these elements, together with well-planned programmes on promoting non-smoking and cessation, provide the supportive environment in which a school's initiatives can take place.

The Norwegian Tobacco Act was very successful in supporting the anti-smoking lobby. The Icelandic Tobacco Act is another exemplary piece of legislation. Sweden, even though it does not have special anti-smoking school programmes, was able to provide an environment that promoted non-smoking and this produced a decline in smoking among the young. The Swedish smoke-free generation movement was also a contributory factor. Finland, often taken as the model country for anti-tobacco use, not only has specific intervention programmes aimed at schools but also a wide range of legislation, including a tobacco priority policy. However, an interesting recent development in Finland has been the upturn in smoking among young people. One Finnish study pointed to the increased purchasing power of the young and the co-relation between pocket money and cigarette smoking. In brief, it was found that in real terms cigarettes were becoming cheaper.

International solidarity is also a vital link to fight tobacco use. In Europe, 12 out of 29 nations prohibit cigarette sales to young people. In the U.S., all but six States also have such legislation. Australia, New Zealand and Japan have similar regulation. Hopefully, many more countries will do likewise. In looking at smoking prevention in schools it is essential to give full play to the major influence of the wider environment, including the example of teachers, parents and public figures in setting a non-smoking fashion.

To be successful it is essential to approach this major health challenge on a well planned, scientific basis. Health system surveys and evaluation are key factors. Their role has often been undervalued and sometimes other more pressing demands push them to one side. This is changing and a broad picture is emerging which validates pupil involvement and a comprehensive approach in schools. There also needs to be a supportive environment providing suitable legislative and fiscal policy, well planned coordinated action and media support against tobacco.

Schools As a Powerhouse

There are two major areas to be tackled as matter of urgency and with innovation and flexibility, using the school as a powerhouse for change. Developing countries must begin with prevention and control programmes on tobacco use.

The Killer Stick

Tobacco causes more death and suffering among adults than any other toxic material in the environment. This has long been known but now it is feared that involuntary exposure to cigarette smoke causes more cancer deaths than any other pollutant. Protecting non-smokers from cigarette smoke will require marked change in society's treatment of tobacco, one that could also help eliminate its direct threat to users themselves.

No country is yet taking action against tobacco commensurate with the cost it imposes. The global use of tobacco has grown nearly 75 per cent over the past two decades. In China, its use has doubled. In only four countries are fewer cigarettes smoked now than in 1964. In the United States, the percentage of adults who smoke has fallen from 43 to 32 per cent but even there 20 per cent more tobacco is used than when an anti-smoking campaign began in 1964 and the country still ranks third in the world in per capita cigarette use.

Killer

The world-wide cost in human lives now approaches 2.5 million per year, almost five per cent of all deaths. Tobacco kills 13 times more Americans than hard drugs and eight times more than automobile accidents. Passive smokers (those who inhale the smoke of others' cigarettes) are perhaps three times likelier to die of lung cancer than they would be otherwise. The smoking by mothers diminishes the physical and mental capabilities of their children and in many countries more than one fifth of the children are exposed to smoke in this way. These statistics add up a cost that is increasingly viewed—in countries where the information is available, at least, not only as unnecessary, but as intolerable.

Though the health consequences of tobacco are now well known, policies to avoid them lag far behind. Most efforts to control tobacco are merely attempts to control or mere information about the product. Government sometimes warn people that tobacco is unhealthy, ban its advertisement or restrict its use in theatres or buses, though often the efforts are no stronger than the Japanese cigarette package warning : 'For your health, let's be careful not to smoke too much'. No national tobacco control efforts have been launched with the vigour of anti-drug campaigns or even of campaigns against toxic chemicals though hard drugs and chemicals claim far fewer victims than tobacco.

Health leaders in government, international organisations and public interest groups have also failed in this fight—particularly because tobacco is tenaciously addictive, partly because both government and industry promote tobacco, but partly because the leadership of health and environmental authorities is weak. This conclusion is borne out not only by the continued high levels of tobacco use in industrial countries but by the explosive growth of cigarette smoking in Eastern block countries and in China. Yet the information campaigns of concerned health leaders have at least succeeded in getting many analysts to recognise that tobacco is a high priority, worldwide public health problem, and one that needs stronger medicine.

Smoking As an epidemic

Smoking is an epidemic growing at 2.1 per cent per year, faster than world population. Growth in tobacco use slowed briefly in the early eighties, primarily for economic reasons but is again increasing rapidly. Over a billion people now smoke, consuming almost five trillion cigarettes per year, an average of more than half a pack a day. Even in the U.S., where smoking is prevalent, the portion of a population who smoke has declined. The 20 per cent increase in tobacco use since 1964 indicates that those who smoke, now smoke more heavily.

Greece leads the world in per capita cigarette consumption. Japanese, Americans, Canadians, Yugoslavs and Poles follow close behind. People in industrial countries smoke twice as much as people in the Third World. Although Chinese men smoke almost as much as Western men do, the negligible amount of smoking by Chinese women means that country does not rank very high in overall per capita consumption. Nevertheless, China uses a quarter of the world's tobacco. Tobacco is increasingly grown near where it is consumed. China is the world's leading producer, using all it grows. The U.S., India, the former Soviet Union and Brazil rank second to fifth.

Change in tobacco use can be measured in two ways : changes in the use of tobacco products, in absolute or per capita terms and changes in smoking prevalence. Measuring the latter is often preferred by health educators as a sign of progress towards their primary goal which is to get smokers to quit and non-smokers not to start. Prevalence also provides an index of a key goal of tobacco control policy—to make smoking socially unacceptable.

But the absolute quantity of tobacco used provides an essential measure of total health costs that a society must bear. Indeed, the total number of cigarettes smoked over a lifetime is a more important health index than used per day at any given time. Health risks increase in proportion to the total amount of tobacco used.

Hazards of Smoking

Doctors report that heavy crack smoking appears to be causing a growing number of addicts to become critically ill with symptoms similar to those of pneumonia. They say, the condition is not easily detected unless the patients reveal their crack habits, but if it is not treated properly, it can lead to death. Emergency room physicians at four inner city hospitals in Detriot are confronting an average of at least one patient a day with what they are calling cracklung. The problem also has been identified in New York and Baltimore. The doctors say that although the patients give outward signs of suffering from pneumonia, most do not respond to antibiotics or other standard pneumonia treatments. They are being successfully treated instead with corticosteroid, anti-inflammatory drugs.

Preliminary studies suggest that the condition is a variation on a number of ailments that result when poisonous compounds invade the lungs. Some scientists say it is also possible that the condition is caused by other drugs used in combination or by the adulterates used to produce crack. Dr. Dana G. Kissner, a Detroit pulmonary specialist who reports on the condition in the journal *Radiology* said the crack patients suffer severe chest pains and breathing problems, often have high temperatures and cough up blood. She said, the condition can lead to death from lack of oxygen or loss of blood.

Because of its similarity to pneumonia and because patients rarely mention the use of crack—either out of failure to associate smoking crack with their symptoms or reluctance to admit drug-abuse—the crack-lung condition is not always rapidly detected. Dr. Kissner is on the staff at the Detroit Medical Centre where three separate hospitals treat victims of crack-lung. In New York, Dr. James T. Kennedy, an associate professor at the New York University School of Medicine and Medical Director of Covenant House, a private agency that shelters young people, said he had seen about two dozen young people with the condition in the last two years.

The ailment usually appears after heavy use or very deep inhaling of crack, the smokable derivative of cocaine, in an effort to maintain a prolonged high effect, the doctors say. Dr. Kennedy said the young people he saw with the condition had been smoking 12 to 25 pipes of crack a day. At the Handy Ford Hospital in Detroit, five doctors did a retrospective evaluation of the chest X-rays of 71 patients who had been treated in the emergency room for the problem to determine how the X-ray might help detect it. Only nine X-rays showed any abnormalities and there was nothing to distinguish these from either pneumonia or asthma or a number of other lung problems, Dr. Daniel W. Eurman, the senior resident in diagnostic radiology at the hospital said. Dr. Kissner, who is also an assistant professor of medicine at Wayne State University, said the distinguishing factors are the lack of evidence of an infectious organism—a virus or bacterium and the ineffectiveness of antibiotics in combating the condition.

No for Nicotine

The world's first "No Tobacco Day" was observed on April 7, 1988 and it was decided that "From this year, it would be observed every year on May 31". May 31 is the World's 'No Tobacco Day'—a day that the World Health Organization urges countries to dedicate to the cause of tobacco-free societies. It is a day marked by an appeal to those who smoke and to those who chew tobacco to quit for at least 24 hours, however, difficult it may be, as a first step to breaking their addiction to a harmful and wasteful habit. While calling for a day free from tobacco use, WHO is also focussing on the threat to the health of women—both girls and adults—and has chosen the slogan—"The Female Smoker : At the Added Risk".

Stop Before You End up in Smoke

According to the World Health Organization study release, an estimated 500 million people will die from smoking in the next 25 years. WHO statistician Dr. Alan Lopez and Jichard Peto said, their study indicated smoking would be the number one cause of death in the world by the turn of century. "Today there will be 8000 deaths a day from smoking but when the kids reach middle age, there will be something like 28000 deaths a day", Peto told reporters at the World Conference on Tobacco and Health. "If present trend continued, 200 million people under the age of 20 and 300 million adults will die of smoking by 2015", Peto said. "Two-thirds of the forecast deaths in the 1990s will occur in wealthy nations where 20 per cent of all deaths are attributed to smoking", the study said. "But developing nations will probably share 70 per cent of all world deaths from smoking by 2025", they said. The study, said to be the first comprehensive world mortality forecast for smoking, rates tobacco consumption a bigger potential killer than AIDS.

People said there was no doubt about the accuracy of their forecast as it was based on more reliable figures than those for AIDS.

The IMA has expressed deep concern at the growing increase in tobacco use in India and its unethical and unregulated promotion by producers and sellers. In India, the menace of tobacco use is increasing at the rate of two per cent while it is falling at the rate of one per cent in the western world. India, with serious health problems like malnutrition and communicable diseases, cannot afford the hazards to tobacco use and the economic loss associated with it, the release said. It calls for health education of the masses and an aggressive anti-tobacco campaign with active co-operation of government and non-government agencies.

Smoking Habits Among Young

Surveys on smoking habits among youths carried out in 20 countries, 16 of them industrialised, between 1980 and 1985, show that females are relegating males in smoking. All those surveyed were in their early or late teens with the exception of Switzerland and Uruguay where the age varies from 15 to 24. WHO officials warn, "While in most industrialised countries the prevalence of smoking among boys is decreasing, that among girls it is increasing." In the industrialised world, the rate of women dying of lung cancer has increased by more than 200 per cent—a rate of increase that far exceeds for men. Unless governments act now, some 'tens of millions of children' will be at risk of addiction to smokeless tobacco, the sales of which have been shot up sharply over the last 20 years, according to a group of experts from nine countries. WHO estimates put at 2.5 million, the number of lives lost every year from tobacco-induced diseases.[3]

Indian Scene

The past and present tobacco consumption in the country is encapsualised in a 23-page document of the Indian Council of Medical Research. It concludes after giving facts and figures of the matter with a hope that India would usher in a 'tobacco-use-free society' in the year 2000. But reality seemed starker than statistics. The ICMR spokesperson, Dr. Usha Luthra admitted that smoking was on the rise. The number of smokers, including women had reached 150 million. She said, "We do not want to burden you with undue figures". The ICMR wanted to initiate an effective debate amongst the people, so as to bring out more ideas to tackle the problem effectively. The emphasis, this time, was more on women smokers, it turned out.

The lively discussion touched upon the whys and wherefore of tobacco consumption in India. Certain areas were to be best left in grey, one felt, after sometime. It was asked why one witnessed an increase in tobacco advertising. Since female smokers were the focal point of the debate, the participants were told about a brand, exclusively for the chic women smoker, soon to hit at the 'Paan' shops.

The ICMR officials preferred to remain at low key on this matter, still maintaining that it was possible to contain the increase in urban smoking. Dr. Luthra, however, conceded that "money is in the other side of the balance". The implications were clear—tobacco industry in India, besides being a money-spinner, thrived on the image of a 'glamorous and urbane' consumer which was being reflected in the gloss of the advertisements. ICMR document conceded that tobacco industry earned revenues to the tune of Rs. 1733 crore yearly. While India ranked third in tobacco producing countries of the world, "there is still scope for improvement in per hectare yield in our country", which meant there would be more tobacco available in India in the years to come.

Cigarettes, 'Beedi' and even chewing tobacco consumption heads the different modes of tobacco intake in the country. The cigarette and *beedi* factories provide direct employment to a little more than 30 lakh people besides the lakhs involved in the supply and distribution. Alternative mode of generation and employment had to be explored if one was to aim for a "tobacco-free society".

The ICMR document pointed out, on the other hand, the cost of treatment of tobacco-related afflictions was more than that of the actual revenue generated in tobacco production in India. It worked out to be Rs. 685 crore more than "revenue and foreign exchange provided by tobacco to the Government". The innocuous statutory warning, "Cigarette smoking is injurious to Health", was now reduced to a mere embellishment on the attractive cigarette packets.

Women too Enticed into Smoke-ring

WHO's Director-General, Dr. Hiroshi Nakajima said, "It is deplorable that while tobacco consumption by men is clearly decreasing in certain countries, women are becoming the principal target of tobacco advertising." "The burden of illness is being shifted", he added and warns that "women who smoke like men will die like men. All told, some 2.5 million people, at a conserative estimate", die each year from tobacco-induced diseases, e.g., lung cancer, chronic bronchitis, emphusema and heart disease. That is, a death every 13 seconds.

Since an increasing number of men are giving up this habit, the tobacco industry's advertising is pushing its products on women, youth and the Third World. Already in developing nations, it is urban women—often university educated, one of society's pace-setters—who are taking to cigarettes. There's a benefit to be gained if they stop, the report says. The risk of heart attack and stroke is reduced within two years, for example and a normal birth results from "giving up at any stage of the pregnancy...though the earlier, the better".

Nearly 40 per cent of the female population in India above the age of ten years use tobacco in one form or the other. In a press release issued on the eve of the second 'No Tobacco Day' organised by the World Health Organization (WHO), the Indian Medical Association (IMA) said that of the estimated population of 800 million in India, nearly 112 million females in the country use tobacco in one form or the other. 'No Tobacco Day' was observed in 1989 with the theme, "The female smoker : at added risk." Female tobacco users, like their male counterparts, are exposed to health hazards and tobacco-related diseases. It is suspected that one out of every 16 deaths in female is due to tobacco use. "Women are exposed to all the risks that apply to men including lung cancer, but in addition run other specific risk to their sex", says Dr. Eileen Crofton, former Director of Scotland's Action on Smoking and Health (ASH).

Stated differently, women are on their way to achieving equality with men in deaths from lung cancer. Their rate of increase in mortality far exceeds that for men. Women are at increased risk of heart attack and stroke if they are also on the pill. "Women should be aware that if they use oral contraceptives, they should not smoke", says a report. In addition, "a mother to-be smoker, she runs an increased risk of an unsuccessful pregnancy. Her child is likely to be born under-weight, weighing less than 2400 grams (5 lbs. 0.8 ozs.). Because carbon monoxide in her blood and nicotine cause a narrowing of blood vessels, the foetus is deprived of oxygen and nutrition."

"Smoking reduces fertility", the report also points out, citing a study of some 50000 women smokers and non-smokers who took pill which showed non-smokers were twice as likely to conceive than smokers. Smoking women also reach menopause some two or three years earlier than non-smokers and they are more susceptible to osteoporosis, a major cause of fractures. Women who smoke are also three times more likely to develop cancer of the cervix as non-smokers.

International Phenomenon

Smoking is an epidemic growing at 2.1 per cent per year, faster than world population. Growth in tobacco use slowed briefly in the early eighties, primarily due to economic reasons, but is again increasing rapidly. Over a billion people now smoke, consuming almost five trillion cigarettes per year, an average of more than half a packet a day. Even in the US, where smoking prevalence—the portion of a population who smoke—has declined, the 20 per cent increase in tobacco use since 1964 indicates that those who smoke now, smoke more heavily.

Greece leads the world in per capita cigarette consumption. Japanese, Americans, Canadians, Yugoslavs and Poles follow close behind. People in industrial countries smoke twice as much as people in the Third World. Although Chinese men smoke almost as much as Western men do, the negligible amount of smoking by Chinese women means that country does not rank very high in overall per capita consumption. Nevertheless, China uses a quarter of the world's tobacco.

Prevalence of Smoking among Men and Women in selected countries, Circa 1994

Country	*Men*	*Women* (*per cent*)	*All Adults*
Poland	70	30	50
Norway	42	30	36
Brazil	63	33	48
Australia	40	31	36
Ireland	64	36	45
U.K.	38	33	36
Canada	44	36	40
Italy	54	17	35
Japan	66	14	40
East Germany	53	17	33
Bangladesh	67	1	37
Soviet Union	65	11	33
Netherlands	41	33	37
United States	35	30	32
France	49	25	37
China	56	1	29
Sweden	31	26	28
India	46	1	24
Greece	41	2	21
Egypt	40	1	21

Tobacco is increasingly grown near where it is consumed. China is the World's leading producer, using all it grows. The U.S., India, former Soviet Union and Brazil rank second to fifth, with all but the former Union being major exporters. Other major Third World exporters include Zimbabwe and Malawi.

Change in tobacco use can be measured in two ways : change in the use of tobacco products, in absolute or per capita terms and changes in smoking prevalence. Measuring the latter is often preferred by health educators as a sign of progress towards their primary goal which is to get smokers to quit and to start.

But the absolute quality of tobacco used provides an essential measure of total health costs that a society must bear. Indeed, the total number of cigarettes smoked over a lifetime is a more important health index than cigarettes used per day at any given time.

Smokeless Tobacco

One ironic result of campaign to reduce smoking in the absence of more general effort to control tobacco has been the marked increase in the use of 'smokeless' tobacco. The use of 'chew' or 'snuff' in the U.S. has increased by over 40 per cent in the last two decades. Much of the new interest in these forms of tobacco comes from teenage boys who like the stimulus of the nicotine, perhaps feel 'grown-up' when they try it and believe that it is safer than smoking. Surveying in some localities show that 20-40 per cent of high school boys chew tobacco or use snuff. Unfortunately, these forms of tobacco are strongly linked to oral cancer, an effect seen in India, where chewing is common, hence high rate of oral cancer.

The Consumption of Cigarettes in selected Countries, 1994

Country	*Cigarette use per capita (per cent of world average)*	*Change in total consumption since 1975 (per cent)*
Greece	237	+24
Japan	232	+6
United States	227	0
Poland	216	+3
Australia	203	+9
South Korea	186	+45
East Germany	167	+23
Italy	162	+17
United Kingdom	152	–27
Soviet Union	150	+8
France	145	+6
Philippines	130	+24
Finland	128	+4
Sweden	124	–3
Egypt	119	+138
Brazil	104	+17
China	102	+85
Mexico	77	+10
India	56	+33
Kenya	37	+48
Zimbabwe	35	–35
Bangladesh	19	+29

Health advocates have generally dismissed stronger measures for dealing with smoking and tobacco. The U.K. Royal College of Physicians, for example, the first governmental body in the world to launch a campaign to save the health of smokers, has conceded that banning tobacco is impractical. The physicians compared such a move to prohibition and expressed fear that it would not only be un-enforceable but would lead to criminality. Yet, there are a few effective ways to strengthen tobacco control policies without an outright ban. The policy questions are, how to prevent the young and the naive from beginning to smoke; how to persuade smokers to quit and how to protect the health of passive or involuntary smokers in the interim. When naive smokers first light up, ignorant of the life-threatening implications of their careless experimentation, they can become addicted and in effect 'involuntary' smokers themselves. Psychology and medicine currently do not know much about how to help these addicts except to recommend that they quit cold turkey.

Ingenious Marketing

The Indian cigarette industry is out in the streets again trying to entice people into smoking its particular brand. If you nodded in the affirmative or even shown a slight interest, you get the usual blandish of the free pack, cigarette lighters and may be T-shirts.

Just as the ban on cigarettes advertising gets extended into more areas, the cigarette industry discovers newer and more ingenious ways of catching the attention of people. It is one of the most sophisticated practitioners of the art of subtle sell. After all, it's worth the trouble. India, like the rest of the Third World, is the new frontier to be conquered. With 76 billion cigarettes of 'sticks' as they call them, going up in smoke in 1988, and the market is growing at a healthy 10 per cent.

In the West, with the growing awareness of the danger of smoking and health being in fashion, cigarette sales are falling. But in India cigarettes still come with an aura of glamour and a very low awareness of the health hazards. Every *bidi* smoker dreams of one day shifting to cigarettes. It still has a touch of defiance in its imagery. The American who smokes is a fool risking his health. The Indian smoker is an independent being rebelling against conventional morality. So, brands from the West are tripping over themselves to enter this burgeoning market of gullible smokers. Chesterfield, Ms, Rothmans and Virginia Slims, are all waiting expectantly for the cash registers to start ringing in India.

Advertisement Image-Craft

It seems most smokers cannot really differentiate between different blends in blind tests. So all that talk of 'Virginia blend' and 'real toasted tobacco' really means very little. Of course, most smokers with their habitual loyalty to a brand would not believe this. What every smoker really buys when he walks up to a *paanwala* and asks for a brand, is a fantasy, a dream of himself in a particular role. The Charminar man thinks he is an intellectual—*Jhola-totting* leftist. The classic chap would usually step out of a fancy car smelling of an expensive smuggled Cologne. The Red & White type identifies with macho filmi heroes. The Four Square man is an adventurer at heart. If you look at cigarette advertising carefully, you'll realise how intelligently every brand has selected its slot, and how imaginatively it is sustained.

Every brand is positioned with surgical precision. After the pricing has been decided upon, the exact smoker to be tapped is fitted to the price. That very unconventionally successful Charms exercise, at a cheap 20 paise per stick, logged on to the denim generation, guitars, T-shirts, and librated girl-friends. And the Spirit of Freedom was born. Red and White, in its most effective excerise, took the young men of the small towns—'B-class' in industry parlance, whose main entertainment is the fantasy world of films. They selected Raj Babbar as the model and created a persona for him of a mysterious rescuer of damsels in distress. Quick with his fists but modestly refusing the congratulations, quietly lighting a fag as the audio breathes out in breathless admiration, *"Red & White peenewalon-ki-baat hi kuchh aur hai"*.

Target Group

Extensive research is done into the personality and life-style of the target group. From his hobbies, favourite film-stars, to dreams and aspirations, attitudes, reactions, prejudices, and a definite target image is created—A psychographic profile, as the MBAs in the business will solemnly tell you. Brand names are chosen from a selection of hundreds. Pack designs are created by the dozen. Charms produced the denim soft pack, Chesterfield tried to look American, Ms liberated but feminine. Everything from brands names and pack to ads and films are extensively tested among the target audience. The ads and films fit the fantasy scenario of our smoker and at times his wife's. The Wills 'Made for Each Other' which originally meant the perfect match of the tobacco blend and the cigarette filter, has now by some mysterious osmosis turned into this trendy couple living in a happy westernised neverland with a casually held cigarette matching the fish tank and Polish Joke Book, very yuppy and consciously casual. While in the category of the rich but bored, Gold Flake goes for the stodgy old money look of silks, marble balustrades and crystal dinnerware.

Films are still the best medium for reaching the smoker, lifestyle can be shown really effectively only on films. The Four Square Man soaring on a Hang Glider or toasting elegant women on a boat looks static in a still photograph. All his *pizazz* turned to a frozen profile and a macho glitter in the eye. However, television is out of reach, behind governmental disapproval. Cinema halls are the only outlet and now-a-days that's all one sees there. Creating such a chaotic melange of images that really nothing remains in the mind. That man turning to look at the viewer with a James Bondish half smile—what was he selling ?—Red and White, Four Square or a Shaving Foam ?

Every other media is used extensively. Hoarding designs get the attention of top designers. And posters, neons, other point of purchase graphics, slots on popular video films. Then young kids walk into offices in identical T-Shirts, glittering smiles and a practised sales patter. Cigarette companies are also turning into czars of culture-sponsor beauty-shows, music festivals, films, plays, books, photography awards. Anything you can think of. They'll be willing to sponsor-tonga races or conventions of Qawwali singer, if it suits their purpose.

Female Fetish

The newest development has been a beeline to the woman smoker. A segment that is growing very fast. Ms took the well-worn route of the Virginia Slims line of 'You've come a long way baby' by trying to equate liberation of women with smoking. Sadly, it's an image women do fall for, just as fast as the men do for the macho male stuff. After matching the price and image the smoker has another bewildering choice of kings, mini-kings, filters, slims, milds, lights, browns, menthol low tar...Only pink for the Barbara Cartland red or green for the gardening enthusiast are left.

Now a suggestion has come up that even in the print media cigarette ads will only be allowed to show the pack without any indication of the life-style involved. Then, as in the liquor advertising, the ingenuity of the advertising agencies will again be challenged. In Britain, a similar ban led to the highly photography oriented campaigns like for Benson & Hedge and Silk Cut. Where the pack, the logo and even just the lettering and coloures are used in nearly abstract, at times bizarre picture. Just the health warning against smoking being there in print.

Fantasy Over Health

For all the indignation expressed and martyrdom claimed by the cigarette industry, they don't really have to worry too much. As long as the Indian smoker chooses fantasy over health, that is, as long as that overweight man stepping out of a chauffeur-driven Mercedes-Benz clutching a 555 pack and a tin of *pan-masala* decides to forget about his last heart attack and ignore lung cancer, cigarettes will sell. Because cigarette advertising is the clickest seller of dreams and we still seem to prefer them to reality. Cigarette sales will continue to soar because the Indian smoker is mentally blind to the health warning on the pack and chooses to live by the rebellion inherent in the statement, "Please don't ask me to smoke".[4]

Horrors of Nicotine Consumption

In a seminar on 'Smokng and Women', Mrs. Maneka Gandhi said that there are some freedoms that are guaranteed to the people : the most important are the right to live and the right to be healthy. What are you subjected to when someone lights up a cigarette ? Everytime you inhale a whole cigarette you lose two minutes of your life. Unfortunnately, smokers refuse to inhale what their cigarettes produce themselves. Video recordings show that an ordinary smoker takes no more than 11 puffs at each cigarette. Each puff lasts two seconds. The rest of the time—300 seconds or more—she (because today more women smoke than men) talks, exhales, sips her drink or waves her hands arounds.

The Poison Stick

What is in this generously released smoke ? Suppose you had a dinner guest who offered to do some magic tricks. She brought in an aluminium cauldron and then dropped metal shaving into it. Then, she poured in caustic soda and sent up a suffocation cloud. She then added paint stripper, the methane and then some ammonia. Before you as host, regain enough of your senses to throw her out, she whipped out the skull and crossbones and added nitric oxide and formaldehyde. Then would come hydrogen cyanide. Your smoking comrade is pouring out exactly this mix of chemicals as she lights up her cigarette. There are reactive metal fragments according to 'The Secret House', in a burning cigarette, as well as ammonia fumes, the paint stripper, chemical acetone, hydrogen sulphide, methane, hydrogen cyanide, nitric oxide, formalydehyde, mosaic virus and over 1000 other delights ranging from irritants, poisonous and nerve gases to motagens and lots of carcinogens. It's not that the cigarette is packed with them in the factory—nicotine, originally a nerve gas, is bad enough—but the lit cigarette which, during inhalation, reaches a temperature of 1700° F makes them itself.

These newly created poison chemicals band together in micro-cosmic balls, about 200 billion cigarettes, released at the rate of 6 billion per second. Some of them are rich in ammonia and acetone, some in embalming liquid formaldehyde. But once your colleague's cigarette has burnt half way, the balls rich in hydrongen cyanide come rolling out. These little balls stick to things. Hair—specially if it's been washed and consequently has a negative charge. Walls, all fabrics, upholstery—all these are excellent landing spots.

Enviromentalists pointed out at the conference that some smoking is exceedingly high factor in city pollution. The doctors pointed out that 76 per cent of those, who came in with heart problems and cancer, smoked. And that they also included in these statistics who had been subjected to constant cigarette smoke. Why are we getting so hot up about AIDS being a communicable disease spread in the world-over ? Five million people die annually either because they have been active or passive smokers.

Banning Smoking of Public Places

The government will first have to ban cigarettes in all public places. They will have to give incentives or disincentives to tobacco growers to switch to other agricultural activity. But most of all is what you, as a passive smoker can do ? Demand your rights. Smoking falls into the category of 'Intent to Murder'. 'Are you going to have two minutes of your life killed every few hours by people you don't even know. So stop them. If someone lights up, protest loudly. It is an unclean thing to do, it is a smelly, uncouth thing to do. Women need not smoke to prove their independence and men are much more macho without it. Despite the ads, people who don't smoke are much more popular than people who do and probably a lot quicker, because nicotine is a nerve gas. If you work in an office or run your own business, stop people smoking in your space, on your time. Smoking is not a private activity. It is a public one that harms everyone around. Stop people at airports, at public meetings, at political conferences. Don't listen to the smoker's argument that it is his/her cigarette—it is smoked by you as well.

Sachets of smokeless tobacco and the sponsorship of sporting events by the industry is to be discouraged. In yet another recommendation, the experts call for the end of duty-free sales of all tobacco products including smokeless tobacco at airports. Studies have shown that smokeless tobacco causes cancer, particularly oral cancer and increases the risk of cancer of the pharynx, larynx, oesopgagus, pancreas and urinary tract'. In addition, "the nicotine in smokeless tobacco, like that in the cigarette smoke produces addiction in users", the experts assert adding that smokeless tobacco use is dangerous to health and should not be viewed as less harmful than or as an alternative to smoking.[5]

Hazards from Crack Smoking

Doctors report that heavy crack smoking appears to be causing a growing number of addicts to become critically ill with symptoms similar to those of pneumonia. They say the condition is not easily detected unless the patients reveal their crack habits but if it is not treated properly it can lead to death. Emergency-room physicians at four inner-city hospitals in Detroit are seeing an average of at least one patient a day with what they are calling crack-lung. The problem has also been identified in New York and Baltimore. The doctors say that although the patients give outward signs of suffering from pneumonia most do not respond to anti-biotics or other standard pneumonia treatments. They are being successfully treated instead with corticostriods anti-inflammatory drugs.

Primary studies suggest that the condition is a variation of a number of ailments that afflict when poisonous compounds invade the lungs. Some scientists say it is also possible that the condition is caused by other drugs used in combination or by the adulterates used to produce crack. Dr. Dana G. Kissner, a Detroit pulmonary specialist who reported on the condition in the journal 'Radiology' said that the crack patients suffer severe chest pain and breathing problems, often have high temperature and cough up blood. She said the condition can lead to death from lack of oxygen or the loss of blood. At the Henry Ford Hospital in Detroit five doctors did a retrospective evaluation of the chest X-rays of 71 patients who had been treated in the emergency room for the problem to determine how the X-rays might help detect it.

Smoking Causes Cancer

Cancer of the bladder, pancreas, lip, mouth, oesophagus and pharynx can also be traced to the use of tobacco, though in the last two types of cancer alcohol has claimed more victims. The use of tobacco may be linked with cervical and stomach cancer as well, although these connections are less clear. Lung cancer is predominantly a disease of smoking. Active smoking habits account for an esimated 85 per cent of lung cancer. The claims of the tobacco industry that some types of people are predisposed to lung cancer and some unknown mechanism unrelated to smoking habits causes this condition are unlikely to prove true because different rates of smoking in men and women for different periods produce different rates of lung cancer. When women in U.S. did not smoke, for example, they rarely developed lung cancer. But as they took up the habit, lung cancer increased in proportion, after the lag of 20 years that it takes for cancer to develop following exposure to mutagens. In fact, in 1981 for the first time, lung cancer was as prevalent as breast cancer in American women over the age of 55 years.

Smoking causes two other serious lung diseases—bronchitis and emphysema, referred to together as chronic obstructive lung disease. Bronchitis is a cndition of secretion in the large air passage of the lung system that reduces the lung's ability to expel germs and can lead to infection. In emphysema, the air sacs in lungs coalesce and become less efficient in absorbing oxygen and releasing carbon dioxide. Smoking kills 52000 Americans each year through chronic obstructive lung disease.

The link between tobacco and other causes of death must not be overlooked. Fires caused by cigarettes kill between 2000 and 4000 Americans each year. And passive smoking may cause 5000 lung cancer deaths each year in U.S. alone. Altogether, smoking causes 10-20 per cent of deaths in Europe and the U.S. Passive smoking has been co-related lung cancer in non-smoking spouses of smokers in more than 10 studies. One particularly important study derived from other research designed to track lung cancer in smokers in Japan. This work focused on passive smoking as careful records were kept of spousal smoking habits. Wives, who did not smoke but who lived with their heavy smoking partners were found to be twice likely to die of lung cancer as wives of men who did not smoke.

Europe 1983-1985 : Total Annual Mortality from Tobacco

(Millions)

Female cancer	0.04 M
Male cancer	0.36 M
Total cancer	0.4 M
Other disease caused by smoking (e.g. heart or chronic lung disease)	0.4 M*
Total	0.8 M

* Estimated as approximately equal to the total number of cancer deaths caused by tobacco (and not, as in UK, by a larger number).

Predicted Mortality Patterns if there is no change in current Smoking Patterns

	Current population of children (i.e. aged under 20 in 1985) Billions	*Predicted annual deaths from tobacco when these reach middle age (in about 2025) Millions*

Europe	0.25	2
Other developed		
(chiefly N. America, Japan)	0.1	1
China	0.4	2
India	0.4	? 1
	0.4	? 1
Other less developed		
(chiefly S. America, Africa)	0.5	? 1
Total	2 B	8 M/year*

* Note : Most new aged 0-19 will die during the second quarter of the next century (2025-2049), when annual mortality from tobacco may, if present smoking habits persists, be about 10 million per year.

Hence, about 200 million of them will be killed by the habit.

Prediction for Europe in 2025 if current Smoking Patterns persists

	Attributed to tobacco	*Other deaths*
Deaths in middle age 40-69	1 M*	2 M
Deaths in old age (70 +)	1 M	8 M

* The people dying in middle age in 2025 are already alive (born 1955-1985) as the children and young adults of today.

Deaths from Tobacco during the period 2025-2050 (i.e. when children born in 1985 are aged 40-65)

	Growth in population of males aged 60 or over (ratio, 2050/2025 population)*	*Predicted** annual deaths due to tobacco (millions)* (a) in about 2025	(b) in about 2050
Europe WHO	1.0	2	2
Other developed			
(chiefly N. America, Japan)	1.0	1	1
China	1.3	2	3
India	2	? 1	2
Other Asia	2	? 1	2
Other less developed			
(chiefly S. America, Africa)	2	? 1	2
World total	1.6	about 8 M	about 12 M
Predicted average rate			
during 2025-2050		about 10 M	

* Infants in the late 1980s will be in their early 60s in 2050, and may be nearing retirement.

** Predicted assuming no large change in the proportions of young adults who become regular smokers (but with continued replacement of traditional forms of tobacco by manufactured cigarettes).

Source for all tables on this page : Richard Peto.

It can be done—A smoke free Europe W.H.O.—United Nations.

A CRISIS IN LEADERSHIP

Given the enormous number of early deaths and the appalling catalogue of disease caused by tobacco use, why has so little been done ? In his opening address, Dr. Asvall put his finger on the problem when he said the tobacco 'pandemic' indicated not only a crisis in health but also a crisis in leadership, in professional responsibility and organizational capacity. He laid the blame directly at the door of the politicians and public health administrators who had failed in their duty to protect the rights and defend the welfare of the people; and at the public health sector that, despite having the information about the dire health consequences of tobacco use, had for the most part failed to force the community to confront the problem. It was also, he said, a "gross ambarrassment" for millions of health personnel who for so many decades had failed to act on their knowledge to advise their patients and communities.

His words were echoed by other speakers. Given the enormity of the tobacco problem and the potential for prevention, one might have expected a comprehensive worldwide public health programme aimed at eradicating this man-made health problem. With few exceptions, the response to this human trgedy has been "characterized by a paralysing passivity on the part of the health profession, bureaucrats and the decision makers".[a]

But the picture is not all grim. In a number of countries control measures have been introduced that have successfully reduced tobacco consumption and that are feasible in all countries.

These measures are clearly meant to respond to different aspects of tobacco control, but one common by-product of the measures is that they all contribute to the erosion of the social acceptability of the tobacco habit. Public health strategy must aim at changing the climate of opinion and reversing the social profile of smoking—from a habit that is perceived as being adult, sophisticated and glamorous to one that gives a more realistic impression of a product responsible for the death of one in four of its regular users. What the success of these individual measures tells us is : It can be done.

Smoking A Leading Cause of Death

Smoking will become the leading cause of death in developing nations in the next decade, a Senate Committee was told. William Forge, former Director of the U.S. Government's Centres for Disease Control, testified at a hearing of the Senate Labour and Human Resources Committee that the aggressive efforts in the name of free trade by American tobacco companies may be blamed for this. Their need for increase in the market for their cigarettes abroad is contributing to an increase in smoking in many countries that will cause million of tobacco related deaths, he said.

Smoking already causes an estimated three million deaths world-wide each year. Smoking is declining at a rate of ten per cent a year in developed countries, Mr. Forge noted, but it is increasing at the rate of two per cent a year in developing countries as the U.S. Government encourages it to create markets. The U.S. is the world's leading tobacco exporter and in the last few years U.S. trade officials have induced such countries as South Korea, Japan and Taiwan to allow the import of American cigarettes and to relax advertising restrictions. Mr. Forge, now executive Director of the Carter Centre at Emory University in Atlanta, said, "The U.S. effect on world health may be negative because of tobacco."

Tobacco industry representatives opposed curbs on their efforts to open up markets abroad saying that the easing of trade barriers against American cigarettes abroad had improved the U.S. trade deficit . They maintained that tobacco export is an economic, not a health issue. Senator Terry Sanford from the home of tobacco industry, North Carolina, agreed with them. He said, "It would be unfair to have persons who desire to abolish smoking (at home) undermine U.S. exports". While health officials in the U.S. are campaigning to reduce teenage smoking; U.S. Television commercials in many Asian countries link smoking with the glamour of American youth culture, Gregory N. Connolly of the Massachusetts Department of Public Health alleged.

In Thailand, where smoking rates are low, the Government banned all forms of cigarette advertising last year, but ads for American cigarettes continue to appear illegally in these magazines, said Prakit Vateesatokit, a Professor of Medicine at Thailand's Mohidol University. "If smoking is a difficult problem for America, America should not make it hard for us", he said.[6]

One of the greatest preventable causes of ill-health is tobacco use. Among the many harmful consequences of such use, mainly through cigarette smoking, is cancer. Tremendous progress has been made in the prevention and cure of cancer, but a major battle has yet to be won against tobacco use and cigarette smoking which is "...the chief, single, available cause of death in our society". The more one learns about the effects of cigarette smoking, the better one realises how pernicious, wide-ranging and long-lived they are. For many years the effects of tobacco smoke were studied mainly in relation to the smoke, since the harm done was more directly observable. But now passive smoking is becoming a cause for concern. Also known as involuntary smoking, it is called "the exposure of subjects to environmental tobacco smoke (ETS)"—the exhaled smoke and that released between puffs by cigarette smokers. This sidestream smoke contains a complex combination of chemicals and has higher concentrations of toxic chemicals than mainstream smoke that is inhaled by the smoker.

Adolescence Attraction

Adolescence is a dynamic period of development when young people have to cope with physical, psychological, social and sexual changes. Adolescents begin to draw away from parents while developing intense relationships with peers from whom they seek approval. The adolescent years are stepping stones to adult life. Sporting prowess and community involvemnet usually boost health, while unprotected sexual activities, drinking and smoking can compromise present and future well-being. Of course the risk of smoking may be a challenge, but by resisting it, young people gain in strength and learn more about themselves.

Making Smoking Difficult

Indian cigarette manufacturers will soon be forced into making radical changes in packaging design, if the tough new anti-smoking legislation currently under formulation, is pushed through Parliament. According to Health Ministry sources, one of the proposed changes will make it mandatory for manufacturers to make the statutory health warning more prominent on the pack. As of now, although cigarette producers are required to display the warning "prominently", the word "prominent" has not been defined unambiguously in legal terms. If the current proposal goes through the size of the lettering of the statutory warning will have to be the same as the size of the largest letter appearing on the pocket, whether as part of the brand name, or elsewhere in the pack. The wording of the health will also be amended, from the bland "Cigarette smoking is injurious to health" currently in use, to more direct ones like "Smoking may lead to lung cancer and heart disease".

The thorny issue of a total ban on cigarette advertising, however, continues to be the subject of a hot debate between the Health, Commerce, Information and Broadcasting and Law Ministries. While cigarettes are on the Central List, *bidis*, tobacco chewing are on the State List. "Obviously, viewpoint differ", says a senior Health Ministry official, adding, "While we might press for a ban on tobacco, my colleague on the Tobacco

Board, also representing the government's interest, is trying to increase tobacco cultivation." This has been the main reason why comprehensive legislation covering all aspects of tobacco control has not been able to come through. Cigarettes, which are controlled by the Centre, have thus become the starting point of the anti-smoking programme.

Although cigarette advertising has already been banned in the government-controlled media, it continued to use cinema, print and outdoor media extensively. While government-sponsored slides highlighting the negative effects of smoking are already being shown in cinema halls, it is learnt that the Ministry of Information and Broadcasting is reluctant to allow a total ban on print advertising. The government might also try some subtler tactics. One of the proposals under consideration is that no character be allowed to smoke in any of the serial on TV or at best allow only the villains or those with negative image attributes to be shown smoking on screen.

Meanwhile, the series of administrative measures announced recently—curbing smoking in public areas like hospital, dispensaries, Indian Airlines flights, air-conditioned trains, etc.—will be made stringent with punitive clauses, so that implementation can be ensured. Clearly, the government has begun to get serious about its anti-smoking drive. But whether all this huffing and puffing will be able to contain the multi-million rupee cigarette industry remains to be seen.[7]

"I read so much against smoking that finally I decided to give up readily."

Graffiti—near a Cigarette Shop

Statutory Warning

Soon smokers won't be able to dismiss anti-smoking warnings so lightly. The Ministry of Environment has banned smoking at hospitals, dispensaries, air-conditioned railway compartments, stadia and Indian Airlines extended its ban on smoking to all its domestic flights. A draft legislation proposes to ban cigarette advertising totally. The government has also asked the industry to publish statutory warnings on *pan-masalas* and fluoride toothpastes; has defended the use of iodised salt in the Bombay High Court; and has forced the soft drinks industry to replace brominate vegetable oil (BV) as an emulsion stabilizer and clouding agent in citrus-based drinks.

Pan-masala, the pan substitute which has grown to a Rs. 100 crore industry, now must carry a warning saying—Consuming more than 4 grams a day is harmful. Argues a manufacturer : "I'll shut down my business if it is proved to be harmful". But dentists say, *pan-masala* is so harmful, it should be banned rather than regulated. Chewing *pan* filled with tobacco, is a major cause of oral cancer, the most widely prevalent one among Indian males. Says Madhu Mantri, President, Mumbai Grahayak Panchayat, a Bombay-based consumer group : "When there's a bomb scare, you cancel a flight. Why can't you ban harmful products ?" The regulations reflect the growth of health consciousness among Indians, who had found an ally in the former Minister of State for Environment, Mrs. Maneka Gandhi.

Despite the onset of the summer—a period in which the soft drinks industry sells nearly 70 per cent of its annual production, sales have been sluggish. Reason : BVO use has scared consumers. The Monopolies and Restrictive Trade Practices Commission pulled up Tripp, Campa, Orange, Rush, Rasna, Limca and Gold Spot, but its replacement, ester gum was hardly the solution. Unlike BVO, it is untested and unproven; it is not even permitted as a additive under the Prevention of Food Adulteration Act. The domestic soft drink manufacturers have been crying foul, claiming the ban is part of a multi-national conspiracy (read Pepsi). Pepsi plot or not, Indian soft drinks manufacturers routinely used BVO-about 45-50 parts per million (PPM)—way above the US permissible norm of 15 PPM. The trouble with BVO is that it causes lesion on arteries and excessive usage may cause heart trouble. The World Health Organization termed it harmful in 1969 and it is banned in 129 countries.[8]

Tobacco habit : A custom
loathsome to the eye,
hateful to the nose, harmful
to the brain and heart,
dangerous to the lungs and
in the black stinking fume
thereof, nearest resembling
the horrible stygian smoke
of the pit that is bottomless

—James I in 1604

The Tobacco Truth

How many of us care about this seventeenth century statement. The World Health Organization has pointed out that tobacco is the greatest villain in the history of mankind, because even according to conservative estimates tobacco accounts for over three million avoidable deaths and 32 million diseases in the world every year. Smoking, chewing and snuffing are widely prevalent throughout nature, has far outrun the control over himself. Otherwise how can we explain the increasing number of tobacco habituates every year despite the knowledge of its deleterious effects ? Or are we all ignorant ?

Irritants in tobacco smoke cause narrowing of the bronchial tubes, increased air-way resistance and mucous production. Irritants have a direct effect on the mucosa of the throat, nose, trachea, pharynx, oesophagus and the mouth. The smoke contains three toxic substances.

Tar, a mixture of 2000 chemicals some of which are carcinogenic, nicotine and carbon monoxide. Nicotine causes narrowing of the blood vessels and reduces blood flow to the peripheral arteries like in coronary arteries. Smokers are more prone to a heart attack than those who do not. Nicotine can also be converted to potent carcinogenic agents like nitrosamines. Carbon monoxide interfaces with oxygen transport and utilisation.

A single death is tragedy, a million deaths is a statistic. Several studies have shown that the use of tobacco is the main etiology in cancer of the upper gastriontestinal tract and respiratory tract. Indian women who chew tobacco increase the risk of still births three-fold. Non-smokers associated with smokers, like wives, children and co-passengers are forced to breath polluted air and they too have an increased risk of gettinglung cancer of the lung. Thus, ignorance and inconsideration are the two great causes of the ruin of the makind.

Tobacco being responsible for one third of the cancer cases in the country, primary prevention of tobacco related cancer is a justifiable aim. Dr. Krishnan Nair, Director of the Regional Cancer Centre at Trivandrum, told a workshop on oral cancer and pre-cancer organised by W.H.O's Regional Cancer Centre last year that any anti-tobacco activity in our country should be pursued with courage to face the sociopolitical issues that will ensue. Keeping this factor in mind, it is indeed extremely gratifying to learn about the anti-tobacco activities carried out in Kerala by the members of the Regional Cancer Institute as also certain voluntary organizations.

Aiming at a World Free of Tobacco

"Childhood and youth without tobacco" will be the theme of 'World No-Tobacco Day' to be observed all over the world on May 31. Observing that 'every child should have the right to grow up without tobacco,' the World Health Organization (WHO) has appealed to the national authorities at all levels, non-governmental organisations and all concerned with health throughout the world, to join forces in a sweeping offensive to find the right answer to the question : "Tobacco or health ?" In a message for the Third World "No Tobacco Day", WHO's South-East Asia Regional Director, Dr. U. Ko Ko said that tobacco related diseases would claim more than 2.5 million lives in the world over in 1990. Dr. Ko Ko said that about 90 per cent of lung cancer cases, 30 per cent of all types of cancer cases and 80 per cent of the case of bronchitis and emphysema are due to tobacco use. In addition, about 25 per cent of coronory heart diseases and stroke deaths are related to this addiction, he said. Dr. Ko Ko further said that a welcome change was discernible in the world with more and more people generally recognising the inherent dangers of tobacco use and its lethal effects on health.

The regional director cautioned against the new strategies of tobacco and tobacco product manufacturers who are now focussing attention on the developing world, particularly on women and youth, whom they see as a market waiting to be tapped, as developed countries had become more conscious about the ill-effects of tobacco use; smoke-free zones, particulary in and around educational institutions, could also prove effective in the fight against tobacco use. He expressed confidence that with the involvement of people and with the efforts of the member countries it will be possible to achieve "Our goal of tobacco-free society".

Proposal to Curb Smoking

Government is considering a proposal for enacting a law to curb smoking. Under the proposed law, the existing statutory warning on cigarette packets—"Smoking is injurious to health" will be expanded to include slogans such as, "Smoking can lead to oral cancer" and "Smoking can cause heart problems".

Ban on smoking, in hospitals and health institutions, aircrafts, trains, buses and schools and universities and control of smoking in work places are some of the other features of the legislation.[9]

References

1. *The Hindu*, Sunday, February 18, 1990, p. 24.
2. *The Times of India*, Snapshots, 23 Aug. 1994.
3. *The Pioneer*, Lucknow, Thursday, June 1, 1989, p. 9.
4. *Indian Express Weekend*, New Delhi, Saturday, September 8, 1990, p. 4.
5. *The Times of India*, New Delhi-May 31, 1989, p. 6 Leisure.
6. *The Pioneer*, Page 5, May 9, 1990.
7. *The Times of India*, New Delhi, Tuesday, June 12, 1990.
8. *India Today*, July 15, 1990, p.142.
9. *The Pioneer*, May 9, 1990, p. 9.

66

HAZARDS OF TOBACCO SMOKING

It has become a billion dollar question with public activists and the tobacco industry crossing sword—the former seeking a tobacco-free world and the latter fighting not only to retain its market but also to expand it.

Alarming Situation

Geographically, Asia had the largest cigarette consumption. In 1995 some 2.6 trillion cigarettes were smoked here. Consumption per adult varies from 120 in Nepal to over 3000 in Japan and the Republic of Korea. The smoking figures are 53 per cent amongst men and 3 per cent amongst women. Cigarette consumption in China is increasing faster than in any other region.

Smoking and Health Hazard

There are now an estimated 1 billion smokers in the world and they smoke over 5 trillion cigarettes a year. Manufactured cigarettes account for about three quarter of the total world consumption of about 7 million tonnes of tobacco a year. In some countries, like India, a major portion of tobacco consumption is accounted for by non-smokers such as by chewing, snuffing, which also affects health. Oral cancer while smoking leads to lung cancer. According to W.H.O. report tobacco causes 3 million deaths a year worldwide and this may reach the calamitous figure of 10 millions a year over the next three decades. "Unless smoking behaviour changes, three decades from now, premature deaths caused by tobacco in the developing world will exceed the expected deaths from AIDS, tuberculosis and complications of child birth combined". Warns the World Body in its 1993 World Development Report.[1]

Among population where cigarette smoking has been widespread for decades, 90-95 per cent cases of lung cancer, 80-85 per cent cases of chronic bronchitis and 20-25 per cent of deaths from heart disease and stroke are attributed to it. There are other risks including respiratory infection and peptic ulcers. For women aged 35 to 39 years in developing countries, smoking is the cause of one death in eight and for men in the same age-group it is one in three. Each of three smokers loses on an average 22 years of life. In fact, 40-50 per cent of male deaths from cancer in the developed countries are traced to tobacco. Every year smoking causes almost two-third of a million deaths from cardio-vascular diseases in industrialised countries.

The non-smokers are affected by ETS—Environmental Tobacco Smoke. It causes lung cancer. The cigarette industry however, asserts that the nexus between lung cancer and smoking has not been established beyond doubt.

Media and Smoking

The part played by the media in helping reduce tobacco use amongst cigarette smokers and influencing the policymakers has been well acknowledged. As a result of general atmosphere created by media blitz smoking curbs have been initiated like banning of smoking in public transport, cinema halls and public places. The Government of India is contemplating to bring about more stringent legislation so as to reduce consumption of tobacco. In the United States more stringent measures have been brought in.

Tobacco Trouble

According to the U.S. Food and Drug Commissioner one of the tobacco companies, Brown and Williams, has manipulated the nicotine content in order to enhance addiction. It is accused of using genetically engineered tobacco which has double the amount of nicotine found in other cigarettes. This was allegedly grown in Brazil and used in five American brands. The company initially denied the allegation but then admitted to having in its warehouse three million pounds of this tobacco named "Y-1".

The tobacco industry has also been accused of adding to tobacco certain amonia compound which increases the delivery of nicotine which serves either as a stimulant or a depressant and is habit forming. Over the years, the nicotine level in the cigarettes came down and the industry is evidently apprehensive that the level may drop too low to satisfy the smoker. It does not come under drugs category. However, the Food and Drug Commissioner had hinted that if the industry manipulates nicotine levels and ammonia compound, cigarette use will be treated as a "drug delivery" product and brought under the purview of regulations governing drugs.

Industry Prepares Itself to Meet 'The Challenge'

Faced with threats from different directions, the tobacco industry has mounted a counter offensive. The tobacco giant Philip Morris of Marlboro fame brys a series of full page advertisements in US dailies carrying an article by Forbes, a Media critic questioning the basis for the conclusion by the Environmental Protection Agency that ETS is "a known human lung carcinogen". The article took the media to task for what it termed "passive reporting on passive smoke" and "how science lost out to politics on second-hand smoke".

The industry has also sought to rely on the present population of smokers, which is round 48 million in the United States, to resist every further curb on their 'right to smoking'. The industry spent a fortune on advertisements and contribution to political parties during the 1992 United States elections. Likewise in India, the industry has sought to counter the official measures against smoking by exposing the cause of tobacco growers. At a meeting of the Parliamentary Consultative Committee attached to the Agriculture Ministry, strong opposition was expressed by some member to the proposed comprehension law to curb tobacco consumption. However, Mr. Balram Jakhar, Agriculture Minister, did not yield. Not only this, on July 25, 1994, Prime Minister P.V. Narasimha Rao, speaking in the Lok Sabha, requested tobacco growers in Andhra Pradesh and Karnataka to reduce gradually the area of cultivation and switch over to other crops in line with the worldwide trend.

Smoking Prevalence Around 1996

(Percentages of adults)

Country	*Men*	*Women*
Argentine	45	30
Brazil	40	25
Chile	41	37
China	61	7
Colombia	42	25
Equador	52	28
India	53	3
Indonesia	65	5
Iraq	40	5
Korea (Republic of)	74	5
Malaysia	41	5
Mexico	43	20
Morocco	40	9
Niger	41	11
Peru	41	13
Philippines	64	19
Saudi Arabia	53	2
South Africa	40	21
Sri Lanka	55	2
Thailand	47	4
Average	49	14

(Courtesy : The W.H.O.)

The Plight of Non-Smokers

Non-smokers are quite likely to have no choice about breathing tobacco smoke. In the U.S., people typically spend 90 per cent of their time indoors. On the job, some 63 per cent of U.S. workers are exposed to tobacco smoke, while at home over 60 per cent of all householders have at least one smoker. Altogether, only 14 per cent of Americans escape being exposed to tobacco smoke in the home or at the place of work. The rest involuntarily smoke on an average, the equivalent of almost one cigarette per day. Some people—a musician, for example, who performs at smoky bars and lives with a chain-smoker—passively smokes the equivalent of 14 cigarettes a day.

Protecting the public from the carcinogens in passive cigarette smoke requires urgent action. Increasing the ventilation in a building appears to be impractical it being prohibitively expensive. Reducing the risk of cancer due to cigarette smoke would require replacing the volume of air in the living space about 250 times more often than is currently the norm and use, therefore, 250 times the heating, cooling and pumping. The only certain way to make indoor air safe from cigarette is to eliminate the source.

Threat to Unborn Children

Tobacco effects on children, beginning with exposure before they are born, deserve special attention. Passive smoking poses a serious risk to unborn children. Nicotine, numerous toxic chemicals and radioactive polonium may all interefere with foetal development and the foetus can receive these substances through the mother's blood whether she smokes or chews tobacco. Furthermore, studies in both industrial and developing countries show that smoking by pregnant women reduces infant's weight at birth by roughly one-tenth. In one U.S. survey, smokers gave birth to underweight babies twice as often as other women did. Research has found a strong inverse relationship between birth weight and levels of cigarette residue (thiocyanate) in infant's umbilical cords. Low birth weight is considered to be the result of tobacco chewing in India (39 per cent of women in India chew tobacco). Because birth weight is a key factor in infant mortality, tobacco use seriously endangers infants' lives. Each year, at least three million new-born—estimated number of live birth to women who smoke—are thus potentially handicapped by their mothers. Children with parents who smoke, experience much higher rates of respiratory illness, including colds, influenza, bronchitis, asthma and pneumonia. One British study undertaken almost 10 years ago showed that children under one year whose mothers smoke more than one pack a day are twice as likely to get bronchitis and pneumonia. This finding has since been repeatedly corroborated.

Smoking During Pregnancy in selected countries

Country	*Share of pregnant women who smoke (per cent)*	*Infants exposed (Number)*
Ireland	36	26000
Sweden	34	33800
West Germany	32	211700
Canada	26	104400
Chile	25	31600
Belgium	25	31600
Venezuela	24	125200
Brazil	20	715800
Yugoslavia	20	73900
United States	19	706800
Colombia	19	150600
Austria	18	15700
Hungary	13	21500
Mexico	9	227300
Japan	8	130800
Philippines	6	91600
Bangladesh	3	135400
Egypt	1	17700
India	1	96900

In addition, the evidence indicated that parental smoking retards child development. One study found that lung capacity in boys was reduced by 7 per cent by their mothers smoking. If the teenage boys also smoked, their lung capacity was reduced by 25 per cent. The effect of the passive smoking in children can last a lifetime because it delays physical and intellectual development and because the longer people are exposed to carcinogens, the more likely they are to deveolp lung cancer.

Parents who smoke may also reduce the intellectual development of their children. One study in Italy found that children whose mothers smoked learned to read more slowly than children of non-smokers. In the U.S., the learning ability of 11-year olds whose mothers smoke has been shown lagging by six months.

Smokers Warned of Bronchitis

More than 30 per cent of smokers in India will eventually end up as cases of chronic bronchitis, a study has revealed. The extensive study on the combined effect of smoking and the rising pollution in India was conducted by Dr. D.D.S. Kulpati, Director-Professor of Medicine at Maulana Azad Medical College and Medical Superintendent of Lok Nayak Jayaprakash Hospital in New Delhi. "Two of seven per cent of the Indian population is suffering from chronic bronchitis", Dr. Kulpati said, "the industrial belts are the worst affected areas where in addition to human suffering, the productivity is also seriously affected. A survey in Delhi showed that the percentage of the population suffering from bronchitis is 14 among textile workers and 6.1 among university employees." Dr. Kulpati said, "Smokers in India, as compared to other countries stood 8.6 times higher chances of getting lung cancer, three times getting heart diseases and six times getting the second heart-attack. Contrary to the popular belief, he said, "The *beedi* smokers suffered more damage to their health than cigarette smokers." Chronic smokers can now instantly know the extent of damage done to their system by breathing into modern respiratory diagnostic instruments. The LNJP Hospital has established the country's most modern respiratory instruments for effective diagnosis of respiratory diseases. The hospital will soon have the country's first combined heart-lung intensive care unit with 16 beds for integrated treatment of patients suffering from chest diseases", he said. "This unit will cost Rs. 2.5 crores", he added.[1]

Smoking Grave Risk to Heart

"Beginning treatment when the disease has already occurred is like starting to prepare arms after the war has been declared", said All-India Institute of Medical Sciences (AIIMS) leading cardiologist Dr. Harbans Singh Wasir, citing the Ayurveda to stress the significance of prevention of heart diseases. Dr. Wasir, who is Chairman of the organising committee of the International Symposium on Preventive Cardiology being held at AIIMS, said, "The preventive aspect of heart diseases was not given much importance in India where it was all the more necessary not only for achieving better health but also because of the high cost of treatment of heart diseases. The 53-year old specialist who has over 25 years of experience with AIIMS Department of Cardiology said, "One common risk factor for heart diseases in younger patients was smoking." In fact, this was greater among those who smoke about 20 cigarettes a day.

Rising Trends in Heart Diseases

The other major risk factor included high blood pressure or hypertension, high blood cholesterol, lack of physical exercise, a stressful life style and presence of diabetes. Dr. Wasir said, "While no date were available, hospital figures and impression of general physicians, however, indicated that there has been a marked rise in number of persons having heart attacks and angina than about two decades ago. The incidence of heart disease among persons between 30 to 40 years age had also gone up." He said, "By and large rheumatic heart diseases and coronary heart diseases were preventable. This could be done by reducing the risk factors." In countries like the United States where smoking, high blood pressure and high cholesterol had been tackled on a war-footing, there had been a fall by about 25 to 33 per cent in the incidence of heart attacks. Regular physical exercises and mental relaxation also decreases coronary heart diseases, he added. Dr. Wasir said, the common cardio-vascular disease in Indians as in any other country was the disorder of hypertension or high blood pressure. Between 3 to 10 per cent of the adult population in India had this problem. Referred to as the silent killer, the disease had no symptoms and if left untreated or uncontrolled, the person may for the first time and at times for the first and last time have a stroke, may be fatal at that." He said, "While the causes of high blood pressure are still not known, in most cases it was true that prevalence of this disease was more in urban population, among people engaged in stressful occupation, obese of over-weight persons, those consuming excessive alcohol, smoking and those leading sedentary life. It was less prevalent among people leading an active life, consumed less salt, whose diets were rich in potassium to include vegetables and fruits like citrus fruits."[2]

Passive Smoking Can Kill

The U.S. Environmental Protection Agency (EPA) has found that every year as many as 3000 non-smokers may die from lung cancer because they inhale the smoke of people they live or work with. William Reilly, EPA administrator, said recently that "smoking is not just a health danger for smokers, but a significant risk for non-smokers, particularly children who are exposed to second hand smoke." Louis Sulivan, Secretary of Health and Human Services and a physician announced the launching of a campaign to urge the public to demand ban on smoking at plants, offices and restaurants because of the "pioson of second hand tobacco smoke".

Stronger Action Against Smoking in Public Places

As with past government studies linking smoking to cancer, the tobacco industry objected to the conclusion, while non-government health organisations urged the federal and local governments to take stronger action against smoking in public places. This latest report, for years in the making, has generated so much heat that its conclusions were leaked to some newspapers two days before its release and tobacco manufacturers and groups had news conferences condemning and praising it the day before its publication.

One fear of tobacco makers is the possibility of an increased number of law suits by persons claiming to be harmed by second hand smoke now that they are armed with a comprehensive 530-page report. An EPA official said in an interview that the possibility of such suits might turn out to be a major reason for private employers and restaurants to ban smoking.

Officials of Philip Major cigarette manufacturers stated that the EPA study did not "convincingly support" the conclusion of increased risk of cancer from second hand smoke. They also argued that the newest and largest study of the effects of tobacco smoke was not considered by EPA. Reilly said that scientists studied 30 reports from around the world to conclude that second hand tobacco smoke produces cancer. EPA analysed 11 U.S. studies, six done in Japan, four in China, four in Hong Kong, two in Greece, two in Sweden, and one each in England and Scotland. The EPA report states that about 434000 persons die each year in the United States from cancer, heart diseases and respiratory ailments caused by smoking. "Smoking is the leading preventable cause of death in our society today", Reilly stated.

Both the smoke rising from cigarettes and that exhaled by smokers contains a complex mix of more than 4000 compounds, a mix that "contains many known or suspected human carcinogens and toxic agents", the report stated.

Dangers of Passive Smoking

Passive smoking is particularly hard on children, the report states, contributing 150000 to 300000 cases annually of respiratory illnesses such as pneumonia, bronchitis and bronchiolitis among children less than 18 months old. That results in 7500 to 15000 hospitalisations a year, it estimates. In addition, the report says that passive smoking worsens the condition of 200000 to one million asthmatic children.

This report did not examine the relationship between passive smoking and heart disease. Some news services articles charge that such a study was killed by EPA officials because EPA was finding that there are more than 10 times as many heart disease as lung cancer deaths caused by passive smoking. Reilly called the charge "wholly unjustified". The heart disease study is continuing and is not ready for publication, he said, adding that "we have far better date and more persuasive data on lung cancer than heart disease". Michael Eriksen, director of the office on smoking and health of the U.S. Centres for Disease Control, said in an interview that smoking has dropped in the past 30 years to a low of 25 per cent of the population. In Eastern Europe, however, the current proportion of smokers is 70 to 80 per cent. Smaller proportion smoke in developing country but the rate is rising.

China has a great problem because its population is large and smoking is rising. In the Far East, smoking may lead to millions of deaths a year. Only a few developed countries—the United States, United Kingdom and Australia have effective anti-smoking programme, he said. Even fewer have begun to worry about the effects of passive smoking. France is an exception but its regulation protecting non-smokers have not yet been effective, he said.

Smoking Can Cause Impotence

Smoking can cause impotence in men, even younger ones, and reduce fertility in women. This was disclosed by Dr. (Col) K. L. Chopra, Chairman, Heart Care Foundation of India at a public discussion on evils of smoking. Dr. Chopra said tobacco consumption caused premature hardening and narrowing of penile (male sex organ) arteries with consequential incidence of impotence. Smoking also constricts the smaller penile arteries. This has been substantiated by detailed studies on the subject. Smoking also reduces the number of spermatozoa and increases their mortality making men less fertile, according to Dr. Chorpa. Persons who smoked more than 15 cigarettes a day for 10 years or more are specially prone to impotence, Dr. Chopra added. If smoking was stopped and other risk factors taken care of these men can regain potence, he said.

Decreased Fertility

Decreased fertility could also occur in women smokers. Smoking may cause menopause. The women who smoke during pregnancy are exposed to the risk of delivering stillborn babies or baby's death immediately after birth. Even where the baby survives there is likelihood of delay in physical and intellectual development till it reaches 11 years of age, Dr. Chopra said. Dr. K.K. Agarwal, Vice-chairman, Heart Care Foundation of India said smoking and tobacco chewing were major risk factors for heart attacks. Smoking also causes chronic bronchitis, peptic ulcer and cancer. Dr. Agarwal said chances of sudden death were twice as much in smokers as non-smokers.

Sudden death occurs because the normal electrical stimulus gets disorganised causing erratic and ineffective contraction and the heart is unable to pump blood into the vital organs of the body. No pulse is felt and heart may stop contracting altogether (cardiac arrest) unless effective resuscitative measures are undertaken.

Dr. Agarwal said high incidence of coronary diseases leading to angina and heart attack is also largely on account of smoking habits. Nicotine content of tobacco increases secretion of chemicals like adrenalin, which increases the heart rate, raises blood pressure and enhances work load of the heart. Nicotine also interferes with the functioning of blood vessels increasing the tendency of blood to clot. All these factors lead to atherosclerosis (narrowing of arteries) of the coronary arteries, causing attack. Smoking is also responsible for paralytic strokes, brain haemorrhage and hypertension. It also causes pain and stiffness in the legs specially while walking. Dr. Chopra recommended regular meditation as an effective technique for giving up smoking.

The Marlboro Man

Wayne McLaren had learnt pool as he had learnt most things : hanging around, quietly working at it so that by the time he got good it seemed as if he had always been good. In this way, he mastered an array of pursuits that were glamorous mostly in their danger, like riding bulls in competition. He knew how to fight and he didn't mind doing it, but he could also play chopin on the piano. 'Of course', an old friend said, 'that combination made him especially attractive to a certain kind of woman'.

For nearly three decades a small, highly selective club of men—that handful of strong silent types who have been chosen as Marlboro men—have influenced a generation in the United States and around the world, not only in what they smoke, but in what they wear and how they act. The Marlboro man image has proven so powerful and seducive that Marlboro's parent company, Philip Morris, guards it zealously. The models most of whom are actual cowboys, 'are sworn to secrecy' about all aspects of their work, a company spokesman said. But just before he died of lung cancer in July at 51, McLaren went very public saying he had worked briefly as a cowboy-model for Marlboro. At the end of a life spent around the edge of celebrity, McLaren's anti-smoking advocacy made him a romantic hero of another sort, "I'm dying proof", he said before his death, "that smooking will kill you."

Cowboy Image

When he came to Hollywood in the early 1960s, McLaren knew that his rodeo career was nearly over : a bull had stomped his knee. But with Westerns so popular then, he thought he could make a living as a stunt man. "It's moslty just falling off horses", he once said. Meanwhile he would try to be an actor.

"Wayne had the looks and the talent", says Jum Myres, a long time assistant director for films and television who helped the fledgling actor get into the Screen Actors Guild, "but he had a flaw. Three flaws, really : beer, cigarettes and women". McLaren's resume listed television credits like 'Paint Your Wagon', 'Junior Bonner' and 'Count your Bullets'. They weren't all Westerns, but mostly. The cowboy roles were easy, "I don't have to develop a character', he said once, "I just walk on".

Between acting jobs in the mid-1970s, McLaren took a job modeling for some print advertising. He said he was one of a number of men who posed as cowboys for advertisements for Philip Morris's most popular cigarette brand, Marlboro. But as he entered his 40s, Wayne McLaren's acting career was fizzling. 'He had plenty of breaks', Myers recalls, 'but he just didn't know how to take advantage of them'. He had lived with a French Actress for some years, but that had broken up and all his other liaisons were short. McLaren left Hollywood and moved to Orange Country, settling around Newport Beach. He took odd jobs to pay the bills and spent a lot of time at a local bar.

Cancer Cover Death

A week before his 49th birthday, he went to a doctor for an examination. Eight months later, he learnt that he had advanced lung cancer. For the next two years, McLaren was in and out of hospitals undergoing chemotherapy surgery to remove his left lungs, and radiations therapy. In June 1990, though the lung cancer had been removed, doctors discovered that the cancer had spread to his brain. "I think once he knew that he was dying he may have felt a little guilt in that promotion he'd done for cigarettes", says his brother Charles, "It may be that this could be a legacy for him. He had never really accomplished anything that was particularly noteworthy and may be in his own mind he felt that this might be his last hurrah and he was going to make it count". McLaren has learnt enough about images in Hollywood to know he'd get more attention for his message if he accentuated a Marlboro connection.

In March, he appeared wearing jeans and boots and a white cowboy hat before the combined houses of the Massachusets Legislature, which was considering a Bill that would add taxes to cigarettes sales to pay for health education. After saying that he had appeared in Marlboro ads, he told the legislators, "I started to somke in my early teens because it seemed to be the thing to do, a rite of passage to adulthood." In April, McLaren showed up in his white hat to speak at the annual Phillip Morris stockholders' meeting in Richmond, Virginia to support a resolution that the company voluntarily limit its advertisting. When he was finished, Michael A. Miles, the company's Chairman, said from the podium, "We are sorry to hear about your medical problem. Without knowing your medical history, I don't think I can comment any further." A Phillip Morris spokesman, Karon Daragan, now concedes that McLaren appeared in retail promotional displays for Marlboro Texan Poker Card, pictured sitting in a group of four cowboys playing cards. But at the time of McLaren's anti-smoking campaign, the company maintained that it could not find 'any records supporting Wayne McLaren's claim that he had appeared in Marlboro advertising'.

McLaren's rugged good looks were hideously transformed by the time he died on July 22. In the last weeks at home, before his final hospitalisation, he often sprawled on a sofa near the framed photos from his rodeo days and his career in Hollywood. Usually, he would switch on a CD by the country. Western singer George John on the stereo and listen over and over to the song "Hell stays open All Night Long".[3]

Smokeless Tobacco

One ironic result of campaigns to reduce smoking in the absence of a more general effort to control tobacco use has been the marked increase in the use of 'smokeless' tobacco. The 'chew' or 'snuff' in the U.S. has increased by over 40 per cent in the last two decades. Much of the new interest in these forms of tobacco is evinced by teenage boys who like the stimulus of the nicotine, perhaps feel 'grown-up' when they try it and believe that it is safer than smoking. Surveys in some localities show that 20-40 per cent of high school boys chew tobacco or use snuff. Unfortunately, these forms of tobacco are strongly linked to oral cancer, an effect seen in India where chewing and oral cancer, are common. Cancer of the bladder, pancreas, lip, mouth, oesophagus and pharynx can also be traced to the use of tobacco though alcohol plays a major role in the last two types. The use of tobacco may be linked with cervical cancer and stomach cancer as well, although these connections are less clear.

Lung Cancer

Lung cancer is predominantly a disease of smoking. Active smoking habit account for an estimated 85 per cent of lung cancer. The claims of the tobacco industry that some types of people are pre-disposed to lung cancer and that some unknown mechanism unrelated to smoking habits causes this condition are unlikely to prove true, because different rates of smoking in men and women over different periods of time produce different rates of lung cancer. When women in the U.S. did not smoke, for example, they rarely developed lung cancer. But as they took up the habit, lung cancer increased in proportion, after the lag of 20 years that it takes for cancer to develop following exposure to mutagens. In fact, in 1981, for the first time lung cancer was as prevalent as breast cancer in American women over age 55.

Smoking causes two other serious lung diseases—bronchitis and emphysema, referred to together as chronic obstructive lung diseases. The link between tobacco and other causes of death must not be overlooked. Fire caused by cigarettes kill between 2000 and 4000 Americans each year. And passive smoking may cause 5000 lung cancer deaths each year in the U.S. alone. Altogether, smoking causes 10-20 per cent of deaths in Europe and the U.S.

Passive smoking has been co-related with lung cancer in non-smoking spouses of smokers in more than ten studies. One particularly important study was derived from other research designed to track lung cancer in smokers in Japan. This work lent itself to a consideration of passive smoking because careful records were kept of spousal smoking habits. Wives who did not smoke but who lived with heavy smokers were found to be almost twice as likely to die of lung cancer as wives of men who did not smoke.

A dangerous drug clearly should—but cannot be banned. The economic strength of the tobacco industry is so great that it can exploit for its own purpose as the safeguard built into democracies to protect legitimate minorities. In non-democracies, governments may lack the credibility—and the motivation—to tackle so insiduous and pernicious a habit. And under both systems the social conditioning and chemical habituation characteristic of tobacco make banning the product a formidable task, one that would take a long time. Yet the current 'informational' campaign to control tobacco is falling behind as worldwide use increases faster than population.

Banishing Tobacco

An alternative approach is inherent in a new movement to protect passive smokers : Banishing Tobacco. This campaign, which stops short of an outright ban on tobacco sales, includes either the prohibition of smoking in the place of work and in public buildings or the strict limitation of smoking to specified areas. The movement may be the single great success of the informational campaign against tobacco. Its leaders insist that despite the continued sale, advertising and use of tobacco, non-smokers—the majority in most societies—have every right not to be exposed to the carcinogens, carbon monoxide and irritants in tobacco smoke. Such a campaign can make three important contributions—

First, by banishing tobacco use in places where innocent people will be exposed and placed at risk, thousands of lives may be saved.

Second, forcing smokers to give up smoking in the presence of non-smokers. If smokers must get through a working day without smoking, then they are more likely to be able to quit completely. This has been the result of bans in Minnesota and California. In any case, their total dosage of carcinogens and carbon monoxide should decline.

And third, by stigmatising tobacco use as dangerous and anti-social, the passive smokers' rights movement can accomplish a goal of all anti-smoking informational campaign: to make smoking socially unattractive.

Restricting Smoking at Public Places

The passive smokers' rights campaign focuses on the place of work, public gathering places and public transportation. Many countries now prohibit smoking on public transportation and in theatres and auditoriums though the impetus for these restrictions has usually been conventional safety concerns. In a few areas, such as the State of Minnesota and cities of San Francisco and Los Angeles, smoking is now prohibited in public buildings (except in restricted areas) and non-smokers must be protected in restaurants and on the job.

Interestingly, non-smokers have an important ally in the place of work : employers. Companies, at least in the U.S. are rapidly realising two things—first, most of their employees do not smoke and do not like to breathe the smoke of others. Second, smokers cost employers money. Survey indicate that the combination of inefficiency and ill-health as a result of smoking wastes seven per cent of a smoker's working time. They also suggest that smokers cost employers at least $ 650 each per year. Smokers add to insurance and clean-up costs and they dampen non-smoker employee's morale.

American industry is responding rapidly to the non-smoker movement. A number of well-known industries have prohibited smoking on the job for most employees. A few even refuse to hire smokers. The predominent trend, however, is towards banishing the practice from the place of work. In 1984, the rate of increase in adoption of policies against smoking for the publishing, insurance, finance, pharmaceutical and scientific equipment industries in the U.S. was between 10 and 25 per cent. That is, one month to one quarter of the top 1000 businesses in this group of five industries implemented new policies that year to banish smoking.

Role of Government

A particular difficulty in banishing tobacco is the role of the governments in promoting tobacco use. The schizophrenic state of affairs persists not just in the market-oriented West, but also in centralised economies. Governments most often own the tobacco industries in these areas. China, the former Soviet Union and India, for example, grow their own tobacco—they are not victims of some cabal of multi-national companies. The state-owned tobacco industry in China is being carefully nurtured and expanded rapidly even as another part of the government is telling the Chinese that smoking is bad for them.

These incompatible policies are also in place in the West. In the U.S., the Department of Agriculture administers a price-support system to protect tobacco producers. West European nations subsidize tobacco farmers with about $ 660 million in price support each year. Ironically, the system protects small inefficient farmers who earn higher prices than they would obtain without the subsidy. More efficient producers, who could underprice the small farmers, are not allowed to compete fully. The result is that tobacco costs the user more than it would without the system. As tobacco use varies negatively in response to price increases, smoking is being directly reduced by price supports.

There is an even more subtle effect, however. The tobacco industry, though it loses the right to compete with small-scale farmers gains the powerful political support of the small farmer. The added political clout helps counter anti-smoking forces. Moreover, it retains the appealing appearance of official tolerance and even endorsement of the use of tobacco, which in turn diminishes the effectiveness of the informational campaign in reducing the social acceptability of tobacco. Any child about to start smoking could be inclined to think that the U.S. government sees tobacco as desirable. This is, of course, the implicit position of any government that promotes tobacco production.

Anti-smoking Efforts

The anti-smoking then is at best a stand-off in industrial countries and a compaign non-starter in developing ones. At the current rate, Western countries will not see a major improvement in the health effects of smoking for many decades, but Eastern and developing countries will see a rapid worsening. Unfortunately, even the World Health Organization, allots less than one per cent of its budget to this problem, though it calls smoking the most prominent preventable health problem in the world. Its current budget for the mid-eighties had no funds for this purpose.

Effective policies fall in four categories—

(i) The continuation of the informational campaign, is worthwhile as a foundation for other stringent measures.

(ii) The second step is for those countries that have low smoking rates, no indigenous tobacco industry and a reliance on imported tobacco products to ban tobacco altogether. It would completely eliminate the epidemic's threat to them, placing these nations in the forefront of the campaign, much as the industrial world spearheaded the campaigns to eradicate smallpox.

These governments have an incentive to act—the reduction in foreign exchange losses for the purchase of a non-productive product. Only a few people in the societies are now heavily addicted, making national withdrawal politically easier. Many African nations fall into this category. Unfortunately, other impoverished nations, such as India, do not fall in this category because cigarette consumption is already high.

(iii) The third approach is for those nations that must be politically pragmatic to at least act to protect the health of the innocent. Experience in the U.S. and in Poland shows that tobacco can be banished from public buildings, from the place of work and from public eating establishments and meeting places. Banishing cigarette use in the presence of non-smokers should be considered a minimum level of protection.

All governments can pass national clean air acts for public buildings, places of work and entertainment establishments. The United Nations Organisation could do well to establish no-smoking policies for their employees, especially those who work with children and the poor, for they are unavoidably going to be viewed as symbols of modernity and success and therefore, should not introduce such a clearly harmful product.

(iv) The fourth level is to use the power of economic tools to eliminate smoking as much as possible. Estimates of the cost of smoking amounts to $. 1.25-$ 3.15 per pack. A tax of this magnitude in the western nations would reduce smoking by as much as 40 per cent over time. Any tax increase, even of five to ten per cent per pack, would rapidly encourage light smokers to quit in order to avoid the high cost, would provide additional pressure on the heavily addicted to bring themselves to the point of enduring withdrawal and most alternatively, would discourage the young (with low income) and poor from ever starting. Additionally, tobacco support system can be dismantled in order to signal that governments now wish to discourage the use of tobacco.

These measures will not be easy nor will they solve the tobacco problem. They will not, for example, assure that children will be protected in the home against the smoke of their parents. They will not protect the newborn from harm as a result of their mothers' smoking. Parents alone can take this responsibility, though in some cases their addiction makes them risk the health and the intellectual development of their children.

New measures will not assure that smokers themselves will be persuaded to quit. Nor will they guarantee that innocent young people do not become addicted. But without more responsible efforts on the part of the health professions and public interest organisations, even these efforts will remain in abeyance.[4]

Tobacco-free Generation

Cambridge University tried to ban the smoking and drinking 380 years ago but today the skill of Tobacco promotion still lies in the way it constantly tries to circumvent laws aimed at helping young people avoid the initial addiction to tobacco. Law-makers are meeting both success and failure. But legislation which supports a public demand such as smoke-free areas and restricts the time and place for smoking, seems to be gaining ground. Price increases may slow consumption but tobacco availability also stimulates sales. So a number of countries are now banning automatic vending machines and trying to enforce laws against the sale of tobacco by shopkeepers to minors.

Prohibiting sales to Minors

But the laws prohibiting the sale of cigarettes and other tobacco products to minors are being widely flouted. No less than 43 out of a random sample of 50 tobacconists' shops in England and Wales sold cigarettes to children. According to a report by the Royal College of Physicians in the U.K. the widespread breaches of bans on tobacco sales to minors have not gone unnoticed. In some countries, associations and youth groups have mobilised to ensure compliance with existing laws and to enact stronger ones. Such laws might :

* require the posting of signs indicating that sales to minors are illegal;
* require purchasers to show proof of age;
* require the licensing of sellers of tobacco; and
* increase the penalties for violations of the laws in question.

Other major legislation aimed at protecting children and youth from this devastating habit include :

* prohibiting or restricting sales of cigarettes from vending machines, and
* banning smoking in schools and other places frequented by the young.

Prohibiting or restricting sales from vending machines makes cigarettes, often one of the most widely available products, less accessible to young people. Cyprus, Hong Kong and Iceland prohibit all sales of tobacco from such machines. Belgium, Finland and Ireland restrict places where they may be installed and the owners of the premises are held responsible for their supervision. In the United States, many local ordinances have required the removal of vending machines.

Banning smoking in schools and other places frequented by the youth conveys a strong message that smoking is not acceptable social behaviour. The Finish Tobacco Act was one of the earliest and most comprehensive laws of this type; it prohibits smoking on the premises of day nurseries, schools and other places intended for use by children at or below the age of compulsory education as well as at indoor, public events for children or young people or to which children are admitted. A Singapore notification of 1988 prohibits smoking in "Fast Food" establishments, indoor roller-skating rinks and roller discotheques. Chewing tobacco, snuff in sachets to place between the cheek and gum or plain sniffing are being promoted now as "Smokeless Tobacco". But they do affect the consumer, causing cancer of the mouth and are as addictive as cigarette smoking, yet some young people use them in the belief that they are less harmful.

Pre-emptive Ban

A WHO Study Group has called in strong terms for a pre-emptive ban, wherever the product has not been introduced. Hong Kong, Ireland, Israel and New Zealand have prohibited importation, sale and use of smokeless tobacco. The United Kingdom has banned the product under the Consumer Protection Act. The U.S. has enacted legislation requiring health warnings on smokeless tobacco products and on advertisements other than on billboards. Emphasis on the importance of legislation in preventing tobacco use by young people does not diminish the absolute necessity for effective educational programmes on the hazards of smoking and other forms of tobacco use. While most countries do have educational programmes aimed at discouraging smoking, not all of them make health education on this subject mandatory and even fewer allocate funds for such programmes.

An effective message to young smokers is conveyed by powerful health warning on cigarette packages. These should change not only in the wording of the message (and thus termed as a rotating warning) but also in the type-face which should stand out visually. Countries requiring this include Australia, Iceland, Ireland, Norway, Sweden, the U.K. and the U.S. Because of the demonstrated linkage between cigarette advertising and the start or continuation of smoking, at least 50 countries have restricted or prohibited advertising of cigarettes and 21 of these have imposed a total ban on all tobacco advertising. But as restrictions have been strengthened, the tobacco industry has increased its indirect promotion of tobacco through sponsorship of sports and cultural events, including rock music groups which are particularly attracting young people. Some governments have moved against this kind of activity contending that it circumvents the existing laws. Also effective is the promotion of a non-smoking atmosphere. As evidence on the harmful effects of passive smoking increases, authorities—both public and private—are prohibiting smoking in a wider range of public places. Of the 68 or more countries with legislation to control smoking, 48 restrict smoking in public places. The non-smoker also feels empowered by the law or the local prohibition to ask the smoker to desist and in the case of children, to tell them off.

Legislation has the power to make cigarettes less attractive and less available to young people. It contributes to an environment in which children are allowed to grow up free from the hazards of smoke. Sweden has set itself the goal of making the next generation of children smoke-free and Canada has adopted the WHO slogan—"Here comes the breakaway generation", the generation that can break away from the dangerous addiction to tobacco.

Howls of Protest Over Anti-smoking Law in France

Some French are delighted, some are disgusted, while others are blatantly flouting a tough new anti-smoking law that went into effect across the country. "It's a clash of the fors versus the againsts'—the battlelines are drawn", said Alain Genestar, writing in a front-page editorial in Le Journal Dimanche, Paris' only Sunday newspaper. "What do I mean, battle ? It's a war."

The law bans smoking in restaurants, offices and public places except in designated smoking areas, and in public transportation, "All French, or almost, are in agreement. For once", Genestar wrote, "The anti-tobacco law going into practice today—at least as a test—is good thing". Polls show up to 84 per cent of the French approve of the law and 65 per cent are prepared to apply pressure to enforce it. Not all French think so. Howls of protest have risen from cafe, restaurant and hotel owners, who fear losing clientele for enforcing the law, or hefty fines of up to 6000 francs ($ 1200)—or prison terms—if they don't. Individual violators are to be fined $ 120-260.

"It's an imbecile law", Mr. Roger Cournut, a burly 40-year old at La Grappe d'Orgueil, a bar in Paris' Garment district, grumbled last week as he lit up. One French lawmaker even refused to conform to the law in the National Assembly, which he said was different than other public places.

Forbidden to Forbid

"It's forbidden to forbid", sniffed Jean Le Garrec, a socialist member of the Assembly. "I plan to abrogate all the idiotic socialist laws, beginning with this one". Dozens of cafes and bars across Paris put up yellow signs on their doors saying, Bar Fumeurs (Smokers' bar). "It's going to be extremely difficult to enforce this", Joel Ro Roubuchon, the famed chef at the three-star restaurant Jamin, one of the swankiest in Paris where diners often enjoy a pungent cigar, said on French radio yesterday, "It's hard to harmonise this with the dinning experience".

Mr. Daniel Estival, owner of L'estival restaurant, also in Garment district, said flatly, "I refuse to let them force me to refuse my own clients...Eighty-five per cent of them smoke." He said he would join the Bar Fumeurs movement to warn customers before they enter. "We'll wait for the police to visit or for a client to complain," Mr. Estival said, "then we'll see."

France has one of the highest rates of smokers in western Europe—40 per cent of all adults, according to 1991 figures. Youths aged 18-24 account for the highest proportion of smokers. Air France announced on Friday that smoking will be banned from January 1 on internal European flights shorter than two hours. The airline had already banned smoking on domestic flights since 1990.

Action Against Addiction

Children can get hooked on nicotine after smoking just three cigarettes. The study undertaken by the London Institute of Psychiatry, has convinced researchers that nicotine can be as addictive as heroin. It also dispels the notion that it is harmless for teenagers to smoke a few cigarettes 'as an experiment'. A report was presented at Europe's first conference on addiction in London. The Government launched a £ 10 million campaign aimed at preventing teenagers from becoming tobacco addicts. The researchers questioned 600 girls aged 11 to 16 attending a south London girls' school for the two-year project funded by the Medical Research Council. They found that nearly a third—188—were smokers and one in three of those had smoked their first cigarette by the age of eight. Of the children smoking 7 out of 10 had tried to give up. Most had failed and suffered the same nicotine withdrawal symptoms as adult smokers, including cravings for a cigarette and irritability when they were unable to obtain one. "Many of the girls reported that they felt calm again once they had a cigarette", said Dr. Ann McNeill, a psychologist at the Institute.

In a Tobacco Trance

The researchers looked closely at the school's 270 girls aged between 11 and 13 and found that 180 of them had tried one or two cigarettes. Of those more than half had continued smoking. For those who had tried three or more cigarettes, the findings were even more striking : two-thirds were smokers two years later. "We don't sufficiently appreciate just how dependent young people are on cigarettes", said Dr. McNeill, "We consider that nicotine addiction is comparable to heroin addiction." The findings have been endorsed by Professor Richard Peto, who heads the Imperial Cancer Research Fund's Cancer studies unit in Oxford. He said that in his experience nine of 10 smokers started while they were still in their teens. "If you can hold out to the end of your teens then you have escaped most of the risk of getting hooked", said Prof. Peto. The study has also been welcomed by Parents against Tobacco, a pressure group. One member Sue Braund, 36, from Covelly, north Devon is trying to persuade her 13-year-old daughter to give up smoking. "I have tried many times but have never successfully given up", Ms Braund said. "There is no doubt that smoking is an addiction. I first started when I was 13. By the time I was 14, I was on 20 cigarettes a day. I want to prevent my daughter doing the same."

An aim of the conference organised by the Action on Addiction was to take the first steps towards setting up a national addiction centre for research and treatment of all addictions, including tobacco. There is particular concern that women have been less influenced than men by anti-smoking campaigns and that teenage girls are more likely to start smoking than boys. A recent opinion poll found that by the age of 15 nearly one in four girls was smoking regularly, compared with only about one in six boys.

The office of the American surgeon-general, whose health warnings appear on every pack of cigarettes in America, has concluded that like other addictive drugs, tobacco causes temporary mood changes and is used compulsively—"despite damage to the individual or to society". Dr. Michael Farr, the conference organiser and a psychiatrist at the addiction research unit at the Maudsley Hospital in South-East London said that in some cases tobacco was even more addictive than illegal drugs. "Studies of heroin addicts have shown that they find giving up cigarettes more difficult than giving up heroin", he said. Such evidence is challenged by the Tobacco Advisory Council, which is funded by the U.K. tobacco industry. "It is a very dangerous precedent talking about heroin and tobacco in the same breath. You have to define what you mean by addiction", said Brendan Barry, the Council's Public Affairs Manager.

He said that even if all advertising were banned, people would still smoke and children would be even more curious to try cigarettes. "We have no interest in the juvenile market whatsoever", said Mr. Barry. For some, the arguments against smoking are far more straight-forward. In a Health Education Authority advertisement launched in teenage magazines, Victoria Smith, 14, from East London asks, "How do you French kiss a boy who smokes ?" Her friend, Amanda Davies from Cleaveland replies, "Quickly". But for Victoria, a pupil at the Italia Conti Academy of Theatre Arts in London, the real reason is far more pressing, "You can't be a dancer and smoke", she said.[5]

Combating Smoking

The fact that tobacco smoking can lead to utterly catastrophic consequences just doesn't seem to strike home. On the occasion of the recently celebrated 'World No Tobacco Day', Dr. J.P. Wali of the All India Institute of Medical Sciences reeled off some disturbing facts about smoking and Third World countries. For instance: for every smoker who kicks the habit in the West, two are initiated in the developing nations. Global tobacco consumption has increased by 75 per cent by the turn of the century. Cigarette consumption in India has increased by 400 per cent between 1970 and 1980. India ranks third in the world tobacco production. Nearly 250 million kgs. of tobacco is consumed locally—78 per cent chewed and two per cent taken in the form of snuff. And, interestingly enough, it is predicted that smoking related diseases will disappear from the developed world even before communicable diseases and malnutrition are controlled in developing countries.

Among measures called for, are a ban on promotion in the mass media, a prohibition of sales to adolescents, tax increases to keep pace with the rise in the cost of living as a disincentive to customers and health warnings on packages.

67

THE EXCISE TANGLE

Enmeshed in diverse litigation cases the industry has fought the Excise Department over the years. But now, just as things were beginning to look brighter, comes the nightmarish rumour that the government is considering reverting to the dreaded ad valorem excise duty structure again. If the government reverts to the ad valorem excise duty structure, the cigarette industry will be in big trouble.

Ad Valorem System

According to an ad valorem system the manufacturer is compelled to pay excise duty on the printed packed price of cigarettes. In simple terms, it means that if the rate of excise is 300 per cent, then a rise of rupee one in the cost of production results in a rise of rupees four in the printed price. Consequently, cigarette companies fought shy of cost increases—they simply compromised by using inferior quality tobacco.

While the ad valorem structure suited the government, the cigarette industry hated it, both because it limited the potential for increasing cost and because the Excise Department pursued tobacco giants, claiming that they were refusing to record price rises on their packets. The famous ITC excise case saw the government claiming that the cigarette company was telling its wholesalers to charge more than the price noted on the packet. After the exit of Shri V.P. Singh from the Finance Ministry in 1987, the government finally accepted the industry's pleas. The budget that year discarded the ad valorem on the grounds that it generated too many controversies. And that too many cases were piling up.

The new length-specific excise payment rules stated that the manufacturers now had to pay excise on the length of the cigarettes sold. Now, cigarettes were categorised into several segments according to the length and quality—the premium, the king size, the plain filter, the regular plain and the mini. The industry welcomed the change and claimed that companies could now begin concentrating more on quality control and exports. And within a year, the industry recorded substantive growth, both in terms of volume as well as the foreign exchange earned.

The vote-on-account, instead of proper budget, came as temporary respite for the industry. But the companies remained scared. They felt that the government, desperate to salvage India's already crippled economy, might still effect a change in structure. For, ad valorem translates into more excise revenue for the government because of its simple co-relation between price and the rate of tax. "For God's sake don't change this present structure", urged Ashok Bhatia, Director of business strategy for ITC Ltd. (turnover in 1990—Rs. 1827 crores, but net turnover after payment of excise and other duties—Rs. 677 crores). "The entire cigarette industry has now become a little more profitable". So frantic were the tobacco giants that (ITC), Vazir Sultan Tobacco (VST) and Godfrey Phillips India (GPI) which together command 90 per cent of the market, sent desperate letters to the concerned officials. In his letter, Ravi Bhoottalingam, Managing Director of VST (an ITC subsidiary), noted that the government could increase collections by increasing the base of production. One way of doing this could be to switch consumer preferences further from *bidis* to cigarettes, suggested ITC Chairman, Jagdish Narain Sapru. Cigarette had to combat *bidis* in the market-place and the only way this was possible by reducing duties on certain low-priced brands of cigarettes to compete with *bidis* in prices.

The ITC argument is best summarised by its Vice-Chairman, Kishen Lal Chugh, "If you look at the consumption of tobacco in totality, then cigarettes constitute a very small percentage. There are *hookahs, bidis,* chewing tobacco, *zarda,* etc. Ours being the only organised sector, the government collects majority of duty from us...."

Increasing Production

Going by figures alone, the excise duty paid by the industry every year has steadily been on the rise, from Rs. 644.96 crores in 1980-81 to Rs. 1850 crores in 1989-90. The figures for 1991 crossed the Rs. 2000 crores mark. In comparison, the excise duty paid by other tobacco products is peanuts. Together these products accounted for a duty of Rs.111.55 crores in 1980-81 and Rs. 165 crores in 1989-90. In 1991, figures totalled the Rs. 200 crores which is about 10 per cent of what the cigarette industry pays.

On the other hand, sales have not been all that bright. Nothing near the record figure of 96072 million sticks in 1984-85. However, after the 1987-88 budget, the sales have been recording a marginal growth of five per cent—from 72162 million sticks in 1987-88 to 79792 million sticks in 1989-90. The figures for 1991 were around 83000 milion sticks.

Sanjay Dalmia Factor

In every tobacco industry problem, the giants always manage to include the name of Golden Tobacco's Sanjay Dalmia. Off the record, the cigarette trade's view is that Dalmia will use his political connections to secure a return to the ad valorem structure. Because nobody is willing to go on the record, such allegations are difficult to rebut, though Dalmia denies that he is conspiring to pin down his competitors. Nevertheless, the giants persist in their view that Golden Tobacco—which would benefit by ad valorem brands in its product-mix is conspiring with the politicians.

The big three remain wary of Golden Tobacco.The 1989-90 sales figures place GTC (Golden Tobacco Company) fourth with 7751 million sticks, behind ITC's 43235 million, VST's 14631 million and GPI's 14371 million sticks. And the company is doing good business, eating into the sales of the top trio. Something that bothers the giants.

There is a history of confrontations. Earlier in the Madras High Court case filed by ITC on May 28, 1990 against GTC's Gold Flake Virgina 59 mm, Dalmia was caught on the wrong foot, trying to push his brand with a packet almost as identical as ITC's Gold Flake. This caused great confusion in the market and ITC promptly took GTC to court. On February 5, 1991, a Division Bench judgement passed an injunction restraining GTC from infringing ITC's Gold Flake brand or mark. Now the trade buzzes with speculation over GTC's recently launched Flair Kings which retails at an unbelievable Rs. 5 per ten. The excise duty on that length of cigarette itself works out to Rs. 5.64 per ten. This suggests that Dalmia is launching a brand at a loss. "What is he trying to do ?" asks a perplexed Bhatia, "probably he has price information that the ad valorem structure is being re-introduced. Otherwise this makes no sense". Attempts to get a response to these allegations from GTC were unsuccessful.

The panic within the industry demonstrates how much it is at the mercy of the government. Already, ITC has tried to ease its dependence on tobacco by diversifying into other businesses including paper boards and hotels. Nevertheless cigarettes remain its core business. Says Sapru, its Chairman, "I am not one of those men who believe in abandoning a car once it has met with an accident. Improve the car, get your road rules better implemented. Make it more competitive. We cannot diversify into new areas just like that forgetting our long tradition as tobacco men." Sapru's pragmatism is echoed by much of the tobacco industry. But the central problem still remains : What do you do when the government sees you as a cash cow and your principal product has no redeeming social value whatsoever ?[1]

Excise Income of Government

Financial Year	*Total Excise*	*Excise from cigarettes Collection*	*Excise from other tobacco products*
1980-81	6479	644.95	111.55
1981-82	7421	722.18	117.12
1982-83	8069	683.08	188.76
1983-84	10222	857.09	102.00
1984-85	11151	861.76	106.00
1985-86	12920	1202.00	114.00
1986-87	14066	1410.00	151.02
1987-88	14194	1470.00	161.26
1988-89	18288	1599.10	165.00
1989-90	18613	1850.00	173.26
1990-91	20973	2070.00	Sic

The Cigarette Wars

On January 21, the court of the District Judge, Sikkim passed an injunction banning the manufacture, marketing and sale of ITC Ltd's cigarette brands which carry the W.D. & H.O. Wills endorsement. The injunction was the result of complaint lodged by three Sikkim residents—Krishna Moktan, Tsering Dorji Bhutia and Lhendop Dorji Lepcha, who accused ITC of "unfair trade practices" by passing off foreign brands as its own under the trademark of W.D. & H.O. Wills. The charge implied that ITC was misleading consumers through fraud and misrepresentation, violating industrial policy guide-lines issued by the Central Government and contravening the provision of the Trade Mark and Merchandise Act of 1958. It also meant that 98 per cent of the cigarettes in ITC's arsenal, which carrry the W.D. & H.O. Wills name—on packets as well as on cigarette sticks—would have to be scrapped.

Predictably, all hell broke loose at ITC's Virgina House headquarters in Calcutta. The company immediately sent a high level team of officials to Gangtok, the State capital. But when they arrived, they found to their horror that the judge was on leave. There, the company pleaded that the W.D. & H.O. Wills name had been ITC's property since 1910, the days when UK's Imperial Tobacco Company ran the show and later, through British American Tobacco Co. The argument worked : On January 31, Supreme Court Chief Justice R.N. Mishra, presiding over a panel with Justice M.H. Kania and Justice Kuldip Singh, granted the company some relief. In a modified order, ITC was allowed to manufacture cigarettes with the controversial brand name, but sales were prohibited. But even with sales restricted, ITC was not complaining because the country's highest court also directed the Sikkim District Court to decide on the case one way or the other by February 7. And it did on February 6. In a final assessment of the case A.B. Subba, the District Judge incharge, noted that the name W.D. & H.O. Wills was owned by ITC and that the company's various products were covered by the Trade Marks and Merchandise Act of 1958. That was that. On the other hand, legal points apart, it definitely was not. The events at Gangtok sent ripples across the country, because say industry observers, the Sikkim saga was just the prelude to what might have been a deadly cigarette war.

The Men Behind Complaint

Though ITC officials insist that the case was more of embarrassment than a serious problem, it has considerably agitated them. So, who was behind the complaint ? The three men who filed the complaint certainly were not the real force, even ITC officials agree on that. Company sources say that the three were simply egged on by someone else. "Who in your opinion could gain by embarrassing ITC ?" asks Ashok Bhatia, the company's director, "For business strategy, it is not for me to say who it is, but I have a very good idea". He adds, "The lawyers who argued the case (for the Gangtok threesome) were heavyweight ad explosive." R.P. Bhatt of New Delhi, for one, who is well known and respected legal brain. What happened was that ITC cigarettes went out of the market and prices—thanks to the instant shortage and some nifty hoarding by traders—sky-rocketed. And they have not come down since then. "We are very concerned about the consumer" says Kishen Lal Chugh, Vice-chairman of ITC. "But I think the persons behind the price hike are someone else and something else."

Bhatia and Chugh decline to name a culprit but the industry grapevine suggests that it could have been Sanjay Dalmia, ITC's major competitor and chief of the Golden Tobacco Company. His alleged strategy : get ITC cigarettes out of the market and push his own to fill the gap, especially during the pre-budget period which regularly witnesses a shortage of cigarettes. (Dalmia could not be reached for comment, but a GTC official in New Delhi, who declines to be quoted, vehemently denies the charge).

The second factor that struck ITC was the timing of the case. the "embarrassment", as the legal fracas has been termed by the industry, came a month before the budget session of Parliament. In August last year, say observers, Chandra Shekhar had berated the then Finance Minister Madhu Dandavate at a Janata Dal Parliamentary Party meeting over the V.P. Singh Government's decision to refund Rs. 10000 crore in excise dues. This created a panic in government circles and soon enough, the then Deputy Finance Minister, Anil Shastry, demanded Rs. 80.3 crores in unpaid excise dues from ITC's Saharanpur factory in Uttar Pradesh. At that time, the demand was perceived to pacify Shekhar. And biased too, because ITC is not the only company to have been mired in excise controversies—the entire cigarette industry is fighting excise cases wih the government.

ITC is contesting the claim and company officials suggest that the Sikkim case came at a time when the whole issue could be placed back on the centre stage. And at a time when the government was strapped for cash, Shekhar—then Prime Minister—was looking for every way to get resources, he may press ITC to pay its dues. That might not work then. Also, as ITC's Director, Bhatia says, "the case is dead". But both the company and the government have evidently paid for it. With its cigarettes stalled in factories and the market-place, ITC officials claim to have lost Rs. 65 crores in revenue. And for that period between January 21 and February 6, the goverment lost Rs. 60 crores in excise collections from the company.

But one thing is certain, say industry observers : the cigarette wars are going to get worse.

Competitive Marketing

It's a battle on many fronts. At one level, it is the tobacco kings *versus* the Government. At another, it is a feud within the family, as the poor cousins slug it out with the affluent elders. And the war cries gain in stridency as the D-day (the budget announcement) draws near.

Why should their industry be subjected to an excise squeeze, totally out of proportion to its domestic growth, ask the cigarette manufactures. The irony, they point out, is that while during the past decade, the excise duty collection from tobacco has increased almost fourfold (to Rs. 2,748 crore in 1991-92) cigarette consumption has actually declined. "From 96 billion cigarettes a year, five to six years ago, it has fallen to a mere 80 billion today," laments Ram A Poddar, chief executive of Godfrey Phillips India (GPI) and chairman of the newly founded Tobacco Institute of India.

Unfair Taxation Policy

What is even more perturbing to the cigarette giants such as the Indian Tobacco Company (ITC), Vazir Sultan Tobacco (VST), GPI and Golden Tobacco Company (GTC), is the fact the India is virtually the only country where close to 80 per cent of the tobacco products have been kept out of the tax net. Cigarettes seem to be the sole target of the Government's revenue-raking exercise, which is blatantly unfair, they charge. But these are not the only anomalies. The Government's present policy of excise collection is also under fire by a certain section of the industry. And this is where the war moves to the home front. While one lobby, consisting of the majority of big wigs such as ITC, VST and GPI, favours the continuance of the specific duty structure that has been in force since 1987, another, led by GTC, would prefer a return to the pre-'87 ad-valorem system. This system, being linked to prices, would ensure that the Government gets a fair share of the windfall profits being reaped by the giants, contends the latter lobby.

The specific duty formula lays down that excise collection should be on the basis of the length of the cigarette. While on the surface, this would appear to be a somewhat irrational criterion to adopt, the Government's contention that switching to the specific mode would put a lid on the plethora of legal disputes had helped the industry to put off paying taxes, thus depriving the exchequer of huge amounts rightfully due to it. While the specific duty system might have resolved the problem of administrative bottlenecks for the Government, and thereby led to an overall increase in revenue in terms of relative percentage growth, which has declined, point out the critics, who favour the ad-valorem structure.

The total number of cases locked up in adjudication amount to approximately Rs. 2,200 crore over 10 years. Which implies that the Government is currently subjected to a loss of Rs. 200 crore per annum, thanks to the turn-around in excise policy. But this is still only fifty per cent of annual loss of Rs. 400 to Rs. 500 crore suffered by the exchequer, after it adopted the present specific excise levy.

Repressive and Discriminatory

And that's not all. The current system of levy favours the brands that the affluent opt for and discriminates against the poor man's cigarettes. According to S.K. Mukherjee, managing director, GTC, at present the structure does not allow scope for an even playing field. Our duty structure gives credence to Myrdal's theory that "the poor of the world will go on getting poorer, and the rich richer," he expostulates a trifle dramatically. He fears that under the present system, the small companies will get smaller while the bigger ones will get even larger. He suggests that the Government rationalise the tax system, so that cheaper brands don't have to cough up steep taxes. Taking the argument a little further, to the realm of specifics, Mukherjee argues, "Is it fair that I pay 47 per cent duty on my brand of cigarettes, while my driver pays 77 per cent? Our main brands are for the poorer segments. Is it fair this segment pays more excise than the affluent?"

Indeed, there is more than a grain of truth in his argument. Unlike in the Western markets, which have also opted for specifics, in India, the price difference in each segment of equal length is 40 to 60 per cent between the lowest and the highest priced. In countries like the US in contrast, the price differential between brands rarely exceeds 10 per cent.

Ad-valorem System

Countering this argument, Poddar contends that a little leg-up given to the upper segment of the market will help prevent debasement of quality. Indeed, under the ad-valorem system, it is this segment which was subjected to the cascading system of pricing, by which, "if I add value or improve quality by Re. 1, the price has to go up by Rs. 5". When this happens, company will be compelled to face rejection by the customer, he says. "The discerning Indian customer may be willing to pay a marginally higher price for a superior quality product, but he will be put off by a prohibitive price tag."

The most persuasive aspect of the argument about upgrading quality, he points out, is the enhanced status India will occupy in the international market. "At present, we are exporting only to the CIS, not to any of the dollar-paying countries, which even China and Brazil are doing. And this is because we are not competitive enough in reaching up to international standards. Moreover, we make no investments in research, making do with antiquated technology." While India is the third largest producer of tobacco after China and USA, yet we rank number ten in the list of cigarette exporters. On this count, there is little divergence of opinion among the cigarette lobbies. Indeed, the ad-valorem system experts hold that exports need to be stepped up so that benefits can accrue equitably to both exchequer and the companies. But there is a catch in this, cautions Poddar. There can be no question of expanding the export potential without simultaneously stabilising the domestic market. "No international buyer will risk bidding for a product that has not been already tried and tested and approved domestically," he points out. Which means, in effect, that the Government loosen its grip of the industry instead of tightening it. That it allow scope for the sector to expand and experiment, all of which is not possible in the volatile tax structure that prevails at present, with shifts in Government policy being the order of the day. As Satish Mehta, former chairman of ITC, puts it, "The cigarette companies do not want the Government to lose excise, nor do they want the Government to desist from raising excise in the forthcoming Budget. The Government is entitled to collect the duty, but it should be a little kinder to the industry. An irrational increase of excise on cigarettes will only push the prices up, which means they will go beyond the reach of at least a section of consumers, thus affecting the sales."

Any further increase in excise rates for cigarettes will only lead to the "Government trying to kill the goose that is laying the golden eggs," is how S. Thirumalai, director and secretary of VST, put it. "Considering that the cigarette industry is a major source of revenue, contributing as much as Rs. 2,449 crore (90 per cent of the total tobacco share in the excise duty), the Government has been increasing the duty in a harsh manner," he says. Between 1987 and 1992, the excise duty on cigarettes was increased by as much as 64 per cent of the lowest level and 112 per cent at the highest level, he points out. According to him, with a stable policy and tax structure, the revenue from cigarettes could rise to as much as Rs. 6,000 crore by 1997-98.

Export Possibility

VST vice-chairman and managing director Raj Sujan further contends that with stability in the excise duty structure, the export potential of the industry would definitely see an upswing, which in turn will help tobacco growers in the country. His view is endorsed by Kamal Ramnath, executive vice-president (marketing and exports), ITC. "We would favour an excise structure that encourages investment and supports revenue earnings by the industry. The Indian cigarette manufacturer has the advantage of low labour and low leaf cost, but an aggressive excise structure is standing in the way of growth," he says. Madhav Mehrotra, area sales manager of GPI could't agree more. Indeed, it would appear, it is thanks to the growing excise rates, which they perceive to be weighted against them, that the GPI is eyeing the bidi sector with unconcealed interest. As Mehrotra admits, "For the Government, the bidi is inefficient for mopping up taxes." Therefore why not exploit the sector to circumvent the excise tax net? The anti-specific duty lobby, however, would not advocate such diversionary tactics. They would prefer to remain in the fray, facing up to the heat and smoke generated in the course of the battle. They found some ammunition in the interim report of the Chelliah Committee on Tax Reforms, released in December which, however, has been negated by the position taken in the final report, which is now out. The interim report has suggested that "for further administrative purposes, some commodities like petroleum products, tobacco products..., etc., may continue to have specific duties." But it also recommended that "where specific rates are retained, there should be a system of revising the rates every year taking into account price increases as represented by the relevant sectoral wholesale price index." The final report makes no mention of the annual increase related to prices, and recommends a unilateral retention of the specific rates formula. This will be a major triumph for the all-powerful multinational lobby in the cigarette industry, for whom any price-related amendment to the excise tax structure is anathema. But the claim of resisting a price-hike in the interest of the consumer is only a smokescreen used by the tobacco kings to garner more profits and make inroads into the international market. In the process, are the needs and tastes of the consumer put on the back burner.

Indeed, in this battle between the Government and the cigarette lobby for a bigger share of the cake, it is inevitably the customer who is the casualty, making do with crumbs.

A New Smoke Strategy

Puff for puff, this will be a fight to the finish. In the country's nearly 80-billion-sticks-a-year cigarette market, years of loyalty will be put to test even as new smokers—including a host of foreign brand names enter the fray. For a start, there's Godfrey Phillips India (GPI) that plans to bring in brands from the Philip Morris stable, including the best selling Marlboro. Arch-rival R.J. Reynolds, too, is warming up to introduce its Camel, even as Japan Tobacco Company plans to bring in its Mild Seven. New launches, line extensions and innovative merchandising strategies are the order of the day. Consider these:

* VST has launched its new Gold Premium in the 50 paise per stick regular filter segment.
* GPI has introduced its Army Club (60 paise) and Time (50 paise) in the regular filter category.
* ITC has unveiled Classic Milds as a line extension of Classic and has launched Scissors filter cigarettes in new markets.
* Both ITC and VST recently introduced their Wills and Charms brands in packs of 20s. Internationally, cigarettes are sold in a 20s pack. The 10s shell-and-slide pack is unique to India.
* ITC is now selling its premium brands through its new Wills shops. In future, retail atmospherics could be the key in ITC's move to tap new sales channels.

Low Price-High Brand Strategy

So far, the market has been moving along predictable lines. Overall growth is declining as price increases have been over 20 per cent in the last two years because of excise hikes. The premium segment which was slowing down around the end of 1991 has seen sales plateau. The decline in the low end of the market continues. Most of the action is now taking place in the regular filter category perceived as the most promising segment. And within the regular size filter category, the top end of the filter segment (70 paise plus) and the 50 paise smokes have started growing. Marketeers are clearly shifting towards the 50 paise popular regular filter category.

The reason for the shift is not difficult to gauge: the 50 paise category is the fastest-growing segment (5 per cent) currently as it offers coinage convenience and value-for-money appeal to the smoker. After all, today popular regular filter smokes are selling at the same stick price as regular plain brands. For instance, a Cavender's Gold Leaf plain brand costs Rs. 4.65 paise per pack and 50 paise per stick and so does a regular filter brand like Capstan which has a pack price of Rs. 4.85 but costs 50 paise a stick. The brands remain pegged to the 50 paise brand. Current volumes, estimated to be 550 million sticks per months, are not very high, but are expected to go up.

While the going is good in this segment for now, most companies know that they cannot bet on this for long. A slight change in the excise structure is all it would take for the segment to go up in smoke. Admitting that the affair with the 50 paise segment is just temporary, Kamal Ramnath, ITC's executive vice-president (marketing and exports), says: "Your abilities to improvise on a Rs. 5.60 pack are limited." He feels that brands like Bristol will be very important to ITC in the future.

New Brands-New Strategy

While GPI is launching new brands, ITC is taking the line extension route and is repositioning most of its brands. "It's easier to create 'extensions' rather than launch a new brand name," says Ramnath. And in anticipation of the invasion of other foreign brand names, ITC launched Classic Milds, a lighter variety of Classic Kings and a new filter variety of Scissors. The move makes sense as Classic sales have been stagnant at 20 million sticks per month for the last two years and Scissors is ITC's largest selling brand in terms of volumes (900 million sticks a month). The new launches are backed by a Rs. 2.5 crore advertising blitzkrieg. While marketing and ad bills are being marked up, players are going in for different spending strategies. Market leader ITC, for one, has cut down its outdoor media spend—which was 15 per cent of the total spend—by whopping 25 per cent. Argues ITC's Ramnath, "All our brands are going upmarket. As hoardings are a downmarket media we are spending more to put up our own exclusive shops."

In fact, ITC is exploring alternate channels of distribution by entering non-traditional outlets like bookshops, restaurants, pubs, department stores, etc. That apart, ITC is now laying greater emphasis on event marketing. The company set-up a separate brand sponsorship group to spearhead its effort and tied up with Mark McCormack's International Management Group in Hong Kong. The World Badminton Cup is the first event to be handled by this group.

In contrast, VST is reducing its sponsorship spend and is pouring in more money into press, and outdoor sale. Says Philipose: "We are looking closely at outdoor technology and are trying to sharp focus our efforts." But why the change in tack? Philipose maintains that whatever you do benefits the most dominant player—ITC. And to prevent this spillover effect for its rival, VST is looking at relationship marketing. For a start, it will reach out to Kingston smokers in Bombay. And the company has also upped its total spend by ten per cent over the last year.

All this doesn't mean that the premium segment is being ignored. In fact, GPI can hardly afford to do that as the segment accounts for over 50 per cent of its total sales. GPI's latest moves in the non-premium segment are clearly an attempt to corner a larger share in this segment. Evidently, GPI's strategy is to have a nationwide network in the premium segment and in the non-premium category it is hoping to develop local or regional brands with strong consumer loyalty. The air is to increase market share by aggressively marketing the non-premium brands. The decision to bring in international brands like Marlboro is aimed at consolidating its position in the premium segment where ITC is likely to pose a major threat.

The premium segment remains the centre-piece of ITC's strategy too. Says Ramnath: "The entry of a whole host of foreign brands will blow up the market." Most marketeers maintain that five years hence, premium kingsize smokes should rule the roost. With the entry of international brands, larger marketing and ad budgets, and a stronger distribution network, the top-end of the cigarette marked should grow exponentially. And the present frenetic activity—both in premium and non-premium brands—is merely a build-up to the big show-down.[2]

Asia's Enlarging Market

The cigarette hawking Cowboy may be under siege in the United States but half a world away, in Asia, he is prospering.

The Marlboro Man has Found Greener Pastures

The Marlboro man and his cigarettes have now been more popular in Asia reaching to sixty per cent of the world population. The tobacco critics who called for extinction of multi-billion tobacco industry now need to take a stroll down the tobacco scented streets of almost any city in Asia where Marlboro man is prospering with his craggy all-American smug slapped upon bill board and flickering across television screens. The ad is the evidence of a booming Asian growth market that promises vast profits for the tobacco industry and deaths to millions !

Changing Scene

In the hazy market of Seoul at lunch-time throngs of fashionably dressed Korean women gather in a fast food restaurant to enjoy a last cigarette before returning to work, a scene that draws disturbed stares from older Koreans who remember a time when it would have been seandalous for women from responsible homes to smoke. Likewise, in Hong Kong, shoppers flock into the Saloon Attired Boutique, picking trendy sports clothes stamped with logo of 'Salem' cigarettes. Even in Phnom Penh the war shattered capital of Cambodia visitors leaving an audience with King Sihanouk are greeted by a giant bill-board right across the street from. It advertises trendy shirts.

The Asian cigarette market is likely to grow by more than a third during the 1996-2000 with most of the bounty going into the pockets of multinational tobacco giants eager for an alternative to the shimmering market in the United States.

Huge Market

However bad the news in the West, the tobacco companies can always find comfort in Asia and throughout the Third World markets. So huge and so promising that they make the once all prominent American markets seem insignificant. Besides Asia, the cigarette consumption is also expected to grow enormously in Africa, Latin America, Eastern Europe and in the nations of the former Soviet Union. Today, China is the main target of the multinational tobacco companies because cigarette consumption has increased in China by 7 per cent: there are 300 million smokers in China and they buy nearly 1.6 trillion cigarettes. In fact, since the mid-1980s, Japan, South Korea, Taiwan and Thailand have all succumbed to pressure from Washington, accordingly the sale of foreign brand cigarettes has increased manifold. According to Mr. James W. Johnston, Chairman of Reynolds Tobacco Woddinde wrote in his company's 1993 annual report "Woddinde, hundreds of millions of smokers prefer American-blend cigarettes.... Today, Reynolds has access to 50 per cent of the world markets, a decade ago, only 40 per cent. Opportunities have never been better".

Markets for 'Made in the USA'

Shares of total American cigarette exports in 1993.

South America 2.2% Canada and Mexico 0.6%

America is pushing tobacco in the manner, Britain pushed opium in China. Yesterday it was opium and today it is tobacco.

However, a strong lobby of health activists and congressmen is opposing the export of Tobacco, "I think it is exporting death. It's not just a wasteful programme, it's a deadly programme", said congressman Peter H. Frostmayer of Pennsylvania. However, American tobacco producers and cigarette manufacturers seem determined to push up American cigarette sale.

Terrifying

Physicians say that the health implications of the tobacco boom in Asia are absolutely terrifying. According to Richard Peto, an Oxford University epidemiologist, the annual worldwide death toll from tobacco-related illness will be more than triple over the next two decades, from about three million a year to 40 million a year by 2050. A fifth of these deaths will be in China as a result of which 50 million Chinese children alive today will eventually die from disease-linked to cigarette smoking.

References

1. *Sunday*, March 10-16, 1991, pp 62-65.
2. *Economic Times*, September 8, 1993.

Appendix-1

INTERNATIONAL CONFERENCE ON DRUG ABUSE AND ILLICIT TRAFFICKING

PART ONE

Declaration of the International Conference on Drug Abuse and Illicit Trafficking

We, the States participating in the International Conference on Drug Abuse and Illicit Trafficking.

Believing in human dignity and the legitimate aspirations of human-kind for a decent life with moral, humanitarian and spiritual values in a healthy, safe environment.

Concerned at the human suffering, loss of life, social disruption, especially the effect on youth who are the wealth of nations, brought about by drug abuse worldwide.

Aware of its effects on States' economic, social, political and cultural structures, and its threat to their sovereignty and security,

— Commit ourselves to vigorous international actions against drug abuse and illicit trafficking as an important goal of our policies;

— Express our determination to strengthen action and co-operation at the national, regional and international levels towards the goal of an international society free of drug abuse;

— Strive for the universal accession to the 1961 Single Convention on Narcotic Drugs or this Convention as amended by the 1972 Protocol and to the 1971 Convention on Psychotropic Substances and their strict implementation as well as the completion and adoption of the draft Convention against Illicit Trafficking in Narcotic Drugs and Psychotropic Substances, at the earliest possible date.

Agree on the following :

1. We express our determination to pursue the goals we have set for ourselves at various levels of government towards combating this scourge and to adopt urgent measures to strengthen international co-operation through a balanced, comprehensive and multidisciplinary approach. In this regard, we emphasize the pivotal role of Governments in developing appropriate national strategies within which such measures could be implemented.

2. In evolving effective action against drug abuse, illicit production and trafficking, we emphasize the need for the international community to adopt measures to treat all aspects and causes of the problem. To be effective, these measures must take into consideration the relevant social, economic and cultural factors and should be conducted in the context of States' policies in this regard. We recognize the collective responsibility of the States to provide appropriate resources for the elimination of illicit production, trafficking and drug abuse.

3. We affirm the importance of and the need for wider adherence to the 1961 Single Convention on Narcotic Drugs or this Convention as amended by the 1972 Protocol and to the 1971 Convention on Psychotropic Substances. We call for the urgent but careful preparation and finalization, taking into account the various aspects of illicit trafficking, of the draft Convention Against Illicit Trafficking in Narcotic Drugs and Psychotropic Substances to ensure its entry into force at the earliest possible date and to complement existing international instruments.

4. We recognize the important role of the United Nations system in the efforts to combat drug abuse and illicit trafficking, and in particular the role of the Secretary-General of the United Nations in facilitating co-ordination and interaction among Member States and within the United Nations system. We attach importance to the role of the United Nations Commission on Narcotic Drugs as the policy-making body of the United Nations on drug control matters. We commend the positive action carried out by the Division of Narcotic Drugs, the International Narcotics Control Board and the United Nations Fund for Drug Abuse Control, and we urge strong national and international support for the Fund so as to enable it to fulfil its mandate.

5. We recognize the constant, determined efforts of Governments at the national, regional and international levels to counter the escalating incidence of drug abuse and illicit trafficking and the growing link between drug trafficking and other forms of international organized criminal activities.

6. We also recognize and welcome the significant role played by non-governmental organizations in the drive against drug abuse, and urge that further initiatives be encouraged to strengthen the efforts made at the national as well as international levels.

7. We welcome the compilation of the Comprehensive Multidisciplinary Outline of Future Activities in Drug Abuse Control as a compendium of possibilities for future action by all concerned.

8. Recognizing the magnitude and extent of the world-wide drug problem, we agree to intensify efforts against drug abuse and illicit trafficking. As an expression of our commitment, we also agree to promote inter-regional and international co-operation in :

(a) Prevention and reduction of demand;

(b) Control of supply;

(c) Suppression of illicit trafficking; and

(d) Treatment and rehabilitation.

For this purpose, we consider that the following, *inter alia*, should guide the development of our actions:

(a) Prevention and reduction of demand:

(i) Develop methodologies and institute systems for assessing prevalence and trends of drug abuse on a comparable basis;

(ii) Develop and implement the necessary measures to reduce drastically illicit demand through adequate techniques and programmes.

(b) Control of supply:

(i) Encourage contributions from international financial institutions and Governments, where possible, for the implementation of programmes and projects for integrated rural development activities, including crop eradication/ substitution schemes, and continue scientific research in related areas;

(ii) Develop and implement the necessary procedures to eliminate the illicit supply of specific precursors and other materials necessary for the manufacture of narcotic drugs and psychotropic substances, and to prevent the diversion of pharmaceuticals to the illicit drug market.

(c) Suppression of illicit trafficking:

(i) Develop bilateral and other instruments or arrangements for mutual legal assistance which might include among other things, if appropriate, extradition and tracing, freezing and forfeiture of assets, and for enhancing international legal or law enforcement co-operation in this field;

(ii) Improve dissemination of information to national and international law enforcement bodies, especially concerning profiles and methods of operation of drug trafficking organizations, and further develop international, financial, technical and operational co-operation in investigation and training for officers and prosecutors.

(d) Treatment and rehabilitation:

(i) Develop, promote and evaluate effective treatment and rehabilitation techniques;

(ii) Provide health professionals and primary health workers with information and training concerning appropriate medical use of narcotic drugs and psychotropic substances.

9. We affirm our determination to continue our efforts and request the Secretary-General of the United Nations to keep under constant review the activities referred to in this Declaration and in the Comprehensive Multidisciplinary Outline. We request the Secretary-General of the United Nations to propose in the context of the United Nations programme and budget and within available resources how the priority attached to the field of drug abuse control can best be carried out. The Commission on Narcotic Drugs should examine the most suitable modalities for following up these activities, as appropriate, at the international level.

PART TWO

Comprehensive Multidisciplinary Outline of Future Activities in Drug Abuse Control

Introduction

1. The Comprehensive Multidisciplinary Outline of Future Activities in Drug Abuse Control is a repertory of recommendations addressed to Governments and to organizations setting forth practical measures which can contribute to the fight against drug abuse and to the suppression of illicit trafficking. At the national level, it is for each Government to determine which of the recommendations could be useful in its country in the light of economic and social conditions and to the extent consistent with national law. The Comprehensive Multidisciplinary Outline is not and was not designed to be a formal legal instrument; it does not create either rights or obligations of an international character. Its purpose will be achieved when the text is used as a handbook by national authorities and by interested organizations as a source of ideas to be selected and translated into action appropriate to local circumstances in the manner considered fit by these authorities and organizations. The text is accordingly drafted in non-mandatory style as a working guide, rather than as a package to be accepted in its entirety.

2. The recommendations have been drafted in terms fully consistent with the principal international instruments concerned with drug abuse control; that is, the Single Convention on Narcotic Drugs, 1961, as amended by the 1972 Protocol Amending the Single Convention on Narcotic Drugs[1], and the 1971 Convention on Psychotropic Substances of 1961[2].

3. In addition, with a view to safeguarding the principle of the sovereignty of the State and the primacy of the fundamental principles of the law and constitution of the State, many recommendations include a proviso concerning respect for these principles.

Background

4. Throughout recorded history, substances to relieve suffering and to alter moods have been known and used in human societies. The ambivalence of these substances—they are indispensable for the relief of pain and suffering, but addictive and destructive when misused or abused—led societies from earliest times to make rules restricting their use to religious or curative purposes and entrusting them only to priests or to healers and doctors.

5. Since the mid-nineteenth century, drug abuse has been spreading in many countries as a result of a number of factors. The causes are many and vary in intensity; they include, in particular, the increased availability of products, the expansion of communications, socio-economic factors, migration and rapid urbanization, changes in attitudes and in the sense of values and the ruthless exploitation of fellow human beings by criminals.

6. In order to deal with the increase in drug abuse, the community of nations has since the early twentieth century gradually evolved global control mechanisms intended to limit the availability of drugs for abuse, because it was quickly realized that no single country could succeed alone in preventing drug abuse and illicit trafficking. Between 1912 and 1972, no less than 12 multilateral drug control treaties were concluded. Under the auspices of the United Nations, the Single Convention on Narcotic Drugs, 1961, as amended by the 1972 Protocol, consolidated most of the earliest instruments, and the 1971 Convention on Psychotropic Substances created a system of international control for a number of previously uncontrolled psychoactive substances. During this period, the main effort was directed to the gradual devising and reinforcing of a network of administrative controls, having as its primary object the regulation of the supply and movement of drugs with a view to limiting their manufacture and import to the quantities required for legitimate medical and scientific purposes. Governments have also increasingly recognized the need to co-operate in the fight against the illicit production and manufacture of and traffic in drugs and

accordingly to furnish to the international control organs, established first under the auspices of the League of Nations and then included in the United Nations system, periodic reports on their application of the international instruments and to submit to international supervision for their common benefit.

7. As gaps in the international control system were progressively closed, it became more generally apparent that the mechanisms originally created for the introduction of international supply control were not in themselves a sufficient response to the current needs of the international community.

8. Without prejudice to the importance of continuing administrative control of narcotic drugs and psychotropic substances and of international co-operation in the fight against the illicit traffic, counter-offensive of another dimension are now needed at the national and international levels to respond to the threat drug abuse poses not only to millions of persons but also to whole population groups and even to societies and economies in some countries. To take up the challenge it is necessary to intensify not only measures and programmes directed against the illicit production of and trafficking in drugs but also the activities undertaken to prevent the illicit demand for drugs and to further the treatment and eventual social reintegration of drug addicts.

9. Many heads of State or Government have recently directed their personal attention to launching such counter-offensive, and the Secretary-General of the United Nations, addressing the Economic and Social Council on 24 May, 1985, noted that the moment had arrived for the international community to expand its efforts in a global undertaking, that would be more concerted and more comprehensive. He envisaged a truly worldwide effort to contain the plague of illicit drugs and thus proposed that a world conference be convened at the ministerial level in 1987 to deal with all aspects of drug abuse. Significant progress has already been achieved by the international community in building up defences against the production of addictive drugs and the illicit traffic in and abuse of those drugs. Building on this foundation of international co-operation developed over nearly 80 years, the international community is now concerting its efforts to expand the United Nations work in the field of drug abuse control and in the fight against illicit trafficking in drugs.

10. The political will on the part of the international community to take urgent, effective and concerted action to deal with the disturbing situation created by drug abuse and the illicit drug traffic was reflected in the decision of the General Assembly to convene the International Conference on Drug Abuse and Illicit Trafficking (hereinafter referred to as "the Conference") in 1987 (General Assembly resolution 40/122 of 13 December 1985). The Assembly gave the Conference an ambitious mandate, encompassing the full range of issues relevant to the fight against drug abuse, illicit trafficking and related criminal activities at the national, regional and international levels. It directed the Conference to adopt a "comprehensive multidisciplinary outline of future activities which focuses on concrete and substantive issues directly relevant to the problems of drug abuse and illicit trafficking" [paragraph 4 (a)]. This mandate transcends the traditional concern with the control of the supply of and the illicit traffic in drugs and calls for a commitment by the Governments of all States to reinforce their individual efforts and to intensify and expand the scope of international co-operation into new areas.

Structure of the Comprehensive Multidisciplinary Outline of Future Activities in Drug Abuse Control

11. The Outline comprises of four chapters, covering the main elements involved in the fight against drug abuse and illicit trafficking and the subjects included in the agenda of the Conference; the prevention and reduction of illicit demand; the control of supply; action against illicit trafficking, and treatment and rehabilitation.

12. Each chapter indicates specific targets, particularizing the objective to be attained and the action to be taken at the national level (by Governments, professional associations, academic institutions, non-governmental organizations, communities, parents and individuals); at the regional level (by regional inter-governmental and non-governmental organizations and bodies); and at the international level (by international organizations, especially those of the United Nations system).

13. The catalogue of proposed actions is not exhaustive, nor does it follow any particular order of priority. At the national level, it could be for each State to determine its own order of priority among the targets, in the light of its own needs and resources. While all the targets are applicable to most countries, the modalities for the suggested courses of action are not necessarily applicable in each instance. Any action should be considered within the socio-economic context of the country concerned and may need to be adapted to the particular cultural, social, political or legal setting in order to be successfully applied. In particular, due account being taken of the constitutional system, legislative and regulatory provisions and court practice may need to be adapted and revised in order to give full effect to national efforts.

General Principles

14. The application of the recommendations contained in the Comprehensive Multidisciplinary Outline in the national setting implies the participation of many branches of the national governmental machinery—leg-

islative organs; the authorities concerned with public health, education, social welfare, the judiciary, law enforcement and economic affairs; and all the many sectors in which the Government exercises responsibilities—and also institutions of higher learning, research and other academic bodies and private sector organizations. It is in this sense that the Outline is "multidisciplinary" in character.

15. If the contributions of these multifarious entities to a combined national effort are to be marshalled in support of the achievement of the objectives of the Outline, it then follows that machinery has to be established, where it does not yet exist, for co-ordinating the activities of these bodies, services, agencies and institutions. Indeed, the Outline, like the 1961 and 1971 Conventions, refers in numerous passages to national co-ordinating machinery. Accordingly, Governments which are determined to take effective action against drug abuse and illicit trafficking may find it advisable, and even indispensable, to set up a co-ordinating mechanism, in so far as they have not already done so, or to strengthen the existing machinery by establishing a nation-wide strategy. The national agency or authority designated to carry out this strategy could in turn be guided by certain general principles, which might include the following :

(a) A clear set of achievable objectives should be defined;

(b) Target groups should be clearly identified, priority being given to initiatives aimed at reducing drug abuse among young people;

(c) A balanced approach should be adopted for dealing with illicit demand, illicit supply and illicit trafficking;

(d) Programmes should be comprehensive and long-term;

(e) The development of the programmes and their implementation should be supported by research into the extent of drug abuse and its aetiology and consequences;

(f) The achievements of the programmes should be evaluated periodically;

(g) The work of all the agencies concerned at the national, regional (cantonal, provincial) and local levels should be part of a co-ordinated plan;

(h) The formulation of the national strategy should take into account in the first place the existing resources and, in the implementation of the strategy, a cost-effective approach should be adopted, with external support if necessary;

(i) The potential of community actions should be recognized and developed;

(j) The national strategy could make the fullest possible use of the experience and attainments of other countries in fighting drug abuse and illicit trafficking, and the co-ordinating agency might offer to share its experiences with corresponding foreign authorities.

16. The recommended measures set out in the CMO are intended to be applicable in the generality of situations where drugs abuse or illicit trafficking, or both, have assumed dimensions that give rise to concern. National authorities are best equipped to judge what action is called for within the country for dealing with the evils of drug abuse and illicit trafficking. The recommended activities are designed to reinforce and not to replace those now undertaken in fulfilment of treaty obligations arising from the international conventions. At the same time, it cannot be too strongly emphasized that any State which is not yet a party to the two principal international instruments relating to drugs—the Single Convention, 1961 (as amended by the Protocol of 1972) and the 1971 Convention on Psychotropic Substances—should ratify or accede to these two Conventions as a matter of urgency, and so join the community of nations that is formally pledged to fight these evils by a common effort.* In addition, States parties to the Conventions should ensure that the provisions of these instruments are strictly applied. They are also urged to designate, in so far as they have not already done so, the special administrations or agencies referred to in articles 17 and 35 of the 1961 Convention and in articles 6 and 21 of the 1971 Convention.

17. The restriction of the scope of the Comprehensive Multidisciplinary Outline to problems arising from the abuse of narcotic drugs and psychotropic substances should not be taken to imply that Governments should not consider taking action to combat the serious health consequences arising from the abuse of other substances, specifically alcohol and tobacco, since many of the drugs which demand reduction strategies in the Comprehensive Multidisciplinary Outline could be integrated into other government substance abuse prevention programmes.

* As of 1 June 1987, 115 States were parties to the 1961 Convention, 83 States were parties to the 1961 Convention as amended by the 1972 Protocol and 86 States were parties to the 1971 Convention.

I

(International Conference on Drug Abuse and Illicit Trafficking)

PREVENTION AND REDUCTION OF THE ILLICIT DEMAND FOR NARCOTIC DRUGS AND PSYCHOTROPIC SUBSTANCES

Introduction

18. Different authors and different Governments take different views as to the best way of fighting drug abuse and illicit trafficking and of dealing with their attendant phenomena. According to one school of thought, it is the sources of the illicit supply of drugs that should form the primary target of concerted and determined action, for—so it is argued—if the supply is halted or at least curtailed, addiction and trafficking would cease or at least diminish. Much of the national and international action taken so far has been aimed at the supply side of the drug economy, in the form of control of the legitimate production and trade (to prevent diversion into illicit channels), bans on cultivation of narcotic plants, and their eradication (physical uprooting or destroying of illicitly cultivated plants) etc. The philosophy underlying the two principal international instruments relating to narcotic and psychotropic substances—the 1961 and 1971 Conventions—is guided by the same considerations, in that they make provision for strict control, by licensing and other means, of the production, manufacture, internal and international trade, prescribing, storage etc. of the substances covered. In certain circumstances the cultivation of narcotic plants may be prohibited by the national authorities, and the two Conventions contain some provisions for penal sanctions to be enacted by States parties, with respect to violations of the provisions of the Conventions, which are treated as punishable offences.

19. The inference to be drawn is that the manifest intention of the Conventions was to curb and prevent the uncontrolled flow of drugs and substances, the non-medical consumption of which was considered dangerous to the health of the individual, to the social fabric and even in some cases to political stability and security.

20. The other school of thought holds that efforts to deal with what has come to be regarded as a social scourge should concentrate on the illicit demand side, that is on the market. In the opinion of this school of thought, the more efficacious way of dealing with the criminal activities associated with drug abuse and illicit trafficking would be to lower the illicit demand for drugs.

21. For the purpose of dealing with the totality of the problems posed by drug abuse and illicit trafficking, both the supply of and the demand for drugs, should be reduced and action should be taken to break the link between demand and supply, that is, the illicit traffic. It should also be recognized that an important deterrent to both the illicit demand and the illicit supply of drugs, as well as drug trafficking, is the existence of effective and enforced legal sanctions.

22. While it is not claimed that the restriction of the illicit demand is the panacea for the social sickness of drug abuse or for breaking the chain of criminal activities of illicit trafficking, various possible methods for reducing the illicit demand are suggested in chapter I. Nor should these suggestions be regarded in isolation or as contrasted with those put forward elsewhere in this Outline for curbing the illicit supply: Rather, they should be thought of as complementary to those other suggestions, for they all have ultimately the same objective of banishing an acknowledged evil and of rescuing human beings from a precarious situation.

23. To give attention to the fundamental causes of

the problem of drug abuse, social, economic and cultural factors must be taken into account.

24. In some regions of the world, popular movements and voluntary organizations play a primary role in drug abuse prevention, education and awareness. The value of non-governmental organizations and government agencies working together on complex national and international problems is nowhere more apparent than in the comprehensive long-term effort to reduce illicit demand for narcotic drugs and psychotropic substances.

25. It will be noted that in chapter I considerable emphasis is laid on information, statistical data and exchanges of experience, for there is no doubt that there are large gaps in the knowledge about the extent of drug abuse and about methods for restraining it. To the extent that countries can learn from each other, and subject to due allowance for differences in resource endowment, social and cultural conditions and economic system, national authorities should be encouraged to ask for and communicate results of measures for reducing the illicit demand for drugs. For their part, the United Nations bodies concerned will wish to take into account, in projects or activities designed to alert the public, to the risks of drug abuse and to diminish the illicit demand for drugs, the special need for protection of certain vulnerable population groups.

TARGET 1. Assessment of the Extent of Drug Misuse and Abuse

The problem

26. Thanks to modern techniques of data gathering and processing, it is now possible in many countries to measure precisely the extent of drug use for medicinal purposes once a drug has been prescribed by doctors. Drug abuse, however, defies precise measurement, owing to its complexity and to its diffusion throughout a neighbourhood, city, region or nation.

27. A first point to consider is the meaning of the terms used: what effects of use are undesirable and which patterns are abusive? Under national law as well as under the international drug control treaties, a "narcotic drug" or a "psychotropic substance" is defined not in either the medical or scientific sense but rather in a legal sense as a substance included in a list determined by national or international law. At the same time, the use and possession of these specific substances are to be limited exclusively to medical and scientific purposes; misuse and abuse are, therefore, defined as illicit and are distinguished from the strictly medical use (i.e., medical administration or dispensing) by the legal status of the drug in question. Data contained in records of law enforcement agencies and of the judiciary, will therefore only relate to unlawful use, whereas those from hospital emergency rooms will also include incidents caused by the use of substances that are not (or not yet) under control. The World Health Organization is developing a standardized nosology for drug abuse. The classification system will be valuable in clarifying the use of drug related terms, fostering communication and allowing for collaborative research.

28. Second, what is to be measured and by what methods? Surveys of illicit drug use can be unreliable, owing both to users' fears of exposure and to the fact that certain user groups cannot be reached by means of survey techniques. Surveys of illicit drug use can be unreliable for a number of reasons, including utilization of inappropriate methodology, under-reporting because of fear of exposure and the inability to sample certain user groups by survey techniques. In addition, owing to differences in statistical capability and approaches between countries, methods appropriate in one country may not be applicable in another. Research into the drug problem should be carried out by means of statistical studies and epidemiological surveys, in accordance with the circumstances.

Suggested Courses of Action

At the national level

29. The responsible authorities, in co-operation with all other ministries and services concerned or the national co-ordinating body where it exists, and with due regard for the national legislation safeguarding individual freedoms and rights to privacy could:

(a) Review current methodology for epidemiological studies of drug abuse and, where necessary, sponsor research leading to the development of more valid and reliable methodologies and instruments for assessing the extent of drug abuse;

(b) Review present methods of data collection by statistical, medical, legal and demographic services;

(c) Provide for the systematic collection of data on subpopulations of drug abusers from records of the police, registrars of deaths, courts (including coroners' courts), hospital emergency rooms, drug treatment centres, prisons, mental hospitals, psychiatric clinics, correctional institutions, social security and welfare organizations, schools and universities, the armed forces, employers, trade unions, as well as community agencies, institutions and associations;

(d) Establish training programmes for personnel involved in the collection and analysis of data;

(e) Establish central records for storing, analysing and evaluating data;

(f) Determine drug abuse patterns in special groups and environments, such as boarding schools, army bases, workplaces, prisons, hospitals;

(g) Determine, in respect of each type of drug of abuse:

(i) The prevalence of active use (e.g., number of current users within a given period; say, the last year or the last month);

(ii) Incidence of first use (e.g., number of new cases per year);

(iii) Age at onset of use and the demography of users;

(iv) Drug of abuse initially used and/or other drugs currently used;

(v) The source of supply of currently abused drugs;

(vi) Prevalence of use by sub-groups (e.g., males and females);

(vii) Extent of use of particular drugs which cause diseases or affect special health conditions (e.g., pregnancy);

(h) Establish machinery for monitoring trends of abuse and evaluating the effectiveness of instruments, preventive policies, facilitate training, and the planning of prevention strategies;

(i) Test periodically the validity of the methodology used.

30. A country lacking a system for estimating the extent of drug abuse could develop such a system by stages, possibly with the assistance of WHO or any other international organization.

31. The appropriate authorities could undertake or sponsor studies to determine the frequency or gravity of drug addiction (or addiction to particular drugs) in specific population groups and the connection (if any) that exists between drug addiction and a variety of social, economic and cultural factors. These studies could include both particular and general factors related to the individual, the family, and society. It is recommended that the results of the studies be brought to the attention of the agency or authority responsible for the social services that need to be mobilized in order to lessen the illicit demand for drugs.

32. Universities, research institutes and other academic institutions should conduct studies to determine the implications of present patterns of drug abuse in terms of predicted demographic developments.

33. If the appropriate authority has reason to suspect, in the light of statistical or other data, that a change has taken place or is about to take place in the pattern of consumption of addictive drugs (e.g. a shift towards a newly "fashionable" drug among drug users, or a marked rise in illicit demand in a particular locality), it should instruct the appropriate agencies to investigate the circumstances and possible causes of the suspected phenomenon promptly and to report back with recommendations for dealing with the situation. In countries where an "early warning" system of this nature does not already exist, the authority concerned may consider the establishment of such machinery if in its opinion, the situation warrants it.

34. Regional organizations should make comparative studies of drug abuse patterns in countries in the region and inquire into reasons for variations in drug abuse (patterns, methods etc.).

35. Formal agreements should be established for international collaboration, using comparable methodology and instrumentation, so that resulting data can provide measures of international drug abuse patterns. Field testing of the common methodology and instrumentation is essential.

36. National focal points of States in the region concerned might share information on drug abuse and on methodologies for determining its incidence and prevalence, as well as exchanging experiences.

37. Ongoing information collection, analysis and dissemination activities within the United Nations system, including in particular the regional commissions, should include data on the full range of drug abuse issues in their areas of responsibility in order to permit an integrated approach to activities touching on drug abuse (e.g. activities concerning social development policy and planning, integrated rural development, popular participation, women and youth). Country focal points could be invited to collaborate closely with the regional commissions in this effort.

38. Regional organizations should establish training programmes for personnel engaged or about to be engaged in epidemiological surveys; expert knowledge available in the region should be used in the first instance for this purpose.

39. The United Nations Secretariat (Division of Narcotic Drugs), in association with WHO and regional and international organizations as well as non-governmental organizations, should disseminate information on the experiences of states in developing an appropriate methodology for assessing drug abuse.

40. The Division of Narcotic Drugs, in co-operation with WHO, should prepare and publish handbooks suggesting methodologies for collecting and analysing data on drug abuse.

41. The United Nations Fund for Drug Abuse Control (UNFDAC), in co-operation with WHO and other international organizations, should include among its priorities providing assistance to States, at their request, in support of their own efforts to establish standardized systems of case registration as well as to carry out surveys of the causes, extent and patterns of drug

abuse and assistance in training specialized personnel to collect, analyse and evaluate data.

42. Technical assistance in the planning and carrying out of epidemiological surveys should be provided to States by international organizatons, such as WHO, and by countries with the necessary expertise.

TARGET 2. Organization of Comprehensive Systems for the Collection and Evaluation of Data

The problem

43. Crude indicators of future trends in drug use and abuse are available from many sources, including the records of medical practitioners, hospitals, social insurance agencies, school administrators, police, customs authorities and courts. Medical associations world-wide should also be urged to co-operate in assembling this information in collaboration with WHO. This information should be taken into account, subject to respect for confidentiality, in the formulation of national policies intended to prevent abuse and reduce demand for the drugs in question.

44. To be useful to policy-makers, this information, which in general exists only in the records of the agency or person concerned, needs to be collected, collated and analysed, preferably by a system or machinery designed for this specific purpose, with particular emphasis on the consistency and comparability of national and regional data.

Suggested Courses of Action

At the national level

45. The appropriate national authority could, as a beginning, appoint a working group to devise simple, easy-to-use, reliable standardized instruments that may:

(a) Facilitate the gathering of common information about drug abuse from agencies concerned with different aspects of the abuse;

(b) Be used by all entities concerned for collecting basic descriptive data; and

(c) Be used nationally as a foundation for epidemiological surveys.

46. At the local and national levels, a simple framework could be created for the collation of data sheets. The data could be made freely available, without disclosure of names or personal particulars to agencies and individuals involved in research and/or planning. Mechanisms could be established for sharing the data among appropriate policy-makers and drug abuse professionals. Sharing encourages common data collection techniques and makes it possible to give early warning of new drug use patterns.

47. Field trials of the reliability and validity of any instruments should be conducted and independently analysed by appropriate researchers.

48. In addition to sponsoring measures and activities designed to reduce the illicit demand for drugs, the ministry concerned (health, education, social welfare, as the case may be) might inquire into the multiple possible causes of drug abuse and, as appropriate, recommend action for eliminating the causes, or at least the most serious of them. The inquiries might, as appropriate, take the form of surveys of the social and family circumstances of drug addicts or groups of addicts or persons at risk to drug abuse, their housing, employment, level of education and any other characteristic that may appear as a contributory cause of drug abuse. Research should be conducted whenever possible to evaluate the effectiveness of the programmes and techniques utilized by states to reduce drug abuse.

49. In view of the constant evolution of the pharmaceutical industry, of continuing intensive research into the properties of newly discovered drugs and their effects on the human organism and of the risk of abuse, the appropriate authority could make regulations or issue orders (in so far as these do not already exist) directing local health authorities, medical practitioners, pharmacists, directors of educational establishments and research institutes to report to it promptly any case that has come to their notice of persons showing symptoms of incipient addiction to a newly or recently discovered drug or substance or to a pharmaceutical product, previously considered inocuous, which is found to possess narcotic or hallucinogenic properties. The authority concerned might consider whether the product or substance in question should be placed under control or, as the case may be, under stricter control.

At the regional and international levels

50. Standardized instruments developed by appropriate national authorities should be field-tested and, if found useful, made available for the collection of comparable data. National authorities should determine what kinds of data are needed to address key issues, e.g. incidence, prevalence, risk factors and aetiology. Governments might collaborate with each other and with appropriate international organizations, e.g. WHO and the Division of Narcotic Drugs, in an effort to identify common risk factors and establish prospective studies to determine if these factors are predictive.

51. Provisional data sheets for pilot studies should be initially drawn up in a small number of States; if found useful, such studies might be carried out in a larger number of countries.

52. Technical assistance might be made available for the purpose of standardized record keeping, making due allowance for the social and cultural setting in which instruments will be used.

53. The committees established in New York and at Vienna by international non-governmental organizations in consultative status with the Economic and Social Council that have an interest in the control of drug abuse and illicit traffic are invited to act as a clearing-house of relevant information for dissemination to national non-governmental organizations concerned with drug abuse acting as focal points.

54. In regions where the dimensions of illicit drug trafficking and drug abuse are very imperfectly known and where the laws and the practices of the agencies concerned with the suppression of these illegal activities vary greatly from country to country, a unit at the headquarters of the regional commission might co-operate with existing specialized bodies of the United Nations system and other international organizations in identifying the needs of their member countries for strengthening drug control. Where there is a need for improved law enforcement to combat illicit drug trafficking, such a unit could usefully co-operate with the International Criminal Police Organization (ICPO/Interpol) and the Customs Co-operation Council (CCC) in their joint regional programme involving, *inter alia*, the holding of regular working meetings of senior police and customs officials, with a view to increasing co-operation and exchange of information between police and customs authorities, both nationally and internationally. If a regional commission originates regional projects for this purpose, consideration might be given to the granting of support for such projects from sources within the United Nations system.

55. The Governments of countries in a region in which the illicit use of and traffic in drugs give rise to common concern, and in which common action is desirable at the regional level for the purpose of warding off drug abuse and illicit trafficking in these countries or in some of them, may wish to consider the establishment of a regional centre for scientific research and education, that would contribute to the fight against drug abuse and illicit traffic (as envisaged in article 38 *bis* of the 1961 Convention).

TARGET 3. Prevention through Education

The problem

56. Comprehensive and effective education programmes are a necessary part of measures for counteracting drug abuse world-wide. In many countries drug abuse has already spread to the various age and population groups. It is essential that all individuals in the formal and general education system, as well as their families, should be enlightened about the risks of drug abuse.

57. Prevention through education should be considered as a continuous process and a lengthy exacting action, the objective of which is to seek and improve understanding of the long-term and immediate causes of recourse to drugs, to help young people and adults to find solutions to their difficulties and to lead their lives without resorting to drugs.

58. Curricula and programmes for enhancing the community's awareness, developed where appropriate as part of an agreed national strategy, should be so constructed as to strengthen the population's motivation to avoid drug abuse. Indications are that the impact of preventive education is greatest when it:

(a) takes place in its appropriate social, economic and cultural setting;

(b) is integrated into the overall framework of academic, social and cultural learning;

(c) promotes a healthy drug-free life-style as a primary goal, as opposed to placing emphasis on abstinence from drugs and on the negative effects of drug abuse;

(d) reaches individuals before they are exposed to the drug sub-culture and other influences that contribute to initial drug use;

(e) does not involve elements that evoke curiosity or the desire to experiment with narcotic drugs (detailed "positive" descriptions of euphoria etc.), but clearly indicates the negative, harmful consequences of drug abuse and emphasizes the positive effects of alternative activities and a life-style free from narcotic drugs and psychotropic substances; and

(f) does not contain details which might make access to illicit drugs easy, such as detailed descriptions of methods and routes of illicit traffic in narcotic drugs, places of origin of illicit production, non-medical uses of narcotic drugs etc.

59. Care should be taken that preventive education in school and out-of-school does not become yet another addition to curricula and that additional costs are avoided. Where advisable, attention should be paid to integrating progressively into school curricula and out-of-school activities an element concerning the prevention of drug abuse within the framework of activities for the improvement of the quality of life and which would be attached to existing disciplines and curricula. Continuity and the development of the prevention of drug abuse will thus be assured, without extending curricula which are already overloaded and without involving additional costs for the educational system.

Suggested Courses of Action

At the national level

60. The appropriate authority could establish a multidisciplinary unit in which teaching personnel who have received training in the subject of prevention must be represented. The unit would perform such functions

as:

(a) Prescribing or recommending for all levels of educational institutions, the development of drug abuse prevention curricula and instructional materials, taking into account cultural values and traditions and emphasizing the benefits of a healthy drug-free life-style;

(b) Preparing training and educational materials for the unemployed and undereducated, especially the young people, to assist them in developing vocational and self-employment skills;

(c) Within the school system, teacher education and motivation are essential to ensure the effective education of children regarding drugs and healthy life-styles. Therefore, it is essential to prepare training materials and to conduct training programmes for teachers and counsellors that will enable them to instruct their students in the advantages of a drug-free life. The teachers themselves must play an important role in the preparation of these training materials and programmes. This prevention instruction must be provided by the teachers themselves and should in no case become a separate course; on the contrary, it should be made an integral part of as many courses as possible;

(d) Preparing basic information publications on drug abuse and the different models of prevention, and acting as a focal point for the collection, collation and dissemination of information about drug abuse and the different models of prevention. Particular attention should be given to ensuring that preventive instruction at school is not instituted as an isolated, one-time measure, but rather that it is designed as a long-term programme accompanied by preventive measures involving students, parents, workers, the clergy, doctors and pharmacists. Only a comprehensive approach at the regional and local levels will have a chance of success;

(e) Undertaking periodic reviews of the curricula characteristics and publicity material in question to determine their effectiveness, and recommending adjustments as necessary that promote, on behalf of the authority responsible for education, training sessions for interested members of the teaching staff.

(f) Preparing programmes and information relative to promoting drug-free sport, cultural and leisure time facilities and activities.

61. The appropriate authorities could give priority to the training of educational personnel for both school and out-of-school activities (e.g. teachers, educators, monitors, counsellors, inspectors, headmasters). At the level of pre-service training, all educational personnel should be familiarized with problems of drug abuse and prevention techniques. In-service courses should be organized in specialized areas for certain categories of personnel.

62. The authorities responsible for public education might formulate a school policy aimed at preventing drug abuse, its implementation to be monitored by school administrators, in consultation, as appropriate, with teachers, students and parents. In particular, student leaders should be encouraged to develop attitudes and activities in their schools and communities aimed at preventing drug abuse. Seminars and training courses should be established to help parents recognize the symptoms of drug abuse at the early stages and to enable them to educate their children to recognize the damaging physical and mental effects of drug abuse.

63. School boards and authorities responsible for private and denominational schools should similarly formulate school policies aimed at preventing drug abuse, consonant with their curricula, and develop training and instructional material for classroom and related community use.

64. Interested members of the training staff and school and guidance counsellors should be given training to advise students and parents about the dangers of drugs abuse.

65. An important milieu for the education of the young is the family. Programmes should be developed to educate parents about the harmful effects of drug use and about methods to develop the competence skills of their own children.

66. The authority responsible at the national or local level, in conjunction with other appropriate authorities, might issue instructions or recommendations or provide consultation and advice to institutions responsible for training persons for a profession in which they will be concerned with the prevention or treatment of drug abuse (social workers, police, medical practitioners, paramedical personnel, health teachers) to include in their curricula or programmes of study a course specially designed to enable such persons to recognize and deal with cases of drug abuse.

67. In view of the vulnerability of young persons to the temptations offered by certain narcotic and psychotropic substances and to the solicitations of "drug pushers", the appropriate authority may consider establishing publicity and educational programmes to alert young people and society in general, particularly in urban areas, to the dangers with which the consumption of dangerous, habit-forming and dependence-producing substances is fraught: disruption of family and social life, health risks, impairment of

intellectual capability, perversion of moral values and antisocial and criminal behaviour.

68. The national co-ordinating agency, where it exists, or the appropriate authority could consider favourably the idea of establishing a small body composed of experts, representatives of agencies and other persons familiar, through their research or professional activity, with the situation regarding drug abuse in the country (e.g. teachers, social workers, members of the police force, representatives of national non-governmental organizations and others) who would be given the mandate to design a programme (where it does not already exist) intended specifically to reduce and ultimately prevent the illicit demand for drugs in a particular population group such as, and especially, youth. The programme would make provision for the training of specialized personnel to be deployed at the national and local level for the purpose of implementing demand-reducing schemes. Prevention programmes should be designed not only in the light of the dangers of drug consumption but on the basis of educational messages inculcating an active sense of responsibility with respect to the quality of life.

At the regional and international levels

69. Regional bodies should consider establishing regional training, resource and information centres for individuals involved in designing curricula, in order to evaluate and develop teaching techniques consonant with the cultural patterns of the region and to disseminate these techniques by means of regional courses and exchange of personnel.

70. The United Nations Educational, Scientific and Cultural Organization (UNESCO) and other international and non-governmental organizations might, at the regional level, encourage the creation of co-operative networks among institutions, programmes and projects dealing especially with preventive action and, at the request of Member States, provide data on strategies, methods and experience in prevention through education, facilitate and encourage the exchange of educational personnel and participate in the identification and development of programmes aiming at the prevention of drug abuse. Specific initiatives should be encouraged to promote exchanges of experience between countries having similar problems.

71. In cases where projects with an educational content are envisaged for the purpose of preventing and reducing the illicit demand for drugs, the Government or regional or international bodies or non-governmental organizations sponsoring or initiating the projects may, if they lack funds, apply for support to the United Nations Fund for Drug Abuse Control which should give priority to such projects.

72. Programmes and projects of United Nations organizations, especially the regional commissions, dealing with vulnerable population groups should include information about the scope and nature of drug abuse problems, their causes, and preventive and remedial actions as they relate to these groups.

73. Where large groups of refugees are receiving care (including health and education care) from an intergovernmental organization and where cases of drug abuse have been noted among them, the organization concerned should investigate such cases and ascertain their frequency and gravity and, in the light of the information, issue warnings of the dangers of drug abuse (especially among the young) within the refugee population and, if possible, undertake an educational, cultural and therapeutic programme for refugee groups at risk. Any relevant activity should be carried out in co-operation and in accord with the State of asylum.

TARGET 4. Prevention of Drug Abuse in the Workplace

The problem

74. The multiple and harmful consequences associated with drug abuse in the workplace and in occupational activities include deterioration of productivity and performance, defective quality of products, loss of qualified employees, accidents, loss of income and professional disqualification, all of which are costly to the individual, the employer, and the community.

75. Where alcohol abuse prevention programmes exist, consideration should be given to applying the experience gained in these programmes to the prevention and reduction of the abuse of drugs in the workplace.

76. Drug abuse by individuals in a wide range of sensitive occupations can result in disaster. Measures for dealing with drug abuse in the workplace and in occupational activities must have a significant prevention component.

Suggested Courses of Action

At the national level

77. With a view to preventing or reducing drug abuse in the workplace, the authority concerned could:

(a) Publicize information in the workplace warning of the risks of drug abuse;

(b) Set up national training workshops for supervisors, programme developers and others.

78. The appropriate authority, after consulting the parties concerned, might issue guidelines drawing the attention of employers and workers and of their organizations to the resources, in the form of publicity material, information through the media etc., that can be made available to them in connection with programmes for preventing or reducing drug abuse in the workplace,

and for treating and rehabilitating drug addicts.

79. Special emphasis should be placed on the urgent need to prevent abuse of drugs by those persons who are often regarded as role models. Professional and occupational associations should accordingly develop drug abuse prevention programmes for their members.

80. Employers' and workers' organizations should develop joint action programmes for their memberships with a view to discouraging drug abuse.

81. Medical authorities and medical laboratories should design reliable testing methods for the purpose of determining the presence of drugs in the human organism, identifying drug users and ensuring the safety of the public.

At the regional and international levels

82. The Governments of countries in a region could, through the ministries concerned or through regional bodies or non-governmental organizations, identify regional resources for training and programme development. They should organize regional seminars, as necessary, for the exchange of experiences and dissemination of relevant information. The International Labour Organisation (ILO) might be invited to initiate and participate actively in the intensification of the exchange of experiences at the regional and international levels.

83. The ILO should be invited, in co-operation with UNESCO, to co-ordinate international efforts to prevent and reduce drug abuse in the workplace, for example by distributing multimedia resource kits for the promotion and implementation of programmes to control, prevent and treat drug abuse in the workplace and to rehabilitate former drug abusers in employment; it might in addition monitor the use of these kits and the results thereof.

84. The Division of Narcotic Drugs, in collaboration with WHO and ILO, should promote and harmonize national efforts by developing internationally acceptable guidelines, criteria and methodologies for national testing programmes; a central source of reference standards of major drug metabolites should be established to serve national laboratories.

TARGET 5. Prevention Programmes by Civic, Community and Special Interest Groups and Law Enforcement Agencies

The problem

85. The support and involvement of community organizations, as well as law enforcement, health, social and educational agencies involved in the prevention of drug abuse, are indispensable in counteracting the harmful factors that foster drug abuse. Community organizations and health and social agencies are well situated to detect drug abuse and its consequences and to identify groups at risk. Moreover, these organizations and law enforcement agencies are aware of the needs, resources and goals of their members.

86. Hence, the local community and law enforcement agencies should not only adopt drug abuse prevention as part of their basic goals but should also be fully conversant with the objectives of national prevention programmes and should initiate or participate in the formulation and implementation of such programmes.

87. Most community activities are voluntary. Effective co-ordination is therefore required to ensure that community projects and activities for preventing drug abuse are consistent with the national plans for drug abuse prevention, and they should be periodically evaluated for efficacy. To the degree possible and when appropriate, representatives of the edu- cational system should be involved in the co-ordination function.

Suggested Courses of Action

At the national level

88. All civic groups, associations and clubs, especially those directly concerned with youth and other groups at risk, should prepare and disseminate to their members information drawing attention to the dangers of drug abuse. Voluntary organizations might, for example, be asked to provide package programmes of counselling and guidance, preventive education, drug abuse awareness, referral, detoxification, aftercare and rehabilitation. To the degree possible these activities should be co-ordinated to ensure consistency with national policy and, where appropriate, with international recommendations for the prevention of drug abuse.

89. National voluntary agencies, religious groups, political parties, civic organizations, parent-teacher associations, and other community-wide associations, should make a point of publicizing the advantages of a drug-free lifestyle and educating the public about the dangers of drug abuse. Professional associations (e.g. pharmaceutical, medical, teaching, social workers) should promote the voluntary participation of their members in prevention programmes, both within their communities and among their members.

90. For the purpose of taking action to prevent the illicit demand for drugs it is essential to impress upon the public the value of good health as a component in ensuring the quality of life. Community activities should emphasize the importance of positive cultural values and a sound social basis enhancing the quality of life without drug abuse.

91. Initiatives at the national, regional and local levels should strive for a co-ordination of actions among the agencies responsible for health, justice, education and other appropriate agencies. This collaborative effort among agencies and groups should promote a

comprehensive approach to the drug problem. Channels of communication should accordingly be established, in accordance with relevant national legislation, by community organizations with representative target groups in order to achieve a better understanding of the effects of drug abuse (e.g. hotlines, peer group counselling).

92. The ministry concerned (justice, interior, education, finance, health, religious or social affairs, as the case may be) might consider, if appropriate in the societal and cultural setting, sponsoring and, if necessary or desirable, supporting with financial or other resources the participation of representatives of certain law enforcement agencies in activities aimed at informing the public or specific sectors of the public about what is being done and could be done to prevent drug abuse. Representatives of the agencies in question (police, customs service, judiciary) might give talks describing their operations against drug abuse, suggesting how the community, the family and educational establishments should contribute to the campaign against drug abuse and indicating the willingness of the agencies to co-operate in local initiatives to enlighten the population about the dangers of drug abuse. Similarly, the national authorities responsible for monitoring the manufacture and distribution of pharmaceutical products, in association with the representative organization of pharmaceutical manufacturers, might participate in such campaigns. These campaigns could include the concept that prescription drugs should be taken only when, and in the manner, prescribed.

At the regional and international levels

93. Regional meetings of various civic groups should be periodically convened to exchange information on action that has improved community activities for the prevention of drug abuse.

94. Regional organizations should encourage increased participation of non-governmental organizations and their own communities in their efforts to combat drug abuse and illicit trafficking. The activities of the non-governmental organizations in the fight against drug abuse should be co-ordinated, as appropriate, in close co-operation with national Governments, other co-ordinating bodies, and relevant agencies in their regions.

95. The agencies concerned, communities and educational systems, including law enforcement agencies of countries in the same region, should be encouraged to co-ordinate their regional activities for the purpose of conducting a campaign against drug abuse at the regional or international levels, for example by participating in regional meetings dealing with drug abuse and its prevention or by undertaking other joint efforts to curb drug abuse and illicit trafficking. Similarly, these agencies should be encouraged to enter into or broaden contacts with counterparts in countries outside the region, with a view to planning world-wide activities to counter drug abuse. International and intergovernmental organizations such as the Inter-Parliamentary Union, ICPO/Interpol and CCC might have a useful role to play in arranging such contacts.

96. International non-governmental organizations should:

(a) Be encouraged to collate and disseminate information on national and regional activities of their constituent organizations, keeping in mind the need to foster more communication between governmental and non-governmental organizations;

(b) Facilitate discourse between their constituent national organizations and other non-governmental organizations concerned on community action that has had some success;

(c) Help their constituent organizations to formulate programmes that would give effect to recommendations by international bodies concerning the prevention of drug abuse;

(d) At their own request, be given the support of the United Nations and the specialized agencies concerned in carrying out programmes for the prevention of drug abuse;

(e) Work out ways in which to undertake co-operative efforts with the United Nations and with Governments.

TARGET 6. Activities at Leisure in the Service of the Continuing Campaign against Drug Abuse

The problem

97. It is in the community's interest to ensure that is used constructively. A wide range of activities, such as community service and sporting and cultural activities, should provide alternatives to drug abuse for the various social groups, particularly for the young, the elderly and the disabled. Studies have demonstrated that substantial benefits can be realized for both groups by fostering contact between the elderly and youth. Moreover, activities at leisure can exert a positive effect on individuals, as they improve social adaptation and tend to develop skills, talent and a sense of responsibility.

98. If alternative activities are to be in any way successful, target groups must be motivated and encouraged to participate in the design, initiation and running of activities of interest to them. This is particularly important in areas of high unemployment and youth unemployment, and for risk groups living in isolation, which are especially vulnerable to the misuse of prescription drugs.

99. While community groups should initiate and

design activities at leisure, government support in the provision of resources, facilities and materials is often necessary. Governmental support for alternatives to drug abuse can serve as an important counterbalance to certain types of commercial entertainment, where drug abuse is idealized or glorified. At the same time, such support tends to ensure a reciprocal flow of information.

Suggested Courses of Action

At the national level

100. Within the resources available, local government authorities should provide facilities, materials and funding to promote drug-free leisure-time activities. Such leisure-time activities, in the service of drug abuse prevention, should reinforce cultural and familial values, enhance health, and encourage positive social behaviour.

101. Community groups and government agencies concerned might review the existing range of leisure-time activities and develop programmes tending to promote wholesome and drug-free cultural and sporting activities. They could, in addition, make sure that those responsible for organizing cultural and sporting events, and the participants themselves, abstain from any action that might, explicitly, or implicity, give the public a misleading impression of the nature of drug abuse.

102. The appropriate authority might develop guidelines and suggested courses of action for use by public and private travel agencies, operators of transport undertakings and other bodies concerned with passenger travel and tourism urging them to refrain from publishing or disseminating material advertising, overtly or surreptitiously, the easy availability of narcotic or psychotropic drugs in any particular country or area. The authority might also consider posting notices at official points of entry into the country, warning travellers of the penalties applicable in respect of illicit dealing in drugs.

At the regional and international levels

103. Sporting and cultural organizations of countries in the region, in co-operation with the national authorities concerned, should promote and expand sporting and cultural exchanges, with an emphasis on drug-free events.

104. Information on local drug laws should be included with information for travellers on tourist activities (available, for example, at international borders and in hotels and youth hostels).

TARGET 7. Role of the Media

The problem

105. The mass media reach a vast audience every day. While the media's potential contribution to the campaign for preventing drug abuse is enormous, their publications and broadcasts can also be damaging and counterproductive. The use of inaccurate or misleading terminology regarding narcotic drugs and psychotropic substances and their properties, such as the artificial distinction between so-called "hard" and "soft" drugs, the advocacy of legalization of the non-medical use of drugs, the glamorizing of drug abuse in songs, movies and other commercial products, the emphasis given in reports of the street value of seizures to the enormous profits to be made from the illicit drug traffic, and the association of drug use with the names of successful or famous persons—all these can lead to false perceptions and can flaw the individual's powers of judgment.

106. Poorly designed campaigns to prevent drug abuse may also have the opposite effect from that intended, by arousing curiosity and inducing, rather than preventing, undesirable behaviour. It is vital, therefore, that there should be continuous research and evaluation of action taken, programmes implemented and the impact of message transmitted by the media. In addition, care should be taken to ensure that the fundamental human rights of freedom of speech and of the press and of other forms of expression are respected.

Suggested Courses of Action

At the national level

107. The appropriate authority concerned with broadcasting, the film industry and other media, if it has not already done so, might consider drawing up guidelines in the form of codes of conduct to be observed voluntarily by the public and private enterprises concerned in portraying or representing any event or incident involving the abuse of drugs. In addition, it may wish to recommend the observance of any guidelines of like nature that have been approved in an international forum.

108. Health agencies and other organizations concerned should more actively seek the support of the media as responsible advocates of a drug-free life-style. This could be achieved by convening, on a regular basis, workshops, briefings, information sessions etc. for media personnel to increase their knowledge of drug abuse and their awareness of the efforts of governmental and non-governmental agencies to counteract the problem.

109. In addition, in so far as it has not already done so, the authority concerned might consider :

(a) Inviting or, as appropriate, instructing local and central agencies responsible for drug control to designate, in the particular territory or area, a person or group of persons to act as liaison with the media and to provide the media with information and advice about the risks of drug abuse;

(b) Encouraging programme planners to consult the media on the portrayal of drugs and drug abuse in the broadcasting of drama and news items, and to provide them with appropriate material so as to ensure that news items are presented in a factual manner which does not, intentionally or unintentionally, mislead the audience or glamorize drug abuse;

(c) Through its agencie enrolling the media in co-operative efforts to uce the illicit demand for drugs.

110. Media executives might be brought into partnership with government and the community and invited: (a) to assist in the formulation of programmes in support of preventive policies and efforts; (b) to provide factual and honest information on drug abuse; and (c) to formulate programmes that promote cultural values and a healthy life-style. To the extent possible, continuous co-operation between educational authorities and the media should be encouraged.

111. Governments and non-governmental organizations concerned might co-operate in selecting research relating, for example, to an understanding of patterns of drug consumption. Such research should be supported by advertisers and public relations and public opinion researchers.

112. All means of communication and, in particular, the mass media, whether in the public or in the private sector, might be invited to participate in a concerted and long-term effort by nation-wide and local authorities and community groups to enhance the public image of a drug-free life, to disparage the drug-taking habit that has spread to certain classes of society, to induce all population groups to become health conscious and to realize the hazards associated with drug abuse, and to urge parents, teachers, community leaders and persons in public life to set an example by abstaining from drug abuse.

113. The appropriate authority could consider establishing channels of communication through which suggestions or recommendations might be addressed, informally and without implying any interference that might smack of censorship, to persons or bodies responsible for the management of radio or television broadcasting or other mass media. The media should voluntarily comply with requests by the authorities concerned with drug control to refrain from publicizing any news relating to ongoing investigations of drug-related offences.

114. In countries where several languages are spoken and understood, the appropriate authority, in conjunction with any other authorities concerned and with the co-operation of the media, could consider ensuring that warnings about the harm caused by drug abuse are broadcasted, disseminated or prominently displayed on public premises and in schools, youth and sports clubs etc. in the various languages used in the country.

115. In countries with a high rate of illiteracy, the authority concerned could consider publicizing such warnings by audio, visual or graphic means, e.g. posters, photographs etc., or by sending representatives of the public health service to rural areas to spread the warning by word of mouth. Anti-drug education should be recommended for inclusion in literacy programmes. The authority may also wish to use the good offices of "grass-roots" organizations for this purpose.

116. In view of the vital importance of information in planning and carrying out a programme to restrain and reduce the illicit demand for controlled drugs, the appropriate authority could prepare television spots, notices, leaflets, pamphlets or other material seeking information from all sources (including those not in the health system) that may lead to improved, more accurate information on illicit demand and also contribute to the formulation by the media of programmes for curbing this demand.

117. Mothers and fathers may need to be alerted to the ways of preventing their children from becoming drug abusers and of weaning themselves from drug abuse if they have contracted the habit.

118. Drug abuse during pregnancy may lead to serious effects for the child. Special emphasis should therefore be placed on informing young women of the specific dangers of drug abuse during pregnancy.

At the regional and international levels

119. UNESCO, in co-operation with specialized non-governmental organizations, could facilitate exchanges of audio-visual material which could be used in preventive action, contribute to the evaluation of information campaigns and implement international co-operative research on the impact of messages transmitted by the media.

120. Periodic meetings of journalists and media executives from different countries might be organized by the professional organizations concerned to exchange views on the portrayal of drug abuse in the media.

121. Regional and international bodies should be invited to offer training courses for journalists who report on drug issues to improve the accuracy of their reports and to avoid conveying false impressions to their readers and viewers.

122. With a view to stimulating the production of audio and visual programmes on the prevention of drug abuse, the creation of an international prize or award to recompensate the best amongst them should be envisaged.

II

(International Conference on Drug Abuse and Illicit Trafficking)

CONTROL OF SUPPLY

Introduction

123. A most important achievement of the international community in controlling the supply of drugs has been the development of a system for the international control of narcotic drugs, which includes control of the cultivation, production, manufacture and use of and international trade in substances controlled under the Single Convention on Narcotic Drugs, 1961, and that Convention as amended by the 1972 Protocol. The object of the system is to achieve a world-wide equilibrium between demand for and supply of narcotic drugs for licit purposes and to prevent their diversion to illicit channels.

124. In view of the huge growth in the illicit demand for opiates in many countries in the past two decades, an illegal heroin "industry" has developed, involving illicit or uncontrolled opium poppy cultivation and the illegal production and distribution of opium, heroin and other opiates. A complicating factor is that opiates have a recognized medical value and that no adequate substitutes have as yet been approved for certain applications. Consequently, the international control system provides for restricted cultivation of limited amounts of the opium poppy carefully calculated to correspond to the estimated annual medical requirements of each State. With a view to strengthening the treaty provisions relating to the control of poppy straw, it has been suggested that the full range of measures of control applicable to opium should be extended to poppy straw. According to another view, however, the provisions of the amended 1961 Convention are adequate. The treaties also provide that licit cultivation of the opium poppy for commercial purposes other than the production of opium or for the production of poppy straw is also authorized, provided that all necessary measures are taken to ensure that no opium is produced from such poppies and that the manufacture of drugs from poppy straw is adequately controlled.

125. The eradication of illicitly cultivated opium poppy, coca bush and cannabis plants, is a complex undertaking, because they are often grown in remote areas, or in areas beyond effective government control. Additional complications arise because in some developing countries these crops sometimes provide supplementary sources of livelihood for the rural population.

126. Cocaine, an alkaloid derived from the coca-leaf, differs from morphine, codeine and some other opiates in that only relatively small amounts are used worldwide in medical practice. Coca-leaf chewing is traditional and deeply rooted among large segments of the Andean population, and its use has been associated for centuries with a rich variety of anthropological, social and cultural features. The amounts of coca-leaves required both for licit purposes (the extraction of cocaine for medical purposes and the extraction of a flavouring agent which does not contain any alkaloid) and for chewing represent only a small fraction of the total amount now being produced. This means that most of the production is clearly destined for the illicit manufacture of cocaine and for the illicit traffic. The illicit or uncontrolled growth of the coca bush has created grave problems for States in the Andean region, and burgeoning demand has given rise to illicit traffic and marketing networks in this and other regions.

127. A further important development in international co-operation has been the evolution of a control system for psychotropic substances. Since the 1940s, advances in organic chemistry have resulted in the development of a wide range of synthetic medicaments acting on the central nervous system, and the abuse of many of them has been found to result in drug addiction. Although the controls introduced for narcotic drugs have demon-

strated their effectiveness even in the case of those drugs that are manufactured in an entirely synthetic manner and whose number is increasing, the situation is quite different with respect to most psychotropic substances. The response of the international community has been to devise progressively a system of control for psychotropic substances similar to that for narcotic drugs. In view of the diversity of these manufactured substances and their wide applicability and also because the 1971 Convention is a relatively recent instrument, this system of international control has not evolved to the same levels of acceptance and enforcement. Some voluntary measures have been designed that are applied by Governments and have strengthened the control system. Nevertheless, the system has only begun to take effect and diversions from the licit manufacture of certain psychotropic substances under international control and the illicit manufacture of some of them pose a serious threat to the well-being of individual and public health and, moreover, to the national security.

TARGET 8. Strengthening of the International System of Control of Narcotic Drugs and Psychotropic Substances

The problem

128. Under article 19 of the 1961 Convention, a State party to the Convention is required to provide to the International Narcotics Control Board (INCB) an annual estimate of the actual amount of narcotic drugs required during the following calendar year. A large number of States have, in addition, undertaken on a voluntary basis to provide similar assessments with respect to psychotropic substances under international control. In some cases estimates have been submitted on the basis of information provided by national pharmaceutical producers or by pharmaceutical companies, which tend to reflect an over-generous appraisal of market potential or excessive prescribing by medical practitioners, or both. Other action is needed to provide more accurate estimates of legitimate medical needs and should include a determination of the therapeutic applications of the narcotic and psychotropic drugs currently being manufactured and prescribed. Within the context of its Drug Action Programme, WHO has prepared guidelines for more accurate forecasting of requirements for therapeutic use of drugs, including narcotic drugs and psychotropic substances. The improvement of prescription, delivery and utilization practices regarding psychoactive drugs requires an intensification of the co-operation between national pharmaceutical services, medical and pharmaceutical bodies, research institutions, the pharmaceutical industry and others, with a view to ensuring congruence between such practices and the requirements of the international system of controls.

Suggested Courses of Action

At the national level

129. The national drug control authority, under the aegis of the appropriate ministry, could gather systematic detailed information and data regarding therapeutic applications of narcotic drugs and psychotropic substances and their preparations from physicians, hospitals, clinics, pharmacists, academic institutions and the pharmaceutical industry as well as from individual manufacturers in order to establish more accurate estimates of the country's legitimate medical needs, work out the annual plan of production and carry out the plan in specific factories assigned by the Government to ensure licit use and prevent abuse.

130. The authority concerned, in association (as appropriate) with social security agencies, could study the incidence or prevalence of specific diseases treated by the preparations in question.

131. Professional associations, private foundations and academic institutions, in co-operation with the pharmaceutical industry, and, where appropriate, the ministry of health, should be encouraged to conduct research towards finding less abusable psychoactive drugs, which are at least of equal effectiveness and research on combination products where the addition of a counteractive substance leads to products with lower dependence-producing potential. Such research should take fully into account the validity of specific narcotic and psychotropic substances in the management of pain.

132. In addition, government and non-governmental organizations should be encouraged to conduct scientific studies for the purpose of determining whether equally effective, non-pharmacologic treatment can be developed.

133. In accordance with the spirit of the Conventions, manufacturers' associations should urge their members to assist the national drug control authority by providing it with data on the quantities actually delivered by manufacturers and distributors to individual bulk consumers.

134. The national drug control authority, in collaboration with the professional association of pharmacists and social security agencies, could, as appropriate, systematically collect the figures of actual sales of drugs by retail pharmacies.

135. The authority concerned with public health, in consultation with other appropriate authorities, and in accordance with national legislation, could institute a system for monitoring by means of computers or other systems, the movement of narcotic drugs and psychotropic substances in the national territory, from the point of manufacture or importation to the point of dispensing in pharmacies, hospitals or offices of medical practitioners.

136. To assist in forecasting national requirements,

the national drug control authority, in co-operation with appropriate bodies, could determine the trends in the use of narcotic drugs and psychotropic substances over preceding years and work out projections of their use for the coming three-year period.

At the regional and international levels

137. International organizations or Governments within a region might carry out comparative studies of therapeutic practice in various countries with similar characteristics (such as population, health service and climate), and investigate the reasons for variations in drug use. Such studies ought to be undertaken in co-ordination with appropriate United Nations entities and agencies, such as INCB, the Division of Narcotic Drugs, and WHO.

138. The United Nations should offer to assist Governments with analytical studies of national experiences with regard to the effect of the national drug supply systems on the distribution and availability of narcotic and psychotropic drugs in their countries. Such analyses should also include the effects that international treaty controls have on the distribution, availability and consumption of controlled drugs.

139. The donors of bilateral and multilateral assistance should consider favourably requests by States for financial and other support (including expert advice) for efforts to improve or strengthen national systems for controlling the distribution of narcotic and psychotropic drugs. Such assistance could in the first instance be obtained by authorities of States through technical consultations with the appropriate United Nations entities and agencies and from other Governments with existing effective control systems.

140. Expert meetings should be organized in various regions, on request, by INCB in association with WHO, for the purpose of exchanging experiences in the methodology of forecasting and record-keeping with respect to drug use.

141. The United Nations should ensure that manuals describing the methodologies for estimating requirements of drugs with very high abuse potential are available to all States and that the model INCB reporting forms are available in the official languages and disseminated.

142. The international bodies concerned, in particular WHO, should consider favourably, requests by national authorities for assistance in strengthening or establishing a pharmaceutical control agency, under the ministry of health or national authority concerned, for monitoring the manufacture, importation, dispensing and distribution of substances under control. Such assistance could also be provided by Governments with relevant experience and resources.

TARGET 9. Rational Use of Pharmaceuticals Containing Narcotic Drugs or Psychotropic Substances

The problem

143. There is evidence that the supply of essential narcotic drugs and psychotropic substances is not always adequately assured in every country. It is incompatible with the spirit and letter of international drug treaties and with the WHO goal "Health for all by the year 2000" to deprive patients of the necessary and appropriate drug therapy. The identification of national needs and the assurance of an adequate supply of essential drugs are a first priority in any system of health care. At the same time, over-use and misuse of narcotics and psychotropic substances should be avoided.

144. In some countries there are indications that most prescriptions for psychotropic substances are made by medical practitioners who may not be fully aware of the psychotropic properties of these drugs and their potential for misuse and abuse. In addition, these drugs are sometimes prescribed, without thorough evaluation and diagnosis, to a large population of patients for long periods of time without periodic re-evaluation of the need for continuation.

145. In some countries, there are indications that patients are not adequately treated with narcotic drugs and psychotropic substances either initially or for adequate periods of time. While these practices vary among countries, they are often dependent upon the practitioners' knowledge of the various indications for use, appropriate treatment strategies and over-concern for the dependence-producing characteristics of these drugs.

146. The national use of narcotic and psychotropic substances is frequently hindered by incomplete and often inaccurate data. National health authorities, academic institutions, health professional organizations and others often do not provide adequate information on proper drug use to health professionals. Promotional activities vary in substance and completeness (manufacturers' brochures, advertisements, personal representatives, etc.) In many countries, the monitoring of the prescribing of these substances is poorly organized. There is a need for better statistics and internationally comparable data for the comparative study of trends in the prescribing of drugs. Also, there is a need for improved information on drug use to be made available to health professionals.

147. In some countries, national or international controls inadvertently affect the supply and distribution of drugs, thereby indirectly affecting the availability of these drugs for legitimate medical purposes. This problem has been brought to the attention of and discussed by the WHO Executive Board (EB 79/6, 7 November 1986, paragraphs 17-19). In a number of WHO regions,

difficulties have been encountered in obtaining controlled pharmaceuticals. In some cases, pharmacies have become reluctant to stock certain drugs because of the extra record-keeping imposed by national authorities or because inclusion in the Convention on Psychotropic Substances has been interpreted as meaning that the substance is prohibited. This has led to serious problems in availability of these drugs, particularly in rural areas; in primary health care settings such drugs are often no longer available for prescription by health-care workers. This kind of restricted availability can have dire consequences for the proper treatment of serious medical problems and such consequences should not be overlooked.

Suggested Courses of Action

At the national level

148. The national drug control authority or, as appropriate, the ministry of health could specify, on the basis of needs, the narcotic drugs and psychotropic substances in Schedules II and III which as a minimum must be available for drug therapy in general and for primary health-care in particular.

149. The authority concerned, in co-operation with the pharmaceutical industry and with the support of the medical and paramedical professions and pharmacists, as well as with law enforcement agencies, as appropriate, should endeavour to work out agreed guidelines for improving prescription and dispensing practices. National health authorities should ensure that accurate and unbiased information concerning the appropriate use of narcotics and psychoactive substances is made available to health professionals and primary health-care workers.

150. National health authorities should also ensure that national and international controls are not misinterpreted so as to limit the availability of narcotic drugs and psychotropic substances for legitimate medical purposes. Where legislation is an impediment to the availability and use of needed drugs, national authorities could give consideration to adopting a more flexible legislative structure.

151. The authority concerned, in co-operation with research and academic institutions, could conduct, or initiate, studies envisaging the monitoring of trends in the prescribing and consumption of psychotropic substances and on their rational use. To improve data collection and analysis, the authority and professional organizations concerned could make use of the WHO Anatomic-Therapeutic-Chemical Drug Classification System and the Defined Daily Doses measurement techniques.

152. For the detection of falsely labelled or low-quality pharmaceutical preparations, either imported or locally manufactured, containing narcotic drugs or psychotropic substances, national pharmaceutical quality control laboratories should be established or strengthened, as necessary. In order to ensure the control of narcotic drugs and psychotropic substances, the activity of national narcotics laboratory services and pharmaceutical control laboratories should be co-ordinated.

153. The ministry of health and, where applicable, other ministries, medical and paramedical professions, manufacturers and other concerned parties, should ensure that the promotion of psychoactive drugs is always based on safety and effectiveness. The ministry of health or any other appropriate authority should regulate the free distribution of samples of narcotic drugs and psychotropic substances.

154. The authority concerned could, as appropriate, review with the pharmaceutical industry or its representative organizations, any legislative or voluntary programme concerning any printed or other material publicizing their products, advertisements published or broadcast through mass media and other methods of promoting sales of their pharmaceutical products. The legislative or voluntary programme could make provision for disallowing any exaggerated claims about the stimulative or restorative properties of pharmaceuticals that might tempt drug addicts or potential drug addicts.

At the regional and international levels

155. WHO should promote collaborative research on an international basis, including field research, aimed at providing data to improve drug prescribing and drug consumption. These research efforts should be undertaken both in developing and in developed countries.

156. The appropriate organs of the United Nations should consider co-ordinating studies which assess the availability and utilization of narcotics and psychotropic substances for legitimate medical purposes and which assess the impact of international control on this availability and use. In the light of the results of these studies, these organs should make recommendations—which may be either of a general character or adapted to the country or region concerned—for improving availability and use and on the need to modify, as appropriate, the international controls of these drugs.

157. WHO should continue to assist national educational authorities to develop training modules for medical and pharmaceutical education and to conduct training courses for medical and other health personnel so as to promote rational prescribing and rational use of pharmaceuticals containing narcotic drugs or psychotropic substances.

TARGET 10. Strengthening the Control of International Movements of Psychotropic Substances

The problem

158. There is evidence that some lawfully manufactured psychotropic substances are frequently diverted into illicit channels, and the provisions of the 1971 Convention are not adequate to prevent this form of illicit traffic. There are cases in which authorizations for the import of Schedule II psychotropic substances have been falsified and doubts have been cast on the efficacy of the declaration system provided for Schedule III substances. Furthermore, the 1971 Convention does not provide for specific control of international trade in Schedule IV substances. In the absence of an "estimates" system, the only mechanism available to States parties to protect themselves against unwanted shipments is the prohibition provided for in article 13 of the Convention. Inadequate implementation of some provisions of the Convention as well as the absence of national controls in some countries often make the control of the movement of psychotropic substances illusory.

Suggested Courses of Action

At the national level

159. To the extent possible, all States should voluntarily extend the system of import and export authorizations provided for in article 12 of the 1971 Convention to cover international trade in substances listed in Schedules III and IV. In addition, immediate action should be taken whenever substances are found to cause significant abuse and/or trafficking problems.

160. Where medicaments containing psychotropic substances under international control are included in donations for emergency relief in disaster areas, the authorities concerned in the recipient and exporting countries should require that such consignments be authorized for all Schedule II substances and at a minimum be notified to the appropriate authorities with regard to substances in Schedules III and IV, in order that action may be taken to forestall the risk of possible diversion to illicit channels.

At the international level

161. The parties to the 1971 Convention should consider the possibility of adding to the Convention provisions contemplating the transformation into mandatory measures of the hitherto voluntary measures endorsed by the Economic and Social Council (resolutions 1981/7 and 1985/15) concerning in particular:

(a) The assessment of legitimate annual needs for Schedule II substances;

(b) The monitoring by exporting countries of shipments of Schedule IV substances.

162. The States parties should consider making more general use of the procedure for notification of a national prohibition of and restrictions on the export and import of specific psychotropic substances provided for in article 13 of the 1971 Convention.

163. Countries that import psychotropic substances, and that need help, should seek assistance from the secretariat of the International Narcotics Control Board, the Division of Narcotic Drugs and the United Nations Fund for Drug Abuse Control in establishing or strengthening national control services.

164. Countries receiving false or fraudulent permits or orders for psychotropic substances should immediately communicate this fact, together with all known particulars, to the appropriate international organizations such as the INCB and ICPO/Interpol.

165. Manufacturing and transit countries should, as far as possible, provide the Secretary-General of the United Nations with basic industrial and commercial information regarding sources of supply and movement of psychotropic substances and chemicals and precursors (see also target 12) utilized in the illicit drug traffic, in order to assemble a manageable set of data. The information could, to the degree possible, be collected, analysed and provided to national enforcement, drug control and customs administrations, taking into consideration the principles of the freedom of trade and the need to safeguard industrial secrets.

TARGET 11. Action related to the Increase in the Number of Controlled Psychotropic Substances

The problem

166. The number of substances in Schedules III and IV of the 1971 Convention that are used as medicines rose from 15 in 1971 to 60 in 1986, a fourfold increase. These 60 substances are used in hundreds of single and combination products under a variety of trade names. The continuing review process engaged in by WHO will no doubt lead to the addition to the schedules of an even greater number of medically used substances, and the marketing of new psychoactive drugs will necessitate a further extension of the scope of control. There is a genuine need for new psychotropic drugs, and this is one of the reasons for the dynamic psychopharmacological research activities being undertaken. About 16 per cent (138) of the 876 drugs launched in 1985 were psychopharmaceuticals, and pharmaceuticals still in research and development were estimated to account for about 21 per cent (851 out of 3,926). Consequently, the list of pharmaceuticals (mainly hypnotics and tranquillizers) will grow even longer in the future. It would be unrealistic to expect that all of these future drugs will be used without the risk of abuse and dependence.

167. Abuse of substances in Schedules III and IV can in some instances be prevented by pharmaceutical

control methods with limited recourse to law enforcement measures. In countries with developed pharmaceutical control services, the control of the entire pharmaceutical supply system can be made more effective. This national pharmaceutical control constitutes the basis for the control of these substances.

Suggested Courses of Action

At the national level

168. In view of the fact that the number of psychotropic substances has been steadily increasing during the past few years, and seems certain to continue increasing, each State could elaborate and implement a pharmaceutical policy as a part of its health policy. This would ensure that only medicaments containing psychotropic substances which are necessary for medical care of the population are put on the market. Existing criteria for drug registration should be reviewed and updated, as necessary.

At the international level

169. National regulations for new drug approval should explicitly state that a test for drug dependence be conducted in the case of a new psychoactive substance. If it is found to be a potential cause of drug dependence, it should be placed under control so as to prevent its abuse, and WHO should be informed accordingly so that the information can be disseminated.

170. Since the minimum requirements of Schedules III and IV of the 1971 Convention are designed to prevent either the abuse or the likelihood of abuse of the controlled substances to avoid their constituting public health and social problems, the Commission on Narcotic Drugs, in consultation with WHO, should ensure that no undue administrative burdens arise from the control of medically used substances in Schedules III and IV of the 1971 Convention.

171. WHO, in close co-operation with the secretariat of INCB, the Division of Narcotic Drugs and the United Nations Fund for Drug Abuse Control, should consider giving appropriate assistance (if requested) to countries whose pharmaceutical control services are inadequate, in order that these services may be capable of ensuring the control of the manufacture, import, distribution, prescribing and dispensing of psychotropic substances, in accordance with national health policy.

172. WHO might provide international non-proprietary names for the substances included in Schedule I of the 1971 Convention and so far designated by acronyms only.

TARGET 12. Control of the Commercial Movement of Precursors, Specific Chemicals and Equipment

The probelm

173. The continuing increase of the amounts of substances being illegally used for the illicit manufacture of narcotic drugs and psychotropic substances has made it all the more necessary to watch suspect movements of such materials and special equipment used by illicit laboratories. Some of the compounds which are used as precursors for the clandestine manufacture of narcotic drugs are subject to control provisions of the 1961 Convention (for example, ecgonine and thebaine). But there are no provisions in the 1971 Convention which would make possible the scheduling of precursors of psychotropic substances (for example, P^{2}P, ergotamine, or other lysergic acid derivatives). The prompt reporting of suspect movements of such items would make it easier to discover and apprehend traffickers. Seizures of illicit consignments would also reduce their availability for illicit manufacture. While the use of some specific chemicals is in some cases limited to the manufacture of narcotic drugs or psychotropic substances, other chemicals, and some materials and equipment (for example, tableting and encapsulating machines) are traded and required legitimately in large volume and the shipments should therefore be monitored in such a way as to permit law enforcement intervention with a minimum of burden on legitimate commerce. Particularly strict control measures should, however, be enforced only with respect to those specific chemicals whose use is practically limited to the manufacture of narcotic drugs and psychotropic substances.

Suggested course of action

At the national level

174. All exports of specifically identified precursors, chemicals and equipment ought to be distinctively, descriptively and accurately labelled and recorded in the exporter's records. A record giving full particulars of the consignee should also be required. Orders which manifestly exceed or are suspected of exceeding normal and legitimate requirements—particularly if the order in question originates in a country where illicit manufacturing is known to take place—should be promptly reported on a voluntary basis to the appropriate drug law enforcement agency by the manufacturer. If reasonable grounds exist for suspecting that the order is illicit, the shipment should be seized by the authorities of the country of origin. Such excessive orders should also be notified promptly to the country of destination and to INCB.

175. As appropriate, in countries where illicit manufacture of narcotic drugs and psychotropic substances occurs, the authority concerned could adopt a system of control over the importation and supply of selected materials and equipment that could be used for such illicit manufacture. If an import order for such items should appear suspect, the said authority could promptly notify the country in which the consignment

originates. For the purpose of giving full effect to such system of control, the authority should arrange special training for customs and other law enforcement personnel. Similarly, countries which are sources of such materials and equipment could adopt measures to detect and prevent suspicious shipments as referred to in the preceding paragraph.

176. States should observe and give full effect to the principles embodied in the 1961 and 1971 Conventions which deal with this subject and implement the resolutions on this subject adopted by the Economic and Social Council and by the Commission on Narcotic Drugs, pending entry into force of appropriate international treaty provisions envisaged for inclusion in the new Convention against illicit trafficking.

At the regional and international levels

177. An internationally recognized customs nomenclature should be established under the auspices of CCC for selected precursors, specific chemicals and equipment, and an export consignment including such items should bear a corresponding distinctive and internationally agreed label describing accurately the contents of the consignment.

178. The government agencies concerned could arrange to exchange information regularly and systematically with counterpart agencies in other countries about suspect movements of or transactions in materials and equipment of the kind referred to above and to keep INCB notified accordingly.

179. Joint training courses for customs and other law enforcement officers should be organized by the Division of Narcotic Drugs and other relevant international organizations such as ICPO/Interpol and CCC to familiarize them to the fullest possible extent with the nature of the substances involved and the techniques and *modus operandi* used by traffickers.

180. Governments may wish to consider entering into bilateral or multilateral agreements for the purpose of facilitating exchanges of information and co-operation in this field. In this connection, those negotiating the new draft convention against illicit traffic in narcotic drugs and psychotropic substances could consider the inclusion in that instrument of appropriate clauses making provision for such exchanges and co-operation at the international and regional levels.

TARGET 13. Control of Analogues of Substances under International Control

The problem

181. Researchers endeavouring to develop a new substance usually identify and study a range of products that have comparable properties but differ slightly in their molecular structure (analogues). They finally select, after a long screening process, the product thought to be the most effective for specific cases and having the least secondary effects. Narcotic drugs and psychotropic substances that are under national and international control are identified by their international or generic names and their chemical formulae. Their analogues, which have not been authorized and are not marketed, though having similar properties are, in many case, not controlled either by national law or under the international drug control treaties.

182. Unscrupulous persons have, in recent years, produced and distributed through irregular channels some very potent analogues of controlled substances (also referred to as "designer" drugs) in order to circumvent drug laws. These analogues not only pose a serious challenge to drug control, but are highly toxic, contain by-products and impurities and entail a serious threat to health and even to life.

Suggested Courses of Action

At the national level

183. Whenever a new analogue is detected on the illicit market, the appropriate authority could be empowered by the legislative authorities to take emergency action to place it under provisional control, with immediate effect.

184. The appropriate authority could prepare draft legislation making it a punishable offence:

(a) To manufacture with the intent to distribute individual analogues of controlled substances for human consumption in non-research environments;

(b) To distribute such analogues.

185. The pharmaceutical control authorities might take action to control analogues of substances under international control and whenever necessary to review the existing criteria for registration and approval of new drugs.

186. The appropriate United Nations agencies should, in conjunction with Governments, review the possible options available for addressing the potential problems of the illicit manufacture and distribution of analogues of controlled substances. Recommendations should be made to the Commission on Narcotic Drugs regarding the most appropriate United Nations mechanism for dealing with this potential problem. Pending such review, any Government having knowledge of a new addictive analogue should promptly notify the Secretary-General accordingly, with a request to communicate the information to other Governments and to WHO.

187. The Commission on Narcotic Drugs may wish to request the Division of Narcotic Drugs to explore the possibility of developing sensitive and specific detection techniques to detect and identify specific analogues of controlled substances.

TARGET 14. Identification of Illicit Narcotic Plant Cultivation

The problem

188. Under the 1961 Convention, article 22, the cultivation of the opium poppy, the coca bush and the cannabis plant is to be prohibited in certain circumstances; this general provision is subject only temporarily and in a few countries to some transitional reservations (article 49). In only one country is opium today produced for licit export. A few countries produce poppy straw for the extraction of opiates for licit use. In a number of countries the opium poppy is cultivated for the production of edible poppy seeds and oil. In some countries there is illicit and uncontrolled production; in certain of the latter countries some production has been traditional. Several States recognize that the successful elimination of this traditional production requires and deserves transitional economic assistance.

189. The coca bush is grown in countries in and near the Andean region. Under the 1961 Convention, the cultivation of the coca bush is subject to control. The traditional use of its leaves in several of those countries dates from time immemorial. In predominantly farming and mining regions, where leaf-chewing is also practised, there exist deep-rooted cultural, religious and medicinal uses. In recent times the leaf has had licit uses, including the use of a flavouring agent obtained from the leaves after the removal of the alkaloid content. In those countries, growers now sell much of their crop to illicit traffickers, and enormous new acreage has been planted to coca in these and in neighbouring States, all the output of coca leaves being intended exclusively for the illicit market.

190. Attempts to eliminate illicit production frequently encounter the difficulty of locating sizeable plantations with precision. To avoid detection, growers often place them in remote areas which are very difficult to reach on the surface, often in mountainous terrain, in narrow, steep-walled canyons that are difficult to reach even by air.

191. In a number of regions the most serious handicap for law enforcement agencies is that planting occurs in areas which are difficult to control. The opium poppy, coca bush and cannabis plants are often grown interspersed with other plants, which makes locating and eliminating them even more difficult.

Suggested Courses of Action

At the national level

192. In order to organize and mount effectively the most appropriate campaign to eliminate the illicit planting of opium poppy, coca bush and cannabis plants, the national authority concerned should make the necessary efforts to obtain maximum international support and use its own internal resources in order:

(a) To identify the map co-ordinates of areas under illicit cultivation;

(b) To gather data concerning wild growth as well as illicit crops being grown, size of holdings, yields per hectare, labour input and prices obtained by the farmer;

(c) To assess land capability (types of soil and soil fertility), climatic factors and land tenure and irrigation system, to permit selection of substitute crops;

(d) To analyse the geographical, political, social and economic conditions of the area in question.

193. At the national level, law enforcement and local government officials, the authority concerned with agriculture and field offices should be consulted, as should organizations and associations of persons likely to have the required information, and aerial photography should be used for spotting clandestine plantations. Each State, subject to respect for the national interest, could report its findings annually to the Commission on Narcotic Drugs in order to facilitate exchange of information among States.

194. In countries in which the cultivation of plants used in the manufacture of narcotics is carried on clandestinely on a large scale, the appropriate authority could, where necessary, conduct a survey with the assistance of the cartographical institute to determine the extent of such cultivation.

195. The survey should, in addition, endeavour to determine how the earnings and mode of life of the rural population concerned would be affected by the discontinuance of illicit cultivation and its replacement by legitimate food or cash crops. Where necessary or desired, assistance might be requested from the United Nations Fund for Drug Abuse Control or other bilateral or multilateral sources for the purpose of carrying out the survey.

196. The appropriate authority in countries where narcotic plants are cultivated illicitly and in areas where rural development assistance is considered appropriate might, after gathering full information on such cultivation, assess the suitability of the land for the planting of legitimate crops, select the crops best adapted to the environment, and likely to provide the farmers with an adequate livelihood, and work out plans for lessening or eliminating the farmers' dependence on earnings from the sale of illicitly grown plants.

At the regional and international levels

197. In cases where it is suspected that a sizeable illicit cultivation of the plants in question is carried on but the Government of the territory does not possess

the cartographical or other equipment for periodic aerial surveys of suspect areas, that Government could seek co-operation for carrying out such surveys, possibly from a regional or international organization financed by one or more States and/or by the United Nations Fund for Drug Abuse Control.

198. In this connection, the Secretary-General of the United Nations could convene an expert group meeting composed of experts from countries having the technological capacity for high-resolution satellite imagery and aerial photagraphy and from countries in which illicit cultivation is known to take place. The object of this meeting would be to determine whether and how such technology could be used to detect illicit plantations at the discretion of the country where cultivation is taking place.

TARGET 15. Elimination of Illicit Plantings

The problem

199. The opium poppy and the cannabis plant can be eradicated by being uprooted or beaten down manually or mechanically, by being sprayed manually (i.e. using back packs) with herbicides and, if the plantations are sufficiently large and the terrain permits, by being sprayed aerially. They may also be eliminated in connection with programmes of transitional economic assistance. The eradication efforts might also be co-ordinated with long-term integrated development projects accompanied by law enforcement activities, as appropriate.

200. Manual or mechanical uprooting is often hampered by the rugged nature or inaccessibility of the terrain; besides, the process is inefficient and slow, especially if many areas are to be cleared, and furthermore in some areas law enforcement officers who are present on the spot may be risking their lives. These considerations apply also to manual or back-pack spraying. In developing countries these eradication techniques may require a larger deployment of scarce resources than does aerial spraying.

201. In certain circumstances, aerial spraying may be considered appropriate: it involves the use of suitable aircraft, adequately qualified pilots, and carefully tested herbicides which do not cause any harm to human beings or to the environment. Aerial spraying may be inadvisable where the plants in question are interspersed with other plantings or are grown in small numbers or near populated areas. Fixed-and rotary-wing aircraft are commonly used in agriculture and are efficient for spraying large fields from the air.

202. Any chemical used in spraying must be proved to be harmless to human beings and to the particular environment. The International Code of Conduct on the distribution and use of pesticides, adopted by resolution 10/85 of the 23rd FAO Conference, could provide useful guidance in this regard. All the herbicides utilized in illicit crop eradication are widely employed in commercial agriculture and are generally registered for use in major agricultural countries. Some herbicides utilized for commercial range and bush control are partially effective against coca. Research is currently being sponsored to develop a herbicide, or herbicide application technique, which is fully effective in all the major growing environments.

203. Some states in the Middle East and Asia had tolerated the cultivation of the opium poppy under circumstances which lead certain States to believe that the growers, if now required to abandon these plantations, should be given transitional economic assistance. Similar considerations may apply also to certain traditional growers of coca in the Andean region. Rural development of the areas, improved roads, schools, medical treatment, adult education and literacy programmes, advice on other crops, fertilizers and the like can help persuade the former growers to comply with the ban on cultivation; appropriate opportunity should be given for an initial phase of voluntary eradication prior to the imposition of compulsory measures.

204. International co-operation for rural development or crop substitution should be conditional upon States adhering to eradication targets, so far as possible, within mutually agreed time schedules, in conformity with the obligations under the 1961 Convention. An effective crop reduction or eradication programme would require:

(a) Surveys of the project area conducted jointly by United Nations executing agencies and interested States to determine the success of the programme;

(b) Monitoring, by the State concerned, of the area where crops have been destroyed and where substitute crops are cultivated, to clarify that there is no illegal replanting of poppy, coca or cannabis.

205. Transitional economic co-operation may also be used by some countries in the context of a programme for the eradication of existing illicit crops and the prevention of the extension of the illicit growth of narcotic plants.

Suggested Courses of Action

At the national level

206. If they have designated the co-ordinating agency referred to in article 35(a) of the 1961 Convention and article 21(a) of the 1971 Convention, States parties to these instruments might as a first step direct the agency to prepare and keep up to date, a comprehensive plan for the elimination of the illicitly cultivated narcotic plants that have been discovered (see target 14):

(a) **Opium poppy.** Where a sizeable illicit growth of opium poppy has been discovered, the extent of the plantings, their physical accessibility and local political conditions might be investigated. A campaign might then be prepared to eliminate them in the manner most appropriate in the circumstances. Aerial spraying is efficient if the fields are large enough, if equipment, pilots and funds are available and if the chemical to be used has been found to be environmentally safe on the terrain. Alternatively, arrangements could be made to have personnel on the spot in good time to uproot or spray the plantations manually. If the country concerned is one of those that believe that such plantings can be eliminated in a reasonably short time through a programme of transitional economic assistance or that eradication should be accompanied by transitional economic assistance in the growing area, such a programme might be planned and appropriate technical and financial assistance sought, if needed. Any transitional economic assistance could be contingent on commitment to the complete abandonment of illicit cultivation. The ban could be imposed in stages;

(b) **Coca bush.** States in whose territories sizeable plantations of coca bush have been discovered could proceed along lines similar to those suggested above for opium poppy. Since the coca bush has been cultivated and coca leaves have been chewed for centuries in some parts of the Andean region, and in some of them the practice continues, transitional economic assistance and phased elimination of coca growing may be important in some areas;

(c) **Cannabis plant.** States in whose territories sizeable plantations of illicit cannabis have been discovered could proceed along the lines outlined above. Much of the illicit cannabis is grown on very small plots that are not easy to find and cannot be aerially sprayed.

At the regional and international levels

207. The Division of Narcotic Drugs, working in close collaboration with the international bodies concerned, should encourage States to report on their efforts to locate and to eliminate the illicit cultivation of narcotic plants in their annual report to the Secretary-General, updated summaries of which should be communicated to the Commission on Narcotic Drugs. The United Nations Fund for Drug Abuse Control should continue to encourage and help States to develop master plans for narcotics control, including the elimination of illicit crops.

208. The Secretary-General, in consultation with States where the illicit cultivation takes place and with other concerned States, should consider convening a study group of specially selected experts, including experts of the States directly concerned to study and recommend environmentally safe methods for eradicating illicit plants. The United Nations Environment Programme (UNEP), WHO, FAO and the United Nations Fund for Drug Abuse Control may be invited to participate in the study. Special attention should be given to appraising herbicides for use in safely eradicating illicit plantations of the three narcotic plants, in the various environments and conditions likely to be encountered. In particular, attention should be paid to determining which herbicides will be safe to human beings and effective against illicit cultivation by spraying from the air and at ground level. In this connection, the resources of the International Register on Potentially Toxic Chemicals (IRPTC)—a UNEP data base located in Geneva—can prove particularly helpful.

209. The authorities concerned of States that have the material, technological and financial capability to offer to the corresponding ministries of other States, particularly developing States, co-operation in carrying out measures of eradicating illicitly cultivated narcotic crops might co-operate with those other States at their request and, if appropriate, offer to supply chemical or like substances for use in the eradication measures. Requests for such co-operation might be transmitted directly to the United Nations Fund for Drug Abuse Control.

TARGET 16. Redevelopment of Areas Formerly under Illicit Drug Crop Cultivation

The problem

210. Narcotic plants can be, and often are, cultivated on marginal land in remote areas; they do not require specialized agricultural skills or costly inputs. The expense of transporting narcotic products is almost irrelevant, because of their high value per unit weight. Although the grower's income from the sale of the narcotic plants represents only a small fraction of the money that changes hands in the illicit trafficking and distribution of the drugs, the grower's earnings from these crops are generally higher than those from any alternative food or cash crops that can be identified as being suitable for cultivation on the same land area.

211. The reduction of illicit cultivation through law enforcement measures may affect the income of the farming communities concerned and in some cases leave the farmers in the short term without any means of subsistence. Accordingly, the Government may wish to consider undertaking in the traditional growing areas a programme that would assist these farmers in raising crops other than narcotic plants and/or in seeking other means of livelihood. Experience shows that such programmes are most effective if they form part of more general comprehensive rural development programmes offering, for example, improved infrastructure, credit, and marketing facilities. Integrated rural development and crop substitution involve a multidisciplinary and integrated approach to change from traditional low-income

and low-input agriculture to a more broadly based economy with higher technology and involving alternative cash crops as well as other sources of income. Any rural redevelopment scheme of this kind should, of course, be closely supervised lest it inadvertently give some benefit or advantage to drug traffickers. Moreover, to be successful these development efforts must be accompanied by effective enforcement, including bans on cultivation and effective interdiction both within the compass of the rural development programme itself and beyond.

Suggested Courses of Action

At the national level

212. The national authorities concerned could:

(a) Carry out market surveys followed by in-depth studies, where necessary, aimed at identifying domestic and export markets for substitute crops and determining the feasibility of setting up product-processing facilities, possibly with the participation of the private sector;

(b) Strengthen agricultural research and extension in areas growing illicit drug crops with a view to determining the short-term and long-term requirements of any crop substitution programmes envisaged, for example as regards input needs, plant population, harvesting techniques and marketing of the harvest. Long-term research would also cover seed-bed preparation, efficient use of irrigation, dry farming techniques etc.;

(c) Develop the infrastructure with a view to creating an improved living environment for the farmer, in support of government efforts to eliminate illicit cultivation, including, for example, improvement of health and sanitary conditions, launching adult education programmes and literacy campaigns, provision of education facilities, road construction and other initiatives, that would take into account the remoteness and isolation of the drug-producing areas in general.

At the regional and international levels

213. Regional bodies should co-ordinate the actions of Government at regional or subregional levels for the purpose of identifying areas formerly under illicit cultivation that cut across national boundaries and that may be suitable for redevelopment, having regard to the different crops grown in the region.

214. An international campaign with the participation of the United Nations family of organizations and other regional and international organizations should be undertaken to increase the flow of resources to programmes that are designed to reduce the supply of drugs through rural development, crop substitution, agricultural and industrial development, educational programmes, research and extension activities.

215. International intergovernmental and non-governmental bodies interested in this field should set up training and information activities and prepare materials to this end, including video films and material based on other advanced techniques, in support of the programmes aimed at recommending alternative crops or other means of livelihood for farmers in areas in which narcotic plants are grown illicitly.

216. International assistance should be granted to the countries concerned in the form of financing provided by the international community through the United Nations Fund for Drug Abuse Control or other channels. Donors to UNFDAC should be encouraged to increase their contributions to enable the Fund to expand the scope of its activities.

217. In cases where an area previously used by the resident rural population for the illicit cultivation of opium poppy or coca plants or cannabis has been transformed into land planted to substitute crops for which there is a potential export market, the Governments of countries that may offer an outlet for such substitute crops might consider favourably the grant of preferential tariff and non-tariff treatment to imports of produce grown on such converted land. For this purpose, the importing and exporting countries concerned may consider it desirable to enter into agreements stipulating the precise conditions and specifying the commodities in question and the nature of the preferences granted.

218. International financing institutions should contribute more extensively to integrated rural development in support of the eradication of illicit plantings and crop substitution programmes, if the countries concerned so request.

219. Where illicit cultivation of opium poppy, coca bush or cannabis plant occurs on a considerable scale, development co-operation in a project should be subject to mutual agreement by the co-operating parties. The provision of co-operation funds should be based on mutually agreed terms which could include crop reduction, crop substitution or other methods designed to eliminate illicit cultivation of the said plants.

220. Government agencies and international institutions should exchange information on progress in crop substitution in the course of deliberations on agricultural and integrated rural development in the international forums concerned.

221. The results of successful integrated rural development (IRD) and crop substitution programmes should be made generally available with a view to encouraging

other countries to undertake similar programmes where needed. Data on both successful and unsuccessful experiences should be gathered and analysed by FAO, with a view to elaborating an improved methodology for integrated rural development and crop substitution. United Nations Organizations, including the regional commissions, which deal with various aspects of IRD (e.g., popular participation, women's and youths' access to productive roles, and intra-governmental and inter-agency co-operation) should give explicit attention to the problems associated with illicit drug production. UNFDAC and the United Nations regional commissions should examine the possibilities of joint co-ordination of such efforts.

222. In view of the market rise in illicit demand for drugs in a number of developing countries where, until recently, drug abuse had not been on a serious scale, the government agencies concerned in donor countries and the governing bodies of intergovernmental financial institutions might take into account in their development assistance programmes the desirability of supporting or promoting in those developing countries, with their concurrence, economic and other projects or activities that would help to curb the rise in drug abuse and/or illicit drug production in those countries. Projects and activities of this kind might include integrated agricultural, industrial or rural development schemes, more comprehensive in scope than mere crop substitution schemes. While the emphasis to be given in bilateral or multilateral development assistance programmes to projects specifically designed to discourage the growth of illicit demand for drugs is bound to vary from case to case and according to the preferences of the Governments of countries where the projects are being carried out, the donors of financial and other assistance should, as far as possible, co-ordinate the flow and composition of assistance in such a way as to make an impact on drug abuse. In this respect, the United Nations Fund for Drug Abuse Control is performing a valuable co-ordinating function.

III

(International Conference on Drug Abuse and Illicit Trafficking)

SUPPRESSION OF ILLICIT TRAFFICKING

Introduction

223. Drug trafficking is sophisticated and complex. A wide variety of drugs is involved, and they may be of external or domestic origin. The illicit traffic in drugs not only violates national drug laws and international conventions, but may in many cases also involve other antisocial activities, such as organized crime, conspirarcy, bribery, corruption and intimidation of public officials, tax evasion, banking law violations, illegal money transfers, criminal violations of import or export regulations, crimes involving firearms, and crimes of violence. Narcotics are now frequently used instead of money as a medium of exchange for trading in weapons and other contraband, and some large drug-trafficking networks have gained virtual control of certain areas. Because of the far-reaching consequences of the illicit drug trade, even the integrity and stability of certain Governments have been threatened. This wide range of illegal activities presents an equally wide range of openings for imaginative law enforcement action, including action recognizing the need to develop and implement law enforcement programmes relating to subjects identified in the draft convention against illicit drug trafficking and to give emphasis to the recommendations of the Interregional Meeting of Heads of National Drug Law Enforcement Agencies, as reflected in the report of the Meeting (A/41/559).

224. Action on the lines suggested in chapter I, concerning the prevention and reduction of illicit demand, and in chapter II, on the elimination of illicit supply, would obviously go a long way towards suppressing illicit trafficking.

225. In addition, however, it is necessary to ensure vigorous enforcement of the law in order to reduce the illicit availability of drugs, deter drug-related crime, and contribute to drug abuse prevention by creating an environment favourable to efforts for reducing illicit supply and demand. The challenge is to overcome the obstacles posed by the complexity of international transactions, the covert nature of the drug traffic and the large sums of money to be made from illicit drugs in proportion to their often low bulk. Co-ordination of activities and co-operation among national agencies within each country and between countries are vital for the achievement of the objective.

226. Many Governments have initiated innovative methods for disrupting drug-trafficking organizations. These successful initiatives may usefully be shared with other Governments, where applicable, and can be adapted to the particular situation in a given area or region.

227. Special emphasis should be placed on supplementing the activities of the police and customs authorities by increasing the effectiveness of the criminal justice system in the arrest, prosecution and appropriate sentencing of convicted traffickers. The support of the full range of non-governmental organizations that have an interest in law enforcement and judicial matters should also be enlisted. Mutual legal and judicial assistance between national jurisdictions should be fostered and facilitated, as should co-operation between law enforcement agencies. Assets gained from the illicit drug trade should be made liable to seizure, as should the instruments used in trafficking. The extradition from one country of persons accused of drug offences in another country should be facilitated to the extent that it is not incompatible with the existing national legislation of the countries concerned.

228. An important step now being taken by the

international community in this regard is the drafting of a new convention against illicit traffic in narcotic drugs and psychotropic substances,[3] which is being prepared under the auspices of the Commission on Narcotic Drugs at the request of the General Assembly and which it is hoped will be adopted in the near future. States are encouraged to take an active part in the elaboration of the new convention and to become parties to the convention once it has been adopted. Pending its entry into force, which may take some time, measures envisaged in the proposed new convention could be taken to the extent possible by, when necessary, the introduction of appropriate provisions into national law.

229. Ratification and effective implementation by all States of the international conventions relating to drug abuse control will greatly enhance the prospects of ridding the world of illicit drug trafficking.

230. The international intergovernmental bodies concerned should bring to the attention of Governments any deficiencies noted in the operation of the international drug control system (e.g. leakages into illicit channels), and invite them to suggest or consider making (as the case may be) efforts to remedy the shortcomings at the national, regional and international levels.

TARGET 17. Disruption of Major Trafficking Networks

The problem

231. Timely information on the criminal activities of traffickers is required in order that they can be identified and caught. Such information is often available in the records of organizations such as banks, air, road, rail and maritime transportation companies, managements of ports and airports, free-port authorities, courier services, money changers and financial and investment houses. Personnel in all organizations concerned should be trained to recognize the value of law enforcement authorities of such information.

232. The object is to channel all pertinent information to the appropriate law enforcement agencies promptly, so that traffickers can be identified and caught. It is therefore necessary to strengthen co-operation between law enforcement agencies within each State and, under bilateral agreements, between these and their counterparts in other States, enhance training of law enforcement personnel (in skill and integrity) and enlist the assistance of the non-governmental sector in gathering information.

233. For this purpose, States parties that have not yet done so are urged to designate the co-ordinating agency envisaged in article 35(a) of the 1961 Convention and article 21(a) of the 1971 Convention, which should be given the necessary authority to co-ordinate the actions set forth below.

Suggested Courses of Action

At the national level

234. Having due regard to the national administrative and legal system, the agency could gather from all government agencies information useful for drug law enforcement and, subject to respect for privacy and confidentiality, ensure that this information is communicated promptly to the appropriate law enforcement agency; this requirement could, if necessary, be spelt out in appropriate laws and regulations.

235. The ministry or authority concerned could be made responsible for arranging training for the personnel of all such agencies to enable them to recognize and to transmit promptly to the appropriate agency any information useful for the purpose of identifying traffickers and detecting their activities.

236. The appropriate agency could approach air and rail transport firms and shipping and trucking firms which operate internationally, and/or the associations of such firms, urging them to review their procedures for the purpose not only of safeguarding their services against misuse by traffickers (see also target 24) but also of ensuring that information about any trafficking operation whatsoever is reported promptly.

237. Legislative bodies could consider enacting legislation applying penalties to transport companies that are aware of such misuse and illicit traffic and do not take prompt and adequate steps to correct and report it or are reckless or grossly negligent in this regard. Legislation providing in such circumstances for the seizure and immobilization of transport equipment used in drug trafficking could be enacted if not already in effect (see also target 23).

238. Subject to the limitations necessitated by the constitutional, legal and administrative system, the appropriate ministry or other national authority responsible for financial matters should ensure that any suspect activity by banks, money changers, financial and investment houses and courier and related services is promptly reported to the ministry or authority concerned and to the drug law enforcement agency. Similarly, in cases where information held by the tax authorities might assist investigations, this information could be made accessible to the drug law enforcement agency. Any amendment needed in banking and corporate secrecy laws should be in terms conducive to the discovery of drug-related offences. With the support or advice of the ministry or authority, training courses might be arranged for the staff of banking and financial institutions, so that they can learn how to recognize suspect transactions. The movement of unusually large amounts of cash or negotiable instruments, the deposit of exceptionally large amounts of cash in banks, the unreported maintenance of accounts abroad by nationals or resi-

dents in cases where such amounts are required to be reported, and large unexplained accumulations of wealth of obviously illicit origin should by law be liable to penalties if there is evidence of "laundering" or concealment of funds connected with illicit drug trafficking.

239. The appropriate ministry or other appropriate authority at the local level could establish standards of conduct and integrity for law enforcement agencies and officers that are involved in drug law enforcement. All States should ensure that an appropriate legal framework with penal sanctions is enacted (if they do not already exist) to support the criminal prosecution and punishment of corruption offences.

240. The appropriate national authorities could, with due regard for the fundamental principles of the national legal system, make use of all modern techniques of investigation in the fight against organized international drug trafficking, including corruption offences by public officials.

241. If in the course of investigating suspect activities, a government agency discovers evidence of a connection between illicit drug trafficking and illicit arms trafficking or international organized criminal activities, it should promptly inform other national authorities accordingly and communicate relevant particulars to the authority concerned in the country that is the probable target of the illicit traffic or organized criminal activity in question.

At the regional and international levels

242. The ministries concerned and national law enforcement agencies should, as appropriate, under international treaties and bilateral agreements co-operate closely with their counterparts in other States and in liaison with ICPO/Interpol when appropriate, with a view to enhancing the effectiveness of their law enforcement action to suppress the illicit drug traffic. For this purpose, they could establish and maintain channels of communication between their respective law enforcement agencies, by means of which information can be exchanged without delay.

243. States should endeavour, directly or through the appropriate international organizations, to establish regional and international agreements strengthening co-operation in the fight against the illicit drug traffic.

244. Bilateral and multilateral assistance should be sought, where needed, for the purpose of financing programmes of technical co-operation and assistance and improving channels of communication for the transmission of information relating to the fight against the illicit drug traffic.

245. With the co-operation of ICPO/Interpol and CCC, when appropriate, the Division of Narcotic Drugs should continue to organize regular regional and interregional training courses to train personnel of law enforcement and related agencies and officials of national tourist agencies and to promote co-operation among these agencies in the fight against the illicit drug traffic.

246. Since the operations of drug trafficking gangs may vary from region to region, information available on profiles and methods of operation could be gathered by national authorities at the regional level in co-operation with ICPO/Interpol and CCC, when appropriate, to be used by States as well as by international agencies and entities concerned.

247. In countries where it is known or suspected that the "informal" or "parallel" sector of the economy accounts for a significant share of the national product and of international trade, the appropriate ministry or authority might investigate the ways in which the illicit production of and traffic in drugs contribute to the "informal" economic activities and take counteraction. The ministry or the authority concerned may wish to communicate the relevant information to other countries, in the region or elsewhere, which are known destinations of the illicit cross-border drug traffic or of the fight of assets representing earnings of traffickers from unlawful transactions.

248. If conclusive evidence comes to light of illicit trafficking being carried on by means of the misuse of the diplomatic bag or of the diplomatic status, or of the consular status, it is open to the Government of the receiving State to take measures for halting this traffic and for dealing with the diplomatic or consular staff involved in strict conformity with the provisions of the Vienna Conventions on Diplomatic and Consular Relations.[4] The Conference draws the attention of the International Law Commission to possible misuse of the diplomatic bag for illicit drug trafficking so that the Commission could study the matter under the topic relating to the status of the diplomatic bag.

TARGET 18. Promoting Use of the Technique of Controlled Delivery

The problem

249. The law enforcement technique of controlled delivery is widely considered to be an efficient tool in identifying and neutralizing major organizers of international drug trafficking. This procedure involves allowing a delivery of illicit drugs, once detected, to proceed, under constant and secret surveillance, to the ultimate destination envisaged by the traffickers, the object of the surveillance being to lead to the discovery and eventual arrest of the trafficking ringleaders. Difficulties have arisen in some national jurisdictions, where legislative provisions require the suspect's immediate arrest upon detection, and also where the responsibility for such surveillance is ill-defined and the strict enforcement of penal law by the country of destination is not guaranteed or where the penalty for trafficking is lax or less stringent. In addition, there is a risk that the shipment may escape into illicit hands. High operational costs, and the non-availability of trained staff, have in some

countries inhibited the use of this most useful method of tracing the delivery of illicit drugs to the ultimate destination.

Suggested Courses of Action

At the national level

250. Unless the constitution of the State concerned rules out the amendment of the law in such a way as to permit the use of the technique of controlled delivery, consideration should be given to amending the law so as to permit its use following prior bilateral agreements or arrangements. The legislature and the appropriate ministry or authorities could take the necessary measures pursuant to domestic law to authorize the appropriate use of the technique of controlled delivery for the purpose of identifying and bringing to justice the individuals, corporations or other organizations involved in the shipment, transportation, delivery, concealment or receipt of an illicit consignment of controlled substances that might not be detected if the intermediaries or couriers were arrested immediately on identification.

At the regional and international levels

251. In order to ensure that controlled delivery is being effectively co-ordinated at both national and international levels, States could, if they consider it appropriate, designate an agency or agencies as responsible for such co-ordination.

252. With the assistance of the United Nations Fund for Drug Abuse Control and in close co-operation with ICPO/Interpol and CCC, the Division of Narcotic Drugs should organize regional training courses for law enforcement and judicial officers for the purpose of drawing up guidelines and instructing them in techniques of surveillance, control, and co-ordination of controlled delivery.

TARGET 19. Facilitation of Extradition

The problem

253. Despite the community of interests and obligations shared by the States parties to the 1961 and 1971 Conventions, the laws and regulations concerning persons accused of drug-related offences vary considerably from one country to another. These differences in legislations provide traffickers with opportunities for evading capture and trial; for example, the traffickers may be resident in a country whose law or administration of justice is so lax that they enjoy virtual immunity from prosecution or from extradition to a country in which they are accused of a criminal offence.

254. It would obviously be a strong deterrent to traffickers if they knew that prosecution, trial and, if convicted, punishment for drug-related offences would be inescapable, and that extradition to the country in which the alleged offence was committed and which asks for extradition would be almost certain, subject to observance of the proper national legal procedure. The enactment of appropriate legislation where it does not already exist, and its strict enforcement, would go a long way towards denying drug traffickers any haven of refuge.

Suggested Courses of Action

At the national level

255. If the appropriate national authority deems it necessary, the national legislation could be reviewed to ensure that each of the acts enumerated in paragraph 273, be defined as extraditable offences and included (by amendment if necessary) in existing and in contemplated extradition treaties.

At the regional and international levels

256. States could consider entering into bilateral or multilateral extradition treaties concerning drug trafficking and other related offences if they deem this to be appropriate.

TARGET 20. Mutual Judicial and Legal Assistance

The problem

257. The multinational aspects of illicit trafficking in drugs greatly complicate law enforcement, investigation and judicial counteraction. Witnesses, documents and other evidence are often scattered in States other than the State in which persons accused of drug-related offences are brought to trial, and the detailed rules concerning the production of evidence can create difficulties for judicial bodies. Subject to the limitations of the constitutional, legal and administrative system, needed mutual legal assistance includes, for example:

(a) Taking evidence, including compelling testimony;

(b) Serving judicial documents;

(c) Executing requests for searches and seizures;

(d) Examining objects, sites and conveyances;

(e) Locating or identifying witnesses or suspects;

(f) Verifying in narcotics laboratories the illegal nature of substances seized;

(g) Exchanging information and objects;

(h) Providing relevant documents and records, including bank, financial, corporate and business records: existing bank secrecy laws are being used in many instances to obstruct co-operation and the provision of information needed for the investigation of allegations of drug-related offences.

Suggested Courses of Action

At the national level

258. The appropriate authorities could suggest that, in conformity with the relevant bilateral and multilateral agreements, the greatest possible measure of

mutual judicial assistance should be provided in judicial proceedings, including investigations and prosecutions relating to illicit trafficking offences, making or proposing any necessary modifications in the legislation, regulations or procedures. Legislative provisions could be enacted as required, granting broad powers to the courts to assist courts in other jurisdictions in gathering evidence in accordance with the laws of the requested State and, to the greatest extent possible, in conformity with the laws of the requesting State.

259. Each State could ensure that the appropriate agency or responsible authority has the capacity to receive requests for mutual legal assistance and to address such requests to other States. The agency or authority of the requested State should have capacity to recommend that requests for mutual legal assistance be executed in accordance with the procedural requirements specified in the request in so far as they are not incompatible with the law of the requested State.

At the regional and international levels

260. The appropriate ministries or authorities could, in co-operation with ministries of foreign affairs, initiate action to enter into regional or international agreements that would serve the purposes described above. A number of States have entered into or are negotiating bilateral and regional agreements for these purposes. Many such agreements relax the rules governing bank secrecy in drug trafficking cases, thus reducing the number of "safe havens" available to traffickers.

261. The Secretary-General of the United Nations should be requested by the Commission on Narcotic Drugs to issue periodically, lists of the national agencies or authorities designated by State parties to facilitate legal and judicial co-operation.

262. The Secretary-General should also be requested to publish a compendium of the bilateral and regional agreements on mutual legal assistance entered into by States, and States should report the conclusion of such agreements to the Secretary-General, if the Parties deem this to be appropriate.

263. In view of the paramount importance of timely intelligence in the fight against illicit trafficking, Governments would be able to intensify their efforts against these illicit activities if they possessed efficient channels of communication enabling them to track movements of traffickers promptly. For this purpose they may find it useful to enter into multilateral, bilateral or regional arrangements providing for the reciprocal exchange of relevant information among law enforcement agencies, including those of transit States. The latter may request assistance from the United Nations Fund for Drug Abuse Control and bilateral and multilateral assistance programmes in establishing or expanding their communications network for this purpose.

264. In cases where they consider it desirable with a view to strengthening international or regional peace and security and building confidence, interested Governments might envisage the conclusion of formal agreements (in so far as these do not already exist) that contain provisions for pursuing the fight against illicit drug trafficking. Such agreements might envisage, *inter alia*, reciprocal training courses for officials, greater ease of communication between authorities, the establishment of direct telex links etc.

265. States whose systems of law and rules of evidence and procedure are much alike may wish to consider entering into agreements for the transfer of criminal proceedings as appropriate and for the reciprocal recognition of judicial decisions concerning drug-related offences. In such cases, the provisions of such an agreement might specify that the order of the court in one State party to the agreement is enforceable in another State party, provided that in cases where a sentence is imposed, the respect of the convicted person's fundamental human rights is guaranteed in the place where the sentence is to be served.

TARGET 21. Admissibility in Evidence of Samples of Bulk Seizures of Drugs

The problem

266. Frequently, law enforcement authorities seize an illicit shipment of drugs of considerable bulk. Laws and regulations in some States require the holding of that entire bulk shipment pending the completion of investigation and trial. During the waiting period there is a risk that the seized drugs might again leak into the illicit traffic. Also, some States lack adequate laboratories for the analysis of the seizure; yet accurate and timely analysis is essential to the successful prosecution of drug-related offences. Technical methodologies need to be established for the safe destruction of bulk seizures of drugs, as well as for the accurate chemical analysis of samples, including the determination of adequate procedures for taking samples from bulk seizures.

Suggested Courses of Action

At the national level

267. Within the scope of the fundamental principles of the national legal system, legislation could be enacted or amended, as appropriate, to authorize the early destruction or other lawful disposal of seizures of narcotic drugs and psychotropic substances after the legally required sample or samples have been taken for analysis and evidentiary purposes. In any event, security procedures should, as necessary, be introduced and maintained for the safe storage and disposal of seized drugs to ensure that no part of the seizure is diverted into the illicit market.

268. In States lacking adequate capabilities for

carrying out chemical or forensic analyses, every effort should be made to establish such laboratories; in the absence of such facilities, the law may empower the judiciary to admit in evidence the analytical findings of foreign laboratories recognized by the Government. Such laboratories may include but should not be limited to those of the appropriate international bodies or those established on a regional basis. Furthermore, in some cases, it may be desirable to permit use of the mails to ship, in a controlled manner, samples of seized narcotics and psychotropic substances to regional or other laboratories for analysis.

At the regional and international levels

269. The Division of Narcotic Drugs should continue to give high priority, with the assistance of UNFDAC, to its programme of scientific and technical assistance and should give full support to the establishment and strengthening of national and regional narcotics laboratories in States having limited resources and affected by the illicit production, trafficking, transit or consumption of drugs of abuse.

270. The Division should work out acceptable standardized methods for carrying out analyses of seized narcotic and psychotropic material, continue to serve as a central source of pure reference standards and establish universally recognized forms for the submission of results of analyses of seizures. In this connection, increased co-operation for the exchange, pooling and dissemination of information, for example relating to methods of analysis, new trends, non-controlled substances of abuse and source of manufacture, should be promoted.

TARGET 22. Adequacy with a View to Improved Efficacy of Penal Provisions

The problem

271. Articles 36 and 22 of the 1961 and 1971 Conventions respectively provide that acts contravening the provisions of those Conventions are to be declared punishable offences by States parties to them. Because definitions of drug-related offences and the penal provisions are not uniform but may vary from country to country, there are loopholes by which drug traffickers and their accomplices can escape prosecution. Sentencing practices also vary widely within and between national jurisdictions, as do policies regarding conditional release after arrest and after sentences involving deprivation of liberty.

Suggested Courses of Action

At the national level

272. The legislature, ministries or other authorities concerned, university faculties of law, research institutions and like academic bodies, might review national law and practice to ensure that illicit drug production and trafficking offences are punishable by adequate measures.

273. Within the scope of the fundamental principles of the national legal system, States might take necessary measures to establish as offences under criminal law :

(a) The illicit production, trafficking, or importation of narcotic drugs or psychotropic substances;

(b) The manufacture, distribution, possession or temporary possession of materials or equipment intended for use in the illicit production or manufacture of narcotic drugs or of psychotropic substances;

(c) The acquisition, possession, temporary possession, transfer or laundering of proceeds or instrumentalities derived from or used in illicit traffic;

(d) Conspiracy to commit, attempts to commit and participation in the commission of the said offences.

274. States could ensure, by legislation if appropriate, that their judicial system takes into consideration, in connection with initial sentencing and eventual parole, aggravating circumstances which may include, *inter alia:*

(a) The involvement of the planners and organizers of organized criminal groups;

(b) The use of firearms or violence;

(c) The fact that the offender holds public office or is a medical practitioner;

(d) Previous drug trafficking offences, wherever committed;

(e) The manifest intent of the offender to infiltrate and introduce drugs into closed institutions such as prisons, military premises, boarding schools, treatment and rehabilitation centres for drug addicts and the like, and to employ minors in furtherance of the commission of the offence;

(f) The victimization of minors.

275. States may wish to ensure that their judicial authorities observe the utmost caution in granting to an alleged offender conditional release on bail or bond, after arrest and pending trial, for traffickers command such large resources that they can easily afford losses arising from any bail or bond forfeiture. As drug trafficking offences are usually part of a continuing criminal enterprise, States could provide a legal basis for denying conditional release when there is evidence that such release would constitute a continuing threat to the community.

276. The appropriate ministry or authority could propose legislation (in so far as it does not already exist)

providing disciplinary or penal measures with respect to any medical prescriber who is proved to have accepted financial or other favours in return for prescriptions. Similar measures would be applicable to any medical practitioners who are proved to have prescribed inappropriate medicaments to persons known to them to be drug addicts or as being likely to divert the drugs prescribed to the illicit traffic. Analogous provisions would be applicable to pharmacists and nurses in cases of improper dispensing.

At the regional and international levels

277. For the purpose of promoting concerted efforts to deter traffickers from exploiting the diversity of laws and sentencing practices and penal prosecution in countries in a particular region, the Governments and legislatures should encourage greater co-operation among judicial, police and customs authorities and should consider to what extent the diverse laws, sentencing practices and penal prosecution concerning illicit trafficking may be harmonized at the regional level and effectively implemented at the national level.

TARGET 23. Forfeiture of the Instruments and Proceeds of Illicit Drug Trafficking

The problem

278. In line with the provision of article 37 of the 1961 Convention and article 22, paragraph 3, of the 1971 Convention referring to objects directly associated with the commission of a drug trafficking offence, most national criminal or civil legal systems make provision for the seizure and forfeiture of the tools and devices actually used in committing the offence. Most of such existing provisions, however, cannot be constructed as being applicable to assets acquired by means of the proceeds resulting from drug trafficking.

279. The volume of the property and money transactions, and especially of cash transfers, related to drug trafficking has increased so greatly that these transactions affect some national economies in their entirety. The increased use by traffickers and their associates of complex corporate structures and intricate business transactions involving banks, trust companies, firms dealing in real estate and other financial institutions has added to the difficulty of seizing assets obtained as a result of trafficking in drugs. Because bank, tax and investment legislation varies from country to country, traffickers and their accomplices can find loopholes in national laws and procedures and can quickly adapt laundering schemes and techniques to hide their ill-gotten gains.

Suggested Courses of Action

At the national level

280. Within the scope of the fundamental principles of the national legal systems, the legislature, ministries or other authorities concerned, university law faculties, research institutes and like academic bodies could review the national legislation and regulations and consider the desirability of proposing any necessary modifications that would facilitate and ensure the seizure, freezing, and forfeiture of the objects knowingly used in trafficking and the proceeds thereof, including objects knowingly acquired with those proceeds.

281. The law could provide that, where it has been determined by appropriate judicial or administrative procedures that specific assets were acquired by means of the proceeds of trafficking, title to all such assets is forfeited. If some of these assets are located in another State, the State in which the action was initiated could assist that other State in seizing those assets.

282. States could in their legislation and regulations authorize their judicial or other authorities concerned to accede to appropriate requests for such actions from other States where the offence may have been committed.

283. Associations of banks, investment houses and like institutions should devise codes of conduct, whereby their members would pledge themselves to assist the authorities in tracing the proceeds of trafficking activities. Subject to respect for the fundamental principles of national laws, the legislation could provide that the personnel and/or management of such firms will be liable to fines or other penalties if they knowingly participate in or facilitate schemes for concealing information relating to such transactions.

284. Subject to the limitations of the constitutional, legal and administrative system, the ministry concerned may wish to carry out, or cause to be carried out for the purpose of detecting the sources of illicit supply of drugs and of reducing that supply, an investigation into the income levels of persons suspected of serving as channels of supply. The investigators should be directed to look for evidence in the records of the tax authorities, motor vehicle licensing authorities, land registry, public register of companies and any other accessible statistical or financial records, that may disclose a gap between declared income and ostentatious expenditure.

285. The ministry or other authority concerned should consider the desirability and possibility of establishing a special fund whose resources would be mobilized to serve the cause of the fight against the illicit drug traffic and drug abuse. The assets of the fund might be constituted from, for example, voluntary contributions, special governmental allocations, monies or property seized in connection with the prosecution and conviction of drug traffickers.

286. The distribution of the value of the proceeds of seizure and forfeiture could be a matter for agreement between the States concerned. Each State could empower an appropriate agency to establish a trust fund for holding

such forfeited property.

At the regional and international levels

287. The Division of Narcotic Drugs in co-operation with ICPO/Interpol and CCC, assisted, as appropriate, by the United Nations Fund for Drug Abuse Control, should encourage the exchange of information about trans-border laundering schemes and techniques and of experience with the training of staff of law enforcement agencies and of financial institutions.

288. In cases where bilateral or multilateral agreements are negotiated for the purpose of promoting international trade, economic co-operation, cultural exchanges or for like purposes, the representatives of the State negotiating such agreements might consider the advisability of inserting in these instruments provisions designed to prevent legitimate transactions under the agreements from being used as vehicles for the laundering of gains from illicit drug trafficking, illicit drug manufacture and illicit cultivation of narcotic plants.

TARGET 24. Tightening of Controls of Movement Through Official Points of Entry

The problem

289. The security of airports, seaports and land border crossings is normally within the scope of responsibility of customs and, to a certain extent, immigration authorities. There are, however, usually gaps in the effective control by law enforcement agencies. For example, without full-time coverage, these entry points are vulnerable to penetration, as are other frontier areas. Even if covered on a full-time basis, the organization and layout of the facilities often offer opportunities for evasion of controls, and few entry points are equipped with modern or appropriate means of detecting illicit movements of drugs, such as sensing devices and sniffing dogs. Service personnel carrying out activities such as maintenance, cleaning, refuelling and catering, and crew members, are not always adequately controlled. Customs services are invariably under the jurisdiction of central government authorities, whereas the management of airports and seaports may be vested in a variety of local government or corporate entities. Organized private messenger and courier services moving across borders pose potential risks.

290. The results obtained at the world level by law enforcement agencies responsible for combating illicit drug trafficking show that the largest seizures are now made in commercial freight, where drugs are concealed either among goods in normal traffic or inside specially fashioned cavities in means of transport. This form of carriage currently appears to be the preferred means of transportation, of the highly organized trafficking networks. The expansion of international trade, the accelerated rotation of international means of transport and the development of containerized traffic, should induce the supervising authorities to initiate action with respect to the establishment of effective national and international means of deterrence and enforcement, which are compatible with a rapid flow of international trade.

Suggested Courses of Action

At the national level

291. Law enforcement agencies, customs services and the appropriate ministries or authorities could jointly analyse the current control systems, organization and layout of all official entry points with a view to recommending legislation and regulations or the redesign of facilities, in order to ensure full security control and appropriate jurisdictional authority over the physical facilities at all official entry points and so to curb illicit trafficking.

292. Analogous measures might be taken with regard to the movement of goods into and out of free-trade zones or free ports, so that countries take the necessary steps to assure that free trade zones are not used to divert specific chemicals, precursors, and controlled drugs.

293. While avoiding the restriction of or interference in international commerce, countries should develop programmes to ensure that all drugs or chemicals entering or transiting their territory are of legitimate origin and destined for legitimate use. Countries should enact appropriate legislation for this purpose, where necessary. Appropriate law enforcement and customs officials should be trained in the identification of drugs and chemicals, suspicious routes and methods of illicit shipments. Manifests and shipping documents should be closely scrutinized for suspicious shipments. Authorities could be empowered to search incoming and outgoing vessels, aircraft and vehicles in order to monitor the movement and trans-shipment of drugs and chemicals in free trade areas. Patrols should be maintained in harbours, airports and other free zones.

294. The appropriate law enforcement agency, usually the customs service, may initiate and implement these measures through the establishment of a joint task force with the agencies that manage the facilities of all official entry points and with the associations of transportation and shipping companies concerned.

295. Appropriate training might be provided by the law enforcement agencies concerned to personnel involved in border controls. Such training could include techniques relating to the gathering and use of intelligence, the development of high-risk profiles and effective control of outward-bound passengers, particularly those departing for known drug source countries or regions.

296. States might provide their appropriate law enforcement agencies with adequate sensing equipment, including trained dogs, and kits for the preliminary iden-

tification of suspect material; they might enlist the assistance of bilateral, multilateral or international agencies, if needed, for this purpose.

297. The responsible authorities could seek, to the extent possible and by whatever means appear the most appropriate, to encourage shipping companies, rail and road transport undertakings and airlines serving international routes to enter into arrangements with the customs service, defining their respective responsibilities. Detailed arrangements may vary according to the degree of risk of drug trafficking and should whenever possible be based upon the Memoranda of Understanding established between CCC and the various international organizations concerned, e.g. the International Chamber of Shipping and the International Air Transport Association (IATA), and any accompanying guidelines detailing the co-operative measures to be taken by both the authorities and the members of those organizations. The law enforcement authority might undertake to provide training for the personnel in order to reduce these risks, while maintaining appropriate confidentiality.

298. Within the scope of the fundamental principles of the national legal system, legislation could be enacted whereby firms whose employees are found to be in possession of large quantities of drugs acquired unlawfully in the course of their business would themselves be liable to penalties, if proved to have been negligent in their management and to criminal penalties if their negligence rises to the level of recklessness.

299. Notices and leaflets prominently displayed at embassies, consulates, airports, seaports and border crossings should warn travellers of the serious consequences of illegal trafficking and of the penal measures to which persons convicted of drug offences are liable.

300. The appropriate ministry or authority could make regulations (in so far as they are not already in force) requiring industrial, manufacturing and trading corporations to report to the ministry or authority any case in which they have grounds for suspecting that goods, substances, equipment or any other item produced, manufactured or traded by them have been or are likely to be diverted to the illicit drug traffic or used for the purpose of making illicit drugs. In particular, the appropriate ministry or authority should instruct the licensing authority of the State, whether party to the 1961 and 1971 Conventions or not, or exercise caution in processing applications for export authorization for controlled drugs and should take steps to ensure the authenticity of the documents presented.

301. The appropriate ministry or authority should issue instructions (if these do not already exist) to the customs service to verify the accuracy of the description given in the transport documents relating to suspect consignments of drugs entering or leaving the country and, in case of discrepancy or of misleading description, to refuse clearance of the goods in question, pending rectification by the consignor.

At the regional and international levels

302. The International Civil Aviation Organization (ICAO), the International Maritime Organization (IMO), the World Tourism Organization (WTO), IATA and the International Chamber of Shipping should consider and adopt (in so far as they have not already done so) standards or codes of conduct to be recommended to their members and designed to improve control of the movement of passengers and goods, with a view to curbing the illicit traffic in drugs. In this respect, and where they have not already done so, they should seek to establish memoranda of understanding with CCC.

303. The United Nations Fund for Drug Abuse Control and regional and bilateral programmes should assist countries that need such assistance in equipping the law enforcement authorities at points of entry with drug sensing devices, trained sniffing dogs, drug identification kits and other means of control. The drug identification kit and related materials prepared by the Division of Narcotic Drugs can be made available in this connection.

304. ICAO, IMO, the Universal Postal Union (UPU), WTO, ICPO/Interpol, CCC, IATA, the International Chamber of Shipping and the International Association of Ports and Harbours should provide, if requested, technical advice and assistance to Governments with respect to modalities for effecting appropriate physical security in standard layout and design of premises at official points of entry.

305. Where a Government has reason to believe that its country is the target of an illicit supply of drugs originating in another country, it could consider entering into an agreement with the Government of that other country whereby it would be authorized to post duly qualified persons in that other country to co-operate with the local law enforcement agency in planning measures for stopping the supply.

TARGET 25. Strengthening of External Border Controls and of Mutual Assistance Machinery within Economic Unions of Sovereign States

The problem

306. Where a number of sovereign States have joined together to form an economic union or community by virtue of a treaty that provides, *inter alia*, for the free movement of goods and persons between the territories of its member States, it may be difficult or even impossible (because there are no intra-community border controls) to detect illicit movements of drugs and traffickers from one State to another.

Suggested Courses of Action

At the national and regional levels

307. Without prejudice to the principle of freedom of movement of goods and persons laid down in the treaty establishing the union, the States members of the union would agree *inter se* to inform each other, and in particular the law enforcement agencies, of any suspect cross-border movement of drugs or of traffickers that comes to the notice of their authorities.

TARGET 26. Surveillance of Land, Water and Air Approaches to the Frontier

The problem

308. Frontiers are particularly difficult to keep under effective surveillance and offer smugglers many opportunities for evasion. Similarly, in some countries possibilities for building private air strips and for effecting parachute deliveries in remote areas have been widely exploited by traffickers. To supplement the controls applied by the police and customs authorities at official points of entry, more complete coverage of frontiers, airspace and remote areas is needed to protect societies from the nefarious activities of the illicit traffickers in drugs.

Suggested Courses of Action

At the national level

309. The forces which are responsible for or have jurisdiction over the controls at point of entry and other related agencies with responsibility in this area could develop, implement and, as appropriate, co-ordinate plans for the surveillance of air and water approaches by appropriate means and equipment in order that suspect movements may be reported promptly to customs and other law enforcement agencies. The coast guard or similar agencies could be authorized, on reasonable grounds of suspicion of illicit carriage of drugs, to stop and search vessels and aircraft within and over their territorial waters, without prejudice to the safety of such vessels and aircraft.

310. The appropriate authorities should strictly enforce the existing domestic and international regulations regarding the registration of all aircrafts—commercial or private—and enforce the obligation of all aircraft operators to operate strictly in accordance with approved flight plans and in conformity with the instructions of the air traffic control agencies.

311. The appropriate ministry or authority could consider making regulations (where they do not already exist) requiring all privately owned boats, including pleasure craft, arriving from abroad outside any official port of entry, to report immediately to the nearest designated authority, giving full details of port of origin, cargo, passengers, owners and master of the ship or skipper, in order to request permission to refuel and obtain supplies. An aircraft entering or leaving the territory of the State should be strictly required to land at, or take off from, a designated customs airport (article 10 of the Convention on International Civil Aviation)[5] the appropriate authorities of each State have an internationally recognized right, without unreasonable delay, to search any aircraft on landing or departure, and to inspect the certificates and other documents prescribed by national law and/or by international conventions. The non-observance of such regulations would be punishable. Persons or companies providing fuel or supplies to such craft without verifying that they have permission would be liable to fines or other penalties.

312. The appropriate ministry or authority should ensure that law enforcement agencies responsible for combating illicit trafficking are provided with efficient communication networks and means of transport and that their staff is trained to deal with drug trafficking between official points of entry. In countries that lack the financial resources to develop the necessary installations, networks, equipment and facilities for training, the Government might propose projects qualifying for multilateral or bilateral assistance or for assistance for UNFDAC for the purpose of obtaining them.

313. Non-governmental associations of amateur pilots, yachtsmen and owners of pleasure craft and owners of private aircraft, boats and ferries, as well as associations of commercial and private fishermen and hunters and their individual members, are urged to co-operate with law enforcement authorities by reporting to these authorities suspected drug trafficking activities.

314. Law enforcement agencies could consider the possibility of establishing telephone "hot lines" that are free of charge and connected to a permanently manned office, so that any person may report suspect drug-related occurrences without fear of reprisal.

315. The appropriate ministry or authority could establish and maintain a system of licensing for private boats and marinas. The appropriate authorities should strictly enforce the existing domestic and international regulations concerning the registration of aircraft, the issuance of operator's permits and the use of properly designated airports or airstrips. Private operators and their organizations should be encouraged to report to law enforcement agencies suspected drug trafficking activity.

316. The appropriate authorities could consider arranging for civic recognition or awards to individuals and non-governmental associations that have made outstanding contributions to the protection of the national frontiers against illicit drug trafficking.

At the regional and international levels

317. The air traffic control agencies and other authorities concerned should strengthen fight control regulations in co-operation with their counterparts in the region and on a world-wide basis.

318. The ministry or authority concerned, together with law enforcement agencies at the national and local levels, may wish to ensure that clear and effective channels of communication with corresponding agencies in other countries are established and maintained.

319. Regional seminars should be organized to facilitate the exchange of ideas and techniques designed to strengthen frontier controls.

320. The ministries or authorities concerned should take full advantage of regional and interregional co-operative mechanisms, of the sessions of the Commission on Narcotic Drugs and its Sub-Commission on Illicit Drug Traffic and Related Matters in the Near and Middle East, regional meetings of Heads of National Drug Law Enforcement Agencies, and of ICAO, IMO, CCC and ICPO/Interpol and IATA, in order to ensure maximum co-operation and consistency of implementation and training methods in safeguarding and strengthening the security of frontiers.

TARGET 27. Controls over the Use of the International Mails for Drug Trafficking

The problem

321. Customs services have determined by various techniques, including sniffing dogs, sensing devices, X-rays and the like, that controlled substances are being sent through the international mails notwithstanding the prohibition of such use of the post by the Universal Postal Convention. If the suspicious item is destined for delivery within the national territory, the customs service may obtain a search warrant and open the item for inspection, if the national laws so provide. However, if the suspicious item is in transit in the mails through the territory of the State whose authorities detect the illicit shipment, article 1 of the Universal Postal Convention and of the Constitution of UPU[6] provides that postal items in transit through a State party may not be opened. When this problem was discussed between CCC and UPU, the latter invited postal administrations:

"(a) To co-operate in combating the traffic in narcotics and psychotropic substances whenever they are legally required to do so by their national authorities responsible for this matter; to ensure respect for the fundamental principles of the international post, in particular, the freedom of transit (article 1 of the Constitution and of the Convention)" ;

"(b) To make all appropriate arrangements with the relevant authorities of their countries to ensure that bags of mail in transit suspected of enclosing items containing narcotics or psychotropic substances are not opened, but to :

(i) Advise by the quickest means, at the request of their customs authorities, the administration of destination so that the suspected bags can easily be identified on arrival ;

(ii) Verify the origin of the mail."

In effect, this provides for a procedure not unlike controlled delivery for postal items in transit, a procedure already available with respect to items addressed to domestic destinations. However, the detection of suspicious items in the huge volumes of mail involved is no easy task, regardless of the destination of the items.

Suggested Courses of Action

At the national level

322. Within the scope of the constitutional order, the ministry responsible for postal communications and the national postal authorities could strengthen controls over the use of the international mails for drug trafficking. Items that are suspect by reason of origin, address, nature or other characteristics may be subjected to a test by a sensing device, sniffing dogs, X-ray or other mode of detection by postal authorities working closely with the appropriate law enforcement agency and :

(a) If the item is addressed to a person located in the national territory, the customs service should obtain a search warrant, if needed, and should inspect the item in accordance with domestic law and procedures;

(b) If the item is in postal transit, the customs service should urgently notify the customs authorities of the State of destination, by the quickest possible means, fully identifying the item and indicating its origin.

323. Local post offices could prominently display notices giving particulars of the penalties that are by law applicable to the use of the postal service for the carriage of controlled substances.

At the regional and international levels

324. UPU should provide States parties to the Universal Postal Convention with models of standard procedures for co-operation of postal authorities with customs.

325. The States parties to the Universal Postal Convention might further consider how to prevent the use of the international mails for drug trafficking; for this purpose they may wish to propose appropriate

amendments to the Convention. A study of this problem and of appropriate action which may be taken to combat it would be useful.

TARGET 28. Controls over Ships on the High Seas and Aircraft in International Airspace

The problem

326. Vessels and aircraft are utilized for the illicit transport of drugs between countries, outside national boundaries, on the high seas and in international airspace. As numerous countries may be affected by the international shipment of drugs, appropriate co-operative procedures for interception need to be devised which do not interfere with legitimate passage and commerce, subject to compliance with existing relevant international conventions.

Suggested Courses of Action

At the national level

327. Should the ministry or authority concerned have reasonable grounds for suspecting that a vessel or aircraft registered under the laws of the State is illicitly carrying drugs, it may request another State to assist in carrying out a search: for example, that other State may be asked to direct its authorities to board and inspect the vessel and, if drugs are found, to seize them and arrest persons involved in the trafficking. In such circumstances, the State's own authorities may board or seize a vessel or aircraft registered under its laws.

328. Subject to the provisions of international law, the law enforcement authorities should, to the fullest extent permitted by national law, undertake to board and seize a vessel unlawfully carrying drugs, provided that the authorization of the State of registry and, when applicable, of a coastal State has been obtained. A State should endeavour to respond promptly when asked for permission to stop, board and search a vessel under its registry for reasons of illicit drug trafficking control. Subject to the same considerations, an aircraft may be subject to search upon landing at a designated airport.

329. The appropriate ministry or authority should, after the seizure of such a vessel or aircraft, deal promptly with illicit drugs and traffickers found thereon under the country's own laws if the conveyance is registered under that country's laws or, if registered under the laws of another State, pursuant to such agreement as is reached with the State of registry without unnecessary delay.

330. States could authorize the appropriate agency or responsible authority to take appropriate action in these matters. This action might include the prompt communication of information indicating whether a particular vessel or aircraft is registered under the laws of the requested State and also authority to empower a requesting State to seize the suspect vessel or aircraft.

At the regional and international levels

331. International bodies and States could consider whether international standards can be established for the identification, seizure and disposition of vessels and aircraft on the surface suspected of carrying drugs illicitly, and of the drugs and traffickers found thereon. States should also make every effort to conclude bilateral, multilateral and regional agreements to strengthen such co-operation between States.

332. Existing intergovernmental forums, including the transport and shipping programmes of the regional commissions, should consider the question of illicit drug movement, the need to co-ordinate efforts to halt it, and the importance of support for the new convention.

IV

(International Conference on Drug Abuse and Illicit Trafficking

TREATMENT AND REHABILITATION

Introduction

333. In many cultural patterns and in many regions, man has had recourse to drugs for various purposes: ritualistic, initiatory, diagnostic, dionysiac or therapeutic. In response to the obvious dangers, societies have made laws or pronounced taboos to prevent abuse and to protect the community.

334. The upsurge of drug addiction since the 1960s represents a previously unknown phenomenon, at least so far as its dimensions are concerned. Addiction has spread over the entire planet, sparing almost no nation, no social class and no age, regardless of sex and race. The damage caused to the physical, psychological and social health of individuals and of communities has made drug addiction a public hazard on the world scale. Addiction has become a matter of serious concern to many Governments, for it affects public and social health and economic resources. In some countries, drug abuse is regarded as an offence and persons are prosecuted for drug abuse and are liable to penalties. Moreover, the stigma attached to drug addiction may lead to a reluctance to apply for assistance and treatment at available services.

335. Because drug addiction is a phenomenon with world dimensions and ramifications, the treatment and rehabilitation of addicts call for collaboration on a global scale and a multidisciplinary context at the national, regional and international levels.

336. Development of a policy for treatment for drug addiction is difficult and complex, because it must include counselling, guidance, motivation, treatment in the medical sense, rehabilitation and social reintegration and should ideally culminate in the drug addict's return to a drug-free life. Great importance should be ascribed to social, cultural and environmental factors. Particular stress is laid on the multidisciplinary aspects of the strategy of treatment and rehabilitation, which involves the participation of a wide range of experts from several specialities. Account must be taken of the establishment and maintenance of parent/family and peer groups designed to support the drug addict throughout the treatment and aftercare process. These psychosocial support structures can greatly enhance the potential for recovery and prevention of relapse. In face, a successful treatment mitigates the health and social consequences of drug abuse and thus reduces *pari passu* the use of drugs, the activities of traffickers and the risks of relapse. A high degree of commitment on the part of the addict plays a crucial role.

337. Developing countries have particular problems where drug addiction becomes an additional burden on limited resources. Special attention should be given to the needs of developing countries and ways in which approaches to treatment and rehabilitation can be developed which take account of the economic, cultural and social constraints upon these countries.

TARGET 29. Towards a Policy of Treatment

The problem

338. The definition of a clear and precise policy is fundamental to the conduct of treatment operations. It is well known that, apart from its health and social implications, drug abuse has economic repercussions which constitute serious handicaps to the development of some countries, impairing the productivity of citizens and representing heavy charges on medical and social support systems. Treatment has often suffered from the lack of guiding principles and a lack of cohesion—each category of experts tended to pursue its own programme

in isolation. For a long time, law enforcement measures played, directly or indirectly, an important part as the only alternative to treatment. Owing to the inconsistency of the results of this combination with those obtained by self-help associations, for example, the concept that eventually prevailed was that drug addiction was a disease amenable to treatment. For this purpose, a coherent policy of action is needed, avoiding dissipation or duplication of effort and making possible the integration of the programme into the general primary health care plan. It must be a policy that permits the judicious choice of objectives, the identification of target groups at risk and the establishment of an order of priorities. Therapeutic measures can operate more effectively with the participation of the community and when such measures ensure that the addict remains within the context of the community.

Suggested Courses of Action

At the national level

339. The competent national authorities concerned, where appropriate, in collaboration with non-governmental organizations, may consider establishing a nation-wide co-ordinating body to be responsible for co-ordinating and giving guidance to the development and maintenance of a comprehensive national treatment programme for drug addiction.

340. It is essential to collect data without prejudice to confidentiality, and then to identify priority targets and groups at risk and to estimate the cost of programmes and the resources available. The medical files of health services, the records of the social service and court documents are all potential sources of data. The data should be supplemented by systematic field surveys, which might be carried out by researchers and institutes in the social sciences.

341. The ministry or authority concerned with health matters, in conjunction with the ministries or authorities responsible for education, law enforcement, security, employment etc., could work out, in the light of the data gathered by statistical services, customs authorities or other agencies concerning drug abuse, illicit trafficking and the incidence of drug-related diseases such as acquired immune deficiency syndrome (AIDS) and hepatitis, a national policy, and a programme for its implementation, for preventing and reducing drug addiction and for reintegrating drug addicts into social and occupational life. The national policy and the implementation programme might, as appropriate, pay attention to intravenous drug abuse and to countering the spread of HIV virus among this group of drug abusers. Provision should be made for the periodic testing of the functioning of the programme.

342. The various targets, methods and priorities should be strictly laid down in order to facilitate implementation of the programme. Work should proceed on the basis of and with the facilities available, improvements being introduced progressively. The start should not be delayed by aspirations to achieve perfection and the programme should not be limited to an "all-or-nothing" approach. Methods which do not take account of the actual local situation should be avoided. Allowance should be made for the changing nature of addiction by making provision for possible adjustments in the light of the demands of the moment.

343. Incentives could be offered to bodies and institutions in the private sector which carry on research into drug abuse or which perform therapeutic functions, social welfare services or services of some other kind in support of the community efforts to combat drug abuse and to treat drug addicts.

At the international level

344. WHO should indicate the principal factors to be taken into account in the formulation of a policy for the treatment of drug addiction.

345. United Nations Organizations, including in particular the regional commissions, which undertake activities in the various aspects of social development should incorporate issues related to the treatment and rehabilitation of drug abusers into their social development programmes.

346. Member States should involve WHO in establishing or strenghtening national policy for the treatment of drug addiction.

347. WHO should be urged to prepare, in co-operation with international and governmental organizations, a plan for disseminating knowledge about the way in which national treatment policies can be established and strengthened.

348. The work of WHO on nosology should be encouraged and expanded, and governments should assist WHO in field testing the validity of the nosology in various locations.

TARGET 30. Inventory of Available Modalities and Techniques of Treatment and Rehabilitation

The problem

349. After an action policy has been adopted, an inventory is needed of the modalities and techniques actually available at the national level for undertaking the treatment, rehabilitation and social reintegration of drug addicts. In most countries, many different services and agencies are involved in the prevention of drug abuse and in the treatment and rehabilitation of drug abusers. Some are specialist, others are generalist. Some are centrally funded, others locally funded, and some depend on voluntary contributions. In the absence of

evaluation and co-ordination this network of services will develop haphazardly and without regard to the manner in which each is expected to complement the others.

350. It is important that the treatment system should be carefully planned and properly co-ordinated in order to make better use of existing resources. Appropriate intervention, possibilities for care and referral mechanisms should be available and be integrated within the primary health care system. Furthermore, it is essential that the provision of services in all of the treatment areas should be monitored and evaluated. Sound monitoring and evaluation can lead to efficient and cost-effective use of resources.

Suggested Courses of Action

At the national level

351. The appropriate authority could:

(a) Draw up an inventory of and evaluate the facilities for treatment actually available at the national level, their capacities and their locations. The inventory should include information regarding the staffing, objectives, methods and other characteristics of the treatment provided at the primary care community level, and as specialist health and social services;

(b) In co-operation with the fiscal, legislative and other appropriate authorities, adjust the apportionment of resources in the light of priority targets and of the groups at risk;

(c) Evaluate the material and manpower in the various professional categories involved, taking into account the resources actually available at the national level.

(d) Evaluate the effectiveness of treatment approaches making use of WHO advice on treatment outcome evaluation.

352. The appropriate authority could evaluate the efficacy of methods of dealing with cases of drug abuse or, in cases where it considers that other, more effective methods could be applied, may recommend a change of methods. From time to time it could critically evaluate the progress made and make recommendations for the future. The persons in charge of the treatment and rehabilitation centres should participate in the evaluation process.

353. Close collaboration should be established or maintained with international organizations (especially with WHO) and with non-governmental organizations so that information on treatment approaches and outcomes can be adapted to specific local and national settings. In the light of this information, national strategies may, as appropriate, be revised.

354. WHO, collaboration with the Division of Narcotic Drugs, regional organizations and international non-governmental organizations, should gather data on modalities and techniques of treatment available, as well as on their application. Such information should be widely disseminated by WHO, in particular to countries that lack relevant experience.

355. Exchanges of experts with other countries should be encouraged under the auspices of WHO. These exchanges should make it possible to form a comparative and objective estimate of the progress of work.

356. International organizations should provide technical assistance as requested to evaluate the efficacy of national programmes of treatment. WHO, in co-operation with Governments, should consider developing comparable evaluation design and methodology for treatment outcome studies. International organizations should assist in disseminating the findings and in comparing results.

357. International organizations (including UNFDAC, WHO and ILO), non-governmental organizations and Governments should be requested to provide technical assistance and, if needed, financial assistance, for epidemiological studies, including research to identify high risk groups and to establish the aetiology of drug addiction.

358. At the request of Governments or intergovernmental organizations, regional seminars on available modalities and techniques of treatment and rehabilitation could be organized for more extensive information and mutual benefit. Experts from ILO, WHO, UNESCO and other bodies whose experience in their respective fields is essential could participate in these seminars. Assistance by the United Nations Fund for Drug Abuse Control may be required.

359. Regional training and retraining courses could be arranged, enabling participants to learn about the latest development in the field of the treatment of drug addiction, rehabilitation and re-integration of drug addicts.

360. The need for a common language calls for the preparation by WHO of a glossary of terms and expression in use in the area of drug addiction. With the aid of such a glossary, experts of various nationalities would be able to understand each other better and more accurately. These terms should be defined operationally and tested for validity before dissemination with the necessary consideration of medical, social and legal connotations, including the provisions of international law.

361. A handbook or compendium of treatment and rehabilitation techniques prepared by WHO, in collaboration with ILO, for use by workers in the field would be of great value. The available manuals on treatment techniques should be reviewed and, if appropriate, revised as necessary for broad applicability and after field testing made readily available.

TARGET 31. Selection of Appropriate Treatment Programmes

The problem

362. The selection of treatment programmes is fraught with various difficulties. It should take account of the local social, cultural and environmental factors for the purpose of mobilizing local resources and energies and facilitating a drug addict's social reintegration. The programme model should be personalized, i.e. its thrust and form should correspond to the nature of the drug and to the personality of the drug abuser. Drugs do not all produce the same kinds of addiction. Some people misuse drugs without being addicted. Also, people who have recourse to drugs do so for different reasons. In some cases the use of drugs masks a pathological condition.

363. Such techniques as continued provision of controlled amounts of addictive substances have not been universally accepted. However, some countries consider that programmes in which substitutes are prescribed are appropriate in reducing illicit drug use. Other modalities of treatment or drug-free therapeutic communities are successful in reducing the rate of recidivism.

364. Detoxification represents only a part of the treatment and rehabilitation process. Detoxification must therefore be succeeded by rehabilitation, which is a long process aimed at teaching the individual to resume civic life free of the chains of the drug. The rehabilitated drug abuser learns to give a meaning to life, to establish a healthy relationship with others and to face the difficulties of day-to-day life without recourse to drugs.

Suggested Courses of Action

At the national level

365. With respect to the technique and course of treatment, a national or regional optimum therapeutic programme should be elaborated in the light of the different conditions of the countries or regions concerned. Scientific research should be strengthened in order to guide the clinical practice of primary health care agencies. WHO and other international organizations can sum up the experience of various countries and promote the exchange of information.

366. The appropriate ministry or authority could issue guidelines addressed to the members of the medical and paramedical professions and to educational establishments, social welfare agencies and others concerned, recommending that, in cases where they have to deal with drug addicts, they should not ignore the possibility that the addiction or abuse had its origin in a psychopathological or psycho-social condition, which may require treatment by psychotherapeutic or chemo-therapeutical means.

367. The ministry or authority concerned could issue guidelines recommending that those administering treatment should adjust the mode of treatment in the light of changing patterns of drug use, e.g. the appearance of more potent drugs, increased poly-drug use, and other factors. In addition, they should take into account situations where the drug habit is aggravated or complicated by psycho-social factors, or by the presence of certain diseases, or other conditions, such as pregnancy.

368. The appropriate ministry or authority, in countries where drug addiction has assumed or threatens to assume serious dimensions, could establish centres (in so far as they do not already exist) for carrying out detoxification operations and for treating serious cases. The availability and possible success of treatment should be made known to the public. Provision should be made for seeking out drug addicts in their customary environment with a view to guiding them towards treatment; this function might be fulfilled by social workers.

369. Where appropriate, it might be desirable to enlist the participation of groups of volunteers and former addicts, subject to appropriate training and satisfactory evidence of their qualifications in the treatment and rehabilitation teams.

370. Treatment programmes should make provision for involving the families of drug addicts. In countries where the resident population is covered by public health insurance schemes it might be useful for the various ministries concerned (e.g. health, finance, social welfare, labour) to obtain precise information on the cost to the nation, for example in terms of public funds, time lost, personnel, material and medicaments, of the treatment of drug addicts that necessitates their absence from school or training courses or from work. In the light of the information and of the success rate of treatment the ministry concerned may make recommendations concerning the circumstances in which out-patient treatment of such persons may be perferable to their in-patient treatment.

371. Where appropriate, the ministry concerned may wish to recommend that health insurance schemes, whether public or private, should offer the insured persons coverage of the expenses of detoxification, treatment for drug abuse and rehabilitation.

372. The persons in charge of assistance centres should be able to carry out "individualized" treatment programmes geared to the drug addict's genuine problems and should involve, as appropriate, the person's family; the treatment should, if necessary, deal with drug-related diseases and include pharmacological or psychotherapeutic treatment, social assistance, participation in a community, etc. The treatment should

be chosen on the basis of the best possible diagnosis and of an evaluation of its true efficacy, as well as on the basis of a cost-benefit evaluation.

373. In urban environments where the risk of addiction is considered to be particularly high in certain social groups or age groups, the civic or municipal authority might be invited by the ministry concerned to establish a permanently staffed emergency aid centre specially designed to respond to urgent appeals for help by addicts or their families.

374. The ministry concerned, taking account of the national legislation, might consider the possibility of supporting private initiatives such as self-help associations which have proved to be useful in providing care for drug abusers or for persons at risk of becoming drug addicts or for persons who have undergone a course of treatment for addiction and who may need to be protected against relapse. Such initiatives may relieve congestion in public health institutions and may be found to be cost-effective.

At the regional and international levels

375. WHO may be asked by Governments to send experts to help in setting up a treatment programme.

376. Regional meetings should be organized by regional bodies to assess the progress of work and consider future projects regarding the treatment of drug addiction. The United Nations regional commissions could provide an appropriate forum for such meetings, organized in collaboration with WHO.

377. International organizations (in particular WHO and ILO) and other appropriate organizations as well as non-governmental organizations should review existing operational manuals available for sponsors of treatment programmes, determine if a need exists for additional or improved manuals, and, if so, revise or prepare them as necessary and distribute them.

378. The appropriate authorities or ministries concerned with treatment and rehabilitation should consider the possibility of seconding to each other senior staff in charge of treatment programmes, in order that they may have an opportunity to broaden their experience.

TARGET 32. Training for Personnel Working with Drug Addicts

The problem

379. Social service and health personnel who deal with drug addicts have undergone a basic university or university-type education designed to provide them with the fundamental technical skills and specific knowledge needed for the performance of their professional activities. This basic training must, however, be supplemented by specific training programmes to enable them to develop their aptitude for an activity, that is essentially interdisciplinary, for treatment can only be effective if it takes into account both the medical and psychological problems of the addict and those that are sociological in nature and related to the surroundings.

380. In view of the fact that numerous other professionals and officials come into contact with addicts, these persons should be provided with appropriate training regarding not only the cultural aspects of drug addiction, but also the treatment centres whose services are available to them in cases of need.

Suggested Courses of Action

At the national level

381. The appropriate authority might consider appointing working groups consisting of experts, specialist workers and other competent people working closely with cases of drug abuse to design specific curricula and training courses for specific groups of professionals, volunteers and community leaders.

382. Where provision for training already exists, the working group should identify and fill gaps in the curriculum. This should be done through close consultation with persons and organizations responsible for the delivery of services. The curriculum for personnel working with drug addicts should contain a syllabus concerning the treatment, management, and rehabilitation of drug dependent persons.

383. Specialized training courses should be established (in so far as they do not already exist) that confer professional and academic status (i.e. recognized by professional bodies or teaching institutions) and degrees or titles entitling graduates of the course to certain advantages (e.g. seniority or higher salary). Wherever possible, training courses should be multidisciplinary bringing together all workers concerned with similar issues, including lawyers, judges and prison officers, teachers and community workers.

384. Professional organizations could include rules of conduct concerning the handling of drug-related cases in their general rules or professional standards.

At the international level

385. Specialized agencies, other international organizations and non-governmental organizations should, on request, provide advice on the content of existing training courses, and suggest general guidelines for States wishing to establish courses *de novo*.

386. In addition, these organizations should review existing materials available for training purposes, revise the material if necessary and make it available for integration within national training programmes for application by local training staff.

387. Training projects conducted by United Nations

organizations, especially the regional commissions, related to social development, primary health care and special population groups should have a training and rehabilitation component which utilizes WHO material and which reports to WHO, the responses to this training and its use.

388. In certain parts of the world it may be cost-effective to establish courses on a regional rather than a national basis. The regional offices of WHO may be consulted in cases where countries wish to initiate regional training courses concentrating on the treatment of drug addiction.

TARGET 33. Reduction of the Incidence of Diseases and the Number of Infections transmitted through Drug-using Habits

The problem

389. A large number of persons misusing drugs suffer from diseases not caused by the drug itself but by secondary factors connected with a certain life style or a special method of drug application (especially intravenous), thus creating health hazards not only to themselves but also to others. While some of those hazards may affect individuals only (e.g. damage to the column of the nose caused by sniffing habits), others may have a wider impact, such as AIDS and certain communicable diseases, such as hepatitis, which may be spread by the use of the same hypodermic syringe by different persons, one of whom has the disease. The adverse health consequences of drug abuse and addiction on the next generation are only partially known. The study of the short- and long-term damage of drug taking to future generations has to be intensified.

Suggested Courses of Action

At the national level

390. The ministry or authority concerned with public health might collect the data available on the frequency of contagious diseases among drug users. For this purpose the ministry or authority might undertake or cause to be carried out (in so far it has not already done so) a scientific investigation to determine, *inter alia:*

(a) The different diseases prevalent among the drug-using population, the number of persons already infected and the number of persons at risk;

(b) The channels of contagion and their possible relation to the drug-using habit;

(c) Possible means of halting the transmission of disease.

391. Especially in cases where possible infections tend to become a health hazard to larger segments of the society and where a certain drug-using habit cannot be stopped immediately but is known to accelerate the spread of the contagion, the ministry or authority concerned with public health should consider inviting experts to study possible prophylactic measures which will not promote or facilitate drug abuse and to make recommendations accordingly. The ministry or authority and other appropriate agencies could publish information about those health hazards and about ways of avoiding them, both among the drug-using population and in the community at large. The ministry or authority and other appropriate agencies could also consider expanding treatment capacity and incorporating treatment outreach programmes for intravenous drug users as a means of interrupting the transmission of disease.

At the international level

392. Information about experience in the field of the treatment of drug-related infectious diseases should be communicated to WHO, which should analyse the data and publish its conclusions.

393. WHO may be asked by the ministry or authority concerned with public health to send experts or provide other information about such diseases to help to identify existing health hazards, groups at risk and possible dangers in the future and in planning strategies for counteracting them.

394. WHO should consider publishing guidelines addressed to a broad range of professionals and authorities as well as family members for dealing with drug use in cases where drug use is associated with contagious diseases. WHO should be requested to study this problem.

395. The scientific community at large and drug manufacturers should continue their efforts to invent and develop vaccines, medicines and other remedies to prevent and cure AIDS and other drug-abuse related infectious diseases.

TARGET 34. Care for Drug-addicted Offenders within the Criminal Justice and Prison System

The problem

396. According to the provisions of the 1961 Convention [article 36, paragraph 1(b) and article 38, paragraph 1] and the 1971 Convention [article 20, paragraph 1, and article 22, paragraph 1(b)] the parties should consider providing, either as an alternative to conviction or punishment or in addition to punishment, that abusers of drugs and/or of psychotropic substances should undergo measures of treatment, education, after-care, rehabilitation and social reintegration.

397. A large number of drug-abusing persons come into contact with the criminal justice system. In some States drug-related offences represent more than half of all the criminal cases dealt with by the courts. As a

consequence, many of the offenders are then channelled through the correctional institutions, and in some States offenders convicted of drug-related offences form the majority of the prison population.

398. Hence the criminal justice and correctional system occupies a prominent place in the handling, treatment and rehabilitation of persons who have committed a criminal offence against the legislation governing narcotic drugs and psychotropic substances or a criminal offence for the purpose of obtaining illicit supplies thereof or while under the influence of such drugs or substances.

Suggested Courses of Action

At the national level

399. The appropriate ministry or authority, in collaboration with the other ministries or authorities concerned, could collect the data available on the incidence of drug addiction among offenders. The ministry or authority might take appropriate actions to ensure that effective care and treatment are provided to offenders who are drug addicts.

400. In countries where it is known or believed that a substantial proportion of the total prison population is accounted for by persons serving sentences imposed for drug-related offences, or being held in custody pending trial for an alleged drug-related offence, the ministry concerned could undertake or cause to be carried out (in so far as it has not already done so) a scientific investigation, conducted with due regard to preserving anonymity, for the purpose of determining, *inter alia*:

(a) The number of persons serving sentences for offences involving drugs or drug-related offences, broken down according to age groups;

(b) The number of persons in custody charged with such an offence and awaiting trial;

(c) The proportion of the total prison population accounted for by persons convicted of or charged with drug-related offences, including offences committed by drug addicts;

(d) The type and number of offences committed by persons in the categories described in the preceding subparagraphs;

(e) The male-female ratio among prisoners convicted of or charged with drug-related offences;

(f) The proportion among prisoners convicted of or charged with such offences that is accounted for by persons previously convicted or accused of such an offence;

(g) The proportion of such prisoners that is accounted for by persons who have previously received therapeutic treatment by reason of drug addiction.

401. In the light of the results of the investigation, the ministry or ministries concerned could consider in what respects the national plan or policy for dealing with drug abuse and illicit trafficking needs to be reviewed, amended or strengthened with a view to reducing the number of persons imprisoned by reason of drug-related offences. If it should decide to publish a report containing the result of the investigation, the ministry could invite research institutes, research workers, sociologists, criminologists, experts in the law, specialized non-governmental organizations and other interested bodies to submit comments on the report and to suggest remedial measures.

402. In countries where the relevant information is not yet available and where the treatment of drug addiction is under study or review, the ministry or authority concerned with public health might consider carrying out or causing to be carried out a statistical survey for the purpose of determining what proportion of the inmates of penal, correctional and other closed institutions is accounted for by drug abusers and former drug addicts. In the light of the data and after consultation with experts, the ministry or authority may wish to make appropriate changes—if needed—in the national policy for the treatment of drug addiction, or in the guidelines for the treatment of drug addicts, or in the conditions governing admission to and discharge from such institutions.

403. In cases where the appropriate national authority is contemplating preparing new legislation that would make drug-related offences punishable by more severe penalties than are applicable in respect of such offences under the existing law, it could invite comments on its proposals from scholars and experts in the law, from youth leaders, social workers and others having direct experience of the behaviour of drug addicted offenders. The principal object of such consultations would be to ensure that the proposed legislation will not conflict with the longer-term objectives of the campaign for reducing the illicit demand for drugs, drug abuse and illicit trafficking and will not tend to add to the number of hardened offenders.

404. The appropriate authority might direct (in so far as it has not already done so) that the necessary medical, educational and support care should be provided for drug addicts held in custody pending trial and for convicted drug abusers.

At the international level

405. The Centre for Social Development and Humanitarian Affairs of the United Nations Office at Vienna should co-ordinate, in co-operation with the United Nations Social Defence Research Institute, the United Nations regional institutes for crime prevention and

control and WHO, research into the interdependence of activities for the prevention of crime and the treatment of offenders.

406. The ministries concerned could arrange exchanges of experience of treatment alternatives and intramural rehabilitation efforts, especially for staff of the judicial system and personnel of the correctional system.

407. WHO and the United Nations Social Defence Research Institute might consider preparing guidelines for the treatment of drug offenders who are drug addicts.

TARGET 35. Social Reintegration of Persons who have undergone Programmes for Treatment and Rehabilitation

The problem

408. A former drug abuser is a fragile being who has passed through a difficult stage. Assistance is needed by such persons in order that they should be able to readjust to social life and its constraints. By helping the individual to take a place in its midst, society is helping itself. The community, which, after all, has stood by the drug addict throughout the period of treatment, should mobilize resources for assisting this person thereafter. Drug addiction should be viewed as a chronic recurring disorder which responds to treatment. Several treatment episodes may be necessary before long-term abstinence is realized. The treatment of a drug addict does not end until the addict is reintegrated into society.

409. In most cases, drug addiction occurs first at an age when the person has not yet acquired mastery of a trade or occupation, or else, if addiction occurs later, it often disorganizes work habits. The adult drug addict who has been treated runs into almost insurmountable difficulties in attempting to resume civic life. Hence, vocational and social rehabilitation techniques could be employed to assist reintegration of addicts into society.

410. Promoting social rehabilitation is an important part of the aid process because both during and after treatment addicts and former addicts often have no realistic prospects in society in the sense of training opportunities and accommodation. The necessary attention should therefore be given at the earliest possible stage to developing genuine social alternatives for drug addicts. This means that such matters as housing (supervised or otherwise), vocational training and appropriate work, training for work and assistance in finding employment are not only important in the after-care stage but also constitute an inseparable element of treatment, to be incorporated from the outset. Only then can addicts be sufficiently motivated to take part in an aid programme.

411. The purpose of treatment is to secure the individual's return to a drug-free life. This is rendered difficult by a series of factors for which allowance must be made: the prejudices that in some social groups stain the reputation of the drug abuser, even after treatment, poor or non-existent occupational qualifications, awkwardness and poor performance at school and at work, resistance to discipline, tendency to instability—all these are negative factors on the road back to a productive drug-free social life. Moreover, some former drug addicts, for fear of revealing their past, are reluctant to offer themselves openly on the labour market; they do not dare to make use of the social niches placed by society at the disposal of all.

412. The guiding principle is that the particular programme should be adopted which is most cost-effective within the community. Attention should be given to on-the-job apprenticeships in various crafts. Master craftsmen in the community may accept apprentices in return for some financial compensation by the community. This policy is sometimes preferable to the establishment of vocational centres or schools, which calls for the investment of resources not always immediately available, particularly in the developing countries.

413. In appropriate settings, consideration might be given to the rehabilitative value of outdoor work for addicts. In the rural environment vocational training might concentrate on agricultural pursuits.

Suggested Courses of Action

At the national level

414. In as much as the success of treatment and rehabilitation of drug addicts can be evaluated by reference to the extent to which the harm caused by drug abuse has been reduced, as well as by reference to the extent to which they have been integrated into society and into a drug-free environment at home as well as at work, and as the risk of recidivism is always present where the facilities and resources for after-care are lacking or inadequate, the appropriate authority could ensure that after-care services, including the assistance of social workers, are provided on a scale commensurate with needs. Also treatment and rehabilitation programmes could include follow-up programmes.

415. In order to prevent a relapse after the return of the ex-addict to his previous environment, the ministry or authority concerned with public health or social welfare could consider establishing "transit" institutions in which the ex-addict could be trained, through work and education, to an appropriate life-style, in a therapeutic environment.

416. Youth movements, sports clubs and like associations, as well as associations of former drugs users, should be encouraged to make their contribution towards the rehabili- tation and social reintegration of former drug users.

417. In appropriate cases, religious organizations might provide assistance to former drug users. Where these exist, ministries of religious affairs might wish to issue guidelines for religious groups to take part in the effort of treatment and rehabilitation.

418. The ministry or authority concerned with education, in conjunction with other ministries or authorities concerned and voluntary bodies (e.g. parent organizations), could initiate or expand programmes for associating families in the process of rehabilitation and reintegration.

419. By reason of the deleterious effect of enforced idleness, particularly among the young, the ministry or other authority concerned with employment might take into account, when shaping the country's employment policy, the desirability of introducing a wider range of occupational training and creating employment opportunities for persons specially vulnerable to the risks of drug abuse, because of their age, social environment, lack of skills or for any other reason. The ministry concerned could consider working out, in consultation with employers' and workers' organizations, special training (including on-the-job training) for the benefit of such persons, and if possible should support such schemes by the grant of financial or other assistance.

420. The ministry or authority concerned with public health could give strict instructions, applicable to both public and private institutions, clinics, hospitals or other treatment centres, to the effect that confidential data disclosed to the centre or institution concerned by a medical practitioner about drug users who have received or are under treatment there, or disclosed by the persons themselves, must not be divulged to any unauthorized person or agency and must not in any case be used in any manner that might prejudice their chances of resuming a position in society and of finding employment.

421. Employers should be invited to contribute to the treatment of addicted employees by keeping the workplace open during the period of treatment or agreeing to the person's reinstatement after discharge. At enterprise level, early detection and assistance under conditions of confidentiality should be encouraged.

422. The ministry or authority concerned with employment questions could urge employers, as part of the rehabilitation process, to provide equal employment opportunities to former drug addicts.

423. Trade union organizations should consider assisting the family of any member who has become addicted and who is under treatment.

424. The support of professional and occupational associations and chambers of commerce and industry should be enlisted for the purpose of opening up possibilities of apprenticeship and placement for former addict. Social workers should help in finding employment for such persons.

425. The ministry or authority concerned might wish to consider inviting a non-governmental organization to suggest a scheme for aiding persons who have been receiving treatment for drug addiction and who have good prospects of being cured to readjust to life in society. Such a scheme might be modelled *mutatis mutandis* on those operated by discharged prisoners' aid societies, and should avoid any hint of moral stigma attached to the person concerned. Similarly, the ministry or authority concerned might consider inviting philanthropic or other non-governmental organizations to offer aid and counselling services to former drug abusers and their families in need of advice or support.

At the regional and international levels

426. Governments and workers' and employers' organizations may wish to ask ILO for technical assistance and expert advice in setting up schemes for the reintegration of former drug addicts in occupational activities or for their vocational training or retraining. ILO might consider publishing guidelines for this purpose, as well as arranging for exchanges of experience with respect to methods of reintegrating drug addicts into society.

427. Regional bodies, including non-governmental organizations, might consider establishing schemes to be operated jointly by two or more of their member countries for the social reintegration of former drug addicts.

References

1. *United Nations Treaty Series*, vol. 976, No. 14152, p. 106.
2. *Ibid.*, vol. 1019, No. 14956, p. 176.
3. See the report on the first session (A/CONF. 133/PC/6).
4. *United Nations Treaty Series*, vol. 500, No. 7310 and vol. 596, No. 8638.
5. *United Nations Treaty Series*, vol. 15, p. 295.
6. *United Nations Treaty Series*, vol. 611, Nos. 8844 and 8845, pp. 14 and 64.

Appendix 2

CONVENTION ON PSYCHOTROPIC SUBSTANCES*

Preamble

The Parties,

Being concerned with the health and welfare of mankind,

Noting with concern the public health and social problems resulting from the abuse of certain psychotropic substances,

Determined to prevent and combat abuse of such substances and the illicit traffic to which it gives rise,

Considering that rigorous measures are necessary to restrict the use of such substances to legitimate purposes,

Recognizing that the use of psychotropic substances for medical and scientific purposes is indispensable and that their availability for such purposes should not be unduly restricted,

Believing that effective measures against abuse of such substances require co-ordination and universal action,

Acknowledging the competence of the United Nations in the field of control of psychotropic substances and desirous that the international organs concerned should be within the framework of that Organization,

Recognizing that an international convention is necessary to achieve these purposes.

Agree as follows:

Article 1

Use of Terms

Except where otherwise expressly indicated, or where the context otherwise requires, the following terms in this Convention have the meanings given below:

(a) "Council" means the Economic and Social Council of the United Nations.

(b) "Commission" means the Commission on Narcotic Drugs of the Council.

(c) "Board" means the International Narcotics Control Board provided for in the Single Convention on Narcotic Drugs, 1961.

(d) "Secretary-General" means the Secretary-General of the United Nations.

(e) "Psychotropic substance" means any substance, natural or synthetic, or any natural material in Schedule I, II, III or IV.

(f) "Preparation" means :

(i) Any solution or mixture, in whatever physical state, containing one or more psychotropic substances, or

(ii) One or more psychotropic substances in dosage form.

(g) "Schedule I", "Schedule II", "Schedule III" and "Schedule IV" mean the correspondingly numbered lists of psychotropic substances annexed to this Convention, as altered in accordance with article 2.

(h) "Export" and "import" mean in their respective connotations the physical transfer of a psychotropic substance from one State to another State.

(i) "Manufacture" means all processes by which psychotropic substances may be obtained, and

---

* *Note by the Secretariat:* In the following text a number of minor corrections are included which were required owing to certain errors and omissions in the English text of the original of the Convention and which were made by a *Procés-Verbal* of Rectification of the Original of the Convention, signed on 15 August 1973 and communicated to Governments by the Office of Legal Affairs of the United Nations in circular notes C.N.169. 1973. TREATIES-5 and C.N.321. 1974. TREATIES-1 dated 30 August 1973 and 9 December 1974 respectively. They affect article 2, para. 7 (a) and the chemical formulae of certain substances in Schedules I, II and IV annexed to the Convention.

includes refining as well as the transformation of psychotropic substances. The term also includes the making of preparations other than those made on prescription in pharmacies.

(j) "Illicit traffic" means manufacture of or trafficking in psychotropic substances contrary to the provisions of this Convention.

(k) "Region" means any part of a State which pursuant to article 28 is treated as a separate entity for the purposes of this Convention.

(l) "Premises" means buildings or parts of buildings, including the appertaining land.

Article 2

Scope of Control of Substances

1. If a Party or the World Health Organization has information relating to a substance not yet under international control which in its opinion may require the addition of that substance to any of the Schedule of this Convention, it shall notify the Secretary-General and furnish him with the information in support of that notification. The foregoing procedure shall also apply when a Party or the World Health Organization has information justifying the transfer of a substance from one Schedule to another among those Schedules, or the deletion of a substance from the Schedules.

2. The Secretary-General shall transmit such notification, and any information which he considers relevant, to the Parties, to the Commission and, when the notification is made by a Party, to the World Health Organization.

3. If the information transmitted with such a notification indicates that the substance is suitable for inclusion in Schedule I or Schedule II pursuant to paragraph 4, the Parties shall examine, in the light of all information available to them, the possibility of the provisional application to the substance of all measures of control applicable to substances in Schedule I or Schedule II, as appropriate.

4. If the World Health Organization finds:

(a) That the substance has the capacity to produce

(i) (1) A state of dependence, and

(2) Central nervous system stimulation or depression, resulting in hallucinations or disturbances in motor function or thinking or behaviour or perception or mood, or

(ii) Similar abuse and similar ill effects as a substance in Schedule I, II, III or IV, and

(a) That there is sufficient evidence that the substance is being or is likely to be abused so as to constitute a public health and social problem warranting the placing of the substance under international control,

the World health Organization shall communicate to the Commission an assessment of the substance, including the extent or likelihood of abuse, the degree of seriousness of the public health and social problem and the degree of usefulness of the substance in medical therapy, together with recommendations on control measures, if any, that would be appropriate in the light of its assessment.

5. The Commission, taking into account the communication from the World Health Organization, whose assessments shall be determinative as to medical and scientific matters, and bearing in mind the economic, social, legal, administrative and other factors it may consider relevant, may add the substance to Schedule I, II, III or IV. The Commission may seek further information from the World Health Organization or from other appropriate sources.

6. If a notification under paragraph 1 relates to a substance already listed in one of the Schedules, the World Health Organization shall communicate to the Commission its new findings, any new assessment of substance it may make in accordance with paragraph 4 and any new recommendations on control measures it may find appropriate in the light of that assessment. The Commission, taking into account the communication from the World Health Organization as under paragraph 5 and bearing in mind the factors referred to in that paragraph, may decide to transfer the substance from one Schedule to another or delete it from the Schedules.

7. Any decision of the Commission taken pursuant to this article shall be communicated by the Secretary-General to all State Members of the United Nations, to non-member States Parties to this Convention, to the World Health Organization and to the Board. Such decision shall become fully effective with respect to each Party 180 days after the date of such communication, except for any Party which, within that period, in respect of a decision adding a substance to a Schedule, has transmitted to the Secretary-General a written notice that, in view of exceptional circumstances, it is not in a position to give effect with respect to that substance to all of the provisions of the Convention applicable to substances in that Schedule. Such notice shall state the reasons for this exceptional action. Notwithstanding its notice, each Party shall apply, as a minimum, the control measures listed below:

(a) A Party having given such notice with respect to a previously uncontrolled substance added to Schedule I shall take into account, as far as possible, the special control measures enumerated in article 7 and, with respect to that substance, shall:

(i) Require licences for manufacture, trade and distribution as provided in article 8 for substances in Schedule II;

(ii) Require medical prescriptions for supply or dispensing as provided in article 9 for substances in Schedule II;

(iii) Comply with the obligations relating to export and import provided in article 12, except in respect to another Party having given such notice for the substance in question;

(iv) Comply with the obligations provided in article 13 for substances in Schedule II in regard to prohibition of and restrictions on export and import;

(v) Furnish statistical reports to the Board in accordance with paragraph 4(a) of article 16; and

(vi) Adopt measures in accordance with article 22 for the repression of acts contrary to laws or regulations adopted pursuant to the foregoing obligations.

(b) A Party having given such notice with regard to a previously uncontrolled substance added to Schedule II shall, with respect to that substance:

(i) Require licences for manufacture, trade and distribution in accordance with article 8;

(ii) Require medical prescriptions for supply or dispensing in accordance with article 9;

(iii) Comply with the obligations relating to export and import provided in article 12, except in respect to another Party having given such notice for the substance in question;

(iv) Comply with the obligations of article 13 in regard to prohibition of and restrictions on export and import;

(v) Furnish statistical reports to the Board in accordance with paragraphs 4 (a), (c) and (d) of article 16; and

(vi) Adopt measures in accordance with article 22 for the repression of acts contrary to laws or regulations adopted pursuant to the foregoing obligations.

(c) A Party having given such notice with regard to a previously uncontrolled substance added to Schedule III shall, with respect to that substance:

(i) Require licences for manufacture, trade and distribution in accordance with article 8;

(ii) Require medical prescriptions for supply or dispensing in accordance with article 9;

(iii) Comply with the obligations relating to export provided in article 12, except in respect to another Party having given such notice for the substance in question;

(iv) Comply with the obligations of article 13 in regard to prohibition of and restrictions on export and import; and

(v) Adopt measures in accordance with article 22 for the repression of acts contrary to laws or regulations adopted pursuant to the foregoing obligations.

(d) A Party having given such notice with regard to a previously uncontrolled substance added to Schedule IV shall, with respect to that substance:

(i) Require licences for manufacture, trade and distribution in accordance with article 8;

(ii) Comply with the obligations of article 13 in regard to prohibition of and restrictions on export and import; and

(v) Adopt measures in accordance with article 22 for the repression of acts contrary to laws or regulations adopted pursuant to the foregoing obligations.

(e) A Party having given such notice with regard to a substance transferred to a Schedule providing stricter controls and obligations shall apply as a minimum all of the provisions of this Convention applicable to the Schedule from which it was transferred.

8. (a) The decisions of the Commission taken under this article shall be subject to review by the Council upon the request of any Party filed within 180 days from receipt of notification of the decision. The request for review shall be sent to the Secretary-General together with all relevant information upon which the request for review is based.

(b) The Secretary-General shall transmit copies of the request for review and the relevant information to the Commission, to the World Health Organization and to all the Parties, inviting them to submit comments within ninety days. All comments received shall be submitted to the Council for consideration.

(c) The Council may confirm, alter or reverse the decision of the Commission. Notification of the Council's decision shall be transmitted to all States Members of the United Nations, to non-member States Parties to this Convention, to the Commission, to the World Health Organization and to the Board.

(d) During pendency of the review, the original decision of the Commission shall, subject to paragraph 7, remain in effect.

9. The Parties shall use their best endeavours to apply to substances which do not fall under this Convention, but which may be used in the illicit manufacture of psychotropic substances, such measures of supervision as may be practicable.

Article 3

Special Provisions regarding the Control of Preparations

1. Except as provided in the following paragraphs of this article, a preparation is subject to the same measures of control as the psychotropic substance which it contains, and, if it contains more than one such substance, to the measures applicable to the most strictly controlled of those substances.

2. If a preparation containing a psychotropic substance other than a substance in Schedule I is compounded in such a way that it presents no, or a negligible risk of abuse and the substance cannot be recovered by readily applicable means in a quantity liable to abuse, so that the preparation does not give rise to a public health and social problem, the preparation may be exempted from certain of the measures of control provided in this Convention in accordance with paragraph 3.

3. If a Party makes a finding under the preceding paragraph regarding a preparation, it may decide to exempt the preparation in its country or in one of its regions, from any or all of the measures of control provided in this Convention except the requirements of:

(a) article 8 (licences), as it applies to manufacture;

(b) article 11 (records), as it applies to exempt preparations;

(c) article 13 (prohibition of and restrictions on export and import);

(d) article 15 (inspection), as it applies to manufacture;

(e) article 16 (reports to be furnished by the Parties), as it applies to exempt preparations; and

(f) article 22 (penal provisions), to the extent necessary for the repression of acts contrary to laws or regulations adopted pursuant to the foregoing obligations.

A Party shall notify the Secretary-General of any such decision, of the name and composition of the exempt preparation, and of the measures of control from which it is exempted. The Secretary-General shall transmit the notification to the other Parties, to the World Health Organization and to the Board.

4. If a Party or the World Health Organization has information regarding a preparation exempted pursuant to paragraph 3 which in its opinion may require the termination, in whole or in part, of the exemption, it shall notify the Secretary-General and furnish him with the information in support of the notification. The Secretary-General shall transmit such notification, and any information which he considers relevant, to the Parties, to the Commission and, when the notification is made by a Party, to the World Health Organization. The World Health Organization shall communicate to the Commission an assessment of the preparation in relation to the matters specified in paragraph 2, together with a recommendation of the control measures, if any, from which the preparation should cease to be exempted. The Commission, taking into account the communication from the World Health Organization, whose assessment shall be determinative as to medical and scientific matters, and bearing in mind the economic, social, legal, administrative and other factors it may consider relevant, may decide to terminate the exemption of the preparation from any or all control measures. Any decision of the Commission taken pursuant to this paragraph shall be communicated by the Secretary-General to all States Members of the United Nations, to non-member States Parties to this Convention, to the World Health Organization and to the Board. All Parties shall take measures to terminate the exemption from the control measure or measures in question within 180 days of the date of the Secretary-General's communication.

Article 4

Other Social Provisions regarding the Scope of Control

In respect of psychotropic substances other than those in Schedule I, the Parties may permit:

(a) The carrying by international travellers of small quantities of preparation for personal use; each Party shall be entitled, however, to satisfy itself that these preparations have been lawfully obtained;

(b) The use of such substances in industry for the manufacture of non-psychotropic substances or products, subject to the application of the measures of control required by this Convention until the psychotropic substances come to be in such a condition that they will not in practice be abused or recovered;

(c) The use of such substances, subject to the application of the measures of control required by this Convention, for the capture of animals by persons specifically authorized by the competent authorities to use such substances for that purpose.

Article 5

Limitation of Use to Medical and Scientific Purposes

1. Each Party shall limit the use of substances in Schedule I as provided in article 7.

2. Each Party shall, except as provided in article 4, limit by such measures as it considers appropriate, the manufacture, export, import, distribution and stocks of, trade in, and use and possession of, substances in Schedules II, III and IV to medical and scientific purposes.

3. It is desirable that the Parties do not permit the possession of substances in Schedules II, III and IV except under legal authority.

Article 6

Special Administration

It is desirable that for the purpose of applying the provisions of this Convention, each Party establish and maintain a special administration, which may with advantage be the same as, or work in close co-operation with, the special administration established pursuant to the provisions of conventions for the control of narcotic drugs.

Article 7

Special Provisions regarding Substances in Schedule I

In respect of substances in Schedule I, the Parties shall :

(a) Prohibit all use except for scientific and very limited medical purposes by duly authorized persons, in medical or scientific establishments which are directly under the control of their Governments or specifically approved by them;

(b) Require that manufacture, trade, distribution and possession be under a special licence or prior authorization;

(c) Provide for close supervision of the activities and acts mentioned in paragraphs (a) and (b);

(d) Restrict the amount supplied to a duly authorized person to the quantity required for his authorized purpose;

(e) Require that persons performing medical or scientific functions keep records concerning the acquisition of the substances and the details of their use, such records to be preserved for at least two years after the last use recorded therein; and

(f) Prohibit export and import except when both the exporter and importer are the competent authorities or agencies of the exporting and importing country or region, respectively, or other persons or enterprises which are specifically authorized by the competent authorities of their country or region for the purpose. The requirements of paragraph 1 of article 12 for export and import authorizations for substances in Schedule II shall also apply to substances in Schedule I.

Article 8

Licences

1. The Parties shall require that the manufacture of, trade (including export and import trade) in, and distribution of substances listed in Schedules II, III and IV be under licence or other similar control measure.

2. The Parties shall:

(a) Control all duly authorized persons and enterprises carrying on or engaged in the manufacture of, trade (including export and import trade) in, or distribution of substances referred to in paragraph 1;

(b) Control under licence or other similar control measure the establishments and premises in which such manufacture, trade or distribution may take place; and

(c) Provide that security measures be taken with regard to such establishments and premises in order to prevent theft or other diversion of stocks.

3. The provisions of paragraphs 1 and 2 of this article relating to licensing or other similar control measures need not apply to persons duly authorized to perform and while performing therapeutic or scientific functions.

4. The Parties shall require that all persons who obtain licences in accordance with this Convention or who are otherwise authorized pursuant to paragraph 1 of this article or sub-paragraph (b) of article 7 shall be adequately qualified for the effective and faithful execution of the provisions of such laws and regulations as are enacted in pursuance of this Convention.

Article 9

Prescriptions

1. The Parties shall require that substances in Schedules II, III and IV be supplied or dispensed for use by individuals pursuant to medical prescription only, except when individuals may lawfully obtain, use, dispense or administer such substances in the duly authorized exercise of therapeutic or scientific functions.

2. The Parties shall take measures to ensure that prescriptions for substances in Schedules II, III and IV are issued in accordance with sound medical practice

and subject to such regulation, particularly as to the number of times they may be refilled and the duration of their validity, as will protect the public health and welfare.

3. Notwithstanding paragraph 1, a Party may, if in its opinion local circumstances so require and under such conditions, including record-keeping, as it may prescribe, authorize licensed pharmacists or other licensed retail distributors designated by the authorities responsible for public health in its country or part thereof to supply, at their discretion and without prescription, for use for medical purposes by individuals in exceptional cases, small quantities, within limits to be defined by the Parties, of substances in Schedules III and IV.

Article 10

Warnings on Packages, and Advertising

1. Each Party shall require, taking into account any relevant regulations or recommendations of the World Health Organization, such directions for use, including cautions and warnings, to be indicated on the labels where practicable and in any case on the accompanying leaflet of retail packages of psychotropic substances, as in its opinion are necessary for the safety of the user.

2. Each Party shall, with due regard to its constitutional provisions, prohibit the advertisement of such substances to the general public.

Article 11

Records

1. The Parties shall require that, in respect of substances in Schedule I, manufactures and all other persons authorized under article 7 to trade in and distribute those substances keep records, as may be determined by each Party, showing details of the quantities manufactured, the quantities held in stock, and, for each acquisition and disposal, details of the quantity, date, supplier and recipient.

2. The Parties shall require that, in respect of substances in Schedules II and III, manufacturers, wholesale distributors, exporters and importers keep records, as may be determined by each Party, showing details of the quantities manufactured and, for each acquisition and disposal, details of the quantity, date, supplier and recipient.

3. The Parties shall require that, in respect of substances in Schedule II, retail distributors, institutions for hospitalization and care and scientific institutions keep records, as may be determined by each Party, showing, for each acquisition and disposal, details of the quantity, date, supplier and recipient.

4. The Parties shall ensure, through appropriate methods and taking into account the professional and trade practices in their countries, that information regarding acquisition and disposal of substances in Schedule III by retail distributors, institutions for hospitalization and care and scientific institutions is readily available.

5. The Parties shall require that, in respect of substances in Schedule IV, manufacturers, exporters and importers keep records, as may be determined by each Party, showing the quantities manufactured, exported and imported.

6. The Parties shall require manufacturers of preparations exempted under paragraph 3 of article 3 to keep records as to the quantity of each psychotropic substance used in the manufacture of an exempt preparation, and as to the nature, total quantity and initial disposal of the exempt preparation manufactured therefrom.

7. The Parties shall ensure that the records and information referred to in this article which are required for purposes of reports under article 16 shall be preserved for at least two years.

Article 12

Provisions relating to International Trade

1. (a) Every Party permitting the export or import of substances in Schedule I or II shall require a separate import or export authorization, on a form to be established by the Commission, to be obtained for each such export or import whether it consists of one or more substances.

(b) Such authorization shall state the international non-proprietary name, or, lacking such a name, the designation of the substance in the Schedule, the quantity to be exported or imported, the pharmaceutical form, the name and address of the exporter and importer, and the period within which the export or import must be effected. If the substance is exported or imported in the form of a preparation, the name of the preparation, if any, shall additionally be furnished. The export authorization shall also state the number and date of the import authorization and the authority by whom it has been issued.

(c) Before issuing an export authorization, the Parties shall require an import authorization, issued by the competent authority of the importing country or region and certifying that the importation of the substance or substances referred to therein is approved, and such an authorization shall be produced by the person or establishment applying for the export authorization.

(d) A copy of the export authorization shall accompany each consignment, and the Government issuing the export authorization shall send a copy to the Government of the importing country or region.

(e) The Government of the importing country or region, when the importation has been effected, shall return the export authorization with an endorsement certifying the amount actually imported, to the Government of the exporting country or region.

2. (a) The Parties shall require that for each export of substances in Schedule III exporters shall draw up a declaration in triplicate, on a form to be established by the Commission, containing the following information:

(i) The name and address of the exporter and importer;

(ii) The international non-proprietary name, or, failing such a name, the designation of the substance in the Schedule;

(iii) The quantity and pharmaceutical form in which the substance is exported, and, if in the form of a preparation, the name of the preparation, if any; and

(iv) The date of despatch.

(b) Exporters shall furnish the competent authorities of their country or region with two copies of the declaration. They shall attach the third copy to their consignment.

(c) A Party from whose territory a substance in Schedule III has been exported shall, as soon as possible but not later than ninety days after the date of despatch, send to the competent authorities of the importing country or region, by registered mail with return of receipt requested, one copy of the declaration received from the exporter.

(d) The Parties may require that, on receipt of the consignment, the importer shall transmit the copy accompanying the consignment, duly endorsed, stating the quantities received and the date of receipt, to the competent authorities of his country or region.

3. In respect of substances in Schedules I and II the following additional provisions shall apply :

(a) The Parties shall exercise in free ports and zones the same supervision and control as in other parts of their territory, provided, however, that they may apply more drastic measures.

(b) Exports of consignments to a post office box, or to a bank to the account of a person other than the person named in the export authorization, shall be prohibited.

(c) Exports to bonded warehouses of consignments of substances in Schedule I are prohibited. Exports of consignments of substances in Schedule II to a bonded warehouse are prohibited unless the Government of the importing country certifies on the import authorization, produced by the person or establishment applying for the export authorization, that it has approved the importation for the purpose of being placed in a bonded warehouse. In such case the export authorization shall certify that the consignment is exported for such purpose. Each withdrawal from the bonded warehouse shall require a permit from the authorities having jurisdiction over the warehouse and, in the case of a foreign destination, shall be treated as if it were a new export within the meaning of this Convention.

(d) Consignments entering or leaving the territory of a Party not accompanied by an export authorization shall be detained by the competent authorities.

(e) A Party shall not permit any substances consigned to another country to pass through its territory, whether or not the consignment is removed from the conveyance in which it is carried, unless a copy of the export authorization for consignment is produced to the competent authorities of such Party.

(f) The competent authorities of any country or region through which a consignment of substances is permitted to pass shall take all due measures to prevent the diversion of the consignment to a destination other than that named in the accompanying copy of the export authorization, unless the Government of the country or region through which the consignment is passing, authorizes the diversion. The Government of the country or region of transit shall treat any requested diversion as if the diversion were an export from the country or region of transit to the country or region of new destination. If the diversion is authorized, the provisions of paragraph 1 (e) shall also apply between the country or region of transit and the country or region which originally exported the consignment.

(g) No consignment of substances, while in transit or whilst being stored in a bonded warehouse, may be subjected to any process which would change the nature of the substance in question.

The packing may not be altered without the permission of the competent authorities.

(h) The provisions of sub-paragraphs (e) to (g) relating to the passage of substances through the territory of a Party do not apply, where the consignment in question is transported by aircraft which does not land in the country or region of transit. If the aircraft lands in any such country or region, those provisions shall be applied so far as circumstances require.

(i) The provisions of this paragraph are without prejudice to the provisions of any international agreements which limit the control which may be exercised by any of the Parties over such substances in transit.

Article 13

Prohibition of and Restrictions on Export and Import

1. A Party may notify all the other Parties through the Secretary-General that it prohibits the import into its country or into one of its regions of one or more substances in Schedules II, III or IV, specified in its notification. Any such notification shall specify the name of the substance as designated in Schedules II, III or IV.

2. If a Party has been notified or a prohibition pursuant to paragraph 1, it shall take measures to ensure that none of the substances specified in the notification is exported to the country or one of the regions of the notifying Party.

3. Notwithstanding the provisions of the preceding paragraphs, a Party which has given notification pursuant to paragraph 1 may authorize by special import licence in each case the import of specified quantities of the substances in question or preparations containing such substances. The issuing authority of the importing country shall send two copies of the special import licence, indicating the name and address of the importer and the exporter, to the competent authority of the exporting country or region, which may then authorize the exporter to make the shipment. One copy of the special import licence, duly endorsed by the competent authority of the exporting country or region, shall accompany the shipment.

Article 14

Special Provisions concerning the carriage of Psychotropic Substances in first-aid kits of ships, aircraft or other forms of public transport engaged in International Traffic

1. The international carriage by ships, aircraft or other forms of international public transport, such as international railway trains and motor coaches, of such limited quantities of substances in Schedules II, III or IV as may be needed during their journey or voyage for first-aid purposes or emergency cases shall not be considered to be export, import or passage through a country within the meaning of this Convention.

2. Appropriate safeguards shall be taken by the country of registry to prevent the improper use of the substances referred to in paragraph 1 or their diversion for illicit purposes. The Commission, in consultation with the appropriate international organizations, shall recommend such safeguards.

3. Substances carried by ships, aircraft or other forms of international public transport, such as international railway trains and motor coaches, in accordance with paragraph 1 shall be subject to the laws, regulations, permits and licences of the country of registry, without prejudice to any rights of the competent local authorities to carry out checks, inspections and other control measures on board these conveyances. The administration of such substances in the case of emergency shall not be considered a violation of the requirements of paragraph 1 or article 9.

Article 15

Inspection

The Parties shall maintain a system of inspection of manufacturers, exporters, importers, and wholesale and retail distributors of psychotropic substances and of medical and scientific institutions which use such substances. They shall provide for inspections, which shall be made as frequently as they consider necessary, of the premises and of stocks and records.

Article 16

Reports to be furnished by the Parties

1. The Parties shall furnish to the Secretary-General such information as the Commission may request as being necessary for the performance of its functions, and in particular an annual report regarding the working of the Convention in their territories including information on:

(a) Important changes in their laws and regulations concerning psychotropic substances; and

(b) Significant developments in the abuse of and the illicit traffic in psychotropic substances within their territories.

2. The Parties shall also notify the Secretary-General of the names and addresses of the governmental authorities referred to in sub-paragraph (f) of article 7, in article 12 and in paragraph 3 of article 13. Such information shall be made available to all Parties by the Secretary-General.

3. The Parties shall furnish, as soon as possible after the event, a report to the Secretary-General in respect of any case of illicit traffic in psychotropic substances or seizure from such illicit traffic which they consider important because of :

(a) New trends disclosed;

(b) The quantities involved;

(c) The light thrown on the sources from which the substances are obtained; or

(d) The methods employed by illicit traffickers.

Copies of the report shall be communicated in accordance with sub-paragraph (b) of article 21.

4. The Parties shall furnish to the Board annual statistical reports in accordance with forms prepared by the Board:

(a) In regard to each substance in Schedules I and II, on quantities manufactured, exported to and imported from each country or region as well as on stocks held by manufacturers;

(b) In regard to each substance in Schedules III and IV, on quantities manufactured, as well as on total quantities exported and imported;

(c) In regard to each substance in Schedules II and III, on quantities used in the manufacture of exempt preparations; and

(d) In regard to each substance other than a substance in Schedule I, on quantities used for industrial purposes in accordance with sub-paragraph (b) of article 4.

The quantities manufactured which are referred to in sub-paragraphs (a) and (b) of this paragraph do not include the quantities of preparations manufactured.

5. A Party shall furnish the Board, on its request, with supplementary statistical information relating to future periods on the quantities of any individual substance in Schedules III and IV exported to and imported from each country or region. That Party may request that the Board treat as confidential, both its request for information and the information given under this paragraph.

6. The Parties shall furnish the information referred to in paragraphs 1 and 4 in such a manner and by such dates as the Commission or the Board may request.

Article 17

Functions of the Commission

1. The Commission may consider all matters pertaining to the aims of this Convention and to the implementation of its provisions, and may make recommendations relating thereto.

2. The decisions of the Commission provided for in articles 2 and 3 shall be taken by two-thirds majority of the members of the Commission.

Article 18

Reports of the Board

1. The Board shall prepare annual reports on its work containing an analysis of the statistical information at its disposal, and, in appropriate cases, an account of the explanations, if any, given by or required of Governments, together with any observations and recommendations which the Board desires to make. The Board may make such additional reports as it considers necessary. The reports shall be submitted to the Council through the Commission, which may make such comments as it sees fit.

2. The reports of the Board shall be communicated to the Parties and subsequently published by the Secretary-General. The Parties shall permit their unrestricted distribution.

Article 19

Measures by the Board to ensure the Execution of the Provisions of the Convention

1. (a) If, on the basis of its examination of information submitted by governments to the Board or of information communicated by United Nations organs, the Board has reason to believe that the aims of this Convention are being seriously endangered by reason of the failure of a country or region to carry out the provisions of this Convention, the Board shall have the right to ask for explanations from the Government of the country or region in question. Subject to the right of the Board to call the attention of the Parties, the Council and the Commission to the matter referred to in sub-paragraph (c) below, it shall treat as confidential, a request for information or an explanation by a government under this sub-paragraph.

(b) After taking action under sub-paragraph (a), the Board, if satisfied that it is necessary to do so, may call upon the Government concerned to adopt such remedial measures as shall seem under the circumstances to be necessary for the execution of the provisions of this Convention.

(c) If the Board finds that the Government concerned has failed to give satisfactory explanations when called upon to do so under sub-paragraph (a), or has failed to adopt any remedial measures which it has been called upon to take under sub-paragraph (b), it may call the attention of the Parties, the Council and the Commission to the matter.

2. The Board, when calling the attention of the Parties, the Council and the Commission to a matter in accordance with paragraph 1 (c), may, if it is satisfied that such a course is necessary, recommend to the Parties that they stop the export, import, or both, of particular psychotropic substances, from or to the country or region concerned, either for a designated period or until the Board shall be satisfied as to the situation in that country or region. The State concerned may bring the matter before the Council.

3. The Board shall have the right to publish a report on any matter dealt with under the provisions of this article, and communicate it to the Council, which shall forward it to all Parties. If the Board publishes in this report a decision taken under this article or any information relating thereto, it shall also publish therein the views of the Government concerned if the latter so requests.

4. If in any case a decision of the Board which is published under this article is not unanimous, the views of the minority shall be stated.

5. Any State shall be invited to be represented at a meeting of the Board, at which a question directly interesting it is considered under this article.

6. Decisions of the Board under this article shall be taken by a two-thirds majority of the whole number of the Board.

7. The provisions of the above paragraphs shall also apply if the Board has reason to believe that the aims of this Convention are being seriously endangered as a result of a decision taken by a Party under paragraph 7 of article 2.

Article 20

Measures against the Abuse of Psychotropic Substances

1. The Parties shall take all practicable measures for the prevention of abuse of psychotropic substances and for the early identification, treatment, education, after-care, rehabilitation and social reintegration of the persons involved, and shall co-ordinate their efforts to these ends.

2. The Parties shall, as far as possible, promote the training of personnel in the treatment, after-care, rehabilitation and social reintegration of abusers of psychotropic substances.

3. The Parties shall assist persons whose work so requires to gain an understanding of the problems of abuse of psychotropic substances and of its prevention, and shall also promote such understanding among the general public if there is a risk that abuse of such substances will become widespread.

Article 21

Action against the Illicit Traffic

Having due regard to their constitutional, legal and administrative systems, the Parties shall:

(a) Make arrangements at the national level for the co-ordination of preventive and repressive action against the illicit traffic; to this end they may usefully designate an appropriate agency responsible for such co-ordination;

(b) Assist each other in the campaign against the illicit traffic in psychotropic substances, and in particular immediately transmit, through the diplomatic channel or the competent authorities designated by the Parties for this purpose, to the other Parties directly concerned, a copy of any report addressed to the Secretary-General under article 16 in connexion with the discovery of a case of illicit traffic or a seizure;

(c) Co-operate closely with each other and with the competent international organizations of which they are members with a view to maintaining a co-ordinated compaign against the illicit traffic;

(d) Ensure that international co-operation between the appropriate agencies be conducted in an expeditious manner; and

(e) Ensure that, where legal papers are transmitted internationally for the purpose of judicial proceedings, the transmittal be effected in an expeditious manner to the bodies designated by the Parties; this requirement shall be without prejudice to the right of a Party to require that legal papers be sent to it through the diplomatic channel.

Article 22

Penal Provisions

1. (a) Subject to its constitutional limitations, each Party shall treat as a punishable offence, when committed intentionally, any action contrary to a law or regulation adopted in pursuance of its obligations under this Convention, and shall ensure that serious offences shall be liable to adequate punishment, particularly by imprisonment or other penalty of deprivation of liberty.

(b) Notwithstanding the preceding sub-paragraph, when abusers of psychotropic substances have committed such offences, the Parties may provide, either as an alternative to conviction or punishment or in addition to punishment, that such abusers undergo measures of treatment, education, after-care, rehabilitation and social reintegration in conformity with paragraph 1 of article 20.

2. Subject to the constitutional limitations of a Party, its legal system and domestic law,

(a) (i) If a series of related actions constituting offences under paragraph 1 has been committed in different countries, each of them shall be treated as a distinct offence;

(ii) International participation in, conspiracy to commit and attempts to commit, any of such offences, and preparatory acts and financial operations in connexion with the offences referred to in this article, shall be punishable offences as provided in paragraph 1;

(iii) Foreign convictions for such offences shall be taken into account for the purpose of establishing recidivism; and

(iv) Serious offences heretofore referred to committed either by nationals or by foreigners shall be prosecuted by the Party in whose territory the offence was committed, or by the Party in whose territory the offender is found if extradition is not acceptable in conformity with the law of the Party to which application is made, and if such offender has not already been prosecuted and judgement given.

(b) It is desirable that the offences referred to in paragraph 1 and paragraph 2 (a) (ii) be included as extradition crimes in any extradition treaty which has been or may hereafter be concluded between any of the Parties, and, as between any of the Parties which do not make extradition conditional on the existence of a treaty or on reciprocity, be recognized as extradition crimes; provided that extradition shall be granted in conformity with the law of the Party to which application is made, and that the Party shall have the right to refuse to effect the arrest or grant the extradition in cases where the competent authorities consider that the offence is not sufficiently serious.

3. Any psychotropic substance or other substance, as well as any equipment, used in or intended for the commission of any of the offences referred to in paragraphs 1 and 2 shall be liable to seizure and confiscation.

4. The provisions of this article shall be subject to the provisions of the domestic law of the Party concerned on questions of jurisdiction.

5. Nothing contained in this article shall affect the principle that the offences to which it refers shall be defined, prosecuted and punished in conformity with the domestic law of a Party.

Article 23

Application of Stricter Control Measures than those required by this Convention

A Party may adopt more strict or severe measures of control than those provided by this Convention if, in its opinion, such measures are desirable or necessary for the protection of the public health and welfare.

Article 24

Expenses of International Organs incurred in Administering the Provisions of the Convention

The expenses of the Commission and the Board in carrying out their respective functions under this Convention shall be borne by the United Nations in such manner as shall be decided by the General Assembly. The Parties which are not Members of the United Nations shall contribute to these expenses such amounts as the General Assembly finds equitable and assesses from time to time after consultation with the Governments of these Parties.

Article 25

Procedure for Admission, Signature, Ratification and Accession

1. Members of the United Nations, States not Members of the United Nations which are members of a specialized agency of the United Nations or of the International Atomic Energy Agency or Parties to the Statute of the International Court of Justice, and any other State invited by the Council, may become Parties to this Convention:

(a) By signing it; or

(b) By ratifying it after signing it subject to ratification; or

(c) By acceding to it.

2. The Convention shall be open for signature until 1 January 1972 inclusive. Thereafter it shall be open for accession.

3. Instruments of ratification or accession shall be deposited with the Secretary-General.

Article 26

Entry into force

1. The Convention shall come into force on the nine-

tieth day after forty of the States referred to in paragraph 1 of article 25 have signed it without reservation of ratification or have deposited their instruments of ratification or accession.

2. For any other State signing without reservation of ratification, or depositing an instrument of ratification or accession after the last signature or deposit referred to in the preceding paragraph, the Convention shall enter into force on the ninetieth day following the date of its signature or deposit of its instrument of ratification or accession.

Article 27

Territorial Application

The Convention shall apply to all non-metropolitan territories for the international relations of which any Party is responsible except where the previous consent of such a territory is required by the Constitution of the Party or of the territory concerned, or required by custom. In such a case the Party shall endeavour to secure the needed consent of the territory within the shortest period possible, and when the consent is obtained the Party shall notify the Secretary-General. The Convention shall apply to the territory or territories named in such a notification from the date of its receipt by the Secretary-General. In those cases where the previous consent of the non-metropolitan territory is not required, the Party concerned shall, at the time of signature, ratification or accession, declare the non-metropolitan territory or territories to which this Convention applies.

Article 28

Regions for the Purposes of this Convention

1. Any Party may notify the Secretary-General that, for the purposes of this Convention, its territory is divided into two or more regions, or that two or more of its regions are consolidated into a single region.

2. Two or more Parties may notify the Secretary-General that, as the result of the establishment of a customs union between them, those Parties constitute a region for the purposes of this Convention.

3. Any notification under paragraph 1 or 2 shall take effect on 1 January of the year following the year in which the notification was made.

Article 29

Denunciation

1. After the expiry of two years from the date of the coming into force of this Convention any Party may, on its own behalf or on behalf of a territory for which it has international responsibility, and which has withdrawn its consent given in accordance with article 27, denounce this Convention by an instrument in writing deposited with the Secretary-General.

2. The denunciation, if received by the Secretary-General on or before the first day of July of any year, shall take effect on the first day of January of the succeeding year, and if received after the first day of July it shall take effect as if it had been received on or before the first day of July in the succeeding year.

3. The Convention shall be terminated if, as a result of denunciations made in accordance with paragraphs 1 and 2, the conditions for its coming into force as laid down in paragraph 1 of article 26 cease to exist.

Article 30

Amendments

1. Any Party may propose an amendment to this Convention. The text of any such amendment and the reasons therefor shall be communicated to the Secretary-General, who shall communicate them to the Parties and to the Council. The Council may decide either:

(a) That a conference shall be called in accordance with paragraph 4 of Article 62 of the Charter of the United Nations to consider the proposed amendment; or

(b) That the Parties shall be asked whether they accept the proposed amendment and also asked to submit to the Council any comments on the proposal.

2. If a proposed amendment circulated under paragraph 1 (b) has not been rejected by any Party within eighteen months after it has been circulated, it shall thereupon enter into force. If however, a proposed amendment is rejected by any Party, the Council may decide, in the light of comments received from Parties, whether a conference shall be called to consider such amendment.

Article 31

Disputes

1. If there should arise between two or more Parties a dispute relating to the interpretation or application of this Convention, the said Parties shall consult together with a view to the settlement of the dispute by negotiation, investigation, mediation, conciliation, arbitration, recourse to regional bodies, judicial process or other peaceful means of their own choice.

2. Any such dispute which cannot be settled in the manner prescribed shall be referred, at the request of any one of the parties to the dispute, to the International Court of Justice for decision.

Article 32

Reservations

1. No reservation other than those made in accor-

dance with paragraphs 2, 3 and 4 of the present article shall be permitted.

2. Any State may at the time of signature, ratification or accession make reservations in respect of the following provisions of the present Convention:

(a) Article 19, paragraphs 1 and 2;

(b) Article 27; and

(c) Article 31.

3. A State which desires to become a Party but wishes to be authorized to make reservations other than those made in accordance with paragraphs 2 and 4 may inform the Secretary-General of such intention. Unless by the end of twelve months after the date of the Secretary-General's communication of the reservation concerned, this reservation has been objected to by one third of the States that have signed without reservation of ratification, ratified or acceded to this Convention before the end of that period, it shall be deemed to be permitted, it being understood however that States which have objected to the reservation need not assume towards the reserving State any legal obligation under this Convention which is affected by the reservation.

4. A State on whose territory there are plants growing wild which contain psychotropic substances from among those in Schedule I and which are traditionally used by certain small, clearly determined groups in magical or religious rites, may, at the time of signature, ratification or accession, make reservations concerning these plants, in respect of the provisions of article 7, except for the provisions relating to international trade.

5. A State which has made reservations may at any time by notification in writing to the Secretary-General withdraw all or part of its reservations.

Article 33

Notifications

The Secretary-General shall notify to all the States referred to in paragraph 1 of article 25:

(a) Signatures, ratifications and accessions in accordance with article 25;

(b) The date upon which this Convention enters into force in accordance with article 26;

(c) Denunciations in accordance with article 29; and

(d) Declarations and notifications under articles 27, 28, 30 and 32.

IN WITNESS WHEREOF, the undersigned, duly authorized, have signed this Convention on behalf of their respective Governments.

DONE at Vienna, this twenty-first day of February one thousand nine hundred and seventy-one, in a single copy in the Chinese, English, French, Russian and Spanish languages, each being equally authentic. The Convention shall be deposited with the Secretary-General of the United Nations, who shall transmit certified true copies thereof to all the Members of the United Nations and to the other States referred to in paragraph 1 of article 25.

LISTS OF SUBSTANCES IN THE SCHEDULES*

List of Substances in Schedule I

	INN	*Other non-proprietary or trivial names*	*Chemical Name*
1.		DET	N, N-diethyltryptamine
2.		DMPH	3-(1, 2-dimethylheptyl)-1-hydroxy-7, 8, 9, 10-tetrahydro-6, 6, 9-trimethyl-6*H*-dibenzo [*b*, *d*] pyran
3.		DMT	*N*, *N*-dimethyltryptamine
4.	(+)-LYSERGIDE	LSD, LSD-25	(+)-*N*, *N*-diethyllysergamide (*d*-lysergic acid diethylamide)
5.		mescaline	3,4,5-trimethoxyphenethyl-amine
6.		parahexyl	3-hexyl-1-hydroxy-7, 8, 9, 10-tetrahydro-6, 6, 9-trimethyl-6*H*-dibenzo [*b*, *d*] pyran
7.		psilocine, psilotsin	3-(2-dimethylaminoethyl)-4-hydroxyindole
8.	PSILOCYBINE		3-(2 dimethylaminoethyl)-indol-4-yl dihydrogen phosphate
9.		STP, DOM	2-amino-1-(2, 5-dimethoxy-4-methyl) phenylpropane
10.		tetrahydro-cannabinols, all isomers	1-hydroxy-3-pentyl-6a, 7, 10, 10a-tetrahydro-6, 6, 9-trimethyl-6-*H*-dibenzo [*b*, *d*] pyran

The salts of the substances listed in this Schedule whenever the existence of such salts is possible.**

List of Substances in Schedule II

1. AMPHETAMINE		(±)-2-amino-1-phenylpropane
2. DEXAMPHETAMINE		(+)-2-amino-1-phenylpropane
3. METHAMPHETAMINE		(+)-2-methylamino-1-phenyl- propane

4. METHYLPHENIDATE — 2-phenyl-2-(2-piperidyl)acetic acid, methyl ester

5. PHENCYCLIDINE — 1-(1-phenylcyclohexyl) piperidine

INN *or trivial names* *Chemical Name*

6. PHENMETRAZINE — 2-methyl-2-phenylmorpholine

The salts of the substances listed in this Schedule whenever the existence of such salts is possible**

List of Substances in Schedule III

Other non-proprietary or trivial names

1. AMOBARBITAL — 5-ethyl-5-(3-methylbutyl) barbituric acid

2. CYCLOBARBITAL — 5-(1-cyclohexen-1-yl)-5-ethylbarbituric acid

3. GLUTETHIMIDE — 2-ethyl-2-phenylglutarimide

4. PENTOBARBITAL — 5-ethyl-5-(1-methylbutyl) barbituric acid

5. SECOBARBITAL — 5-allyl-5-(1-methylbutyl) barbituric acid

The salts of the substances listed in this Schedule whenever the existence of such salts is possible.**

List of Substances in Schedule IV other non-proprietory or trival names

1. AMFEPRAMONE — 2-(diethylamino) propiophenone

2. BARBITAL — 5, 5-diethylbarbituric acid

3. ethchlorvynol — ethyl-2-chlorovinylethinylcarbinol

4. ETHINAMATE — 1-ethynylcyclohexanol-carbamate

5. MEPROBAMATE — 2-methyl-2-propyl-1, 3-propanediol dicarbamate

6. METHAQUALONE — 2-methyl-3-*o*-tolyl-4 (3*H*)-quinazolinone

7. METHYLPHENOBARBITAL — 5-ethyl-1-methyl-5-phenylbarbituric acid

8. METHYPRYLON — 3, 3-diethyl-5-methyl-2, 4-piperidine-dione

9. PHENOBARBITAL — 5-ethyl-5-phenylbarbituric acid

10. PIPRADROL — 1,1-diphenyl-1-(2-piperidyl)-methanol

11. SPA — (–)-1-dimethylamino-1, 2-diphenylethane

The salts of the substances listed in this Schedule whenever the existence of such salts is possible.**

Notes

* The names printed in capitals in the left-hand column are the International Non-Proprietary Names (INN). With one exception [(+)-LYSERGIDE], other non-proprietary or trival names are given only where no INN has yet been proposed.

** *Note by the Secretariat:* The Commission on Narcotic Drugs decided through a vote by correspondence, pursuant to its decision 6 (XXVII) of 24 February 1977, to include this sentence at the end of each Schedule.

AMENDMENT/REV. 4
Ref. : Sales No. : E. 78. XI. 3
1 March 1989
Original : English

CONVENTION ON PSYCHOTROPIC SUBSTANCES

Revised Schedules including all amendments made by the Commission on Narcotic Drugs in force as of 12 October 1988

Amendment

Replace the Schedules by the following text.

LIST OF SUBSTANCES IN THE SCHEDULES

List of Substances in Schedule I

International nonproprietary name	*Other nonproprietary or trivial names*	*Chemical Name*
1. BROLAMFETAMINE	DOB	(±)-4-bromo-2, 5-dimethoxy-α -methylphenethylamine
2. CATHINONE		(—)-(S)-2-aminopropiophenone
3.	DET	3-[2-(diethylamino) ethyl] indole
4.	DMA	(±)-2, 5-dimethoxy- α-methyl-phenethylamine
5.	DMHP	3-(1, 2- dimethylheptyl)-7, 8, 9, 10-tetrahydro-6, 6, 9-trimethyl-6 *H*-dibenzo [*b, d*] pyran-l-ol
6.	DMT	3-[2-(dimethylamino) ethyl] indole
7.	DOET	(±)-4-ethyl-2, 5-dimethoxy- α-phenethylamine
8. ETICYCLIDINE	PCE	*N*-ethyl-phenylcyclohexylamine
9. (+)-LYSERGIDE	LSD, LSD-25	9, 10-didehydro-*N, N*-diethyl -6, methylergoline-8 *β*-carboxamide
10. TENAMFETAMINE	MDA	α-methyl-3, 4-(methylenedioxy) phenethylamine
11.	mescaline	3, 4, 5-trimethoxyphenethyla-mine
12.	MMDA	2-methoxy- α-methyl-4, 5-(methylenedioxy) phenethylamine
13.	parahexyl	3-hexyl-7, 8, 9, 10-tetrahydro- 6, 6, 9-trimethyl-6*H*-dibenzo [*b,d*] pyran-1-ol
14.	PMA	*p*-methoxy- α-methylphen-ethylam ine
15.	psilocine, psilotsin	3-[2-(dimethylamino) ethyl] indol -4- ol
16. PSILOCYBINE		3-[2-(dimethylamino) ethyl] indol-4-yl dihydrogen phosphate
17. ROLICYCLIDINE	PHP, PCPY	1-(1-phenylcyclohexyl) pyrrolidine
18.	STP, DOM	2, 5-dimethoxy- α, 4-dimethyl-phenethylamine
19.	MDMA	(±)-*N*, α-dimethyl-3, 4-(methylenedioxy) phenethylamine
20. TENOCYCLIDINE	TCP	1-[1-(2-thienyl) cyclohexyl] piperidine
21.		tetrahydrocannabinol, the following isomers and their stereochemical variants
		7, 8, 9, 10-tetrahydro-6, 6, 9-trimethyl 3-pentyl-6*H*-dibenzo [*b, d*) pyran-1-ol
		(9*R*, 10a*R*)- 8, 9, 10, 10a-tetrahydro-6, 6, 9-trimethyl-3-pentyl-6*H*- dibenzo (*b, d*) pyran-l-ol
		(6a*R*, 9*R*, 10a*R*)-6a, 9, 10, 10a-tetrahydro-6, 6, 9-trimethyl-3-pentyl-6*H*-dibenzo [*b, d*) pyran-l-ol
		(6a*R*, 10a*R*)-6a, 7,10, 10a-

			tetrahydro-6,6,9-trimethyl-3-pentyl-6H-dibenzo [b, d] pyran-1-ol
			(6a*R*, 10a*R*)-6a, 7, 8, 10a-tetrahydro-6, 6, 9-trimethyl-3-pentyl-6*H*-dibenzo [*b*, *d*] pyran-1-ol
			6a, 7, 8, 9-tetrahydro-6, 6, 9-trimethyl-3-pentyl-6*H*-dibenzo [*b*, *d*) pyran-1-ol
			(6a*R*, 10a*R*)-6a, 7, 8, 9, 10, 10a-hexahydro-6,6-dimethyl-9-methylene-3-pentyl-6*H*-dibenzo [*b*,*d*] pyran-1-ol
22.	TMA		(±)-3, 4, 5-trimethoxy- α-methylphen-ethylamine

The salts of the substances listed in this Schedule whenever the existence of such salts is possible.

List of Substances in Schedule II

1. AMEFETAMINE	amphetamine	(±)-α-methylphenethylamine
2. DEXAMFETAMINE	dexamphetamine	(+)-α-methylphenethylamine
3. FENETYLINE		7-[2-[(α-methylphenethyl) amino] ethyl] theophylline
4. LEVAMFETAMINE	levamphetamine	(—)-(*R*)-α-methylphenethylamine
5.	levometham-phetamine	(—)-*N*, α dimethylphenethyla mine
6. MECLOQUALONE		3-(*o*-chlorophenyl)-2-methyl-4 (3*H*)-quinazolinone
7. METAMFETAMINE	methamphetamine	(+)-(S)-*N*- α-dimethylphenethylamine
8. METAMFETAMINE RACEMATE	methamphetamine racemate	(±)-*N*, α-dimethylphenethylamine
9. METHAQUALONE		2-methyl-3-*o*-tolyl-4 (3*H*)-quinazolinone
10. METHYLPHENIDATE		Methyl α-phenyl-2-piperidinea- cetate
11. PHENCYCLIDINE	PCP	1-(1-phenylcyclohexyl) piperidine
12. PHENMETRAZINE		3-methyl-2-phenylmorpholine
13. SECOBARBITAL		5-allyl-5-(1-methylbutyl) barbituric acid

The salts of the substances listed in this Schedule whenever the existence of such salts is possible.

List of Substances in Schedule III

International nonproprietary name	*Other nonproprietary or trivial names*	*Chemical name*
1. AMOBARBITAL		5-ethyl-5-isopentylbarbituric acid
2. BUTALBITAL		5-allyl-5-isobutylbartituric acid
3. CATHINE	(+)-norpseudo-ephedrine	(+)-(*R*)- α-[(*R*)-1-aminoethyl] benzyl alcohol
4. CYCLOBARBITAL		5-(1-cyclohexen-1-yl)-5-ethylbar bituric acid
5. GLUTETHIMIDE		2-ethyl-2-phenylglutarimide
6. PENTAZOCINE		(2*R**, 6*R**, 11*R**)-1, 2, 3, 4, 5, 6-hexahydro-6, 11-dimethyl-3-(3-methyl-2-butenyl)-2, 6-methano- 3-benzazocin-8-ol
7. PENTOBARBITAL		5-ethyl-5-(1-methylbutyl) barbituric acid

The salts of the substances listed in this Schedule whenever the existence of such salts is possible.

List of Substances in Schedule IV

1. ALLOBARBITAL		5, 5-diallylbarbituric acid
2. ALPRAZOLAM		8-chloro-1-methyl-6-phenyl-4*H*-*s*-triazolo [4, 3-*a*] [1,4] benzodiazepine
3. AMFEPRAMONE		2-(diethylamino) propiophenone
4. BARBITAL		5, 5-diethylbarbituric acid
5. BENZFETAMINE	benzphetamine	*N*-benzyl-*N*, α-dimethylphenethylamine
6. BROMAZEPAM		7-bromo-1, 3-dihydro-5- (2-pyridyl)-2*H*-1, 4- benzodiazepin-2-one
7. BUTOBARBITAL		5-butyl-5-ethylbarbituric acid
8. CAMAZEPAM		7-chloro-1, 3-dihydro-3-hydroxy-1-methyl-5- phenyl-2 *H*-1, 4- benzo-diazepin-2-one dimethyl-carbamate (ester).
9. CHLORDIAZEPOXIDE		7-chloro-2- (methylamino)-5-phenyl-3*H*-1, 4-benzodiazepine-4-oxide
10. CLOBAZAM		7-chloro-1-methyl-5-phenyl-1*H*-1, 5-benzodiazepine-2, 4 (3*H*, 5*H*)-dione
11. CLONAZEPAM		5-(*o*-chlorophenyl)-1, 3-dihydro-7- nitro-2*H*-1, 4-benzodiazepin-2-one
12. CLORAZEPATE		7-chloro-2, 3-dihydro-2-oxo-5-phenyl-1*H*-1, 4-benzodiazepine- 3-carboxylic acid
13. CLOTIAZEPAM		5-(*o*-chlorophenyl)-7-ethyl-1, 3-dihydro-1-methyl-2*H*-thieno [2, 3-*e*]-1, 4-diazepin-2-one

International nonproprietary name	Other nonproprietary or trivial names	Chemical name
14. CLOXAZOLAM		10-chloro-11b-(*o*-chlorophenyl)- 2, 3, 7, 11b-tetrahydro oxazolo- [3, 2-*d*] [1, 4] benzodiazepin- 6 (5*H*)-one
15. DELORAZEPAM		7-chloro-5-(*o*-chlorophenyl)-1, 3-dihydro-2*H*-1, 4-benzodiazepin-2- one
16. DIAZEPAM		7-chloro-1, 3-dihydro-1-methyl-5-phenyl-2*H*-1, 4-benzodiazepin-2-one
17. ESTAZOLAM		8-chloro-6-phenyl-4*H*-*s*-triazolo [4, 3-*a*] [1, 4] benzodiazepine
18. ETHCHLORVYNOL		1-chloro-3-ethyl-1-penten-4-yn-3-ol
19. ETHINAMATE		1-ethynylcyclohexanol carba- mate
20. ETHYL LOFLAZEPATE		ethyl 7-chloro-5- (*o*-fluorophenyl)-2, 3-dihydro-2-oxo-1*H*-1, 4-benzodiazepine-3-carboxylate
21. ETILAMFETAMINE	N-ethylamphetamine	*N*-ethyl- α-methylphenethylamine
22. FENCAMFAMIN		*N*-ethyl-3-phenyl-2-norborna- namine
23. FENPROPOREX		(±)-3-[(α-methylphenethyl) amino] propionitrile
24. FLUDIAZEPAM		7-chloro-5-(*o*-fluorophenyl)-1, 3-dihydro-1-methyl-2*H*-1, 4-benzodiazepin-2-one
25. FLUNITRAZEPAM		5-(*o*-fluorophenyl)-1, 3-dihydro-1-methyl-7-nitro-2*H*-1, 4-benzodiazepin-2-one
26. FLURAZEPAM		7-chloro-1-[2-(diethylamino) ethyl]-5- (*o*-fluorophenyl)-1, 3-dihydro-2*H*-1, 4-benzodiazepin-2-one
27. HALAZEPAM		7-chloro-1, 3-dihydro-5-phenyl-1- (2, 2, 2-trifluoroethyl)- 2*H*-1, 4-benzodiazepin-2-one
28. HALOXAZOLAM		10-bromo-11b-(*o*-fluorophenyl)-2, 3, 7, 11b-tetrahydrooxazolo [3, 2-*d*] [1, 4] benzodiazepin-6 (5*H*)-one
29. KETAZOLAM		11-chloro-8, 12b- dihydro-2,8-dimethyl-12b-phenyl-4*H*- [1, 3] oxazino [3, 2-*d*] [1, 4] benzodiazepine-4, 7 (6*H*)-dione
30. LEFETAMINE	SPA	(—), *N,N*-dimethyl-1, 2-diphenyl-ethylamine

International nonproprietary name	Other nonproprietary or trivial names	Chemical name
31. LOPRAZOLAM		6-(*o*-chlorophenyl)-2, 4-dihydro-2-[(4-methyl-1-piperazinyl) methylene]-8-nitro-1 *H*-imidazo [1, 2-*a*] [1, 4] benzodiazepin-1-one
32. LORAZEPAM		7-chloro-5-(*o*-chlorophenyl)-1, 3-dihydro-3-hydroxy-2*H*-1, 4-benzodiazepin-2-one
33. LORMETAZEPAM		7-chloro-5-(*o*-chlorophenyl)-1, 3-dihydro-3-hydroxy-1-methyl-2*H*-1, 4-benzodiazepin-2-one
34. MAZINDOL		5-(*p*-chlorophenyl)-2, 5-dihydro-3*H*-imidazo [2, 1-*a*] isoindol-5-ol
35. MEDAZEPAM		7-chloro-2, 3-dihydro-1-methyl-5-phenyl-1*H*-1, 4-benezodia- zepine
36. MEFENOREX		*N*-(3-chloropropyl)-α-methylphenethylamine
37. MEPROBAMATE		2-methyl-2-propyl-1, 3-propanediol, dicarbamate
38. METHYLPHENOBARBITAL		5-ethyl-1-methyl-5-phenylbar- bituric acid
39. METHYPRYLON		3, 3-diethyl-5-methyl-2, 4-piperidinedione
40. NIMETAZEPAM		1, 3-dihydro-1-methyl-7-nitro-5-phenyl-2*H*-1, 4-benzodiazepin -2-one
41. NITRAZEPAM		1, 3-dihydro-7-nitro-5-phenyl-2*H*-1, 4-benzodiazepin-2-one
42. NORDAZEPAM		7-chloro-1,3-dihydro-5-phenyl-2*H*-1, 4-benzodiazepin-2-one
43. OXAZEPAM		7-chloro-1, 3-dihydro-3-hydroxy-5- phenyl-2*H*-1, 4-benzodiazepin-2-one
44. OXAZOLAM		10-chloro-2, 3, 7, 11b-tetrahydro-2-methyl-11b-phenyloxazolo [3, 2-*d*] [1, 4] benzodiazepin-6 (5*H*)-one
45. PHENDIMETRAZINE		(+)-(2*S*, 3*S*)-3, 4-dimethyl-2-phenylmorpholine
46. PHENOBARBITAL		5-ethyl-5-phenylbarbituric acid
47. PHENTERMINE		α, α-dimethylphenethylamine
48. PINAZEPAM		7-chloro-1, 3-dihydro-5-phenyl-1- (2-propynyl)-2*H*-1, 4-benzodiazepin-2-one
49. PIPRADROL		α, α-diphenyl-2-piperidinemethanol

International nonproprietary name	Other nonproprietary or trivial names	Chemical name
50. PRAZEPAM		7-chloro-1-(cyclopropylmethyl)-1, 3-dihydro-5-phenyl-2*H*-1, 4-benzodiazepin-2-one
51. PROPYLHEXIDRINE		*N*, α-dimethylcyclohexane ethylamine
52. PYROVALERONE		4'-methyl-2-(1-pyrrolidinyl) valerophenone
53. SECBUTABARBITAL		5-*sec*-butyl-5-ethylbarbituric acid
54. TEMAZEPAM		7-chloro-1, 3-dihydro-3-hydroxy-1-methyl-5-phenyl-2*H*-1, 4-benzodiazepin-2-one
55. TETRAZEPAM		7-chloro-5-(1-cyclohexen-1-yl)-1, 3-dihydro-1-methyl-2*H*-1, 4-benzodiazepin-2-one
56. TRIAZOLAM		8-chloro-6-(*o*-chlorophenyl)-1-methyl-4*H*-*s*-triazolo [4, 3*a*] [1, 4] benzodiazepine
57. VINYLBITAL		5-(1-methylbutyl)-5-vinylbarbituric acid

The salts of the substances listed in this Schedule whenever the existence of such salts is possible.

Appendix-3

TABULAR SUMMARY OF LEGISLATION

Information on legislation is presented by WHO region, for each of which the following tables are given:

1. Control of tobacco advertising and promotion.
2. Health warnings and statement and control of tar and nicotine (T/N) content.
3. Restrictions on places of sale.
4. Restrictions on smoking in public places.
5. Restrictions on smoking in the workplace.
6. Preventing young people from smoking.
7. Health education on tobacco.
8. Legislation establishing a national organization for policy development and coordination.

A. African Region

Table A 1. Control of tobacco advertising and promotion

Country	*Total ban*	*Restrictions on the use of mass media (television, radio, press, billboards)*	*Restrictions on the promotion of tobacco through sponsorship of sports and cultural events*	*Restrictions on the content, format, or location of advertising*
Algeria	X			
Burkina FASO		X		
Gambia[1]		X		
Mozambique	X			
Nigeria			X	W
Senegal		X		X
Sudan[3]	X		X	

* Text of legislation not available to author.

W = health warning required.

Notes to Table A1

1. **Gambia**. The Tobacco Products (Control of Advertisements) Act, 1985 prohibits the advertising of tobacco products on television and radio and by photographs and films, as well as by the display or distribution of notices, signs, posters, models, placards, labels, or showcards. Advertising on billboards and by lotteries or games is banned.
2. **Nigeria**. The Tobacco Smoking (Control) Decree 1990 provides that a health warning is required on tobacco advertising in newspapers, magazines, radio, television, cinemas, billboards, and handbills. Sponsorship or promotion of sports events by the tobacco industry is prohibited.
3. **Senegal**. Law No. 85-23 of 25 February, 1985 prohibits the advertising of tobacco products on television. Where the promotion or advertising of tobacco products is authorized, it may be effected only under conditions specified by decree. Under Decree No. 85-1375 of 28 December, 1985, advertising announcements may not employ the voice of or depict a person under 21 years of age, and films advertising tobacco may not be projected during programmes for young persons.

Table A 2. Health warnings and statement and control of tar and nicotine (T/N) Content

Country	*Familiar warning*	*Rotating/strong warnings*	*Smokeless tobacco warnings*	*Statements of T/N content*	*Control of T/N content*
Algeria	X			X	
Burkina Faso	X			X	
Ghana*	X				
Kenya	X				
Mauritius	X				
Nigeria		X		X	
Senegal	X			X	

* Text of legislation not available to author.

Notes to Table A 2

1. Nigeria. The Tobacco Smoking (Control) Decree 1990 requires the following rotating warnings on cigarette packages and on advertising :

   The Federal Ministry of Health warns that tobacco smoking is dangerous to health.

   Smokers are liable to die young.

   The tar and nicotine content of each unit of the product must be stated on the package.

Table A 3. Restrictions on places of sale

Country	*Government buildings*	*Health establishments*	*Schools and places where young people congregate*	*Other*
No information available				

Note : Restriction on the use of vending machines are presented in Table A-6.

Table A 4. Restrictions on smoking in public places

Country	*Government agencies*	*Health establishments*	*Schools & places where young people congregate*	*Public transport*	*Lifts*	*Indoor public places*	*Markets and food storage places*	*Restaurants*
Algeria								
Benin*		X		X				
Botswana*		X	X	X				
Burkina Faso		X	X					
Cape Verde*		X	X	X				
Chad*		X	X					
Central African Republic*		X	X					
Cote d'Ivoire		X	X	X	X	X		
Kenya*								
Lesotho*		X	X	X				
Madagascar*			X	X				
Mauritius		X	X		X	X	X	
Mozambique*								
Niger*		X	X	X				
Nigeria	X	X	X	X	X	X		
(Benue State)		X						

Country	Government agencies	Health establishments	Schools & places where young people congregate	Public transport	Lifts	Indoor public places	Markets and food storage places	Restaurants
Senegal								
Togo*		X	X	X				
Uganda*		X	X	X				
United Republic of Tanzania*			X					

* Text of legislation not available to author.

Notes to Table A4

1. **Algeria** (Law No. 85-05 of 16 February, 1985) and Senegal (Law No. 85-23 of 25 February, 1985) provide that the public places in which smoking is prohibited are to be specified by administrative regulations.

2. **Mauritius**. Under Regulations effective 1 July, 1990, smoking is prohibited in any ward or room earmarked for treatment or examination of patients, including a casualty room for patients in a public hospital, dispensary, clinic, or other health institution; in any classroom, library, and student's workshop in a preprimary, primary, secondary, or tertiary educational institution; and in any sports hall, gymnasium and swimming pool to which the public has access. No person may smoke while engaged in the preparation, serving, or sale of food in a public place.

3. **Nigeria**. The Tobacco Smoking (Control) Decree 1990 bans smoking in a wide variety of public places; cinemas, theatres, and stadiums; offices; public transportation; lifts; medical establishments; schools, and nursery institutions.

Table A5. Restrictions on smoking in the workplace

Country	*General laws restricting smoking in the workplace*	*Government agencies*	*Health establishments*	*Schools*	*Authorization for administrative regulations or accommodation of smokers and non-smokers*	*Other*
Algeria					X	
Mauritius[1]						X
Senegal					X	

Note to Table A5

1. **Mauritius**. The Public Health (Prohibition of Smoking) Regulations of 1990 prohibit smoking while engaged in the preparation, serving, or sale of foodstuffs in a public place.

Table A6. Preventing young people from smoking

Country sales to minors frequented by young people	*Prohibiting from vending machines young people*	*Restricting sales in schools and other places*	*Prohibiting sales advertising that influences*	*Restricting*	*Other*
Mauritius	X				
Senegal				X	

Table A 7. Health Education on Tobacco

Country	*Public education*	*Anti-smoking education in schools*	*Allocation of funds*
Algeria	X		

Table A8. Legislation establishing a national organization for policy development and coordination to control tobacco use

Country	*National organization*
Kenya	Kenya Action on Smoking and Health (KASH) (voluntary Organization)

B. Region of the Americas[*]

Table B1. Control of tobacco advertising and promotion

Country or territory	*Total ban*	*Restrictions on the use of mass media (television, radio, press, billboards)*	*Restrictions on the promotion of tobacco through sponsorship of sports and cultural events*	*Restrictions on the content, format, or location of advertising*
Argentina		X	X	XW
Bahamas				W
Bermuda				W
Bolivia		X	X	XW
Brazil		X	X	XW
Canada	X	X	X	W
Cayman Islands		X		
Chile		X		W
Colombia		X		XW
Costa Rica[7]		X		X
Cuba	X			
Ecuador		X		W
El Salvador		X		
Mexico		X		XW
Panama		X		W
Paraguay		X		X
Peru		X		W
Trinidad and Tobago		X		XW
United States of America		X		W
Uruguay		X		W
Venezuela		X		W

W = health warning required.

[*]Minor discrepancies between the tables given here and those in the 1992 report of the Surgeon General (US Department of Health and Human Services, 1992) are due mainly to differences in availability of information.

Notes to Table B1

1. **Argentina.** Under Law No. 23344 of 31 July, 1986, advertising on television and radio is prohibited from 08 : 00 to 22 : 00 except that the brand name can be stated. Advertising directed to young people is prohibited and in theatres and cinemas to which persons under 18 are admitted. The distribution and promotion of samples at colleges and universities is prohibited. Advertising associated with the practice of sports is prohibited. Featuring personalities from artistic and cultural circles whose public consists essentially of young persons is prohibited. The use of expressions or terms characteristic of young persons is prohibited. Young persons may not be shown as models in tobacco advertising, and the representation of persons smoking excessively is prohibited. Low-tar and low-nicotine cigarettes may not be represented as beneficial to health.

2. **Bolivia.** Under the Regulation of 15 March, 1982, advertising of tobacco is restricted to the format of "tombstone" advertising in which only the name, brand, symbol, tar and nicotine contents are given. Only products directly associated with smoking may be depicted. The content of advertising is restricted to statements concerning the quality, origin, and purity of tobacco products. Persons inhaling or exhaling cigarette smoke, adolescents, pregnant women, and children may not be shown. Tobacco advertisements may not be associated with sporting, domestic, or occupational activities. Labels and advertisements for tobacco must be licensed by the Health Authority to prevent indiscriminate promotion of tobacco comsumption.

3. **Brazil.** Under Order No. 490 of 25 August, 1988, advertising of tobacco products is permitted on television only between 21 : 00 and 06 : 00. Advertising in theatres before 20 : 00 is prohibited if persons under 18 are admitted. Tobacco advertisements must meet certain requirements; the advertisement must not incite excessive or irresponsible consumption; it must not allude to health, holidays, or say that tobacco has soothing

properties; it must not associate tobacco products with sexuality, virility, or femininity; reference to children and adolescents is prohibited; and tobacco advertising may not be addressed to young people. The size and frequency of health warnings are prescribed. Announcement of sponsorship of events by tobacco companies is limited to the presentation of the name and logo of the company, and such announcement is not permitted as part of the programme of the event.

4. **Canada.** The Tobacco Products Control Act of 1988 prohibits the advertising of any tobacco product offered for sale in Canada. Until 1 January, 1991, a tobacco manufacturer or importer may advertise the product by signs, subject to a limitation on the amount to be expended on their preparation and presentation. The Act limits the amount of a tobacco manufacturer's or importer's contributions to cultural or sporting activities or events where brand names of tobacco products are used to the value of contributions to such events in 1987. Regulations under the Act specify the health warnings that must appear on signs, the number and size of signs at a retailer's place of business, and on vending machines.

   The Act is coming into force in stages. Print media advertising was prohibited as of 1989, signs as of January 1991, and point of sale advertising will be prohibited in 1993.

   Legislation restricting sponsorship prohibits merely the use of brand names but not corporate name sponsorship. The cigarette companies have created limited companies named after the brand names to circumvent the ban on sponsorship.

5. **Chile.** Under Decree No. 106 of 8 April, 1981, a health warning is required on advertisements for tobacco products in print media, on television and radio, and in cinemas, where the warning must remain on the screen for at least 5 seconds after the advertisement is shown. No direct or indirect reference to minors may be made, and young people may not be depicted in tobacco advertising.

6. **Colombia.** Tobacco advertising is restricted to the presentation of brand, quality, price and system of marketing. Depiction of minors and the act of smoking is prohibited. Advertising of tobacco is allowed on television only after 23 : 00 and is limited to 30 seconds for each brand. All commercials advertising cigarettes must devote 20% of transmission time to the warning "tobacco is harmful to health". The Council of the District of Bogota prohibits tobacco advertising at children's sports and in scientific publications; on murals, posters, or signs in sports, cultural, educational, or residential places; and in public transportation vehicles.

7. **Costa Rica.** All advertising or promotion of cigarettes in newspapers, radio, television, and cinemas must be approved by the Ministry of Health so as to avoid publicity detrimental to the public health. Advertising of tobacco is prohibited on radio and television programmes intended for children.

8. **Ecuador.** Tobacco advertising directed at or referring to minors is prohibited. Also prohibited are broadcasting tobacco advertisements before 19 : 30, inserting such advertisements in programmes for children, using minors as models, advertising near schools and colleges or in comics, depicting sports figures or people who have contributed to the literature and history of Ecuador.

9. **El Salvador.** Tobacco advertising on radio, television, and in cinemas in programmes directed to children is prohibited, but advertising in programmes not directed to children is allowed.

10. **Mexico.** The legislation prohibits advertising of tobacco that asserts that the product enhances social prestige, or induces consumption by asserting that it is a sedative or reduces fatigue or tension, or by attributing to it stimulant qualities leading to success, or induces persons to consume the product on health grounds, or associates tobacco with sports, domestic, or professional activities, or emphasizes femininity or virility, or suggests greater success in sexual relations, or depicts children or adolescents, or attributes an effect of well-being, or depicts persons smoking in public. Tobacco advertising is limited to information on the characteristics, quality, and techniques of preparation of these products.

11. **Panama.** All advertising of tobacco must be approved by the Ministry of Health. Advertising that shows people smoking is prohibited.

12. **Paraguay.** Under Law No. 836/80 of 15 December, 1980, tobacco advertisements may refer only to the quality and origin of the tobacco and must not encourage consumption. The use of figures or characters representing children or adolescents is prohibited, as is the association of tobacco with sports, work, study, or home. No televised promotion of tobacco may occur before 19 : 00 except for presentations of international meetings, whether produced locally or abroad.

13. **Peru.** Advertising of cigarettes is prohibited before 20 : 00 on radio and television and in places of entertainment during shows suitable for minors.

14. **Trinidad and Tobago.** The Bureau of Standards, established under the Standards Act, No. 38 of 1972, has enunciated standards based on the Code of Advertising Practice, 1969 of the Advertisers' Association of Trinidad and Tobago, which was developed in cooperation with the Advertising Standards Authority and other agencies. These standards require the traditional health warning on tobacco advertisements; permit advertisements for free samples only in the trade press; prohibit the inclusion of coupons or trading stamps in cigarette packages; and forbid the directing of tobacco advertisements and promotion at audiences including children.

15. **United States of America.** Federal legislation prohibits the advertising of cigarettes, little cigars, and smokeless tobacco on television and radio. Health warnings are required on print advertising and on billboard advertising.

16. **Uruguay.** Legible health warnings are required on written tobacco advertising, and they must remain on a screen long enough to be read. Oral advertising must refer to the health warning once for every five references to tobacco products. No promotion of tobacco products, direct or indirect, may be undertaken in schools or other educational institutions, whether public or private. Legislation proposed in 1988 would lay down standards for the advertising of tobacco products, including the prohibition of directing advertising to young people, the depiction of young people and of low-tar and low-nicotine cigarettes as beneficial to health, and the association of smoking with sports, physical strength, social prestige, virility, or femininity; advertising would be restricted to objective facts on the characteristics of the product, its price, and quality.

17. **Venezuela.** Under Decree No. 849 of 21 November, 1980 and Decree No. 996 of 19 March, 1981, all advertising of tobacco products by television or radio that directly or indirectly encourages the consumption of cigarettes and tobacco products is prohibited.

Table B2. Health warnings and statement and control of tar and nicotine (T/N) contents

Country or territory	*Familiar warning*	*Rotating/strong warnings*	*Smokeless tobacco warnings*	*Statements of T/N content*	*Control of T/N content*
Argentina	X				
Bahamas		X			
Barbados	X				
Bermuda		X		X	
Bolivia	X				
Brazil	X				
Canada		X	X	X	
Chile		X			
Colombia	X				
Costa Rica		X			
Ecuador	X			X	
El Salvador	X				
Mexico	X			X	
Panama	X				
Paraguay	X				
Peru	X				
Trinidad and Tobago	X				
United States of America		X	X	(X)	
Uruguay	X			X	
Venezuela	X				

(X) = statements appear on some packages, by decision of the tobacco company.

Table B3. Restrictions on places of sale

Country	*Government buildings*	*Health establishments*	*Schools and places where young people congregate*	*Other*
Brazil				
Rio Grande do Sul	X			
Chile		X		
Cuba		X		X

Note: Restrictions on the use of vending machines are presented in Table B6.

Notes to Table B3

1. **Brazil.** Under Law No. 7813 of 21 September, 1983, the State of Rio Grande do Sul prohibits the sale of cigarettes in any establishment that is administered or subsidized by the state, including buildings of the Legislative Assembly.

2. **Chile.** The Ministry of Health prohibits the sale of tobacco products in all establishments of the Chilean National Health Service. The Ministry of the Interior in 1989 recommended restrictions on the sale of tobacco products in kiosks and other places in all government services.

Table B4. Restrictions on smoking in public places

Country or territory	*Government agencies*	*Health establishments*	*Schools and places where young people congregate*	*Public transport*	*Lifts*	*Indoor public places*	*Markets and food storage places*	*Restaurants*
Argentina								
Province of Cordoba	X		X	X		X	X	
Province of Jujuy	X	X	X	X	X			
Province of Mendoza	X	X	X	X	X			
San Fernando del Valle de Catamarca	X			X	X			
Valle Viejo	X			X	X			
Municipality of Buenos Aires			X	X		X	X	
Bolivia		X	X	X	X	X		
City of Cochabamba				X		X		
City of Oruro				X		X		
Brazil	X	X				X		
Rio Grande do Sul		X	X	X	X	X		
Sao Paulo		X	X	X	X		X	
Brazil (continued)								
Rio de Janeiro		X	X	X	X	X	X	
Porto Alegre			X	X	X	X		X
Curitiba						X		
Florianopolis						X		
Canada	X	X	X	X	X	X	X	
Province of Manitoba		X	X	X		X		X
Province of Quebec	X	X	X	X		X		
Chile[3]	X	X	X	X	X			
Colombia[4]								
Bogota	X	X	X	X	X	X		
Costa Rica[5]	X			X		X		
Cuba[6]		X		X				
Guatemala[7]	X			X				
Honduras[8]	X	X	X	X		X		
Mexico[9]		X						
Federal District	X	X	X	X	X	X		
Panama[10]				X				
Paraguay[11]		X						
Asuncion				X				
Peru[12]	X	X						
Trinidad and Tobago[13]				X				
United States of America	X	X	X	X	X	X	X	X
Uruguay[14]	X	X	X					
Montevideo		X	X			X		
Venezuela[15]	X	X	X	X		X		

Notes to Table B4

1. **Bolivia.** Under the Regulations of 15 March, 1982, smoking is prohibited in schools, health premises, indoor public places, and public transport vehicles. Separate smoking areas are to be provided in indoor public places.

2. **Brazil.** On 31 May, 1990, the Ministry of Health of Brazil adopted Regulation No. 731 prohibiting smoking in any health institution, public or private. It recommends that the states, the Federal District, and the municipalities adopt measures restricting smoking in public premises, public transportation vehicles, elevators, auditoriums, cinemas and theatres, libraries, and premises for use by the public. Smoking is prohibited on all flights of two hours or less, and on flights exceeding two hours, space must be reserved for smokers in the rear of the aircraft.

   In 1988, Order No. 490 of 25 August, 1988 recommended that Federal, State, and municipal levels of government adopt or encourage limitations on smoking in enclosed public places that lack adequate ventilation.

   In 1980, the Secretary of State for Health and the Environment prohibited cigarette smoking in places where services are provided to the public in health units, hospitals, and other agencies of the Secretariat.

   At the State level, Rio Grande do Sul (Law No. 7813 of 21 September, 1983) prohibits smoking in public educational establishments, halls used for meetings, entertainment events and lectures, and museums and libraries; in public health establishments, gymnasia or other closed premises used for sports activities, either maintained or subsidized by the State; and intercity passenger transport vehicles.

   In Sao Paulo, smoking is prohibited on intercity buses, in schools, hospitals, health centres, and other local public health buildings. Both the smoker and the person in charge of the facility are subject to fines for violation of the legislation. Since July 1990, all restaurants of more than 100 square metres must reserve 50% of the space for non-smokers.

   In the State of Rio de Janeiro smoking is prohibited in meetings of the Federal Council on Medicine.

   At the municipal level, the city of Rio de Janeiro has enacted legislation prohibiting smoking on buses, in elevators, cinemas and theatres, stores and supermarkets, hospitals and health services, museums, schools, garages, and taxis.

   Porto Alegre prohibits smoking in businesses, cinemas, theatres, schools, elevators, buses, and places where explosives or flammable materials are processed or stored.

   The municipalities of Curitiba and Florianopolis prohibit smoking in enclosed public places and businesses.

3. **Chile.** The Ministry of Education in 1981 issued Circular No. 601/81 of 11 May, 1981 requiring professors, whenever possible, to refrain from smoking in class and while complying with their obligations to students. The head of the institution is responsible for the enforcement of this regulation.

   In 1981, the Ministry of Health prohibited smoking by staff in the premises of the National Health System, in patients' waiting rooms, administrative offices, elevators, auditoriums, and meeting rooms.

   Acting on the recommendation of WHO, the Minister of the Interior recommends that smoking be prohibited in the waiting rooms, offices, anterooms, and places of public service in government organisations.

   Law No. 18290 of 1985 prohibits smoking in public transport vehicles.

4. **Colombia.** The District of Bogota prohibits smoking in cinemas, theatres, public libraries, museums, and other buildings to which the public is admitted or that are devoted to cultural or sports activities; in buses and taxis; in enclosed buildings and schools; in enclosed areas of hospitals, sanatoria, and health centres; and in government offices where the public is served.

5. **Costa Rica.** Smoking by employees and visitors is prohibited in government buildings, with the exception of persons incarcerated in buildings of the National Penitentiary System; however in each public institution a smoking area is to be provided.

   Smoking is also prohibited in places for public entertainment, cinemas and theatres throughout the country. The owners or managers of these facilities are responsible for enforcement.

   Under Decree No. 18248-MOPT-S of 23 June, 1988, smoking is also prohibited in all means of public transport. Drivers are responsible for enforcement and may refuse to continue the service and seek help from the authorities.

6. **Cuba.** Under Ministerial Resolution No. 165 of 17 August, 1981, smoking by the staff, patients, and visitors of the National Health System is prohibited. Another Ministerial Resolution of 1981 prohibits smoking on all means of public transport.

7. **Guatemala.** Government Accord No. 681 of 3 August, 1990 prohibits smoking in public transport and in public areas in government and private offices.

8. **Honduras.** The Law of the Honduran Institute for the Prevention of Alcoholism and Drug Addiction, Decree No. 136-89, dated 14 October, 1989, prohibits smoking in public and private schools, cinemas and theatres, collective ground, air, and sea transport, public and private hospitals, government offices and workplaces, sports centres, and sessions of the National Congress.

9. **Mexico.** A Decree of the Secretary of Health of 17 April, 1990 restricts smoking in the medical facilities of the Secretary of Health and the National Institutes of Health, including areas for preventive, curative, and rehabilitative care, auditoria and places for group meetings, lectures, teaching, and other areas.

   In the Federal District of Mexico, a Regulation for the protection of non-smokers dated 5 July, 1990 restricts smoking in a wide range of indoor public places, including public transport vehicles, kindergartens, primary, secondary, and high schools, hospitals, clinics, and waiting rooms of health facilities, libraries, cinemas, theatres, auditoriums, government offices, and shops and business providing a service to the public, such as automobile service shops, banks, and financial, industrial, and commercial offices.

10. **Panama.** A Decree of 1978 prohibits smoking in buses.

11. **Paraguay.** A Resolution of the Ministry of Public Health and Social Welfare, issued on 23 January, 1990, prohibits smoking in clinics and waiting rooms as well as in other offices and buildings of the Ministry.

    In Asuncion, the Municipal Council has prohibited smoking in vehicles of the public transport system.

12. **Peru.** A Ministerial Resolution prohibits smoking in buildings and offices of the Ministry of Health and its decentralized agencies. The text of the resolution must be posted at the entrance to, and in visible places in the buildings and offices of the Ministry of Health. Managers and staff are required to ensure strict compliance with the ban on smoking.

13. **Trinidad and Tobago.** Although there is no national legislation restricting smoking on aircraft, British West Indian Airway, in compliance with regulations of the International Airline Transport Association, prohibits smoking on flights of less than 1½ hours.

14. **Uruguay.** Since 1976 the Faculty of Medicine of the University of the Republic of Uruguay has prohibited smoking by physicians, students, staff, patients, and visitors in hospitals of the Faculty.

    Special Order of the Ministry of Public Health No. 3904 prohibits smoking in all Ministry hospitals. This ban applies to administrative personnel while on duty and in contact with patients and their visitors.

    Smoking is prohibited in plenary sessions and working committee sessions of the Chamber of Deputies.

    Legislation proposed on 16 June, 1987 would prohibit smoking in public offices; health centres, public and private schools, and public transport.

    Municipal legislation in Montevideo prohibits smoking in theatres, cinemas, circuses, and in all other places where public performances are presented, although a 1979 Decree permits the sale of cigarettes in theatres. Smoking is also prohibited by personnel of companies engaged in the storage, sale, and transportation of flammable liquids; in storage places of microcontainers of "supergas", in storage places of cylinders and equipment for respiratory therapy; and in storage places of bulk liquid petroleum gas.

    Montevideo also prohibits smoking in city buses and on short trips (less than 110 kilometres) of interdepartmental buses. A 1975 Decree prohibits smoking by drivers of school buses. On long-distance lines, including national and international tourism buses, smoking is permitted in the last three rows of seats.

15. **Venezuela.** Regulations of 1979 under the Law of 13 September, 1978 ban smoking in public transportation vehicles, in buildings where groups of people gather, such as waiting rooms in theatres and cinemas, hospitals and other health facilities, sports arenas, and other places that may be designated. Smoking areas may be set aside in these facilities. No-smoking signs must be posted. Managers of these public places are responsible for compliance with the ban on smoking.

    In 1985, the Venezuelan Institute of Social Security prohibited smoking in all the administrative service units of the Institute, and the Ministry of Education prohibited all smoking in school buildings.

Table B5. Restrictions on smoking in the workplace

Country	*General laws restricting smoking in the workplace*	*Government agencies*	*Health establishments*	*Schools*	*Authorization for administrative regulations or accommodation of smokers and non-smokers*	*Other*
Brazil[1]		X	X			
Rio Grande do Sul		X	X	X		
Canada[2]		X	X	X	X	
Province of Ontario		X	X	X		
Chile[1]			X			
Costa Rica[3]	X					
Cuba[1]			X			
Mexico[1]			X			
Paraguay[1]			X			
Peru[1]		X	X			
United States of America[4]		X	X	X	X	
Uruguay[1]			X			
Venezuela[1]		X	X	X		

Notes to Table B5

1. In Brazil, Chile, Cuba, Mexico, Peru, Uruguay, and Venezuela, smoking by employees in health premises is banned.

2. **Canada.** The Non-smokers' Health Act of 1990, as amended, restricts smoking in all workplaces under federal jurisdiction. The Smoking in the Workplace Act of 26 July, 1989 of the Province of Ontario restricts smoking in all workplaces under provincial jurisdiction and establishes minimal standards that limit exposure to tobacco smoke in the workplace. In addition to provincial legislation, municipal by-laws regulate smoking at the workplace.

3. **Costa Rica.** The Decree of 23 June imposes a general ban on smoking in workplaces and requires the director of the workplace, or his representative, to ensure compliance with this provision. An area may be reserved at the workplace for smokers, to the extent that this is possible.

4. **United States of America.** Regulatory polices of federal agencies generally restrict smoking at the workplace. At the state level, 31 states restrict smoking at public worksites, and 13 have restricted smoking at private worksites. Local ordinances provide for accommodation of the interests of smokers and non-smokers.

Table B6. Preventing young people from smoking

Country	*Prohibiting sales to minors*	*Restricting sales from vending machines*	*Prohibiting sales in schools and other places frequented by young people*	*Restricting advertising that influences young people*	*Others*
Agrentina[1]			X	X	X
Bahamas					X
Bolivia[2]			X	X	X
Brazil[3]	X		X	X	X
Canada	X	X	X	X	X
Chile[4]			X	X	X
Colombia[5]	X		X	X	
Costa Rica[6]	X		X		
Cuba	X			X	
Ecuador[7]	X			X	X
El Salvador[8]				X	X
Mexico[9]	X				X
Panama[10]				X	X
Paraguay[11]				X	X
Peru[12]				X	X
Trinidad and Tobago				X	X
United States of America	X	X	X	X	
Uruguay[13]	X			X	X
Venezuela[14]				X	X

Notes to Table B6

1. **Argentina.** Law No. 23344 of 31 July, 1986 prohibits tobacco advertising on radio and television from 08.00 to 22.00 except that the name of the brand may be presented. Tobacco advertising is prohibited in publications intended for young people and in theatres and cinemas to which persons under 18 are admitted. Distribution and promotion of samples of cigarettes at colleges and universities is prohibited. Young people may not be used as models in advertisements for tobacco. Advertising directed at young people or associated with sports is prohibited.

2. **Bolivia.** The Regulations of 1982 ban smoking in schools as "premeditated harm" because it exposes persons of low resistance to the polluting effects of tobacco and because minors are susceptible to a bad example. Tobacco advertising must not depict children or adolescents, nor may it associate tobacco with sports. In 1984, the Minister of Education and Culture prohibited students, professors, and parents of students from smoking in public and private educational institutions throughout the country.

3. **Brazil.** The order of 25 August lays down the requirements that advertising of tobacco must meet, including a specification that no reference to children may be made and that advertising must not be addressed to them. No tobacco advertising may be presented in theatres before 20:00 if persons under 18 may attend. Advertising on television is allowed only betwen 21:00 and 06:00. An order of the Ministry of Health of 1990 prohibits the sale of cigarettes to minors and promotion of tobacco products through the distribution of free samples at public events. The municipalities of Rio de Janeiro, Sao Paulo, and Porto Alegre prohibit smoking in schools.

4. **Chile.** Any presentation of tobacco advertising on radio or television before 21:30 is prohibited. Young people may not be depicted in tobacco advertising. Circular No. 601/81 of the Ministry of Education of 11 May, 1981 prohibits smoking

in schools and by teachers during classes.

5. **Colombia.** Sales to minors under 14 are prohibited, as is smoking in schools.

6. **Costa Rica.** The Decree of 4 February, 1983 prohibits the sale of cigarettes to minors in all commercial establishments. It is the responsibility of the administators or managers of the establishments to assure compliance with the decree. Violators will be sentenced under the General Health Law, which provides for a penalty of 5—30 days in jail.

7. **Ecuador.** The distribution of samples of cigarettes to minors is prohibited, as is tobacco advertising aimed at children or referring to them. Tobacco advertisements may not be presented on television before 19:30 or included in programmes intended exclusively for children. Tobacco may not be advertised in or near schools, on school buses, in sports centres, or in comic books. Sports stars and young artists may not be depicted using or smoking cigarettes in posters, films, or on record albums. A similar ban applies to the use of historical figures and members of the learned professions. Encouraging smoking as a means of improving concentration or performance is prohibited.

8. **El Salvador.** Tobacco advertising is permitted on radio, television, and in cinemas, but only in programmes not intended for children.

9. **Mexico.** The General Law on Health of 1983 sets forth the objectives of the Programme against Smoking, including the education of the family, children, and adolescents about the effects of tobacco on health through individual methods and mass communication. The Law contains no specific ban on advertising directed at children.

10. **Panama.** All advertising of tobacco must be approved by the Ministry of Health. Such advertising may not depict persons smoking.

11. **Paraguay.** Tobacco advertising that depicts persons representing children or adolescents or that associates tobacco with sports is prohibited.

12. **Peru.** Cigarette advertising may be presented on radio and television only after 20:00. It is an offence to present tobacco advertising in performances suitable for minors before 19:00.

13. **Uruguay.** The sale of cigarettes, cigars, and tobacco products to minors under 18 years of age is prohibited. The sale of single or loose cigarettes is prohibited. Advertising of cigarettes is allowed on radio and television only after 21:00. It is the responsibility of television stations to aviod guests smoking on programmes between 06:00 and 24:00.

14. **Venezuela.** Stating that tobacco advertising leads to intemperate use of tobacco, especially by young people. Decrees of 1980 and 1981 prohibit television and radio broadcasting, respectively, of commercial advertising that leads to the use of cigarettes and tobacco products. Violation of these decrees is punishable by suspension or revocation of the broadcasting permit.

Table B7. Health education on tabacco

Country	*Public education*	*Anti-smoking education in schools Argentina*	*Allocation of funds*
San Fernando del Valle de Catamarca	X		
Bolivia	X	X	
Brazil	X		
Sao Paulo	X	X	
Canada	X	X	
Chile	X	X	
X			
Colombia	X		
Costa Rica	X		
Cuba	X		
Ecuador	X		
El Salvador	X	X	
Mexico	X	X	
Peru			
United States of America	X	X	
Uruguay	X	X	

Notes to Table B7

1. **Bolivia.** The Regulations of 1982 require the Health Authority (the Ministry of Social Welfare and Public Health) to draw up mass education programmes to counter the harmful effects of tobacco and to supervise the use of the media for tobacco advertising. An Intersectoral Council for health training and education, created by joint action of the Ministries of Social Welfare and Public Health and of Education and Culture, is charged with analysing the educational programmes, including compulsory anti-smoking education, for systematic and programmed teaching of health education.

2. **Brazil.** The Law of 11 June, 1986 provides for a National Anti-Smoking Day to be held each year on 29 August, with a national campaign in the preceding week to alert the people to the dangers of tobacco.

3. **Chile.** The National Commission for the Control of Smoking established by a 1986 decree, is charged with designing and evaluating a programme for smoking control in the medium and long term in the fields of education, information, regulation, and evaluation. The Commission is required to identify resources in the public and private sectors for informational, educational, and cessation activities. The functions of the Intersectoral Commission for Primary Prevention of Alcoholism in Schools, established in 1980, have been expanded to cover prevention of the use of drugs and tobacco. In 1984, the Decree on the Advisory Joint Commission of Education was modified to strengthen joint activities of the Ministries of Health and Education and their constituent bodies and to increase support to structures at the local level.

4. **Colombia.** A decree enacted in 1986 provides for educational programmes and preventive campaigns on tobacco use.

5. **Costa Rica.** A 1988 decree urges that campaigns and activities should be undertaken to mark No-Smoking Day, established by WHO, with emphasis on the injury to health caused by smoking.

6. **Cuba.** The Ministerial Resolution No. 165 of 17 August, 1981 requires the staff of the National Health System to take advantage of all opportunities to provide information on the harmfulness of tobacco and to persuade the people of this.

7. **Ecuador.** The 1979 Constitution of Ecuador recognizes the right to welfare of the population, which includes the protection of health and requires programmes aimed at eliminating alcoholism and other addictions.

8. **El Salvador.** The Decree of 11 May 1988, requires the Ministry of Public Health and Social Welfare to draw up programmes on the effects of the consumption of drugs and tobacco and to encourage cultural and sports activity to prevent such consumption.

9. **Mexico.** The General Law on Health of 1983 lays down the objectives of the Programme against Smoking, which include education of the family, children, and adolescents through individual methods and mass communications, with emphasis on the education of the family to prevent tobacco use by children and adolescents. Coordination agreements between the Secretariat for Health and Welfare and the States provide for implementing smoking-control programmes in higher educational institutions and for preventing smoking by children and adolescents.

10. **Peru.** Under the Law of 20 October, 1982, a tax increase of 9% on cigarettes made from blond tobacco is allocated to the construction and equipping of a cancer hospital, to the development of cancer detection centres throughout the country, and to research on neoplastic diseases.

11. **Uruguay.** Concerned about the increase in smoking among young people of both sexes, the Ministry of Public Health, with the active participation of the Ministry of Education, has organized No-Tobacco Day, involving educational councils at the primary, secondary, and teacher training levels. Legislation proposed in 1988 would authorize a Commission for the Control for Smoking to coordinate educational programmes on tobacco with the National Administration of Public Education, the University of the Republic, and other educational organizations.

Table B8. Legislation establishing a national organization for policy development and coordination to control tobacco use

Country	*National organization*
Bolivia	Ministry of Health and Public Welfare
Brazil	Ministry of Health
Canada	Department of National Health and Welfare
Chile	National Commission for the Control of Smoking
Ecuador	Interinstitutional Committee against Smoking
Mexico	National Council against Addictions, with support from the Secretariat for Health and Welfare in coordination with the Governments of the various States
Panama	National Commission to Study Tobacco Use in Panama
Peru	Permanent National Commission against Tobacco
United States of America	Office on Smoking and Health, Centers for Disease Control, US Department of Health and Human Services
Uruguay	Ministry of Health acting in collaboration with the Ministry of Education and Culture
Venezuela	National Council of the Ministry of Health and Welfare, composed of representatives of the Ministry of Health and the Ministries of Agriculture, Labour, Transportation and Communications, Justice, Environment and Natural Resources, Information and Tourism, Youth Affairs, the Venezuelan Institute of Social Secu-

rity, the National Academy of Medicine, the Cancer Society, and the Venezuelan Medical Federation.

C. Eastern Mediterranean Region

Table Cl. Control of tobacco advertising and promotion

Country	*Total ban*	*Restrictions on the use of mass media (television, radio, press, billboards)*	*Restrictions on the promotion of tobacco through sponsorship of sports and cultural events*	*Restrictions on the content, format, or location of advertising*
Afghanistan*	X			
Bahrain		X		X
Cyprus				W
Egypt		X		W
Iraq	X			
Jordan	X			
Kuwait		X		
Lebanon				W
Morocco		X		
Pakistan		X		
Saudi Arabia		X		
Sudan		X		

* Text of legislation not available to author.

W = health warning required.

Notes to Table C1

1. **Bahrain.** By order of the Minister of Information, smoking is banned in television serials, plays, and other theatrical presentations.
2. **Egypt.** The advertising of tobacco products on radio and television is prohibited.
3. **Iraq**. All advertising of tobacco is banned.
4. **Pakistan.** In 1977, the advertising of tobacco products and smoking in schools and public places were banned.
5. **Saudi Arabia.** By order of the Minister of Information, smoking in television serials, plays, and other theatrical presentations is banned.
6. **Sudan.** The distribution and use of any form of poster, picture, or publication for the advertising of cigarettes is prohibited.

Table C2. Health warnings and statement and control of tar and nicotine (T/N) content

Country	*Familiar warning*	*Rotating/ strong warnings*	*Smokeless tobacco warnings*	*Statements of T/N content*	*Control of T/N content*
Bahrain		X		X	X
Cyprus	X				
Egypt	X			X	X
Iraq	X				X
Jordan	X			X	
Kuwait	X	X		X	X
Lebanon	X				
Oman		X		X	X
Pakistan	X				
Qatar		X		X	X
Saudi Arabia	X		*		
Sudan	X				
United Arab Emirates		X		X	X

* Total ban.

Notes to Table C2

1. **Arab Gulf States.** By Resolution 24 of the Health Ministers of the Arab Gulf States, the warning in all these States is: "Health warning: Smoking is a leading cause of lung cancer and of pulmonary and cardiovascular diseases."
2. **Kuwait.** The warning "Smoking is a leading cause of lung cancer and of pulmonary and cardiovascular disease" was required by a Ministerial Resolution in 1980. In 1988, a Ministerial Decree modified the warning and reduced the maximum premitted levels of tar and nicotine in cigarettes offered for sale on the Kuwaiti market.
3. **Saudi Arabia.** The Central Laboratory in Riyadh analyses samples of cigarettes sold on the market in the Arab Gulf States to determine their nicotine and tar content.

Table C3. Restrictions on places of sale

Country	*Government buildings*	*Health establishments*	*Schools and places where young people congregate*	*Other*
Iraq				X

Note : Restrictions on the use of vending machines are presented in Table C6

Table C4. Restrictions on smoking in public places

Country	*Government agencies*	*Health establishments*	*Schools and places where young people congregate*	*Public transport*	*Lifts*	*Indoor public places*	*Markets and food storage places*	*Restaurants*
Bahrain		X	X					
Cyprus[1]		X	X	X	X	X	X	X
Egypt	X			X		X		
Alexandria[2]						X		
Jordan				X		X		
Lebanon*								
Libyan Arab Jamahiriya*[3]								
Morocco*[4]								
Oman*[5]		X	X					
Pakistan[6]			X			X		
Saudi Arabia[7]	X	X	X	X				
Sudan				X	X			

* Text of legislation not available to author.

Notes to Table C4

1. Cyrus. Under the Regulations of 27 May 1988, the following health warning is required to be posted in cafes and restaurants. "CAUTION. Smoking is seriously harmful to your health : Ministry of Health."
2. Egypt. Under Law No. 52 of 20 June 1981, smoking is prohibited in public transport vehicles and enclosed public places defined by the Minister. The Alexandria Municipal Ordinance of 20 July 1904 prohibited smoking in theaters. It was amended on 29 December 1938 and by Law No. 372 of 29 October 1956 and again in 1964 to ban smoking in public places and advertising.
3. Libyan Arab Jamahiriya. An administrative order of 1989 imposed a ban on smoking in designated public places.
4. Morocco. Legislative measures to ban smoking in public places are being reviewed by the National Parliament.
5. Oman. A decree was issued banning smoking in hospitals and health institutions except in designated places.
6. Pakistan In 1977, the Govenment banned smoking in schools and public places.
7. Saudi Arabia. Smoking is prohibited in ministerial offices, government agencies and public institutions. Domestic flights by the national airline are smoke-free.

Table C5. Restrictions on smoking in the work place

Country	*General laws restricting smoking in the workplace*	*Government agencies*	*Health establishments*	*Schools*	*Authorization for administrative regulations or of smokers and non-smokers*	*Other*
Saudi Arabia[1]						*X*

Note to Take C5

1. Saudi Arabia Smoking in public institutions is prohibited.

Note to Table C3

1. In conformity with resolution 24 adopted at the Eighth Meeting of the Health Ministers of the Arab States of the Gulf Area, held in Baghdad on 20-26 January 1980, Iraq requested airlines not to sell cigarettes and tobacco on board their aircraft.

Notes to Table C4

1. **Cyprus.** Under the Regulations of 27 May, 1988, the following health warning is required to be posted in cafes and restaurants. "CAUTION. Smoking is seriously harmful to your health: Ministry of Health."
2. **Egypt.** Under Law No. 52 of 20 June, 1981, smoking is prohibited in public transport vehicles and enclosed public places defined by the Minister of Health. The Alexandria Municipal Ordinance of 20 July, 1904 prohibited smoking in theatres. It was amended on 29 December, 1938 and by Law No. 372 of 29 October, 1956 and again in 1964 to ban smoking in public places and advertising.
3. **Libyan Arab Jamahiriya.** An administrative order of 1989 imposed a ban on smoking in designated public places.
4. **Morocco.** Legislative measures to ban smoking in public places are being reviewed by the National Parliament.
5. **Oman.** A decree was issued banning smoking in hospitals and health institutions except in designated places.
6. **Pakistan.** In 1977, the Government banned smoking in schools and public places.
7. **Saudi Arabia.** Smoking is prohibited in ministerial offices, government agencies, and public institutions. Domestic flights by the national airline are smoke-free.

Note to Table C5

1. **Saudi Arabia.** Smoking in public institutions is prohibited.

Table C6. Preventing young people from smoking

Country	*Prohibiting sales to minors*	*Restricting sales from vending machines*	*Prohibiting sales in schools and other places frequented by young people*	*Restricting advertising that influences young people*	*Other*
Bahrain		X			
Cyprus[1]		X			

Note to Table C6

1. **Cyprus.** The sale of cigarettes from vending machines is prohibited under Law No. 51 of 1980.

Table C7. Health education on tobacco

Country	*Public education*	*Anti-smoking education in schools*	*Allocation of funds*
Bahrain	X		
Iraq	X	X	
Saudi Arabia	X	X	

Table C8. Legislation establishing a national organization for policy development and coordination to control tobacco use

Country	*National organization*
Bahrain	Anti-Smoking Society, chaired by a laywer
Saudi Arabia	Smoking Control Society, chaired by the Minister of Health.

D. European Region

Table D1. Control of tobacco advertising and promotion

Country	*Total ban*	*Restrictions on the use of mass media (television, radio, press, billboards*	*Restrictions on the promotion of tobacco through sponsorship of sports and cultural events*	*Restrictions on the content, format, or location of advertising*
Austria		X		
Belgium		X		XW
Bulgaria	X			
Czechoslovakia	X			
Denmark		X		
Finland	X		X	
France	X		X	
Germany		X		
Greece		X		
Hungary	X			
Iceland	X		X	
Ireland		X	X	XW
Italy	X			
Luxembourg[9]		X	X	X

Malta		X		W
Netherlands		X		
Norway	X		X	
Poland	X			
Portugal	X		X	XW
Romania	X			
Spain		X		
Sweden		X		XW
Former USSR	X	X		
United Kindgom		(X)	(X)	(W)
Yugoslavia	X			
European Communities		X		

W = health warning required.

(X) = provisions laid down by voluntary agreement.

Notes to Table D1

1. **Belgium.** The Crown Order of 10 April, 1990 strengthens the controls on tobacco advertising specified in the Crown Order of 20 December, 1982. Under these two orders, advertising of tobacco products on radio and television and in newspapers, reviews, and other publications aimed mainly at minors is prohibited, as is advertising:

   — by means of aircraft, boats or vehicles, except for those taking part in competitions or being used to transport tobacco, tobacco products, or similar products;

   — by using films, videotapes, slides, or other types of visual presentations;

   — by the distribution or door-to-door delivery of stickers or promotional leaflets;

   — by means of public speeches promoting tobacco products;

   — by means of free samples of tobacco, tobacco products, and other products in any form whatsoever, even under the pretext of carrying out market research, or a public opinion or consumer survey;

   — by using the brand name or symbol of tobacco, tobacco products, or similar products, or by using any other image or reference usually associated with such products, on, or close to everyday objects or by any other means other than those which are part of the personal equipment of participants in sporting competitions or of the accompanying technical team;

   — by referring to, or mentioning or representing the act of smoking or the representation of everyday products or articles linked with tobacco use;

   — by using illuminated signs, except within or at the entrance to places where tobacco, tobacco products, or similar products are available for sale.

   Where tobacco advertising is permitted, its contents are limited to the name and address of the manufacturer, importer, or distributor, the name of the product, its brand name and symbol, representations of products directly associated with smoking, tar and nicotine levels, and information on price and quality.

2. **Czechoslovakia.** On 28 March, 1989, the Czech National Council passed Law No. 37/89 on protection against alcoholism and other addictions, which prohibits publicity, promotion, and advertising of smoking in the press and other information media. Inconsistent with this total ban on tobacco advertising is the Law of 30 October, 1991 on radio and television broadcasting, adopted by the Czech and Slovak Federal Republic, which provides that advertising of tobacco products must indicate the tar and nicotine content and contain the health warning that smoking is harmful. A new Tobacco Act is in preparation.

3. **Denmark.** There has long been a voluntary agreement with the tobacco industry. Legislation was enacted in 1988 prohibiting the advertising on radio and television of tobacco products or goods primarily used in connection with tobacco smoking.

4. **Finland.** One of the first countries to enact a total ban on tobacco advertising, its Law No. 693 of 13 August, 1976 banned advertising of tobacco, tobacco products and imitations, and smokers' accessories, and other sales promotion activities directed at the consumer, as well as their association with advertising for other products or services, or other sales promotion activity.

5. **France.** From 1976 to 1990, French legislation authorized tobacco advertising only in the "tombstone" format, in which only the name of the product, the package, and the lettering may be shown. Advertising of products using the name or emblem of tobacco products was required to conform to the same restrictions as those applying to tobacco products. Sponsorship of sports events by the tobacco industry was prohibited, with the

exception of motor car racing.

In January 1991, the National Assembly repealed the 1976 law and enacted a new statute, effective 1 January, 1993, banning all advertising, direct and indirect, of tobacco and tobacco products, all free distribution of such products, and all sponsorship of tobacco that has the aim or purpose of advertising tobacco or tobacco products directly or indirectly. Indirect advertising of tobacco includes all advertising for an agency, service, activity, product or article other than tobacco or a tobacco product, when that advertising, on account of its logo, presentation, use of a trade mark, symbol, or any other distinctive sign, is suggestive of tobacco or a tobacco product.

Until 31 December, 1992 all advertising of tobacco and tobacco products, allowed under transitional provisions, must carry the health warning "Abuse is dangerous".

6. **Germany.** The use of advertising material is prohibited if it is likely to have an effect on physical performance or well-being, to induce adolescents to smoke, to make it appear that inhaling tobacco smoke is something to be imitated, or if it suggests that tobacco products are pure.

7. **Iceland.** A total ban on all advertising and promotion of tobacco products has been in force since 1971. It is reported that, overall, the ban has been closely observed, even though indirect advertising occurs on a small scale.

8. **Ireland.** The Tobacco (Health Promotion and Protection) Act, 1988, restricts the media that may be used for tobacco advertising. Advertising is allowed only in the print media and at points of sale. The content of advertisements is restricted to the brand name, emblem, representation of the package, and health warning. Rotating health warnings on packages and advertisements are compulsory. Expenditures on advertising and sponsorship are limited to the actual expenditures for such activities in 1985, with an increase each year to allow for inflation. Regulations under the Act restrict sponsorship of events by the tobacco industry. An event which was not sponsored prior to 1 May, 1996 cannot now be sponsored. Events or activities in which the participants are aged 18 years or under cannot be sponsored. Sponsored events can be advertised only in newspapers and magazines, at the place where the event is taking place, and internally in premises which are points of retail sale. Such events cannot be advertised in comics or other publications aimed primarily at persons 18 years or under or along a public roadway. An advertisement relating to a sponsored event may contain only the name of the event, a written description of the event, the name of the sponsor, and the corporate emblem. An advertisement about a sponsored event may not appear alongside an advertisement for a tobacco product. Participants at a sponsored event or their assistants are not permitted to display the sponsor's emblem, brand name, or symbol of a tobacco product on themselves or their equipment, such as racing cars or golf bags. The restrictions relating to participants at sponsored events were designed to defeat the industry's objective of using tobacco advertisements at such events in order to circumvent the ban on television advertising of tobacco products.

9. **Luxembourg.** The Law of 24 March, 1989 prohibits advertising of tobacco products on television, radio, in theatres and other places open to the public, by means of boats and aircraft, by distribution of free samples of tobacco products, by using the emblem or name of the brand of tobacco products on products other than those directly related to the use of tobacco, in magazines and periodicals designed to reach minors, in stadiums and public and private sports arenas (although the name of the product on a billboard or vehicle not surrounded by text or graphic representation is allowed). The content of advertising is restricted; it may not be addressed to minors; it may not make use of arguments based on health; and it may not contain any representation of public figures. The tobacco industry is prohibited from sponsoring events for children or minors, and the organizers of such events may not accept their patronage.

10. **Norway.** All forms of tobacco promotion are prohibited. Tobacco products include cigarette paper, cigarette rollers, and pipes. Tobacco products may not be included in advertisements for other products, nor may a person smoking be depicted in advertisements of other products. The name or logo of a tobacco company may not be used in the sponsorship of sports or cultural events. Advertising low-price tobacco products is forbidden. Inside a sales outlet, tobacco products may be displayed only to the extent that this is necessary to serve customers in a rational way; special displays with product markers or penants are forbidden. Window displays of tobacco products are forbidden. The secondary use of tobacco names, pictures of tobacco products, etc., on such things as playing cards, matchboxes, ashtrays, napkins, wardrobe tags, and sunshades is forbidden. The ban on tobacco advertising applies to pamphlets, catalogues,

Table D2. Health warnings and statement and control of tar and nicotine (T/N) content

Country	*Familiar warning*	*Rotating/strong warnings*	*Smokeless tobacco warnings*	*Statements of T/N content*	*Control of T/N content*
Austria	X			X	
Belgium	X		X	X	X
Bulgaria	X				
Czechoslovakia	X				
Denmark		X		X	X
Finland		X	X	X	X
France	X			X	
Germany	X		X	X	X
Greece	X				
Hungary	X				
Iceland		X	X		
Ireland			*		X
Israel			*		
Luxembourg		X	X	X	X
Malta		X	*	X	X
Netherlands		X		X	X
Norway		X	X	X	X
Portugal	X			X	X
Spain		X		X	X
Sweden		X	X	X	
Switzerland	X			X	
Former USSR	X				X
United Kingdom		(X)	*		(X)
European Communities		X		X	X

(X) = provisions laid down by voluntary agreement.

* Total ban.

samples distributed to the consumer or other groups, whether sent through the mail, distributed at trade fairs, in shops, or in public or private places. Selling or supplying imitations for tobacco products (candy cigarettes) to persons under 16 years of age is forbidden. In order to restrict promotion of tobacco in 1989 Norway adopted regulations (No. 1044 of 13 October, 1989) prohibiting the importation, sale, and supply of new forms of products containing tobacco or nicotine. The Ministry of Social Affairs and the Directorate of Health are authorized to demand that manufacturers, importers or dealers submit representative samples of the product, or institute such investigations as are necessary to evaluate the product's characteristics and effects. The manufacturer or importer shall bear the cost of such investigations. The above-mentioned authorities are authorized to have free access to buildings, means of transport, warehouses, installation areas, etc., where products governed by these regulations are to be found.

11. **Poland.** Tobacco advertising on television and radio is banned. The Ordinance of 4 June, 1974 also provides for a reduction in smoking on television programmes and an increase in popular scientific broadcasts on the harmful effects of smoking.

12. **Portugal.** Advertising is prohibited in the national media and by apposition of the name, brand, or symbol of a tobacco product to consumer goods that are not directly concerned with the use of tobacco.

13. **Spain.** Under Crown Decree No. 709/1982, informational advertising of new tobacco products with low tar or nicotine contents is permitted for two years following their introduction on the market. Legislation introduced in the Spanish Parliament would ban all tobacco advertising, except on radio from 22:00 to 08:00.

14. **Sweden.** Tobacco advertising is restricted to the "tombstone" format. Under the 1979 Guidelines, advertising of tobacco products in the sports pages of daily newspapers, in sports newspapers, and in publications aimed at people under 20 years of age is prohibited. Direct mail advertising and billboard advertising of tobacco products are prohibited. There is no commercial advertising on Swedish radio or television. Advertising in films, in hospitals, and in places frequented by young people is prohibited. No human figures or beautiful scenery may be shown.

    In March 1990, the Committee for Tobacco Control, appointed by the Minister of Health and Social Security recommended the adoption of a comprehensive tobacco act, which would impose a total ban on tobacco advertising.

15. **Former USSR.** The laws in the states of the former Soviet Union are in flux. In Lithuania, for example, a draft tobacco control act would regulate cigarette production, sales, and advertising, provide for increased taxes and prices, allocate tobacco tax revenue for health education, and ban smoking in schools, health premises, and other public places.

16. **United Kingdom.** By voluntary agreement with the industry, all advertisements in newspapers and magazines, on posters and cinema screens, in leaflets and brochures, in consumer catalogues and circular letters must carry a reference to the government warning that appears on cigarette packets. Certain restrictions on the content of advertisements are also imposed by voluntary agreement, but much more than "tombstone" advertising is permitted. On 24 March, 1986 the voluntary agreement banned advertising of cigarettes in cinemas.

17. **European Communities.** A Directive of 10 March, 1989 prohibits tobacco advertising on television. A directive on advertising on radio is under consideration. On 12 February, 1992, the European Parliament voted to support a total ban on tobacco advertising. The Council of Health Ministers will make the final decision.

Notes to Table D2

1. **Belgium.** The Crown Order of 19 January, 1990 limits the tar and nicotine content of cigarettes. As of 31 December, 1992, the maximum tar yield permitted is 15 mg, and the maximum nicotine yield is 1.5 mg. These limits will be lowered to 12 mg for tar and 1.2 mg for nicotine on 31 December, 1997.

2. **Denmark.** On 28 June, 1990, the Ministry of Health issued Order No. 507 requiring all packages of tobacco products to carry the general warning, "Extremely harmful to health. National Board of Health" and also a specific warning selected from a specified list. The contents of tar and nicotine, measured according to the ISO methods indicated, must be stated on the package. Law No. 426 of 13 June, 1990 on the labelling of tobacco products and the tar content of cigarettes authorizes the Ministry of Health to issue provisions to implement Directives of the European Communities on the labelling of tobacco products and the tar content of cigarettes and to supervise and monitor compliance with these requirements.

3. **Finland.** In 1984, Decree No. 770, the following maximum permissible levels of harmful substances were laid down for factory-manufactured cigarettes: tar (condensates)—15 mg (previously 18 mg) per cigarette smoked; carbon monoxide 12mg (previously 14 mg) per cigarette smoked; nicotine—1.2 mg (previously 1.3 mg) per cigarette smoked.

4. **France.** The additives (flavouring agents, texture agents, preservatives, and colouring matters) permitted to be added in the manufacture of tobacco products and the permitted levels and purity criteria for such additives are specified in an order of 16 July, 1984. Under the Law of 10 January, 1991, the full composition of every package of cigarettes (except filters) and the average tar and nicotine content must be indicated on the cigarette package.

5. **Germany.** The substances that may be used in tobacco products are subject to regulation.

6. **Iceland.** Under Regulations No. 499 of 28 December, 1984, rotating warnings with illustrative designs are required. The warning for snuff and chewing tobacco states: "Snuff and chewing tobacco may damage the mucous membranes."

8. **Luxembourg.** The Regulations of 19 June, 1990 under the Law of 24 March, 1989 restricting advertising of tobacco products and prohibiting smoking in certain places provide for rotating health warnings and a maximum tar yield of 15 mg per cigarette, to become effective two years after the publication of the Regulations.

9. **Malta.** Under the Tobacco (Smoking Control) Act, 1986, health warnings must be displayed in premises where tobacco products are sold so as to attract the attention of customers or other persons on the premises and to be easily seen and read by them. Under the 1987 Regulations made under the Act, rotating warnings are required on cigarette packages, in cinemas, and in advertisements.

10. **Norway**. The Norwegian Government has decided to strengthen the 12 rotating warnings adopted in 1984 and to add illustrative designs, as in Iceland.

11. **Spain.** The Crown Decree of 4 March, 1988 requires rotating health warnings and the display of the tar and nicotine contents. As from 31 December, 1992, the maximum tar content for cigarettes is 15 mg. and the maximum nicotine content 1.3 mg. Cigarettes classified as "low nicotine and tar", including "light", "mild", or similar designations, must have, six months after the effective date of the Decree, a tar content not exceeding 13 mg and a nicotine content not exceeding 0.9 mg. After 31 December, 1992, the corresponding figures for such cigarettes are 10 mg and 0.8 mg. respectively.

12. **Sweden.** The legislation requires not only a declaration of the number of milligrams of tar, nicotine, and carbon monoxide but also comparison figures (the arithmetical average for all brands marketed in Sweden) determined each year by the National Board of Health and Welfare. Under an Order of 10 April, 1986, one warning used for snuff and chewing tobacco states that these products contain nicotine and therefore may be as addictive as tobacco.

13. **United Kingdom.** On 24 March, 1986, the Government reached an agreement with the tobacco industry requiring six rotating health warnings on tobacco packages, posters, and press advertisements, as follows:

— Smoking can cause fatal diseases.

— Smoking can cause heart disease.

— Smoking when pregnant can injure your baby and cause premature birth.

— Stopping smoking reduces the risk of serious diseases.

— Smoking can cause lung cancer, bronchitis and other chest diseases.

— More than 30,000 people die each year in the UK from lung cancer.

The sale of oral tobacco to children is prohibited by the Protection of Children (Tobacco) Act 1986. In 1990, the Oral Snuff (Safety) Regulations 1989 banned the sale of moist snuff in the following terms: "goods which consist mainly or wholly of tobacco in fine cut, ground or particulate form or in any combination of those forms and which are for oral use other than smoking." This is interpreted to exclude dry snuff and chewing tobacco. These regulations were overturned because of a legal technicality, and the Government has appealed against this ruling. The European Community, however, is expected to approve amendments to its Directive 89/622/EEC on the labelling of tobacco, which will impose a Community-wide ban on oral moist snuff. This will require some slight amendments to the United Kingdom Regulations and it is expected that the Government may introduce new regulations and drop the appeal against the existing ones.

14. **European Communities.** The Directive of 13 November, 1989 requires all packages of tobacco products to carry the general warning, "Tobacco

seriously damages health", and alternating specific warnings taken from those listed in the annex to the Directive (see p. 49) so as to guarantee the appearance of each warning on an equal quantity of unit packages, with a tolerance of around 5%. The Directive also requires tar and nicotine yields to be indicated on cigarette packages and such yields to be measured and their accuracy verified in accordance with the relevant ISO standards. The directive of 17 May, 1990 sets the maximum tar yield of cigarettes marketed as 15 mg as from 31 December, 1992 and as 12 mg from 31 December, 1997.

Table D3. Restrictions on places of sale

Country	*Government buildings*	*Health establishments*	*Schools and places where young people congregate*	*Other*
Bulgaria			X	
Malta		X		
Netherlands	X	X	X	X
Spain		X	X	
United Kingdom			X	

Note : Restrictions on the use of vending machines are presented in Table D6.

Notes to Table D3

1. **Netherlands.** The sale of tobacco products in health care, social welfare, sports, sociocultural, and educational establishments administered by the State is prohibited under the Tobacco Law of 1988.
2. **Spain.** The sale of tobacco products in health establishments, educational establishments, and establishments primarily intended for the care of children and adolescents is prohibited under the Crown Decree of 1988.

Notes to Table D4

1. **Austria.** Reports indicate that the control of smoking in public places and public transport is effective to some extent, but legislation on the subject is not available to the author.
2. **Belgium.** The Crown Order of 15 May, 1990 strengthens the legislation banning smoking in public places enacted in 1987. The new legislation prohibits smoking in enclosed places used by the public which form part of premises where:
   - services are available to the public, with or without charge, including places serving food and/or drinks for public consumption:
   - ill or elderly people are admitted or treated;
   - health care for prevention or treatment is provided;
   - children or young people or school age are admitted, treated, or are in residence;
   - education and/or professional training are provided;
   - entertainment is offered;
   - exhibitions are mounted;
   - sports are practised.

   An exception is made only for places specially designated for the provision and consumption of food and/or drinks (restaurants, snackbars, cafes, etc.) which do not exceed 30 square metres in area. Clearly defined areas can be set aside for smokers, but they must be established so as to minimize the nuisance of smoke to non-smokers. Until December 1992, these special areas must measure less than two-thirds of the total area of the enclosed premises and, after that date, they cannot exceed half the total area of the enclosed premises. In places where smoking is premitted, a smoke extractor and/or smoke-reducing ventilation system must be installed.
3. **Czechoslovakia.** The legislation provides that, in public services, smoking is permitted only in places specially reserved for the purpose. This provision does not apply to restaurants, but a minimum of one section must be set aside for non-smokers, at least during normal meal hours. Smoking is prohibited in passenger trains (except in smoking compartments), in meetings and conferences in enclosed places, and in all health, school, cultural and enclosed sports premises.
4. **Denmark.** A Circular issued in March 1988 by the Ministry of Health bans smoking in all workplaces under its jurisdiction and in meetings of public councils, boards, and commissions unless all participants agree to permit smoking. The Circular also provides that individual Ministers shall introduce non-smoking environments on public premises under their jurisdiction. A Circular of the Ministry of Social Affairs dated 22 December, 1988 bans smoking on premises frequented by children. Order No.

Table D4. Restrictions on smoking in public places

Country	*Government agencies*	*Health establishments*	*Schools and places where young people congregate*	*Public transport*	*Lifts*	*Indoor public places*	*Markets and food storage places*	*Restaurants*
Austria				X				
Belgium	X	X	X	X	X	X	X	X
Bulgaria		X	X	X		X	X	X
Czechoslovakia		X	X	X		X	X	X
Denmark	X	X	X	X		X	X	
Finland	X	X	X	X	X	X		
France	X	X	X	X	X	X	X	X
Germany		X	X	X		X	X	
Greece		X			X	X		
Hungary		X		X				
Iceland	X	X	X	X	X	X	X	X
Ireland	X	X	X	X		X	X	
Israel		X	X	X	X	X	X	X
Italy		X	X	X		X		
Luxembourg	X	X	X	X		X		
Malta		X	X	X		X	X	
Netherlands	X							
Norway	X	X	X	X	X	X	X	X
Poland	X	X	X	X		X	X	
Portugal		X	X	X		X		X
Romania		X	X	X		X		
Spain	X	X	X	X	X	X	X	
Sweden	X	X	X	X	X	X	X	X
Turkey				X				
Former USSR	X	X	X	X		X		
United Kingdom								
European Communities	X	X	X	X		X		X

121 of the Ministry of the Environment dated 28 March, 1980 regulates tobacco smoking and retail selling of food. A Circular of the veterinary authorities dated October 1974 forbids smoking on premises where food is being prepared for resale.

5. **Finland.** Smoking is prohibited in all public places unless specifically allowed.

6. **France.** Legislation has long prohibited smoking in places intended for use by groups of people if the premises do not meet specified ventilation requirements. Law No. 91—32 of 10 January, 1991 prohibits smoking in places intended for collective use and in particular scholastic use and in collective means of transport, except in areas specifically reserved for smokers. A decree (No. 92-478 of 29 May, 1992) under this Law, adopted by the Council of State in 1992, prohibits smoking in all public places, including businesses, restaurants, schools, workplaces, and public transport, with areas reserved for smokers. Smoking is totally prohibited in theatres, exhibition halls, sports arenas, places where food is prepared or presented for sale, lifts, taxis, aircraft on flights of less than two hours, and dining cars of trains.

7. **Germany.** As in the United Kingdom, there are voluntary restrictions by institutions, government agencies, and transport companies on smoking in public places.

8. **Greece.** The Ministerial Decision of 25 April, 1980 prohibits smoking in all enclosed public places (public waiting-rooms, conference and meeting halls, lifts, etc.) belonging to State agencies, public or private companies and organizations, and in other establishments. Under the Ministerial Decision of 12 April, 1979, smoking is also prohibited in public and private hospitals and private nursing homes; a smoking room, reserved for hospital personnel and visitors, is required on every floor of an establishment that has an area of 200m^2 or more, and smokers may smoke only in these rooms.

9. **Hungary.** MALEV, the Hungarian airline, has prohibited smoking on all its short-distance flights.

10. **Iceland.** Smoking is prohibited under the Law of 28 May, 1984 in specific institutions (primary schools, nurseries, and buildings used for leisure activities for children and adolescents) and in those parts of public and private institutions rendering service to, and accessible to, the public. Restaurants and places of entertainment are required to reserve a certain number of tables for non-smokers. In general, smoking is allowed only in places where it will not inconvenience non-smokers.

11. **Ireland.** Under the Tobacco (Health Promotion and Protection) Act, 1988, the Minister for Health may, by regulations, prohibit or restrict the consumption of tobacco products in a designated area or a designated facility, including aircraft, trains, public service vehicles or other facilities; in such parts of health premises and schools as may be specified; in such parts of buildings to which the public has access and which belong to or are occupied by the State or a body established under an Act of the Oireachtas or in any other area designated by the Minister; and in cinemas, theatres, concert halls and such other places of indoor public entertainment as may be specified. Violation of these regulations is punishable by a fine not exceeding £Ir 100, and the owner, manager or other person in charge of a designated area or a designated facility who fails, neglects, or refuses to ensure compliance is subject to a fine not exceeding £Ir500 and/or imprisonment for six months unless he has taken all reasonable steps to ensure full compliance with the regulations. The Regulations under this Act, issued in 1990, set forth the following designated areas where smoking is prohibited: State and semi-State buildings, public offices, primary and secondary schools (a separate smoking facility may be provided for teachers), universities (specific smoking facilities may be provided), food preparation areas, supermarkets and grocery stores, bus and railway stations, sports centres, cinemas, theatres, concert halls, art galleries, museums, and public libraries. (In the last three areas, smoking facilities may be provided in staff rooms or restaurants.) Smoking is prohibited on all buses used by the public.

In another list of designated areas and facilities, smoking is restricted, and clearly defined no-smoking areas must be provided. These include health premises, hospitals, nursing and maternity homes, centres for the mentally and physically handicapped, restaurants, canteens, snack bars, trains, aircraft and the seating areas in the arrival and departure sections of aerodromes, harbours, ferryports, and airports (in the last two facilities, smoking is prohibited in at least one-third of the seating area).

12. **Israel.** A procedure has been established, under an Order of 1986, for serving "tickets" on violators of the law banning smoking in public places, similar to the tickets served on those committing traffic offences.

13. **Italy.** An exemption to the ban on smoking in

specified premises imposed by the Law of 11 November, 1975 is made under the Ministerial Decree of 18 May, 1976 for premises in which an air-conditioning or ventilation system meeting prescribed conditions has been installed. Failure of such systems to operate within their prescribed limits must cause illuminated "No smoking" signs to switch on automatically.

14. **Luxembourg.** The Law of 24 March, 1989 prohibits smoking in hospitals, waiting rooms of physicians and dentists, pharmacies, educational establishments, places that receive or house persons under 16 years of age, indoor sports halls, cinemas and theatres, museums, art galleries, libraries, and reading rooms open to the public, all halls and rooms of public buildings where a sign prohibits smoking, buses and non-smoking railroad cars, and non-smoking sections of airports.

15. **Netherlands.** The Law of 10 March, 1988 requires the competent authorities to take measures to ensure that establishments, services, and companies administered by the State and public bodies can be used and activities carried on there without any nuisance arising out of the use of tobacco products.

16. **Norway.** Law No. 24 of 6 May, 1988 amending Law No. 14 of 9 March, 1973 restricts smoking in premises and means of transport to which the public has access, and in meeting rooms and work premises where two or more people are present. Under certain circumstances, premises or means of transport may be divided into smoke-free and smoking areas provided that it is not possible for the smoke to pass into the smoke-free area and that the smoke-free premises and areas are not smaller or of lower standard than the premises or areas where smoking is permitted.

    Restaurants, hotels and other establishments that serve food and/or offer overnight accommodation are exempted from these restrictions. In Regulations of 8 July, 1988 issued pursuant to the above-mentioned Law, the Ministry of Social Affairs urges restaurants, hotels, etc., to arrange for smoke-free environments as soon as possible. As of 1993, not less than one-third of the seats in restaurants and of hotel rooms are to be smoke-free. Responsibility for enforcing this provision is shared by the Directorate of Labour Inspection and the Health and Social Affairs Board.

17. **Poland.** Smoking is banned in all organizational units under the jurisdiction of the health and welfare services, and it is recommended that other ministries, central agencies, and local governments should prohibit smoking in cafeterias and canteens for workers and students, at meetings and conferences, and that a separate room should be designated for non-smokers in waiting-rooms and clubs with several rooms. In addition to the Decree of the Ministry of Health and Social Welfare of 4 June, 1974, banning smoking in organizational units of the Department of Health, the Decree of the Minister of Education of 17 October, 1974 deals with reducing smoking at schools; the Decree of the Minister of Transport of 4 December, 1974 concerns the reduction of smoking in the Department to Transport, including means of transport, railway and bus stations, and airports; and the Decision of the Minister of Internal Trade and Services of 6 July, 1974 concerns the reduction of smoking in organizational units of the Department, including catering units, groceries, and trade shops (T. Gorski, personal communication, 1988).

18. **Portugal.** In 1988, smoking was banned in public places and, in 1989, the ban was extended to provide a partial ban on smoking in restaurants and other catering establishments, such as tea rooms, coffee houses, sandwich bars, pubs and cake shops, where food is consumed on the premises. Managers of these establishments are authorized to designate a certain area within their establishments as smoke-free.

19. **Spain.** Smoking is not permitted in welfare centres for persons under age 16, health centres, teaching centres, public administration premises to which the public has access, auditoria, libraries, theatres, cinemas, enclosed premises for sports or entertainment, and enclosed commercial premises where people frequently assemble. If it is possible to delimit separate areas for smokers, such areas may be designated. In enclosed commercial establishments where considerable numbers of people are present, such areas must be designated with particular clarity. Smoking is also prohibited on all urban and long-distance vehicles for collective transport, on school vehicles, and in rail and sea transport, where whole carriages or cabins may be allocated for smokers. Smoking is prohibited in premises where foodstuffs are prepared or sold (Crown Decree 192/1988 of 4 March, 1988). Smoking by food handlers is prohibited by Crown Decree 2505-1983.

20. **Sweden.** In 1983, the General Recommendations (AFS 1983:10) concerning smoking restrictions were issued by the National Board of Occupational Safety and Health and the National Board of Health and Welfare. The main aim of the

Recommendations is that no one should against his will be subjected to discomfort or to health hazards caused by tobacco smoke in public places and workplaces. The intention of the Recommendations is to accommodate both smokers and non-smokers, priority being given to the rights of non-smokers. An Interdepartmental Committee for Tobacco Control appointed by the Minister of Health and Social Affairs recommended in 1990 that all indoor public places (including public transport and workplaces) should, in principle, be made smoke-free. Separate smoking rooms or areas may be designated under certain circumstances. Restaurants with seating for more than 50 people must have smoke-free sections comprising at least half the seats. In hotels and similar premises at least half the number of rooms must be smoke-free.

21. Turkey. As of 25 January, 1988, the Ministry of Transport has banned smoking on domestic flights. On international flights, a section at the front of the aircraft including 60% of the entire seating capacity has been designated for non-smokers; on outward-bound flights, smoking is allowed only in the rear part of the aircraft.

22. Former USSR. In Resolution No. 706 of 12 June, 1980, the Central Committee of the Communist Party and the Council of Ministers called on government agencies to restrict or prohibit smoking in official and work premises. By Regulations of the Ministry of Public Health and the Ministry of Education, smoking is banned in medical institutions and schools, respectively.

23. United Kingdom. There is no national legislation governing smoking in public places, which is controlled either by rules imposed by the owners or operators of the accommodation or transport or, in some cases, by by-laws. Smoking in health institutions and government buildings is regulated by the management of those institutions and buildings.

24. European Communities. The Resolution of 18 July, 1989 of the Council and the Ministers for Health of the Member States invites Member States to ban smoking in enclosed premises open to the public that form part of public or private establishments and to extend the ban to all forms of public transport. Public places include:

- establishments where services are provided to the public, whether for a charge or free, including the sale of goods;
- hospitals and health care establishments;
- establishments where elderly persons are received;
- schools and other premises where children or young people are received or housed;
- establishments where higher education and vocational training are given;
- enclosed establishments used for entertainment (cinemas, theatres) and radio and television studios open to the public;
- enclosed establishments where exhibitions are held;
- establishments and enclosed places where sports are practised;
- enclosed premises of underground and railway stations, ports, and airports.

Notes to Table D5

1. Bulgaria. Under the Sanitary Rules published on 21 January, 1969, smoking at work places where any non-smoker is employed is prohibited, except by written agreement of the non-smoker, and without exception where expectant or nursing mothers are employed, even if they agree.

2. Denmark. By the Circular of 23 March, 1988 of the Ministry of Health, smoking is prohibited in workplaces under the Ministry's jurisdiction that constitute a place of work for more than one person in rooms for general occupation unless the employees having places of work on the same premises declare smoking to be acceptable or if effective measures are adopted to ensure that no employee is inconvenienced. Other ministers are required to introduce non-smoking environments in public premises and means of transport under their jurisdiction.

3. France. A 1992 decree of the Council of State restricts smoking in workplaces to areas designated by the head of the enterprise after consultation with representatives of the workers, but recreation rooms may not be chosen as places where smoking is allowed.

4. Norway. As from 1 July, 1988, a new Section 6 of the Law of 9 March, 1973 introduced by the Law of 6 May, 1988 provides that the air shall be smoke-free in work premises where two or more persons are present.

5. Spain. Crown Decree 192/1988 prohibits smoking in any place where a greater risk to the health of workers exists through the combination of the harm caused by tobacco and the adverse effects of industrial contamination. Smoking is prohibited in any occupational area where pregnant women work. Smoking by food handlers is prohibited by Crown Decree 2505-1983.

6. Sweden. In 1983, detailed General Recommen-

Table D5. Restrictions on smoking in the workplace

Country	General laws restricting smoking in the workplace	Government agencies	Health establishments	Schools	Authorization for administrative regulation or accommodation of smokers and non-smokers	Other
Austria	X					
Belgium	X					
Bulgaria[1]	X					
Czechoslovakia	X					
Denmark[2]		X	X	X	X	
France[3]	X					
Germany	X					
Iceland	X				X	
Ireland	X					
Norway[4]	X					
Poland	X					
Romania	X					
Spain[5]	X					
Sweden[6]	X				X	

Notes to Table D5

1. Bulgaria. Bulgaria. Under the Sanitary Rules published on 21 January 1969, smoking at work places where any non-smoker is employed is prohibited, except by written agreement of the non-smoker, and without exception where expectant or nursing mothers are employed, even if they agree.
2. Denmark. By the Circular of 23 March 1988 of the Ministry of Health, smoking is prohibited in workplaces under the Ministry's jurisdiction that constitute a place of work for more than one person in rooms for general occupation unless the employees having places of work on the same premises declare smoking to be acceptable or if effective measures are adopted to ensure that no employee is inconvenienced. Other ministers are required to introduce non-smoking environments in public premises and means of transport under their jurisdiction.
3. France. A 1992 decree of the Council of State restricts smoking in work places to areas desginated by the head of the enterprise after consultation with representatives of the workers, but recreation rooms may not be chosen as places where smoking is allowed.
4. Norway. As from 1 July 1988, a new Section 6 of the Law of 9 march 1973 introduced by the Law of 6 May 1988 provides that the air shall be smoke free in work premises where two or more persons are present.
5. Spain. Crown Decree 192/1988 prohibits smoking in any place where a greater risk to the health of workers exists through the combination of the harm caused by tobacco and the adverse effects of industrial contamination . Smoking is prohibited in any occupational area where pregnant women work. Smoking by food bandlers is prohibited by Crown Decree 2505-1983.
6. Sweden. In 1983, detailed General Recommendations (AFS 1983:10) concerning smoking restrictions were issued by the National Board of Occupational Safety and Health and the National Board of Health and Welfare. The essential principle is that a smoker's desire to smoke should take second place to other peoples desire for a smoke free environment. Since 1988, 25 country councils have banned smoking on country council premises.

In 1990, an Interdepartmental Committee for Tobacco Control, appointed by the Minister of Health and Social Affairs recommended that all workplaes should in principle, be made smoke free, Separate smoking rooms or areas could be designated under crtain circumstances. In workplaes, smoking would be premited only in special smoking rooms and private offices if the ventilation is sufficient to prevent smoke from spreading into smoke-free rooms.

dations (AFS 1983:10) concerning smoking restrictions were issued by the National Board of Occupational Safety and Health and the National Board of Health and Welfare. The essential principle is that a smoker's desire to smoke should take second place to other people's desire for a smoke-free environment. Since 1988, 25 county councils have banned smoking on county council premises.

In 1990, an Interdepartmental Committee for Tobacco Control, appointed by the Minister of Health and Social Affairs recommended that all workplaces should, in principle, be made smoke-free. Separate smoking rooms or areas could be designated under certain circumstances. In workplaces, smoking would be permitted only in special smoking rooms and private offices if the ventilation is sufficient to prevent smoke from spreading into smoke-free rooms.

Table D6. Preventing young people from smoking

Country	*Prohibiting sales to minors*	*Restricting sales from vending machines*	*Prohibiting sales in schools and other places frequented by young people*	*Restricting advertising that influences young people*	*Other*
Belgium		X		X	
Czechoslovakia[1]	X				
Finland		X	X		
France[2]					X
Hungary					X
Iceland	X	X			
Ireland[3]	X	X		X	
Luxembourg				X	
Malta	X	X			
Norway	X				
Spain[4]	X	X	X		
Sweden[5]				X	
Former USSR	X		X		
United Kingdom	X				

Notes to Table D6

1. **Czechoslovakia.** Law No. 37/89 of the Czech National Council concerning protection against alcoholism and other addictions, dated 28 March, 1989, prohibits the sale of tobacco to persons under 16 years of age.
2. **France.** Law No. 91-32 of 10 January, 1991 on measures to combat tobacco use and alcoholism provides that, on the occasion of medical examinations, required once every two years for staff or public and private educational institutions, including all persons in regular contact with pupils and students, the staff shall be provided with information by the school physician concerning tobacco use and alcoholism.
3. **Ireland.** The Tobacco Products (Control of Advertising. Sponsorshop and Sales Promotion) Regulations, 1986, provide that tobacco products may not be advertised in comics, comic supplements, or any other publications directed primarily to persons under 18 years of age.
4. **Spain.** Under Crown Decree No. 192/1988, the sale of tobacco products is prohibited in health establishments, educational establishments, and establishments primarily intended for the care of children and adolescents. An order of 7 November, 1989 prohibits the sale and distribution of tobacco and alcoholic beverages at public school centres under the authority of the Ministry of Education and Science and directs the provincial departments of the Ministry of Education and Science to adopt the measures necessary to implement this order. Tobacco products may be sold from automatic vending machines only on enclosed premises, and the machine is to display a non-removable health warning of area not less than 20 cm². The sale or supply of tobacco products to persons under 16 years of age is prohibited. Minors under age 16 are prohibited from using automatic vending machines.
5. **Sweden.** Tobacco advertising aimed at people under 20 years of age is restricted. Under the proposed comprehensive tobacco act recommended by the Committee for Tobacco Control, sale of tobacco products to persons under 18 years of age or from vending machines would be prohibited.

Table D7. Health education on tobacco

Country	*Public education*	*Anti-smoking education in schools*	*Allocation of funds*
Belgium	X	X	
Finland	X		X
France		X	
Iceland	X	X	X
Luxembourg		X	
Norway	X		
Portugal	X		
Spain		X	
Valencia	X	X	
Sweden	X	X	
Former USSR		X	

Notes to Table D7

1. **Iceland.** Since 1985, 0.2% of the gross sales of tobacco have been allocated to anti-smoking activities. This amounts to US $ 0.41 annually for each inhabitant.
2. **Sweden.** State-funded information and education on tobacco and health began in 1964. Health education on tobacco is mandatory in all schools. The county councils are responsible for public health education in their regions. The Committee on Tobacco Control has recommended increased funding at (least doubled) for health education on tobacco.

Table D8. Legislation establishing a national organization for policy development and coordination to control tobacco use

Country	*National organization*
Denmark	Council on Prevention and Council on Tobacco-induced Damage to Health
Iceland	Committee for Tobacco Use Prevention, Ministry of Health and Social Security
Malta	Committee on Smoking and Health to advise the Minister of Health on measures, including legislation, to reduce the health risks from tobacco and to coordinate activities for the prevention and control of smoking.
Norway	National Council on Smoking and Health
Portugal	Smoking Prevention Council
Spain	
Valencia	Technical Commission to Reduce Tobacco Consumption, composed of citizens' groups, scientific associations, and institutions
Former USSR	Department of Sanitation and Prevention, Ministry of Public Healh

E. South-East Asia Region

Table E1. Control of tobacco advertising and promotion

Country	*Total ban*	*Restrictions on the use of mass media (television, radio, press, billboards)*	*Restrictions on the promotion of tobacco through sponsorship of sports and cultural events*	*Restrictions on the content, format, or location of advertising*
Bangladesh*		X		W
India		X		W
Nepal		X		
Sri Lanka		X		
Thailand[1]	X			

*Text of legislation not available to author.

W = health warning required.

Notes to Table E1

1. **Thailand.** A resolution of the Thai Cabinet of 26 April, 1988 directed the Ministry of Finance to order the Thailand Tobacco Monopoly to refrain from advertising. Foreign cigarettes continued to be advertised, although they were not imported legally at that time. Accordingly, by cabinet resolution dated 20 December, 1988, a total ban was imposed on all foreign cigarette advertising in all forms, both direct and indirect, in all media. On 13 March, 1989 the Council of Ministers of Social Affairs adopted a resolution directing the Administrative Committee of Radio and Television Broadcasting to control strictly indirect advertising of tobacco in radio and television programmes e.g. the expression of emotion portrayed by the action of smoking.

   On 13 March, 1992 Thailand adopted the Tobacco Products Control Act banning tobacco advertising in the electronic and print media and on billboards. Two exceptions permit advertising in international magazines and on live television shows from abroad. "Advertising" is defined in the Thai law as any act that enables people to see or to know the wording for commercial purposes, such as logos or wording on T-shirts. Advertising of other products with the same name is prohibited. The production, import, advertising, or sale of products that imitate tobacco products, such as candy or savings banks for children, are banned. The free distribution of cigarettes for commercial purposes is prohibited, as are free samples, exchanges, and gifts of tobacco products for commercial purposes.

Table E2. Health warnings and statement and control of tar and nicotine (T/N) content

Country	*Familiar warning*	*Rotating/strong warnings*	*Smokeless tobacco warnings*	*Statements of T/N content*	*Control of T/N content*
Bangladesh	X				
India	X			X	
Indonesia	X				
Sri Lanaka	X				X
Thailand		X		X	

Notes to Table E2

1. **India.** Legislation enacted in 1975 requires a health warning on packages of cigarettes and on advertising. The requirement does not apply to exported cigarettes unless the country to which cigarettes are exported requires the same or a similar health warning.

Table E4. Restrictions on smoking in public places

Country	*Government agencies*	*Health establishments*	*Schools and places where young people congregate*	*Public transport*	*Lifts*	*Indoor public places*	*Markets and food storage places*	*Restaurants*
India[1]	X	X	X	X				
Indonesia[2]		X		X				
Jakarta			X					
Sri Lanka[3]		X		X				
Colombo*						X		
Thailand[4]	X	X		X		X		
Bangkok				X		X		

* Text of legislation not available to author.

2. **Thailand.** A cabinet resolution of 11 July, 1989 requires the printing of six different rotating health warnings on the front of cigarette packages as follows:
   - — Smoking causes lung cancer and emphysema.
   - — Smoking causes ischaemic heart disease.
   - — Smoking is dangerous for babies in the womb.
   - — Please respect the rights of non-smokers by refraining from smoking in public places.
   - — Stopping smoking decreases the risk to health.
   - — For your beloved children and grandchildren, please give up smoking.

   Article 11 of the Tobacco Products Control Act of 13 March, 1992 requires producers and importers of tobacco products to disclose the ingredients of their products in accordance with the standards of the Ministry of Health. The Ministry of Health has the right to determine what should be disclosed. Article 12 provides that the labelling of tobacco products shall be determined by the Ministry of Health.

Table E3. Restrictions on places of sale

Country	*Government buildings*	*Health establishments*	*Schools and places where young people congregate*	*Other*
India[1]	X			
Thailand[2]		X		

Note : Restrictions on the use of vending machines are presented in Table E6.

Note to Table E3

1. **India.** As a component of its ban on smoking in public places, the Government of India in 1990 ordered the removal of all cigarette shops from government buildings or compounds where smoking is prohibited.
2. **Thailand.** The Ministry of Health has directed all trading units in health premises to refrain from selling cigarettes.

Notes to Table E4

1. **India.** On 7 May, 1990, the Government of India issued an Office Memorandum, "Prohibition of tobacco smoking in public places". It was decided to prohibit tobacco smoking, to start with, in a few selected places, namely, hospitals, dispensaries and other health care establishments, educational institutions, conference rooms, domestic flights, air-conditioned chaircars and air-conditioned sleeper coaches in trains, and suburban trains and air-conditioned buses. All domestic flights of Indian Airlines have been made smoke-free.
2. **Indonesia.** The Minister of Internal Affairs has instructed the governors of all provinces to ban smoking in educational institutions.
3. **Sri Lanka.** The Transport Act prohibits smoking in buses. The Ministry of Health prohibits smoking in hospitals and prohibits doctors from smoking in front of patients. The municipality of Colombo prohibits smoking in cinemas
4. **Thailand.** Smoking is prohibited at cabinet meetings and all other meetings in Government House, in the House of Representatives, in all types of air-conditioned public transport, and on flights of less than two hours. All buildings of the Ministry of Health are required to be smoke-free, with provision of designated smoking areas, if necessary. Health personnel must refrain from smoking in public, during work, or outside designated smoking areas. On 10 March, 1989 the Minister of Public Health issued a circular to all Provincial Governors urging the adoption of similar smoke-free policies. The Nonsmokers' Health Protection Act of 13 March, 1992 bans smoking in all public places designated by the Ministry of Health.

Table E5. Restrictions on smoking in the workplace

Country	*General laws restricting smoking in the workplace*	*Government agencies*	*Health establishments*	*Schools*	*Authorization for administrative regulation or accommodation of smokers and non-smokers*	*Other*
No information available						

Table E6. Preventing young people from smoking

Country	*Prohibiting sales to minors*	*Restricting sales from vending machines*	*Prohibiting sales in schools and other places frequented by young people*	*Restricting advertising that influences young people*	*Other*
Thailand[1]	X	X		X	

Note to Table E6

1. **Thailand.** Under its Tobacco Products Control Act of 13 March, 1992, Thailand bans sales of tobacco products to persons under 18 and bans sales of cigarettes from vending machines.

Table E7. Health education on tobacco

Country	*Public education*	*Anti-smoking education in schools*	*Allocation of funds*
Indonesia	X		
Thailand	X		

Table E8. Legislation establishing a national organization for policy development and coordination to control tobacco use

Country	*National organization*
Sri Lanka	Sri Lanka National Federation on Smoking and Health
Thailand	National Committee for the Control of Tobacco Use

F. Western Pacific Region

Table F1. Control of tobacco advertising and promotion

Country or territory	*Total ban*	*Restrictions on the use of mass media (television, radio, press, billboards)*	*Restrictions on the promotion of tobacco through sponsorship of sports and cultural events*	*Restrictions on the content, format, or location of advertising*
Australia[1]		X	X	
Australian Capital Territory		X	X	
South Australia			X	W
Victoria		X	X	
Western Australia	X		X	
China[2]	X			
French Polynesia	X		X	
Hong Kong		X	X	XW
Macao		X		X
Malaysia		X		X
New Zealand	X		X	
Papua New Guinea	X			
Singapore	X			

W = health warning required.

Notes to Table F1

1. **Australia.** The Broadcasting and Television Act 1972 imposed restrictions on tobacco advertising on the electronic media. Because of a loophole in the law, television stations have been permitted to show billboards, banners, and other paraphernalia that advertise tobacco products during sporting events. Legislation has been introduced in the Australian Parliament to ban such incidental television advertising of tobacco. On 28 December, 1989, under the Smoking and Tobacco Products Advertisements (Prohibition) Act, a total ban was introduced on tobacco advertising in the print media (newspapers and magazines). On 1 April, 1992 the Federal Cabinet resolved to phase out tobacco advertising and sponsorship throughout the country. This resolution was passed by the Parliament in August, 1992, and complements the legislation of the various states and territories.

   The Australian Capital Territory established in 1989 the ACT Health Promotion Fund with an additional 3% levy on the wholesale price of tobacco products. The income from this fund is expected to be A$900 000 a year, and it will be used in grants and sponsorships for events formerly sponsored by the tobacco industry.

   The Tobacco Products (Licensing) Act 1986 of the State of South Australia provides for a licence fee to be paid by tobacco merchants as a contribution to State revenues in place of a consumption licence to be taken out by a consumer of tobaccoo products. The Tobacco Products Control Amendment Act 1988 provides for an increase in the tobacco tax to replace industry sponsorship of sports events.

   In the State of Victoria, the Tobacco Act 1987 provides for the establishment of the Victorian Health Promotion Foundation, created by an increase in the tobacco tax, to replace industry sponsorship of sports events and to support health promotion and research.

   In the State of Western Australia, the Tobacco Control Act 1990 phases in a ban on all tobacco advertising and promotion, prohibits sponsorship by a tobacco manufacturer or distributor promoting a tobacco product or brand name, and establishes the Western Australian Health Promotion Foundation to fund health promotion activities and to offer an alternative source of funds for sporting and arts activities currently supported by the tobacco industry.

2. **China.** On 29 July, 1991, the Standing Committee of the 7th Plenary Session of the National People's Congress. China's highest legislative body, announced approval of China's first tobacco law, effective 1 January, 1992. The Law regulates many aspects of the State national tobacco monopoly—growth, distribution, licensing, manufacture and sales, import and export of tobacco products and machinery, measures against smuggling, etc. It also contains sev-

Table F2. Health warnings and statement and control of tar and nicotine (T/N) content

Country or territory	*Familiar warning*	*Rotating/strong warnings*	*Smokeless tobacco warnings*	*Statements of T/N content*	*Control of T/N content*
Australia		X	*	(X)	(X)
Australian Capital Territory		X	*		
South Australia		X	*	X	
Tasmania		X	*		
Victoria	X	X	*	X	
Western Australia		X	*	X	
China	X			X	X
French Polynesia				X	
Hong Kong	X		*	X	X
Japan			*		
Macao	X			X	
Malaysia	X				
New Zealan		X	*	X	X
Papua New Guinea	X				
Singapore		X	*	X	X

* Total ban.

(X) = provisions laid down by voluntary agreement.

eral important provisions aimed at protecting the health of the people. Section 19 bans tobacco advertising on radio, television, and in newspapers and magazines.

3. **Hong Kong.** As of December, 1990, amendments to the Television Code of Practice on Advertising Standards arising from the ban on tobacco advertising and sponsorship specify the conditions under which a sponsored event may be broadcast: the television station may not be paid for the broadcast: no undue emphasis may be given to the sponsorship; and references to the sponsor may not obtrude on the programme or entertainment. Advertising of a product containing nicotine is prohibited. Advertising of a non-tobacco product with the corporate name or image of a tobacco company is allowed only if the advertisement clearly demonstrates that it has no association with any tobacco products, e.g., by referring only to the corporation or the non-tobacco products of the corporation. Advertisements of events sponsored by tobacco companies are permitted if no reference is made to the attributes of the product or the slogan of the company, if the lettering is uniform and the sponsor's name does not stand out, and if the air time is not paid for by tobacco advertisers. Tobacco products may not be presented as prizes or gifts for television contests. Advertisements for tobacco accessories may be directed only to adult audiences.

4. **Malaysia.** The Ministry of Information, which controls the broadcast media, has banned advertisements of cigarettes, but cigarette brand names have been permitted on other products and services, e.g. Kent Fashion, Camel Adventure Gear, Marlboro Travel, Salem High Country Travel, and Mid Seven Fishing Tackle. The Censor Board under the Home Ministry is authorized to restrict the content of advertisements in the mass media.

5. **New Zealand.** The Smoke-free Environments Act of 1990 imposes a total ban on all advertising of tobacco products in New Zealand. The only exceptions are publications printed outside New Zealand and films or radio or television transmissions originating outside New Zealand unless their principal purpose is the promotion of tobacco use or the targeting of a New Zealand audience. Retailers may display tobacco products inside their place of business provided no such product is visible from outside that place and may display inside their place of business notices identifying tobacco products for sale and their price. A tobacco product sold through an automatic vending machine and its price may be depicted on the exterior of the vending machine. The use of tobacco trade marks on goods other than tobacco products or in relation to sponsored events is prohibited. The tobacco industry is prohibited from organizing or promoting or making any financial contribution towards any organized activity (cultural, educational, sporting, or recreational) in New Zealand where the activity involves the use of a tobacco trade mark, except that the Minister of Health may grant an exception for craft participating in an international race under certain circumstances. The free distribution of tobacco products and the offering of rewards to purchasers of tobacco products are prohibited.

   The Act establishes a Health Sponsorship Council to provide alternative sponsorship of sports and cultural events to persons and organizations that formerly received financial assistance from the tobacco industry and, when that assistance was provided, had a reasonable expectation of receiving future financial or other assistance. Funding for the Council shall consist of appropriations by Parliament, gifts and bequests, and interest earned on these funds.

6. **Singapore.** Legislation of 7 November, 1989 imposes a total ban on all forms of tobacco advertising and sales promotion. All advertising of tobacco products on television and radio, in printed material, and on billboards is prohibited, except that on souvenirs or mementos given free of charge or as gifts, on vehicles of tobacco retailers and distributors, and at premises where tobacco products are sold. It is prohibited to sell or offer for sale tobacco products together with any other product as a free gift or vice versa, as is the selling or offering for sale of tobacco products packaged or labelled together with any other product. Distribution of free samples of tobacco products to the public is prohibited. Sponsorship of an event or promotion of goods unconnected with tobacco products is restricted in that these advertisements must not encourage or promote smoking and must be approved by the Minister of Health.

   Penalties for offences under the Act have been increased from a fine of S$2000 for a first offence to S$10000 and from a fine of S$4000 or each subsequent offence to S$20 000.

Notes to Table F2

1. **Australia.** Under the Trade Practices Act 1974, a "Notice of Declaration of Unsafe Goods and of Reasons for Decision" dated 12 October, 1989 prohibits the manufacture, import, and sale of oral snuff and chewing tobacco.

Table F4. Restrictions on smoking in public places

Country or territory	Government agencies	Health establishments	Schools and places where young people congregate	Public transport	Lifts	Indoor public places	Markets and food storage places	Restaurants
Australia[1]	X			X				
Australian Capital Territory				X	X			
Queensland				X				
South Australia				X	X			
Western Australia	X							
China	X	X	X	X	X			
Hong Kong	X	X	X	X	X	X	X	
Japan						X		
Macao		X	X	X				
Malaysia	X	X	X	X		X		
New Zealand	X	X	X	X	X	X	X	X
Wellington		X		X		X		
Papua New Guinea				X				
Republic of Korea		X				X		
Singapore	X	X	X	X	X	X	X	X
VietNam		X				X		

A voluntary agreement between the Federal Government of Australia and the Australian tobacco manufacturers and importers, first entered into in 1982 and periodically renegotiated, provides for the indication of tar and nicotine levels on cigarette packages and specifies approved maximum yields. The 1988 agreement provides that the maximum yield for corrected particulate matter (tar) is 14 mg per cigarette and for nicotine 1.4 mg per cigarette (Winstanley, 1989).

The required health warnings in all the states of Australia are:

Smoking Causes Lung Cancer
Health Authority Warning.

Smoking Causes Heart Disease
Health Authority Warning

Smoking Damages Your Lungs
Health Authority Warning

Smoking Reduces Your Fitness
Health Authority Warning.

The States of South Australia and Victoria have banned smokeless tobacco. The Governments of Tasmania and Western Australia have added smokeless tobacco products to Section 4 of their Poisons Lists, thus restricting advertising and making these products obtainable only on medical prescription.

2. **China.** Section 18 of the Law of 29 June, 1991 requires the printing of the health warning, "Smoking is hazardous to health", and tar levels on cigarette packages. Section 5 requires the reduction of the levels of tar and other harmful substances in tobacco products.

3. **Hong Kong.** The Government Chemist is required to determine the tar group to which each brand of cigarettes belongs, and the Customs and Excise Service has the authority to take samples of cigarettes for analysis by the Government Chemist.

4. **Malaysia.** Under the Trade Descriptive Act (under the Ministry of Trade), all cigarette packages sold in Malaysia must contain the health warning in Malay "Warning from the Government of Malaysia: Smoking is dangerous to your health." Tar and nicotine levels are not controlled by legislation, but the Ministry of Health has directed cigarette manufacturers to keep the tar below 19 mg and nicotine below 2 mg.

5. **New Zealand.** The Smoke-free Environments Act 1990 requires that each tobacco package contain a health warning, a list of the harmful constituents of the product and, for products intended for smoking, a list of the harmful constituents, and their respective quantities, present in the smoke. In addition, a leaflet containing information relating to the effects of the use of the product on health must be placed inside the package with the product, if required by regulations under the Act.

Every manufacturer and importer of tobacco products is required to conduct annual tests for the constituents of each brand of tobacco product sold and the respective quantities of those constituents and to report the results of these tests and specified information concerning the weight of tobacco and of all additives used to the Director-General of Health, who may require further testing at the expense of the manufacturer or importer.

A manufacturer, importer, or distributor who publishes a tobacco advertisement without reasonable excuse is liable to a fine not exceeding NZ$50,000; in any other case to a fine not exceeding NZ$10,000. Similar fines are imposed for violations of other provisions of the Act.

6. **Singapore.** The Consumer Protection Regulations of 11 January, 1989 require the following rotating warning on tobacco product containers:

— Smoking causes heart disease.

— Smoking causes cancer.

— Smoking damages your lungs.

— Smoking harms those around us.

A statement of the tar and nicotine content is also required.

The Poisons (Amendment) (No. 2) Rules provide that any cigarette containing 1.3 mg of nicotine and/or 15 mg of tar will be classified as poison under the Poisons Act, so that the sale of cigarettes containing more than these levels is prohibited.

Table F3. Restrictions on places of sale

Country or territory	*Government buildings*	*Health establishments*	*Schools and places where young people congregate*	*Other*
No information available				

Note : Restrictions on the use of vending machines are presented in Table F6.

Notes to Table F4

1. **Australia.** Smoking on domestic commercial flights was banned on 1 December, 1987 by regulations under the Air Navigation Act. Smoking was banned on 1 September, 1990 on sched-

uled international flights operating between cities in Australia. Administrative regulations require all Australian Government offices to be smoke-free. As of 1 July, 1988, all inter-state buses are smoke-free. State legislation and voluntary action restrict smoking in other public places.

In the Australian Capital Territory, many government and private institutions have adopted voluntary smoke-free policies. Administrative orders generally ban smoking in hospitals and government schools. By decision of local management, smoking is restricted in airports, railway stations, and bus terminals.

Information provided by Michael M. Daube of the Health Department of Western Australia indicates that there is little legislation specifically restricting smoking in public places and the workplace, but substantial occupational health, safety, and welfare legislation may be interpreted to apply to public places and workplaces, obviating the need for tobacco-specific legislation. Moreover, several state governments have decided that all government offices and buildings shall be smoke-free.

2. **China.** The Law of 29 June, 1991 prohibits smoking on public transport and in public places; the latter have not yet been defined.

3. **Hong Kong.** Regulations issued in 1983 restrict smoking in schools, hospitals, clinics, public areas of government offices, and food business premises.

4. **Japan.** Legislation prohibiting smoking in theatres and subway stations has been introduced for fire-prevention purposes. There is no legislation to prevent pollution of indoor areas by tobacco smoke.

5. **Malaysia.** Smoking is banned in cinemas, theatres, and indoor standiums. Under the Road Transport Rules, smoking is banned in air-conditioned buses. In other buses, the driver and conductor are not allowed to smoke while on duty, and taxi drivers are not allowed to smoke while driving passengers. The Ministry of Transport has directed Malaysian Airlines to prohibit smoking on all domestic flights as well as flights to Singapore. By Government directive, smoking is banned in all government offices, schools, and vehicles except in the toilet or in special rooms allocated to smokers. By directive of the Ministry of Health, smoking is banned in all hospitals and government clinics. The Prisons Department has banned smoking in all prisons (because narcotic drugs have been smuggled in through cigarettes).

6. **New Zealand.** The Smoke-free Environments Act of 1990 restricts smoking in hospitals, rest homes and prisons; prohibits smoking on internal flights of airlines; restricts smoking in passenger service vehicles, ships, and trains, in passenger lounges, and waiting rooms; and in restaurants. In restaurants with only one room, half the seating must be designated as non-smoking and must be separated, as far as practicable, from seating where smoking is permitted. Where there is more than one room in a restaurant, at least one room shall be designated as non-smoking. Area health boards are responsible for enforcement. The requirement that every employer prepare a written policy on smoking for every workplace of that employer (see Table F5, note 3) has the effect of achieving a general restriction on smoking in public places.

7. **Papua New Guinea.** Under the Tobacco Products (Health Control) Act 1987, the Minister of Health is authorized to limit or prohibit smoking in public places, including public transport vehicles.

8. **Republic of Korea.** Legislation passed by the National Assembly in January, 1990 provides that persons who own, occupy, or manage facilities that are subject to the control of the Ministry of Health and Social Welfare shall designate certain places within the facility as smoking areas, and smoking shall be prohibited outside these designated areas.

9. **Singapore.** In 1970, under the Prohibition of Smoking in Certain Places Act, smoking was prohibited in cinemas, theatres, other specified buildings, and buses. In 1973, the prohibition was extended to lifts, in 1982, to amusement centres, in 1985, to government hospitals and clinics, in 1987, to all public buildings, in 1988, under the Smoking (Prohibition in Certain Places) Notification, to hospitals, maternity homes, medical clinics and nursing homes, fast-food restaurants, indoor roller-skating rinks and roller discotheques, and in 1989, to air-conditioned restaurants, department stores, supermarkets, mini-supermarkets, indoor stadiums, bowling alleys, billiard saloons, gymnasiums, aerobic and fitness centres, convent halls, and multipurpose halls used as meeting places.

10. **Viet Nam** The Law of 30 June, 1989 on the protection of public health prohibits smoking in meeting-halls, cinemas, theatres, and other stipulated places.

Notes to Table F5

1. **Australia.** See Table F4, note 1; although Australia has little legislation specifically

Table F5. Restrictions on smoking in the workplace

Country or territory	*General laws restricting smoking in the workplace*	*Government agencies*	*Health establish-ments*	*Schools*	*Authorization for administrative regulations or accommodation of smokers and non-smokers*	*Other*
Australia						
Victoria	X					
Hong Kong		X	X	X		
Malaysia		X	X	X		X
New Zealand	X	X	X	X		X
Singapore						X

Table F6. Preventing young people from smoking

Country or territory	*Prohibiting sales to minors*	*Restricting sales from vending machines*	*Prohibiting sales in schools and other places frequented by young people*	*Restricting advertising that influences young people*	*Other*
Australia					
Australian Capital Territory	X	X			X
New South Wales	X				
South Australia	X	X			
Tasmania	X				
Victoria	X	X			
Western Australia	X	X			
China			X		
Hong Kong				X	X
New Zealand	X				
Papua New Guinea	X				
Republic of Korea				X	
Singapore					X

restricting smoking in the workplace, implementation of existing occupational health, safety, and welfare legislation will control worksite smoking.

2. **Malaysia.** By government directives, smoking is banned in all government offices, schools and government vehicles, except in the toilet or in special rooms allocated for smokers.
3. **New Zealand.** The Smoke-free Environments Act of 1990 requires every employer to prepare a written policy on smoking for every workplace of that employer. Every written policy must, at the minimum, prohibit smoking in lifts or in offices where two or more people work in a common air space, in at least half of any workplace lunchroom, or in any part of the workplace to which the public normally has access. Special provisions are laid down for certain institutions, e.g. hospitals, rest homes, and prisons. Area health boards are responsible for enforcement. Violation of the statute by a corporation is subject to a maximum fine of NZ$4000, and violation by an individual to a maximum fine of NZ$400.
4. **Singapore.** Section 7 of the Environmental Public Health (Food Handlers) Regulations 1973 provides that "no person shall... smoke or use tobacco for snuff or any other preparation or chew tobacco or betel nuts while engaged in the sale or preparation for sale of any food".

Notes of Table F6

1. **Australia.** All six states of Australia prohibit sales of cigarettes to minors. In Western Australia, it is illegal to sell cigarettes to persons under age 18; in the other states to persons under 16. The Western Australia Tobacco Control Act 1990 restricts vending machines containing tobacco products to premises licensed to sell liquor or in areas set aside as staff amenities. The penalty for violation of the ban on the sale of tobacco products to minors has been raised from A$4 to A$5000.
2. **China.** The Law of 29 June, 1991 on the exclusive sale of tobacco prohibits smoking by elementary and secondary school students.
3. **Hong Kong.** Tobacco products may not be sold in schools or institutions for children and young people.
4. **Singapore.** The distribution or giving of any free sample of cigarettes, cigars or any other form of tobacco to the public or any section of the public, other than to persons who are associated or concerned with the manufacture, distribution, or sale of tobacco products, is prohibited.

Table F7. Health education on tobacco

Country or territory	*Public education*	*Anti-smoking education in schools*	*Allocation of funds*
Australia			
Queensland	X		X
South Australia			X
Victoria	X	X	X
Western Australia			X
China	X	X	
French Polynesia	X		X
Singapore	X	X	

Table F8. Legislation establishing a national organization for policy development and coordination to control tobacco use

Country or territory	*National organization*
China	Chinese Association on Smoking and Health
Hong Kong	Hong Kong Council on Smoking and Health
Viet Nam	National Campaigning Committee on Tobacco or Health, chaired by Minister of Health

Summary Table A. Number of countries and territories with legislation to control the production, sale, and promotion of tobacco, by WHO region

Type of legislation	*Number of countries and territories*						
	All regions	*Africa*	*Americas*	*Eastern Mediterranean*	*Europe*	*South-East Asia*	*Western Pacific*
Total ban on tobacco advertising	27	3	2	3	13	1	5
Some restrictions on advertising	77	7	19	9	26	5	11
Restriction on sponsorship	18	2	4	—	8	—	4
Rotating/strong warnings on cigarette packages	29	1	6	6	11	1	4
Familiar warnings on cigarette packages	48	6	14	7	12	4	5
Statement of tar and nicotine content	40	4	6	8	15	—	7
Restrictions on places of sale	11		3	1	5	2	—

Summary Table B. Number of countries and territories with legislation to influence smoking practices, by WHO region

Type of legislation	*Number of countries and territories*						
	All regions	Africa	Americas	Eastern Mediterranean	Europe	South-East Asia	Western Pacific
Restrictions on smoking in public places	90	20	17	10	28	4	11
Restrictions on smoking in the workplace	34	3	11	1	14	—	5
Preventing young people from smoking	42	2	18	2	12	1	7
Controlling smokeless tobacco	19	—	2	1	11	1	4
Health education	36	1	14	3	12	2	4

Appendix 4

Resolutions of the World Health Assembly

Resolution WHA23.32

The Twenty-third World Health Assembly,

Having considered the report of the Director-General;

Recalling the resolutions on this subject adopted by the Executive Board, the Directing Council of the Pan American Health Organization/Regional Committee for the Americas, and the Regional Committee for Europe;

Conscious of the serious effects of smoking in promoting the development of pulmonary and cardiac disease, including bronchopulmonary cancer, chronic bronchitis, emphysema and ischaemic heart disease;

Being aware that bronchopulmonary cancer is at present increasing in all countries of the world where records are available in a form which permits assessment;

Holding that health agencies must now demonstrate their concern for the reduction of the main causal factor in diseases related to smoking; and

Considering that smoking of tobacco during meetings may constitute a nuisance to non-smokers.

Resolves that:

(1) all those present at meeting of the Assembly and its committees be requested to refrain from smoking in the rooms where such meetings are held;

(2) the Director-General be requested:

(a) to consider the desirability of making the subject for World Health Day "The health consequences of smoking," on the earliest possible occasion;

(b) to call the attention of all Members and Associate Members to the report on limitation of smoking and to suggest that the advantages of applying the recommendations on pages 19 and 20 of that report[1] should be considered in all countries;

(c) to consider convening an expert group to recommend further action that might be taken to discourage smoking;

(d) to examine to what extent and by what educational methods young people might be persuaded not to begin smoking;

(e) to bring to the attention of FAO the need for studying crop substitution in tobacco-producing countries;

(f) to report to the Executive Board at its forty-seventh session and to the Twenty-fourth World Health Assembly on the action proposed and the financial consequences for the Organization.

May 1970

Resolution WHA24.48

The Twenty-fourth World Health Assembly,

Having considered the report of the Director-General;

Recalling the resolution on this subject adopted by the Twenty-third World Health Assembly, the Executive Board, and the respective Regional Committees;

Recognizing the relationship between smoking and the development of pulmonary and cardiac disease, including lung cancer, ischaemic heart disease, chronic bronchitis and emphysema; and

Believing that a sustained effort by health and education authorities and others is needed to reduce tobacco smoking and to prevent the extension of the habit, with special attention to young people and pregnant women;

1. THANKS the Director-General for his report;
2. ENDORSES the recommendations contained therein;

---

[1]*See WHO chronicle*, 24,365

3. CALLS UPON all Member States and Associate Members to give all possible consideration to putting these recommendations into effect; and

4. REQUESTS the Director-General;

   (i) to continue to assemble information on the health effects of tobacco smoking and the action being taken by countries to reduce the habit;

   (ii) to place emphasis on the control and prevention of smoking as an integral part of operating programmes as and when feasible;

   (iii) to continue in co-operation with the United Nations, the specialized agencies and the appropriate non-governmental organizations to foster a greater awareness of the health hazards of smoking and to take whatever action is deemed necessary to reduce them, and particularly to draw the attention of the Food and Agriculture Organization of the United Nations to the necessity of undertaking a study on crop diversification in tobacco-growing areas in view of the expected decrease in tobacco consumption;

   (iv) to stimulate the strengthening of health education activities, including the production, dissemination and exchange of educational materials to discourage the habit of smoking; and

   (v) to produce a code of practice that can guide governments in the formulation of legislative action relevant to the health consequences of smoking.

May 1971

Resolution WHA29.55

The Twenty-ninth World Assembly,

Recalling resolutions EB45.R9, WHA23.32, EB47.R42 and WHA24.48 concerning the health hazards of smoking and ways towards its limitation;

Noting with satisfaction that the recent WHO Expert Committee report on smoking and its effects on health,[1] prepared in accordance with resolution EB53.R31 and reviewed favourably by the Executive Board at its fifty-seventh session, provides a thorough and authoritative summary of currrent knowledge in the field and contains a number of important recommendations for WHO and the Member States;

Considering that the results of the Third World Conference on Smoking and Health, held in New York in June 1975, gave further support to the evidence and proposals presented by the WHO Expert Committee;

Recognizing the indisputable scientific evidence showing that tobacco smoking is a major cause of chronic bronchitis,emphysema and lung cancer as well as a major risk factor for myocardial infarction, certain pregnancy-related and neonatal disorders and a number of other serious health problems, and also has harmful effects on those who are involuntarily exposed to tobacco smoke;

Seriously concerned about the alarming worldwide trends in smoking-related mortality and morbidity and the rapidly increasing consumption of tobacco, especially in cigarettes, in countries in which it was not previously widespread, and about the growing number of young people and women who are now smoking;

Recognizing that an effective strategy to tackle the problem requires a concerted effort consisting of educational, restrictive and legislative measures, combined with coherent taxation and price policies, and supported by continuous research and evaluation on a multidisciplinary basis;

Noting that very few countries have thus far taken effective steps to combat smoking;

Believing that no organization devoted to the promotion of health can be indifferent in this matter, and that WHO has an important role to play in promoting effective policies against smoking, as envisaged in the Sixth General Programme of Work of WHO covering the period 1978-1983;

1. URGES governments of Member States to identify the actual or anticipated health problems associated with smoking in their countries;

2. RECOMMENDS governments of Member States:

   (1) to create and to develop effective machinery to coordinate and supervise programmes for control and prevention of smoking on a planned, continuous and long-term basis;

   (2) to strengthen health education concerning smoking, as a part of general health education and through close collaboration with health and school authorities, mass media, voluntary organizations, employers' and employers' organizations and other relevant agencies, taking into account the different needs of various target groups, laying emphasis on the positive aspects of non-smoking, and supporting individuals wishing to stop smoking;

   (3) to consider steps which can be taken towards ensuring that non-smokers receive protection, to which they are entitled, from an environment polluted by tobacco smoke;

[1]WHO Technical Report Series, No. 568 1975.

(4) to give serious consideration to the legislative and other measures suggested by the WHO Expert Committee in its recent report on smoking and its effects on health;

3. REQUESTS the Director-General:

(1) to continue, and intensify, WHO's anti-smoking activities;

(2) to collate and disseminate information on smoking habits, smoking-related health problems and smoking control activities in Member States;

(3) to give assistance and encouragement to research in smoking and health, with particular emphasis on studies that are directly relevant to the assessment and improvement of the effectiveness of anti-smoking activities;

(4) to promote the standardization of:

(a) definitions, measurement methods and statistics concerning smoking behaviour, tobacco consumption and the occurrence of smoking-related morbidity and mortality;

(b) laboratory techniques used for the quantitative analysis of the harmful substances in tobacco products;

(5) to give assistance, upon request, to governments in the formulation, implementation and evaluation of their policies and programmes to combat smoking;

(6) to continue, in cooperation with the United Nations, the specialized agencies and appropriate nongovernmental organizations, to make all efforts deemed necessary to reduce smoking; and particularly to work out with the Food and Agriculture Organization of the United Nations and with the United Nations a joint strategy for crop-diversification in tobacco-growing areas with a view to avoiding the anticipated economic consequences of reducing tobacco consumption in the world as a whole for public health reasons;

(7) to convene an expert committee in 1977 or 1978 to review and evaluate the world situation in regard to smoking control;

(8) to report to a future Health Assembly on developments in this field;

May 1976

Resolution WHA31.56

The Thirty-first World Health Assembly,

Recalling resolutions EB45.R9, WHA23.32, EB47.R42, WHA24-48, EB53.R31 and WHA29.55 concerning the health hazards of tobacco smoking and ways towards its limitation;

Recognizing the increasing and indisputable scientific evidence showing that tobacco smoking is a major cause of chronic bronchitis, emphysema and lung cancer, as well as a major risk factor for myocardial infarction, certain pregnancy-related and neonatal disorders and a number of other serious health problems, and that it also has harmful effects on those who are involuntarily exposed to tobacco smoke;

Seriously concerned at the alarming increase in production and consumption of cigarettes during the last two decades in some of the countries, particularly developing countries, in which it was previously not widespread, and at the extensive promotional drive for the sale of cigarettes being carried out on radio and television, in newspapers and other news media, and through association with sporting and cultural events, often inducing young people to smoke tobacco;

Noting that few countries have so far taken comprehensive action to effectively combat smoking through educational, restrictive and legislative measures for the control of publicity and advertisements in the news media, combined with coherent taxation and price policies for tobacco cultivation and cigarette production;

Believing that WHO has an important role in promoting effective policies against smoking, as envisaged in the Sixth General Programme of Work covering the period 1978-1983 inclusive;

1. URGES Member States:

(1) to strengthen health education programmes concerning tobacco smoking as a part of general education, through close collaboration among health and education authorities and other relevant agencies, taking into account the different needs of the various target groups;

(2) to adopt comprehensive measures to control tobacco smoking, *inter alia* by providing for increased taxation on the sale of cigarettes and restricting as far as possible all forms of publicity for promotion of smoking;

(3) to protect the rights of non-smokers to enjoy an atmosphere unpolluted by tobacco smoke;

(4) to seek economically sound alternative undertakings to replace tobacco growing and processing, where appropriate;

2. REQUESTS the Director-General:

(1) to continue to intensity WHO's activities in connexion with control of tobacco smoking;

(2) to collaborate with Member States, the United

Nations, the specialized agencies and appropriate nongovernmental organizations as required, in the formulation, implementation and evaluation of programmes to combat smoking, including studying possibilities for crop diversification in tobacco-growing areas;

(3) to cooperate with Member States upon request in developing measures for the control of publicity with regard to smoking through the news media, especially newspapers, radio and television;

(4) to give urgent consideration to having non-smoking as a theme for World Health Day as soon as possible, and in this and other ways to give maximum publicity to an anti-smoking campaign;

(5) to encourage research as to the cause of tobacco smoking;

(6) to report on progress in this field not later than the Thirty-third World Health Assembly.

May 1978

Resolution WHA33.35

The Thirty-third World Health Assembly,

Recalling resolutions EB45.R9, WHA23.32, EB47.R42, WHA24.48, EB53.R31, WHA29.55, and WHA31.56 concerning the health hazards of tobacco smoking and WHO's role in the limitation of this harmful habit;

Noting the report of the WHO Expert Committee on Smoking Control;[1]

Reiterating its firm conviction that the effect of tobacco smoking is now a major public health problem in all industrialized countries and in many developing countries and that it will become so in the near future in all other developing countries unless action is taken now;

Mindful of the ill-effects of smoking, particularly on risk groups such as pregnant women, lactating mothers and children;

Seriously concerned about the aggressive promotional drives for the sale of cigarettes that occur in developing as well as developed countries, thus inducing the new generations to take up the habit of smoking;

Alarmed by the fact that advertising practices using psychological means in both industrialized and developing countries have the effect of inducing and perpetuating smoking habits, especially among youth;

Encouraged by the existence of total bans, restrictions or limitations on tobacco advertising in several countries;

Noting encouraging signs of expanded national activities and of increasing public awareness of the harmful health effects of cigarette smoking in many countries, partly as a result of WHO's efforts and of this year's World Health Day on "Smoking or health: the choice is yours";

Realizing that national and international strategies to combat the spreading of the habit of smoking must be carried out on a continuous, long-term basis;

Believing that WHO has an essential role to play in promoting effective smoking control policies;

1. URGES Member States:

(1) to strengthen, and to initiate where lacking, the smoking control strategies outlined in the above-mentioned resolutions, laying special emphasis on educational approaches, particularly with respect to youth, and on measures to ban, restrict or limit advertising of tobacco products;

(2) to support WHO's action in the field of smoking and health;

2. REQUESTS the Director-General:

(1) to further develop an effective WHO action programme on smoking and health, clearly defining lines of responsibility and priority areas, and taking into account the multidisciplinary and intersectoral character of the relationship between smoking and health;

(2) to ensure that WHO plays a leading role in coordinating international activities and to strengthen collaboration with other United Nations agencies and with relevant nongovernmental organizations, and, particularly, to pursue the study on crop diversification in tobacco-growing areas in collaboration with FAO;

(3) to collaborate with Member States in their efforts to reduce smoking;

(4) to consider problems caused by the marketing and consumption of tobacco, particularly in developing countries;

(5) to mobilize financial and other resources for the implementation of the programme;

(6) to report on progress of this programme to the Thirty-fifth World Health Assembly.

May 1980

Resolution WHA39.14

The Thirty-ninth World Health Assembly,

[1]WHO Technical Report Series, No. 636, 1979.

Recalling resolutions WHA31.56 and WHA33.35 on the health hazards of tobacco smoking and the WHO action programme on smoking and health;

Deeply concerned by the current pandemic of smoking and other forms of tobacco use, which results in the loss of the lives of at least one million human beings every year and in illness and suffering for many more;

Believing that the battle between health and tobacco must and can be won for the sake of human health;

Encouraged by the Existence of total bans, restrictions or limitations on tobacco advertising in several countries;

1. AFFIRMS;

(1) that tobacco smoking and the use of tobacco in all its forms is incompatible with the attainment of health for all by the year 2000;

(2) that the presence of carcinogens and other toxic substances in tobacco smoke and other tobacco products is a known fact; and that the causal link between tobacco and a range of fatal and disabling diseases has been scientifically proven;

(3) that passive, enforced or involuntary smoking violates the right to health of non-smokers, who must be protected against this noxious form of environmental pollution;

2. CALLS for a global public health approach and action *now* to combat the tobacco pandemic;

3. DEPLORES all direct and indirect practices the aim of which is to promote the use of tobacco, as this product is addictive and dangerous even when used as promoted;

4. URGES those Member States which have not yet done so to implement smoking control strategies; these, as a minimum, should contain the following:

(1) measures to ensure that non-smokers receive effective protection, to which they are entitled, from involuntary exposure to tobacco smoke, in enclosed public places, restaurants, transport, and places of work and entertainment;

(2) measures to promote abstention from the use of tobacco so as to protect children and young people from becoming addicted;

(3) measures to ensure that a good example is set in all health-related premises and by all health personnel;

(4) measures leading to the progressive elimination of those socioeconomic, behavioural, and other incentives which maintain and promote the use of tobacco;

(5) prominent health warnings, which might include the statement that tobacco is addictive, on cigarette packets, and containers of all types of tobacco products;

(6) the establishment of programmes of education and public information on tobacco and health issues, including smoking cessation programmes, with active involvement of the health professions and the media;

(7) monitoring of trends in smoking and other forms of tobacco use, tobacco-related diseases, and effectiveness of national smoking control action;

(8) the promotion of viable economic alternatives to tobacco production, trade and taxation;

(9) the establishment of a national focal point to stimulate, support, and coordinate all the above activities;

5. APPEALS to other organizations of the United Nations system:

(1) to support WHO in all ways possible within their fields of competence;

(2) to show solidarity with WHO's efforts to stem the spread of tobacco-induced diseases by protecting the health of non-smokers on their premises, as this action would have a major exemplar role;

(3) to help Member States in identifying and implementing economic alternatives to tobacco cultivation, production and trade;

6. REQUESTS the Director-General:

(1) to strengthen the present programme on smoking and health without waiting for its official introduction in the Eighth General Programme of Work, as a visible and resolute attitude on the part of WHO would provide Member States with encouragement and support, which are necessary prerequisites to abating the smoking pandemic before the year 2000;

(2) to mobilize support for the present programme on smoking and health in terms of funds and manpower which would ensure adequate programme continuity on a long-term basis;

(3) to coordinate activities in support of WHO's action on smoking and health with other organizations of the United Nations system at the highest executive level;

(4) to continue and strengthen collaboration with

nongovernmental organizations as appropriate;

(5) to ensure that WHO plays an effective global advocacy role in tobacco and health issues and that, in common with other health institutions, it plays an exemplar role in non-smoking practices;

(6) to provide support to national smoking control efforts;

(7) to report on progress to the Executive Board at its eighty-first session and to the Forty-first World Health Assembly.

May 1986

Resolution WHA40.38

The Fortieth World Health Assembly,

Bearing in mind the objective of the World Health Organization contained in article 1 of the Constitution, "the attainment by all peoples of the highest possible level of health".

Recalling resolution WHA39.14 on "Tobacco or health";

Expressing its satisfaction at the measures increasingly being taken by Member States to reduce smoking;

Gratified at the decision taken by the Director-General to declare the WHO premises a smoking-free area;

Aware that the consumption and use of tobacco result in serious health consequences, as well as economic and social problems, notably in developing countries;

Noting that on 7 April, 1988 the World Health Organization will celebrate its fortieth anniversary;

1. CALLS UPON all Member States, as part of their continuing efforts to reduce the smoking pandemic, through all appropriate means including, where applicable, legislative and regulatory measures;

   (1) to celebrate 7 April, 1988 as a world no-smoking day;

   (2) to encourage the population, by all appropriate means, to desist from smoking and from using tobacco in all other forms on that day;

   (3) in conjunction with governmental and nongovernmental organizations, to use the occasion to launch, or strengthen existing, anti-smoking campaigns and health-promoting initiatives;

   (4) to encourage vendors to refrain voluntarily from selling all forms of tobacco on that day;

   (5) to inform the Director-General on actions taken in response to this resolution;

2. APPEALS to all manufacturers of tobacco and those who promote its consumption, in the spirit of this resolution and of resolution WHA39.14, to refrain voluntarily from all publicity activities in all countries, especially in developing countries, and calls upon the press and the other media in each country voluntarily to do likewise;

3. REQUESTS the Director-General to report to a subsequent Health Assembly on the action taken in this regard.

May 1987

Resolution WHA42.19

The Forty-second World Health Assembly,

Recalling resolution WHA39.14 and resolution WHA41.25 requesting the Director-General to draw up a plan of action on tobacco or health for submission through the Programme Committee to the eighty-third session of the Executive Board;

Recognizing that the use of tobacco is responsible worldwide for more than two million premature deaths annually;

Recalling that active efforts are needed to resolve the economic issues involved in reducing tobacco production;

Concerned at the fact that, while tobacco consumption is decreasing in developed countries as a result of effective health promotion supported by appropriate legislation and regulations, the developing countries are registering increases in tobacco consumption;

Reaffirming that the health services should clearly and unequivocally publicize the health risks connected with the use of tobacco and actively support all efforts to prevent the associated diseases;

1. THANKS the Director-General for having already accelerated implementation of the WHO programme on tobacco or health;

2. APPROVES the plan of action for the WHO programme on tobacco or health for 1988-1995 as proposed by the Director-General and endorsed by the Executive Board;

3. REQUESTS the Director-General:

   (1) to continue to support this programme as outlined in the plan of action and to mobilize extrabudgetary funds for its implementation;

   (2) to support national authorities, at their request, in taking measures to disseminate information on the health risks of tobacco, to promote life-styles without tobacco, and

to control the promotion of tobacco consumption;

(3) to work, in close collaboration with national health authorities, with organizations of the United Nations system, and with relevant nongovernmental organizations in official relations with those organizations, to ensure that both health and economic aspects are fully taken into account;

(4) to review the impact of tobacco production on the economy, environment and health of the populations in developing countries which depend upon tobacco production as a major source of income, and to report on this issue to the Forty-third World Health Assembly;

(5) to collaborate actively with FAO and other relevant United Nations agencies with a view to developing agricultural projects that demonstrate how crop substitution programmes can be implemented in countries whose economies depend heavily upon tobacco production and to encouraging such countries to implement these programmes;

4. RESOLVES that each year 31 May shall be World No-Tobacco Day.

May 1989

Resolution WHA43.16

The Forty-third World Health Assembly,

Recalling the strong statement on the issue of smoking and health made by the President in opening the Forty-third World Health Assembly;

Recalling resolution WHA33.35, WHA39.14, WHA41.25 and WHA42.19 on the health hazards of tobacco smoking and the WHO programme on tobacco or health;

Recalling the requirement contained in resolution WHA42.19 concerning a review of crop substitution and the health and economic aspects of tobacco production and consumption;

Recalling further that resolution WHA39.14 urged Member States to implement a comprehensive nine-point smoking control strategy;

Encouraged by:

(a) the significant progress made by Member States in the implementation of this strategy;

(b) the continuing decline in tobacco consumption in Member States that have adopted comprehensive smoking control policies;

(c) recent information demonstrating the effectiveness of tobacco control strategies, and in particular:

— legislation or other measures to provide protection from involuntary exposure to tobacco smoke in workplaces, public places and public transport;

— policies to achieve progressive increases in the real price of tobacco;

— comprehensive bans and other legislative restrictive measures to control effectively direct and indirect advertising, promotion and sponsorship concerning tobacco;

Deeply concerned by increasing evidence of the dangers to health of passive smoking and by a new WHO estimate that, unless current smoking rates decrease, there will be 3 million tobacco-related deaths per year during the 1990s, and that this figure will rise sharply to 10 million deaths per year by the 2020s;

Believing that millions of future premature deaths can be avoided if current smoking rates are quickly and substantially reduced;

1. URGES all Member States:

(1) to implement multisectoral comprehensive tobacco strategies which, at a minimum, contain the nine elements outlined in resolution WHA39.14;

(2) to consider including in their tobacco control strategies plans for legislation or other effective measures at the appropriate government level providing for:

(a) effective protection from involuntary exposure to tobacco smoke in indoor workplaces, enclosed public places and public transport, with special attention to risk groups such as pregnant women and children;

(b) progressive financial measures aimed at discouraging the use of tobacco;

(c) progressive restrictions and concerted action to eliminate eventually all direct and indirect advertising, promotion and sponsorship concerning tobacco;

2. NOTES that, in countries where more than one level of government exists, national authorities may not have complete jurisdiction over these issues;

3. REQUESTS the Director-General:

(1) to intensify support for the 1988-1995 plan of action for the WHO programme on tobacco or health;

(2) to ensure the provision of sufficient budgetary resources to assist Member States in

implementing comprehensive tobacco control programmes;

(3) to ensure that the report requested in resolution WHA42.19 is presented to the Forty-fourth World Health Assembly;

(4) to monitor and report biennially to the Health Assembly on the progress and effectiveness of Member States' comprehensive tobacco control programmes;

(5) to report to the Forty-fourth World Health Assembly on the progress made in assistance to countries that depend on tobacco production as a major source of financial resources for health and development, with emphasis on measurement of efficacy of such assistance.

May 1990

Resolution WHA44.26

The Forty-fourth World Health Assembly,

Recalling resolutions WHA33.35, WHA39.14, WHA41.25 and WHA42.19 on the health consequences to tobacco consumption and the WHO "tobacco or health" programme, formerly the action programme on smoking and health;

Recalling in particular resolution WHA43.16, which urges all Member States to adopt effective measures to prevent involuntary exposure to tobacco smoke in enclosed public places and public transport;

Recognizing that there is no safe level of exposure to tobacco smoke;

Aware of the technical problems of ensuring a smoke-free environment in many public conveyances, especially trains and aircraft;

Congratulating the transport authorities and companies that have adopted measures to offer their passengers a smoke-free environment and encouraging all those responsible for public transport to do likewise;

Deeply concerned about the dangers to the health, and the violation of the right to health, of non-smokers caused by enforced, or passive, smoking and about the WHO-approved estimates that the annual number of deaths in the world attributable to smoking will be about three million in the 1990s,

1. URGES all Member States:

(1) to adopt appropriate measures for effective protection from involuntary exposure to tobacco smoke in public transport;

(2) to ban smoking in public conveyances where protection against involuntary exposure to tobacco smoke cannot be ensured, and to adopt effective measures of protection wherever possible;

(3) to promote educational activities necessary to make people aware of the importance of protecting themselves and their families, especially children, against passive smoking, for example, while travelling in private cars;

2. REQUESTS the Director-General:

(1) to collaborate with the International Civil Aviation Organization and all competent international and national agencies in developing guidelines and recommendations for a smoke-free travel environment in all types of public conveyances;

(2) to support Member States at their request in implementing effective measures to protect people against involuntary exposure to tobacco smoke in public transport;

(3) to keep the Executive Board and the Health Assembly informed of the progress made in implementing this resolution as an element of the WHO "tobacco or health" programme.

May 1991

Appendix 5

List of Legislation Reviewed[1]

Afghanistan

Order of the Cabinet of 28 December, 1971 (prohibiting all advertising of cigarettes)*

Algeria

Law No. 85-05 of 16 February, 1985 on health protection and promotion (Sections 63-66) (IDHL, 1985, 36(4): 909, 916)

Argentina

Law No. 18604 of 1970 (prohibiting certain forms of advertising of cigarettes)*

Order No. 33.266 prohibiting drivers of school buses from smoking and prohibiting smoking on vehicles transporting dangerous substances

Order No. 22.900 prohibiting smoking on public transport vehicles

Order No. 09-12-910 prohibiting smoking in theatres including interior vestibules and corridors

Law No. 23344 of 31 July, 1986 on restrictions on the advertising of tobacco, cigars, cigarettes, and other products intended for smoking and their packaging (IDHL, 1986, 37(4): 796-797)

Parliamentary Decree No. 226 of 27 April, 1988 requiring health warnings that smoking is prejudicial to health in all advertising and promotion of tobacco

Argentine Food Code, Article 18 (prohibiting use of tobacco in food establishments and places where food products are handled)

Argentina (Buenos Aires)

Order No. 6762-DOC5-84 of 5 December, 1984 concerning smoking in public transport, stations of the underground, school buses, vehicles transporting dangerous substances, theatres, and food establishments

Law No. 10.600 of 12 November, 1987 prohibiting smoking in public transport vehicles

Argentina (Cordoba)

Order No. 8425 of 11 October, 1988 prohibiting smoking in the offices of the municipal government that serve the public

Law No. 7827 of 20 September, 1989 prohibiting smoking in enclosed places of the executive, legislative, and judicial branches of the government

Argentina (Jujuy)

Law No. 4292 of 17 June, 1987 prohibiting smoking in public buildings, school rooms of all levels, hospitals, and means of urban and suburban transport

Argentina (Mendoza)

Law of 3 December, 1988 prohibiting smoking in indoor public places, elevators, and public offices, in hospitals and health centres, in official banks and educational establishments of all levels

Argentina (San Fernando Del Valle De Catamarca)

Order No. 565-C-89 prohibiting smoking in enclosed places of the municipal government and ordering a campaign against smoking with the objective of extending the prohibition to all public and private places

---

[1] Texts marked with an asterisk were not available to the author. Texts marked with two asterisks are due to be published in forthcoming issues of the *International digest of health legislation* (abbreviated to IDHL in the list).

Argentina (Valle Viejo)

Order of 25 October, 1988 prohibiting smoking in government offices, indoor public places, and means of transport

Australia

The Broadcasting and Television Act 1972 (IDHL, 1973, 24(3): 453) Notice of Declaration of Unsafe Goods and of Reasons for Decision of

12 October, 1989 under the Trade Practices Act of 1974 prohibiting the manufacture, import, and sale of oral snuff and chewing tobacco.

Air Navigation Regulations (Amendment) 25 November, 1987

Air Navigation Regulations (Amendment) 13 September, 1990

The Smoking and Tobacco Products Advertisements (Prohibition) Act 1989.

28 December, 1989 (IDHL. 1991, 42(1): 48)

Australia (Capital Territory)

The Tobacco Products (Health Warnings) Ordinance 1986 requiring the printing of health warnings on containers in which tobacco products are packaged (IDHL, 1987, 38(3): 541-542)

Australia (New South Wales)

The Cigarettes (Labelling) Act, 1972 (IDHL, 1973, 24(3): 459)

Australia (South Australia)

The Cigarettes (Labelling) Act, 1971-1972 (IDHL, 1973, 24(3): 468)

The Cigarettes (Labelling) Regulations. Dated 3 May, 1973 (IDHL, 1974, 25(4): 714-715)

An Act to prohibit the sale or supply of certain tobacco products to children, dated 15 November, 1984 (IDHL, 1986, 37(1): 45) (repealed by the Tobacco Products Control Act, 1986).

The Tobacco Products Control Act, 1986 (IDHL, 1988. 39(2): 392-394)

The Tobacco Products (Licensing) Act, 1986 (IDHL, 1988, 39(2): 394-395)

The Tobacco Products Control Amendment Act 1988

Australia (Tasmania)

The Cigarettes (Labelling) Act, 1972, IDHL, 1974, 25(4): 721)

The Cigarettes (Labelling) Regulations 1973 (IDHL, 1976, 27(1): 20-21)

The Tobacco Products (Labelling) Act 1987 to prohibit the sale of tobacco products except in packages marked with health warnings (IDHL, 1988, 39(2): 395)

Australia (Victoria)

The Health (Amendment) Act 1969 (IDHL, 1972, 23(1): 22-23)

The Tobacco Act 1987 creating a State-wide programme to reduce the prevalence of smoking in Victoria, restricting various forms of advertising, prohibiting the supply of tobacco to persons under 16 years of age and restricting the siting of vending machines and establishing the Victorian Health Promotion Foundation (IDHL, 1988, 39(3): 652-653)

Australia (Western Australia)

The Tobacco (Warning Labels) Regulations, 1987 (IDHL, 1988, 39(2): 395-396)

The Tobacco Control Act 1990 (IDHL, 1991, 42(2): 273-275)

Austria

Decree of 15 February, 1979 of the Federal Ministry of Health and Environmental Protection on smoking in hospitals (IDHL, 1980, 31(1): 19)

The Employees' Protection Law of 1972, as amended by the Federal Law of 20 October, 1982 (IDHL, 1986, 37(1): 114-115)

Bahamas

The Health Services (Amendment) Act of 1976, cited as the Health Services Rules, requiring a health warning on tobacco advertising and cigarette packages.

Belgium

Crown Order of 28 December, 1979 on the manufacture and marketing of tobacco, tobacco products, and similar products (IDHL, 1981, 32(1): 85-86) as amended by the Crown Order of 30 March, 1981 (IDHL, 1982, 33(1): 32-33)

Crown Order of 5 March, 1980 on advertising for tobacco, tobacco products, and similar products (IDHL, 1981, 32 (1): 86-87), as amended by Crown Order of 22 September, 1980 (IDHL, 1982, 33(1): 32)

Crown Order of 21 January, 1982 amending the Crown Order of 5 March, 1980 on advertising for tobacco, tobacco products, and similar products (IDHL, 1983, 34(4): 767-768)

Decree of 2 December, 1982 on the control of smoking and tobacco consumption (IDHL, 1986, 37(1):

45-48)

Crown Order of 20 December, 1982 on the advertising of tobacco, tobacco products, and similar products (IDHL, 1985, 36(1): 62-63)

Crown Order of 20 December, 1982 amending the Crown Order of 28 December, 1979 on the manufacture and marketing of tobacco, tobacco products and similar products (IDHL, 1985, 36(1): 63)

Crown Order of 31 March, 1987 prohibiting smoking in certain public places (IDHL, 1987, 38(3): 542-543)

Crown Order of 19 January, 1990 modifying the Crown Order of 29 December, 1979 on the manufacture and marketing of tobacco and tobacco products**

Crown Order of 10 April, 1990 modifying the Crown Order of 20 December, 1982 on advertising of tobacco and tobacco products**

Crown Order of 15 May, 1990 prohibiting smoking in certain public places**

Bermuda

The Tobacco Products (Public Health) Act 1987 requiring warnings on packages and advertisements for tobacco products (IDHL. 1989, 40(1): 100)

The Tobacco Products (Public Health) Regulations 1988 requiring health warnings on cigarette packages (IDHL, 1989, 40(1): 100-101)

Bolivia

Decree-Law No. 15629 of 18 July, 1978 promulgating the Health Code (including Chapter IX. Tobacco) (IDHL, 1983, 34(2): 232)

Ministerial Resolution (Ministry of Social Welfare and Public Health) of 7 April, 1980 (establishing a health warning on cigarette packets)

Regulations of 15 March, 1982 on the use of tobacco (IDHL, 1983, 34(3): 538-539)

Ministerial Resolution No. 883 of 12 October, 1984 prohibiting smoking in any educational establishment, private or public, throughout Bolivia

Brazil

Law No. 7488 of 11 June, 1986 establishing a National Anti-Smoking Day (IDHL, 1989, 40(2): 406)

Order No. 490 of 25 August, 1988 restricting smoking in public places, requiring health warnings on tobacco packages, restricting advertising (IDHL, 1989, 40(2): 406)

Regulation No. 731 of the Ministry of Health dated 31 May, 1990 restricting advertising of tobacco products, requiring a health warning on packages and advertising, regulating smoking in health institutions and on airline flights, encouraging federal districts and municipalities to restrict smoking in public places, and forbidding sale of tobacco products to persons under 18 years of age (Resolution No. 490 of 25 August, 1988 is repealed)

Brazil (Rio Grande Do Sul)

Order No. 1/80-SSMA of 8 April, 1980 concerning smoking in the workplace, smoking in health institutions, restrictions on sales in health institutions (IDHL, 1981, 32(1): 87)

Law No. 7813 of 21 September, 1983 laying down provisions on smoking, and other provisions (IDHL, 1983, 34(4): 768)

Brazil (Sao Paulo)

Law No. 3.938 of 8 September, 1950 prohibiting smoking on public transportation vehicles, in elevators, and in places of public entertainment

Law No. 8.421, of 14 July, 1976 prohibiting smoking in indoor supermarkets and other stores

Law No. 9.032 of 27 March, 1980 concerning educational programmes in the schools on the harmful consequences of tobacco and alcohol consumption

Law No. 9.120 of 8 October, 1980 prohibiting smoking in public places, health establishments, elementary and secondary schools, and public urban transportation vehicles.

Law No. 2.845 of 20 May, 1981 prohibiting smoking on school premises, sports grounds, and public health establishments

Decree No. 17.451 of 22 July, 1981 regulating Law No. 9.120 of 8 October, 1980 prohibiting smoking in public places, hospitals, elementary and secondary schools

Bulgaria

Sanitary Rules approved by the Chief State Sanitary Inspector to restrict smoking in working premises and workplaces (IDHL, 1970, 21(3): 536)

Decree No. 2431 of 2 November, 1973 embodying the Law on public health (Section 58) (IDHL, 1974, 25(3): 502, 512-513)

Ordinance No. 2 on health requirements in connection with smoking (IDHL., 1980, 31(4): 752-755)

Burkina Faso

Order No. ANV-81 FP. SAN CAPRO. DP of 29 February, 1988 containing measures aimed at the control of smoking (IDHL, 1990, 41(2): 273-274)

Canada

The Tobacco Restraint Act (Chapter 266 of the Revised Statutes of Canada) Aeronautics Act: Air Regulations, amendment, SOR/87-554, dated 10 September, 1987 (IDHL, 1988 39(1): 86)

The Tobacco Products Control Act. An Act (C-51) to prohibit the advertising and promotion and respecting the labelling and manufacturing of tobacco products, dated 31 May, 1988 (IDHL, 1988, 39(4): 858-859)

The Non-smokers' Health Act. An Act (C-204) to regulate smoking in the federal workplace and on common carriers and to amend the Hazardous Products Act in relation to cigarette advertising, dated 31 May, 1988 (IDHL, 1988, 39(4): 859-860)

The Non-smokers' Health Act, 1988, Chapter 21, Revised Statutes of Canada, as amended by Chapter 7, Revised Statutes of Canada, 1988 (IDHL, 1990, 41(1): 83-84)

Tobacco Products Control Regulations, Statutes of Canada 1988, Chapter 20

The Non-smokers' Health Regulations, Dated 14 December, 1989 (IDHL, 1990, 41(1): 84-85)

Canada (British Columbia)

The Tobacco Products Act, Dated 27 October, 1972 (IDHL, 1975, 26(2): 268)

The Tobacco Products Regulations. British Columbia Regulation 258/72.

Regulation made by Order in Council 3941, approved 2 November, 1972 (IDHL, 1974, 25(1): 52)

Canada (Manitoba)

An Act to Protect the Health of Non-Smokers. 15 March, 1990

Canada (Nova Scotia)

The Drug Dependency Act. Dated 30 March, 1972 (IDHL, 1973 24(1): 66)

Canada (Ontario)

The Smoking in the Workplace Act. Dated 26 July, 1989 (IDHL, 1990, 41(3): 450-451)

Canada (Quebec)

An Act respecting the protection of non-smokers in certain public places, dated 18 June, 1986 (IDHL, 1987 38(1): 65-66)

Cayman Islands

The Tobacco Product and Intoxicating Liquor Advertising Law, 1986, making provision for the display of a health warning in connection with the advertisement of tobacco products and providing for the control of advertising of tobacco products and intoxicating liquor (IDHL, 1989, 40(3): 602)

Chile

Decree Law No. 2763 of 11 July, 1979 (IDHL, 1982, 33(4): 671) and Supreme Decree No. 395 of the Ministry of Health creating a National Commission for the Control of Smoking

Decree No. 106 of 8 April, 1981 prescribing a warning in connection with the marketing and advertising of tobacco (IDHL, 1982, 33(4): 732)

Circular No. 601/81 of the Ministry of Education, dated 11 May, 1981, restricting smoking by teachers and in the schools

Circular No. 3H/95 of 23 June, 1982 of the Ministry of Health prohibiting smoking by professionals, officials, and the general public in hospital rooms, clinics, waiting rooms, administrative offices serving the public, elevators, auditoriums and waiting rooms of the National Health Service

Law No. 18290 of February 1985 concerning the public transport of passengers and prohibiting smoking in the interior of public vehicles

Decree No. 1 of 2 January, 1986 establishing the National Commission for the Control of Smoking (IDHL, 1987, 38(4): 797-798)

Decree No. 164 of 4 June, 1986 prescribing a new warning for use in the marketing and advertising of tobacco (IDHL, 1987, 38(4): 787)

Circular No. 3F/123 of 13 August, 1986 of the Ministry of Health restricting smoking in the health facilities of the National Health Service

Circular No. 1-27 of the Ministry of Health of July 1989 concerning promotion of the anti-tobacco campaign in the community and in the schools of the municipal education system:

Circular No. 27 of 4 July, 1989 of the Ministry of the Interior recommending restrictions on smoking in government services and on the sale of tobacco products in kiosks and other places of the government services

China

Law of 29 June, 1991 on the exclusive sale of tobacco

of the People's Republic of China (IDHL, 1991, 42(4): 666)

Colombia

Decree No. 1188 of 25 June, 1974 promulgating the National Statute on Narcotics in which Section 20 restricts tobacco advertising in cinemas and the broadcast media (IDHL, 1978, 29: 23-26)

Resolution No. 001974 of 7 May, 1975 prescribing Regulations for the implementation of Sections 10 and 20 of Decree No. 1188 of 1974

Decree No. 3430 of 26 November, 1982 concerning restrictions on advertising of tobacco

Resolution No. 4063 of 1982 regulating Decree No. 3430 of 26 November concerning restrictions on advertising

Resolution No. 07559 of 12 June, 1984 creating the National Board on Tobacco and Health

Decree No. 3788 of 1986 concerning educational campaigns against tobacco

Colombia (Bogota)

Accord No. 3 of 1983 concerning smoking in public places, public vehicles, schools, health establishments, and government offices

Costa Rica

Decree No. 1520-SPPS of 24 February, 1971 requiring warnings on cigarette packages (IDHL, 1974, 24: 61)

Decree No. 17377-S of 13 January, 1987 concerning health warnings on packages (IDHL, 1987, 38(4): 787)

Decree No. 17967-S of 4 February, 1988 concerning restrictions on sales to minors (IDHL, 1989, 40(1):101)

Decree No. 17969-S of 4 February, 1988 concerning information programmes (IDHL, 1989, 40(1): 101)

Decree No. 18216-S-TSS of 23 June, 1988 concerning smoking in the workplace (IDHL, 1989, 40(1): 101)

Decree No. 18248-MOPT-S of 23 June, 1988 concerning smoking on public transport vehicles (IDHL, 1989, 40(1): 101-102)

Cote D'ivoire

Decree No. 79-477 of 6 June, 1979 prohibiting smoking in certain public places (IDHL, 1981, 32(2): 258)

Cuba

Ministerial Resolution No. 165 of 17 August, 1981 concerning smoking in health institutions, smoking in the workplace (IDHL, 1989, 40(2): 407)

Cyprus

The Health Protection (Smoking Control) Law of 1980 (IDHL, 1981, 32(4): 731-732)

The Health Protection (Smoking Control) Regulations of 1988 (IDHL, 1989, 40(1): 102)

Czechoslovakia[1]

Order No. 45 of 13 June 1966 of the Ministry of Health concerning the establishment and protection of healthy living conditions (Section 22) (IDHL, 1967, 18(2): 327)

Instructions No. 12 of 1972 of the Ministry of Health of the Czech Socialist Republic (sub-section 3 of Section 21)*

Instructions No. 13 of 1973 of the Ministry of Education of the Czech Socialist Republic (sub-section 2 of Section 3)*

Regulations of the Slovak Ministry of Health of 1 September, 1976 (prohibiting smoking in all health establishments)

Standard CSN 569560, Dated 1 November, 1979

Law No. 37/89 of 28 March, 1989 of the Czech National Council on protection against alcoholism and other addictions

Denmark

Law No. 398 of 10 June, 1987 on the Council on Prevention and the Council on Tobacco-induced Damage to Health (IDHL, 1988, 39(2): 396)

Order No. 416 of the Ministry of Culture prohibiting advertising of tobacco on television. Dated 18 June, 1987

Circular of 23 March, 1988 of the Ministry of Health on the provision of nonsmoking environments in public premises, means of transport, etc. (IDHL, 1989, 40(2): 407-408)

Order No. 441 of 26 July, 1988 on advertisements through local radio services (IDHL, 1989, 40(1): 603)

Law No. 426 of 13 June, 1990 on the labelling of tobacco products and the tar content of cigarettes (IDHL, 1991, 42(1): 49-50)

Order No. 507 of 28 June, 1990 on the labelling of tobacco products and the tar content of cigarettes (IDHL, 1991, 42(1): 50)

---

[1]There are a number of other items of legislation in Czehoslovakia restricting smoking in public transport vehicles, etc.

Ecuador

Supreme Decree No. 965 of 24 August, 1973 promulgating Regulations governing manufacturing, sales, and advertising activities associated with the use and consumption of cigarettes and alcoholic beverages (IDHL, 1978, 29(1): 64-65)

Egypt

Order No. 386 of 1977 of the Ministry of Communications and Culture prohibiting the advertising of cigarettes on radio and television

Law No. 52 of 20 June, 1981 on protection against the harmful effects of smoking (IDHL, 1981, 32(3): 459)

Egypt (Alexandria)

Alexandria Municipal Ordinance of 20 July, 1904 prohibiting smoking in theatres, amended on 29 December, 1938 and by Law No. 372 of 29 October, 1956 and again in 1964 to ban smoking in public places and advertising

El Salvador

Decree No. 955 of 11 May, 1988 promulgating the Health Code (IDHL, 1990, 41(1): 1-15)

European Communities

Council Directive No. 77/805/EEC of 19 December, 1977 amending Directive 72/464/EEC on taxes, other than turnover taxes, which affect the consumption of manufactured tobacco (IDHL, 1978, 29(2): 330-331)

Resolution of 18 July, 1989 of the Council and the Ministers of Health of the Member States, meeting within the Council, inviting the Member States to ban smoking in enclosed premises open to the public which form part of public or private establishments and to extend the ban on smoking to all forms of public transport (IDHL, 1989, 40(4): 890-891)

Council Directive 89/552/EEC of 3 October, 1989 on the coordination of certain provisions laid down by law, regulation or administrative action in Member States concerning the pursuit of television broadcasting activities

Council Directive 89/622/EEC of 13 November, 1989 on the approximation of the laws, regulations and administrative provisions of the Member States concerning the labelling of tobacco products, and requiring statements of tar and nicotine yields in specified form and health warnings on tobacco packages (IDHL, 1990, 41(1): 85-88)

Council Directive 90/239/EEC of 17 May, 1990 on the approximation of the laws, regulations and administrative provisions of the Member States concerning the maximum tar yield of cigarettes (IDHL, 1990, 41(3): 451-452)

Finland

Law No. 693 of 13 August, 1976 on measures to restrict smoking (IDHL, 1977, 28(3): 486-489)

Law No. 1147 of 31 December, 1976 concerning allocation of revenue from tobacco taxes to anti-smoking activities

Ordinance No. 225 of 25 February, 1977 on measures to restrict smoking (IDHL, 1977 28(3): 489-491)

Decree No. 227 of 25 February, 1977 of the Council of State on the maximum permitted limits for substances dangerous or harmful to health that are contained in tobacco products approved for sale or that are released when such products are smoked (IDHL, 1977, 28(3): 491)

Decree No. 254 of 25 February, 1977 of the Ministry of Social Affairs and Health on smokers' accessories (IDHL, 1977, 28(3): 491-492)

General Directive on Health Education (1978) DNO 3113/02/78

Decree No. 747 of 26 September, 1979 of the Council of State on the maximum levels of harmful substances in tobacco products approved for sale and the classification of self-burning factory-manufactured cigarettes (IDHL, 1980, 31(2): 307-308)

Ordinance No. 125 of 22 February, 1980 amending the Ordinance on measures to restrict smoking (IDHL, 1980, 31(2): 308)

Decree No. 770 of 15 November, 1984 of the Council of State on the maximum levels of harmful substances in tobacco products approved for sale and the classification of factory-manufactured cigarettes (IDHL, 1985, 36(4): 992-993)

Law No. 914 of 12 December, 1986 on product safety (IDHL, 1991, 42(1): 100-101)

Law No. 1037 of 11 December, 1987 concerning allocation of revenue from tobacco taxes to anti-smoking activities

France

Law No. 76-616 of 9 July, 1976 concerning measures to combat smoking (IDHL, 1976, 27(4): 732-734)

Decree No. 77-1042 of 12 September, 1977 prohibiting smoking in certain places intended for use by groups of people where this practice may have harmful effects upon health (IDHL, 1978, 29(2): 352-353)

Decree No. 77-1273 of 17 November, 1977 for the implementation of Law No. 76-616 of 9 July, 1976 concerning measures to combat smoking (IDHL, 1978, 29(2): 361)

Order of 30 January, 1978 establishing the list of substances which must be indicated on cigarette packaging units and the conditions for determining the presence of such substances (IDHL; 1979, 30(1): 44)

Order of 1 March, 1978 on the limits of the space devoted to advertising of tobacco in the printed press (IDHL, 1979, 30(1): 44)

Circular No. 519 of 1 March, 1978 on the opening of tobacconists' shops in hospital establishments (IDHL, 1979, 30(1): 44)

Order of 9 March, 1978 for the implementation of the third and fourth paragraphs of Section 10 of Law No. 76-616 of 9 July, 1976 concerning measures to combat smoking (IDHL, 1979, 30(1): 44)

Order of 2 October, 1978 concerning conditions for the implementation of the prohibition of smoking in aircraft (IDHL, 1979, 30(1): 45)

Decree No. 87-1108 of 23 November, 1978 for the implementation of the Law of 1 August, 1905 on fraudulent practices and misbranding with regard to products or services in respect of tobacco, tobacco products, and tobacco substitutes (IDHL, 1979, 30(4): 785-786)

Order of 5 June, 1979 on the additives premitted in the manufacture of tobacco and tobacco products and their substitutes (IDHL, 1979, 30(4): 786)

Law No. 83-25 of 19 January, 1983 prescribing various measures relating to social security (IDHL, 1983 34(4): 767)

Order of 4 March, 1983 determining the form to be taken by the distinctive mark to be apposed to the packaging of alcoholic beverages and tobacco (IDHL, 1983, 34(4): 767)

Order of 16 July, 1984 on the additives permitted in the manufacture of tobacco products and their substitutes (IDHL, 1985, 36(3): 648)

Circular No. 231 of 13 January, 1988 on the control of tobacco comsumption in hospital establishments (IDHL, 1989, 40(1): 102)

Law No. 89-18 of 13 January, 1989 amending the law of 9 July, 1976 concerning the fight against smoking by requiring advertising of products bearing the name or emblem of tobacco products to comply with the restrictions on advertising of tobacco

Law No. 91-32 of 10 January, 1991 on measures to combat tobacco use and alcoholism (IDHL, 1991, 42(1): 44-47)

Decree of 26 April, 1991 laying down the maximum tar content of cigarettes

Decree of 26 April, 1991 on the methods of analysis of the content of nicotine and tar and the methods of checking the accuracy of the information given on packages, as well as the modalities for showing health messages and the obligatory statements on packs of tobacco and tobacco products

Decree No. 92-478 of 29 May, 1992 establishing the conditions for application of the prohibition of smoking inpremises assigned for community use and modifying the Public Health Code.

French Polynesia

Deliberation No. 82-11 of 18 February, 1982 of the Territorial Assembly of French Polynesia on the orgnization of the control of tobacco abuse and tobaoco dependence in the Territory (IDHL, 1982, 33(2): 283)

Gambia

The Tobacco Products (Control of Advertisements) Act, 1985 (IDHL, 1986, 37(3): 570-571)

Germany[1]

Law on the protection of juveniles in public (version dated 27 July, 1957)

Law of 15 August, 1974 for the overall reform of foodstuffs law (IDHL, 1975, 26(3): 524-526)

Workplace Ordinance of 20 March, 1975 (Section 5)

The Tobacco Ordinance. Dated 20 December, 1977 (IDHL, 1978, 29(4): 737)

Greece

Decision No. A2a/5648 of 20 November, 1976 introducing the Child Health Booklet (IDHL, 1978, 29(2): 379-380)

Ministerial Decision No. A2g/Ec 1989 of 2 April, 1979 prohibiting smoking in hospital establishments and private nursing homes (IDHL, 1981, 32(3):459-460)

[1]The German Democratic Republic had a number of statutes restricting smoking and access to cigarettes. The terms of the unification of the two Germanys provide that the laws of the Federal Republic of Germany shall prevail.

Ministerial Decision No. A2g/EC 3051 of 25 April, 1980 prohibiting smoking in enclosed public places (IDHL, 1981, 32(3): 460)

Ministerial Decision No. A2/ik. 6139 of 23 December, 1988 on the labelling of tobacco products with regard to the harmful effect of tobacco on health (IDHL, 1989, 40(4): 844-845)

Guatemala

Decree No. 80-74 of 13 September, 1974 establishing a health warning on cigarette packets

Government Accord No. 681 of 3 August, 1990, prohibiting smoking in public transportation vehicles and in public places in government and private offices

Honduras

Law of the Honduran Institute for the Prevention of Alcoholism and Drug Addiction, Decree No. 136-89 of 14 October, 1989 providing for control of smoking in public places

Hong Kong

The Smoking (Public Health) (Notices) (Amendment) Order 1983 (IDHL, 1983, 34(4): 768-769)

The Smoking (Public Health) (Amendment) Ordinance 1984 (IDHL, 1984, 35(4): 771-772)

The Smoking (Public Health) (Amendment) Regulations 1984 (IDHL, 1984, 35(4): 772)

The Smoking (Public Health) (Specification of Newspaper Circulation) Notice 1984 (IDHL, 1984, 35(4): 772)

The Public Health and Municipal Services (Amendment) (No. 2) Ordinance 1986 containing a definition of "smokeless tobacco products" (IDHL, 1987, 38(1): 66-67)

The Smokeless Tobacco Products (Prohibition) Regulations 1987 (IDHL, 1987, 38(1): 67)

The Hong Kong Council on Smoking and Health Ordinance 1987 (IDHL, 1988, 39(1): 86-87)

Hungary

Regulations No. 118244 of 1972 of the Minister of Posts and Transport concerning the transportation of passengers, baggage, and urgent goods, taking into account all subsequent amendments and regulations (Sections 7 and 8)

Policy Statement No. 127128 of 1973 of the Minister of Culture concerning the internal regulations of schools

Policy Statement No. 4 of 1976 of the Ministry of Internal Trade concerning smoking in commercial and hotel establishments

Instructions No. 500551 of 1976 introducing provisions intended to restrict smoking in buses

Instruction No. 10 of 1977 of the Minister of Agriculture and Food concerning the warning notice that must appear on confectionery products and on tobacco products

Ordinance No. 10 of 14 May, 1977 of the Minister of Internal Trade amending Ordinance No. 13 of 12 March, 1968 concerning the marketing of tobacco products for purposes of consumption

Recommendation of the Ministry of Internal Trade of 1979 concerning abstention from smoking during children's mealtimes

Policy Statement No. 31 of 1980 of the Ministry of Health on measures to control smoking (IDHL, 1980, 31(4): 839)

Iceland

Law No. 59 of 18 April, 1971 amending Law No. 63 of 28 May, 1969 on State trade in alcoholic beverages, tobacco, and medicaments (IDHL, 1978, 29(3): 581)

Law No. 27 of 11 May, 1977 prescribing measures to reduce smoking (IDHL, 1979, 30(3): 570)

Law No. 74 of 28 May, 1984 on the prevention of the use of tobacco (IDHL, 1984, 35(4): 772)

Regulation No. 499 of 28 December, 1984 concerning warnings on tobacco products (IDHL, 1986, 37(1): 47-48)

India

The Cigarettes (Regulation of Production, Supply and Distribution) Act, 1975 (IDHL, 1977, 28(4): 996-999)

The Cigarettes (Regulation of Production, Supply and Distribution) Rules, 1976 (IDHL, 1977, 28(4): 999)

Prohibition of tobacco smoking in public places. Office Memorandum No. 27/1/3/90 Cab. Government of India, Cabinet Secretariat, New Delhi, 7 May, 1990

India (State of New Delhi)

The Delhi (Place of Public Entertainment) Prohibition of Smoking Act, 1953

Ireland

The Tobacco Products (Control of Advertising, Sponsorship and Sales Promotion) Act, 1978 (IDHL, 1979, 30(4): 806-807)

The Tobacco Products (Control of Advertising, Sponsorship and Sales Promotion) Regulations, 1979 (IDHL, 1980, 31(3): 537-538)

The Health (Restricted Article) Order, 1985 (IDHL, 1986, 37(1): 48)

The Tobacco Products (Control of Advertising, Sponsorship and Sales Promotion) Regulations, 1986 (IDHL, 1986, 37: 48-49, 571)

The Tobacco (Health Promotion and Protection) Act, 1988 (IDHL, 1989, 40(1): 103-104)

The Tobacco (Health Promotion and Protection) Regulations, 1990. Dated 23 February, 1990 (IDHL, 1990, 41(3): 452)

Israel

The Restriction of Smoking (Public Places) Law, 1983 (IDHL, 1984, 35(3): 617)

The Aircraft (Operation of Aircraft and Flight Rules) (Amendment) Regulations, 1987 (IDHL, 1988, 39(4): 861)

The Criminal Procedure (Violations Subject to a Fine—Smoking Restrictions in Public Places) Order, 1986 (IDHL, 1988, 39(4): 861)

Order restricting smoking in public places (change in Appendix to the Law of 1983), 1988

Italy

Law No. 165 of 10 April, 1962 prohibiting advertisements for tobacco products

Law No. 584 of 11 November, 1975 prohibiting smoking in specified premises and on means of public transport (IDHL, 1977, 28(3): 621)

Ministerial Decree of 18 May, 1976 laying down provisions governing airconditioning or ventilation systems as referred to in Law No. 584 of 11 November, 1975 prohibiting smoking in specified premises and in means of public transport (IDHL, 1977, 28(4): 1018)

Japan

Law No. 33 of 7 March, 1900 prohibiting smoking by minors, as amended by Law No. 251 of 18 December, 1948

Jordan

Law No. 64 of 1977 for the protection of public health against health hazards caused by smoking (IDHL, 1983, 34(2): 283)

Decision No. 1 of 1980 of the Ministry of Health establishing a health warning on cigarette packets (IDHL, 1985, 36(4): 993)

Kenya

The Public Health (Warning on Cigarette Smoking) Rules, 1984 (IDHL, 1985, 36(4): 993)

Kuwait

Ministerial Resolution No. 25 of 1980 on the particulars to be stated on cigarette packages (IDHL, 1981, 32(2): 258)

Ministerial Decree No. 180 of 28 April, 1988 laying down detailed provisions on the analysis of the components of imported cigarettes (IDHL, 1988, 39(4): 861)

Kuwait (Municipality of Kuwait)

Resolution No. 981 of 1980 of the Mayor of the Municipality of Kuwait (cancelling licences for advertisements for tobacco products within the municipality) (IDHL, 1981, 32(2): 258)

Lebanon

Decree-Law No. 101 of 16 September, 1983 on the warning of citizens of the harmful effects of tobacco (IDHL, 1985, 36(1): 63-64)

Luxembourg

Law of 24 March, 1989 relating to the restriction of advertising of tobacco products and prohibiting smoking in certain places (IDHL, 1990, 40(4): 845)

Law of June, 1990 implementing the Law of 24 March, 1989 restricting advertising of tobacco products and prohibiting smoking in certain places (IDHL, 1990, 41(4): 636)

Macao

Law No. 3/83/M of 11 June, 1983 on the prevention and control of smoking (IDHL, 1985, 36(3): 648-649)

Malta

The Tobacco (Control of Advertisement) Act, 1970 (IDHL, 1973, 24(1): 183)

The Prohibition of Advertisement of Tobacco on Television and in Cinemas Regulations, 1970 (IDHL, 1973, 24(1): 183)

The Tobacco (Smoking Control) Act, 1986 (IDHL, 1987, 38(3): 544-547)

The Health Warning (On Smoking) Regulations, 1987 (IDHL, 1989, 40(2): 408)

Smokeless Tobacco (Ban) Regulations 1988, to prohibit the importation, manufacture, preparations, storage, sale or supply of smokeless tobacco

Mauritius

Public Health (Prohibition of Smoking) Regulations 1990, requiring health warnings, prohibiting sales of cigarettes to minors, and restricting smoking in public places

Mexico

Decree of 26 February, 1973 promulgating the Health Code of the United Mexican States (Sections 249-252) (IDHL, 1974, 25(1): 123-141)

Regulations of 16 December, 1974 on advertising for foodstuffs, beverages, and medicaments (IDHL, 1976, 27(1): 163-168)

The Geneal Law on Health, dated 26 December, 1983, including Chapter II, Programme Against Smoking, Chapter XI, Tobacco, and Title 13, Advertising (IDHL, 1986, 37(3): 493, 498, 501)

Coordination Agreement of 10 November, 1986 between the Federal Executive and the Executive of the State of Tabasco, for the purpose of supporting the Smoking Control Programme (IDHL, 1987, 38(4): 787-788)

Decree of the Secretary of Health of 17, April, 1990 restricting smoking in medical facilities of the Secretary of Health and in the National Institutes of Health

Mexico (Federal District)

Regulation for the protection of non-smokers dated 5 July, 1990 prohibiting smoking in indoor public places, public transport vehicles, public and private schools, hospitals and clinics, government offices, cinemas, theatres, and shops and business places where the public is being served

Netherlands

The Cigarette and Cigarette Tobacco Labelling Decree (Commodities Law) of 29 April, 1981 (IDHL, 1982, 33(1): 33)

Decree of 9 December, 1986 amending the Cigarette and Cigarette Tobacco Labelling Decree (IDHL, 1988, 39(1): 87-88)

Law of 10 March, 1988 prescribing measures for the restriction of tobacco use, and in particular for the protection of non-smokers (IDHL, 1989, 40(2): 408-410)

New Zealand

The Police Offences Act 1927 (reprint as of 1 January, 1974) (IDHL, 1975, 26(2): 398-399)

The Toxic Substances Act 1979 (Sections 72-74) (IDHL, 1980, 31(4): 855-856)

The Toxic Substances Regulations 1988. Amendment No. 2. Dated 21 March, 1988

The Smoke-free Environments Act 1990 (IDHL, 1990, 41(4): 636-640)

The Smoke-free Environments Regulations 1990

Nicaragua

Decree of 30 June, 1976 (establishing a health warning on cigarette packets)*

Nigeria

The Tobacco Smoking (Control) Decree 1990, Decree No. 20. Dated 25 June, 1990 (IDHL, 1990, 41(4): 640-641)

Norway

Law of 19 April, 1899 for the protection of children against the use of tobacco

Law No. 14 of 9 March, 1973 on restrictive measures in trade in tobacco products, etc. (IDHL, 1975, 26(3): 571-572)

Regulations of 25 October, 1974 on the labelling of packets of cigarettes, packets of smoking tobacco, and packets of cigarette paper (IDHL, 1975, 26(3): 572), as amended on 5 June, 1975 (IDHL, 1976, 27(3): 593-594)

Regulations of 25 October, 1974 on the prohibition of tobacco advertising (IDHL, 1975, 26(3): 572)

Regulations of 29 November, 1977 on conditions governing exemptions from the prohibition on advertising in the law on restrictive measures in trade in tobacco products, etc. (IDHL, 1979, 30(1): 125-126)

Regulation No. 830 of 25 March, 1983 on the labelling of packets of cigarettes, packets of smoking tobacco, and packets of cigarette paper (IDHL, 1984, 35(2): 344-345)

Regulation No. 2145 of 14 December, 1984 on asbestos (IDHL, 1985, 36(4): 1060)

Law No. 24 of 6 May, 1988 amending Law No. 14 of 9 March, 1973 on restrictive measures in trade in tobacco products (IDHL, 1988, 39(4): 862)

Regulation No. 563 of 8 July, 1988 on protection against tobacco-induced injury (IDHL, 1988, 39(4): 862-863)

Regulation No. 925 of 4 September, 1989 amending the Regulations concerning exceptions to the ban on advertising contained in Section 2 of the Law on protection against tobacco-induced injury (IDHL, 1990, 41(2): 274)

Regulation No. 1044 of 13 October, 1989 on the

prohibition of new products containing tobacco and nicotine (IDHL, 1990, 41(2): 275)

Pakistan

The Cigarettes (Printing of Warning) Ordinance, 1979 (IDHL, 1984, 35(1): 74-75)

The Cigarettes (Printing of Warning) Rules, 1982 (IDHL, 1984, 35(1): 75)

Panama[1]

Cabinet Decree No. 56 of 17 March, 1970 prescribing measures against cigarettes (IDHL, 1973, 24(3): 581)

Decree No. 129 of 19 June, 1978 prescribing measures on the advertising of alcoholic beverages, non-alcoholic beverages, cigarettes, and tobacco

Papua New Guinea

The Tobacco Products (Health Control) Act 1987 (IDHL, 1988, 39(4): 863-865)

Paraguay

Law No. 836/80 promulgating the Health Code, 15 December, 1980 in which Sec. 202 restricts advertising of tobacco and authorizing the Ministry of Health to require a health warning on tobacco products (IDHL, 1981, 32(4): 624-634)

Resolution S.G. No. 20 of the Ministry of Public Health and Social Welfare, 23 January, 1990, prohibiting smoking in the facilities of the Ministry of Public Health and Social Welfare and setting forth means of control

Decree-Law No. 4012 regulating Articles 202-205 of the Sanitary Code on Advertising of Tobacco and Alcohol

Paraguay (Asuncion)

Capital Municipality Transit Rule 298 of August, 1981 prohibiting smoking in urban passenger vehicles

Capital Municipality Ordinance 15.381 dated 2 February, 1984 prohibiting smoking in cinemas, theatres, and other similar public places

Order of the Municipal Council, Article 298 in relation to World No-Tobacco Day, 1991 prohibiting smoking in collective public transportation vehicles

Peru

Supreme Decree No. DS-0079-70-SA of April 1970 requiring health warnings on cigarette packages and advertising and restricting advertising (IDHL, 1977, 28(3): 689)

Law No. 23482 of 20 October, 1982 on the selective consumption tax on cigarettes made from blond tobacco (IDHL, 1987, 38(1): 67)

Poland

Ordinance of 4 June, 1974 of the Minister of Health and Social Welfare concerning restrictions on smoking based on health grounds (IDHL, 1975, 26(1): 200-201)

Portugal[2]

Order No. 212 of 18 April, 1978 prohibiting smoking in urban public transport as well as in inter-urban public transport on journeys lasting up to one hour (IDHL, 1979, 30(3): 638-639)

Order No. 375/78 of 11 July, 1978 prescribing Regulations for the implementation of Order No. 212/78 of 18 April, 1978 concerning the prohibition of smoking in public transport, prescribing the manner in which the prohibition and, in certain cases, permission to smoke are to be indicated, and determining the right to carry out inspection and collect fines and the manner in which this is to be done (IDHL, 1979, 30(3): 639)

Decree-Law No. 421/80 of 30 September, 1980 prescribing rules governing advertising (IDHL, 1981, 32(3): 481-482)

Law No. 22/82 of 17 August, 1982 on the control of smoking (IDHL, 1985, 36(1): 64)

Decre-Law No. 226/83 of 27 May, 1983 for the implementation of Law No. 22/82 of 17 August, 1982 on the control of smoking, and establishing the Smoking Control Council (CPT) (IDHL, 1985, 36(1): 64-65)

Romania

Law of 6 July, 1978 on the safeguarding of the health of the population (Section 84) (IDHL, 1979, 30(2): 263)

Saudi Arabia

Royal Decree M/10 dated 1392 (H)-03-03 (1982) requiring health warnings in Arabic and English on cigarette packages as well as tar and nicotine contents, and specifying maximum tar and nicotine contents

---

[1]There are a number of other provisions in Panama restricting smoking in buses, cinemas, etc.

[2]According to a report in the *Daily Telegraph* (London) of 11 June 1981, the Portuguese Parliament has approved a ban on smoking by children under 16 years of age.

Circular of 18 October, 1983 prohibiting smoking in ministerial offices, Government agencies, and public institutions (IDHL, 1986, 37(1): 49)

Senegal

Law No. 81-58 of 9 November, 1981 prohibiting advertising for tobacco and smoking in certain public places (IDHL, 1983, 34(1): 90)

Ministerial Order No. 8236 M.S.P.-D.P.H. of 30 July, 1982 prescribing the substances that must be indicated on the packaging units of tobacco or tobacco products and the conditions for determining the presence of such substances (IDHL, 1983, 34(1): 90)

Law No. 85-23 of 25 February, 1985 amending Law No. 81-58 of 9 December, 1981 prohibiting the advertising of tobaco and smoking in certain public places (IDHL, 1986, 37(4): 798)

Decree No. 85-1375 of 28 December, 1985 specifying the conditions governing the promotion or advertising of tobacco or tobacco products (IDHL, 1986, 37(4): 798-799)

Singapore

The Prohibition of Smoking in Certain Places Act, 1970 (IDHL, 1973, 24(1): 217)

The Prohibition of Advertisements relating to Smoking Act, 1970 (IDHL, 1973, 24(1): 218)

The Environmental Public Health (Food Handlers) Regulations 1973

The Consumer Protection (Warning Against Danger of Smoking) Regulations, 1980 (IDHL, 1981, 32(3): 460)

The Smoking (Prohibition in Certain Places) Notification 1988. Dated 15 June, 1988 (IDHL, 1990, 41(2): 276)

The Consumer Protection (Labelling of Tobacco Product Containers) Regulations 1989, Dated 11 January, 1989 (IDHL, 1990, 41(2): 276)

An Act to amend the Smoking (Prohibition of Advertisements) Act 1989, Dated 7 November, 1989

Spain

Crown Decree No. 1100 of 12 May, 1978 regulating advertising for tobacco and alcoholic beverages by State broadcasting media (IDHL, 1978, 29(4): 817)

Crown Decree No. 1259 of 4 April, 1979 on the designation of cigarette products as "low-nicotine" and "low-tar" (IDHL, 1981, 32(4): 732)

Order of 23 May, 1980 approving the technological standards necessary in order to perform the chemical analysis of cigarette products as referred to in Crown Decree No. 1259 of 4 April, 1979 (IDHL, 1981, 32(4): 732)

Crown Decree No. 709/1982 of 5 March, 1982 regulating the advertising and use of tobacco (IDHL, 1985, 36(2): 403-404)

Crown Decree No. 192/1988 of 4 March, 1988 laying down restrictions on the sale of tobacco, for the protection of the health of the population (IDHL, 1988, 39(3): 653-656)

Order of 8 June, 1988 for the implementation of Crown Decree No. 192/1988 laying down restrictions on the sale of tobacco, for the protection of the health of the population (IDHL, 1989, 40(3): 603)

Order of 7 November, 1989 prohibiting the sale and distribution of tobacco and alcoholic beverages at public school centres under the authority of the Ministry of Education and Science (IDHL, 1991, 42(3): 479-480)

Spain (Catalonia)

Law No. 20 of 25 July, 1985 on prevention and care in regard to potentially dependence-producing substances

Spain (Valencia)

Order of 20 June, 1988 of the Council on Health and Consumer Affairs approving the Programme on the Reduction of Tobacco Consumption (IDHL, 1989, 40(3): 603-605)

Sri Lanka[1]

Direction 14 under Section 6(1)(c) of the Consumer Protection Act, No. 1 of 1979 (IDHL, 1980, 31(2): 383-384)

Sudan

The Regulation of Cigarette Smoking Act, 1982. Dated 12 April, 1983 (IDHL, 1983, 34(4): 769)

Regulation No. 16 of 1983 on the control of smoking (IDHL, 1986, 37(1): 49)

Sweden

Law of 11 December, 1975 concerning the warning notices and declarations of content to be displayed on tobacco products (IDHL, 1976, 27(4): 868-869)

Order No. 32 of 31 May, 1976 of the National Board of Health and Welfare concerning the warning notices and declaration of content to be displayed on tobacco products (IDHL, 1976, 27(4): 869-870) as amended by Order No. 13 of 24 February, 1978 (IDHL, 1979, 30(2): 320-321) and Order No. 17 of 26 February, 1981 (IDHL, 1981, 32(3): 460-461)

---

[1]There are also a number of other provisions in Sri Lanka restricting smoking in public transport vehicles.

Law No. 764 of 30 November, 1978 prescribing certain provisions concerning the marketing of tobacco products (IDHL, 1980, 31(3): 556)

Guidelines No. 7 of the National Board for Consumer Policies on the marketing of tobacco products (IDHL, 1980, 31(3): 556-558)

Order No. 18 of 26 February, 1980 of the National Board of Health and Welfare on the mean values determined in 1980 for the content of cigarette smoke in accordance with Annex B to Order No. 32 of 31 May, 1976 of the National Board of Health and Welfare concerning the warning notices and declaration of content to be displayed on tobacco products (IDHL, 1980, 31(3): 558-559)

Order No. 17 of 26 February, 1981 of the National Board of Health and Welfare amending Order No. 32 of 31 May, 1976 concerning the warning notices and the declaration of content to be displayed on tobacco products (IDHL, 1981, 32(3): 460-461)

General Recommendations (AFS 1893: 10) concerning smoking restrictions, issued by the National Board of Occupational Safety and Health and the National Board of Health and Welfare (IDHL, 1988, 39(1): 88)

Order No. 10 of 10 April, 1986 of the National Board of Health and Welfare amending Order No. 32 of 1976 of the National Board of Health and Welfare concerning the warning notices and declarations of contents to be displayed on tobacco products (IDHL, 1988), 39(1): 88-89)

Order No. 6 of 5 February, 1988 of the National Board of Health and Welfare concerning the warning notice and declaration of contents on tobacco products (IDHL, 1989, 40(1): 104)

Switzerland

Ordinance on foodstuffs. Amendments of 18 October, 1978 (IDHL, 1979, 30(2): 327-328)

Switzerland (Canton of Vaud)

Law of 29 May, 1985 on public health (Sections 52-54) (IDHL, 1985, 36(4): 942)

Thailand

Order of the Commission on Advertisement of the Consumer Protection Board providing for a total ban on tobacco advertising in accordance with Cabinet resolutions of 26 April, 1989 and 20 December, 1989 approving the principles of tobacco control. *Royal Gazette.* Vol. 106, Section 25, 10 February, 1989

Tobacco Products Control Act 1992. *Royal Gazette*, Vol. 109, Section 38. 5 April, 1992

Nonsmokers' Health Protection Act 1992 *Royal Gazette*, Vol. 109, Section 40, 7 April, 1992

Trinidad and Tobago

An Act relating to the protection of children and young persons, industrial schools, and orphanages, and juvenile offenders (the Children Act), Chap. 46: 01. Laws of Trinidad and Tobago, 17 March, 1925

Compulsory standard. Requirements for labelling: Part II—Labelling of retail packages of cigarettes. TTS 2110500 Part II, 10 March, 1989

(Former) Union of Soviet Socialist Republics

Resolution No. 706 of 12 June, 1980 of the Central Committee of the Communist Party of the Soviet Union and the USSR Council of Ministers on measures to intensify the campaign against smoking (IDHL, 1981, 32(1): 87-90)

United Kingdom

The Children and Young Persons Act 1933 (Section 7)

The Television Act 1964*

The Finance Act 1978 (IDHL, 1979, 30(3): 653)

The Tobacco Products (Higher Tar Cigarettes) Regulations 1978 (IDHL, 1979, 30(3): 653)

An Act (Chapter 34) to amend the Children and Young Persons Act 1933, and the Children and Young Persons (Scotland) Act 1937, to make it an offence to sell any tobacco product to persons under the age of sixteen, and for connected purposes (IDHL, 1986, 37(4): 799-800)

The Oral Snuff (Safety) Regulations 1989. Dated 13 December, 1989 (IDHL, 1990, 41(3): 452-453)

United States of America[1]

The Federal Cigarette Labeling and Advertising Act, 1965 as amended by the Public Health Cigarette Smoking Act, 1969 and the Comprehensive Smoking Education Act, 1984 (IDHL, 1971 22: 998: IDHL, 1985, 36(3): 649)

The Comprehensive Smoking Education Act concerning information programmes, warnings on packages, evaluation of smoking control programmes, advertising restrictions (IDHL, 1985, 36(3): 649-652)

The Comprehensive Smokeless Tobacco Health Education Act of 1986 concerning information programmes, smokeless tobacco, restrictions on sales to minors, health warnings on pack

[1]There are many other items of legislation relating to smoking control at the State, county, and municipal levels.

ages, advertising restrictions, levels of toxic constituents, evaluation of smoking control programmes (IDHL, 1987, 38(1): 67-70)

Regulations under the Comprehensive Smokeless Tobacco Health Education Act of 1986 (IDHL, 1987, 38(3): 547)

Smoking Regulations Part 101-20 Management of Buildings and Grounds) of Title 41 (Public Contracts and Property Management) of the US Code of Federal Regulations (IDHL, 1987, 38(3): 547-548)

The Department of Transportation and Related Agencies Appropriations Act 1988 concerning smoking on board aircraft (IDHL, 1988, 39(4): 865)

Smoking aboard aircraft. Parts 121 and 135 of Title 14 (Aeronautics and Space) of the US Code of Federal Regulations (IDHL, 1989, 40(1): 104)

An Act (Public Law 101-164) making appropriations for the Department of Transportation and related agencies for the fiscal year ending 30 September, 1990 and concerning the permanent prohibition against smoking on scheduled airline flights. Dated 21 November, 1989 (IDHL, 1990, 41(3): 453)

Prohibition against smoking; final rule. Parts 121, 129, and 135 of Title 14 (Aeronautics and Space) of the US Code of Federal Regulations. Dated 28 February, 1990 (IDHL, 1990, 41(3): 453)

United States of America (California)

Chapter 139 of the Laws of 1980...relating to smoking (IDHL, 1981, 32(1): 90)

United States of America (Kentucky)

Chapter 438 of the Human Resources Law of Kentucky, 1976 Supplement (IDHL, 1977, 28(2): 394-395)

United States of America (Minnesota)

The Minnesota Clean Indoor Air Act. Approved 2 June, 1975 (IDHL, 1976, 27(3): 704-705)

United States of America (New York)

An Act to amend the penal law, in relation to a label upon cigarette packages. Approved 24 June, 1965 (IDHL, 1967, 18(2): 442)

An Act to amend chapter seven hundred eighty-seven of the laws of nineteen hundred sixty-seven, entitled "An Act directing the commissioner of education to establish a five-year programme for critical health problems, and making an appropriation therefore", in relation to the continuing of such programme. Approved 8 May, 1970 (IDHL, 1974, 25(2): 458)

An Act to amend the public health law, in relation to smoking restrictions and to repeal article 13-E of such law relating thereto. Approved by the Governor: 5 July, 1989 (IDHL, 1990, 41(1): 88)

United States of America (Utah)

Chapter 10 of the Laws of 1976 (IDHL, 1979, 30: 146)

Uruguay

Resolution No. 1150/970 of 21 July, 1970 assigning the Ministry of Health the task of studying the effects of smoking and disseminating information thereon through a special commission (IDHL, 1973, 24(3): 680)

Resolution 765602 adopted 23 September, 1976 prohibiting smoking in the clinics and hospital of the Faculty of Medicine by physicians, students, and technical and administrative personnel; requiring inclusion of smoking histories in patient charts; establishing smoking cessation programmes in the hospital; intensifying education against tobacco in the maternal and child clinics; and increasing information on smoking and its risks at all levels of instruction—professional, middle-level, and primary education

Decree No. 407/981 of 17 December, 1980 prohibiting the smoking of tobacco products in any form in buses used for interdepartmental transport of passengers

Law No. 15361 of 24 December, 1982 adopting provisions on the advertising and marketing of cigarettes, cigars, and tobacco products (IDHL, 1983, 34(3): 539)

Decree No. 263983 of 22 July, 1983 regulating the marketing and advertising of tobacco products

Law No. 15656 of 10 October, 1984 extending the interval for publishing the maximum contents of nicotine and tar by cigarette manufacturers and importers (IDHL, 1988, 39(2): 396)

Resolution of the Chamber of Deputies dated 9 May, 1989 prohibiting smoking in the plenary sessions and working committee meetings of the Chamber of Deputies

Ministry of Public Health, Special Order No. 3904 (undated) prohibiting smoking in the hospitals of the Ministry of Public Health by patients and their visitors, and by physicians, students, and technical and administrative personnel while on duty and in contact with patients, and calling for intensified education on tobacco especially in the maternal and child health clinics, and requiring inclusion of information on smoking in clinical histories in the hospital

Uruguay (Montevideo)

Decree No. 16.750 of 21 March, 1975 prohibiting smoking by bus drivers for school children

Decree No. 19.067 of March 1979 concerning requirements for theatrical performances, including authorization for the sale of non-alcoholic drinks, cigarettes, and other items in theatres

Decree 407/981 of 12 August, 1981 concerning smoking on interdepartmental passenger transportation

Venezuela

Law of 13 September, 1978 prescribing the tax on cigarettes and tobacco products (IDHL, 1979, 30(4): 925)

Decree No. 3007 of 2 January, 1979 prescribing regulations for the implementation of the law prescribing the tax on cigarettes and tobacco products (IDHL, 1979, 30(4): 925)

Decree No. 849 of 21 November, 1980 prohibiting the transmission by television stations of any commercial advertising which directly or indirectly encourages the consumption of cigarettes and other products derived from tobacco manufacture (IDHL, 1982, 33(3): 499)

Decree No. 996 of 19 March, 1981 prohibiting the transmission by radio stations of any commercial advertising which directly or indirectly encourages the consumption of cigarettes and other products derived from tobacco (IDHL, 1982, 33(3): 499)

Resolution of 23 October, 1984 establishing a Standing Honorary National Council attached to the Division of Chronic Disease of the Ministry of Health and Social Welfare for studying health problems associated with smoking and with a view to formulating policies for the prevention of smoking and the organic diseases resulting therefrom (IDHL, 1986, 37(2): 276-277)

Resolution of 1985 of the Venezuelan Institute of Social Security prohibiting smoking in administrative and medical care facilities.

Viet Nam

Law of 30 June, 1989 on the protection of public health (IDHL, 1990, 41(1): 15-27)

Yugoslavia

Decree of 29 November, 1972 to promulgate the Law on the wholesomeness of foodstuffs and consumer goods (IDHL, 1974, 25(1): 249)

Appendix 7

Effect of Legislation on Tobacco Consumption

Table A4.1 Annual average rates of change in the percentage of adults smoking according to tobacco advertising restrictions in force, 1970-86

Country and year of ban	*Age group studied*	*Years of study*	*% who smoke*		*Annual percentage change over study*	
			beginning	*end*	*per country*	*per group*
Enforced tobacco advertising ban						
Iceland, 1972	18-69	1985-86	40.0	36.2	—9.5	
Finland, 1978	15-64	1978-86	25.8	25.5	—0.1	average
Norway, 1975	16-74	1973-86	41.5	34.5	—1.1	—3.6
Portugal, 1983	na	1983-84	na	23.5	na	
Tobacco promotion in few media						
Belgium	15 +	1980-87	41.5	32.0	— 3.3	
France	15 +	1976-83	44	39	— 1.6	
Italy (enforced ban)	14 +	1980-83	34.9	31.1	— 3.6	average
New Zealand	15 +	1976-86	35	30	— 1.4	—2.5
Singapore	15 +	1977-84	23	19	— 2.5	
Sweden	16 +	1976-85	38.5	29.6	— 2.6	
Tobacco promotion in most media						
Australia	16 +	1972-83	38.0	33.5	— 1.1	
Austria	15 +	1972-81	28	27	— 1.4	
Belgium	15 +	1970-79	20.2	21.3	+ 0.6	
Canada	15 +	1970-86	45	32	— 1.8	
Denmark	15 +	1970-79	57.9	48.9	— 1.7	
France	15 +	1970-76	53.1	46.7	— 2.0	
Federal Republic of Germany	15 +	1970-84	41.7	36	— 1.0	average —1.2
Ireland	15 +	1973-82	43	35	— 2.1	
Netherlands	15 +	1970-86	58.5	38.0	— 2.2	
Switzerland	15 +	1972-81	34.6	37.5	+ 0.9	
United Kingdom	16 +	1972-86	46	33	— 2.0	
United States	17 +	1970-86	36.6	26.5	— 1.7	
Tobacco promotion in all media						
Greece	15 +	1985	na	21.5	na	
Japan	20 +	1970-86	46.6	37.6	— 1.2	— 1.2
Spain	15 +	1970-87	na	40.5	na	

Source: *Health or tobacco. An end to tobacco advertising and promotion.* Wellington, Toxic Substances Board, 1989.
na = data not available.

Health warnings on Tobacco Products in Sweden

- Do not expose your colleagues to tobacco smoke. It is harmful and irritating. *National Board of Health and Welfare.*
- Tobacco smoke contains many carcinogens. The smoke in passive smoking, which affects those in the vicinity of the smoker, contains the highest level of carcinogens. *National Board of Health and Welfare.*
- Tobacco smoking lowers resistance to infections, including those of the respiratory tract. *National Board of Health and Welfare.*
- Do not smoke when children are present. Smoke irritates their respiratory tract. *National Board of Health and Welfare.*
- If you are pregnant or breast-feeding, do not smoke; both you and your child may be harmed. *National Board of Health and Welfare.*
- In 1983, 779 persons died in traffic accidents and at least 8000 from tobacco smoking. *National Board of Health and Welfare.*
- Lung cancer among women will soon be commoner than breast cancer. Smoking is the cauase. *National Board of Health and Welfare.*
- Almost all persons who suffer from arteriosclerosis affecting the blood vessels of the legs and have difficulty in walking are smokers. *National Board of Health and Welfare.*
- It is practically only smokers who have a heart attack before they are 50. *National Board of Health and Welfare.*
- Smoking heightens the risk of inflammation of the gums. This can lead to loss of teeth. *National Board of Health and Welfare.*
- If one person smokes, everyone smokes. Most tobacco smoke enters the air breathed by everyone. Your smoking may harm others. *National Board of Health and Welfare.*
- The combination of smoking and contraceptive pills heightens the risk of a heart attack for women over 30. *National Board of Health and Welfare.*

References

Adriaanse H. Van Reek J (1992) Adults in the European community favour a tobacco advertising ban. *Tobacco control. 1: 57.*

Aho v. Rettig Oy and Suomen Tupakka Oy, 5.1 TPLR 2.4, 5.3 TPLR 2.119 (1990). *Alexander v. California Unemployment Insurance Appeals Bd.* 162 Cal. Rptr 411 (1980).

Amos A (1990) Women's magazines and tobacco—the preliminary findings of a survey on the tobacco policies of the top women's magazines in Europe. In Durston B. Jamrozik K. eds. *Tobacco and health 1990. The global war, Proceedings of the Seventh World Conference on Tobacco and Health*, Perth, Health Department of Western Australia, pp. 912-916.

Appeal of Joan Clay, Case no. 2:11:1935, Comm'r's File CI/364/1989 (Social Security Comm'r. United Kindgom, 16 July, 1990).

Asian advertising review 88-89 (1989) *Asian advertising and marketing*, April: 24-76.

Atkinson AB. Skegg (Townsend) JL (1973) Antismoking publicity and the demand for tobacco in the UK. *Manchester school*, 41: 265.

Australian Federation of Consumer Organisations v. *Tobacco Institute of Australia*, 6.1 A TPLR 2.77 (Federal Court of Australia, 1991).

Bagdikian BH (1987) *The media monopoly, 2nd ed*, Boston, Beacon Press.

Bal DG et al. (1990) Reducing tobacco consumption in California. *Journal of the American Medical Association, 264: 1570-1574.*

Ball K (1986) Smoking spells death for millions. *World health forum*, 7: 211-216.

Ball K (1989) Tobacco: a doctor speaks. *African business*, June 12-14.

BASP Newsletter (1990a) No. 8, March.

BASP Newsletter (1990b) Smoking prevention. No. 9, June.

BASP Newsletter (1990c) Smoking prevention. No. 11, December.

Beaglehole R (1991) Science, advocacy and health policy: lessons from the New Zealand tobacco wars. *Journal of public health policy*, 12 (2): 175-183.

Benowitz NL et al. (1983) Smokers of low-yield cigarettes do not consume less nicotine. *New England journal of medicine*, 309 (3): 139-142.

Bjartveit K (1978a) The Norwegian Tobacco Act. *Scandinavian journal of respiratory diseases* (Suppl. 102): 142-146.

Bjartveit K (1978b) Governmental action on smoking and health. *Bulletin of the International Union against Tuberculosis*, 53 (4): 334-339.

Bjartveit K (1990) Fifteen years of comprehensive legislation: results and conclusions. In Durston B, Jamrozik K, eds. *Tobacco and health 1990. The global war. Proceedings of the Seventh World Conference on Tobacco and Health.* Perth, Health Department of Western Australia, pp. 71-80.

Bjartveit K, Lund KE (1987) *Smoking control in Norway.* Oslo, National Council on Smoking and Health.

British medical journal. (1980) Call for stronger government action against smoking, 2: 1574.

Brown ER et al. (1988) Workplace smoking policies: attitudes of union members in a high-risk industry. *Journal of occupational medicine.* 30: 312-320.

Buchholz P (1980) Legal aspects of the control of tobacco. *Legal medical quarterly, 4: 14-34.*

Calgary Health Services (1988) *Smoking by-laws in Canada.* Calgary.

Canadian Cancer Society et al. (1988) *Tobacco taxation and public health. A submission to the Honourable Robert Nixon. Treasurer of Ontario.* Toronto.

Canadian Department of National Health and Welfare (1989) Tobacco products control regulations. *Canada gazette,* Part I, 18 January.

Canadian Department of National Health and Welfare (1990) *Health protection branch information letter # 776.1 March.*

Carlisle v. Philip Morris. Inc. 6.1 TPLR 2.1 (Texas Ct. of Appeals, 3d Dist., 1991).

Carroll v. Melbourne Metropolitan Transit Authority, Workcare claim. Victoria Accident Compensation Tribunal. noted in 3.7 TPLR 1.92 (1988).

Carroll v. Tennessee Valley Authority, 697 F. Supp. 508 (D.D.C. 1988).

Centers for Disease Control (1990a) Cigarette advertising—United States—1988. *Morbidity and mortality weekly report,* 39 (16): 261-265.

Centers for Disease Control (1990b) State laws restricting minors' access to tobacco. *Morbidity and mortality weekly report.* 39 (21).

Chapman S (1985) *Cigarette advertising and smoking: a review of the evidence.* London, British Medical Association, Professional Division.

Chapman S, Fahey P (1989) *Cigarette smoking and economic prosperity: the relationship of GNP per capita and cigarette consumption in 103 countries.* Sydney. Department of Community Medicine. University of Sydney.

Chapman S, Leng WW (1990) Tobacco control in the Third World: a resource atlas. Penang. International Organization of Consumers Unions.

Chapman S. Richardson J (1990) Tobacco excise and declining tobacco consumption: the case of Papua New Guinea. *American journal of public health,* 80 (5): 537-540.

Chen TTL. Winder AE (1990) The opium wars revisited as US forces tobacco exports in Asia. *American journal of public health,* 80 (6): 659-662.

Choi WS et al. (1992) Restricting minors' access to tobacco—a review of state legislation 1991. *American journal of preventive medicine,* 8 (1): 19-22.

Cipollone v. Liggett Group Inc. 789 F.2d 181 (3d Cir), cert. denied, 107 S. Ct. 907 (1986).

Cipollone v. Liggett Group Inc., 683 F. Supp. 1487 (D.N.J. 1988)

Cipollone v. Liggett Group Inc., 893 F. 2d (3d Cir, 1990).

Cipollone v. Liggett Group Inc., 112 S. Ct. 2608. 120 L. Ed. 2d 407. 60 U.S.L.W. 703 (24 June, 1992).

Cockburn A (1989) Light up on Asia. *The Nation,* 30 October, p. 483.

Collishaw NE (1990) *Monitoring effectiveness of Canada's health-oriented tobacco policies.* Paper presented to the 1990 National Workshop on Smoking and Health. Halifax, Nova Scotia, 21 September, 1990.

Collishaw NE et al. (1990) *Monitoring effectiveness of Canada's health-oriented tobacco policies.* Paper presented to the Seventh World Conference on Tobacco and Health. Perth, Western Australia.

Collishaw NE et al. (1985) An estimate of the demand for tobacco in Canada 1950-1982. In: *Proceedings of the 34th International Congress on Alcoholism and Drug Dependence, Calgary, Alberta, 4-10 August,* pp. 201-203.

Commission des Affaires Culturelles. Familiales et Sociales (1990) *Rapport fait a l'Assemblee Nationale sur le projet de loi (no 1418) relatif a la lutte contre le tabagisme et a la lutte contre l' alcoolisme. [Report to the National Assembly of France on the proposed law, No. 1418, concerning the struggle against smoking and the struggle against alcoholism.]* Document No. 1482, distributed 22 June, 1990.

Connolly GN (1991) Tobacco trade. GATT and Eastern Europe. *Tobacco on trial, 15 July* pp. 1-2.

Connolly GN (1988) *World health implications of United States tobacco trade policies.* Presentation to the US Interagency Committee on Smoking and Health, 18 February, 1988.

Correa P et al. (1983) Passive smoking and lung cancer. *Lancet*, 2: 595-597.

Council of Europe Consultative Assembly (1973) *Recommendation 716 on the control of tobacco and alcohol advertising and on measures to curb consumption of these products.*

Council of the European Communities and the representatives of the governments of the Member States (1986) *Resolution on a programme of action of the European Communities against cancer* (86/C184/05).

County of Fresno v. Fair Employment and Housing Comm'n of the State of California, noted in 6.1 TPLR 1.4 (1991).

Cox H. Smith R (1984) Political approaches to smoking control: a comprehensive analysis. *Applied economics*, 16: 569-582.

Currie K. Ray L (1984) Going up in smoke: the case of British American tobacco in Kenya. *Social science and medicine*, 19 (11): 1131-1139.

Dalla-Vorgia P et al. (1990) An evaluation of the effectiveness of tobacco-control legislative policies in European Community countries. *Scandinavian journal of social medicine*, 18: 81-89.

Daube MM (1975) Towards an advertising ban. In: *Proceedings of the 3rd World Conference on Smoking and Health*, Bethesda, MD. National Institute of Health, pp. 841-850 (DHEW publication No. (NIH) 77-1413).

Daube MM (1990) Call for cigarette chemical warning. *West Australian*, 3 July.

Davis RM, Jason LA (1988) The distribution of free cigarette samples to minors. *American journal of preventive medicine*, 4 (1): 21-24.

Davis RM et al. (1990) 'Common courtesy' and the elimination of passive smoking. *Journal of the American Medical Association*, 263 (16): 2208-2210.

Davis RM et al. (1990) Information on tar and nicotine yields on cigarette packages. *American journal of public health*, 80 (5): 551-553.

De Beni Souza v. Kallweit, case no. 807516 (Sacramento Cty. Calif. 1990).

De Givry Y (1984) The impact of the 1976 antismoking legislation in France. *International digest of health legislation*, 35(2): 463-467.

Denoix P (1979) Tobacco—current French legislation. In: *Carcinogenic—risks strategies for intervention*. Lyon, International Agency for Research on Cancer, p.14 (Scientific Publications Series, No.25).

Department of the Secretary of State of Canada (1987) *The Charter of Rights and Freedoms: a guide for Canadians*. Ottawa.

Dewey v. R. J. Reynolds Tobacco Co., 577 A. 2d 1239 (N.J. 1990).

DiFranza JR et al. (1987) Legislative efforts to protect children from tobacco. *Journal of the American Medical Association*, 257:3387-3389.

DiFranza JR et al. (1991) RJR Nabisco's cartoon camel promotes Camel cigarettes to children. *Journal of the American Medical Association*, 266: 3149-3153.

Doll R, Hill AB (1950) Smoking and carcinoma of the lung. *British medicaL journal*, 2: 729-748.

Doll R, Hill AB(1956) Lung cancer and other causes of death in relation to smoking. *British medical journal*, 2: 1071-1081.

Doll R, Hill AB (1964a) Mortality in relation to smoking: ten years' observations of British doctors. *British medical journal*, 1: 1399-1410.

Doll R, Hill AB (1964b) Mortality in relation to smoking: ten years' observations of British doctors (conclusion). *British medical journal*, 1: 1460-1467.

Dumas A (1978) Cigarette ads—conscience versus profits. *Broadcasting and television weekly*, 10 August.

Durston B, Jamrozik K, eds. (1990) *Tobacco and health 1990. The global war. Proceedings of the Seventh World Conference on Tobacco and Health*. Perth, Health Department of Western Australia.

Economist Intelligence Unit (1983) *Tobacco and food crops in the Third World*. London (Studies in Agriculture and commodities, No. 3).

Erickson AC et al. (1990) Past lessons and new uses of the mass media in reducing tobacco consumption. *Public health reports*, 105 (3): 239-244 (May-June).

European Bureau for Action on Smoking Prevention (1991) Smoking prevention. *BASP newsletter*, No. 12, April.

Federal Trade Commission (1986) *Report to Congress pursuant to the Federal Cigarette Labeling and Advertising Act, 1984.*

Femi- Pearse D (1983) Aspects of smoking in developing countries in Africa. *New York State journal of medicine*, 83 (13): 1312-1313.

Fielding JE (1986) Banning worksite smoking. *American journal of public health*, 76 (8): 957-959.

Fielding JE (1991) Smoking control at the workplace. *Annual review of public health*, 12: 209-234.

Fielding JE, Phenow KJ (1988) Health effects of involuntary smoking. *New England journal of medicine*, 319: 1452-1460.

Finland, Advisory Committee on Health Education (1985) *An evaluation of the effects of an increase in the price of tobacco and a proposal for the tobacco price policy in Finland in 1985-87*. Helsinki. National Board of Health.

Finland, National Board of Health (1977) *Legislative and administrative action for control of tobacco-smoking in Finland.* Helsinki.

Finnish Parliamentary Report, 1990 (II La VM 4/119 1989 v. p. - Motion of the II Law Committee. *Applying the law of product liability to tobacco products*, 5.3 TPLR 7.15).

Fire Safe Cigarette Act of 1990, Public Law 101-352. August 10, 1990, 101st Congress. 5.4 TPRL 7.17 (1990).

Foote E (1981) Advertising and tobacco. *Journal of the American Medical Association*, 245 (16): 1667-1668.

Gallagher v. Midalco/CSR (Supreme Court, West Australia), noted in 5.4 TPLR 1.153 (1990), 6.1 TPLR 1.8 (1991).

Garfinkel L et al. (1985) Involuntary smoking and lung cancer: a case-control study. *Journal of the National Cancer Institute*. 75 (3): 463-469.

Geizerova H, Masironi R (1990) Smoking restrictions in public transport and public places. WHO survey. In: Durston B. Jamrozik K. eds. *Tobacco and health 1990. The global war. Proceedings of the Seventh World Conference on Tobacco and Health.* Perth, Health Department of Western Australia, pp. 759-762.

General Agreement on Tariffs and Trade (1990) *Thailand—restrictions on importation of and internal taxes on cigarettes. Report of the panel*, Geneva.

Glantz S. Parmley WW (1991) Passive smoking and heart disease: epidemiology, physiology, and biochemistry. Circulation, 83 (1): 1-12.

Globe and Mail (1991) The Canadian economy, 17 June, p. B3.

Goss J (1990) An economist's rationale for government intervention to reduce the consumption of cigarettes. In: Durston. B Jamrozik K. eds. *Tobacco and health 1990. The global war. Proceedings of the Seventh World Conference on Tobacco and Health.* Perth, Health Department of Western Australia. pp. 789-793.

Government of Victoria (1987) *Submission by the Government of Victoria to the Industries Assistance Commission Inquiry on tobacco growing and manufacturing industries.*

Gray N (1984) *Low tar cigarettes—bane or benefit.* Paper presented to the 6th International Symposium on Prevention and Detection of Cancer. Vienna, 26-29 November, 1984.

Gray N, Daube M, eds (1980) *Guidelines for smoking control*, 2nd ed. Geneva, International Union against Cancer.

Gupta PC (1988) Health consequences of tobacco use in India. *World smoking and health*, 5-10.

Gupta PC (1989) An assessment of excess mortality caused by tobacco usage in India. In: Sanghvi LD, Notani P, eds. *Tobacco and health: the Indian scene.* Bombay, Tata Memorial Centre, pp. 57-62.

Gupta PC, Ball K (1990) India: tobacco tragedy. *Lancet*, 335-594.

Haglund M (1990) *Tobacco control in Sweden-some remarks on recent developments.* Stockholm, Department of Public Health and Epidemiology, National Board of Health and Welfare.

Hall J (1989) Law on cigarette sales to teens called outdated. *The Toronto Star*, 26 January.

Hamilton JL (1972) The demand for cigarettes: advertising, the health scare, and the cigarette advertising ban. *Review of economics and statistics*, 54: 401.

Heise L (1988) Unhealthy alliance. With U.S. government help, tobacco firms push their goods overseas. *World-Watch*, 1 (5): 19-28.

Hentzel v Singer Co., 188 Cal. Rptr 159 (1982).

Hirayama T (1981) Non-smoking wives of heavy smokers have a higher risk of lung cancer: a study from Japan. *British medical journal*, 282: 183-185.

Hong Kong Council on Smoking and Health (1990) *Annual report*, 1989-1990.

Horovitz B (1990) 'Freedom' campaign under siege. *Los Angeles Times*, 8 May, pp. D1, D7.

Humble C et al. (1990) Passive smoking and 20-year cardiovascular disease mortality among nonsmoking wives, Evans County, Georgia. *American journal of public health*, 80 (5): 599-601.

Ierardi v. Lorillard. Inc., U.S.D.Ct. E.D.Pa. Civ. Action No. 90-7049, 6.1 TPLR 3.11 (complaint) (1991).

Industries Assistance Commission (1981) *Manufactured tobacco.* Canberra, Australian Government Publishing Service (Report No. 256).

International Agency for Research on Cancer (1986) *Tobacco habits other than smoking.* Lyon (IARC monographs on the evaluation of the carcinogenic risk of chemicals to humans, Volume 37).

International Agency for Research on Cancer (1987a) *Tobacco smoking.* Lyon (IARC monographs on the evaluation of the carcinogenic risk of chemicals to humans. Volume 38).

International Agency for Research on Cancer (1987b) *Environmental carcinogens: methods of analysis and exposure measurement. Volume 9. Passive smoking.* Lyon (Scientific Publications Series, No. 81).

International digest of health legislation (1981) Four resolutions on smoking and health adopted by Health Ministers of the Arab States of the Gulf Area. 32 (2): 327-328.

Islam N et al. (1990) Smoking in Bangladesh—a pilot study. In: Durston B, Jamrozik K, eds. *Tobacco and health 1990. The global war. Proceedings of the Seventh World Conference on Tobacco and Health.* Perth, Health Department of Western Australia, pp. 242-244.

Janerich DT et al. (1990) Lung cancer and exposure to tobacco smoke in the household. *New England journal of medicine,* 323 (10): 632-636.

Jardeleza v. R. J. Reynolds Tobacco Co., Quezon Trial Ct., 2.7 TPRL 3.409 (complaint) (1987).

Johnston LD et al. (1987) *National trends in drug use and related factors among American high school students and young adults, 1975-1986.* US Department of Health and Human Services, Public Health Service, Alcohol, Drug Abuse, and Mental Health Administration, National Institute on Drug Abuse (DHHS Publication No. (ADM) 87-1535).

Joossens L, Raw M (1991) Tobacco and the European common agricultural policy, *British journal of addiction,* 86: 1191-1202.

Karaoglu A, Joossens L, eds. (1990) *Report on a new form of smokeless tobacco: moist snuff.* Brussels, European Bureau for Action on Smoking Prevention.

Kyle KL (1990) *Tobacco advertising ban in Canada.* A brief presented to the Subcommittee on Health and the Environment. Committee on Energy and Commerce, US House of Representatives, 12 July.

Kyte v. Philip Morris. Inc. and Store 24. Inc., 2.9 TPLR 2.208 (Super. Ct. Mass. 1987).

Lancet (1990) Europe for cancer, 336: 1036.

Lando HA (1990) The movement to restrict vending machines in Minnesota: the fight against pre-emption. In: Durston B. Jamrozik K, eds, *Tobacco and health 1990. The global war. Proceedings of the Seventh World Conference on Tobacco and Health,* Perth, Health Department of Western Australia, pp. 741-744.

Laugesen M. Meads C (1990) Do tobacco advertising bans lower consumption? Tobacco advertising restrictions, price, income and tobacco consumption in OECD countries, 1960-86. In: Durston B. Jamrozik K, eds. *Tobacco and health 1990. The global war: Proceedings of the Seventh World Conference on Tobacco and Health,* Perth, Health Department of Western Australia, pp. 126-129.

Ledwith F (1984) Does tobacco sports sponsorship on television act as advertising to children? *Health education journal.* 43 (4): 85-88.

Lee v. Massachusetts Department of Public Welfare, 1.2 TPRL 2.82 (Mass. Super. Ct. 1983).

Legislative Clearinghouse (1989) *State legislated actions on tobacco issues.* Washington, DC.

Lenfant C (1983) Are low-yield cigarettes really safer? *New England journal of medicine,* 309 (3): 181-182.

Lev M (1991) Tobacco and alcohol lose more ground. *The New York Times,* 24 January p. C4.

Lewit EM (1989) U. S. tobacco taxes: behavioural effects and policy implications. *British journal of addiction.* 84: 1217-1235.

Lewit EM (1990) Economics of tobacco in developing countries: telling it like it is. In: Durston B. Jamrozik K, eds. *Tobacco and health 1990. The global war. Proceedings of the Seventh World Conference on Tobacco and Health.* Perth. Health Department of Western Australia, pp. 785-787.

Lewit EM, Coate D (1982) The potential for using excise taxes to reduce smoking. *Journal of health economics,* 1: 121-145.

Lewit EM et al. (1981) The effects of government regulation on teenage smoking. *Journal of law and economics,* 24: 545-569.

Lokschin FL, Barros FC (1983) Smoking or health: the Brazilian option. *New York State journal of medicine,* 83 (13): 1314-1315.

The Times (1991) Tobacco aid is to end, 12 February.

McCarthy v. State of Washington Dep't of Social and Health Serv., 759 p. 2d 351 (Wash. Sup. Ct. 1988).

Mackay JM (1986) The tobacco industry and tobacco advertising in Hong Kong. In: *Proceedings, Conference on Strategies for Smoking Control,* Hong Kong.

Malhotra SP (1988) The economic significance of tobacco and the future outlook. Paper presented at the First European Conference on Tobacco Policy—*Tobacco or health: the way ahead,* Copenhagen. WHO Regional Office for Europe.

Maron DJ. Fortmann SP (1987) Nicotine yield and measures of cigarette smoke exposure in a large population: are lower-yield cigarettes safer? *American journal of public health,* 77 (5): 546-549.

Marongiu P (1980) FAO activities in tobacco. In: Ramstrom LM, ed, *The smoking epidemic, a matter of worldwide concern.* Stockholm, Almquist & Wiksell International, pp. 320-321.

Mills PK et al. (1990) History of cigarette smoking and risk of leukemia and myeloma: results from the Adventist Health Study. *Journal of the National Cancer Institute,* 82 (23): 1823-1836.

Muller M (1978) *Tobacco and the Third World: tomorrow's epidemic?* London, War on Want

Murphy J (1990) International trade body sends a health message on tobacco in battle over US forcing Thailand to accept US tobacco. *The nation's health,* November.

Naett C. Joossens L (1992) Steps towards a tobacco advertising ban in the European Community. *Tobacco control,* 1: 57.

Nakajima H (1990) WHO—Part of a global strategy for a tobacco-free society. In: Durston B. Jamrozik K, eds. *Tobacco and health 1990. The global war. Proceedings of the Seventh World Conference on Tobacco and Health,* Perth, Health Department of Western Australia, pp. 9-10.

Nakajima H (1991) Presentation to Economic and Social Council, Second regular session of 1991. Geneva, 3-26 July.

Nath UR (1986) *Smoking: Third World alert.* Oxford, Oxford University Press.

National Campaign for Action on Tobacco (1991) *News release.* Ottawa. 1 August.

National Clearinghouse on Tobacco and Health (1990) *Children and tobacco.*

Legislation in Canada—the current status. Ottawa (Fact Sheet Series, No. 9).

National Insurance Office v. Estate of Gun Palm, case no. 2722/83: 8, 6.1 TPLR 2.65 (Ins. Ct of Appeal, Sweden, 1985).

New Straits Times (1991) Ban on duty-free cigarette sales, 1 January.

New Zealand Toxic Substances Board (1989) *Health or tobacco: an end to tobacco advertising and promotion,* Wellington.

Nordgren P, ed. (1983) *Smoking control in Sweden.* Stockholm, Swedish National Smoking and Health Association.

Nordic Council of Ministers (1972) *Recommendation number 12,* Helsingfors, 22 February.

Norwegian National Council on Tobacco and Health (1990) *New political initiatives 1988-90. Controlling tobacco,* Oslo.

Novello A (1991) Can we prevent the Virginia Slims woman from catching up with the Marlboro man? *World smoking and health,* 16 (2): 2.

O'Connell DL et al. (1981) Cigarette smoking and drug use in school children. II. Factors associated with smoking. *International journal of epidemiology,* 10 (3): 223-231.

O'Connor M (1990) Voluntary agreements between the government and the tobacco industry and the control of tobacco promotion in the United Kingdom. In: Durston B. Jamrozik K, eds. *Tobacco and health 1990. The global war. Proceedings of the Seventh World Conference on Tobacco and Health,* Perth, Health Department of Western Australia, pp. 739-740.

Office of Inspector General (1990) *Youth access to cigarettes.* Washington, DC, Department of Health and Human Services.

Office of Technology Assessment (1985) *Staff memo: smoking-related deaths and financial costs.* Washington, DC.

Parkin DM et al. (1984) Estimates of the worldwide frequency of twelve major cancers. *Bulletin of the World Health Organization,* 62 (2): 163-182.

Pearl R (1938) Tobacco smoking and longevity. *Science,* 87: 216-217.

People v. Cook. 34 N. Y. 2d 100, 312 N. E. 2d 452 (1974) upholding Cigarette Tax Law Regulations of New York City Finance Administrator, Article 2-a.

Pertschuk M, Shopland DR. eds. (1989) *Major local smoking ordinances in the United States. A detailed matrix of the provisions of workplace, restaurant, and public places smoking ordinances.* Bethesda, MD, US Department of Health and Human Services, Public Health Service, National Institutes of Health (NIH publication No. 90-479).

Peto R (1990) Women and tobacco. *World health forum,* 11 (1): 3-13.

Peto R. Lopez AD (1990) Worldwide mortality from current smoking patterns: WHO consultative group on statistical aspects of tobacco-related mortality. In: Durston B. Jamrozik K, eds. *Tobacco and health 1990. The global war, Proceedings of the Seventh World Conference on Tobacco and Health,* Perth, Health Department of Western Australia, pp. 66-68.

Philip Morris Companies Inc. (1989) The story of a government that almost didn't happen. *The New York Times,* 3 November, p. A18.

Piha T (1990) The Finnish experience of tar regulation: "light cigarettes." *BASP newsletter, No. 8,* 3-5.

Posadas de Puerto Rico Associates, DBA Condado Holiday Inn v. Tourism Company of Puerto Reco et al., 478 U.S. 328. 106 S. Ct. 2968 (1986).

Ramstrom L. ed. (1980) *The smoking epidemic, a matter of worldwide concern. Proceedings of the Fourth World Conference on Smoking and Health, Stockholm, 1979.* Stockholm, Almquist & Wiksell International.

Ramstrom L (1983) Sweden's programme to eliminate smoking—activities and success to date. In: *Smoking control in Sweden.* Stockholm, Swedish National Smoking and Health Association.

Ramstrom L (1984) Population scale evidence that smoking control can prevent lung cancer. In: *Proceedings of the Sixth International Symposium for Prevention and Detection of Cancer, Vienna, 26-29 November, 1984.*

Raw M et al. (1990) *Clearing the air. A guide for action on tobacco.* London, British Medical Association.

Reid D, Seymour L (1987) *How two million people gave up smoking: the effects of tax increases and the U.K. national health education programme. 1980-86.* Paper contributed to the WHO European Action on Tobacco Programme, March 1987.

Revill. J. Drury CG (1980) An assessment of the incidence of cigarette smoking in fourth year school children and factors leading to establishment. *Public health (London),* 94: 243-260.

Riding A (1990) What price tobacco ads? A countdown in France. *The New York Times,* 9 April.

Rimpela M (1978) Cigarette smoking and public policy. *Scandinavian journal of respiratory diseases.* (Suppl. 102): 139-141.

RJR Macdonald and Imperial Tobacco Limited v. Attorney-General of Canada, Quebec Superior Court, Montreal, 1989.

RJR-Macdonald Inc. v. The Attorney General of Canada, No. 500-05-009755-833 and *Imperial Tobacco Ltd. v. The Attorney General of Canada,* No. 500-05-08-760-883, Superior Court, Province of Quebec, District of Montreal, 26 July, 1991.

Roemer R (1982) *Legislative action to combat the world smoking epidemic.* Geneva, World Health Organization.

Roemer R (1986a) *Recent developments in legislation to combat the world smoking epidemic.* Geneva, World Health Organization (unpublished WHO document WHO/SMO/HLE/86.1).

Roemer R (1986b) legislative action to control smoking—accomplishments and challenges. In: Skirrow J. ed. *Strategies for a smoke-free world. Selected resource documents from the International Workshop on Smoking and Health.* Calgary, Alberta Alcohol and Drug Abuse Commission.

Roemer R (1988) *Legislative strategies for a smoke-free Europe.* Copenhagen, WHO Regional Office for Europe and the Commission of the European Communities.

Rothstein M (1990) Uneasy partners: arts and Philip Morris. *The New York Times.* "The Living Arts": 18 December, p. B1. 10.

Royal College of Physicians of London (1962) *Smoking and health, Summary and report of the Royal College of Physicians of London on smoking in relation to cancer of the lung and other diseases.* London, Pitman Medical Publishing.

Royal College of Physicians of London (1971) *Smoking and health now.* London. Pitman Medical Publishing.

Royal College of Physicians of London (1977) *Smoking and health (the Third Report from the Royal College of Physicians of London).* London. Pitman Medical Publishing.

Royal College of Physicians of London (1983) *Health or smoking? Follow-up report.* London. Pitman Medical.

Satalino v. Satalino, case no. 11440/86, 34 ATLA L.R. 16 (Nassau Cty, N. Y. 1990). Schmeisser P (1988) Pushing cigarettes overseas. *The New York Times Magazine,* 10 July.

Schmidt F (1990) 'Light' smoking is no expedient. In: Durston B. Jamrozik K, eds, *Tobacco and health 1990. The global war. Proceedings of the Seventh World Conference on Tobacco and Health.* Perth, Health Department of Western Australia. pp. 428-429.

Schober v. Mountain Bell Telephone Co., 630 p. 2d 1231 (N. M. Sup. Ct. 1980).

Scull R (1986) Bright future predicted for Asia Pacific. *World tobacco,* 9: 34-41.

Shepherd PL (1985) Transnational corporations and the international cigarette industry. In: Newfarmer R. ed. *Profits, progress and poverty.* Notre Dame. Indiana, University of Notre Dame Press.

Shimp v. N. J. Telephone Co., 368 A. 2d 408 (N. J. Super. Ct. 1976).

Simpson D (1988) The public health and public debate. *Health and hygiene,* 9: 29-34.

Slattery ML et al. (1989) Cigarette smoking and exposure to passive smoke are risk factors for cervical cancer. *Journal of the American Medical Association,* 261 (11): 1593-1598.

Smith RC (1978) The magazines' smoking habit. *Columbia journalism review.* January/February: 29-31.

Smoking and health reporter (1985) Thumbs up/thumbs down. 2: 3.

South China Morning Post (1991) 7 March.

Stebbins KR (1987) Tobacco or health in the Third World: a political economy perspective with emphasis on Mexico. *International journal of health services,* 17 (3): 521-536.

Sweda E (1991) Summary of legal decisions regarding smoking in the workplace and other places, 6.2 TPLR 4.11.

Swedish Committee for Tobacco Control (1990) *Tobacco Act. Report of the Commitee for Tobacco Control.* Presented to the Minister of Health and Social Affairs, Stockholm.

Taylor P (1984) *The smoke ring: the politics of tobacco;* London. The Bodley Head;

Taylor SA (1989) Tobacco and economic growth in developing nations. *Business in the contemporary world.* 55-70.

Tobacco alert (1984) 1 (2): 4. June.

Tobacco alert (1985) 2 (1): 2. March.

Tobacco alert (1986) Study confirms high tar levels in developing countries. 15/16.

Tobacco alert (1990) The European common agricultural policy in the tobacco sector: in need of review.

Tobacco alert (1991) January.

Tobacco on trial (1990) 17 December.

Tobacco on trial (1991) 15 February.

Tobacco Products Liability Project (1988). Selected exhibits from the *Cipollone* case.[1]

Tobacco Products Liability Project (1989), Selected exhibits from the *Palmer* case.[1]

Tobacco reporter (1987) 114 (8): 34.

Townsend J (1987) Economic and health consequences of reduced smoking. In: Williams A. ed. *Health and economics. Proceedings of Section F (Economics) of the British Association for the Advancement of Science, Bristol, 1986.* London, Macmillan Press, pp. 139-161.

Townsend J (1990) *Price and income elasticity of demand, taxation and the smoking epidemic.* Paper presented to the Seventh World Conference on Tobacco and Health, Perth, Western Australia.

Trichopoulos D et al. (1981) Lung cancer and passive smoking. *International journal of cancer,* 27: 1-4.

Tye JB et al., (1987) Tobacco advertising and consumption: evidence of a causal relationship. *Journal of public health policy*, 8 (4): 492-508.

United Nations (1978) *Marketing and distribution of tobacco*. Geneva (Report No. TD/B/C.1/205).

US Department of Health, Education, and Welfare (1979) *Smoking and health: Report of the Surgeon General*. Rockville, MD, Office on Smoking and Health, Office of the Assistant Secretary for Health, Public Health Service (DHEW Publication No. (PHS) 79-50066).

US Department of Health and Human Services (1980) *The health consequences of smoking for women: a report of the Surgeon General*. Rockville. MD. Office on Smoking and Health, Office of the Assistant Secretary for Health.

US Department of Health and Human Services (1981) *The health consequences of smoking—the changing cigarette: a report of the Surgeon General*. Washington, DC, Office on Smoking and Health (DHHS Publication No. (PHS) 81-50156).

US Department of Health and Human Services (1984) *The health consequences of smoking: chronic obstructive lung disease. A report of the Surgeon General*. Office on Smoking and Health (DHHS Publication No. (PHS) 84-50205).

US Department of Health and Human Services (1985) *The health consequences of smoking. Cancer and chronic lung disease in the workplace: a report of the Surgeon General*. Public Health Service, Office on Smoking and Health (DHHS Publication (PHS) 85-50207).

US Department of Health and Human Services (1986) *The health consequences of involuntary smoking: a report of the Surgeon General*. Public Health Service, Centers for Disease Control (DHHS Publication No. (CDC) 87-8398).

US Department of Health and Human Services (1988) *The health consequences of smoking: nicotine addiction. A report of the Surgeon General. 1988*. Public Health Service, Centers for Disease Control, Center for Health Promotion and Education, Office on Smoking and Health (DHHS Publication No. (CDC) 88-8406).

US Department of Health and Human Services (1989) *Reducing the health consequences of smoking: 25 years of progress. A report of the Surgeon General*. Rock ville. MD. Public Health Service, Centers for Disease Control, Center for Chronic Disease Prevention and Health Promotion. Office on Smoking and Health (DHHS Publication No. (CDC) 89-8411).

US Department of Health and Human Services (1990a) *Highlights of state of Tobacco and health data*. Program Services Activity, Office on Smoking and Health, Center for Chronic Disease Prevention and Health Promotion. Centers for Disease Control, Public Health Service.

US Department of Health and Human Services (1990b) *Model Sale of Tobacco Products to Minors Control Act. A model law recommended for adoption by states or localities to prevent the sale of tobacco products to minors*,

[1]These and other documents are available from: Professor R. A. Daynard. Chairman, Tobacco Products Liability Project, Northeastern University School of Law, 400 Huntington Avenue, Boston, MA02115, USA.

US Department of Health and Human Services (1992) *Smoking and health in the Americas. A 1992 report of the Surgeon General, in collaboration with the Pan American Health Organization*. Atlanta, GA, Public Health Service, Centers for Disease Control, National Center for Chronic Disease Prevention and Health Promotion, Office on Smoking and Health (DHHS Publication No. (CDC) 92-8419).

US Federal Trade Commission (1981) *Report of "tar", nicotine and carbon monoxide of the smoke of* 187 varieties of cigarettes. Washington. DC.

US National Research Council (1986) *Environmental tobacco smoke: measuring exposures and assessing health effects*. Washington., DC. National Academy Press.

US Public Health Service (1964) *Smoking and health: report of the Advisory Committee to the Surgeon General of the Public Health Service*. Washington, DC. US Department of Health, Education, and Welfare (PHS Publication No. 1103).

US Public Health Service (1967) *The health consequences of smoking; a Public Health Service review*. Washington, DC, US Department of Health. Education and Welfare (PHS Publication No. 1696, revised 1968a).

Vaithinathan BR (1990) *Measures taken to discourage smoking in Singapore*. Singapore, Ministry of Health.

Wald N et al. (1981) Trends in tar, nicotine and carbon monoxide yields of UK cigarettes manufactured since 1934. *British medical journal*, 282: 763-765.

Waldron I et al. (1988) Gender differences in tobacco use in Africa, Asia, the Pacific and Latin America. *Social science and medicine*, 27 (11): 1269-1275.

Warner KE (1977) The effects of the anti-smoking campaign on cigarette consumption. *American journal of public health*, 67: 645-650.

Warner KE (1979) Clearing the airwaves: the cigarette ban revisited. *Policy analysis*, 5 (4): 435-450.

Warner KE (1984) Cigarette taxation: doing good by doing well. *Journal of public health policy*, 5 (3): 312-318.

Warner KE (1986a) *Selling smoke: cigarette advertising and public health*. Washington, DC. American Public Health Association.

Warner KE (1986b) Smoking and health implications of a change in the federal cigarette excise tax. *Journal of the American Medical Association*, 255 (8): 1028-1032.

Warner KE (1988) The tobacco subsidy: does it matter? *Journal of the National Cancer Institute*, 80 (2): 81-83.

Warner KE (1990a) Tobacco taxation as health policy in the Third World. *American journal of public health*, 80 (5): 529-531.

Warner KE (1990b) Tobacco taxation and economic effects of declining tobacco consumption. In: Durston B. Jamrozik K, eds. *Tobacco and health 1990. The global war. Proceedings of the Seventh World Conference on Tobacco and Health*, Perth, Health Department of Western Australia, pp. 81-87.

Warner KE et al. (1986) Public policy on smoking and health: toward a smoke-free generation by the year 2000. A statement of a working group to the Subcommittee on Smoking of the American Heart Association. *Circulation*, 73 (2): 381A-395A.

Whelan E (1984) *A smoking gun:how the tobacco industry gets away with murder*, Philadelphia, George F. Stickley.

Wilson K (1990) Western Australia Tobacco Control Bill 1990. Second Reading Speech.

Winstanley M. ed. (1989) *Tobacco in Australia: facts and issues.* Surrey Hills. NSW. Action on Smoking and Health (ASH Australia).

World Health Organization (1975) *Smoking and its effects on health: report of a WHO Expert Committee.* Geneva (WHO Technical Report Series, No. 568).

World Health Organization (1976) *Legislative action to combat smoking around the world: a survey of existing legislation*, Geneva.

World Health Organization (1979) *Controlling the smoking epidemic: report of the WHO Expert Committee on Smoking Control.* Geneva (WHO Technical Report Series, No. 636).

World Health Organization (1982) *Prevention of coronary heart disease: report of a WHO Expert Committee.* Geneva (WHO Technical Report Series, No. 678).

World Health Organization (1983) *Smoking control strategies in developing countries: report of a WHO Expert Committee.* Geneva (WHO Technical Report Series, No. 695).

World Health Organization (1985) *Tobacco or health.* Geneva (unpublished document EB77/1986/rec/1).

World Health Organization (1986) *Community prevention and control of cardiovascular disease: report of a WHO Expert Committee.* Geneva (WHO Technical Report Series. No. 732).

World Health Organization (1988) *Smokeless tobacco control: report of a WHO Study Group.* Geneva (WHO Technical Report Series, No. 773).

World Health Organization (1989) *Report of a WHO Consultation on Statistical Aspects of Tobacco-related Mortality.* Geneva (unpublished document, WHO/TOH/CLH/90.2).

World Health Organization (1990a) *Prevention in childhood and youth of adult cardiovascular diseases: time for action: report of a WHO Expert Committee*, Geneva (WHO Technical Report Series. No. 792).

World Health Organization (1990b) *It can be done : a smoke-free Europe. Report of the first European Conference on Tobacco Policy Madrid. 7-11 November, 1988.* Copenhagen, WHO Regional Office for Europe (European Series. No. 30).

World Health Organization (1991a) *WHO Programme on Tobacco or Health: implementation of resolutions WHA42.19 and WHA43.16* Geneva (unpublished document WHA44/1991/REC/1).

World Health Organization (1991b) *Legislative responses to the tobacco epidemic*, Amsterdam, Martinus Nijhoff.

Wynder EL, Graham EA (1950) Tobacco smoking as a possible etiological factor in bronchiogenic carcinoma. *Journal of the American Medical Association*, 143: 329-336.

Wynder EL, Kabat GC (1988) The effect of low-yield cigarette smoking on lung cancer risk. *Cancer*, 62: 1223-1230.

Yach D (1986) The impact of smoking in developing countries with special reference to Africa. *International journal of health services*, 16 (2): 279-292.

Appendix-VI

THE DELHI PROHIBITION OF SMOKING AND NON SMOKERS HEALTH PROTECTION ACT, 1996[1]

(Delhi Act No. 1 of 1997)

[25th September, 1996]

An Act to provide for prohibiting of smoking in places of public work or use and in public service vehicles in the National Capital Territory of Delhi and to make provision for other matters connected therewith.

BE it enacted by the Legislative Assembly of the National Capital Territory of Delhi in the Forty-Seventh Year of the Republic of India as follows :

1. Short title, extent and commencement— (1) This Act may be called the Delhi Prohibition of Smoking and Non-smokers Health Protection Act, 1996.

(2) It extends to the whole of the National Capital Territory of Delhi.

(3) It shall come into force on such date[2] as the Government may by Notification in the Official Gazette appoint :

Provided that different dates may be appointed for different provisions of this Act and any reference in any such provision to the commencement of this Act shall be construed as a reference to the coming into force of that provision.

2. Definitions—In this Act, unless the context otherwise require—

(a) "advertisement means and includes any notice, circular, wall paper, pamphlet, display on hoarding, or any visible representation made by means of any other means of any light, sound, smoke, gas or any other means which has the effect of promoting smoking and the expression 'advertise' shall be construed acordingly :

(b) "authorised officer" means a person authorised under section 4;

(c) "Delhi" means the National Capital Territory of Delhi;

(d) "Government" means the Government of the National Capital Territory of Delhi;'

(e) "Legislative Assembly" means the Legislative Assembly for Delhi;

(f) "place of public work or use" means a place declared as such under section 3 and includes auditoria, hospital buildings, health institutions, amusement centres, restaurants, public offices, court buildings, educational Institutions, libraries and the like which are visited by general public but does not include any open place;

(g) "Official Gazette" means the Delhi Gazette;

(h) "public service vehicle" means a vehicle as defined under clause (35) of section 2 of the Motor Vehicles Act, 1988 (59 of 1988);

(i) "rule" means the rule made under this Act;

(j) "smoking" means smoking of tobacco in any form whether in the form of cigarette, cigar, beedis or otherwise with the aid of a pipe, wraper or any other instruments.

COMMENTS

(i) Authorised officers may be appointed by the Government by notification in the Official Gazette.

(ii) For the convenience of readers for understanding the meaning of 'public service vehicle' the clause (35) of section 2 of the Motor Vehicles Act, 1988 is reproduced below :

> "2(35)—'public service vehicle' means any motor vehicle used or adapted to be used for the carriage of passengers for hire or reward, and includes the maxicab, a motor cab, contract carriage, stage carriage.'

3. Declaration no-smoking places of public work or use—As soon as may be after the commencement of this Act and thereafter from time to time, the Government may, by notification in the Official Gazette, declare any place of public work or use in the Delhi to be no-smoking place for the purpose of this Act.

COMMENTS

In exercise of the powers conferred by section 3, the Government of the National Capital Territory of Delhi declared the following places of public work or use in Delhi and public service vehicles to be "non-smoking places" with effect from 26th January, 1997.

1. All places of public work or use as defined in section 2(f), banquet halls, monuments and stadia (closed area only), cinema halls, hotels and restaurants (provided there is separate place for smokers and non-smokers in hotels and restaurants).
2. All public service vehicles as defined in section 2(h) read with section 2(35) of the Motor Vehicles Act, 1988.

4. Power of Government to authorise officers to Act under this Act. : (1) The Government may, by notification in the Official Gazette, authorise one or more persons who shall be competent to act under this Act.

(2) Every person authorised under sub-section (1) shall be deemed to be a public servant within the meaning of section 21 of the Indian Penal Code, 1860 (45of 1860).

COMMENTS

(i) Section 21 of the Indian Penal Code, 1860 is reproduced below for reference :

21. "Public servant' : The words "public servant" denote a person falling under any of the descriptions hereinafter following, namely:

First ***

Second : Every Commissioned Officer in the Military, Naval or Air Force of India;

Third : Every Judge including any person empowered by law to discharge, whether by himself or as a member of any body of persons, any adjudicatory functions,

Fourth : Every officer of a Court of Justice (including a liquidator, receiver or commissioner) whose duty it is, as such officer, to investigate or report on any matter of law or fact, or to make, authenticate, or keep any document or to take charge or dispose of any property, or to execute any judicial process, or to administer any oath, or to interpret, or to preserve order in the court, and every person specially authorized by a Court of Justice to perform any of such duties;

Fifth: Every juryman, assessor, or member of a panchayat assisting a Court of Justice or public servant;

Sixth: Every arbitrator or other person to whom any cause or matter has been referred for decision or report by any Court of Justice, or by any other competent public authority.

Seventh : Every person who holds any office by virtue of which he is empowered to place or keep any person in confinement;

Eighth : Every officer of the Government whose duty it is, as such officer to prevent offences, to give information of offences, to bring offenders to justice or to protect the public health, safety or convenience;

Ninth : Every officer whose duty it is, as such officer, to take, receive, keep or extend any property on behalf of the Government or to make any survey, assessment or contract on behalf of the Government or to execute any revenue-process, or to investigate, or to report, on any matter affecting the pecuniary interests of the Government or to make, authenticate or keep any document relating to the pecuniary interests of the Government or to prevent the infraction of any law for the protection of the pecuniary interests of the Government.

Tenth : Every officer whose duty it is, as such officer, to take, receive, keep or expand any property, to make any survey or assessment or to levy any rate or tax for any secular common purpose of any village, town or district, or to make authenticate or keep any document for the ascertaining of the rights of the people of any village, town or district :

Eleventh : Every person who holds any office in virtue of which he is empowered to prepare, publish, maintain or revise an electoral roll or to conduct an election or part of an election.

Twelfth : Every person

(a) in the service or pay of the Government or remunerated by fees or commission for the performance of any public duty by the Government;

(b) in the service or pay of a local authority, a corporation established by or under a Central Provincial or State Act or a Government company as defiend in section 617 or the Companies Act, 1956 (1 of 1956).

Illustration

A Municipal Commissioner is a public servant.

Explanation 1 : Persons falling under any of the above descriptions are public servants, whether apoined by the Government or not.

Explanation 2. : Wherever the words "public servant" occur, they shall be understood of every person who is in actual possession of the situation of a public servant. Whatever legal defect there may be in his right to hold that situation.

Explanation 3 : (i) The word 'election' denotes election for the purpose of selecting members of any legislative, municipal or other public authority, of whatever character, the method of selection to which is by, or under, any law prescribed as by election.

(ii) In exercise of the powers conferred by section 4, the Government of the National Capital Territory of Delhi authorised the officers mentioned in coloumn 2 of the Table given below, who shall be the competent officers to act under the said Act in respect of places of public work or use and declared to be no-smoking places and public service vehicles mentioned in column 3 of the said Table with effect from 26th January 1997:

Sl. No.	*Designation*	*Name of Place of Public work or use*
1	2	3
1.	All gazetted officers under the Govt. of NCT of Delhi, Govt of India, State Govt. officers in Delhi and all officers holding Category "B" posts in autonomous corporations and local bodies.	All places of public work or use as defined in section 2(f) of the Delhi Prohibition of Smoking and Non smokers' Health Protection Act, 1996 and banquet halls, monuments and stadia (closed area only), cinema halls, hotles, and restaurants (excluding separate place, if any, provided for smokers) and all public service vehicles as defined in section 2 (h) of the said Act.
2.	Manager/Asstt. Manager in respect of Cinema halls.	
3.	Driver/ conductor in respect of public service vehicles.	

5. Prohibition of smoking in places of public work or use : No person shall smoke in any place of public work or use.

6. Prohibition of smoking in public service vehicles: Without prejudice to the provisions of the Motor Vehicles Act, 1988 (59 of 1988) no person shall smoke in a public service vehicle.

7. Prohibition on advertisement of cigarettes etc. : Notwithstanding anything contained in any other law for the time being force, no person shall advertise in any place and any public service vehicle which may promote smoking, or the sale of cigarettes and beedis etc.

8. Prohibition of sale of cigarettes, etc, to minors: No person shall sell cigarettes, beedis or any other such smoking substance to any who is below the age of eighteen years.

COMMENTS

This section has not been brought into force on 26th January, 1997.

9. Prohibition of storage, sale and distribution of cigarettes, etc. in the vicinity of educational institutions : No person shall himself or by any person on this behalf, store, sell or distribute cigarettes or beedis or any other such smoking substance within an area of one hundred metre around any college, school or educational institution.

COMMENTS

This section has not been brought into force on 26th January.

10. Display and Exhibition of Board : The owner or manager or incharge of affairs of every place of public work or use shall display and exhibit a board at a conspicuous place or places in and outside the premises visited or used by general public prominently starting that the place is a "No Smoking Zone" and that "Smoking is an Offence".

11. Penalties : Any person, who contravenes the provisions of :

(i) section 5, 6 or 10 shall be punishable with fine which may extend to one hundred rupees and in case of second or subsequent offence, shall be punishable with a minimum fine of two hundred rupees, but which may extend to five hundred rupees.

(ii) section 7, 8 or 9 shall be punishable with fine which may extend to five hundred rupees and in case of second or subsequent offence, shall be punishable with imprisonment which may extend to three months, or with a minimum fine of five hundred rupees, but which may extend to one thousand rupees or with both.

12. Ejection of violators of this Act from the place of public work or use.: Any authorized offence or any police officer, not below the rank of sub-inspector, may eject any person from the place of public work or use who contravenes the provisions of this Act.

13. Court competent to try offences under this Act and take cognizance of offence : (1) No court other than the court of a Metropolitan Magistrate shall take cognizance of, any try an offence under this Act.

(2) No court shall take cognizance of any offence under this Act except on a complaint in writing of an authorised officer with respect to offence under sections 5, 6 and 10 and on a report in writing of a police officer, not below the rank of sub-inspector, with respect to the offences under sections 7, 8 and 9.

14. Certain offence to be cognizable and bailable : Notwithstanding anything contained in the Code of Criminal Procedure, 1973 (2 of 1974) offences under sections 7, 8 and 9 shall be cognizable and bailable.

15. Offences under the Act to be tried summarily : All offences under this Act shall be tried summarily in the manner provided for summary trial under the Code of Criminal Procedure, 1973 (2 of 1974).

16. Power to delegate : The Government may, by notification in the Official Gazette, direct that any power exercisable by it, under this Act, may also be exercised by such officer as may be mentioned therein, subject to such conditions, if any as may be specified therein.

17. Composition of offences : The Government or any person authorised by it by general or special order in this behalf, may either before or after the institution of the proceedings compound any offences made punishable by or under this Act.

18. Power to make rules : (1) The Government may make rule to provide for or regulate any matter in respect of which this Act makes no provision or makes insufficient provision and provision is, in its opinion, necessary.

(2) Any rule made by the Government shall be subject to previous publication thereof in the Official Gazette.

(3) Every rule made under this Act shall be laid as soon as may be after it is made before the Legislative Assembly and if the House agrees in making any modification in the rule or the House agrees that the rule should not be made, the rule shall thereafter have effect only in such modified form or be of no effect, as the case may be, so, however that any such, modification or annulment shall be without prejudice to the validity of anything previously done under that rule.

NOTIFICATIONS

Notification dated, the 24th January, 1997

No. F.342/63/94-M&PH/Vol. II. In exercise of the powers conferred by section 3 of the Delhi Prohibition of Smoking and Non-smokers' Health Protection Act, 1996 (1 of 1997), the Government of the National Capital Territory of Delhi hereby declares the following places of public Work or use in Delhi and public service vehicles to be "non-smoking places" for the purpose of the said Act, namely :

Sl. No.	*Name of place of public work or use (with full description)*
1.	All places of public work or use as defined in section 2 (f) On the Delhi Prohibition of Smoking and Non-smokers Health Protection Act, 1996, banquet halls, monuments and stadia (closed area only), cinema halls, hotels and restaurants (provided there is separate place for smokers and non smokers in hotels and restaurants).
2.	All public service vehicles as defined in section 2 (h) of the Delhi Prohibition of Smoking and Non-smokers, Health Protection Act, 1996 read with section 2 (35) of the Motor vehicles Act, 1988.

This notification shall come into force with effect from the forenoon of 26th January, 1997.

Notification dated, the 24th January, 1997

No. F, 342/63/94-M&PH/Vol. II. In exercise of the powers conferred by section 4 of the Delhi Prohibition of Smoking and Non-smokers' Health Protection Act, 1996 (1 of 1997), the Government of the National Capital Territory of Delhi hereby authorises the officers mentioned in column 2 of the Table given below, who shall be the competent officers to act under the said Act in respect of places of public work or use and declared to be no-smoking places and public service vehicles mentioned in column 3 of the said Table :

Sl. No.	*Designation*	*Name of Place of Public work or use*
1.	2	3
1.	All gazetted officers under the Govt. of NCT of Delhi, Govt. of India, State Govt officers in Delhi and all officers holding Category 'B' posts in autonomous corprations and local bodies.	All places of public work or use a defined in section 2(f) of the Delhi Prohibition of Smoking and Non-smokers' Health Protection Act, 1996 and banquet halls, monuments and stadia (closed area only), cinema, halls, hotels and restaurants (excluding separate place, if any, provided for smokers) and all all public service vehicles as defined in section 2 (h) of the said Act.
2.	Manger/Asstt. Manager in respect of Cinema halls.	
3.	Driver/conductor in respect of public service vehicles.	

This notification shall come into effect with effect from the forenoon of 26th January, 1997.

Notification dated, the 24th January 1997

No. F. 342/63/94-M&PH/Vol. II. In exercise of the powers conferred by sub-section (3) of section 1 of the Delhi Prohibition of Smoking and Non-smokers Health Protection Act 1996 (1 of 1997), the Government of the National Capital Territory of Delhi hereby appoints the 26th day of January, 1997as the date on which the provisions of sections, 2, 3, 4, 5, 6, 7, 10, 11, 12, 13, 14,15, 16, 17 and 18 of the said Act shall come into force.

References

1. Published in the Delhi Gazette Extra, Pt. IV, dated 1st January, 1997.
2. 26th January, 1997 except the provisions of sections 8 and 9.